The Development of Evolutionary Genetics

“Kette der Wesen”, Scala Naturae, in Oeuvres d’histoire naturelle, 4 (Neufchâtel, 1781). The Scala Naturae or Ladder of Creation arranges species along a linear scale, with the higher forms of life at the top and lower forms at the bottom. The concept is due to Plato and Aristotle. (From: https://upload.wikimedia.org/wikipedia/commons/c/c5/Kette_der_Wesen%2C_Scala_Naturae%2C_Charles_Bonnet_%281781%29.jpeg)

Prakash Gorroochurn

The Development of Evolutionary Genetics

From Early Ideas on Evolution to the Modern Synthesis

Springer

Prakash Gorroochurn
Department of Biostatistics
Columbia University
New York, NY, USA

ISBN 978-3-031-69376-2 ISBN 978-3-031-69374-8 (eBook)
https://doi.org/10.1007/978-3-031-69374-8

This Springer imprint is published by the registered company Springer Nature Switzerland AG
The registered company address is: Gewerbestrasse 11, 6330 Cham, Switzerland

Non Omnis Moriar

To Nishi, Premal, Geetesh, and Chetand

Preface

The current book attempts to present a history of the development of evolutionary genetics, starting from when ideas on evolution were first contemplated until the 1930s, when the so-called modern synthesis took place.

A historical account cannot possibly be an account of events that occurred in the past. At first, this statement may seem surprising, but a little thought should convince us that it is true. Only an objective, omnipresent being would be able to recount what really happened. At best, a historical account is the most accurate *recorded* version of events that took place in the past. Moreover, few would deny that an accurate version may be obtained if heard from the "horse's mouth," i.e. from the protagonists themselves. Great pain has therefore been taken to trace the primary sources in the discussion of topics. The hope is that the reader will be able to get the flavors, feelings, and motivations of past thoughts. However, many secondary sources have been provided both in the text and as a list at the start of each chapter, usually to be found in the footnotes.

There are some points that perhaps merit clarification and may help anyone wishing to read the book:

(a) *Evolutionary vs. Population Genetics.* Evolutionary and population genetics are often used interchangeably. Population genetics is the study of the genetic composition of a population; when the focus is on changes of the genetic composition across generations and on the forces responsible for such change, we are in the realm of evolutionary genetics. Thus, the square-law component of the Hardy-Weinberg law is intrinsically a population genetics result; the equilibrium component of that law, referring to the constancy of the genetic composition across generations, is more appropriately an evolutionary genetics result. However, this distinction is most often unnecessary since questions of evolutionary importance require results from population genetics, and vice versa. In most cases, the two terms may therefore be used interchangeably.

(b) *Theoretical vs. Mathematical Evolutionary Genetics.* The reader will often come across the terms "theoretical evolutionary (or population) genetics" and "mathematical evolutionary (or population) genetics." Theoretical means pertaining to a body of laws (based on certain assumptions) that attempt to understand a particular phenomenon and is different from mathematical. Thus, Lamarck's work on evolution is theoretical but non-mathematical; Fisher's work is both.

(c) *A Scientific Theory.* A hallmark of a scientific theory is its ability to be tested and potentially falsified. Thus, both Lamarck's theory of evolution and Darwin's theory of evolution by natural selection are testable and therefore scientific. On the other hand, the biblical theory of special creation is not testable and is therefore not scientific. This, of course, does not mean this theory *must* be false. What it means is that it does not fall within the purview of the scientific apparatus. At the same time, confidence in a theory is reduced if it cannot be tested.

(d) *The Fact of a Phenomenon and a Theory to Explain It.* Gravity is a fact and Newton proposed the theory of universal gravitation to explain it. However, this theory was later found to be deficient in certain cases and was subsumed by Einstein's general theory of relativity. However, even when the latter theory was proposed, gravity remained a fact. It is thus important to decouple what is factual from the theory put forward to explain it. If most scientists view evolution as a fact, it remains no less a fact even when those scientists propose competing theories to explain it.

It is hoped that everyone who reads the book will find at least one surprising fact in it, hitherto unknown to them. There are plenty of these in the book, but to name a few: (i) Georges Cuvier, one of the greatest anti-evolutionists, co-wrote a pro-evolutionist paper early in his career; (ii) Charles Darwin did *not* become an evolutionist while visiting the Galapagos islands; (iii) Darwin used Lamarckian ideas in his theory of natural selection; (iv) at least three contemporaries had published ideas "somewhat" related to natural selection before Darwin; (v) like Gregor Mendel, Darwin also performed hybridization experiments and obtained a segregation ratio of 3:1 (or close to it) in some of his experiments; (vi) August Weismann, who made a big contribution in the eventual demise of conventional Lamarckism, was earlier a Lamarckian himself; (vii) Hugo de Vries, one of the co-rediscoverers of Mendelism, went on to espouse an anti-Mendelian theory; (viii) William Castle, who had discovered the Hardy-Weinberg law five years before Hardy and Weinberg, held some anti-Mendelian views until 1919; (ix) Ronald Fisher was not the first to observe that some of the ratios obtained by Mendel were too close to expected ones; and (x) John Haldane was not the only one who responded to Ernst Mayr's charge of "beanbag genetics." Of course, several of these will be well known to many, but a few may be surprising to some, since they are not usually mentioned in textbooks.

In the author's experience, the evolutionary works of non-English-speaking scholars have not been sufficiently treated (if at all) in either textbooks or historical texts. One of the aims of the current book therefore is to adequately treat the contributions of biologists such as Carl Nägeli, August Weismann, Hugo de Vries,

Wilhelm Weinberg, Sten Wahlund, Lucien Cuénot, Wilhelm Johannsen, Herman Nilsson-Ehle, Gustave Malécot, and Sergei Chetverikov.

Some readers will find that certain topics which should have been part of a work on the history of evolutionary genetics were omitted. The reason for such omissions is in large part due to the size of the book. However, it is hoped that whatever was able to be covered will be an enriching reading experience.

New York, NY, USA
June 7, 2024

Prakash Gorroochurn

Acknowledgments

I am grateful to Warren Ewens and Anthony Edwards for their guidance throughout the writing of the book. The following also provided helpful feedback: Tom Dickins, Carlos Herrera, Kostas Kampourakis, Kyung-Man Kim, Alan Stark, and Peter van Dijk. A special thank you goes to the following for help with the proofreading: Premal Gorroochurn, Lauren Lazaro, Joseph LaRocca, Huanyu Chen, Felix Liu, and Haotian Tang. Kenneth Teng, Arun Siva Shanmugam, Karthiga Barath, and others from the team at Springer were a pleasure to work with. Finally, I am forever indebted to my late father and mother.

Contents

Part I
The Problem of Evolution

Chapter 1
Before Darwin's Origin of Species

Abstract The origin of evolutionary thoughts is traced. It is argued that two powerful ideas stood against a possible theory of species transformation, namely, the biblical doctrine of special creation and the fixity of species and Plato's theory of Forms. Evolution was initially used in the sense of unfolding of the preexisting forms. A complete theory of evolution was first proposed by Jean-Baptiste Lamarck, who advocated the inheritance of acquired characters through continued use and disuse. The works of other pre-Darwinian scholars are also presented.

Keywords Special creation · Fixity of species · Plato's theory of Forms · Typological thinking · Epigenesis · Preformation · Chain of Being · Uniformitarianism · Catastrophism · Inheritance of acquired characters · Struggle for existence

1.1 Two Powerful Ideas Against Evolution

1.1.1 Biblical Influence: Special Creation and the Fixity of Species[1]

Discussion about evolution did not start with Darwin; in fact, they date back to Biblical times. According to the Judeo-Christian tradition, creation took place through God over 6 days. Genesis 1 (written around 1400 BC) states that God created Heaven, Earth, and light on Day 1. In the subsequent days, the creation of the firmament, waters, and vegetation took place. On Day 5, life in the waters was created. Then, on Day 6:

> **QUOT. 1** Then God said, "Let the earth bring forth the living creature according to its kind: cattle and creeping thing and beast of the earth, each according to its kind"; and it was so. (Genesis 1:24)

---

[1] The term "species" was first used in a biological sense in 1630 by William Prynne to denote a class of individuals having some common qualities or characteristics: "This kinde of argument from euery [*sic*] indiuiduall to the species will not hold" (Prynne, 1630, p. 180).

P. Gorroochurn, *The Development of Evolutionary Genetics*,
https://doi.org/10.1007/978-3-031-69374-8_1

Fig. 1.1 James Ussher (1581–1656). (From: https://commons.wikimedia.org/wiki/File:James_Ussher_001.jpg)

QUOT. 2 Then God said, "Let Us make man in Our image, according to Our likeness; let them have dominion over the fish of the sea, over the birds of the air, and over the cattle, over all the earth and over every creeping thing that creeps on the earth." (Genesis 1:26)

For centuries, scholars were influenced by the notions of special creation and the fixity of species, as depicted in Genesis 1 above. It is noteworthy that man, being created in the image of God, was believed to have a special status compared to other forms of life.

James Ussher (Fig. 1.1) was the influential Archbishop of Armagh, Ireland. In 1650–1654, he drew a timetable of biblical events in the *Annals of the Old and New Testament* and dated the creation of man at 4004 B.C. Ussher's timetable was widely adopted and later formed the basis of creationist belief that the Earth was only a few thousand years old.[2]

Genesis relates an important event involving Noah, who was of the tenth generation after the first man, Adam. This occurred around 2350 B.C. when Noah was approximately 600 years old:

QUOT. 3 Then the LORD saw that the wickedness of man was great in the earth, and that every intent of the thoughts of his heart was only evil continually.

And the LORD was sorry that He had made man on the earth, and He was grieved in His heart.

[2]Ussher was by no means the first to estimate the age of Creation, but his calculations were the most influential. See the books by Dalrymple (1994) and Jackson (2006), which include descriptions of attempts at estimation from both creationist and non-creationist perspectives. See also Sect. 2.14.4.4 in the current book.

> So the LORD said, "I will destroy man whom I have created from the face of the earth, both man and beast, creeping thing and birds of the air, for I am sorry that I have made them." (Genesis 6:5–7)

However, since Noah "walked with God," he was instructed by the latter as follows:

> **QUOT. 4** "Make yourself an ark of gopherwood; make rooms in the ark, and cover it inside and outside with pitch. (Genesis 6:14)

> **QUOT. 5** "...I will establish My covenant with you; and you shall go into the ark—you, your sons, your wife, and your sons' wives with you.
>
> "And of every living thing of all flesh, you shall bring two of every sort into the ark to keep them alive with you; they shall be male and female. (Genesis 6:18–19)

> **QUOT. 6** "...after seven more days I will cause it to rain on the earth forty days and forty nights, and I will destroy from the face of the earth all living things that I have made." (Genesis 7:4)

Noah's flood holds sway even today. Before Darwin's time, it was embraced by the neptunists and, during Darwin's time, by the catastrophists (Sect. 1.10).

1.1.2 Greek Influence: Plato's Theory of Forms

A second powerful idea that stood in the way of later evolutionary ideas was Plato's *Theory of Forms*. Influenced by Socrates, Plato (427–347 BC) believed that every real world object (i.e., every "particular") had an ideal or eternally perfect form. Thus, humans all looked different but were all manifestations of an ideal form.[3] Plato's Theory of Forms was illustrated in the famous *allegory of the cave* (see Fig. 1.2) in the *Republic*:

> **QUOT. 7** Picture men dwelling in a sort of subterranean cavern with a long entrance open to the light on its entire width. Conceive them as having their legs and necks fettered from childhood, so that they remain in the same spot, able to look forward only, and prevented by the fetters from turning their heads. Picture further the light from a fire burning higher up and at a distance behind them, and between the fire and the prisoners and above them a road along which a low wall has been built, as the exhibitors of puppet shows have partitions before the men themselves, above which they show the puppets. (Hamilton and Cairns, 1961, Republic, VII: 514b)

Plato's prisoners were, therefore, only able to see, right from their childhood, shadows of objects projected on the wall before them, i.e., "in every way such prisoners would deem reality to be nothing else than the shadows of the artificial objects." Now, suppose one prisoner was freed:

> **QUOT. 8** When one was freed from his fetters and compelled to stand up suddenly and turn his head around and walk and to lift up his eyes to the light, and in doing all this, felt pain and, because of the dazzle and glitter of the light, was unable to discern the objects whose shadows he formerly saw, what do you suppose would be his answer if someone told him

---

[3] In the Bible, that ideal form is God.

Fig. 1.2 Depiction of Plato's *allegory of the cave* (Republic VII). The prisoners on the right are chained to the head and, since childhood, can only see the shadows of real objects projected on the wall. (From: https://commons.wikimedia.org/wiki/File:Plato_-_Allegory_of_the_Cave.png)

> that what he had seen before was all a cheat and an illusion, but that now, being nearer to reality and turned toward more real things, he saw more truly? And if also one should point out to him each of the passing objects and constrain him by questions to say what it is, do you not think that he would be at a loss and that he would regard what he formerly saw as more real than the things now pointed out to him? (*ibid.*, VII: 515c)

Plato indicated that the prisoner's reality would still be the shadows, not the (real) objects pointed out to them.

Similarly, according to Plato, the "things" around us are like the shadows in the cave. We, like Plato's prisoners, take them to be reality while the true reality is their form.

The influence of Plato's Theory of Forms cannot be overstated. It is the root of essentialist (or typological) thinking that prevailed throughout centuries. The form gives objects around us their essence. Variations within species are superficial and, therefore, irrelevant since it is the form that is eternal. Plato's Theory of Forms thus had powerful anti-evolutionary implications. See also Sect. 2.9.

In Plato's scheme, the forms were arranged in what was later called the *Ladder of Creation* (*or scala naturae*[4]), with lower forms of life arranged hierarchically at the bottom and higher forms of life at the top. The ladder was based on the principle of continuity,[5] according to which there should always exist a third life form between any two in the ladder. Nature did not tolerate any gaps, as that would indicate an imperfection in God.[6]

Although Heraclitus (540–480 BC) believed in a world that was in a constant state of flux, Aristotle (384–322 BC) and his powerful school put down this idea. Aristotle, who was Plato's student, argued for a static world in which God had created living things, the nature of which remained fixed.

Among the Greeks, Empedocles (490–430 BC) was the one who advanced an idea vaguely reminiscent of Darwin's theory of evolution. According to Empedocles, life arose spontaneously on Earth by the random combination of limbs, heads, and other organs. The fortuitous nature of this process implied that some beings were poorly adapted and thus disappeared. Only the best-adapted organisms survived.

1.2 Epigenesis vs. Preformation: First Use of Evolution in a Biological Sense (1670)

To better understand the sense in which biological evolution first came to be used, it is necessary to review two competing theories of embryonic development in earlier times, namely, the theories of epigenesis and preformation.[7] Epigenesis holds that an organism develops by the gradual appearance of new features, while preformation holds that development occurs by the unfolding and growth of preexisting features in the germline.

Both theories can be traced to Aristotle's *On the Generation of Animals*. Concerning preformation, Aristotle said:

> **QUOT. 9**...if the parts of the future animal are separated in the semen, how do they live? and if they are connected, they would form a small animal. (Ross, 1952b, On the Generation of Animals, I: 722b 5)

---

[4]A term first used by Charles Bonnet in 1764. Plato's *Ladder of Creation* is a consequence of his principle of plenitude, according to which "every possible place [of the Universe] should be filled" (Lovejoy, 1936, p. 54).

[5]First introduced by Aristotle in *Metaphysics*: "I call two things continuous when the limits of each, with which they touch and by which they are kept together, become one and the same, so that plainly the continuous is found in the things out of which a unity naturally arises in virtue of their contact." (Ross, 1952a, Metaphysics, X: 1069a 5).

[6]The same principle of continuity would later be used by Darwin (see QUOT. 176) in the form *Natura non facit saltum*, a phrase attributed to Leibniz: "...*Nature never makes leaps*. I call this *the law of continuity*..." (Leibniz, 1765, p. 11).

[7]The theories of epigenesis and preformation are also discussed in Cobb (2006), Cole (1930), Needham (1959), Roe (1981), Richards (1992, Chap. 2), Pinto-Correia (1997), and Meyer (1939).

Fig. 1.3 William Harvey (1578–1657). (From: https://commons.wikimedia.org/wiki/File:Lives_of_British_Physicians_-_William_Harvey_MD.png)

The existence of a miniature form of an animal in the germ fluids is the essence of the doctrine of preformation. Aristotle debated this theory with that of epigenesis, resulting in his endorsement of the latter:

> **QUOT. 10** Yet again, if the whole animal or plant is formed from semen or seed, it is impossible that any part of it should exist ready made in the semen or seed, whether that part be able to make the other parts or no. For it is plain that, if it exists in it from the first, it was made by that which made the semen. But semen must be made first, and that is the function of the generating parent. So, then, it is not possible that any part should exist in it, and therefore it has not within itself that which makes the parts. (*ibid.*, II, 734a 35)

Aristotle's views greatly influenced William Harvey (Fig. 1.3), the English physician who discovered the circulation of blood, when the latter proposed the theory of (and also coined the name) epigenesis in his *Exercitationes de generatione animalium* (Harvey, 1651, 1653). Epigenesis is so strongly associated with Harvey's name that it is perhaps little known that he actually proposed two different theories of development, namely *metamorphosis* and *epigenesis*. Concerning the first process:

> **QUOT. 11** ...in the *Generation*[8] of *Animals;* some are formed and transfigured, out of *matter* already concocted and grown; and all the parts are made and distinguished together *per metamorphosin*, by a *metamorphosis*, so that a complete *Animal* is the result of that *Generation*. (*ibid.*, p. 222)

[8] All italics are Harvey's.

By metamorphosis, Harvey thus had in mind the coming together at the same time of all body parts "already concocted and grown" so that a complete animal is suddenly formed. He thought this process applied to insects, "as when by a *Metamorphosis a Worm* is made of an *Egge.*"

However, when it came to "the more perfect *Animals* which have blood," Harvey proposed the second process, namely, epigenesis:

> **QUOT. 12** ...but some [animals] again, having one part made before another, are *afterwards* nourished, augmented, and formed out of the same *matter*; that is, they have parts, whereof some are before, and some after other...and therefore we say are made *per Epigenesin*, by a post-generation, or after-production; that is to say, by degrees, part after part. (*ibid.*, pp. 222–223)

The doctrine of epigenesis was later revived by Caspar Friedrich Wolff (1759). He advocated that the embryo originated by the secretion of parental fluids and their solidification through an inherent property they possessed, which Wolff called the *vis essentialis*.

Other adherents to epigenesis include the Comte de Buffon, Erasmus Darwin, Jean-Baptiste Lamarck, and Geoffroy Saint-Hilaire. In 1942, Conrad Hal Waddington changed *epigenesis* to *epigenetics* and gave a new meaning to the term.[9]

A few years after Harvey had proposed his theory of epigenesis, the Dutch biologist Jan Swammerdam (Fig. 1.4) published his *Historia insectorum generalis* (Swammerdam, 1669, 1682). In his book, Swammerdam wished to oppose Harvey's theory of metamorphosis. However, many of his observations were interpreted to support preformation. For example:

> **QUOT. 13** We are now going to pass to the comparison of plants which we see grow from a seed which already contains some leaves in its germ, or at least some buds: & it's in the same manner that we find insects come from a seed, which contains not only all the parts of the animal, but which is effectively the animal itself, enclosed in a new membrane. (*ibid.*, p. 59)

Swammerdam's version of preformation is properly called the doctrine of preexistence (Roger, 1971, pp. 325–326). It was more extreme than the original theory since it postulated that each preformed embryo carried a seed containing a second embryo, which in turn carried another seed, and so on, so that:

> **QUOT. 14** [f]inally, we could even [,] following the thoughts of a very learned man to whom we have revealed the secrets of our experience [,] conclude from this principle the origin of our natural corruption [i.e. the original sin] by conceiving that all creatures were eclosed in the loins of our first fathers. (Swammerdam, 1669, 1682, p. 48)

Swammerdam thus explained the original sin by postulating that all future mankind was already contained in the organs of Adam and Eve. The "very learned man" in the quotation above was the French philosopher and priest Nicolas Malebranche (1638–1715), who expounded the theory of preexistence in his *Recherche de la Vérité* (Malebranche, 1674).

---

[9] See the end of Sect. 1.9.2.

Fig. 1.4 Jan Swammerdam (1637–1680). (From: https://upload.wikimedia.org/wikipedia/commons/4/41/Jan_Swammerdam_%28CardioNetworks_ECGpedia%29.jpg)

In 1670, an anonymous English reviewer explained Swammerdam's *Historia insectorum generalis* as follows (Fig. 1.5):

> **QUOT. 15** *First*, It [Swammerdam's work] lays down the Ground of all Natural *Changes* in Insects; declaring, that by the word *Change*, is nothing else to be understood but a gradual and natural Evolution and Growth of the parts; not any Metamorphosis or Transformation of them. (Anonymous, 1670, p. 2078)

This was the first mention of (biological) evolution. It was thus *first used and recognized as the unfolding of preexisting forms, i.e., an equivalent of preformation.*

The use of evolution in the sense of preformation was repeated by several biologists. For example:

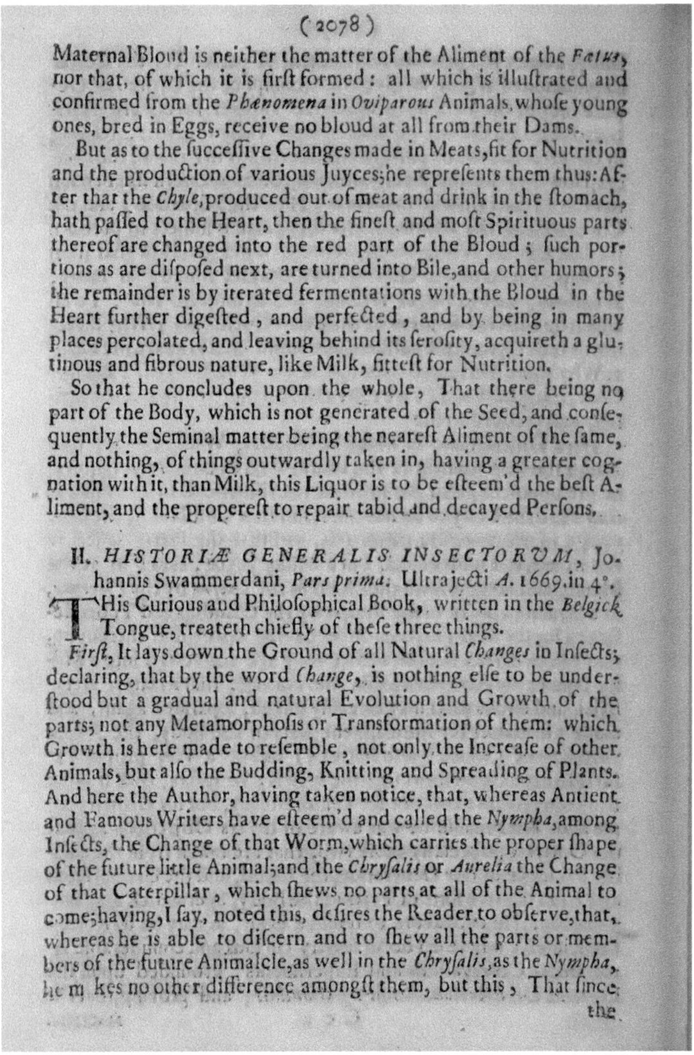
(2078)

Maternal Bloud is neither the matter of the Aliment of the *Fœtus*, nor that, of which it is firſt formed: all which is illuſtrated and confirmed from the *Phænomena* in *Oviparous* Animals, whoſe young ones, bred in Eggs, receive no bloud at all from their Dams.

But as to the ſucceſſive Changes made in Meats, fit for Nutrition and the production of various Juyces; he repreſents them thus: After that the *Chyle*, produced out of meat and drink in the ſtomach, hath paſſed to the Heart, then the fineſt and moſt Spirituous parts thereof are changed into the red part of the Bloud; ſuch portions as are diſpoſed next, are turned into Bile, and other humors; the remainder is by iterated fermentations with the Bloud in the Heart further digeſted, and perfected, and by being in many places percolated, and leaving behind its ſeroſity, acquireth a glutinous and fibrous nature, like Milk, fitteſt for Nutrition.

So that he concludes upon the whole, That there being no part of the Body, which is not generated of the Seed, and conſequently the Seminal matter being the neareſt Aliment of the ſame, and nothing, of things outwardly taken in, having a greater cognation with it, than Milk, this Liquor is to be eſteem'd the beſt Aliment, and the propereſt to repair tabid and decayed Perſons.

II. *HISTORIÆ GENERALIS INSECTORUM*, Johannis Swammerdani, *Pars prima*; Ultrajecti *A.* 1669. in 4°.

THis Curious and Philoſophical Book, written in the *Belgick* Tongue, treateth chiefly of theſe three things.

Firſt, It lays down the Ground of all Natural *Changes* in Inſects; declaring, that by the word *Change*, is nothing elſe to be underſtood but a gradual and natural Evolution and Growth of the parts; not any Metamorphoſis or Transformation of them: which Growth is here made to reſemble, not only the Increaſe of other Animals, but alſo the Budding, Knitting and Spreading of Plants. And here the Author, having taken notice, that, whereas Antient and Famous Writers have eſteem'd and called the *Nympha*, among Inſects, the Change of that Worm, which carries the proper ſhape of the future little Animal; and the *Chryſalis* or *Aurelia* the Change of that Caterpillar, which ſhews no parts at all of the Animal to come; having, I ſay, noted this, deſires the Reader to obſerve, that, whereas he is able to diſcern and to ſhew all the parts or members of the future Animalcle, as well in the *Chryſalis*, as the *Nympha*, he makes no other difference amongſt them, but this, That ſince

the

Fig. 1.5 First Use of "Evolution" in 1670 in a biological sense by an anonymous English reviewer of Swammerdam's *Historia insectorum generalis* (S. Swammerdam, 1669a, b)

(a) In 1744, the Swiss anatomist Albrecht von Haller (1708–1777) edited a book by his teacher Hermann Boerhaave (1668–1738). In reviewing the opposing theories of epigenesis and preformation, von Haller stated:

> **QUOT. 16** But the theory of evolution ['evolutionum theoria'] almost everywhere obtains from Swammerdam and Malphigi. (Boerhaave, 1744, p. 498, Vol. 5, part 2)

By referring to the regeneration of lost parts and the development of the heart of the chick from a simple tube, Haller rejected preformation. However, by 1758, his position had changed in favor of the latter theory, leading to a heated debate with Wolff.[10]

(b) In the book *Considérations sur les Corps Organisés* (Bonnet, 1762), the Swiss naturalist and philosopher Charles Bonnet (1720–1793) wished to develop a theory of unfolding of preexisting forms in the germline:

QUOT. 17 I have just put in front of my Readers several interesting Facts, which seem to unite to make *Evolution* a general Law of an organic System. This Law obviously assumes the pre-existence of Germs; nothing can develop which has not been preformed. (*ibid.*, p. 168)

It was only later that "evolution" began to be used in the sense of organic transmutation, first by Lyell in 1832[11] while explaining Lamarck's views:

QUOT. 18 The testacea of the ocean existed first, until some of them by gradual evolution were improved into those inhabiting the land. (Lyell, 1832, p. 11)

1.3 Gottfried Leibniz: The Chain of Being (1753)

During the Renaissance, Plato's Ladder of Creation developed into the *Chain of Being* (Fig. 1.6). The *Chain* contained a large number of hierarchical links, from the most basic at the bottom to the most perfect, namely God, at the top. The concept was further developed and refined by the German philosopher and mathematician Gottfried Leibniz (Fig. 1.7), reaching its apotheosis in the eighteenth century. In a letter that was posthumously published in 1753, Leibniz said:

QUOT. 19 All the different classes of beings which taken together make up the universe are, in the ideas of God who knows distinctly their essential gradations, only so many ordinates of a single curve so closely united that it would be impossible to place others between any two of them, since that would imply disorder and imperfection. Thus men are linked with the animals, these with the plants and these with the fossils, which in turn merge with those bodies which our senses and our imagination represent to us as absolutely inanimate. (Buchenau and Cassirer, 1924, p. 558 Vol. II; see also Lovejoy, 1936, pp. 144–145)

One infers plenitude (the existence of every type of life form), continuity (the existence of a third life form between any two life forms in the chain) and gradation (the progression of life forms from the simplest to the more complex along the chain) in Leibniz's concept of the Chain of Being.

---

[10] See Roe (1981).

[11] Lyell used evolution only once in the book, preferring the word "transmutation" in other instances. As for C. Darwin, he used "evolution" only in the sixth edition of *The Origin of Species* (Darwin, 1872), preferring instead "descent with modification" in the earlier editions. See later in the text.

Fig. 1.6 The chain of being, from "Rhetorica Christiana" by Fray Diego de Valades (1579). (From: https://commons.wikimedia.org/w/index.php?search=Great+Chain+of+Being&title=Special%3ASearch&go=Go#/media/File:Great_Chain_of_Being_2.png)

In the posthumous *Nouveaux Essais sur l'Entendement Humain*,[12] Leibniz also provided the following definition of a species:

[12]The *Nouveaux Essais* was meant to be a reply to Locke's *An Essay Concerning Human Understanding* (1690) and was written in 1703. Owing to Locke's death in 1704, Leibniz refused to publish his *Nouveaux Essais* in his own lifetime.

Fig. 1.7 Gottfried Leibniz (1646–1716). (From: https://upload.wikimedia.org/wikipedia/commons/3/3b/Gottfried_Wilhelm_Leibniz.jpg)

> **QUOT. 20** . . .we shall define the species by the generation, such that this similarity, which comes or might come from a common origin or seed, would belong to the same species. (Leibniz, 1765, pp. 268–269)

He next contemplated the possibility of species transformation:

> **QUOT. 21** Moreover, the intermixing of species, & even the changes within a species often take place with much success in plants. Perhaps at some time at some place in the Universe, the species of animals are or were or will be more liable to change, than they are presently among us. . . (*ibid.*, p. 277)

The Chain of Being was later popularized by Charles Bonnet (1720–1793).

1.4 Carl Linnaeus: Fixity of Species (1736), Transmutation by Hybridization (1762)

The enlightenment was a time of change and progress in both science and society. Although it originated in England and occurred mainly in France, personalities from other parts of Europe were also involved. One such figure was the Swedish botanist

Fig. 1.8 Carl Linnaeus (1707–1778). (From: https://commons.wikimedia.org/wiki/File:Carl_von_Linn%C3%A9,_1707-1778,_botanist,_professor_(Alexander_Roslin)_-_Nationalmuseum_-_15723.tif)

Carl Linnaeus[13] (Fig. 1.8). In his epochal *Systema Natura* (Linnaeus, 1735), Linnaeus introduced the binomial nomenclature, which is still used today. He classified animals and plants into species based on structural characteristics. The species were then further grouped into genera (or groups of similar species). Although Linnaeus did not work much on the question of evolution, his early writings show belief in special creation and the fixity of species:

> **QUOT. 22** There are as many species as the Infinite Being produced diverse forms in the beginning. (Linnaeus, 1736, No. 157)

However, in his later years, he came to believe that, through hybridization, species could change between genera and that it was the latter that were created in fixed forms by God. In the *Fundamentum fructificationis*, Linnaeus stated:

> **QUOT. 23** I have long cherished the suspicion, and no longer dare to sell it as an unquestionable truth, I present it by way of hypothesis: all species from one genus have from the beginning been one species, and propagated through hybridization, so that all species [within a genus] are born from a single mother, but from a different father they are made into different species. (Linnaeus, 1762, p. 16)

---

[13] Carl Linnaeus's life and work are also discussed in Stafleu (1971), Blunt (1971), Frängsmyr (1983), Koerner (1999), Barber (1980, Chap. 3), Ley (1968, pp. 178–182), Adams (1969, Chap. 3), and Kastner (1977, pp. 27–35).

Fig. 1.9 George-Louis Leclerc, Comte de Buffon (1707–1788). (From: https://commons.wikimedia.org/w/index.php?search=buffon&title=Special%3ASearch&go=Go&ns0=1&ns6=1&ns12=1&ns14=1&ns100=1&ns106=1#/media/File:Georges-Louis_Leclerc,_Comte_de_Buffon.jpg)

1.5 George-Louis Leclerc, Comte de Buffon (1707–1788)

1.5.1 Uniformitarianism (1749)

The Frenchmen George-Louis Leclerc, Comte de Buffon[14] (Fig. 1.9) was another major figure of the Enlightenment. He held the position of intendant of the Jardin du Roi in Paris. Buffon was one of the greatest naturalists that ever lived, having authored a monumental 44-volume treatise, *Histoire Naturelle,* on every aspect of the natural world that was known at the time.[15]

[14]Comte de Buffon's life and work are also discussed in Roger (1997), Loveland (2001), Gayon (1992), Hanks (1966), Adams (1969, Chap. 4), and Fellows & Milliken (1972). His evolutionary views are discussed in Rostand (1945, Chap. IV), Wilkie (1956a, 1956b, 1956c), Ostoya (1951, pp. 47–57), Roger (1997, pp. 320–329), Gaudant & Gaudant (1971, Chap. 5), and Fellows & Milliken (1972, Chap. 7).

[15]Except for the first three volumes, the other volumes were written with the help of collaborators, especially L.J.M. Daubenton (1716–1800).

Buffon was one of the first to reject biblical catastrophism in favor of a uniformitarian doctrine.[16] Referring to the separation of the "old" (Europe, Africa, and Asia) and "new" (Americas) worlds, in *Théorie de la Terre* (T. 1), he stated:

QUOT. 24 . . .this change could then have taken place suddenly. . .by a universal flood, or else this change did not take place suddenly, & much time was needed, but finally it happened, & I even think it happened naturally; for to judge what has happened & even what will happen, we only have to examine what is happening. (Buffon, 1749a, p. 96)

The above is a quintessential uniformitarian statement and implies an ancient Earth.[17] Its age was later estimated by Buffon (1775, p. 404) to be 74,832 years. Such statements brought the wrath of the Sorbonne, the infamous theological faculty in Paris. However, Buffon was tactful in satisfactorily responding to the authorities without having to retract any of his writings.

1.5.2 Epigenesis (1749)

Buffon also rejected the theory of preformation and favored epigenesis. Concerning the former theory, one can read in the *Histoire des Animaux* (T. 2):

QUOT. 25 When we ask how we can imagine the reproduction of beings is done, & that we are answered that in the first being this reproduction was already done, it is not only to confess that we ignore how it is done, but also to renounce the will to conceive of it. (Buffon, 1749b, p. 28)

Buffon's epigenesis was based on a fundamental distinction he made concerning all matter in the universe, namely, that it consisted of either organic or nonorganic particles. The former were vital (i.e., endowed with an inner force) and identical. Through the action of "penetrating forces," the organic particles tended to unite and form a shape according to the "internal mold" (*moule intérieur*) (*ibid.*, p. 35) of the organism. The internal mold was an innate property of each organism. The embryo was formed simply by the mixing of male and female fluids (*ibid.*, p. 65) and then developed by the assimilation of organic particles through nutrition, the shape of the developing organism being determined by its internal mold. Buffon's internal mold

[16]Uniformitarianism holds that all geologic change can be accounted for by existing processes acting in the same manner through extended periods, while catastrophism posits that such change is due to supernatural catastrophes, such as the biblical flood. The words 'uniformitarian' and 'catastrophist' were introduced by William Whewell in the 1832 review of the second volume of Lyell's Principles of Geology (Lyell, 1832): "These two opinions will probably for some little time divide the geological world into two sects, which may perhaps be designated as the *Uniformitarians* and the *Catastrophists*" (Whewell, 1832, p. 216). "Uniformitarianism" was first used in 1865 by T. Marsden (1865, p. 113). See also Sects. 1.6 and 1.14.2. "Catastrophism" was first used in 1845 in French (Richard, 1845, p. 55). Its first English use occurred in 1869 by T. H. Huxley (1869, p. xxxix). See also Sect. 1.10.1.

[17]Buffon was probably influenced by Benoit de Maillet (1656–1738), who was the first to have proposed an ancient Earth in the posthumous *Telliamed* (de Maillet, 1748).

thus replaced the preformationists' preexisting germs. His epigenesis, being based on the concept of forces acting on particles, was mechanical and thus different from that of Wolff.[18]

1.5.3 From Fixism (1749) to Limited Transformism (1766)

Buffon's views on evolution will next be discussed. His writings on the topic are nuanced and need careful interpretation. In the *Histoire des Animaux* (T. 2), Buffon declared that all life forms (animal and plant) were essentially the same:

> **QUOT. 26** ...there is absolutely no essential & general difference between animals & plants... Nature descends by degree & by imperceptible nuances from one animal which seems the most perfect to us to another which is less, and from that one to the plant. (*ibid.*, p. 8)

Note that, in the quotation above, Buffon spoke of a *descent* from a perfect organism to a less perfect one. The concept of descent was formalized to *degeneration* later in the text and formed an integral part of Buffon's evolutionary thinking. *Whenever Buffon spoke of transmutation of species, it was always in the direction of degeneration.*

However, the more important question is, does QUOT. 26 imply that, at the very start (in 1749) of his *Histoire Naturelle*, Buffon endorsed some kind of descent with modification? This question was examined in Buffon's article *L'Asne* (T. 4):

> **QUOT. 27** ...if we once admit that there are *families* in plants and animals, that the donkey is part of the horse *family*, & that it only differs because it has degenerated, we could also say that the monkey is of the *family* of man, that it is a degenerate man, that man & the monkey had a common origin like the horse & the donkey, that each *family*, both in animals as in plants, had only one branch, & even that all animals came from one animal, which, in the succession of times, produced by perfecting and degenerating, all the races of other animals. (Buffon, 1753, p. 382)

Instead of endorsing descent with modification, Buffon ridiculed the statement above and condemned "naturalists who so lightly establish families in animals and plants, do not seem to have felt the whole extent of their consequences, which will reduce the immediate product of Creation to a number as small as we wish." *The* naturalist whom Buffon had in mind was no doubt Linnaeus (Buffon's archrival), who had classified the horse and donkey under the same genus. While Linnaeus' classification was based on *morphology*, Buffon insisted on *fecundity*:

> **QUOT. 28** The resemblance, both external and internal, even if in some animals it is even greater than it is in the horse & the donkey, should therefore not lead us to confuse these animals in the family, any more than to give them a common origin; because if they came from the same stock, if they were indeed from the same family, we could bring them together, ally them again. (*ibid.*, p. 401)

[18] See the paragraph after QUOT. 12.

Thus, there is little doubt that Buffon's initial reticence to transmutation between species was largely due to his opposition to the Linnaean classification system. Another reason was explicitly stated by Buffon as follows:

> **QUOT. 29** ...it is certain, through revelation, that all animals equally participated in the grace of Creation, that the first two of each species and of all species have emerged completely formed through the hands of the Creator, & we must believe that they were such then, more or less, that they make us represented today by their descendants. (*ibid.*, p. 401)

The above is simply the biblical concept of special creation and the fixity of species.[19]

In the 1761 article *Lion* (T. 9), Buffon dropped some of his reservations and accepted that some species could be classified under the same genus. However, this applied only to species that were not *noble.* By noble, Buffon meant those species that were "constant, invariable, & that we cannot suspect of having been degraded" (Buffon, 1761, p. 9). The human species was the most noble since "this species is unique...no animal belong to man neither closely nor from a distant natural parentage." On the other hand, "the horse species is not as noble as a man since it has the donkey species as a neighbor." Furthermore:

> **QUOT. 30** In the dog, the species is perhaps even less noble since it seems to be closely related to those of the wolf, the fox and the jackal, which can be considered as degenerate branches of the same family. (*ibidem*)

In the above, Buffon all but conceded degeneration between (less noble) species within certain genera.

This was made even more explicit in the article *De la Dégénération des Animaux* (T. 14). After having considered degeneration within species, Buffon stated:

> **QUOT. 31** ...a more important and extended consideration presents itself; it is that of the change in species themselves, it is that more ancient & since time immemorial degeneration, which seems to have taken place in each family, or if we wish, in each genus... (Buffon, 1766, p. 335)

By considering all such possible degenerations:

> **QUOT. 32** ...we will find that the two hundred species whose history we have given, may be reduced to a rather small number of families or principal groups, from which it is not impossible that all others may have been derived. (*ibid.*, p. 358)

The two quotations above are Buffon's clearest endorsements of the evolutionary viewpoint. However, his evolution was always of a limited kind—always in the direction of degeneration and not applicable to "noble" species.

---

[19] See QUOT. 1.

1.5.4 The Inheritance of Acquired Characters

Buffon also believed in the inheritance of acquired characters, which would later stand (in the form of Lamarckian evolution[20]) as a major competitor to Darwinian selection. In the article *De la Dégénération des Animaux* (T. 14), one can read:

QUOT. 33 Dogs whose ears & tail have been cut generation after generation, transmit these defects completely or in part to their descendants. (*ibid.*, p. 322)

Buffon believed that flood and climate were able to produce heritable changes in animals. Concerning skin color, in the article *Histoire Naturelle de l'Homme* (T. 3), he said:

QUOT. 34 ...To me, it has always seemed that the same cause which darkens our skin when we are exposed to the open air & to the sun's heat, that cause which makes the Spanish have darker skin than the French, & the Moors darker than the Spanish, also cause Negros to be darker than the Moors. (Buffon, 1749c, pp. 525–526)

The above (Lamarckian) explanation for the inheritance of skin color through the repeated action of the sun's radiation may be compared with the modern (Darwinian) explanation in terms of better adaptation to the sun's radiation by individuals who possess the melanin pigment.

1.6 James Hutton: Uniformitarianism (1785, 1795)

Buffon's uniformitarianism was first developed by the Scottish geologist James Hutton[21] (Fig. 1.10). Uniformitarianism holds that all geologic change can be accounted for by existing processes acting in the same manner through extended periods.[22] The physical and chemical laws that govern nature are thus uniform. In explaining the current conditions on Earth, one only has to consider these natural laws, rather than supernatural, catastrophic occurrences. This was first stated in Hutton's 30-page "Abstract of a Dissertation Read in the Royal Society of Edinburgh, upon the Seventh of March and Fourth of April MDCCLXXV Concerning the System of the Earth, Its Duration, and Stability":

QUOT. 35 Hence we are led to conclude, that the greater part of our land, if not the whole, had been produced by operations natural to this globe. (Hutton, 1785, p. 6)

In the 1788 paper "Theory of the Earth" (Hutton, 1788) (which was also read in 1785), one can further read:

[20] See Sect. 1.8.

[21] And later more fully by Charles Lyell (see Sect. 1.14.2). Hutton's life and work are described in Dean (1992), Repcheck (2003), Hooykaas (1963, pp. 17–24), and Schneer (1969, pp. 122–141).

[22] See footnote 16.

Fig. 1.10 James Hutton (1726–1797). (From: https://commons.wikimedia.org/wiki/File:Hutton_James_portrait_Raeburn.jpg)

> **QUOT. 36** Therefore, upon the supposition that the operations of nature are equable and steady, we find, in natural appearances, mean for concluding a certain portion of time to have necessarily elapsed, in the production of those events of which we see the effects... (*ibid.*, p. 217)

These ideas were later more fully developed in Hutton's two-volume *Theory of the Earth* (Hutton, 1795).

Aligned with Hutton's uniformitarianism is also his Plutonism,[23] the theory that Earth's internal heating was responsible for the cyclical creations of rocks and continents. On opposite ends, one finds the theory of Neptunism[24] (which holds

[23]Named after Pluto, God of the underworld. The word 'plutonic' was first used by Richard Kirwan as follows: "There is another system which attributes not only to basalts, but to all stony substances, an igneous origin... This may be called the *Plutonic* system" (Kirwan, 1794, p. 455, Vol. I). The words "volcanist" and "vulcanist" have also been used interchangeably with "plutonist."

[24]This idea was most clearly developed by Abraham Gottlob Werner (1749–1817). Named after Neptune, the God of the sea. The word "Neptunian" was first used by Richard Kirwan as follows: "Mountains of Neptunian origin are distinguished by their materials, and often by the arrangement of those materials" (Kirwan, 1794, p. 425, Vol. I).

Fig. 1.11 Erasmus Darwin (1731–1802). (From: https://commons.wikimedia.org/wiki/Erasmus_Darwin#/media/File:Portrait_of_Erasmus_Darwin_by_Joseph_Wright_of_Derby(1792).jpg)

such processes were the products of water action), which is aligned with the theory of catastrophism (which holds that Earth's history was a succession of catastrophic or supernatural events such as the biblical flood).

One notable implication of Hutton's uniformitarianism was an extremely old age for the Earth. This was in direct contrast with the Neptunists' and catastrophists' young-Earth view.

1.7 Erasmus Darwin's Ideas on Evolution (1794)

A notable figure inspired by Buffon's evolutionary views was none other than Darwin's grandfather, the English physician and naturalist Erasmus Darwin[25] (Fig. 1.11). E. Darwin favored the spontaneous generation of life, and, like Buffon,

[25] Erasmus Darwin's views on evolution are also discussed in Harrison (1971), King-Hele (1963a, 1963b, Chap. V), McNeil (1987, pp. 92–105, 115–119), Gaudant & Gaudant (1971, Chap. 6), Fothergill (1952, pp. 63–68), Bardell (2001), and Krumbhaar (1931).

he explicitly rejected the concept of fixity of species. In the poem[26] *The Temple of Nature* (Darwin, 1803), one can read lines that are quite close to today's beliefs:

QUOT. 37 Hence without parent by spontaneous birth.
Rise the first specks of animated earth;
From Nature's womb the plant or insect swims,
And buds or breathes, with microscopic limbs. (*ibid.*, Canto I, lines 247–250)

. . . .

QUOT. 38 In branching cones the living web expands,
Lymphatic ducts, and convoluted glands;
Aortal tubes propel the nascent blood,
And lengthening veins absorb the refluent flood;
Leaves, lungs, and gills, the vital ether breath
On the earth's green surface, or the waves beneath. (*ibid.*, Canto I, lines 259–265)

The poem also describes a struggle for existence among humans:

QUOT. 39 So human progenies, if unrestrain'd;
By climate friended, and by food sustain'd,
O'er seas and soils, prolific hordes! would spread
Erelong, and deluge their terraqueous bed;
But war and pestilence, disease, and death
Sweep the superfluous myriads from the earth. (*ibid.*, Canto IV, lines 369–374)

Again, from E. Darwin's two-volume book *Zoonomia*, largely devoted to medicine, one can read:

QUOT. 40 . . .the whole is one family of one parent. (Darwin, 1794, p. 1, Vol. I)

QUOT. 41 . . .shall we conjecture that one and the same kind of living filaments is and has been the cause of all organic life. (*ibid.*, p. 507)

Remarkably, E. Darwin also adduced the concept of sexual selection (involving competition between members of a given sex for the other), a topic that would later be treated at length by his grandson Charles Darwin:

QUOT. 42 The birds, which do not carry food to their young, and do not therefore marry, are armed with furs for the purpose of fighting for the exclusive possession of the females, as cocks and quails. It is certain that these weapons are not provided for their defence against other adversaries, because the females of these species are without this armour. The final cause of this contest amongst the males seems to be, that the strongest and most active animal should propagate the species, which should thence become improved. (*ibid.*, p. 503)

The quotations above provide a glimpse of E. Darwin's views on evolution, namely some kind of descent with modification (although he never used that exact phrase). These views were not different from those of his grandson, Charles Darwin. The break between the two, however, first arises when one tries to answer the following question: *what exactly is the modus operandi of evolution?* E. Darwin believed that a gradual transformation of the members of a species occurred when their "irritations,

[26]E. Darwin also wrote prose but chose this form for some of his works because poetry could conjure up visual images.

sensations, volitions, and associations" made them interact with their environment:[27]

> **QUOT. 43** ...would it be too bold to imagine, that all warm-blooded animals have arisen from one living filament, which the GREAT FIRST CAUSE endued with animality, with the power of acquiring new parts, attended with new propensities, directed by irritations, sensations, volitions, and associations; and thus possessing the faculty of continuing to improve by its own inherent activity, and of delivering down those improvements by generation to its posterity, world without end! (Darwin, 1794, p. 505)

E. Darwin's emphasis on acquired characteristics in the development of *new* features is thus clear. His doctrine that animals changed through their own "volition" would later find a parallel in Lamarck's concept of "inner feeling" (Sect. 1.8). On the other hand, from C. Darwin's viewpoint, evolution took place primarily through the process of natural selection acting on fortuitous, *but already existing* variation. However, this in no way implies that C. Darwin had no place for the doctrine of the inheritance of acquired characteristics in his theory. On the contrary, he believed in it and incorporated it even more fully in the fifth edition of the *Origin of Species*, after criticism from a certain Fleeming Jenkin. These points will be addressed in more detail later (Sect. 2.14.4).

1.8 Jean-Baptiste Lamarck: First Complete Theory of Evolution (1809)

Despite E. Darwin's trailblazing writings, it is with the (slightly) later Frenchman, Jean-Baptiste Lamarck[28] (Fig. 1.12), that the doctrine of the inheritance of acquired characteristics[29] is usually associated. There is good reason for this because Lamarck was the one who thus fully developed a complete theory of evolution.[30]

---

[27] The theory of inheritance, according to which the hereditary material is not constant from generation to generation but may be modified either through the effects of the environment or use/disuse, is known as *soft* or *Lamarckian* inheritance (as opposed to *hard* or *non-Lamarckian* inheritance). See the next section.

[28] Lamarck's life and work are also discussed in Burkhardt (1977), Corsi (1988), Cannon (1959), Barthélemy-Madaule (1982), Greene (1959, pp. 155–166), Ostoya (1951, Chap. III), Pichot (2011, pp. 434–470), Gaudant & Gaudant (1971, Chap. 7), Guyénot (1941, pp. 412–439), Adams (1969, Chap. 5), Gillispie (1959), Jordanova (1984), Delange (1984), Szyfman (1982), and Packard (1901).

[29] The inheritance of acquired characters is also discussed in Zirkle (1946, Chap. I), Cannon (1958, Chap. IV), Blacher (1982), Fothergill (1952, pp. 69–78), Darlington (1964, pp. 50–55), and Wool (2020, Chap. 6).

[30] Lamarck was also one of the first to use the word biology. According to Littré's dictionary (also simply known as Littré), the German naturalist Gottfried Treviranus (1776–1837) (Treviranus, 1802) first used "Biologie," which was then adopted by Lamarck in his *Hydrogeologie* (Lamarck, 1802; Carozzi, 1964). However, in a short note to *Nature*, Müller (1983) has indicated that "biologie" was first used by the German anatomist Theodor Georg August Roose (1771–1803) in 1797 in the book *Grundzüge von der Lehre von der Lebenskraft*. The first English use occurs in James Stanfield's *Biography* (Stanfield, 1813, p. xii).

Fig. 1.12 Jean-Baptiste Pierre Antoine de Monet, chevalier de Lamarck (1744–1829). (From: https://upload.wikimedia.org/wikipedia/commons/6/68/Jean-baptiste_lamarck.jpg)

Lamarck was a student of Buffon, and his main vocation was botany. He believed in the fixity of species until 1797, at which time he started to believe in evolution. This was first displayed in 1800 in an introductory lecture for his new position[31] as Professor of Lower Animals in the newly founded *Muséum d'Histoire Naturelle* in Paris. The opening lecture was published as a preface to Lamarck's *Système de Animaux Sans Vertèbres* (Lamarck, 1801) and contained all the elements of his theory of evolution, namely:

(1) The availability of vast geological times: "We know that time has no limit for her [nature], and consequently she always has it at her disposition" (*ibid.*, p. 13)
(2) Evolution of "degraded" animals to more complex ones: "the most imperfect animals, the most simply organized, those in a word suspected of hardly any animality, those perhaps from which nature has started, when aided by much time and favorable circumstances, she had formed all the others." (*ibid.*, p. 12)

[31] *Discours d'ouverture du Cours de Zoologie donné dans le Muséum national d'Histoire naturelle, le 21 floréal, an 8 de la République (1800).*

(3) Evolutionary changes brought about when an animal, by its own needs, interacted with its environment: "The bird which necessity drives to water to find the prey which makes it live, stretches the toes of its feet when it wishes to strike the water and move on its surface. . .Thus with time, the broad membranes which unite the toes of ducks, geese, &c., are formed in the manner we see them." (*ibid.*, p. 13)

All these elements of Lamarck's theory would later be expanded.

In 1802, Lamarck explicitly rejected catastrophism[32] in favor of a doctrine of uniformitarianism,[33] which allowed for vast geological times (Lamarck, 1802). In the *Philosophie Zoologique* (Lamarck, 1809, 1984), one finds a full elaboration of Lamarck's theory of evolution. Like Charles Darwin later, Lamarck viewed evolution very much as a gradual process of descent with modification. However, in Lamarck's theory alone, one finds a tendency for the organism to reach increasing complexity and perfection.[34] Drawing on Buffon's theory of degeneration (which was now changed to "degradation"), Lamarck stated:

> **QUOT. 44** . . .That it is a positive fact that on following the animal chain in the usual direction from the most perfect to the most imperfect animals, we observe an increasing degradation and simplification of organisation; that, consequently on traversing the animal scale in the opposite direction, that is to say, in the same order as Nature's, we shall find an increasing complexity in the organization of animals, a complexity which would advance with evenness and regularity, if the environmental conditions, mode of life, etc., did not occasion many anomalies in it. (*ibid.*, p. vi)

According to Lamarck, the simplest form of life is the *monad*, which is "nothing but an *animated point*" (*ibid.*, p. 416) and which arises by spontaneous generation. In the chain of increasing complexity, the monad is at the bottom and man is at the top.

Regarding the evolution from simple to more complex life forms, Lamarck proposed two principles (*ibid.*, p. 2):

(a) The continued use of an organ "leads to its development, strengthens it" whilst disuse "is injurious to its development, causes it to deteriorate and ultimately disappear" if the disuse is sustained over an extended period.
(b) When a body accelerates, fluids inside the body "modify the cellular tissue in which they move, open passages in them, form various canals, and finally create different organs, according to the state of the organization in which they are placed."

The first principle tells us why evolution occurs; the second, how.

An important component of Lamarck's theory of evolution, one to which he returned repeatedly in the *Philosophie Zoologique*, was the concept of an organism's "inner feeling" (*sentiment intérieur*). The inner feeling of need was the ultimate driver of change:

---

[32] See Sect. 1.10.

[33] See Sect. 1.6.

[34] Lamarck's theory thus has elements of *orthogenesis* (or directed evolution).

QUOT. 45 In the frequent fits of anger to which the males especially are subject, the efforts of their inner feeling cause the fluids to flow more strongly towards that part of their head; in some there is hence deposited a secretion of horny matter, and in others of bony matter mixed with horny matter, which gives rise to solid protuberances: thus we have the origin of horns and antlers, with which the head of most of these animals is armed. (*ibid.*, p. 122)

According to Lamarck's theory, inner feelings of need were brought about by interactions with the environment. This resulted in the use and disuse of organs, giving rise to new characteristics (or loss of old ones), which were then passed on to future generations. The two principles, namely, (i) that of *use and disuse* and (ii) that of the *inheritance of acquired characteristics*, were so pivotal that Lamarck gave them the status of laws:

QUOT. 46
FIRST LAW.

In every animal which has not passed the limit of its development, a more frequent and continuous use of any organ gradually strengthens, develops and enlarges that organ, and gives it a power proportional to the length of time it has been so used; while the permanent disuse of any organ imperceptibly weakens and deteriorates it, and progressively diminishes its functional capacity, until it finally disappears.

SECOND LAW.

All the acquisitions or losses wrought by nature on individuals, through the influence of the environment in which their race has long been placed, and hence through the influence of the predominant use or permanent disuse of any organ; all these are preserved by reproduction to the new individuals which arise, provided that the acquired modifications are common to both sexes, or at least to the individuals which produce the young. (*ibid.*, p.113)

Lamarck's first law is generally agreed with: use and disuse indeed do effect changes. The controversy concerns the second law: does continued use and disease produce changes that are ultimately inherited? There has been no evidence to indicate that this could be the case.

Although Lamarck's name is usually associated with the theory of inheritance of acquired characteristics, it is seen that there is more to his theory, namely, the concepts of continued use and disuse, of fluids modifying the cellular tissue inside the body, and, above all, of the existence of an inner feeling as a driver of evolutionary change. The theory may be schematically represented as follows:

inner feeling of need → use and disuse → somatic modifications

→ transmission of somatic modifications to offspring

Lamarck's most famous example of the power of the environment to effect heritable change in the organism was that of the giraffe:

QUOT. 47 It is interesting to observe the result of habit in the peculiar shape and size of the giraffe (*Camelo-pardalis*): this animal, the largest of the mammals, is known to live in the interior of Africa in places where the soil is nearly always arid and barren, so that it is obliged to browse on the leaves of trees and to make constant efforts to reach them. From this habit long maintained in all its race, it has resulted that the animal's fore-legs have become longer than its hind legs, and that its neck is lengthened to such a degree that the giraffe, without standing up on its hind legs, attains a height of six metres (nearly 20 feet). (*ibidem*)

Fig. 1.13 Lamarckian Inheritance of Acquired Characters vs. Darwinian Selection (Biological Sciences Curriculum Study, 1963, p. 595)

The example of the giraffe has become a standard way to illustrate Lamarck's doctrine of the inheritance of acquired characters and distinguish it from Darwinian evolution—see, for example, Fig. 1.13.

There are two aspects of Lamarck's theory that have usually been misunderstood and merit clarification:

(a) *Needs vs. desires*. Lamarck's theory has often been ridiculed for allegedly stating that permanent changes can be brought about simply by the occurrence of a particular *desire* in an organism. The great French zoologist Georges Cuvier was the first to make such a charge (see QUOT. 54), which was then often repeated by others (e.g., Holmes (1916, p. 23)). However, a careful reading of Lamarck indicates that the latter emphasized the primacy of *needs* rather than *desires*:

> **QUOT. 48** . . .needs arouse the individual's inner feeling in proportion as he feels them, and this feeling promptly sets in movement the nervous fluid so as to produce actions, either physical or moral, suitable for satisfying them. (*ibid*., p. 380)

(b) *Evolution through inherent forces in Lamarck's theory*. Lamarckian evolution (or Lamarckism) is often equated with the inheritance of acquired characters (e.g., see the dictionaries by King et al. (2006, p. 247) and Lincoln et al. (1982, p. 138)). However, are external forces the only cause of evolution in Lamarck's theory? A careful reading of Lamarck shows that "a continuous gradation in the organism" of animals already occurs through *inherent forces* in the organism (and not by external environmental factors):

> **QUOT. 49** It is obvious that, if nature had given existence to none but aquatic animals and if all these animals had always lived in the same climate, the same kind of water, the same depth, etc., etc., we should then no doubt have found a regular and even continuous gradation in the organisation of these animals. (Lamarck, 1809, 1984, p. 69)

The effect of external environmental forces is to *disturb* this "continuous gradation," leading to deviations and speciation:[35]

> **QUOT. 50** After having produced aquatic animals of all ranks and having caused extensive variations in them by the different environments provided by the waters, nature led them little by little to the habit of living in the air, first by the water's edge and afterwards on all the dry parts of the globe. These animals have in course of time been profoundly altered by such novel conditions; which so greatly influenced their habits and organs that the regular gradation which they should have exhibited in complexity of organisation is often scarcely recognisable. (*ibid*., pp. 69–70)

1.9 Reception of Erasmus Darwin's Ideas and Lamarck's Theory

1.9.1 Reception of Erasmus Darwin's Ideas

The reader is urged to compare Lamarck's concept of "inner feeling" to E. Darwin's "irritations, sensations, volitions, and associations." Indeed, Lamarck's ideas were so close to those of E. Darwin that a few speculated that the former might have plagiarized the latter. The name of C. Darwin should be included among those who have been suspicious. In a letter to his dear friend T.H. Huxley, he once remarked:

---

[35] See also Rostand (1945, pp. 100–101).

QUOT. 51 ... it is curious to observe how exactly and accurately my grandfather (in *Zoonomia*, Vol. I., p. 504, 1794) gives Lamarck's theory. . .Lamarck published *Hist. Zoolog*. In 1809. The *Zoonomia* was translated into many languages. (Darwin and Seward, 1903a, p. 125)

However, there is no evidence that C. Darwin could be right. E. Darwin's *Zoonomia* was translated into French only in 1810, one year after Lamarck's *Philosophie Zoologique*. Lamarck possessed neither the French nor the English editions. Burkhardt (1977, p. 225) has further pointed out that if Lamarck had drawn his ideas from E. Darwin, then Cuvier (who was fiercely opposed to Lamarck's transformism) would have certainly pointed it out.

A more plausible explanation for the proximity of E. Darwin's and Lamarck's ideas seems to be Buffon's influence on both.[36] As mentioned earlier, Lamarck was a student of Buffon, and E. Darwin was also aware of the latter's writings, having referred to them several times in his work.

History has been relatively kind to both E. Darwin and Lamarck, but their contemporaries did not bestow such kindness on them. For example, the poet and theologian Samuel Taylor Coleridge (1772–1834) once remarked:

QUOT. 52 What! did Sir W. R. believe that a male and female ounce (and, if so, why not two tigers and lions, etc.?) would have produced, in course of generations, a cat, or a cat a lion? This is Darwinizing with a vengeance. By this mode of reasoning he might have reduced Noah's stowage to at most half-a-dozen, so beautiful is the gradation of the species and genera of animals from men to mice. (Coleridge, 1875, p. 423)

Apart from its hostility to E. Darwin's evolutionary views, the above also shows that Darwinism was already used during E. Darwin's times.[37] Elsewhere, Coleridge denounced E. Darwin for being a believer "of man's having progressed from an orang-outang State" (Griggs, 1956, pp. 574–575, Vol. IV), foreshadowing later Victorian criticism of evolution.

Regarding his grandfather's evolutionary views, C. Darwin denied much influence on his *Origin of Species* on several occasions. In his own words,

QUOT. 53 . . .I had previously read the *Zoönomia* of my grandfather. . .but without producing any effect on me. Nevertheless it is probable that the hearing rather early in life such views maintained and praised may have favoured my upholding them under a different form in my *Origin of Species*. At this time I admired greatly the *Zoönomia*; but on reading it a

[36]See Sect. 1.5.4.

[37]In fact, it was first used in relation to E. Darwin in 1856 by the Rev. J.B. Reade: "[In September 1856] Mr. Sopwith described the hypothesis of the development of living things from a primordial center. That, said Reade, is rank Darwinism. It was the first time I had heard that word used. It had no reference to Charles Darwin, whose name at that period was not connected with the subject; but it had reference to Erasmus Darwin" (Richardson, 1891, p. 256). After C. Darwin's *Origin of Species*, Darwinism came to be associated with his name. It was first used in this sense by T.H. Huxley in his 1864 review of the *Origin*: ". . .and from none more thoroughly than from those in which he seeks to define what we may term the philosophical position of Darwinism" (Huxley, 1864, p. 567). As for the term "Darwinian," it was first used with reference to E. Darwin in 1794 by W. B. Stevens (Galbraith, 1965, p. 161) and with reference to C. Darwin in 1860 by T.H. Huxley (1860, p. 566).

second time after an interval of ten or fifteen years, I was much disappointed, the proportion of speculation being so large to the facts given. (Barlow, 1958, p. 49)

C. Darwin was undoubtedly right about the speculative nature of E. Darwin's theories, the latter having himself admitted in the *Botanic Garden* that "[e]xtravagant theories, however, in those parts of philosophy where our knowledge is yet imperfect are not without their uses." (Darwin, 1791, p. vii). Equally doubtless is the fact that C. Darwin was influenced by the views of his grandfather, even to the point of entitling his B notebook of 1837 "Zoonomia."

1.9.2 Reception of Lamarck's Theory

Unsurprisingly, Lamarck's evolutionary views did not receive a better reception than those of E. Darwin. Among the fiercest critics of Lamarck was the highly influential Georges Cuvier (1769–1832).[38] A letter (Plan, 1909, pp. lxxxi–lxxxii) from the Italian naturalist Marzari Pencati (1779–1836) to the Swiss pharmacist Henri-Albert Gosse (1753–1816) revealed that as early as 1804, during a course given in Paris, Cuvier ridiculed Lamarck's *Recherches sur l'Organisation des Corps Vivants* (Lamarck & Drouin, 1755) and the latter's views that organs were formed by needs. However, the brunt of Cuvier's criticism was reserved for the eulogy he wrote on Lamarck's death. Cuvier's eulogy was so over-the-top that the committee concerned refused to allow it unless he made drastic changes to it. Cuvier would not rescind and a eulogy was instead delivered by Pierre Latreille (1762–1833). Only after his death was Cuvier's eulogy, in truncated form, read in 1832 by M. le Baron Silvestre. In one of the more critical passages of the eulogy, one can read:

QUOT. 54 Other needs, other desires, resulting by circumstances, will lead to other efforts, which will produce more organs; for, by one more hypothesis added to the others, it is not the organs, that is the nature and form of parts, which give rise to habits and faculties; it is the habits, the manner of living, which, with time, give birth to the organs: it is by the will to swim that membranes form in the legs of water birds... (Cuvier, 1835, p. xix)

By stressing the importance of "desires" and "will," rather than "needs," as the main agent of change, Cuvier had exaggerated Lamarck's position.[39] However, Cuvier's authority ensured that Lamarck's theory did not gain much traction.

In England, Lamarck was first mentioned in the first volume of Charles Lyell's *Principles of Geology* (Lyell, 1830). Previously, in a letter dated March 2, 1827 (Lyell, 1881, p. 168, Vol. I), Lyell had thanked his colleague Gideon Mantell (1790–1852) for having introduced him to the Frenchman's writings. In Lamarck's arguments for transmutation, Lyell had seen "an advocate on the wrong side." Nevertheless, he admired Lamarck for his courage and logic, and the latter's writings had delighted him.

---

[38] See Sect. 1.10.

[39] See QUOT. 48.

However, Lyell's lukewarm tone toward Lamarck changed in the *Principles of Geology* (Lyell, 1830). While he mentioned Lamarck only once in the first volume, Lyell mounted a full-frontal attack against the Frenchman in the second volume (Lyell, 1832). Lyell's attacks were not surprising, though, since he strongly believed in the fixity of species at that time.[40] Lyell characterized the major components of Lamarck's theory as fictitious:

> **QUOT. 55** ...when Lamarck talks 'of the efforts of internal sentiment,' 'the influence of subtle fluids,' and the 'acts of organization,' as causes whereby animals and plants may acquire new organs, he gives us names for things, and with a disregard to the strict rules of induction, resorts to fictions, as ideal as the 'plastic virtue,' and other phantoms of the middle ages. (*ibid.*, p. 8)

Thereupon, Lyell essentially reiterated Cuvier's previous criticism, stating that this aspect (i.e., the ability of one's "needs" to effect change) of Lamarck's theory was "absurd and staggering."

C. Darwin was first introduced to Lamarck's ideas in 1827 through Robert Grant (1793–1874), a Lamarckian.[41] By his own admission, Darwin was left indifferent by Grant's praise for Lamarck:

> **QUOT. 56** ...I knew him [Grant] well; he was dry and formal in manner, but with much enthusiasm beneath this outer crust. He one day, when we were walking together burst forth in high admiration of Lamarck and his views on evolution. I listened in silent astonishment, and as far as I can judge, without any effect on my mind. (Barlow, 1958, p. 49)

Later in 1844, however, Darwin was far from indifferent in a Jan 11 letter to Joseph Hooker (1817–1911), when he said: "Heaven defend me from Lamarck's nonsense" (Darwin, 1888a, p. 23).

To say that Darwin was indifferent to Lamarck or that his beliefs were completely different from those of Lamarck (or E. Darwin) would be inaccurate. Darwin very much believed that the environment influenced evolution (but not to the extent that one's inner feelings of need would lead to newly acquired characteristics). Darwin's 1868 provisional hypothesis of pangenesis (Darwin, 1868b, Vol. II, Chap. XXVII) was very Lamarckian in spirit. According to this hypothesis, cells threw off granules (called "gemmules"), which were representative of organs and were then collected in the germ cells. Furthermore:

> **QUOT. 57** ...in the cases in which the organisation has been modified by changed conditions, the increased use or disuse of parts, or any other cause, the gemmules cast off from the modified units of the body will be themselves modified, and, when sufficiently multiplied, will be developed into new and changed structures. (*ibid.*, p. 397)

---

[40] Lyell changed to transmutationism only in the 1860s, after Darwin's *Origin of Species* (Darwin, 1859) was published. See Sect. 1.14.

[41] The word "Lamarckian" was first used (as an adjective) in 1846 by James Dwight Dana as follows: "These remarks are intended to support no monad or Lamarckian theory, but only to elucidate the established principle that there are in nature certain distinct systems or types of development" (Dana, 1853, p. 107). "Lamarckism" was first used by Ray Lankester in 1884: "Lamarckism looks very well on paper, but...when put to the test of observation and experiment it collapses absolutely" (Lankester, 1884, p. 412).

In the above, Darwin all but embraced (albeit "provisionally") Lamarck's views on the inheritance of acquired characteristics. See also the latter part of Sect. 2.14.4.3.

Darwin's embrace of Lamarckian concepts was one of the major reasons his *Origin of Species* was criticized by several biologists. Indeed, although Lamarckism occasionally gained adherents,[42] several key contributors to evolutionary genetics (e. g., Galton, Weismann, Johannsen) explicitly rejected it. A culmination point was reached with the modern synthesis in the 1930s when Lamarckism was all but dealt a death blow.

However, in 1942, an important event occurred when Conrad Hal Waddington (1905–1975), a British geneticist, published a short paper entitled "The epigenotype" (Waddington, 1942). Waddington's thesis was that the relationship between an individual's genotype and phenotype was mediated by a chain of developmental processes and their interactions, which he called the *epigenotype*:

> **QUOT. 58** We certainly need to remember that between genotype and phenotype, and connecting them to each other, there lies a whole complex of developmental processes. It is convenient to have a name for this complex: 'epigenotype' seems suitable (*ibid.*, p. 19)

Waddington designated the study of the epigenome by the name "epigenetics" (= epigenesis + genetics). Epigenetics[43] is now applied not only to the developmental processes that Waddington had in mind but also to any heritable change in genome function that is not associated with changes in the DNA sequence. For example, histone acetylation and DNA methylation have been shown to be associated with changes in DNA transcription and lead to heritable changes in an individual. Since histone acetylation and DNA methylation are both brought about by environmental changes, epigenetics may appear to be a throwback to Lamarckism. However, this is not exactly the case since epigenetic changes are usually reversible and short-term—unlike the permanent, long-term changes Lamarck had in mind.

---

[42] An infamous example is that of the Austrian biologist Paul Kammerer (1880–1926), who, in the early 1900s, experimented with various amphibians to prove the inheritance of acquired characters. Kammerer's most famous experiments were with the midwife toad. The male midwife toads lack the "nuptial pads" (pigmented patches on the hand), which male toads of other species use to grasp the back of the female while mating in water. Kammerer claimed that, by forcing midwife toads to mate in water for several generations, he was able to obtain male toads with nuptial pads. Kammerer published his findings, which drew the suspicion of the scientific community. In 1926, G. K. Noble, who was Curator of Reptiles at the American Museum of Natural History, visited the Biological Experimental Institute in Vienna and examined the pad in one of the remaining specimens. Shortly after, in a *Nature* article, Noble (1926) declared that the black coloring on the pad was Indian ink. Kammerer committed suicide shortly after.

[43] There is an enormous literature on epigenetics; see, for example, Jablonka & Lamb (1995, 2020), Carey (2012), Ward (2018), van Speybroeck et al. (2002), Cabej (2012), Turner (2001), and Kovalchuk & Kovalchuk (2012).

1.10 Georges Cuvier (1769–1832)

1.10.1 Catastrophism (1792, 1799)

Uniformitarianism[44] posits that the causes that have affected all geological change are *actual or present-day causes*. On the opposite end is the doctrine of *catastrophism*,[45] which holds that such change is due to sudden catastrophic events (such as the biblical flood). The concept was proposed to explain the differences in fossil forms that exist at various stratigraphic levels. Catastrophism postulates that the history of Earth has been marked by a succession of worlds, each separated by sudden catastrophic and supernatural events (such as violent floods). Cataclysmic occurrences account for mass extinctions. The last upheaval is believed to be the biblical (i.e., Noah's) flood. Catastrophism is thus, in essence, creationist and fixist. It is invariably associated with Georges Cuvier[46] (Fig. 1.14). Cuvier was one of the greatest zoologists that ever lived. In 1795, under the recommendation of Geoffroy Saint-Hilaire, he was invited to assist Mertrud, who was Professor of Comparative Anatomy at the *Muséum d'Histoire Naturelle* in Paris. From then on, Cuvier's career took off and he became Professor at the *Jardin des Plantes* in 1802.

Cuvier's first statement of the catastrophist idea seems to have occurred in a March 11, 1792, letter to his younger friend, Christoph Heinrich Pfaff (1773–1852), where he spoke of "several catastrophes" and "revolutions" to explain the existence of different fossils in rock layers (Marchant, Cuvier and Pfaff, 1858, p. 264). The idea was later more clearly put forward in a memoir that was read in 1796 (and published in 1799):

> **QUOT. 59** Let it be asked why one finds so many remains of unknown animals, whilst one finds almost none that can be said to belong to known species, and one finds how much probable it is that they belonged to a world prior to ours, to beings destroyed by some revolutions [catastrophes]; beings from which those that exist today have filled up the place, to be perhaps also destroyed and replaced by others. (Cuvier, 1799, p. 21)

In 1812, Cuvier published his *Recherches sur les Ossemens Fossiles de Quadrupèdes* (Cuvier, 1812) in four volumes. The first volume was introduced by a 120-page *Discours Préliminaire*, where Cuvier's catastrophist theory was further expounded. In 1825, Cuvier published his famous *Discours sur les Révolutions de la Surface du Globe* (Cuvier, 1821), which was the third edition of the *Discours Préliminaire* and where the definite statement of Cuvier's catastrophism may be found.

Cuvier believed that, after each cataclysm, new life forms occurred due to *migration* from other regions. However, some of his disciples (such as Alcide

---

[44] See Sects. 1.6 and 1.14.

[45] See footnote 16.

[46] Georges Cuvier's life and work are also discussed in Coleman (1964), Outram (1984), Appel (1987), Daudin (1926), Pichot (2011, pp. 471–485), Blainville (1890), Gaudant & Gaudant (1971, Chap. 8), and Adams (1969, Chap. 6).

Fig. 1.14 Jean Léopold Nicolas Frédéric, Baron Cuvier (1769–1832). (From: https://commons.wikimedia.org/wiki/Georges_Cuvier#/media/File:Georges_Cuvier_large.jpg)

d'Orbigny, Adam Sedgwick, William Buckland, and William Whewell) instead argued that each catastrophe was accompanied by the *creation* of new and more developed life forms, the latter being a function of the changed conditions of the world. Eventually, mankind was created. The doctrine of successive creations of better life forms is called *progressionism*.[47]

1.10.2 Fixism (1792)

Regarding Cuvier's fixist beliefs, a first insight may be gleaned from the March 11, 1792, letter to Pfaff:

> **QUOT. 60** I believe, I see that aquatic animals have been created for water and others for air. But that they are the branches or roots, or at least the parts of the same stem, is what I cannot understand. (Marchant et al., 1858, p. 270)

---

[47] First used by H. Spencer as follows: "Sir R. Murchison, who is a Progressionist, calls the lowest fossiliferous strata, 'Protozoic'" (Spencer, 1859, p. 81, Vol. II).

Despite Cuvier's early fixist beliefs, it is noteworthy that just a few years later, he coauthored articles with Geoffroy St Hilaire where views favorable to the transmutation of species were expressed (Geoffroy Saint-Hilaire and Cuvier, 1795a, b). However, such views were almost certainly due to St Hilaire. Indeed, in his 1799 paper, Cuvier expressed his fixist beliefs even more categorically:

> **QUOT. 61** Whatever the influence of the climate in varying animals, it surely does go that far: and to say that it can change all the proportions of the bone structure and intimate construction of the teeth, it is to put forward that all quadrupeds can be derived from a single species; that the differences they present are only successive degenerations: in a word, it is to reduce entire natural history to nothing, since its object will consist only of variable forms and transient types. (Cuvier, 1799, p. 12)

The authority and influence of Cuvier ensured that his catastrophist and fixist beliefs won approval beyond French circles. A good portion of that approval was due to Cuvier's perceived support of the Church through his "scientific" interpretation of Noah's flood.

A strong adept of Cuvier was the renowned Swiss-born geologist Jean Louis Rodolphe Agassiz (1807–1873), who moved to the United States in 1847. Agassiz was a believer in the fixity of species, and his anti-evolutionary views helped galvanize resistance to Darwinism (i.e., evolution by natural selection)[48] for some time in the United States. Similarly, the British Richard Owen (1804–1892), Adam Sedgwick (1785–1873), and William Whewell (1794–1866), all disciples of Cuvier, voiced strong opposition to Darwinism.

1.10.3 Empiricism and Positivism (1788)

Cuvier's fixist beliefs invariably put him at odds with Lamarck's evolutionary theory.[49] However, there is a deeper reason why Cuvier opposed Lamarck. It is that Cuvier was an absolute empiricist and was guided only by observed facts. On the other hand, it cannot be denied that Lamarck's views on transmutation contained several elements of speculation. As early as 1788, Cuvier expressed his philosophy towards science as follows:

> **QUOT. 62** There is a necessary thing for each science, it is that everything should be studied in depth. I would like everything that experience shows us be carefully separated from any hypothesis. (Marchant et al., 1858, p. 73)

Cuvier's disdain of speculative hypotheses is an echo of Newton's famous *hypotheses non fingo*. The view that science should rely on only what is positively observed would soon be the backbone of Auguste Comte's influential *positivist* doctrine.

---

[48] Natural selection was thus the primary component of Darwin's theory; evolution was only a secondary component.

[49] See Sect. 1.9.2.

1.11 Geoffroy Saint-Hilaire (1772–1844)

1.11.1 Unity of Plan (1796)

Any discussion of Lamarck and Cuvier should also mention Etienne Geoffroy Saint-Hilaire[50] (Fig. 1.15). Saint-Hilaire believed in the unity of structure of all animals and has been called "the father of homology." The unity of plan was strongly suggestive of an evolutionary relationship between different life forms. As early as 1795, when Georges Cuvier was staying with him after the latter had just moved to Paris, Saint-Hilaire expressed the idea of descent from a common type (in a paper coauthored with Cuvier!):

> **QUOT. 63** Now, it is a sufficiently general law of nature, that the number of species of a genus is more or less due to the fertility of each one; let it be that what we call species, are only various degenerations of the same type... (Geoffroy Saint-Hilaire and Cuvier, 1795a, p. 452)

The term "degenerations" above was used previously by his mentor Buffon (Sect. 1.5). The notion of a common structure ("*plan unique*")[51] was explained the next year, as well as the idea of the effect of use/disuse on development:

> **QUOT. 64** ...it seems that nature has restricted itself to certain limits, and has formed all living beings according to a unique plan, essentially the same in its principle, but which she has varied in thousands of ways in all its accessory parts...it is enough for her to change some of the proportions of organs to make them appropriate for new functions, or to extend or restrict their use. (Geoffroy Saint-Hilaire, 1796, p. 20)

In the above, Saint-Hilaire, in effect, suggested an evolutionary process involving the primacy of a unique body plan from which other life forms evolved, determined by functionality and use. This principle of evolutionary change would soon be developed in more detail by his colleague Lamarck.

Saint-Hilaire's emphasis on structure over function would soon also put him at odds with Cuvier, for whom function determined structure.

---

[50] Geoffroy Saint-Hilaire's life and work are also discussed in Le Guyader (2004), Appel (1987), Pichot (2011, pp. 485–525), Gaudant & Gaudant (1971, Chap. 9), Cahn (1962), Nordenskiöld (1928, pp. 296–298, 341–343), and Blainville (1890).

[51] Saint-Hilaire's emphasis on analogies and unities of plan makes him the French representative of *Naturphilosophie* (Philosophy of Nature). This philosophical and biological trend, developed in Germany, was characterized by a rejection of a purely mechanistic view of the world and the adoption of a romantic viewpoint whereby the material and spiritual worlds were harmonized. The adherents to this philosophy included Johann Wolfgang von Goethe (1749–1832), Friedrich von Schelling (1775–1854), and Lorenz Oken (1779–1851).

Fig. 1.15 Etienne Geoffroy Saint-Hilaire (1772–1844). (From: https://commons.wikimedia.org/wiki/File:Geoffroy_Saint-Hilaire_%C3%89tienne_1772-1844.jpg)

1.11.2 Start of Controversy with Cuvier (1820)

Cuvier expressed his disapproval of Saint-Hilaire's views but, to the latter's irritation, only privately. However, in a series of papers in the 1820s, Saint-Hilaire pushed his views on analogy to the limit. In the first of his "Mémoires sur l'Organisations des Insectes" (Geoffroy Saint-Hilaire, 1820b, p. 345), he claimed that "insects are vertebrate animals" since their exoskeletons were, in fact, the vertebrae! Indeed, according to Saint-Hilaire, Cuvier's class of *Articulata*[52] (e.g., insects, crustaceans) should be considered vertebrates and only his *Mollusca* (e.g., cephalopods, barnacles), and *Radiata* (e.g., echinoderms, jellyfish) should be classified as invertebrates. Saint-Hilaire was aware of the controversial nature of his views, and his second memoir on insects sounded more defensive than expository (Geoffroy Saint-Hilaire, 1820a). This memoir also reported an incident where Cuvier allegedly bashed Saint-Hilaire's first memoir:

[52]This was one of Cuvier's four branches ("*embranchements*"), each with a distinct plan of structure, in his classification of the animal kingdom.

QUOT. 65 'Your memoir on the skeleton of insects lacks logic from start to finish,' M. Cuvier said, expressing himself on my first work, with extreme vivacity, before the professors of the Muséum. 'You compare things which are totally incapable of comparison. Nothing common, absolutely nothing between insects and vertebrates: at most a single point, animality.' (*ibid.*, p. 34n)

However, Cuvier initially refused to attack Saint-Hilaire in print, despite the latter beseeching him to do so in the third memoir on insects (Geoffroy Saint-Hilaire, 1820c, p. 168).

1.11.3 Transformism (1825)

It probably has not escaped the reader that transformism had hitherto been only a necessary by-product of Saint-Hilaire's theory of the unity of plan (although he hinted at it in 1795—see Sect. 1.11.1). It was not until 1825 that Saint-Hilaire stated his allegiance to transformism in explicit terms. A more definite statement of such views would later occur in 1833 (Geoffroy Saint-Hilaire, 1833).

To better understand these events, it is necessary to go back to 1817, when Eudes-Delongchamps, a young medical student, found bone fragments of what seemed to be a crocodile in Caen. The fossils were eventually sent to Cuvier, who hastily classified the skull as that of an extinct species of gavial. This was directly contradicted by Saint-Hilaire, who instead thought that the skull had mammalian features. Saint-Hilaire went even further and interpreted the fossils in evolutionary terms:

QUOT. 66...by pushing the consequences of these undeniable facts, I may propose that it is not contrary to reason, that is to physiological principles, the Crocodiles of the present times could have descended through an uninterrupted succession from antediluvian species, found today in fossil form on our territory. (Geoffroy Saint-Hilaire, 1825, pp. 152–153)

He stated that transformations occurred according to Lamarck's theory, whereby changes in the environment had a transmissible effect on the organs of these animals. However, in contrast to Lamarck's extremely slow, gradual evolution, Saint-Hilaire's was sudden (or saltationist). Thus, while explaining the transformation of birds:

QUOT. 67 Obviously, it was not by an imperceptible change that the lower types of oviparous animals gave rise to the higher degree of organization, or the group of birds. All that was required was a possible accident which was not very considerable in the original production, but of incalculable importance as to its effects... (Geoffroy Saint-Hilaire, 1831, p. 80)

Saint-Hilaire's 1825 paper on the "crocodiles" of Caen dealt a double blow to Cuvier: first to his expertise in paleontology, second to his fixism. Cuvier was now forced to reply through an article entitled "Nature" for the *Dictionnaires des Sciences Naturelles* (Blainville, 1825), where he officially condemned Saint-Hilaire's views:

QUOT. 68 . . .to the hypothesis of the continuous ladder of beings, other philosophers have added the one, that all beings are modifications of a single one, or that they have been produced successively, and through the development of a first germ, and it is on the latter that is based that of a unity of composition in all. . .by seizing some partial similarities, while overlooking differences, it sees in the worm the embryo of a vertebrate; in the warm-blooded vertebrate, the embryo of the cold-blooded animal. . . (*ibid.*, p. 266)

In October 1829, the full battle lines between Cuvier and Saint-Hilaire were drawn in the *Académie des Sciences*, when the young naturalists Meyranx and Laurencet presented a report entitled "Quelques considérations sur l'organisation des mollusques"[53] likening the cephalopod to a vertebrate bent back at the umbilicus. Saint-Hilaire (together with Latreille) welcomed the article since it corresponded to his own views. Thereafter, a back-and-forth between Cuvier and Saint-Hilaire ensued in the *Académie* and gradually reached both the media and the public at large. The details of the deliberation may be found in the book by Appel (1987), but it would not be inaccurate to say that Cuvier got the better of Saint-Hilaire, especially in his 1830 article (Cuvier, 1830).

1.12 Malthus' Essay on the Principle of Population (1798): The Struggle for Existence

Darwin was considerably affected by the political economist Thomas Robert Malthus[54] (Fig. 1.16).[55] In the anonymous (but highly influential) *Essay on the Principle of Population* (Malthus, 1798; Appleman, 1976), Malthus stated that "the power of the population is indefinitely greater than the power of the earth to produce subsistence for man," because:

QUOT. 69 [p]opulation, when unchecked, increases in a geometrical ratio. Subsistence increases only in an arithmetical ratio. (*ibid.*, p. 20)

Unchecked populations thus increased at an exponential rate, whilst food production increased only at an arithmetical rate. To cope with this great inequality, Malthus concluded that internal mechanisms (such as epidemics and famine) operated to keep populations in check. Moreover, these internal mechanisms usually affected the poor, who made up the majority of the population.

Further along the *Essay*, while describing how members of a hypothetical tribe would search for new food resources, Malthus made use of a most interesting phrase when he said:

[53]This report is no longer extant.

[54]Malthus and his connection to Darwin are also discussed in Herbert (1971), Bowler (1976), Manier (1978, pp. 74–85), Ospovat (1979), Remoff (2016), Vorzimmer (1969), Schwartz (1974), Levin (1966), and Flew (1997, pp. 74–83).

[55]Malthus' influence on Darwin was second only to Lyell's. See further in the text.

Fig. 1.16 Thomas Robert Malthus (1766–1834). (From: https://commons.wikimedia.org/wiki/Thomas_Malthus#/media/File:Thomas_Robert_Malthus.jpg)

QUOT. 70 And when they fell in with any tribes like their own, the contest was a *struggle for existence,*[56] and they fought with a desperate courage, inspired by the reflection that death was the punishment of defeat and life the prize of victory. (*ibid*., p. 29)

The "struggle for existence" was a key concept that Darwin later admitted was key to building his theory:

QUOT. 71 In October 1838, that is, fifteen months after I had begun my systematic enquiry, I happened to read for amusement Malthus on *Population*, and being well prepared to appreciate the struggle for existence which everywhere goes on from long-continued observation of the habits of animals and plants, it at once struck me that under these circumstances favourable variations would tend to be preserved, and unfavourable ones to be destroyed. The result of this would be the formation of new species. Here, then, I had at last got a theory by which to work. (Barlow, 1958, p. 120)

Darwin later argued for evolution through natural selection. The "struggle for existence" provided a rationale for natural selection that was both fundamental and intuitive.

There has been considerable debate on the exact influence of Malthus on Darwin. How decisive was Malthus' influence, given that Darwin had already referred to

[56] Italics are our own.

some kind of struggle for existence in his notebooks written before October 1838? For example, in the B Notebook, which was probably written between July 1937 and March 1938, one can read:

> **QUOT. 72** With respect to extinction we can easily see that variety of ostrich, Petise may not be well adapted, & thus perish out, or on other hand like Orpheus, being favourable many might be produced. (Barrett et al., 1987, B: 37–38)

Eiseley (1958, p. 180) has thus contended that Darwin did not receive his complete inspiration for the selective nature of the struggle for existence from Malthus since both William Paley and Charles Lyell used the idea. For example, in Lyell's *Principles of Geology*, one can read:

> **QUOT. 73** ...In the universal struggle for existence, the right of the strongest eventually prevails; and the strength and durability of a race depends mainly on its prolificness. (Lyell, 1832, p. 56)

De Beer (1967, p. 99) conceded that Darwin was aware of the principle of struggle for existence before reading Malthus. However, only when he read the latter did Darwin realize the universality of this principle and the inexorable pressure it exerted on populations.

Darwin first referred to Malthus on September 28, 1838 (Barrett et al., 1987, D: 134). Therefore, his October 1838 date (mentioned in QUOT. 71) is not quite accurate. The September 28, 1838, date for his actual reading of Malthus would also explain the statement he made at the start of the D Notebook (probably written between July 15, 1838, and Oct 2, 1839): "Towards close [of the D Notebook] I first thought of selection owing to struggle." Most likely, this is when Darwin appreciated the universality of the struggle for existence.

1.13 William Paley: Natural Theology and Teleology (1802)

In addition to Malthus, one other person who greatly influenced Darwin was the theologian William Paley[57] (Fig. 1.17), whose name was mentioned a few sentences back. Not only did both Malthus and Paley greatly influence Darwin, but it is also the case that both men influenced each other. Malthus' *Essay* contained several points that had previously been raised by Paley in his *Principles of Political and Moral Philosophy* (Paley, 1835). Likewise, after reading the *Essay on the Principle of Population* (Malthus, 1798), Paley became convinced of Malthus' argument on arithmetical and geometrical ratios.

[57]Paley and his connection to Darwin are further discussed in Nuovo (1992), Manier (1978, pp. 69–75), Phipps (2002, pp. 6–9), Ayala (2004, 2012), Gould (1990), Topham (2010), and Brunnander (2013).

Fig. 1.17 William Paley (1743–1805). (From: https://commons.wikimedia.org/w/index.php?search=william+paley&title=Special%3ASearch&go=Go#/media/File:William-Paley.png)

In his most important book, *Natural Theology* (Paley, 1802), Paley made a forceful argument for the existence of God from design. Using the watch as an example, Paley's theological[58] argument started as follows:

> **QUOT. 74** . . .When we come to inspect the watch, we perceive. . .several parts are framed and put together for a purpose, e.g. that they are so formed and adjusted as to produce motion, and that motion so regulated as to point out the hour of the day. . .This mechanism being observed. . .the inference, we think, is inevitable; that the watch must have had a maker (*ibid.*, pp. 7–8)

Given that the universe abounded with natural entities (from the structure of the eye to that of the universe itself) whose complexity surpassed that of a watch, Paley then argued that the entire universe and all natural entities associated with it must also have a "maker," namely, the Creator. Moreover, all the parts are combined for a *final purpose*. The latter is the essence of the Aristotelian *causa finalis* or *teleological argument.*[59]

In *Natural Theology*, Paley made no explicit statements about evolution. However, it is reasonable to assume that his belief in special creation would be

[58]The word "theologia" was first used c. 1270.

[59]The word "teleologia" was first used by Christian van Wolff in 1728 (Wolff, 1728, Chap. III, Sect. 85).

incompatible with any allowance for descent with modification. In several instances, Paley referred indirectly to Erasmus Darwin, only to disapprove of the latter's views on acquired characteristics.

Paley's *Natural Theology* was strongly influential. Charles Darwin had read Paley since his university days and had at first been convinced by the latter's teleological argument:

QUOT. 75 In order to pass the B.A. examination, it was, also, necessary to get up Paley's *Evidences of Christianity*, and his *Moral Philosophy*. This was done in a thorough manner, and I am convinced that I could have written out the whole of the Evidences with perfect correctness, but not of course in the clear language of Paley. The logic of this book and as I may add of his Natural Theology gave me as much delight as did Euclid. (Barlow, 1958, p. 59)

Later, when he proposed the theory of evolution by natural selection, Darwin saw no need to invoke a special Creator to explain the life forms around him:

QUOT. 76 The old argument of design in nature, as given by Paley, which formerly seemed to me so conclusive, fails, now that the law of natural selection has been discovered. We can no longer argue that, for instance, the beautiful hinge of a bivalve shell must have been made by an intelligent being, like the hinge of a door by man. There seems to be no more design in the variability of organic beings and in the action of natural selection, than in the course which the wind blows. Everything in nature is the result of fixed laws. (*ibid.*, p. 87)

The above certainly does not imply that Darwin became an atheist after developing his theory. This only means that Darwin no longer thought that a Creator *directly* affected the life forms around us.

Malthus' pessimistic undertones (as explained before) should be contrasted with Paley's optimism. Paley's view was that of a happy and harmonious world (e.g., Paley, 1802, p. 238), where every structure was adapted to fulfill a particular purpose, thanks to the Creator's work. For example,

QUOT. 77 The arm is an instrument of motion, principally, if not solely. Accordingly the shallowness of the socket at the shoulder, and the yieldingness of the cartilaginous substance with which its edge is set round, and which in fact composes a considerable part of its concavity, are excellently *adapted*[60] for the allowance of a freer motion and a wider range; both which the arm wants. (*ibid.*, p. 65)

It is a reflection of Darwin's genius that he was able to successfully synthesize Malthus' pessimism and Paley's optimism into one meaningful theory. He used Malthus' struggle for existence as a rationale for his theory of natural selection and Paley's doctrine of harmony to explain the resulting adaptation that occurs because of the struggle.

Two final comments need to be made regarding William Paley's *Natural Theology* and the concept of teleology.

(a) *Natural Theology*. Paley's work undoubtedly influenced Frank Henry Egerton (1756–1829), the 8th Earl of Bridgewater, to bequeath £8000 for the publication

[60] Italics are ours.

of eight books (the so-called *Bridgewater Treatises*) "On the Power, Wisdom, and Goodness of God as manifested in the Creation". The eight volumes were subsequently published between 1833 and 1836, each by a different author, but all from a theological perspective. William Whewell authored the third volume (Whewell, 1833), while William Buckland authored the sixth volume (Buckland, 1836).

(b) *On teleology*. Although the concept of teleology has been criticized by scientists, it is difficult to claim that the development of several organs is devoid of a final purpose. Thus, the eyes of vertebrates and invertebrates may develop by different mechanisms, but they all have one purpose, namely, the ability to see. The property of organisms to develop structures for the purpose of performing a particular *function* is called *teleonomy* (Pittendrigh, 1958). In teleonomic explanations, the final purpose is not decided by a purposive Creator but is dictated by a particular function that increases the adaptation of the organism.

1.14 Charles Lyell (1797–1875)

1.14.1 Influence on Darwin

Of all the figures hitherto mentioned, the one who exerted the most influence on Darwin's *Origin of Species* was undoubtedly the Scottish geologist Sir Charles Lyell[61] (Fig. 1.18). By Darwin's own admission:

> **QUOT. 78** I always feel as if my books came out of Lyell's brain, and that I never acknowledge this sufficiently. (Darwin and Seward, 1903a, p. 117, Letter to L. Horner, Letter 480)

Indeed, during the *Beagle* voyage, Darwin had brought the first volume of Lyell's *Principles of Geology* (Lyell, 1830), which he "studied attentively" and found of the "highest service" (Barlow, 1958, p. 77). He received the second volume of the book in November 1832, when the Beagle stopped at Montevideo in South America, and the third volume in April 1834. When the voyage was over, Darwin met Lyell on Oct 29, 1839, at the latter's house, and they became good friends.

1.14.2 Uniformitarianism (1830)

Lyell's three-volume opus, where he argued for uniformitarianism, propelled him to the forefront of the geological field. The uniformitarian doctrine posits that natural processes have been operating throughout geological time with essentially the same

[61] Charles Lyell's life and work are also discussed in Bartholomew (1973), Bailey (1962), Blundell & Scott (1998), Adams (1969, Chap. 9), Barber (1980, pp. 219–224), and Wilson (1972).

Fig. 1.18 Sir Charles Lyell (1797–1875). (From: https://upload.wikimedia.org/wikipedia/commons/f/f6/SS-lyell.jpg)

intensity. Thus, the causes that have affected all geological changes are *actual (or present-day) causes*. The theory was delineated, especially in the first volume of the *Principles of Geology*. The subtitle of the book spelt out the task of uniformitarianism: *Being an Attempt to Explain the Former Changes of the Earth's Surface by Reference to Causes Now in Operation*. A consequence of this theory was that geological forces operated slowly over long periods, leading to gradual changes. The latter offshoot of uniformitarianism is known as *gradualism*.

As mentioned in Sect. 1.6, uniformitarianism had previously been most clearly stated by James Hutton (1726–1797). Lyell had first learned of Hutton's work in 1816 when he read Robert Bakewell's *Introduction to Geology* (Bakewell, 1815). In 1823, Lyell met the French geologist Constant Prévost (1787–1856) in Paris and was considerably influenced by the latter's uniformitarian views. Finally, and perhaps oddly, Lyell also seems to have received much inspiration from Jean-Baptiste

Lamarck (1744–1829). On the same March 2, 1827, letter[62] (Lyell, 1881) where Lyell had qualified Lamarck's transmutationism as views from "an advocate on the wrong side", Lyell had also said:

> **QUOT. 79** That the earth is quite as old as he [Lamarck] supposes, has long been my creed, and I will try before six months are over to convert the readers of the Quarterly to that heterodox opinion...I am going to write in confirmation of ancient causes having been the same as modern, and to show that those plants and animals which we know are becoming preserved now, are the same as were formerly. (*ibid.*, pp. 168–169)

The competency with which Lyell later made the case for uniformitarianism in his book had the inevitable consequence that Cuvier's opposite doctrine of catastrophism[63] was dealt a serious blow. For Darwin, Lyell's uniformitarianism implied that a vast geological time was available, during which small variations could accumulate because of environmental changes, resulting in the transmutation of species. Lyell's geological gradualism was turned into biological evolution by Darwin.

1.14.3 Rejection of Evolution (1830)

The second volume of the *Principles of Geology* (Lyell, 1832) was specifically devoted to the question of species. The third volume dealt with applications. Although the three volumes contained many of the elements of a viable theory of evolutionary change, Lyell himself believed in the fixity of species at the time of his writing. A famous sentence in the second volume, one that would be reiterated later by Fleeming Jenkin,[64] reads as follows:

> **QUOT. 80** ...there are fixed limits beyond which the descendants from common parents can never deviate from a certain type. (*ibid.*, p. 23)

Lyell's belief stemmed from his uniformitarianism, which was antithetical to both catastrophism and the progressionism associated with it. Thus, as described before, Lyell castigated Lamarck's theory of evolution in the second volume of his book.

To make his claim against the transmutation of species, Lyell used numerous examples in his book. In one instance, he noted the discovery of mammalian remains in the Stonesfield slate, near Oxford. This led him to conclude:

> **QUOT. 81** The occurrence of one individual of the higher classes of mammalia, whether marine or terrestrial, in these ancient strata, is as fatal to the theory of successive development, as if several hundreds had been discovered. (Lyell, 1830, p. 150)

---

[62] See Sect. 1.9.2.

[63] See Sect. 1.10.

[64] See Sect. 2.14.4.2.

Lyell's religious beliefs implied that man was created in God's image and precluded that he could have descended from a less civilized life form. Man's uniqueness rested on his mental and moral powers. Such a stance was very much in accord with prevalent scientific opinion in Victorian Britain.

1.14.4 Later Conversion (1863)

Lyell first fully learned of Darwin's theory of evolution by natural selection on April 16, 1856, when Darwin invited him to his residence at Down House.[65] Lyell knew since the mid-1840s that Darwin was an evolutionist but had also become recently aware of Alfred Wallace's Sarawak Law paper of 1855 (Wallace, 1855), which clearly showed the latter's thought along evolutionary lines. During his visit to Darwin, he informed the latter of Wallace's paper, and once he was exposed to Darwin's theory, he soon wrote to the latter urging him to publish it lest somebody else would do it first. This was despite Lyell's religious views and the realization that Darwin's doctrines were at odds with such views. Lyell was also the one who, together with Hooker, presented the joint paper by Darwin and Wallace on July 1, 1858, to the Linnean Society of London,[66] where the theory of evolution by natural selection was first presented to the world.

Lyell's conversion to descent with modification was slow and qualified. It was first publicly expressed in the book *The Geological Evidences of the Antiquity of Man* (Lyell, 1863), four years after Darwin's *Origin*. Lyell was converted on two fronts. First, he now expressed his belief in fossil man[67] (fossil remains of past man-like creatures). Second, in the last four chapters of his book, he declared his allegiance to Darwin's ideas. As Lyell explained, his conversion was motivated by the ability of Darwin's theory to explain the following phenomena:

(a) Morphological: "the unity of type which runs through the whole organic world, and why there is sometimes a fundamental agreement in structure in the same class of beings which is quite independent of their habits of life." (*ibid*., p. 413)
(b) Physiological: "rudimentary organs are so useful in classification, being the remnants preserved by inheritance of organs which the present species once used—as in the case of the rudiments of eyes in insects and reptiles inhabiting dark caverns, or of the wings of birds and beetles which have lost all power of flight." (*ibidem*)
(c) Biogeographical:

[65]The first person to whom Darwin communicated his new transmutationist viewpoint was his friend Joseph Hooker on Jan 11, 1844.

[66]See Sect. 2.13.4.

[67]The possibility was rejected by Cuvier and by himself until 1853.

(i) "distinct geographical provinces of species of animals and plants, for, after long isolation by physical barriers, each fauna and flora, by varying continually, must become distinct from its ancestral type, and from the new forms assumed by other descendants which have diverged from the same stock." (*ibidem*)

(ii) "no mammalia in islands far from continents, except bats, which can reach them by flying; and also why the birds, insects, plants, and other inhabitants of islands, even when specifically unlike, usually agree generically with those of the nearest continent, it being assumed that the original stock of such species came by migration from the nearest land." (*ibid.*, p. 414)

Nevertheless, Lyell had important reservations about certain aspects of Darwin's theory. He refused to admit that natural selection could explain the whole of evolution:

> **QUOT. 82** In our attempts to account for the origin of species, we find ourselves still sooner brought face to face with the working of a law of development of so high an order as to stand nearly in the same relation as the Deity himself . . . If we confound "Variation" or "Natural Selection" with such creational laws, we deify secondary causes or immeasurably exaggerate their influence.
>
> Yet we ought by no means to undervalue the importance of the step which will have been made, should it ever become highly probable that the past changes of the organic world have been brought about by the subordinate agency of such causes as "Variation" and "Natural Selection." (*ibid.*, p. 469)

While Darwin was grateful that Lyell had retrenched on the immutability of species, he was disappointed that Lyell's new stance seemed to be only half-hearted (especially when Lyell used the phrase "should it ever become highly probable").[68]

For Lyell, evolution had a spiritual component that could not be explained by natural selection alone. He believed in the uniqueness of man, especially concerning intellectual abilities. This was characteristically a vitalistic[69] approach whereby life was endowed with a special force that could not be explained by physicochemical properties. Lyell's vitalism was earlier stated by Alfred Wallace but stood in sharp contrast to Darwin and Huxley's materialist views of evolution. In Lyell's own words:

> **QUOT. 83** So far from having a materialistic tendency, the supposed introduction into the earth at successive geological periods of life – sensation –instinct – the intelligence of the higher mammalia bordering on reason – and, lastly the improvable reason of Man himself, presents us with a picture of the ever-increasing dominion of mind over matter. (Lyell, 1863, p. 394)

Lyell's adoption of Darwinian ideas was also described in the revised, 10th edition of the *Principles of Geology*. The "fixed limits beyond which the descendants from

[68] See *The Life and Letters of Charles Darwin* (Darwin, 1888b, Vol. III) for Darwin's Feb 24, 1863, letter (p. 8) to Hooker and his Mar 6, 1863, letter (p. 11) to Lyell on this topic.

[69] But now rejected by most scientists.

common parents can never deviate from a certain type" in QUOT. 80 was then dropped.

A final note needs to be made regarding the earlier statement in this section that Lyell's convincing exposition of uniformitarianism dealt a serious blow to Cuvier's catastrophism. While it is true that evolutionists and geologists subsequently rallied on Lyell's side for the larger part of the twentieth century, an important event occurred in 1980. The geologist Walter Alvarez (b. 1940) was studying the boundary between the Cretaceous and Tertiary periods (the so-called K-T boundary). He enlisted the help of his father, physicist Luis Alvarez (1911–1988), to analyze a clay layer that had formed at the K-T boundary. The clay showed elevated levels of iridium, and this finding was replicated for samples of the K-T boundary from around the world. The Alvarezes postulated that the iridium must have been carried by an asteroid or comet that crashed into Earth approximately 65 million years ago, creating a dust cloud lasting for several years. This led to several mass extinctions, including those of dinosaurs. The Alvarez hypothesis gained acceptance by most scientists and marked a return to catastrophism (without, however, the supernatural elements in Cuvier's hypothesis).

1.15 "Just Before Darwin": Robert Chambers (1802–1871)

1.15.1 Chambers' Vestiges of the Natural History of Creation (1844)

In the period "just before Darwin",[70] namely, 1844, a book by the suggestive title *Vestiges of the Natural History of Creation* (Chambers, 1844) was published. Its author was Robert Chambers (Fig. 1.19), who had wished to remain anonymous while alive[71] for reasons that would soon become apparent. Chambers was by no means a biologist, although he was the editor of the Chambers Journal and had authored several books on history, literature, and biography.

In the *Vestiges*, Chambers pictured a universe governed by natural laws that were under divine command. He propounded a law of progressive development that was not directly regulated by God and that may be summarized as follows:

> **QUOT. 84** The idea, then, which I form of the progress of organic life upon the globe – and the hypothesis is applicable to all similar theatres of vital being – is, *that the simplest and most primitive type, under a law to which that of like-production is subordinate, gave birth to the type next above it, that this again produced the next higher, and so on to the very*

[70]This is the title Millhauser (1959) has given to his book on the subject matter that will be described next.

[71]The name was revealed only in the 12th edition of the book, in 1884, thirteen years after the author's death.

Fig. 1.19 Robert Chambers (1802–1871). (From: https://commons.wikimedia.org/wiki/File:Robert_Chambers,_publisher,_ca1863.jpg)

> *highest,*[72] the stages of advance being in all cases very small – namely, from one species only to another; so that the phenomenon has always been of a simple and modest character. (*ibid.*, p. 222)

Chambers thus believed that the simplest form of life gave birth to the next form above it. The order was based on the stages of embryological development. When the latter process was extended, a new species was formed.

Chambers illustrated the process through a diagram (Fig. 1.20). Suppose that the embryos of four classes of vertebrates (fish, reptiles, birds, and mammals) advance to an identical stage of development at A. The fish embryo usually diverges at A, giving rise to a mature fish at F. Similarly, the reptile embryo diverges at C, while the mammal goes on the highest point M. However, "if the fish embryo should not diverge at A, but go on to C before it diverges, in which case the progeny will be, not a fish, but a reptile" (*ibid.*, p. 213). According to Chambers, it was thus possible for a fish embryo to develop into a human being!

What could prompt the type of changes that Chambers advocated? Chambers' answer was Lamarckian in spirit:

---

[72]Italics are Chambers'.

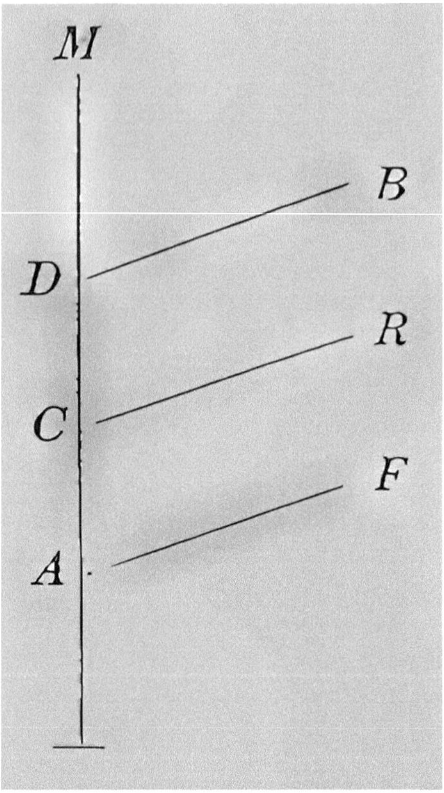

Fig. 1.20 Chambers' diagram of embryological divergence (Chambers, 1844, p. 212)

QUOT. 85 This might be done by the force of certain external conditions operating upon the parturient system. (*ibidem*)

However, he admitted that "the nature of these external conditions we can only conjecture" (*ibidem*).

1.15.2 Reception of Chambers' Work

As Chambers had rightly foreseen (to the point of withholding his name from the book), the *Vestiges* received heavy criticism from all sides. The general scientific opinion in Victorian Britain was fiercely opposed to Chambers' book. Among the critics was the progressionist Adam Sedgwick, who launched a virulent attack on the scientific basis of Chambers' work:

QUOT. 86 The author is intensely hypothetical, and builds his castles in the air, misconceiving the principles of science, or misunderstanding the facts with which it has to deal; or, what is worse still, distorting them to serve his purpose. He does all this, apparently, without having any just conception of the methods by which men, after the toil of many generations, have ascended, step by step, to the higher elevations of physical knowledge—without any even glimmering conception of what men mean when they tell us of Inductive Science and its sober truths. (Sedgwick, 1845, p. 2)

Similar criticisms were added in the fifth edition of Sedgwick's *Discourse on the Studies of the University of Cambridge* (Sedgwick, 1850).

Chambers' model of evolution (Fig. 1.20), especially regarding the implications for man, was scoffed by many. This is how Benjamin Disraeli (1804–1881), an influential British politician, ridiculed it:

QUOT. 87 ...what is most interesting is the way in which man has been developed. You know, all is development. The principle is perpetually going on. First, there was nothing, then there was something; then, I forget the next, I think we were shells, then fishes; then we come, let me see, did we come next? Never mind that; we came at last. And the next change there will be something very superior to us, something with wings. Ah! That's it; we were fishes, and I believe we shall be crows. (Disraeli, 1847, p. 124)

Huxley was not more favorable than Sedgwick in his opinion of the *Vestiges*. In his review of the tenth edition, Huxley accused it of being "a mass of pretentious nonsense"[73] (Huxley, 1854, p. 425).

Darwin's view of the *Vestiges* was more nuanced than Huxley's. In the first edition of the *Origin of Species*, he summarily dismissed Chambers' model of evolution:

QUOT. 88 The author of the 'Vestiges of Creation' would, I presume say that, after a certain number of unknown generations, some bird had given birth to a woodpecker, and some plant to a misseltoe [*isic*], and that these had been produced perfect as we now see them. (Darwin, 1859, pp. 3–4)

However, in a Sep 2, 1854, letter to Huxley, he had previously been more sympathetic:

QUOT. 89 ...I cannot think but that you are rather hard on the poor author. I must think that such a book, if it does no other good, spreads the taste for Natural Science.

But I am perhaps no fair judge, for I am almost as unorthodox about species as the Vestiges itself, though I hope not quite so unphilosophical. (Darwin and Seward, 1903a, p. 75)

Darwin's sympathy was not surprising since he shared Chambers' view of the transmutability of species.

Despite all the criticism, Chambers' *Vestiges* was widely read and went through twelve editions, the last one being in 1884, when the name of the author was revealed.

[73]This was despite two major similarities between Huxley's and Chambers' viewpoints: (i) uniformitarianism and (ii) a saltationist view of evolution.

Chapter 2
Darwin and the Origin of Species

Abstract Darwin's development of the theory of evolution by natural selection is described. When Darwin started his Beagle voyage of 1831, he believed in the fixity of species. Contrary to the opinion of the population, he was not converted to transformism when he visited the Galapagos Islands during his voyage. Conversion took place soon after his return to England in 1836. Darwin was influenced especially by the writings of Lyell, Paley, and Malthus when constructing his theory. Darwin was reluctant to publish his theory, but when he discovered that Wallace had independently discovered an almost identical theory to his, a joint Darwin–Wallace paper was promptly presented to the Linnean Society in 1858. Darwin's Origin of Species was published the following year. Of all the reviews of his groundbreaking book, the one written by the engineer Jenkin in 1867 was the most significant.

Keywords Beagle voyage · Galapagos islands · Finches · Red Notebook · Population thinking · Short sketch of 1842 · Expanded essay od 1844 · Alfred Russel Wallace · Sarawak Law paper · "Ternate" essay · Fleeming Jenkin

2.1 Darwin's Early Education: Religiosity

Charles Darwin[1] (Fig. 2.1) was born in 1809 in England. Darwin's father (Robert Darwin), who was a physician, had at first hoped his son would follow in his footsteps. Charles was thus sent to Edinburgh to study medicine at the age of 16.

---

[1] The literature on Darwin is, of course, enormous; see, for example, Desmond and Moore (1991), Browne (1995, 1996), Eiseley (1958), Gayon (1998), Ghiselin (1969), Limoges (1970), Himmelfarb (1959), Tort (1996), Freeman (1978), van Helvert and van Wyhe (2021), Bowler (1996), Bowlby (1992), Brent (1981), Vorzimmer (1970), Hull (1973), Ospovat (1981), Depew and Weber (1994), Herbert (2005), Porter and Graham (2015), Oldroyd (1980), Darlington (1959), de Beer (1967), Dennett (1996), Manier (1978), Keith (1955), Rose (1998), Ruse (1999), Gale (1982), Ellegard (1958), Caudill (1989), Glick (1974), Norman (2013), Phipps (2002), Clark (1984), Peckham (1959), Barlow (1958), George (1982), Wichler (1961), Jones (2000), Duzdevich (2014), and Kohn (1985).

P. Gorroochurn, *The Development of Evolutionary Genetics*,
https://doi.org/10.1007/978-3-031-69374-8_2

Fig. 2.1 Charles Darwin (1809–1882). (From: https://commons.wikimedia.org/wiki/Charles_Darwin#/media/File:Charles_Darwin_by_G._Richmond.jpg)

In addition to the medical courses, Charles also attended courses on zoology and geology in Edinburgh. After a little time, Robert realized that his son would not be able to make it to medicine and recommended that Charles become a minister instead. The latter soon persuaded himself that this would be his correct vocation:

> **QUOT. 90** . . .I did not then in the least doubt the strict and literal truth of every word in the Bible, I soon persuaded myself that our Creed must be fully accepted. (Barlow, 1958, p. 37)

To join the Church, Charles had to first get a degree from an English university, and then he went to Cambridge in early 1828.

Life at Cambridge was not exactly to his liking; nonetheless, he was able to make the acquaintance of the botanist professor John Henslow and the geologist professor Adam Sedgwick (to whom several references were made in Chap. 1). Furthermore, as mentioned before, Charles absorbed Paley's writings, in particular, the latter's *Natural Theology* (Paley, 1802).[2]

[2]See Sect. 1.13.

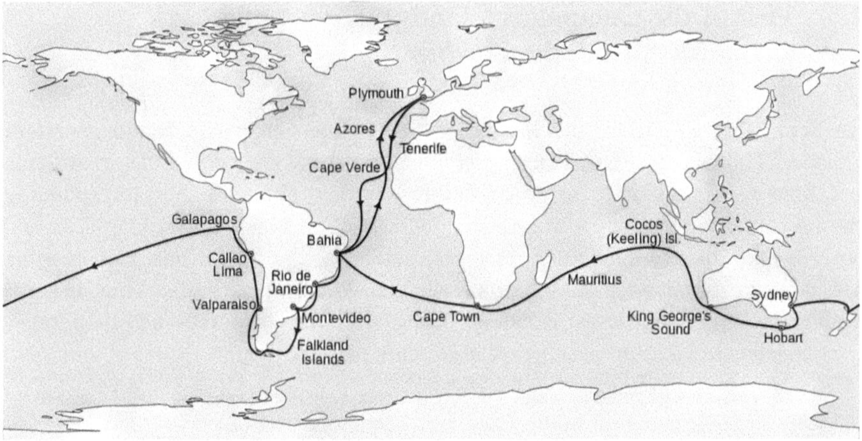

Fig. 2.2 Beagle's (second) voyage (Dec 27, 1831 to Oct 2, 1836). (From: https://commons.wikimedia.org/wiki/File:Voyage_of_the_Beagle-en.svg)

2.2 The Beagle Voyage (1831): Belief in the Fixity of Species

In 1831, upon his return from an excursion with Sedgwick, Charles received a letter from Henslow inviting him to join Captain Robert Fitzroy (1805–1865) of the Royal Navy on a survey voyage aboard the H.M.S. Beagle[3] (Fig. 2.2). Robert had initial doubts about the suitability of the trip but was eventually convinced by Charles' uncle, Josiah Wedgwood, to allow Charles to join the Beagle.

The Beagle voyage proved to be a landmark event in Charles' life, as the latter himself later admitted:

> **QUOT. 91** . . .The voyage of the Beagle has been by far the most important event in my life and has determined my whole career. (Barlow, 1958, p. 76)

Charles may best be described as a budding naturalist just before his voyage. More importantly, as he later recalled in a letter to Zacharias:

> **QUOT. 92** When I was on board the *Beagle* I believed in the permanence of species, but as far as I can remember, vague doubts occasionally flitted across my mind. (Darwin & Seward, 1903a, p. 1903, Letter 278)

Darwin put the first couple of years of his trip to very good use by reading Lyell's *Principles of Geology*.[4] He was very receptive to Lyell's ideas and accepted his uniformitarianism, at the expense of Sedgwick's catastrophism. At the ports of call, Darwin made several excursions both on foot and on horseback.

He also kept scientific notes and a more personal diary.

---

[3]This would be the *second* voyage of the Beagle, the first having taken place in 1826–1830 under Captain Pringle Stokes.

[4]See Sect. 1.14.

2.3 Visit of the Galapagos Islands (1835): Landmark in Darwin's Life, the Finches

On September 17, 1835, the Beagle arrived at the Galapagos Islands, west of Ecuador. The islands are of volcanic origins, and Darwin observed that their fauna and flora were similar to those of South America, although the corresponding environments were vastly different. He collected finches and mockingbirds and observed that the climatic conditions were similar across the islands. However, he noticed that each island might have its particular forms of these birds.[5] Thus, in some islands, some of the finches had small beaks, while in others they had thick ones. Darwin later declared in his *Journal of Researches* that:

> **QUOT. 93** Seeing this gradation and diversity of structure in one of small, intimately related groups of birds, one might really fancy that, from an original paucity of birds in this archipelago, one species had been taken and modified for different ends. (Darwin, 1846, p. 148, Vol. II)

However, despite the feeling conveyed by the quotation above, anybody looking for a eureka moment in the Galapagos Islands when Darwin was converted to evolution would be sadly disappointed. There was no such moment, not least in the Galapagos Islands or on the Beagle later. Darwin's conversion was a slow, gradual process. As Sulloway (1982a, b, 1985) has convincingly argued, Darwin was unable to see evolution in either the finches, the mockingbirds, or the huge tortoises whose structures varied across the islands. This was partly due to his lack of expertise and his orthodox, creationist beliefs at the time. The statement in QUOT. 93 was thus very likely the result of Darwin looking retrospectively years later. No eureka moment is seen in Darwin's contemporaneous (and therefore more reliable) diary and notes. Himmelfarb has summarized the situation as follows:

> **QUOT. 94** There is, in fact, no real continuity between the Beagle and the Origin. Between the two there intervened an idea. It was in the light of that idea that the experiences on the Beagle were re-ordered and re-interpreted by Darwin until they were ready to stand witness for the idea. (Himmelfarb, 1959, p. 123)

Sulloway (1985) further concluded that if there was any development in Darwin's thinking during the Beagle voyage, it was mainly due to his "growing self-assurance over time" as a natural scientist, especially in geology. For example, during the voyage, Darwin proposed a theory of coral reefs that essentially still stands today and provided a novel explanation of species extinction that was different from Lyell's climate-based theory.

---

[5]However, Darwin did not correctly identify some of the birds. He thought that the cactus finch was a blackbird and that the warbler finch was a kind of wren. He also did not carefully note which finches came from which island but was later able to correct his notes from those of others in the expedition, including Captain Fitzroy.

2.4 Back in England (1836): Decisive Evolutionary Evidence from Gould, Owen, and Waterhouse

Although the Beagle voyage had produced no evolutionary epiphany, Darwin's doubts about the fixity of species increased once he reached England. In his own words (which more accurately describe Darwin's recollections than the statement in QUOT. 93):

> **QUOT. 95** I was also informed that many of the islands possess trees and plants which do not occur on the others...Unfortunately, I was not aware of these facts till my collection was nearly completed: it never occurred to me, that the productions of islands only a few miles apart, and placed under the same physical conditions, would be dissimilar. I therefore did not attempt to make a series of specimens from the separate islands. It is the fate of every voyager, when he has just discovered what object in any place is more particularly worthy of his attention, to be hurried from it. (Darwin, 1840, p. 474)

Indeed, back in England, Darwin was faced with three pieces of incontrovertible evidence:

(a) In December 1836 (2 months after the Beagle's return), the anatomist Richard Owen (1804–1892) analyzed the leg and neck bones Darwin had taken from Port Julian in January 1834 and identified them as the "fragments of a Gigantic Llama!"[6]
(b) In January 1837, he showed the finches to the ornithologist John Gould (1804–1881). The latter classified them into 13 different species (rather than simply variations within species or "varieties," as Darwin had at first thought), which were then divided into four subgroups (genera) (Fig. 2.3). Furthermore, Gould concluded that the Galapagos birds were a new group of birds closely related to those on the mainland.
(c) In February 1837, the naturalist George Robert Waterhouse (1810–1888) classified several of the mice brought by Darwin from South America into different species.

Owen's result was an instance of Lyell's law of succession of types (living mammals replaced extinct mammals of the same kind) and suggested that vast geological times were available for species transmutation. The last two results, especially those of Gould, seriously undermined Darwin's belief in the immutability of species. He had already drawn a line in the sand back in June 1836, in the last leg of his voyage, in the following famous passage of his *Ornithological Notes*:

> **QUOT. 96** When I see these Islands in sight of each other, & [but *del.*] possessed of but a scanty stock of animals, tenanted by these birds, but slightly differing in structure & filling the same place in Nature, I must suspect they are only varieties. The only Galapagos fact of a similar kind of which I am aware, is the constant asserted difference—between the wolf-like

[6] This was called *Macrauchemia* by Owen later.

Fig. 2.3 Darwin's finches (Darwin, 1846, p. 148)

> Fox of East & West Falkland—If there is the slightest foundation for these remarks the zoology of Archipelagoes—will be well worth examining; for such facts [would *inserted*] undermine the stability of Species. (Barlow, 1963, p. 262)

In retrospect, it might indeed have been the finches that changed Darwin into an evolutionist. However, this was not during the Beagle voyage but rather when they were analyzed by Gould.

2.5 Darwin's Red Notebook (1837): Changed Viewpoint, the Coral Tree in the B Notebook

Having been exposed to the results of Gould, Owen, and Waterhouse, Darwin returned to his "Red Notebook" (RN)[7] around March 1837 (RN had been started in the last months of the Beagle voyage). However, note the choice of words now:

> **QUOT. 97** Speculate on neutral ground of 2. ostriches; bigger one encroaches on smaller.—3 change not progressif‹e›: produced at one blow. if one species altered... (Barrett et al., 1987, RN: 130)

[7]For a list of Darwin's notebooks, see Table 2.1.

Table 2.1 Darwin's notebooks

1835–1837	Ornithological notes
~May 1836 to mid-June 1837	Red Notebook
Mid-1837 to end of 1839	Notebook A: Geology
~July 1837 to March 1838	Notebook B: Transmutation
1837 to March 1842	Zoology Notes, Edinburgh Notebook
1838	Glen Roy Notebook
After mid-March 1838 to July 1838	Notebook C: Transmutation
July 15, 1838, to Oct 2, 1839	Notebook D: Transmutation
Oct 2, 1838, to July 10, 1839	Notebook E: Transmutation
Jul 15, 1838, to Oct 1, 1838	Notebook M: Metaphysics on morals & expression
Oct 2, 1838, to Aug 1, 1839	Notebook N: Metaphysics & expression
1838–1840	Old & Useless
1839–1841	Torn Apart Notebook
1839–1844	Questions & Experiments
1842	Summer 1842

> **QUOT. 98**. . .if one species does change into another it must be per saltum—or species may perish. (*ibidem*)

Instead of his previous "immutability of species," Darwin now refers to species alteration and changing one species into another! Darwin's conversion to evolution was well underway.

When Darwin started to write the B Notebook[8] (which was his first Transmutation Notebook) around July 1837, he had all but completely espoused the transmutationist viewpoint. For example:

> **QUOT. 99** Countries longest separated greatest differences—if separated from immense ages possibly two distinct type, but each having its representatives—"as in Australia" . . . This view supposes that in course of ages & therefore changes every animal has tendency to change.—this difficult to prove cats &c from Egypt no answer because time short & no great change has happened—I look at two ostriches as *strong argument of possibility of such change,—as we see them in space, so might they in time.*[9] (Barrett et al., 1987, B: 15–17)

Most evocative was Darwin's drawing of a coral tree (or "Tree of Life")[10] (Fig. 2.4), which aptly represented his new view of the interconnectedness of species. Next to the tree, Darwin made the following notes:

> **QUOT. 100** Case must be that one generation then should be as many living as now **To do this & to have many species in same genus (as is). REQUIRES extinction.**[11]—Thus between A & B immens gap of relation C & B the finest gradation, B & D rather greater

[8]This was headed "Zoonomia," an indication by Darwin that he was about to tread on the same evolutionary path as his grandfather (p. 15).

[9]Italics are ours.

[10]This tree was reproduced in a more sophisticated form later in the *Origin of Species* (Darwin, 1859, b/w pp. 117 and 118) and was the only diagram in that book.

[11]The bold type is in the original.

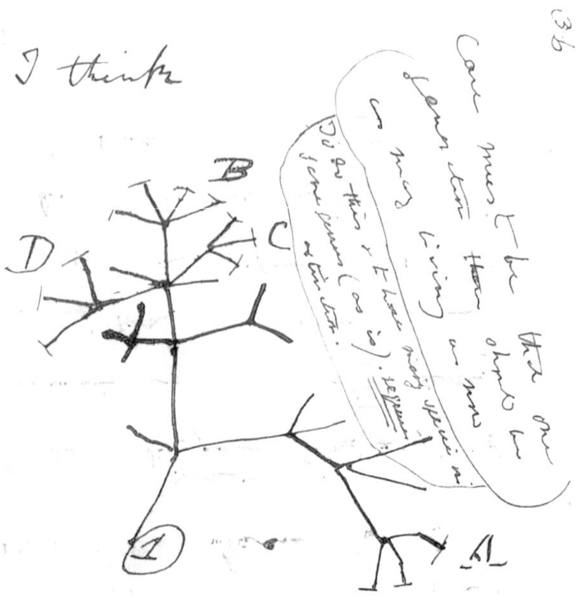

Fig. 2.4 Darwin's first tree, as it appears in the B Notebook (B:36). (From: https://commons.wikimedia.org/wiki/File:Darwins_first_tree.jpg)

> distinction. Thus genera would be formed—bearing relation to ancient types—with several extinct forms, for if each species «an ancient (1)» is capable of making, 13 recent forms.—Twelve of the contemporarys must have left no offspring at all, so as to keep number of species constant.—With respect to extinction we can easily see that variety of ostrich, Petise may not be well adapted, & thus perish out, or on other hand like Orpheus, being favourable many might be produced.—This requires principle that the permanent varieties produced by ‹inter› confined breeding & changing circumstances are continued & produce according to the adaptation of such circumstances & therefore that death of species is a consequence (contrary to what would appear from America) of non-adaptation of circumstances. (*ibid.*, B:36–39)

2.6 Darwin's First Evolutionary Theory Based on Monads: Rejection of the Monad Theory

It is noteworthy that Darwin's "tree of life" in Fig. 2.4 was drawn in the context of his first theory of evolution. This was more than a year before he read Malthus' *Essay on the Principle of Population* (Malthus, 1798) and realized that natural selection was simply due to the universal struggle for existence.[12]

Gruber (1974, Chap. 7) has pointed out that Darwin's first (pre-Malthus) theory of evolution was based on the notion of monads.[13] The latter are simple living

[12]See Sect. 1.12.

[13]Monads were first introduced in a different context by Leibniz.

particles that are spontaneously generated[14] and had previously formed an integral part of Lamarck's evolutionary theory.[15] Darwin used monads to explain the birth and extinction of species. In Darwin's pre-Malthus theory of evolution, each monad developed into a whole group of related species as a result of environmental influences. Since it was assumed that the number of species was approximately constant, species were born at approximately the same rate as they died. Each monad had a limited life span, and once it died, so did all species associated with it. Species thus behaved as individuals.

Yet, Darwin soon distanced himself from the monad theory. Having introduced the monad theory, in the next pages of the B Notebook, he stated:

> **QUOT. 101** If we ‹suppose› «grant» similarity of animals in one country owing to springing from one branch, & the monucle [monad] has definite life, then all die at one period, which is not case, ∴ MONUCULE NOT DEFINITE LIFE. (Barrett et al., 1987, B: 35)

In the above, Darwin had realized the implications of the monad theory: if whole species were to disappear suddenly, this would be consistent with some kind of catastrophism, a doctrine he had already abandoned during the Beagle voyage (upon reading Lyell's *Principles of Geology*). Fortunately, although Darwin rejected the monad theory, he retained the idea of the coral tree.

2.7 From July 1837 to September 1838: All the Elements of a Viable Evolutionary Theory (Except for the Theory)

From July 1837 onward, Darwin contemplated several aspects of his new evolutionary viewpoint. He started studying the work of animal breeders and observed that selection[16] produced variation within species. However, as Bowler has pointed out (2003, p. 160), Darwin's C and D Notebooks show that he did not directly borrow from the breeder's animal model.[17] Darwin was also aware of the concept of natural selection; for example, in the C Notebook (written after mid-March 1838), one can read:

> **QUOT. 102** The infertility of crosse & cross, is method of nature to prevent the picking of monstrosities as Man does.—One is tempted to exclaim. (Barrett et al., 1987, C: 52)

In the above, Darwin's "picking by nature" would later be called natural selection.

---

[14] Spontaneous generation was famously disproved by Pasteur in 1861.

[15] See paragraph below QUOT. 44.

[16] Selection by animal breeders means "artificial selection," and it is in this sense that selection was used at the time.

[17] Despite Darwin's statement to Wallace: "I came to the conclusion that selection was the principle of change from the study of domesticated productions; and then, reading Malthus, I saw at once how to apply this principle" (Darwin & Seward, 1903a, p. 118, Vol. I, April 1869 letter).

In his Notebooks, Darwin frequently used the concept of buds (small copies of parental tissues[18]) as the hereditary material that was passed from parents to offspring. These buds were blended (e.g., Barrett et al., 1987, B:33) during sexual reproduction. Darwin also spoke of mutants, which he called "monsters" or "monstrosities" (e.g., *ibid.*, B:111, C:4). Darwin was aware of species adaptation, as evidenced in the C Notebook:

> **QUOT. 103** Ermine, ptarmigan hare becoming white in winter of Arctic countries few will say it is direct effect, ‹of› according to Physical laws, as sulphuric acid disorganizes wood, but adaptation.—albino however is monster. yet albino may so far be considered an adaptation as best attempt of nature, colouring matter being absent.—again dwarf plant in alpine district & dwarf plants from seed, one adaptation, other monster. (*ibid.*, C:84–85)

The above shows that Darwin, like Lamarck, saw changed environmental conditions as the direct cause of adaptive variation. Last but not least, Darwin was also cognizant of the concept of the struggle for existence (e.g., *ibid.*, B:37–38).

All the elements of a viable theory of evolution were there, except for the theory itself.

2.8 September 28, 1838: Darwin's Reading of Malthus, "A Theory by Which to Work"

Only after he had read Malthus' *Essay on the Principle of Population* (Malthus, 1798)[19] on September 28, 1838, was Darwin able to connect the dots, and a meaningful theory of evolution was crystallized. In his *Autobiography*, as explained at the end of Sect. 1.12, Darwin wrongly gave the date of his reading as October 1838 but then went on to add that he finally "got a theory by which to work" (Barlow, 1958, p. 120).

In his book, Malthus convincingly argued that populations increased in a geometrical ratio, while food production increased only in an arithmetical ratio, resulting in a struggle for existence:

> **QUOT. 104** [p]opulation, when unchecked, increases in a geometrical ratio. Subsistence increases only in an arithmetical ratio. (Malthus, 1798, p. 20)

> **QUOT. 105** And when they fell in with any tribes like their own, the contest was a *struggle for existence*,[20] and they fought with a desperate courage, inspired by the reflection that death was the punishment of defeat and life the prize of victory. (*ibid.*, p. 29)

Darwin was able to see a similar struggle for existence between *species* (as he first thought). In this struggle, any heritable variation that conveyed an adaptive advantage would result in the variation being preserved and less favorable heritable

[18] Darwin would later call these buds "gemmules," in his provisional hypothesis of pangenesis.

[19] See Sect. 1.12.

[20] Italics are our own.

variation being eliminated. Natural selection[21] was the latter process of "discrimination" by nature, based on the adaptive value of variation. Darwin wrote the elements of his new theory in the D Notebook:

> **QUOT. 106** We ought to be far from wondering of changes in number of species, from small changes in nature of locality. Even the energetic language of ‹Malthus› "Decandoelle" does not convey the warring of the species as inference from Malthus—"increase of brutes, must be prevented solely by positive checks, excepting that famine may stop desire" in Nature production does not increase, whilst no checks prevail, but the positive check of famine &consequently death...
>
> Population in increase at geometrical ratio in FAR SHORTER than 25 years—yet until the one sentence of Malthus no one clearly perceived the great check amongst men...One may say there is a force like a hundred thousand wedges trying force ‹into› every kind of adapted structure into the gaps ‹of› in the oeconomy of Nature, or rather forming gaps by thrusting out weaker ones. The final cause of all this wedging, must be to sort out proper structure & adapt it to change. (Barrett et al., 1987, D: 134e–135e)

2.9 A Second Dimension to Malthus' Influence on Darwin: "Population Thinking" versus "Typological Thinking"

QUOT. 106 also shows that Darwin at first thought in terms of struggle between *species*. However, in the E Notebook, he realized that the selection pressure was on the *individual members*, rather than on the species itself.

Following Mayr (1959a), several authors (e.g., Hull, 1965a, b; Ghiselin, 1969; Herbert, 1971) have argued that the shift in Darwin's focus from species to the individual members of the same group (population) was crucial and added a further dimension to Malthus' influence on Darwin. The focus on species harks back to Plato's Theory of Forms (Sect. 1.1.2), in which each individual in a classification is the manifestation (or the "essence") of an ideal type and any variation between the individuals is an illusion. This viewpoint is known as *essentialism* or *typological*. By initially thinking in terms of species, Darwin was following Lyell, who had previously referred to the struggle between species.[22] His new way of thinking, in terms of the individuals themselves, considered the ideal type (average) as an abstraction, and the variation between the individuals as a reality. Mayr (1959a) has aptly called this new way of thinking "population thinking" (in contrast to the traditional "typological thinking").

Walsh (2015, Sect. 2.2) has further elaborated on the issue of "population thinking." In particular, he has differentiated between Grade 1 and Grade 2 population thinking. In Grade 1 (or Malthusian) population thinking, which Darwin adopted, although the conceptualization is in terms of the population, the activities of the component individuals are still relevant. On the other hand, in Grade 2 pop-

[21] Darwin did not use the term "natural selection" yet.

[22] See QUOT. 73.

ulation thinking, which Galton later introduced in the context of regression, the conceptualization is also at the population level, but the activities of the individuals do not matter—only the statistical distribution of the properties (e.g., height, weight, and so on) counts. The difference between the two Grades is important since, by focusing on individual (rather than trait) fitness, Grade 1 thinking does not lend itself to an evolutionary analysis. In Walsh's own words:

> **QUOT. 107** . . .the sort of Malthusian population thinking employed by Darwin, does not render a criterion of evolutionary change. Nor does it allow us to predict or quantify the degree of evolutionary change. Its deficiency lies in the very metaphysical modesty that makes it so intuitively compelling. Evolution may be the consequence of individuals struggling to survive, but citing these individual-level causes of population change is insufficient to explain the object of natural selection theory: change in trait structure. In this respect, Darwin's conception of natural selection is inadequate as a theory of the dynamics of populations. The second moral of these simple models is that in order to explain the dynamics of populations, we must invoke something else: the distribution of trait fitnesses. Invoking trait fitnesses, in turn, requires us to ascend to an ontology of abstract entities, trait types and the statistical properties of a population. (*ibid.*, p.54)

One can equate Galton's (Grade 2) population thinking with the birth of modern statistics, where deviations from the mean are not just accidental errors of no interest but essential and informative aspects of the data.

2.10 The Short Sketch of 1842: "Natural Means of Selection," and Philosophical Influences of John Herschel and William Whewell

2.10.1 The Short Sketch of 1842: Natural Means of Selection

Coming back to September 1838, having discovered his new theory, Darwin was in no hurry to communicate it to the outside world. In his Autobiography, he later explained:

> **QUOT. 108** I was so anxious to avoid prejudice, that I determined not for some time to write even the briefest sketch of it. (Barlow, 1958, p. 120)

In the meantime, Darwin married his cousin Emma Wedgwood in 1839 and became involved in various writing projects, including books on the Beagle Voyage, coral reefs, and volcanic islands (Table 2.2).

In mid-1842, Darwin used his Transmutation Notebooks to sketch a 35-page outline of his theory (Darwin, 1909). The sketch was a rudimentary version of what would later become *The Origin of Species*, but the gist of Darwin's theory was there. The sequence of topics adopted was more or less repeated in both the later 1844 essay and 1859 *Origin* and was as follows:

Table 2.2 Darwin's major publications

1835	Letters on Geology [Extracts from letters addressed to Professor Henslow]
1838–1843	The Zoology of the Voyage of H.M.S. Beagle. Edited and superintended by Charles Darwin (5 Parts). London: Colburn.
1839	The Narrative of the Voyages of H.M. Ships Adventure and Beagle. London: Colburn, 3 vols & appendix
1839	Journal of researches into the geology and natural history of the various countries visited by H.M.S. Beagle. London: Colburn
1839	Questions about the breeding of animals
1842	The structure and distribution of coral reefs
1844	Geological observations on the volcanic islands visited during the voyage of H.M.S. Beagle.
1845	Journal of researches into the natural history and geology of the countries visited during the voyage of H.M.S. Beagle round the world. London: Murray. 2nd ed
1846	Geological observations on South America
1851	Geology of the voyage of the Beagle
1851	A monograph on the fossil Lepadidae, or, pedunculated cirripedes of Great Britain [Vol. 1]
1851	A monograph on the sub-class Cirripedia, with figures of all the species: The Lepadidae; or pedunculated cirripedes [Vol. 1]
1854	A monograph on the sub-class Cirripedia, with figures of all the species: The Balanidae (or sessile cirripedes); the Verrucidae, etc. [Vol. 2]
1854	A monograph on the fossil Balanidae and Verrucidae of Great Britain [Vol. 2]
1859	On the origin of species by means of natural selection; or the preservation of favored races in the struggle for life. London: Murray
1862	On the various contrivances by which British and foreign orchids are fertilized by insects, and on the good effects of intercrossing
1865	On the movements and habits of climbing plants [Reprinted from Journal of the Linnean Society, vol. 9, pp. 1–118]
1868	The Variation of Animals and Plants Under Domestication [vols. 1 & 2]
1871	The Descent of Man, and Selection in Relation to Sex [vols. 1 & 2]
1872	The expression of the emotions in man and animals
1872	The origin of species by means of natural selection; or the Preservation of Favoured races in the struggle for life. London: Murray. 6th ed.
1874	The structure and distribution of coral reefs, 2nd ed
1875	Insectivorous Plants
1875	The movements and habits of climbing plants, 2nd ed
1875	The variation of animals and plants under domestication [vols. 1 and 2], 2nd ed
1876	The effects of cross and self fertilization in the vegetable kingdom
1877	The different forms of flowers on plants of the same species
1878	The effects of cross and self fertilization in the vegetable kingdom, 2nd ed
1880	The power of movement in plants
1881	The formation of vegetable mold through the action of worms, with observations on their habits
1887	The Autobiography of Darwin (published in Vol. 1 of *The Life and Letters of Darwin, including an Autobiographical Chapter*, ed. by Francis Darwin)
1888	Insectivorous plants, 2nd ed

[a]A complete list may be found in Freeman (1977)

PART I

§ I. On Variation Under Domestication, and on the Principles of Selection
§ II. On Variation in a State of Nature and on the Natural Means of Selection
§ III. On Variation in Instincts and Other Mental Attributes

PART II

§ IV. and V. On the Evidence from Geology
§ VI. Geographical Distribution
§ VII. Affinities and Classification
§ VIII. Unity of Type in the Great Classes
§ IX. Abortive Organs
§ X. Recapitulation and Conclusion

The first chapter considered variation under domestication, and artificial selection was referred to as "man's selection." The latter proceeds by first recognizing the presence of variation in organisms. Variation is assumed to be small in magnitude, and the importance of its transmission is recognized. The variety favored by the breeder is then augmented by crossbreeding similar varieties. Concerning the variation that is inherited in an offspring, Darwin stated:

QUOT. 109 Free crossing [is a] great agent in producing uniformity in any breed. (*Ibid.*, p. 2)

In the above, Darwin adopted the prevalent notion of blending inheritance, whereby a fusion of maternal and paternal hereditary material formed the offspring, who then went on to develop an external appearance midway between the two parents.[23]

After introducing the principle of selection under domestication in Chapter 1, the sketch considered natural selection in Chapter 2. "Natural selection" was not used—instead, Darwin referred to "natural means of selection." Darwin recognized that variation in wild animals was exceedingly small but had time to accumulate, given the large geological history. He then asked the following question:

QUOT. 110 . . .But is there any means of selecting those offspring which vary in the same manner, crossing them and keeping their offspring separate and thus producing selected races: otherwise as the wild animals freely cross, so must such small heterogeneous varieties be constantly counter-balanced and lost, and a uniformity of character [kept up] preserved. (*ibid.*, p. 5)

Darwin invoked the Malthusian struggle for existence to explain the process of natural selection:

QUOT. 111 De Candolle's war of nature,—seeing contented face of nature,—may be well at first doubted; we see it on borders of perpetual cold. But considering the enormous geometrical power of increase in every organism and as ⟨?⟩ every country. . .reflection will show that this is the case. Malthus on man,—in animals no moral [check] restraint ⟨?⟩—they breed in time of year when provision most abundant, or season most favourable, every country has its seasons. . .If proof were wanted let any singular change of climate ⟨occur⟩

[23] Hardin later used more colorful words to describe blending inheritance: "the paint-pot theory of heredity" (Hardin, 1959, p. 115).

here ⟨?⟩, how astoundingly some tribes ⟨?⟩ increase, also introduced animals, the pressure is always ready,—capacity of alpine plants to endure other climates,—think of endless seeds scattered abroad,—forests regaining their percentage,—a thousand wedges are being forced into the oeconomy [*sic*] of nature.

The unavoidable effect of this ⟨is⟩ that many of every species are destroyed either in egg or [young or mature (the former state the more common)]. In the course of a thousand generations, infinitesimally small differences must inevitably tell; when unusually cold winter or hot or dry summer comes, then out of the whole body of individuals of any species, if there be the smallest differences in their structure, habits, instincts [senses], health &c., ⟨it⟩ will on an average tell; as conditions change a rather larger proportion will be preserved. (*ibid.*, pp. 7–8)

Darwin's prompt statement of the principle of selection shows the type of savviness that characterized him throughout life. Given the mainstream belief in the fixity of species at the time, there was little chance that any reader would be convinced by Darwin's evolutionism unless, right at the start, a viable mechanism was introduced by which such evolution could occur.

In the 1842 sketch, the causes of variation were attributed to the direct effect of the environment, especially on the organs:

QUOT. 112 The nature of the external conditions tends to effect some definite change in all or greater part of offspring,—little food, small size certain foods harmless &c. &c. organs affected and diseases—extent unknown. (*ibid.*, p. 1)

He next emphasized the indirect influence of the environment on the "reproductive system":

QUOT. 113 But more important is that simple generation ⟨?⟩, especially under new conditions [when no crossing] (causes) infinite variation and not direct effect of external conditions, but only in as much as it affects the reproductive functions. There seems to be no part (beau ideal of liver) of body, internal or external, or mind or habits, or instincts which does not vary in some small degree and [often] some ⟨?⟩ to a great amount. (*ibid.*, p. 2)

The above is a purely Lamarckian principle (see Sect. 1.8). Darwin did not invoke Lamarck's concept of "inner feeling" (*sentiment intérieur*) in his exposition. He simply assumed that the direct influence of the environment was obvious. Moreover, an explicit mechanism of heredity was not specified.

In the 1842 sketch (as well as later in the *Origin*), Darwin repeatedly invoked an omniscient Creator[24] who was responsible for the "final causes" in nature. The "final causes" were achieved by means of natural laws (which were "intermediate" or "secondary" causes). To Darwin, evolution was the province of these natural laws and was not directly acted upon by the Creator. Thus, in of the concluding statements in the sketch, Darwin stated:[25]

QUOT. 114...It accords with what we know of the law impressed on matter by the Creator, that the creation and extinction of forms, like the birth and death of individuals should be the effect of secondary [laws] means. (*ibid.*, p. 51)

[24]Darwin later declared himself an agnostic (Fordyce, 1883, p. 190).

[25]The same statement was made in both the 1844 essay and the various editions of the *Origin*.

The 1842 sketch (and the later 1844 essay) was not published during Darwin's lifetime. It was found in a staircase cupboard after Emma's (Darwin's wife) death and published under the editorship of Francis (Darwin's son) in 1909 (Darwin, 1909).

2.10.2 Philosophical Influences of John Herschel and William Whewell

The structure of the sketch is noteworthy in that Part I concentrates on the enunciation of the theory, while Part II details a series of observations that can be explained by such a theory. The same pattern of exposition may be found in both the later essay and *Origin*. Darwin's approach reflects the inductive influence of both John Herschel's[26] *Preliminary Discourse on the Study of Natural Philosophy* (Herschel, 1830) and William Whewell's *Philosophy of the Inductive Sciences* (Whewell, 1837). Darwin had previously read and been impressed by the first book. Herschel (Fig. 2.5) stressed how a theory could gain support when it was able to explain new phenomena:

> **QUOT. 115** Unexpected and peculiarly striking confirmations of inductive laws frequently occur in the form of residual phenomena, in the course of investigations of a widely different nature from those which gave rise to the inductions themselves. (Herschel, 1830, p. 171)

By using his theory to explain a variety of observational phenomena in Part II of the sketch, Darwin made effective use of this tenet.

In the second book, Whewell (Fig. 2.6) had stressed two criteria:

- *Consilience of inductions*: This is the same as Herschel's principle mentioned earlier, which Whewell defined as follows: "[T]he evidence in favour of our induction is of a much higher and more forcible character when it enables us to explain and determine cases of a kind different from those which were contemplated in the formation of our hypothesis. The instances in which this has occurred, indeed, impress us with a conviction that the truth of our hypothesis is certain" (Whewell, 1840, p. 234, Vol. II).
- *Analogy of sciences*: Whewell drew on the analogy between sciences to explain a new phenomenon in one by a known cause in the other: "The analogy of all other sciences compels us. . .to explain phenomenon by known, not unknown, causes" (*ibid.*, p. 130) Darwin made use of this principle, for example, by drawing on the analogy between domesticated species and species in nature. In the former,

[26]Whom Darwin met in June 1836 at the Cape of Good Hope during the Beagle voyage and who, in an 1836 letter to Lyell, referred to the replacement of extinct species by new ones as the "mystery of mysteries" (Cannon, 1961). The famous phrase later featured in Darwin's 1845 edition of the *Zoology of the Voyage of H.M.S. Beagle* and as the second sentence in Darwin's *Origin*: "These facts seemed to me to throw some light on the origin of species—that mystery of mysteries, as it has been called by one of our greatest philosophers" (Darwin, 1859, p. 1).

Fig. 2.5 Sir John Herschel (1792–1871). (From: https://commons.wikimedia.org/wiki/File:History_of_the_Royal_Astronomical_Society_-_Sir_John_Herschel.png)

changes in organisms occurred through selection (by humans). Darwin was thus led to posit selection (by nature) as the process of change in natural species.

Herschel and Whewell's influence can be seen throughout Darwin's later writings, especially in the *Origin*.[27] Despite their influence on Darwin, both men later rejected Darwin's theory of evolution, owing to their essentialist, fixist beliefs (Hull, 1973, p. 1973, Chap. 4).[28]

[27]Whewell's influence on science is especially notable. He invented many scientific terms, as has been indicated throughout this book. He was the first to use the word "scientist." In an 1834 anonymous review, he suggested the word scientist be used in analogy with...an artist!: "...science, even mere physical science, loses all traces of unity. A curious illustration of this result may be observed in the want of any name by which we can designate the students of the knowledge of the material world collectively...*Philosophers* was felt to be too wide and too lofty a term...*savans* was rather assuming, besides being French instead of English; some ingenious gentleman proposed that, by analogy with *artist*, they might form *scientist*..." (Whewell, 1834, p. 59). Later, in 1840, he said: "We need very much a name to describe a cultivator of science in general. I should incline to call him a Scientist" (Whewell, 1840, p. cxiii, Vol. I).

[28]See also Honenberger (2018).

Fig. 2.6 William Whewell (1794–1866). (From: https://commons.wikimedia.org/wiki/File:PSM_V07_D008_William_Whewell.jpg)

2.11 1844: Confession of "Murder" to Hooker, the Expanded Essay

2.11.1 Confession of "Murder" to Hooker in 1844

Following his 1842 sketch, Darwin admitted his transmutationist views for the first time on January 11, 1844, to his friend, the English botanist Joseph Hooker (Fig. 2.7)[29] (and even likening his confession to that of murder!):

> **QUOT. 116** . . .I have been now ever since my return engaged in a very presumptuous work, and I know no one individual who would not say a foolish one. I was so struck with the distribution of the Galapagos organisms, &c. &c., and with the character of the American fossil mammifers, &c. &c., that I determined to collect blindly every sort of fact, which could bear any way on what are species. I have read heaps of agricultural and horticultural

[29]Darwin first met Hooker in Jan 1839, in the company of Asa Gray at the Hunterian Museum, Royal College of Surgeons. Darwin's friendship with him would prove to be one of the most enduring and fruitful relationships he had. Joseph Hooker's life and interactions with Darwin are also discussed in Endersby (2020), Turrill (1963), Porter (1993), McCalman (2010), Gale (1982, pp. 55–78), and Bower (1919).

Fig. 2.7 Joseph Dalton Hooker (1817–1911) in 1855. (From: https://commons.wikimedia.org/wiki/File:Joseph_Dalton_Hooker_by_George_Richmond_1855.png)

> books, and have never ceased collecting facts. At last gleams of light have come, and I am almost convinced (quite contrary to the opinion I started with) that species are not (it is like confessing a murder) immutable. Heaven forfend me from Lamarck nonsense of a "tendency to progression", "adaptations from the slow willing of animals", &c.! But the conclusions I am led to are not widely different from his; though the means of change are wholly so. I think I have found out (here's presumption!) the simple way by which species become exquisitely adapted to various ends. You will now groan, and think to yourself, "on what a man have I been wasting my time and writing to". I should, five years ago, have thought so... (Darwin, 1888a, p. 384)

Darwin's trepidation in disclosing his theory is an indication of the anti-evolutionist views in Victorian Britain at the time. The anti-evolutionism would be in full display later in October 1844, when Chambers' *Vestiges of the Natural History of Creation* (Chambers, 1844) was published.[30]

Hooker was not convinced by Darwin's letter, but his reply shows that he kept an open mind:

[30] See Sect. 1.15.

QUOT. 117 There may in my opinion have been a series of productions on different spots, & also a gradual change of species. I shall be delighted to hear how you think that this change may have taken place, as no presently conceived opinion satisfy me on the subject. (Burkhardt & Smith, 1987, p. 5)

Even in 1853, Hooker was still arguing for the fixity of species (Hooker, 1853). Desmond and Moore (1991, p. 416) have suggested that Hooker's conversion to evolution took place around July 1855 (Burkhardt & Smith, 1989, p. 372). Hooker's first public statement of his evolutionist views occurred in the *Flora Tasmania* (Hooker, 1860), which started in mid-December 1857 and finished in mid-March 1859.

Despite Hooker's initial reluctance to espouse Darwin's evolutionist views, Darwin maintained a solid friendship with Hooker and benefited immensely from the latter's vast knowledge of botany. Huxley might later have been Darwin's self-proclaimed "bulldog," but Hooker was Darwin's closest friend and one of the earliest converts to transmutationism.

2.11.2 The Expanded Essay of 1844

Darwin's 1842 sketch was expanded to a 231-page essay in mid-1844.[31] This followed the same order as the earlier sketch and the later *Origin*. The principle of selection was explained more succinctly than before, as follows:

QUOT. 118 Where the external conditions are constantly tending to give some character, a race possessing this character will be formed with far greater ease by selecting and breeding together the individuals most affected. (*ibid.*, p. 64)

Chapter 1 again starts with artificial selection, which is denoted as "selection" only. This is followed, as in the 1842 sketch, by natural selection, which is still referred to as "natural means of selection." Although essentially the same chapters figured as before, their contents were substantially increased.

Variation was again attributed either to the direct effect of the environment or to the indirect effect of the environment working through the reproductive system, but the indirect causes were given even more importance than before.

There were, however, a few ideas that were to be later abandoned. In 1844 (and in the 1842 sketch), Darwin seemed to think that:

(a) natural selection acted only when organisms "must cease to be so perfectly adapted to the changed conditions as they originally were" (Darwin, 1909, p. 91).[32]

[31] Hooker was invited to read the essay in Nov 1845.

[32] See also Ospovat (1981, pp. 73–83).

(b) variation in organisms occurred only when they were placed in a new environment (*ibid.*, pp. 84–85) (and, as we explained before, variation was mainly the result of the indirect effect of the environment working on the reproductive system).
(c) variation in nature was "exceedingly little" (*ibid.*, p. 4).

On the other hand, in the 1859 *Origin*:

(A) natural selection acted continually: "...natural selection is daily and hourly scrutinising, throughout the world, every variation, even the slightest" (Darwin, 1859, p. 84). Perfect adaptation was changed to *relative* adaptation: "Natural selection tends only to make each organic being as perfect as, or slightly more perfect than, the other inhabitants of the same country with which it has to struggle for existence" (*ibid.*, p. 201).
(B) variation always occurred and was not contingent on the environment: "The more diversified in habits and structure the descendants of our carnivorous animal became, the more places they would be enabled to occupy" (*ibid.*, p. 113).
(C) variation in nature was common: "I am convinced that the most experienced naturalist would be surprised at the number of the cases of variability, even in important parts of structure, which he could collect on good authority, as I have collected, during a course of years" (*ibid.*, p. 45).

Once the essay was finished, Darwin wrote a letter to his wife, Emma, to be opened in case of his sudden death. This letter contained his "most solemn and last request" (Darwin, 1887, p. 44) that his essay be published after his death, in the hands of a competent editor such as Lyell with the aid of Hooker.

For all the progress Darwin had made in formulating a theory of evolution, he was in no hurry to publish it. Two major reasons may be advanced to explain this:

(i) Darwin may not have felt that all the elements of his theory were coherent enough to be convincing. For example, his belief in 1844 that variation in nature was sparse (Sect. 2.11) implied that the force of natural selection would not be strong enough to drive substantive evolution.
(ii) Darwin may have been dissuaded by the highly critical reception of the *Vestiges* (Chambers, 1844, see Sect. 1.12), which dealt with evolutionary ideas.

To make natural selection a more tenable hypothesis, Darwin collected more data. For example, he initiated an extensive study of barnacles during the period 1846–1854, resulting in a 4-volume work.[33] Darwin's studies revealed that no two barnacles[34] were identical. This suggested to him that the variation in nature was not only vast but also that natural selection would be able to act on these minute, essentially continuous differences.

---

[33] Which Milner has described as Darwin's "most tedious monograph" (Milner, 2009, p. 33).

[34] These may occur in concentrations of 30,000 or more in a square meter (Armstrong, 2007, p. 26, Vol. I).

2.12 Period Leading to the "Origin": Asa Gray and Thomas Huxley

2.12.1 *Asa Gray (1810–1888)*

On another front, Darwin began to engage with some of the foremost scientists of his time so as to convince them of the validity of his theory. Charles Lyell and John Hooker have already been mentioned in this respect. Two other names are those of the American botanist Asa Gray[35] (Fig. 2.8) and the English biologist Thomas Henry Huxley (1825–1895).

Gray first met Darwin in 1839 while visiting England. At the time, Darwin was relatively unknown, and the acquaintance was made only incidentally as Gray accompanied Hooker to meet the famous anatomist Richard Owen. However, in 1855, Gray started a serious correspondence with Darwin (the latter wrote the first letter[36]) and later became Darwin's most stalwart American supporter. In an 1856 letter to his colleague James Dwight Dana (1813–1895), he stated his evolutionary agnosticism as follows:

> **QUOT. 119** On the subject of species, their nature, distribution, what systems in natural history is, etc., certain inferences are slowly settling themselves in my mind, or taking shape; but in some of the most vexed questions I have as yet no opinion whatever and no very strong bias . . . (Gilman, 1899, p. 298)

In a July 20, 1857, letter, Darwin entrusted Gray with a summary of his evolutionary views:[37]

> **QUOT. 120** . . .It is not a little egotistical, but I should like to tell you, (& I do not think I have) how I view my work. Nineteen years (!) ago it occurred to me that whilst otherwise employed on Nat. Hist., I might perhaps do good if I noted any sort of facts bearing on the question of the origin of species; & this I have since been doing. Either species have been independently created, or they have descended from other species, like varieties from one species. I think it can be shown to be probable that man gets his most distinct varieties by preserving such as arise best worth keeping, & destroying the others,—but I should fill a quire if I were to go on. To be brief I assume that species arise like our domestic varieties with much extinction; & then test this hypothesis by comparison with as many general & pretty well established propositions as I can find made out,—in geograph. distribution, geological history—affinities &c. &c. &c. And it seems to me, that supposing that such hypothesis were to explain such general propositions, we ought, in accordance with common way of following all sciences, to admit it, till some better hypothesis be found out. For to my mind to say that species were created so & so is no scientific explanation, only a reverent way of saying it is so & so. But it is nonsensical trying to show how I try to proceed in compass of a note. But as an honest man I must tell you that I have come to the heterodox conclusion that there are no such things as independently created species—that species are only strongly defined varieties. (Dupree, 1959, pp. 244–245)

---

[35] Asa Gray's life, work, and interactions with Darwin are also discussed in Dupree (1959), Porter (1993), Hoeveler (2007, pp. 62–76), and Browne (2010).

[36] "Asa Gray did not discover Darwin. On the contrary, Darwin discovered Asa Gray" (Dupree, 1959, p. 238).

[37] Much in the same fashion he had earlier done with Hooker (Sect. 2.11.1).

Fig. 2.8 Asa Gray (1810–1888) in 1864. (From: https://commons.wikimedia.org/wiki/File:Asa_Gray_by_John_Whipple,_1864.jpg)

Darwin's next letter to Gray indicated that the latter had objected to the possibility of species transmutation (*ibid.*, p. 245). However, their friendly exchange of letters continued throughout the years. In September 1857, Darwin sent the botanist a sketch of his theory, which would later prove to be a key piece of evidence in establishing his priority on the theory of evolution by natural selection.[38]

The first indication of Gray's somewhat tentative conversion to transmutationism occurred on Jan 11, 1859 (11 months before the release of the *Origin*), when he presented his paper "Diagnostic characters of new species of phaenogamous plants" (Gray, 1859) to the American Academy of Arts and Sciences. In Vol. IV of selections from the records for the Jan 11 meeting, Gray declared:

QUOT. 121 . . .the idea of descent of all similar and conspecific individuals from a common stock is so natural, and so inevitably suggested by common observation, that it must need be first tried upon the problem [of distribution], and if the trial be satisfactory, its adoption would follow as a matter of course. (American Academy, 1860, p. 132)

[38] See Sect. 2.13.

The January meeting took place in the presence of Gray's archrival Agassiz,[39] who was then the foremost zoologist in America and an influential proponent of special creation. Several other meetings occurred in the following months, with back-and-fourths between Gray and Agassiz. These were the American versions of the more contentious British debate (see Sect. 2.12.2) that would occur in 1860 between Thomas Huxley (1825–1895) and Samuel Wilberforce (1805–1873).

In a footnote to the published version of the last section of the Japan paper, dated April 25, 1859, Gray went even further than his previous statement:

> **QUOT. 122** The only noteworthy attempt at a scientific solution of the problem, aiming to bring the variety as well as the geographic association of existing species more within the domain of cause and effect, is that of Mr Darwin and (later) of Mr Wallace. . .It will hardly be doubted that the tendencies and causes are really operative; the questions as to their extent of operation. (Sargent, 1889, p. 133, Vol. II)

The above still shows some qualification in Gray's acceptance of Darwin's theory. Indeed, even in later years, Gray's thinking was more aligned with Paley's theology and its teleological implications, than with the random component in Darwin's evolution. In 1861, Gray published a paper entitled "Natural selection not inconsistent with natural theology" (Gray, 1861). He has been aptly called "Darwin's retriever" (because he retrieved Darwin's theory for Christians).

2.12.2 *Thomas Huxley (1825–1895)*

If Gray was Darwin's retriever, then Huxley was Darwin's bulldog. Thomas Henry Huxley[40] (Fig. 2.9) unabashedly bestowed this title unto himself[41] and used all the wits he possessed to become the fiercest defendant of Darwin's ideas. He first met Darwin at the Geological Society in April 1853, although they had exchanged letters earlier. He was the quintessential scientist and adopted a critical outlook on all scientific matters. He was thus opposed to transmutation before reading the *Origin*, even writing an extremely critical review[42] of the tenth edition of the *Vestiges*. However, once Huxley had received a copy of the *Origin* from Darwin himself and had read it, he wrote to Darwin on November 23, 1859:

> **QUOT. 123** Since I read Von Bär's essays, nine years ago, no work on Natural History Science I have met with has made so great an impression upon me, and I do most heartily thank you for the great store of new views you have given me. . .As for your doctrine, I am prepared to go to the stake, if requisite. (Darwin, 1888a, p. 230)

---

[39] See the end of Sect. 1.10.2.

[40] Thomas Huxley's life, work, and interactions with Darwin are also discussed in Desmond (1997), Irvine (1955), di Gregorio (1984), and Bibby (1972).

[41] First used in Osborn (1896, p. 32): "You know I have to take care of him—in fact, I have always been Darwin's bull-dog." In a November 1871 letter to Haeckel, he further added: "The dogs have been barking at his [Darwin's] heels too much of late." See p. 391 of Huxley (1900, Vol. I) for more.

[42] The ferocity of which he later regretted: "The only review I ever had qualms of conscience about, on the grounds of needless savagery, is the one on the *Vestiges*" (Darwin, 1887, p. 189).

Fig. 2.9 Thomas Huxley in 1874 (1825–1895). (From: https://commons.wikimedia.org/wiki/File:Thomas_Henry_Huxley.jpg)

However, he went on to add one important reservation:

> **QUOT. 124** . . .you have loaded yourself with an unnecessary difficulty in adopting *Natura non facit saltum* [Nature does not make any jump] so unreservedly. . . (*ibidem*)

In the above, Huxley disagreed with Darwin's assumption that variation was essentially continuous. He instead believed that variation could be large and that there was no need for transitional forms in the fossil record. Indeed, in a June 25, 1859, letter to Charles Lyell, he had said:

> **QUOT. 125** Suppose that external conditions acting on species A give rise to a new species, B; the difference between the two species is a certain definable amount which may be called A-B. Now I know of no evidence to show that the interval between the two species must *necessarily* be bridged over by a series of forms, each of which shall occupy, as it occurs, a fraction of the distance between A and B. . .
>
> In an organic compound, having a precise and definite composition, you may effect all sorts of transmutations by substituting an atom of one element for an atom of another element. You may in this way produce a vast series of modifications—but each modification is definite in its composition, and there are no transitional or intermediate steps between one definite compound and another. . . (Huxley, 1900, p. 250, Vol. I)

A similar sentiment was expressed in Huxley's *Westminster* review of the *Origin* (1860, pp. 558–559).

Huxley's saltationist views may be seen as precursors to similar views later held by Galton, Bateson, de Vries, and others.

Huxley also had misgivings regarding natural selection. He could not see how the latter could be reconciled with the fact that some varieties could become infertile with their closest relatives. In the *Westminster Review*, he admitted that "it is not absolutely proven that a group of animals, having all the characters exhibited by species in nature, has ever been originated by selection, whether artificial and selection" (*ibid.*, p. 567).

One aspect of Darwinian evolution about which Huxley had few, if any, qualms has to do with his belief that humans evolved from lower forms of life not only in physical traits but also in *mental* ones. In the essay, *On the Relation of Man to the Lower Animals*¸ one can read:

QUOT. 126 I have endeavored to show that no absolute structural line of demarcation, wider than that between the animals which immediately succeed us in the scale, can be drawn between the animal world and ourselves; and may I add the expression of my belief that the attempt to draw a physical distinction is equally futile, and that even the highest faculties of feeling and of intellect begin to germinate in lower forms of life. (Huxley, 1863, p. 129)

Huxley's (and Darwin's) belief in an all-pervasive evolutionary force was the kind of mechanistic approach that stood in contrast to the vitalistic convictions of other friends of Darwin. Thus, both Lyell[43] and Wallace[44] thought it better to leave man's mental capacities outside the domain of Darwinian evolution. On the other hand, the English biologist St George Jackson Mivart found in Huxley's views a source of his alienation from Darwin's views to the point of a total break in 1871.[45] As for Darwin, he mostly abstained from the "man question" in the first edition of the *Origin*,[46] but later in his *Descent of Man* (Darwin, 1871b), he more or less espoused Huxley's position.

Huxley's status as Darwin's "bulldog" was first demonstrated during his confrontation with Samuel Wilberforce[47] (1870–1873), who was the Bishop of Oxford and a fierce critic of Darwin's *Origin*. The occasion was the 30th meeting of the British Association for the Advancement of Science, held in Oxford from June 27 to July 4, 1860.[48] Huxley was a self-declared agnostic[49] and had been persuaded to attend the meeting by none other than Robert Chambers, the (anonymous) author of the *Vestiges*. On June 28, Huxley already had an argument with his archrival Owen regarding the human skull. Owen had claimed that there was a greater difference

[43] See Sect. 1.14.

[44] See Sect. 2.13.5.8.

[45] See Sect. 2.14.5.

[46] Except for the somewhat cryptic: "In the distant future I see open fields for far more important researches. Psychology will be based on a new foundation, that of the necessary acquirement of each mental power and capacity by gradation. Light will be thrown on the origin of man and his history" (Darwin, 1859, p. 488).

[47] Nicknamed "Soapy Sam" because of his slick oratory.

[48] This was approximately 7 months after the release of the *Origin*. The book was now already in its second edition.

[49] A term that, in fact, Huxley himself coined in 1869, although the concept had been most clearly developed earlier by Kant in his *Critique of Pure Reason* (Kant, 1781).

between the skulls of humans and gorillas than between those of gorillas and other primates.

The June 30 meeting was chaired by John Henslow (a catastrophist) and was attended by 700–1000 people, of whom we should mention Joseph Hooker, who was on Huxley's side, and Robert Fitzroy, of the Beagle fame. One notable absentee was Darwin himself, presumably due to poor health. At one point during the meeting, John W. Draper (1811–1882), who was a historian of science, talked about the effect of Darwin's ideas on intellectual development in Europe. Henslow then invited several anti-Darwinians to respond. Wilberforce seized the opportunity and launched a virulent attack on Darwin's theory. Then, turning to Huxley, he asked tauntingly:

> **QUOT. 127** Is it on his [Huxley's] grandfather's or his grandmother's side that the ape ancestry comes in? (Huxley, 1900, p. 270)[50]

Apparently, Huxley had anticipated such an attack and retorted:

> **QUOT. 128** I should feel it no shame to have arisen from such an origin; but I should feel it a shame of have sprung from one who prostituted the gifts of culture and eloquence to the service of prejudice and of falsehood. (*ibidem*)

This exchange reportedly created a moment of intense drama, with several other speakers chiming in, one Lady Brewster fainting, and Fitzroy parading a Bible on his head.

A few myths about the debate have persisted and need to be addressed:

(a) *The belief Huxley landed a big victory over Wilberforce.* For example, in his biography of Darwin, de Beer condemned "the bishop's ignorance, prejudice, and incompetence to deal with the subjects at issue" (de Beer, 1967, p. 166); similarly, in an article on the reception of Darwinism in England and Scotland, Burkhardt spoke of Huxley's "famous rhetoric victory over Bishop Samuel Wilberforce" (Burkhardt, 1974, p. 42). However, more recent authors have questioned the veracity of this depiction. Desmond's description is perhaps more accurate when he said, "[p]erceptions of the event differed so wildly that talk of a 'victor' is ridiculous." (Desmond, 1997, p. 280).

(b) *The belief that the Huxley–Wilberforce debate was a duel between science and religion.* A vivid description of the debate was provided in Andrew Dickson White's monumental *A History of the Warfare of Science with Theology with Christendom* (White, 1897, pp. 70–71, Vol. I), and the authority with which the author wrote invariably shaped much of twentieth-century thinking on the matter. Thus, in a recent book, Seybold (2016, p. 65) declared that "[t]he second famous example of 'conflict' between science and religion was the debate on Darwin's evolutionary theory between Samuel Wilberforce and Thomas

[50] Darwin's theory, of course, does not say that humans are descended from apes. It only says that humans and apes are both descended from a common ancestor.

Huxley." However, this statement cannot be completely true, given that some of the foremost scientists[51] of the time were aligned much more with Wilberforce than with Huxley. Similarly, but to a lesser extent, clergymen such as Charles Kingsley and Frederick Temple were not hostile to Huxley's views. The Huxley–Wilberforce debate was thus more *a clash between two different visions of the origin of species* than a fight between science and religion.

2.13 Just Before the "Origin": Alfred Wallace

2.13.1 Alfred Russel Wallace (1823–1913)

It is now time to consider one of the most important figures in the history of evolutionary thinking, namely Alfred Russel Wallace[52] (Fig. 2.10). Asa Gray might have been Darwin's retriever and Thomas Huxley his bulldog, but Alfred Wallace was his "true knight."[53] In his early youth, Wallace attended courses on agnosticism and secularism. He had virtually no knowledge of natural history until, at the age of 15, he read the fourth edition of John Lindley's *Elements of Botany* (Lindley, 1841), followed by Loudon's *Encyclopedia of Plants* (Loudon, 1829). In the mid-40s, Wallace notably also read Darwin's *Journal of the Voyage of the Beagle*[54] (Darwin, 1840), Malthus' *An Essay on the Principle of Population* (Malthus, 1798), and Lyell's *Principles of Geology* (Lyell, 1830). However, most noteworthy was Wallace's reading of the *Vestiges of the Natural History of Creation* (Chambers, 1844). As explained in Sect. 1.15, the book was ridiculed by virtually the whole scientific community (including Darwin). An important exception was Wallace, who, in a November 1845 letter to his good friend Henry Walter Bates[55] (1825–1892), declared:

> **QUOT. 129** I have rather a more favorable opinion of the "Vestiges" than you appear to have...
>
> I do not consider it as a hasty generalisation, but rather as an ingenious hypothesis strongly supported by some striking facts and analogies but which remains to be proved by

[51] Such as the geologist Adam Sedgwick (1785–1873), the zoologist Louis Agassiz (1873), and the philosopher William Whewell (1794–1866). Moreover, as mentioned, Wilberforce had been coached by the anatomist Owen.

[52] Wallace's life, work, and interactions with Darwin are also discussed in McKinney (1972), Berry (2003), van Wyhe (2013), Beddall (1968, 1988b), Shermer (2002), Slotten (2004), Fichman (2010), Smith and Beccaloni (2010), Costa (2014), Flannery (2018), Raby (2001), George (1964), Brooks (1984), Mayer (1987), Gaudant and Gaudant (1971, Chap. 11), Adams (1969, Chap. 12), and McCalman (2010).

[53] The phrase was first used by Hooker on April 19, 1868, in his Presidential address to the British Association for the Advancement of Science in Norwich (Hooker, 1868, p. lxxi).

[54] Which inspired Wallace's later travels.

[55] Who later gave the first scientific account of mimicry in animals.

Fig. 2.10 Alfred Russel Wallace (1823–1913). (From: https://commons.wikimedia.org/wiki/File:Alfred-Russel-Wallace-c1895.jpg)

> more facts & the additional light which future researches may throw upon the subject—It at all events furnishes a subject for every observer of nature to turn his attention to; every fact he observes must make either for or against it, and it thus furnishes both an incitement to the collection of facts & an object to which to apply them when collected. (Marchant, 1916, pp. 73–74)

Wallace went on an expedition in South America during 1848–1852 with Bates. During his return to England, Wallace's ship caught fire, and all his collections and notes were burnt. Consequently, his first book, *Palm Trees of the Amazon* (Wallace, 1853b) suffered from a lack of detail and did not sell well. His second book, *A Narrative of Travels on the Amazon and Rio Negro* (Wallace, 1853a), fared only slightly better. Wallace briefly met Darwin[56] for the first time in the British Museum in early 1854 (or late 1853).

[56]The meeting is not mentioned in Darwin's autobiography (Barlow, 1958), and Wallace later recollected that "nothing of importance" was exchanged between them during their brief conversation (Wallace, 1903, p. 78).

2.13.2 *The Sarawak Law Paper (1855)*

In March 1854, still a virtual unknown in the scientific community, Wallace left for the Malay Archipelago on a second expedition. In February 1855, while he was in Sarawak, northern Borneo, Wallace was struck with malaria. This gave him time to ponder over the question of the origin of species and a recent paper (Forbes, 1854) by the English naturalist Edward Forbes (1815–1854). In his paper, Forbes argued that more genera were produced during the early parts of the Paleozoic period and the later parts of the Neozoic period (= Mesozoic + Cenozoic) than during any other time span. He attributed his polarity theory to the "regulating law of the divinely originating scheme of creation" (*ibid.*, p. 432).

Wallace "was annoyed to see such an ideal absurdity put forth" (Marchant, 1916, p. 54) and wrote his first consequential paper, "On the law which has regulated the introduction of new species" (Wallace, 1855) (Fig. 2.11). This was a few months before, as Wallace learned later, Darwin started writing his Big Species book (in May 1856). Wallace's paper introduced four major ideas:

(a) "no group or species has come into existence twice" (*ibid.*, p. 186);
(b) (Sarawak Law): "every species has come into existence coincident both in space and time with a preexisting closely allied species" (*ibidem*);
(c) ". . .species being so numerous and the modifications of form and structure so varied, arising probably from the immense number of species which have served as antitypes for the existing species, and thus produced *a complicated branching of the lines of affinity, as intricate as the twigs of a gnarled oak or the vascular system of the human body*[57]" (*ibid.*, p. 187);
(d) "in the long series of changes the earth has undergone, the process of peopling it with organic beings has been continually going on, and whenever any of the higher groups have become nearly or quite extinct, *the lower forms which have better resisted the modified physical conditions have served as the antitypes on which to found the new races*[58]" (*ibid.*, 192).

Statement (a) above restates in time what is known in geography. In analogy with the fact that no species are found in distant localities without being also found in intermediate places, Wallace pointed out that the life of a species cannot be interrupted.

Statement (b) is called the Sarawak Law and is highly suggestive of a process of evolution or transmutation, although neither word is mentioned in the paper. Wallace's repeated use of "creation" and "created" in the paper misled some to believe that he was using creationist arguments like Chambers had done in the *Vestiges*. Thus, Darwin dismissed the paper because it has "nothing very new," "uses my [Darwin's] simile tree," and "it seems all creation with him [Wallace]" (Beddall, 1988a, p. 13). It is noteworthy that Lamarck had earlier made a statement

[57] Italics are ours.

[58] Italics are ours.

XVIII.—*On the Law which has regulated the Introduction of New Species.* By Alfred R. Wallace, F.R.G.S.

Every naturalist who has directed his attention to the subject of the geographical distribution of animals and plants, must have been interested in the singular facts which it presents. Many of these facts are quite different from what would have been anticipated, and have hitherto been considered as highly curious, but quite inexplicable. None of the explanations attempted from the time of Linnæus are now considered at all satisfactory; none of them have given a cause sufficient to account for the facts known at the time, or comprehensive enough to include all the new facts which have since been, and are daily being added. Of late years, however, a great light has been thrown upon the subject by geological investigations, which have shown that the present state of the earth, and the organisms now inhabiting it, are but the last stage of a long and uninterrupted series of changes which it has undergone, and consequently, that to endeavour to explain and account for its present condition without any reference to those changes (as has frequently been done) must lead to very imperfect and erroneous conclusions.

The facts proved by geology are briefly these:—That during an immense, but unknown period, the surface of the earth has undergone successive changes; land has sunk beneath the ocean, while fresh land has risen up from it; mountain chains have been elevated; islands have been formed into continents, and continents submerged till they have become islands; and these changes have taken place, not once merely, but perhaps hundreds, perhaps thousands of times:—That all these operations have been more or less continuous, but unequal in their progress, and during the whole series the organic life of the earth has undergone a corresponding alteration. This alteration also has been gradual, but complete; after a certain interval not a single species existing which had lived at the commencement of the period. This complete renewal of the forms of life also appears to have occurred several times:—That from the last of the Geological epochs to the present or Historical epoch, the change of organic life has been gradual: the first appearance of animals now existing can in many cases be traced, their numbers gradually increasing in the more recent formations, while other species continually die out and disappear, so that the present condition of the organic world is clearly derived by a natural process of gradual extinction and creation of species from that of the latest geological periods. We may therefore safely infer a like gradation and natural sequence from one geological epoch to another.

Fig. 2.11 First page of Wallace's "Sarawak Law" paper (Wallace, 1855) similar to Wallace's: "every individual possessing life always resembles very closely those from which it sprang" (Lamarck, 1809, 1984, p. 35). Similarly, Chambers had previously stated: "The idea, then, which I form of the progress of organic life upon the globe. . .is, *that the simplest and most primitive type, under a law to which that of like-production is subordinate, gave birth to the type next above it, that this again produced the next higher, and so on to the very highest*[59] . . ." (Chambers, 1844, p. 222). However, these authors used theories that were either overly speculative (Lamarck) or supernatural (Chambers). Only Wallace's version was cogent and based on observable facts of geography and geology.

Statement (c) refers to the concept of divergence, which Darwin had previously recognized as a vital component of his theory of evolution. Divergence is the accumulation of different characters in related species as they evolve and thus become increasingly different. There has been some controversy regarding who first discovered divergence and whether Darwin borrowed the idea from Wallace's 1855 paper. This topic will soon be revisited.[60]

Statement (d) suggests some form of struggle for existence: only species that can resist modified physical conditions lead to new, modified species.

It is thus seen that although Wallace's Sarawak Law paper explicitly mentions neither evolution nor a mechanism for it (natural selection), both concepts are more or less strongly suggested, albeit indirectly.

The Sarawak Law paper was published in the *Annals & Magazine of Natural History* in September 1855. In the next few years, Wallace published several papers (e.g., Wallace, 1856a, b) addressing the species question. He was also anxious to know the reaction to his 1855 paper and wrote to Darwin on September 27, 1857. The latter replied thus on December 22, 1857:

> **QUOT. 130** You say that you have been somewhat surprised at no notice having been taken of your paper in the Annals. I cannot say that I am, for so very few naturalists care for anything beyond the mere description of species. But you must not suppose that your paper has not been attended to: two very good men, Sir C. Lyell, and Mr. E. Blyth at Calcutta, specially called my attention to it. Though agreeing with you on your conclusions in the paper, I believe I go much further than you; but it is too long a subject to enter on my speculative notions. (Marchant, 1916, p. 109)

Indeed, Lyell had read Wallace's paper on November 26, 1855, and was quite impressed by it, even opening the first of seven volumes of a species notebook. More importantly, Lyell had seen in the paper perhaps much more than Darwin had, namely that Wallace was edging toward a theory of evolution. A few months later, on April 16, 1856, Lyell visited Darwin at Down House, at which point he was informed of Darwin's theory.[61] Thereupon, Lyell informed Darwin of the Sarawak Law paper, urging him to publish it at once. However, Darwin was not ready. A few weeks later, Lyell wrote to Darwin, urging him again to publish:

---

[59] Italics are Chambers'.

[60] See Sect. 2.13.5.1.

[61] See Sect. 1.14.4.

QUOT. 131 I wish you would publish some small fragment of your data, pigeons if you please & so out with the theory & let it take date & be cited & understood. (Wilson, 1970, p. xlvii)

Darwin was still reticent:

QUOT. 132 To give a fair sketch would be absolutely impossible, for every proposition requires such an array of facts. . .I do not know what to think. . . . I certainly should be vexed if any one were to publish my doctrines before me. (*ibid.*, p. xlvi)

Darwin would soon regret his reluctance.

2.13.3 The "Ternate" Essay (1858)

Before he received Darwin's response to his September 27 letter, Wallace traveled toward Amboyna on November 19, 1857. He stayed in Amboyna until early January 1858, at which point he traveled to Ternate, reaching there on January 8, 1858. On approximately January 20, Wallace left for the island of Gilolo (Halmahera). It was most likely from Gilolo[62] that Wallace wrote his breakthrough paper, "On the Tendency of Varieties to Depart Indefinitely from the Original Type." This is how he later recollected his discovery of the mechanism of natural selection[63] for evolution:

QUOT. 133 I was then living at Ternate in the Moluccas, and was suffering from a rather severe attack of intermittent fever, which prostrated me for several hours every day during the cold and succeeding hot fits. During one of these fits, while again considering the problem of the origin of species, something led me to think of Malthus' Essay on Population (which I had read about ten years before), and the "positive checks"—war, disease, famine, accidents, etc.—which he adduced as keeping all savage populations nearly stationary. It then occurred to me that these checks must also act upon animals, and keep down their

[62] Although Wallace himself later maintained that the paper was written in Ternate (e.g., Marchant, 1916, pp. 86, 88, 295), his itinerary suggests he was in Gilolo. Slotten (2004) has described this as an "incongruity that Wallace never clarified, and that remains a minor mystery." McKinney has gone even further and has accused Wallace of deception: "What difference did it really make if he was on Gilolo rather than Ternate? A distance of ten or twenty miles and a mere three to five hours of travel time probably had little influence on his actual discovery of natural selection, and catching Wallace in his fib does not alter the fact and importance of his discovery. Nevertheless, Wallace's mendacity does reveal a great deal about his character. While we can never perhaps fully understand his motives, we do know that he was a young man extremely anxious to establish a name for himself. He had pondered upon the question of how species evolve for about thirteen years, and he was acutely aware of the far-reaching significance of his discovery. No doubt, he realized that such momentous events tend to be recounted time and again in the pages of history books. It simply would not do to make such a discovery on what he described as the 'large and little known island of Gilolo', a veritable 'terra incognita'" (McKinney, 1972, p. 137). Nevertheless, the 1858 paper is still referred to as the "Ternate paper" by most authors.

[63] Wallace used "natural selection" only later, for the first time in 1860 (Wallace, 1860, p. 144).

numbers; and as they increase so much faster than man does, while their numbers are always very nearly or quite stationary, it was clear that these checks in their case must be far more powerful, since a number equal to the whole increase must be cut off by them every year. While vaguely thinking how this would affect any species, there suddenly flashed upon me the idea of *the survival of the fittest*[64]—that the individuals removed by these checks must be, on the whole, *inferior* to those that survived. Then, considering the *variations* continually occurring in every fresh generation of animals or plants, and the changes of climate, of food, of enemies always in progress, the whole method of specific modification became clear to me, and in the two hours of my fit I had thought out the main points of the theory. (Wallace, 1898, pp. 139–140)

Wallace wrote the whole theory over the next two evenings in a landmark essay. In the latter, he declared confidently:

QUOT. 134 The life of wild animals is a struggle for existence. The full exertion of all their faculties and all their energies is required to preserve their own existence and provide for that of their infant offspring. The possibility of procuring food during the least favourable seasons, and of escaping the attacks of their most dangerous enemies, are the primary conditions which determine the existence both of individuals and of entire species. (Darwin & Wallace, 1858, p. 54)

By winning the struggle, the species would ensure its prolonged existence and preponderance in future generations:

QUOT. 135 It is also evident that most changes would affect, either favourably or adversely, the powers of prolonging existence. If. . .any species should produce a variety having slightly increased powers of preserving existence, that variety must inevitably in time acquire a superiority in numbers. (*ibidem*)

Thereupon, Wallace sent the paper to none other than Darwin,[65] together with a letter in which, as he later recalled:

QUOT. 136 . . .I [Wallace] said that I hoped the idea would be as new to him as it was to me, and that it would supply the missing factor to explain the origin of species. I asked him if he thought it sufficiently important to show to Sir Charles Lyell, who had thought so highly of my former paper. (Wallace, 1905, p. 363)

[64] All italics in this quotation are Wallace's.

[65] Darwin was busy writing his massive Big Species book at the time. Wallace chose to write to Darwin rather than send the manuscript for publication directly (as he had done for the Sarawak Law paper), most probably because he knew Darwin was working on similar matters. In his Dec 22, 1857, letter to Wallace, Darwin said: "You ask whether I shall discuss Man; I think I shall avoid the whole subject, as so surrounded with prejudices, though I fully admit that it is the highest and most interesting problem for the naturalist. My work, on which I have now been at work more or less for twenty years, will not fix or settle anything; but I hope it will aid by giving a large collection of facts with one definite end. I get on very slowly, partly from ill-health, partly from being a very slow worker. I have got about half written; but I do not suppose I shall publish under a couple of years. I have now been three whole months on one chapter on hybridism" (Marchant, 1916, p. 110).

2.13.4 *Presentation of the Joint Darwin–Wallace Paper to the Linnean Society in 1858*

The whereabouts of both the original Wallace manuscript and the accompanying letter are no longer known, but what is known is that they came to Darwin on June 18, 1858, "like a bolt from the blue."[66] Darwin was understandably distraught at the prospect of being forestalled by Wallace. In a July 18, 1858, letter to Lyell, he expressed his full dismay:

> **QUOT. 137** My dear Lyell,—Some year or so ago you recommended me to read a paper by Wallace in the "Annals," which had interested you and as I was writing to him, I knew this would please him much, so I told him. He has to-day sent me the enclosed, and asked me to forward it to you. It seems to me well worth reading. Your words have come true with a vengeance—that I should be forestalled. You said this, when I explained to you here very briefly my views of "Natural Selection" depending on the struggle for existence. I never saw a more striking coincidence; if Wallace had my MS. sketch written out in 1842, he could not have made a better short abstract! Even his terms now stand as heads of my chapters. Please return me the MS which he does not say he wishes me to publish, but I shall, of course, at once write and offer to send to any Journal. So all my originality, whatever it may amount to, will be smashed, though my book, if it will ever have any value, will not be deteriorated; as all the labour consists in the application of the theory.
>
> I hope you will approve of Wallace's sketch, that I may tell him what you say.
>
> My Dear Lyell, yours most truly,
> C. DARWIN
> (Darwin, 1888a, pp. 116–117)

Since Lyell did not reply immediately, Darwin wrote a further letter to him on June 25, 1858. In that letter, whose every word showed Darwin's anguish, he volunteered to publish his 1844 essay (which Hooker had read years before) and to mention that he was prompted to do so by Wallace's paper. Furthermore, he suggested sending Wallace a copy of the letter he had previously sent to Gray to show that he "had not stolen his [Wallace's] doctrine." Finally, Darwin asked Lyell to send everything to Hooker so that he could have the opinions of his "two best and kindest friends." Another postscript was sent to Lyell the next day. The latter eventually contacted Hooker, who, on June 29, urged Darwin to send any material that could be presented together with Wallace's manuscript to the Linnean Society in 3 days. However, Darwin's tenth and youngest child (Charles Waring Darwin) had just died (on June 28), and the only material he could send to Hooker was a handwritten copy of his previous September 1857 letter to Asa Gray[67] and his 1844 essay.[68] On July 1, 1858, a combined paper (Fig. 2.12) was presented by Lyell and Hooker to the Linnean Society, consisting of parts in the following order:

---

[66] In 1891, Leonard Huxley wrote: "Again, when the Abstract [Origin of Species] had been going after Wallace's paper had come like a bolt from the blue, he [Darwin to Hooker] cries, 'in how many ways have you aided me'" (Huxley, 1900, pp. 499–500, Vol. I).

[67] See Sect. 2.12.

[68] See Sect. 2.11.

- An introductory note by Lyell and Hooker.
- Extracts from Darwin's 1844 Essay.[69]
- Extracts from Darwin's 1857 letter to Asa Gray.[70]
- Wallace's 1859 manuscript.

It may be noted that Darwin's and Wallace's contributions were arranged chronologically, a fact that was explicitly stated by Lyell and Hooker in the introductory note.

The small audience attending the presentation was not ready for the radical ideas in the Darwin–Wallace paper and seemed to have been underwhelmed. Thomas Bell, who was the new president of the society, later even claimed:

> The year which has passed [1858]. . .has not, indeed, marked by any of those striking discoveries which at once revolutionize, so to speak, the department of science on which they bear. (Bell, 1859, p. viii)

Bell's declaration would soon prove to be one of the greatest understatements of the nineteenth century.

The actions undertaken by Lyell and Hooker leading to the publication of the Darwin–Wallace paper have been called by a few as "a delicate arrangement."[71] There are certainly some gray areas in the whole affair, and this has prompted a few authors (e.g., Brackman, 1980; Brooks, 1984; Davies, 2008) to level some charges, namely:

(a) Darwin's colleagues conspired to give him priority by failing to seek Wallace's approval before the 1858 paper was presented.
(b) The disappearance of several key documents, including Wallace's original "Ternate" essay and the accompanying letter, was due to Darwin or his friends wishing to eliminate any "incriminating" evidence, such as Darwin borrowing ideas from Wallace.
(c) Darwin received the "Ternate" essay at least 2 weeks before the alleged June 18, 1859, date. During these weeks, he used Wallace's essay to refine his conception of natural selection.
(d) Darwin borrowed the principle of divergence[72] from Wallace's 1855 and/or 1858 papers.

However, none of the authors cited above has been able to convincingly prove their case, and several authors have rebutted the arguments above (e.g., Shermer, 2002, pp. 129–140; Costa, 2014, pp. 157–162, Chap. 6). Most importantly, Wallace never

[69] Darwin added a footnote next to his Essay in the joint paper: "This MS. work was never intended for publication, and therefore was not written with care."

[70] The reason Darwin sent a portion of his letter to Asa Gray was that it included a discussion of the principle of divergence (Sect. 2.13.5).

[71] The phrase was first used by Leonard Huxley.

[72] See Sect. 2.13.5.1.

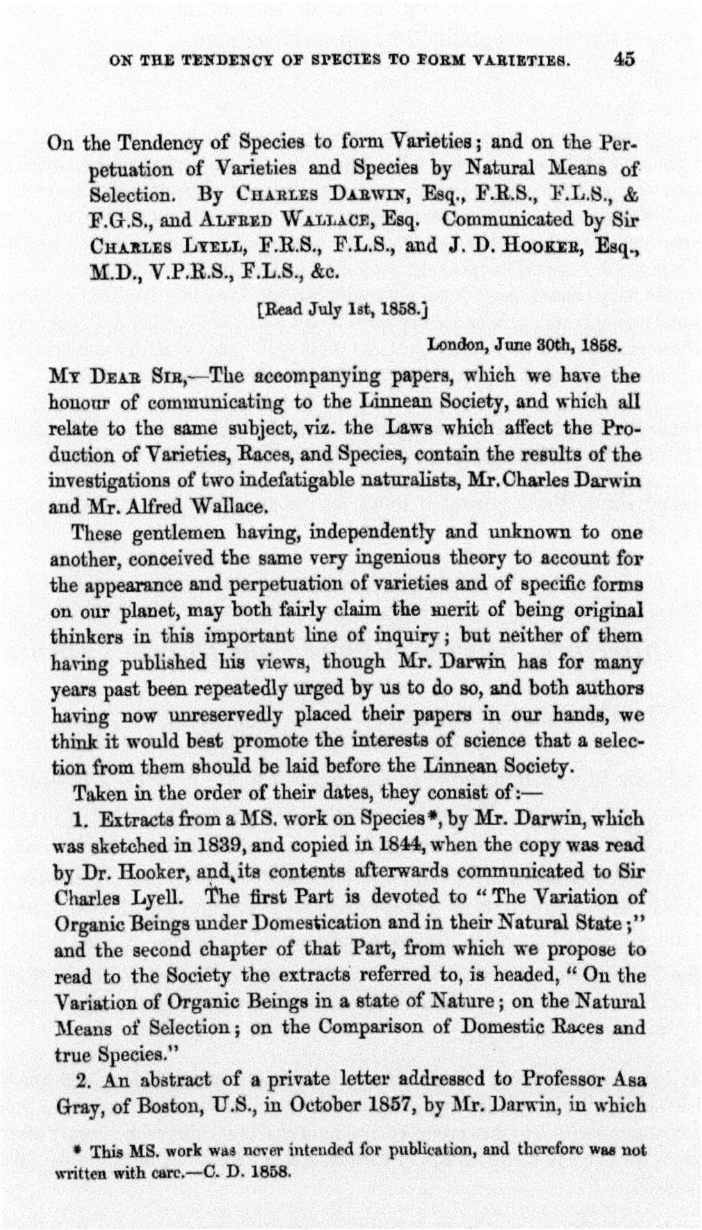

ON THE TENDENCY OF SPECIES TO FORM VARIETIES. 45

On the Tendency of Species to form Varieties; and on the Perpetuation of Varieties and Species by Natural Means of Selection. By CHARLES DARWIN, Esq., F.R.S., F.L.S., & F.G.S., and ALFRED WALLACE, Esq. Communicated by Sir CHARLES LYELL, F.R.S., F.L.S., and J. D. HOOKER, Esq., M.D., V.P.R.S., F.L.S., &c.

[Read July 1st, 1858.]

London, June 30th, 1858.

MY DEAR SIR,—The accompanying papers, which we have the honour of communicating to the Linnean Society, and which all relate to the same subject, viz. the Laws which affect the Production of Varieties, Races, and Species, contain the results of the investigations of two indefatigable naturalists, Mr. Charles Darwin and Mr. Alfred Wallace.

These gentlemen having, independently and unknown to one another, conceived the same very ingenious theory to account for the appearance and perpetuation of varieties and of specific forms on our planet, may both fairly claim the merit of being original thinkers in this important line of inquiry; but neither of them having published his views, though Mr. Darwin has for many years past been repeatedly urged by us to do so, and both authors having now unreservedly placed their papers in our hands, we think it would best promote the interests of science that a selection from them should be laid before the Linnean Society.

Taken in the order of their dates, they consist of:—

1. Extracts from a MS. work on Species*, by Mr. Darwin, which was sketched in 1839, and copied in 1844, when the copy was read by Dr. Hooker, and its contents afterwards communicated to Sir Charles Lyell. The first Part is devoted to "The Variation of Organic Beings under Domestication and in their Natural State;" and the second chapter of that Part, from which we propose to read to the Society the extracts referred to, is headed, "On the Variation of Organic Beings in a state of Nature; on the Natural Means of Selection; on the Comparison of Domestic Races and true Species."

2. An abstract of a private letter addressed to Professor Asa Gray, of Boston, U.S., in October 1857, by Mr. Darwin, in which

* This MS. work was never intended for publication, and therefore was not written with care.—C. D. 1858.

Fig. 2.12 First page of the Darwin–Wallace 1858 Linnean paper (Darwin & Wallace, 1858) gave any hint of slight from the Darwin camp. In fact, his letter to Hooker on Oct 6, 1858, shows that he was gratified by the arrangement:

QUOT. 138
My dear Sir
I beg leave to acknowledge the receipt of your letter of July last, sent me by Mr. Darwin, & informing me of the steps you had taken with reference to a paper I had communicated to that gentleman. Allow me in the first place sincerely to thank yourself & Sir Charles Lyell for your kind offices on this occasion, & to assure you of the gratification afforded me both by the course you have pursued & the favourable opinions of my essay which you have so kindly expressed. I cannot but consider myself a favoured party in this matter...
It would have caused me such pain & regret had Mr. Darwin's excess of generosity led him to make public my paper unaccompanied by his own much earlier & I doubt not much more complete views on the same subject, & I must again thank you for the course you have adopted, which while strictly just to both parties, is so favourable to myself.
...
I remain/My dear Sir/Yours very sincerely/Alfred R. Wallace
(Burkhardt & Smith, 1991, p. 166)

On the same day, Wallace sent a letter to his mother, too, expressing a similar sentiment.

2.13.5 Differences Between Wallace's and Darwin's Theories

2.13.5.1 The Principle of Divergence

As stated in the last section, none of the charges made in Sect. 2.13.4(a)–(d) have been convincingly proven. Yet, it is important to address the issue of divergence in (d) since it is a crucial element in the theories of both Darwin and Wallace. In both theories, divergence arose from considerations of *classification*. Already in the B Notebook of 1837, Darwin drew a coral tree (Sect. 2.5 and Fig. 2.4) that represented different daughter species branching out from a parent species. Kohn (1981, p. 1105) called this classification *taxonomic divergence*, which is simply the arrangement of taxa in a branched, diverging scheme. Taxonomic divergence was also referred to in Wallace's Sarawak Law paper:[73]

QUOT. 139 ...species being so numerous and the modifications of form and structure so varied, arising probably from the immense number of species which have served as antitypes for the existing species, and thus produced a complicated branching of the lines of affinity, as intricate as the twigs of a gnarled oak or the vascular system of the human body. (Wallace, 1855, p. 167)

Given that both authors were aware of taxonomic divergence, it remained for them to explain the *modus operandi* behind it, i.e., it remained to formulate a *principle of divergence*. Darwin was able to discover the principle in November 1854 by

[73] See Sect. 2.13.2.

adapting Henri Milne-Edwards' concept of the "physiological division of labor"[74] (Milne-Edwards, 1851, p. 35, Chap. III) into an "ecological division of labor." In a September 23, 1856, note, Darwin used the concept of the division of labor as follows:

> **QUOT. 140** The advantage in each group becoming as different as possible, may be compared to the fact that by division of ~~land~~ labour most people can be supported in each country—Not only do the individuals of each group strive one against the other, but each group itself with all its members, some more numerous, some less, are struggling against all other groups, as indeed follows from each individual struggling.[75]

The principle of divergence was listed in the table of contents of Darwin's Big Species book in March 1857. It was then mentioned by Darwin in his Aug 22, 1857, letter to Hooker and first expressed in full in Darwin's Sep 5, 1857, letter to Asa Gray:

> **QUOT. 141** . . .One other principle, which may be called the principle of divergence, plays, I believe, an important part in the origin of species. The same spot will support more life if occupied by very diverse forms: we see this in the many generic forms in a square yard of turf (I have counted twenty species belonging to eighteen genera), or in the plants and insects, on any little uniform islet, belonging to almost as many genera and families as to species. We can understand this with the higher animals, whose habits we best understand. We know that it has been experimentally shown that a plot of land will yield a greater weight, if cropped with several species of grasses, than with two or three species. Now every single organic being, by propagating rapidly, may be said to be striving its utmost to increase in numbers. So it will be with the offspring of any species after it has broken into varieties, or sub-species, or true species. And it follows, I think, from the foregoing facts, that the varying offspring of each species will try (only few will succeed) to seize on as many and as diverse places in the economy of nature as possible. Each new variety or species when formed will generally take the place of, and so exterminate its less well-fitted parent. This, I believe, to be the origin of the classification or arrangement of all organic beings at all times. These always seem to branch and sub-branch like a tree from a common trunk; the flourishing twigs destroying the less vigorous-the dead and lost branches rudely representing extinct genera and families. (Darwin, 1888a, pp. 124–125)

Darwin's principle of divergence was thus based on the creation of new evolutionary opportunities through the existence of diverse environmental conditions.

It is now possible to see why Darwin chose to include the portion above in the material he sent for presentation to the Linnean Society: Darwin regarded the principle of divergence as crucial[76] to his whole theory, and this material had not been developed in his 1844 *Essay*.

---

[74] According to this, different parts of the body are specialized for different tasks.

[75] Charles Darwin, Manuscript Notes, held in Special Collections, Cambridge University Library, DAR 205.5.171.

[76] Darwin to Hooker on June 8, 1858: "...the 'Principle of Divergence' which with 'Natural Selection' is the key-stone of my Book..." (Burkhardt & Smith, 1991, p. 102, Vol. VII).

Did Wallace formulate a similar principle of divergence? On reading his portion of the joint 1858 paper, it is seen that his version of divergence was more restricted and was not based on an explanatory principle that had the coherence of Darwin's:

QUOT. 142 The *variety*[77] would now have replaced the *species*, of which it would be a more perfectly developed and more highly organized form. It would be in all respects better adapted to secure its safety, and to prolong its individual existence and that of the race. Such It variety *could not* return to the original form; for that form is an inferior one, and could never compete with it for existence. . . .But this new, improved, and populous race might itself, in course of time, give rise to new varieties, exhibiting several diverging modifications of forms, any of which, tending to increase the facilities for preserving existence, must, by the same general law, in their turn become predominant. Here, then, we have *progression and continued divergence* deduced from the general laws which regulate the existence of animals in a state of nature, and from the undisputed fact that varieties do frequently occur. (Darwin & Wallace, 1858, pp. 58–59)

Thus, Wallace's conception of divergence involved *the formation of new varieties through the extinction of a parent species*. Subsequently, the new varieties were replaced by newer ones, and so on, thus giving rise to continued "progression and divergence." Wallace's explanation for divergences seems to involve nothing more than the process of natural selection. In Darwin's superior version, the concept of the division of labor was a cogent explanation as to how several varieties branching from a single species can coexist, each adapted to a slightly different environment.

The foregoing does not lend any credence to the charge that Darwin had usurped the idea of divergence from Wallace.

2.13.5.2 Domesticated versus Natural Species

As was common belief at the time, Wallace drew a sharp contrast between domesticated varieties and those in nature. In fact, the major thesis of his portion of the 1858 paper was that:

QUOT. 143 . . .there is a general principle in nature which will cause many varieties to survive the parent species, and to give rise to successive variations departing further and further from the original type, and which also produces, in domesticated animals, the tendency of varieties to return to the parent form. (*ibid.*, p. 54)

On the other hand, Darwin thought that the model of selection in domesticated species was as relevant as the model of selection in natural species.[78] He questioned

---

[77] All italics in this quotation are Wallace's.

[78] This does not mean that Darwin's discovery of natural selection proceeded from domesticated to natural species, as has been pointed out by Limoges (1970, pp. 83–85, 99–100) and Gayon (1998, p. 36). Limoges notes that, in his June 25, 1858, letter to Lyell, Darwin had said: "We [Darwin and Wallace] differ only [in] that I was led to my views from what artificial selection has done for domestic animals." However, Limoges then points out that: ". . .Darwin did not proceed in this way: the Notebooks show that the relationship between domestication and natural means of species production was realized only on December 16, 1838, that is, about ten weeks after the solution [of September 28, 1838, see Sect. 2.8] had been discovered" (Limoges, 1970, p. 100).

the common belief that domesticated animals tended to revert to their wild ancestors when left in nature:

> **QUOT. 144** Having alluded to the subject of reversion, I may here refer to a statement often made by naturalists namely, that our domestic varieties, when run wild, gradually but certainly revert in character to their aboriginal stocks. Hence it has been argued that no deductions can be drawn from domestic races to species in a state of nature. I have in vain endeavoured to discover on what decisive facts the above statement has so often and so boldly been made. (Darwin, 1859, p. 14)

2.13.5.3 Lamarckian Evolution

Wallace categorically rejected Lamarckian evolution. In the 1858 joint paper, one can read:

> **QUOT. 145** The hypothesis of Lamarck—that progressive changes in species have been produced by the attempts of animals to increase the development of their own organs, and thus modify their structure and habits... [is] quite unnecessary, by showing that similar results must be produced by the action of principles constantly at work in nature. (Darwin & Wallace, 1858, p. 61)

On the other hand, Darwin accepted the Lamarckian concept of the effect of the environment on the reproductive system of the organism:

> **QUOT. 146** ...changes of external conditions would, from their acting on the reproductive system, probably cause the organization of those beings which were most affected to become, as under domestication, plastic. (*ibid.*, p. 49)

2.13.5.4 Units of Selection

Darwin's unit of selection was the individual, and natural selection operated through slight variations between individuals. In the 1858 joint paper, we can read:

> **QUOT. 147** Now, can it be doubted, from the struggle each individual has to obtain subsistence, that any minute variation in structure, habits, or instincts, adapting that individual better to the new conditions, would tell upon its vigour and health? (*ibidem*)

On the other hand, Wallace's units of selection were the fully formed varieties, i.e., organisms that had departed from the parental type:

> **QUOT. 148** ...there is a general principle in nature which will cause many varieties to survive the parent species, and to give rise to successive variations departing further and further from the original type, and which also produces, in domesticated animals, the tendency of varieties to return to the parent form. (*ibid.*, p. 54)

However, in an 1896 letter to the biologist Edward Poulton, Wallace denied the charge that he had originally thought of natural selection acting only at the level of varieties:

QUOT. 149 I used the term "varieties" because "varieties" were alone recognised at that time, individ[l] variability being ignored or thought of no importance. My "varieties" therefore included "individual variations." (Poulton, 1896, p. 80)

2.13.5.5 Heredity

In Darwin's theory, the concept of heredity was essential. In the 1858 joint paper, he stressed that:

QUOT. 150 . . .any minute variation in structure, habits, or instincts, adapting that individual better to the new conditions. . .would have a better chance of surviving; and those of its offspring which inherited the variation, be it ever so slight, would also have a better chance. (Darwin & Wallace, 1858, p. 49)

On the other hand, in Wallace's theory, heredity was not explicitly stated but only implied:

QUOT. 151 If . . . any species should produce a variety having slightly increased powers of preserving existence, that variety must inevitably in time acquire a superiority in numbers. (*ibid.*, p. 58)

Furthermore, Wallace (wrongly) believed that an understanding of the laws of inheritance was unnecessary for a consistent theory of evolution. Thus, in his reply to Bennett's article on natural selection (Bennett, 1870), Wallace said:

QUOT. 152 Mr. Bennett next returns to the laws of variation, and, because Mr. Darwin says that we are profoundly ignorant of these (although he himself has done so much to elucidate them), maintains that we cannot really know anything of the origin of species. As well might it be said that, because we are ignorant of the laws by which metals are produced and trees developed, we cannot know anything of the origin of steamships and railways. Spontaneous "variations" are but the materials out of which "species" are formed, and we do not require to know how the former are produced in order to learn the origin of the latter. (Wallace, 1870, p. 50)

2.13.5.6 Sexual Selection

Darwin's portion of the 1858 paper contained an explicit reference to, as Darwin originally conceived the idea, a *secondary* means of selection, namely "sexual selection." Darwin had been aware of sexual preferences since 1838.[79] In his 1842 sketch, his ideas about sexual selection were more definite, and he described it as follows:

---

[79] For example, immediately after his Malthusian insight (Sect. 2.8), Darwin noted, "Dr Smith every baboon & monkey, big & little that ever he saw knew women—he has repeatedly seen them try to pull up petticoats., & if woman not afraid clasp them round waist & look in their faces & Mak the st. st noise.—The Cercopithecus chinensis: (or bonnet faced monkey he has seen do this.—These Monkeys had no curiosity to pull up trousers of men. Evidently knew ‹men› women, thinks perhaps by smell.—but monkeys examine sexes of every. . ." (Barrett et al., 1987).

> **QUOT. 153** Besides selection by death, in bisexual animals (illegible) the selection in time of fullest vigour, namely struggle of males; even in animals which pair there seems a surplus (?) and a battle, possibly as in many more males produced than females, struggle of war or charms. Hence that male which at that time is in fullest vigour, or best armed with arms or ornaments of its species, will gain in hundreds of generations some small advantage and transmit such characters to its offspring. So in female rearing its young, the most vigorous and skilful and industrious, (whose) instincts (are) best developed, will rear more young, probably possessing her good qualities, and a greater number will thus (be) prepared for the struggle of nature. (Darwin, 1909, p. 10)

Similar ideas were repeated in the 1844 Essay (and Darwin's section of the joint 1858 paper), the *Origin*, and finally (and much more extensively) in *The Descent of Man, and Selection in Relation to Sex* (Darwin, 1871b).

In the *Origin*, Darwin alluded to two modes of sexual selection: (i) sexual selection involving *male rivalry*, i.e., a "struggle between the males for possession of the females" (Darwin, 1859, p. 88); (ii) sexual selection involving the *female choice* of males who, for example, in the case of birds, "display their gorgeous plumage and perform strange antics before the females, which standing by as spectators, at last, choose the most attractive partner" (*ibidem*).

Wallace generally believed in the overarching power of natural selection and rejected other forms of selection. There was no mention of sexual selection in his portion of the 1858 paper. He conceded that sexual selection involving male rivalry might have some evolutionary role but regarded it as just another "form of natural selection, which increases the vigour and fighting power of the male animal, since, in every case, the weaker are either killed, wounded, or driven away" (Wallace, 1889, p. 282). Concerning sexual selection involving female choice, Wallace was firmer in rejecting it. Drawing on the research of his friend William Bates on mimicry, Wallace argued that the colorations and external "ornaments" of animals were adaptations based on the evolutionary advantage gained by these animals when they resemble other less risk-prone animals (mimicry) or their environment (camouflage).[80]

2.13.5.7 Terminology

To illustrate Nature's "picking" of favorable variations to be transmitted to the next generation, Darwin used "natural means of selection" in the 1842 sketch, the 1844 essay, and the 1858 joint paper. This was changed to "natural selection" in the *Origin*. Darwin's metaphorical use of the term was meant to contrast it with artificial selection by animal breeders.[81] On the other hand, owing to his approach, which

[80] However, in a stunning reversal, Wallace espoused female choice after 1890, owing to his embracement of socialism. See Fichman (2010, pp. 268–270) for more details.

[81] Hodge reports that "by 1800 the use of the verb 'to select' was commonplace in writers on animal and plant breeding. A pamphlet by John Sebright—later to be a decisive source for Darwin's knowledge of such subjects—was entitled *The Art of Improving the Breeds of Domestic Animals*, and in it Sebright declared: 'Were I to define what is called the art of breeding, I should say, that it consisted in the selection of males and females, intended to breed together, in reference to each other's merits and defects' (Sebright, 1809, p. 5). As the definition indicates explicitly, this selection was an intentional action by men, acting on judgments of merit and defect" (Hodge, 1994, p. 213).

drew no analogy to domestication, there was no mention of natural selection in Wallace's portion of the 1858 paper. Indeed, Wallace was skeptical that "natural selection" was an apt word due to its personification of nature and implied teleology. In a July 2, 1866, letter to Darwin, he tried to persuade the latter to drop the term in favor of Herbert Spencer's "survival of the fittest":'[82]

QUOT. 154 I think this [misunderstanding of natural selection] arises almost entirely from your choice of the term Natural Selection, and so constantly comparing it in its effects to man's selection, and also to your so frequently personifying nature as "selecting," as "preferring," as "seeking only the good of the species," etc., etc... I wish, therefore, to suggest to you the possibility of entirely avoiding this source of misconception in your great work (if not now too late)...by adopting Spencer's term (which he generally uses in preference to Natural Selection), viz. "Survival of the Fittest." This term is the plain expression of the fact, Natural Selection is a metaphorical expression of it, and to a certain degree indirect and incorrect, since, even personifying Nature, she does not so much select special variation as exterminate the most unfavourable ones. (Marchant, 1916, pp. 140–141)

In his reply to Wallace's letter, Darwin objected to the use of "survival of the fittest" because it could not be used as a noun. He further declared that:

QUOT. 155 Natural Selection has now been so largely used abroad and at home that I doubt whether it could be given up, and with all its faults I should be sorry to see the attempt made. Whether it will be rejected must now depend "on the survival of the fittest." (*ibid.*, p. 144)

However, he did use Spencer's "survival of the fittest" in 1868 in the *Variation of Animals and Plants Under Domestication*[83] (Darwin, 1868a, p. 6, Vol. I) and in 1869 in the fifth edition of the *Origin* (Darwin, 1869, p. 72), but not at the expense of "natural selection." Likewise, despite his initial doubts about the term, Wallace used it as the subtitle of his 1989 book *Darwinism* (Wallace, 1889).

[82]This expression was coined by the philosopher biologist Herbert Spencer (1820–1903) in his *Principles of Biology*: "This survival of the fittest implies multiplication of the fittest" (Spencer, 1864, pp. 444–445, Vol. I). Then on the next page: "This survival of the fittest, which I have here sought to express in mechanical terms, is that which Mr Darwin has called 'natural selection' or the Preservation of favoured races in the struggle for life". It is inaccurate to equate natural selection with the survival of the fittest because the fittest is, by definition, those who survive, and this makes natural selection a tautology. But natural selection is much more than the survival of the fittest. Natural selection posits that organisms possessing variations that make them best adapt to their environment will, on average, win the struggle for survival and reproduction; furthermore, if the variations are heritable, they will increase from generation to generation, leading to evolution. Thus, natural selection is *not* a tautology—it is experimentally falsifiable and, thus, a genuine scientific theory.

[83]"The preservation, during the battle for life, of varieties which possess any advantage in structure, constitution, or instinct, I have called Natural Selection; and Mr Herbert Spencer has well expressed the same idea by the Survival of the Fittest. The term 'natural selection' is in some respects a bad one, as it seems to imply conscious choice; but this will be disregarded after a little familiarity" (Darwin, 1868a, p. 6, Vol. I).

2.13.5.8 Human Evolution and Vitalism

There was a further difference between Darwin and Wallace, but this was not yet present in the joint 1858 paper. Schwartz (1984) has argued that, after the publication of the 1858 paper, Wallace had started to believe that humans evolved differently from other species. His view on evolution became one of vitalism,[84] whereby life was endowed with a special force that was distinct from mechanistic forces. In an 1864 essay, Wallace thus argued that, in the evolutionary history of man,

> **QUOT. 156** . . .from the moment that his body became stationary, his mind would become subject to those very influences from which his body had escaped; every slight variation in his mental and moral nature which should enable him better to guard against adverse circumstances, and combine for mutual comfort and protection, would be preserved and accumulated. (Wallace, 1864, p. clxiv)

That is, after a while, natural selection stopped acting on man's physical abilities, but his mental evolution continued. This led Wallace to infer the "true grandeur and dignity of man. . .he is, indeed, a being apart, since he is not influenced by the great laws which irresistibly modify all other organic beings" (*ibid.*, p. clxviii).

In 1865, Wallace converted to spiritualism. One year later, he published *The Scientific Aspects of the Supernatural* (Wallace, 1866).

In a famous 1869 article (Wallace, 1869), Wallace went even further, resulting in an initial break between himself and Darwin on the issue of human evolution.[85] In the article, which reviewed the tenth edition of Charles Lyell's *Principles of Geology* and the sixth edition of Lyell's *Elements of Geology*, Wallace demoted the role of natural selection:

> **QUOT. 157** We may even go further, and maintain that there are certain purely physical characteristics of the human race which are not explicable on the theory of variation and survival of the fittest. The brain, the organs of speech, the hand, and the external form of man, offer some special difficulties in this respect. (*ibid.*, p. 391)

Referring to the brain in particular, Wallace argued that it was only slightly inferior in savages compared to those in the most advanced races of man. If natural selection had indeed acted on the brain, the latter should have been much inferior in savages, who had little use of this organ. Hence, the brain of man must have been acted upon by an "Overruling Intelligence":

> **QUOT. 158** In the brain of the lowest savages, and, as far as we yet know, of the pre-historic races, we have an organ so little inferior in size and complexity to that of the highest types (such as the average European). . . But the mental requirements of the lowest savages, such as the Australians or the Andaman islanders, are very little above those of many

[84] Which was later shared by Lyell (end of Sect. 1.14) but has now been rejected by most scientists.

[85] Shortly before the article where this break took place was published, Wallace gave a hint to Darwin in a March 24, 1869, letter: "In my forthcoming article in the 'Quarterly,' I venture for the first time on some limitations to the power of natural selection." To this, Darwin replied: "I shall be intensely curious to read the Quarterly: I hope you have not murdered too completely your own and my child" (Marchant, 1916, p. 197).

animals...Natural selection could only have endowed the savage with a brain a little superior to that of an ape, whereas he actually possesses one but very little inferior to that of the average members of our learned societies... (*ibid.*, pp. 391–392)

QUOT. 159 ...let us not shut our eyes to the evidence that an Overruling Intelligence has watched over the action of those laws, so directing variations and so determining their accumulation, as finally to produce an organization sufficiently perfect to admit of, and even to aid in, the indefinite advancement of our mental and moral nature. (*ibid.*, p. 394)

Darwin's view of evolution gave overarching importance to natural selection and was mainly mechanistic. He did not see any need for an "Overruling Intelligence" regarding the development of man. After reading Wallace's review, he wrote to the latter on April 14, 1869:

QUOT. 160 ...Altogether, I look at your article as appearing in the Quarterly as an immense triumph for our cause. I presume that your remarks on Man are those to which you alluded in your note.

If you had not told me I should have thought that they had been added by someone else. As you expected, I differ grievously from you, and I am very sorry for it.

I can see no necessity for calling in an additional and proximate cause in regard to Man. But the subject is too long for a letter. (Marchant, 1916, p. 199)

2.14 The Origin of Species (1859): Anatomy and Reception

2.14.1 Anatomy

As previously mentioned,[86] Darwin was in the process of writing his Big Species book when he received Wallace's "Ternate" paper. After the rushed presentation of the joint 1858 paper, he started writing a condensed version of his unfinished work in the form of an abstract. Francis Darwin (Charles Darwin's son) further clarified the planned task as follows:

QUOT. 161 The "Abstract" ... was in fact the *Origin of Species*, on which he [Charles Darwin] now set to work. In his *Autobiography* (p. 41) he speaks of beginning to write in September, but in his Diary he wrote, "July 20 to Aug. 12, at Sandown, began Abstract of Species book." "Sep. 16, Recommenced Abstract." The book was begun with the idea that it would be published as a paper, or series of papers, by the Linnean Society, and it was only in the late autumn that it became clear that it must take the form of an independent volume. (Stauffer, 1975, p. 204)

Darwin had wished to call his book *An Abstract of an Essay on the Origin of Species and Varieties Through Natural Selection*. John Murray, the publisher, refused to call the work an abstract[87] since it consisted of several hundred pages. The book was thus

[86] See footnote 65.

[87] On another occasion, Darwin described his book simply as "one long argument from the beginning to the end" (Barlow, 1958, p. 140). The format of an abstract enabled him to cite several names in his book without giving any references.

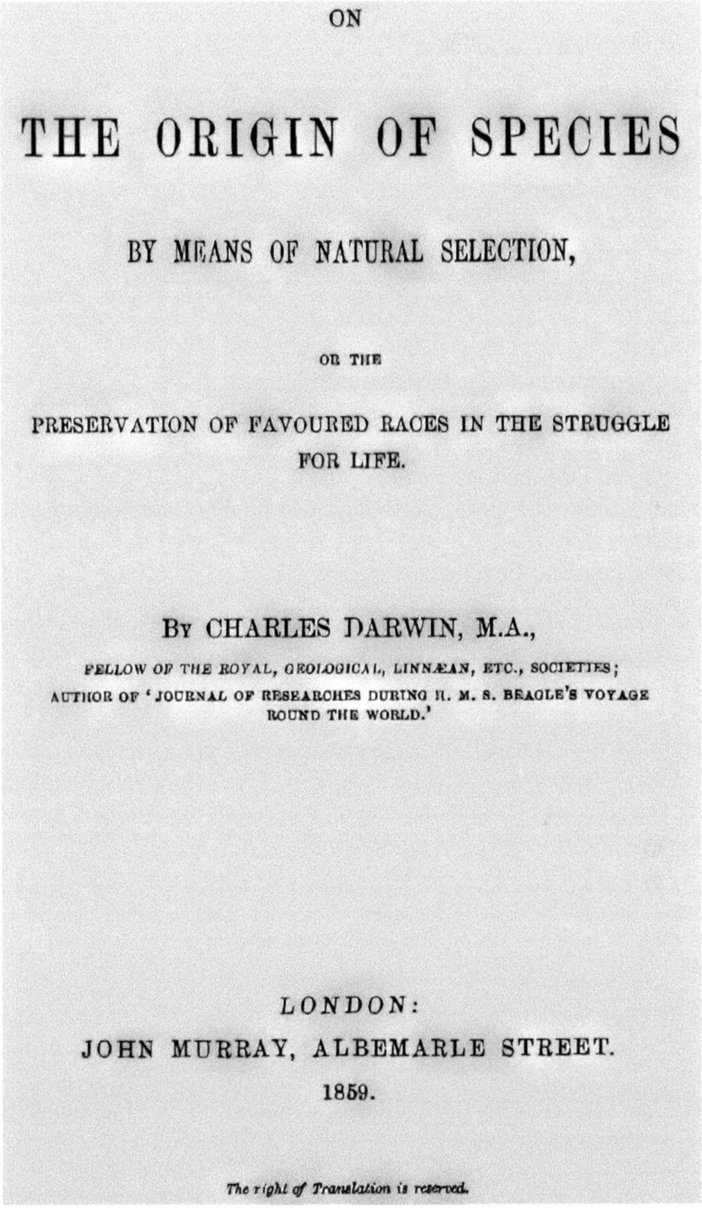

ON

THE ORIGIN OF SPECIES

BY MEANS OF NATURAL SELECTION,

OR THE

PRESERVATION OF FAVOURED RACES IN THE STRUGGLE FOR LIFE.

BY CHARLES DARWIN, M.A.,

FELLOW OF THE ROYAL, GEOLOGICAL, LINNÆAN, ETC., SOCIETIES;
AUTHOR OF 'JOURNAL OF RESEARCHES DURING H. M. S. BEAGLE'S VOYAGE ROUND THE WORLD.'

LONDON:
JOHN MURRAY, ALBEMARLE STREET.
1859.

The right of Translation is reserved.

Fig. 2.13 Title page of the first edition of the *Origin of Species* (Darwin, 1859)

entitled *On the Origin of Species*[88] *by Means of Natural Selection; or the Preservation of Favored Races in the Struggle for Life* (Darwin, 1859) (Fig. 2.13) and was

[88]Despite its tantalizing title, the *Origin of Species* was not about the origin of *life* but rather about how new species formed by transmutation from previous species.

officially published on November 24, 1859. In addition to an Introduction, it consisted of 14 chapters as follows:

Introduction
I. Variation Under Domestication
II. Variation Under Nature
III. Struggle for Existence
IV. Natural Selection
V. Laws of Variation
VI. Difficulties on Theory
VII. Instinct
VIII. Hybridism
IX. On the Imperfections of the Geographical Record
X. On the Geological Succession of Organic Beings
XI. Geographical Distributions
XII. Geographical Distributions-contd
XIII. Mutual Affinities of Quadratic Beings: Morphology: Embryology: Rudimentary Organs
XIV. Recapitulation and Conclusion

Thus, the contents of the *Origin* mirrored that of both the sketch of 1844 (Sect. 2.10) and the essay of 1844 (Sect. 2.11.2). Since the audience would most likely be skeptical, Darwin's exposition was meant to convince, and his writing style was rhetorical.

In the Introduction, Darwin wished to clarify several points, a major one being the issue of priority. He explained that he had started his evolutionary investigations as early as 1837 and acknowledged the role of Wallace in prompting him to write the book (which he still referred to as an *Abstract*):

> **QUOT. 162** I have more especially been induced to do this [i.e. write this book], as Mr. Wallace, who is now studying the natural history of the Malay archipelago, has arrived at almost exactly the same general conclusions that I have on the origin of species. (*ibid.*, pp. 1–2)

A special mention was made of Hooker, "who for the last 15 years has aided me in every possible way by his large stores of knowledge and his excellent judgment" (*ibid.*, p. 3).

The Introduction also provided brief descriptions of the individual chapters. These can be classified as follows:

- Chapters I–V: an exposition of the theory of evolution by natural selection.
- Chapters VI–IX: a description of difficulties of the theory together with attempts at resolving them.
- Chapters X–XIII: an application of the theory to various related fields such as geology, geography, and anatomy.
- Chapter XIV: a brief recapitulation and concluding remarks.

The major keywords in Darwin's theory were (almost) all mentioned and briefly explained: variation, struggle for existence, natural selection, divergence, and modification (of species)[89] (Darwin, 1859).

Natural selection was succinctly described as follows:

> **QUOT. 163** As many more individuals of each species are born than can possibly survive; and as, consequently, there is a frequently recurring struggle for existence, it follows that any being, if it vary however slightly in any manner profitable to itself, under the complex and sometimes varying conditions of life, will have a better chance of surviving, and thus be naturally selected. From the strong principle of inheritance, any selected variety will tend to propagate its new and modified form. (*ibid.*, p. 6)

The last sentence of the Introduction was one of Darwin's main theses:

> **QUOT. 164** I am convinced that Natural Selection has been the main but not exclusive means of modification. (*ibidem*)

In Chapter I, Darwin wished to use artificial selection in domesticated species as a launch pad for his theory about how nature shaped organisms through the process of natural selection. Darwin pointed to the greater amount of variation in domesticated animals than in animals in nature:

> **QUOT. 165** WHEN we look to the individuals of the same variety or sub-variety of our older cultivated plants and animals, one of the first points which strikes us, is, that they generally differ much more from each other, than do the individuals of anyone species or variety in a state of nature. (*ibid.*, p. 7)

He then took issue with naturalists who insisted that nothing could be learned about animals in nature from domesticated species. He attributed this viewpoint to the common belief (which he believed to be erroneous) that when domesticated animals were left in the wild, they reverted to their wild ancestors (Sect. 2.13.5.2). This argument was crucial for Darwin since it enabled him to lay the ground for a transition from artificial to natural selection based on the same underlying principles.

Darwin was fully aware of the importance of variations that were inherited. However, as to the laws of inheritance themselves, he correctly summed up the (pre-Mendelian) state of affairs in 1859 as follows:

> **QUOT. 166** The laws governing inheritance are quite unknown; no one can say why the same peculiarity in different individuals of the same species, and in individuals of different species, is sometimes inherited and sometimes not so. (*ibid.*, p. 13)

He also conjectured that variation was due to environmental conditions:

> **QUOT. 167** When we reflect on the vast diversity of the plants and animals which have been cultivated, and which have varied during all ages under the most different climates and treatment, I think we are driven to conclude that this greater variability is simply due to our domestic productions having been raised under conditions of life not so uniform as, and

---

[89] However, the word "evolution" is absent in the whole book, except as a verb in the very last, powerful sentence: "...whilst this planet has gone cycling on according to the fixed law of gravity, from so simple a beginning endless forms most beautiful and most wonderful have been, and are being, evolved" (Darwin, 1859, p. 492).

somewhat different from, those to which the parent-species have been exposed under nature. (*ibid.*, p. 7)

To the extent that these variations are inherited, the above is all but Lamarck's inheritance of acquired characteristics (see Sect. 1.8).

In Chapter II, Darwin wanted to extend the ideas of the previous chapters to organisms in nature. He went over the concept of species and varieties and called the first major type of variation "monstrosities" (or "sports,"[90] defined as follows:

QUOT. 168 ...some considerable deviation of structure in one part, either injurious to or not useful to the species, and not generally propagated. (*ibid.*, p. 44)

However, it was the second type of variation, namely individual differences, that was more important to Darwin:

QUOT. 169 Again, we have many slight differences which may be called individual differences, such as are known frequently to appear in the offspring from the same parents, or which may be presumed to have thus arisen, from being frequently observed in the individuals of the same species inhabiting the same confined locality...These individual differences are highly important for us, as they afford materials for natural selection to accumulate. (*ibid.*, p. 45)

Sports are discontinuous variations, whereas individual differences are continuous. The question of the type of variation (discontinuous or continuous) upon which natural selection acted would later be the topic of a major debate between biometricians and Mendelians and would shape the whole history of population genetics.

Chapters III and IV proceeded to the main exposition of Darwin's theory, starting with the concept of a struggle for existence in nature and culminating in the process of natural selection as a result of the struggle. The struggle for existence was explained in Chapter III by invoking Thomas Malthus (Sect. 1.12), the very person from whom Darwin got the idea and was able to complete his theory:

QUOT. 170 A struggle for existence inevitably follows from the high rate at which all organic beings tend to increase. Every being, which during its natural lifetime produces several eggs or seeds, must suffer destruction during some period of its life, and during some season or occasional year, otherwise, on the principle of geometrical increase, its numbers would quickly become so inordinately great that no country could support the product. Hence, as more individuals are produced than can possibly survive, there must in every case be a struggle for existence, either one individual with another of the same species, or with the individuals of distinct species, or with the physical conditions of life. It is the doctrine of Malthus applied with manifold force to the whole animal and vegetable kingdoms. (*ibid.*, p. 63)

Chapter IV made an immediate connection between a universal struggle for existence and the resulting process of natural selection. As before, the latter was introduced via selection in domestic animals:

QUOT. 171 How will the struggle for existence, discussed too briefly in the last chapter, act in regard to variation? Can the principle of selection, which we have seen is so potent in the

[90] See Sect. 7.1.3.

> hands of man, apply in nature? I think we shall see that it can act most effectually. Let it be borne in mind in what an endless number of strange peculiarities our domestic productions, and, in a lesser degree, those under nature, vary; and how strong the hereditary tendency is. Under domestication, it may be truly said that the whole organisation becomes in some degree plastic. Let it be borne in mind how infinitely complex and close-fitting are the mutual relations of all organic beings to each other and to their physical conditions of life. Can it, then, be thought improbable, seeing that variations useful to man have undoubtedly occurred, that other variations useful in some way to each being in the great and complex battle of life, should sometimes occur in the course of thousands of generations? If such do occur, can we doubt (remembering that many more individuals are born than can possibly survive) that individuals having any advantage, however slight, over others, would have the best chance of surviving and of procreating their kind? On the other hand, we may feel sure that any variation in the least degree injurious would be rigidly destroyed. This preservation of favourable variations and the rejection of injurious variations, I call Natural Selection. (*ibid.*, pp. 80–81)

As in Chapter II, Darwin stressed that natural selection acted on minute, essentially continuous variations. But how could nature have produced new species by acting on such tiny differences? Darwin's answer lay in the large geological time available for such changes to take place:

> **QUOT. 172** ...as man can certainly produce great results by adding up in any given direction mere individual differences, so could Nature, but far more easily, from having incomparably longer time at her disposal. (*ibid.*, p. 82)

An important working assumption in Darwin's work was that of blending inheritance:[91]

> **QUOT. 173** Intercrossing plays a very important part in nature in keeping the individuals of the same species, or of the same variety, true and uniform in character. (*ibid.*, p. 103)

Chapter IV also addressed the topic of sexual selection (Sect. 2.13.5.6) and contained a single diagram (illustrating the divergence of species and varieties) in the book (Fig. 2.14).

Chapter V concerned the laws of variation and, in retrospect, was the least satisfactory chapter. Darwin stated that he had assumed that variations were essentially random, but this assumption served "to acknowledge plainly our ignorance of the cause of each particular variation." (*ibid.*, p. 131). However, the much greater variability under domestication than in nature had led him to believe that:

> **QUOT. 174** ...deviations of structure are in some way due to the nature of the conditions of life, to which the parents and their more remote ancestors have been exposed during several generations. (*ibid.*, p. 131)

The above is similar to Lamarck's ideas. Another Lamarckian theme that was repeated by Darwin is the inheritance of use and disuse:

> **QUOT. 175** I think there can be little doubt that use in our domestic animals strengthens and enlarges certain parts, and disuse diminishes them; and that such modifications are inherited. (*ibid.*, p. 134)

---

[91] See also Sect. 2.10.1.

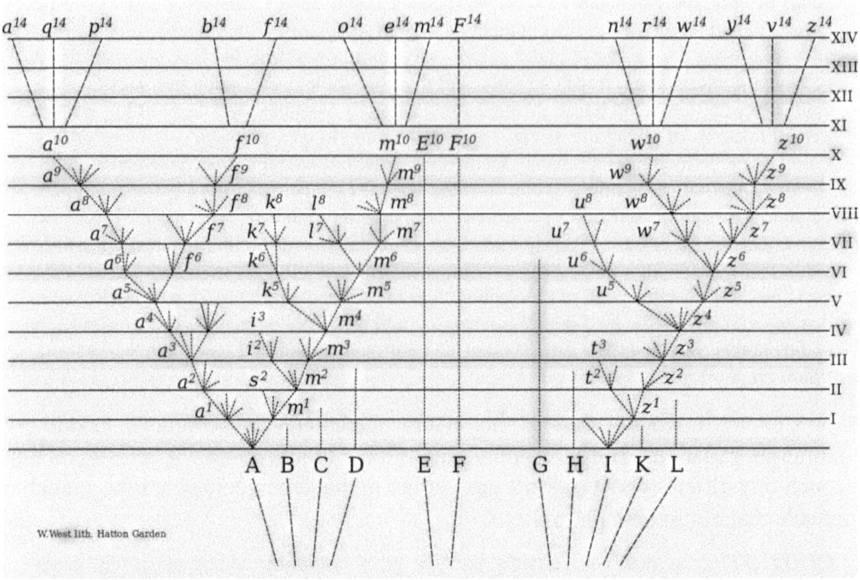

Fig. 2.14 Only diagram in the *Origin of Species* (b/w pp. 117 and 118), showing the divergence of varieties and species from bottom to top

In Chapter VI, Darwin wished to confront the main difficulties of his theory, namely:

(a) *Absence of transitional forms*: "[W]hy, if species have descended from other species by insensibly fine gradations, do we not everywhere see innumerable transitional forms?" (*ibid.*, p. 171).
(b) *Species with peculiar habits or structures*: Darwin here wished to address the issue of how a species with particular structures and habits could have evolved from one with completely different structures and habits. For example, if a land animal evolved from a bat, how could natural selection account for intermediate forms since the latter would surely have been adapted for neither air nor land? Darwin also raised the issue of both perfect and seemingly useless organs: "Can we believe that natural selection could produce, on the one hand, organs of trifling importance, such as the tail of a giraffe, which serves as a fly-flapper, and, on the other hand, organs of such wonderful structure, as the eye, of which we hardly as yet fully understand the inimitable perfection?" (*ibid.*, pp. 171–172).
(c) *Instincts*: "[C]an instincts be acquired and modified through natural selection? What shall we say to so marvelous an instinct as that which leads the bee to make cells, which have practically anticipated the discoveries of profound mathematicians?" (*ibid.*, p. 172).
(d) *Hybridism*: A consequence of the prevailing belief in special creation was that different species were reproductively isolated; therefore, if two different species were crossed, the resulting offspring (hybrid) should be sterile. Darwin was aware that this viewpoint was fatal to his theory and attempted to show that it

was inaccurate. He, therefore, asked, "how can we account for species, when crossed, being sterile and producing sterile offspring, whereas, when varieties are crossed, their fertility is unimpaired?" (*ibidem*).

Darwin was unable to refute each of the problems above conclusively, but he provided the best explanations that were available to him at the time, as follows.

Regarding problem (a), Darwin explained that natural selection acts only to preserve the most profitable forms. This implies that both the parent and intermediate forms are exterminated. Darwin acknowledged that this explanation resulted in another problem: if the intermediate forms were indeed exterminated, why were their fossils missing? Darwin here invoked the imperfection of the geological record, the remains of intermediate organisms not being "embedded and preserved to a future age only in masses of sediment sufficiently thick and extensive to withstand an enormous amount of future degradation" (*ibid.*, p. 173). Finally, Darwin also observed that only in rare cases were transitional forms unknown. This observation supported his working assumption of continuity, that is

QUOT. 176 Natura non facit saltum. (*ibid.*, p. 194)[92]

For (b), referring to the adaptability of a transitional form between an aquatic and an earth animal, Darwin replied that "[i]t would be easy to show that within the same group, carnivorous animals exist having every intermediate grade between truly aquatic and strictly terrestrial habits; and as each exists by a struggle for life, it is clear that each is well adapted in its habits to its place in nature" (*ibid.*, p. 179). Thereupon, he gave several examples where intermediate forms were useful in the environment in which they existed. Concerning the existence of highly complex organs such as the eye, Darwin explained that one only needed to look for various gradations in eye complexity and show how each gradation could be useful. He gave several instances of such cases (e.g., in crustaceans). Furthermore, he found no case where "any complex organ existed, which could not possibly have been formed by numerous, successive, slight modification" (*ibid.*, p. 189). Regarding the existence of seemingly useless organs, Darwin warned that "we are much too ignorant in regard to the whole economy of any one organic being, to say what slight modifications would be of importance or not" (*ibid*). For example, the giraffe's tail is used as a fly-tapper. This may seem to be "so trifling" until one realizes that, without the ability to easily ward off flies, many animals in Latin America are incessantly harassed, have their strength reduced, and become vulnerable to disease.

Regarding (c), Darwin devoted the whole of Chap. VII to the issue of instinct. He argued that natural selection could act on instincts, just like it acted on physical traits. Just as there are variations in traits and natural selection ensures the transmission of the more favorable ones, so too we can find variation in instinctive abilities and natural selection will transmit the more profitable ones: "Hence, as in the case of corporeal structures, we ought to find...in the collateral lines of descent some evidence of such gradations; or we ought at least to be able to show that gradations

[92] See footnote 6 in Chap. 1.

of some kind are possible; and this we certainly can do" (*ibid.*, p. 210). Thereupon, Darwin gave several examples of such gradations (e.g., migratory instincts, nesting of birds, etc.).

Finally, concerning problem (d), Darwin here, too, dedicated a whole chapter (Chap. VIII) to hybridism. He had two major aims. First, he wished to counter the argument that the offspring of intercrossed species were completely sterile and thus show that species were not fundamentally different entities. He cited examples from the works of two researchers (Kölreuter[93] and Gärtner[94]), who showed that crossings between different species did produce some fertile offspring. Darwin's second aim was to show that varieties and species did not differ in kind but only in degree so that domesticated animals were not radically different from wild animals. Therefore, he stated that "neither sterility nor fertility affords any clear distinction between species and varieties; but that the evidence from this source graduates away, and is doubtful in the same degree as is the evidence derived from other constitutional and structural differences[95]" (*ibid.*, p. 248). Thus Darwin preferred to call well-marked varieties "incipient species" (*ibid.*, p. 52) and argued that all of them behaved similarly under the force of natural selection.

Chapter IX was wholly devoted to the issue of the imperfect geological records, which had been used in the first problem of Chapter VI to explain the absence of intermediate forms in fossil records.

In Chapter X, Darwin used his theory to explain various observations made in the available fossil record despite the latter being imperfect. For example, it was known that

> **QUOT. 177** [w]hen a species has once disappeared from the face of the earth, we have reason to believe that the same identical form never reappears.[96] (*ibid.*, p. 313)

Darwin easily explained this observation:

> **QUOT. 178** . . .though the offspring of one species might be adapted. . .to fill the exact place of another species in the economy of nature, and thus supplant it; yet the two forms—the old and the new—would not be identically the same; for both would almost certainly inherit different characters from their distinct progenitors. (*ibid.*, 315)

As another example, Darwin cited the intriguing observation made by Richard Owen that extinct animals often fall in between existing groups. Once again, Darwin used his theory to explain this in a simple fashion. Referring to the only diagram in the book (Fig. 2.14), Darwin showed that two (existing) groups could evolve and

---

[93] See Sect. 3.3.

[94] See Sect. 3.5.

[95] A view on species and varieties that is more or less espoused by modern biologists. Some organisms are morphologically very different yet belong to the same species, whilst others are morphologically very similar yet belong to different species. In addition, whilst hybrids obtained by crossing different species are usually sterile, this is not always the case. Thus, researchers in the past often could not distinguish between different species and varieties within the same species.

[96] Even though Lyell at first believed in cycles of extinction and appearance, only to eventually endorse directional change.

diverge in opposite directions (with respect to specific traits) from a single (extinct) group.[97]

Chapters XI and XII addressed geographical distributions and were partly inspired by Darwin's own travel experience. At the very start, he dispelled a common misconception that the relationships between various species and varieties were due to climate:

> **QUOT. 179** In considering the distribution of organic beings over the face of the globe, the first great fact which strikes us is, that neither the similarity nor the dissimilarity of the inhabitants of various regions can be accounted for by their climatic and other physical conditions. (*ibid.*, p. 346)

Darwin also noted the importance of "obstacles to free migration" in determining the relationships between organisms in various geographical regions. Both of these facts were in accord with his theory.

Chapter XIII addressed the issue of classification. Darwin challenged an important aspect of the Linnaean system of classification (as described in his *Systema Naturae* (Linnaeus, 1735)):

> **QUOT. 180** Naturalists try to arrange the species, genera, and families in each class, on what is called the Natural System. But what is meant by this system? Some authors look at it merely as a scheme for arranging together those living objects which are most alike, and for separating those which are most unlike; or as an artificial means for enunciating, as briefly as possible, general propositions,—that is, by one sentence to give the characters common, for instance, to all mammals, by another those common to all carnivora, by another those common to the dog-genus, and then by adding a single sentence, a full description is given of each kind of dog. The ingenuity and utility of this system are indisputable. But many naturalists think that something more is meant by the Natural System; they believe that it reveals the plan of the Creator; but unless it be specified whether order in time or space, or what else is meant by the plan of the Creator, it seems to me that nothing is thus added to our knowledge. (*ibid.*, p. 413)

Darwin thus questioned the real meaning of the Linnaean system and took issue with the suggestion of a divine plan. He suggested that his theory of common descent with modification provided a better rationale:

> **QUOT. 181** I believe that something more is included; and that propinquity of descent,-the only known cause of the similarity of organic beings,—is the bond, hidden as it is by various degrees of modification, which is partially revealed to us by our classifications. (*ibid.*, pp. 413–414)

Issues related to embryology and morphology were also treated in Chap. XIII.

The final chapter, Chap. XIV, first surveyed the main difficulties of Darwin's theory. He then reiterated the major themes of the theory and summarized his position as follows:

> **QUOT. 182** . . .all the organic beings which have ever lived on this earth have descended from some one primordial form, into which life was first breathed. (*ibid.*, p. 484)

[97]Thus, in Fig. 2.14, b^{14} and f^{14} could represent the existing species, and f^{10} could represent the extinct species.

Darwin predicted that, if his theory was accepted, he could “dimly foresee that there will be a revolution in natural history” (*ibidem*). In retrospect, this prediction was still an understatement.

Concerning matters of the Creator, it was no surprise that, as in his previous writings (see, for example, Sect. 2.10), Darwin denied special creation since his whole theory was antithetical to such a belief. However, he did not reject the existence of a Creator, ascribing the “final causes” to Him. Evolution by natural selection, though, was the purview of “secondary” or “intermediary” cases (natural laws):

> **QUOT. 183** Authors of the highest eminence seem to be fully satisfied with the view that each species has been independently created. To my mind it accords better with what we know of the laws impressed on matter by the Creator, that the production and extinction of the past and present inhabitants of the world should have been due to secondary causes, like those determining the birth and death of the individual. (*ibid.*, p. 488)

This concludes our description of the *Origin.* There is no denying that Darwin’s writing was very compelling, and he buttressed his arguments through a large amount of *observational* evidence. However, no *experimental* evidence was provided to support the theory.

As mentioned at the start of this section, Darwin was in the process of writing his Big Species book (which he had begun in May 1856) when he received Wallace’s “Ternate” paper and was forced to quickly write the *Origin*. After the latter was released, Darwin stopped the Big Species project, and the book was never finished. However, *Variation of Plants and Animals under Domestication* was published in two volumes in 1968 and formed the first part of the Species Book. The second part was transcribed from manuscript notes by R.C. Stauffer and published in 1975 as *Charles Darwin’s Natural Selection Being the Second Part of his Big Species Book Written from 1856 to 1858* (Stauffer, 1975).

The first print of the *Origin* was for 1250 copies, of which 12 were sent to Darwin, 41 for review, five for the Stationers’ Hall copyright, and the rest (1192) for sale. Darwin’s diary entry for Oct 1 read “all copies sold first day,”[98] but that presumably meant that *booksellers* (rather than the *public*) had, on the first day, bought all copies that were made available by the publisher. In all, there were six editions of the *Origin:* First edition 1859, second edition 1860, third edition 1861, fourth edition 1866, fifth edition 1869, and sixth edition 1872.[99] The third edition contained several additional features: (i) a table of differences between it and the second edition (pp. xi–xii), (ii) a seven-page Historical Sketch (pp. xiii–xix),[100] (iii) a postscript (p. xii), which was a notice about a review by Asa Gray of the *Origin*.

[98] This entry would have been made on or after Nov 24, 1859, the date of release of the *Origin.*

[99] For the sixth edition, the title of the book was changed from *On the Origin of Species* to *The Origin of Species.*

[100] However, the first German edition and the fourth American printing, both released in 1860, already contained a shorter version of the historical sketch. The book by Johnson (2019) is completely devoted to Darwin’s historical sketch.

The variorum text by Peckham (1959) provides a sentence-by-sentence comparison of the various editions. The changes made by Darwin in successive editions were often substantive. Peckham has reported that:

> **QUOT. 184** [of] the 3,878 sentences in the first edition, nearly 3,000, about 75 per cent, were rewritten from one to five times each. Over 1,500 sentences were added, and of the original sentences plus these, nearly 325 were dropped. Of the original and added sentences there are nearly 7,500 variants of all kinds. In terms of net added sentences, the sixth edition is nearly a third as long again as the first. (*ibid.*, p. 9)

2.14.2 Reception, Part I: Bibliography of Reviews

In the last chapter of the *Origin*, Darwin admitted that little did he "expect to convince experienced naturalists whose minds are stocked with a multitude of facts all viewed, during a long course of years, from a point of view directly opposite to mine" (Darwin, 1859, p. 481). However, he was hopeful that "[a] few naturalists, endowed with much flexibility of mind, and who have already begun to doubt on the immutability of species, may be influenced by this volume" (*ibid.*, p. 482). The events that followed the publication of the *Origin* made Darwin's predictions quite accurate: a few embraced Darwin's theory, but many voiced strong opposition. Below is a list of reviews and letters, the volume of which should give the reader an indication of the sensational nature of the *Origin:*

1. *The Theory of Natural Selection (Anon.)* (1859) in *The Critic* (Nov 26), 19: 528–530.
2. *[Review of Origin of Species], (Anon.)* (1859), in *English Churchman*, p. 1152.
3. *On the Origin of Species by Means of Natural Selection (Anon.)* (1859), in *John Bull and Britannia* (Dec 24).
4. *[Review of Origin of Species], (Anon.)* (1859), in *The Press* (Dec 10): 1243.
5. *[Review of Origin of Species], (Anon.)* (1859), in *The Daily News* (Dec 26).
6. *Darwin's Origin of Species (Anon.)* (1859), in *Saturday Review* (Dec. 24), 8: 775–776.
7. *Darwin on the Origin of Species (Anon.)* (1859), in *Spectator*, 32(1638): 1210–1211.
8. *Charles Darwin on the Origin of Species* by R. Chambers (*Anon.*), in *Chambers' Journal* (1859), 12:388–391.
9. *On the Origin of Species, by Means of Natural Selection; or the Preservation of Favoured Races in the Struggle for Life* by John Crawfurd (*Anon.*) (Dec 3, 1859), in *Examiner*, pp. 772–773.
10. Letter to Darwin by Erasmus Darwin [Charles Darwin's brother] (Nov 23, 1859), Correspondence, VII (1992), pp. 389–390.
11. *[Review of Origin of Species]* by Joseph Hooker (*Anon.*) (Dec 31, 1859), in *The Gardeners' Chronicle and Agricultural Gazette*: pp. 1051–1052, contd. in

1860, 303–304. First part reprinted in *Darwin and his Critics*, by David Hull (Cambridge, MA: Harvard University Press, 1973), pp. 82–83.
12. *Time and Life: Mr. Darwin's "Origin of Species"* by T.H. Huxley (1859), in *Macmillan's Magazine* 1:142–148.
13. *Darwin on the Origin of Species* by T.H. Huxley (*Anon.*) (1859), in *The Times* (Dec 26).
14. Letter to Darwin by Charles Kingsley (Nov 18, 1859), in *The Life and Letters of Charles Darwin, Including an Autobiographical Chapter* by F. Darwin, 3 vols. (London: John Murray), 2: 287–288.
15. *On the Origin of Species, by Means of Natural Selection; or the Preservation of Favoured Races in the Struggle for Life* by J.R. Leifchild (*Anon.*) (1859), in *Athenaeum* (Nov 19), no.1673: 659–660.
16. *Darwin's Origin of Species* by John Phillips (*Anon.*) (1859), in *Saturday Review*, 8:775–776.
17. Letter to Darwin by Adam Sedgwick (Nov 23, 1859), Correspondence, VII (1992), 396–398.
18. *[Review of Origin of Species]*, by Senex [Robert Fitzroy] (1859), in *The Times* (Dec 1).
19. *Prof. Agassiz On the Origin of Species* by L.J.R. Agassiz (1860), in *American Journal of Science and Arts*, 30:142–154.
20. *The Origin of Species, by Means of Natural Selection; or the Preservation of Favoured Races in the Struggle for Life (Anon.)* (1860), in *British Quarterly Review*, 31:398–421; same article in *Eclectic Magazine*, 50:831–345.
21. *On the Origin of Species, by Means of Natural Selection; or the Preservation of Favoured Races in the Struggle for Life (Anon.)* (1860), in *British and Foreign Evangelical Review* (April), 9:413–441.
22. *On the Origin of Species, by Means of Natural Selection; or the Preservation of Favoured Races in the Struggle for Life (Anon.)* (1860), in *The Christian Observer*, 60:561–574.
23. *The Origin of Species Controversy (Anon.)* (1860), in *Weldon's Register of Facts and Occurrences Relating to Literature, the Sciences and the Arts* (Aust), pp. 1–7.
24. *[Review of Origin of Species]*, *(Anon.)* (1860), in *News of the World* (Jan 8).
25. *[Review of Origin of Species]*, *(Anon.)* (1860), in *The Morning Post* (Jan 10).
26. *[Review of Origin of Species]*, *(Anon.)* (1860), in *St. James Chronicle* (Jan 10).
27. *[Review of Origin of Species]*, *(Anon.)* (1860), in *The Patriot* (Jan 19).
28. *On the Origin of Species (Anon.)* (1860), in *New York Times* (Mar 28).
29. *[Review of the Origin of Species] (Anon.)* (1860), in *The Freeman* (Jan 18), pp. 45–46.
30. *On the Origin of Species (Anon.)* (1860), in *The Eclectic Review* (Mar), 3: 217–242.
31. *The Origin of Species (Anon.)* (1860), *The London Review* (July), 14 (28): 281–308.
32. *Darwin on the Origin of Species (Anon.)* (1860), in *New Englander*, 18(70): 516–519.

33. *On the Origin of Species, by Means of Natural Selection; or the Preservation of Favoured Races in the Struggle for Life (Anon.)* (1860), in *North American Review* (1860) 90:474–506.
34. *On the Origin of Species by Means of Natural Selection; or the Preservation of Favoured Races in the Struggle for Life* by F. Bowen (*Anon.*) (1860), in *North American Review*, 90:474–506.
35. *Ch. Darwin: On the Origin of Species by Means of Natural Selection; or the Preservation of Favoured Races in the Struggle for Life* by H.G. Bronn (1860), in *Neues Jahrbuch für Mineralogie, Geognosie, Geologie und Petrefaktenkunde*, pp. 112–16.
36. *Vestiges of the Natural History of Creation* 11th ed. by Robert Chambers (*Anon.*) (1860), pp. lxiii—lxiv.
37. *The Theory of Development in Nature* by W.B. Carpenter (*Anon.*) (1860), in *British and Foreign Medico-Chirurgical Review*, 25:367–404.
38. *Darwin On the Origin of Species* by W.B. Carpenter (*Anon.*) (1860), in *National Review*, 10:188–214.
39. *Review of On the Origin of Species by Means of Natural Selection* by E.J. Chapman (1860), in *Canadian Journal of Industry, Science and Art*, 5: 367–387.
40. *On the Origin of Species* by W.R. Church (*Anon.*) (1860), in *The Guardian* (Feb 8), pp. 134–135.
41. *Darwin On the Origin of Species by Means of Natural Selection* by John W. Dawson (1860), in *Canadian Naturalist and Geologist* 5: 100–120.
42. *Species* by Charles Dickens (ed.), in *All the Year Round* (1860), 3(58): 174–178.
43. *Natural Selection* by Charles Dickens (ed.) (1860), in *All the Year Round*, 3(63):293–299.
44. *Transmutation of Species* by Charles Dickens (ed.), in *All the Year Round* (1860/1), 4(98): 519–521.
45. *Paleontology* by D.T.A., in *Dublin University Magazine* (1860), 55:712–722.
46. *On the Origin of Species* by Rev. Dunns (*Anon.*) (1860), in *North British Review* 32:455–486.
47. *Popular Exposition of Mr Darwin on the Origin of Species* by H. Fawcett (1860), in *Macmillan's Magazine* (Dec), 8:81–92.
48. *Review of Darwin's Theory on the Origin of Species by Means of Natural Selection* by Asa Gray (*Anon.*) (1860), in *American Journal of Science*, 29(86): 153–184.
49. *Darwin on the Origin of Species* (1860) by Asa Gray (*Anon.*), *Atlantic Monthly*, vol. 6, pp. 109–116, and 229–239. Also published as *Natural Selection not Inconsistent with Natural Theology* (1861). London.
50. *Darwin and His Reviewers* by Asa Gray (*Anon.*) (1860), *Atlantic Monthly*, 6: 406–425.
51. *The Origin of Species* by Asa Gray (*Anon.*) (1860), in *Athenaeum*, p. 161. Extract from Proceedings of the American Academy of Arts and Science, 4: 411–415.

52. *Discussion Between Two Readers of Darwin's Treatise on the Origin of Species, upon Its Natural Theology* by Asa Gray (*Anon.*) (1860), in *American Journal of Science and Arts*, 30(89):226–239.
53. *Darwin on the Origin of Species* by Asa Gray (1860), in *Annals and Magazine of Natural History*, 6:373–386.
54. *Darwin on the Origin of Species* by W.H. Harvey (1860), in *Gardener's Chronicle*, 20:145–146.
55. *Biogenesis* by Samuel Haughton (*Anon.*) (1860), in *Natural History Review*, 7: 23–732. Reprinted with omissions in the *Annals and Magazine of Natural History* (1863), 11:420–429.
56. *Physical Theories of the Phenomenon of Life* by W. Hopkins (1860), in *Fraser Magazine*, Part I in 61:739–752, Part II in 62:74–90.
57. *On the Origin of Species* by F.W. Hutton (*Anon.*) (1860), in *Geologist*, 3(12): 464–472.
58. *On the Origin of Species Origin of Species by Means of Natural Selection; or the Preservation of Favoured Races in the Struggle for Life* by T.H. Huxley (*Anon.*) (1860), in *Westminster Review*, 17 (N.S.): 541–570.
59. *On Species and Races, and Their Origin* by T.H. Huxley (1860), in *Medical Circular* (Mar 7), 401:149–150.
60. *On Species, Races and Their Origin* by T.H. Huxley (1860), in *Journal Proceedings of the Royal Institution of Great Britain*, 3: 195–200.
61. *Report of the Oxford meeting of the British Association for the Advancement of Science in the Summer of 1860* (Jul 7, 1860), London Weekly.
62. *On the Origin of Species* by W. Jardine (*Anon.*) (1860), in *Edinburgh New Philosophical Journal*, 10: 280–289.
63. Letter to Darwin by Leonard Jenyns (Jan 4, 1860), Correspondence, VIII (1993), p. 14.
64. *Nouvelle Théorie d'Histoire Naturelle: L'Origine des Espèces*, by A. Laugel (1860), in *Revue des Deux Mondes*, 26(3):644–671.
65. *Studies in Animal Life* by G.H. Lewes (1860), in *Cornhill Magazine*, 1:438–447.
66. *Darwin's Origin of Species Origin of Species by Means of Natural Selection; or the Preservation of Favoured Races in the Struggle for Life* by J.A. Lowell (*Anon.*) (1860), in *Christian Examiner and Theological Review*, 6:449–464.
67. *Nature's Law of Selection* by Patrick Matthew (1860), in *Gardeners' Chronicle and Agricultural Gazette* (7 April): 312–13.
68. *The Origin of Species* by Patrick Matthew (1860), in *Gardeners' Chronicle and Agricultural Gazette* (12 May): 433.
69. *On the Origin of Species by Means of Natural Selection; or the Preservation of Favoured Races in the Struggle for Life* by Canon Morris (*Anon.*) (1860), in *Dublin Review*, 48: 50–81.
70. *On Mr Darwin's Theory of the Origin of Species* by A. Murray (1860), in *Proceedings of the Royal Society of Edinburgh*, 4: 274–291.

71. *Review of Origin and other works* by Richard Owen (*Anon.*) (1860), in *Edinburgh Review*, 111:487–532. Also in *The Living Age* (1860), 66(840): 3–26.
72. *On the Origin of Species* by T. Parsons (1860), in *American Journal of Science and Arts* 30:1–13.
73. *Eine neue Lehre über die Schöpfungsgeschichte der organischen Welt* by Oscar Perschel *(Anon.)* (1860), in *Das Ausland*, 33:97–101, 135–140.
74. *Sur l'Origine de l'Espèce par Charles Darwin* by Fançois J. Pictet (1860), in *Bibliothèque Universelle. Revue Suisse et Etrangère*, 7: 233–255.
75. *Objections to Mr. Darwin's Theory of the Origin of Species* by Adam Sedgwick (1860) (*Anon.*), in *Spectator* (24 March), 33:285–286, revised on 7 April: 334–335.
76. *[Review of Origin of Species]*, by Richard Simpson (*Anon.)* (1860), in *Rambler* (Mar), pp. 361–376.
77. Letter to Darwin by Herbert Spencer (Feb 22, 1860), Correspondence, VIII (1993), pp. 98–99.
78. *Species not transmutable, nor the result of secondary causes; being a critical examination of Mr. Darwin's work entitled "Origin and Variation of Species"* by C.R. Bree (1860). London and Edinburgh.
79. *President's Address* by H.B. Tristram (1860), in *Transactions of the Tyneside Naturalists' Field Club*, 4: 191–228 (Discussion of *Origin* on pp. 218–228).
80. *On the Origin of Species, by Means of Natural Selection; or the Preservation of Favoured Races in the Struggle for Life* by Samuel Wilberforce (*Anon.*) (1860), in *Quarterly Review*, 108:225–264. Also in *London Quarterly Review* (American Ed., July-Oct) 107:118–138.
81. *On the Origin of Species, by Means of Natural Selection; or the Preservation of Favoured Races in the Struggle for Life* by T.V. Wollaston (*Anon.*) (1860), in *Annals and Magazine of Natural History*, 5:132–143.
82. *M. Darwin et sa théorie de la formation des espèces* by E. Claperède (1861), in *Revue Germanique Française & étrangère* (1861), 16:523–559, and 17:232–263.
83. *On Agassiz' Views of Darwin's Theory of Species* by C. Collingwood (1861), Proceedings of Literary and Philosophical Society of Liverpool, 15:81–99.
84. *Remarks on the Final Causes of the Sexuality of Plants, with Particular Reference to Mr. Darwin's Work "On the Origin of Species by Natural Selection"* by Charles G.B. Daubeny (1861), *in Report of the British Association for the Advancement of Science 30th Meeting (1861), Transactions of the Sections*: 109–10.
85. *The Genesis of Organic Forms* by S. Hibberd (1861), in *Recreative Science*, 2: 268–277.
86. *On Darwin's Theory of the "Origin of Species"* by H. H. Higgins (1861), in *Proceedings of the Literary and Philosophical Society of Liverpool*, no. 15, pp. 42–49, and pp. 135–140.

87. *Some Remarks on Darwin's Theory* by F.W. Hutton (1861), in *Geologist* 3: 464–472.
88. *The Origin of Species, by Means of Natural Selection; or the Preservation of Favoured Races in the Struggle for Life* [3rd ed.] by Edward Newman (1861), *Zoologist*, 19:7577–7611.
89. Report of the first of a series of debates at the Boston Society of Natural History, *Proceedings of the Boston Society of Natural History* (1861), 15 Feb. 1860, 7:231–5.
90. *Darwin on the Origin of Species* by W.C. Wilson (*Anon.*) (1861), *Methodist Quarterly Review*, 43:605–627.
91. *De l'origine des espèces, ou des lois du progrès chez les êtres organisés* by Clémence-Auguste Royer (1862), [in translator's preface] pp. lxiii–lxiv.
92. *Darwin Among the Machines* by Samuel Butler (*Anon.*) (1863), in *New Zealand Press* (June 13); also in *A First Year in Canterbury Settlement with Other Early Essays* (London: A. C. Fifield, 1914), 179–185.
93. *On the Origin of Species, or the Causes of the Phenomenon of Organic nature: A Course of Lectures to Working Men*, by T. H. Huxley (1863), D. Appleton, New York. Reprinted in Huxley's *Darwiniana*, pp. 303–475 as *On our Knowledge of the Causes of the Phenomenon of Organic Nature.*
94. *Review of On the Origin of Species, or the Causes of the Phenomena of Organic Nature: A Course of Lectures to Working Men, by T. H. Huxley* by Rev. W. Hincks (1863), in *Canadian Journal of Industry, Science and Art*, 8(47): 390–404.
95. Letter to Rev. D Brown by William Whewell (Oct 26, 1863), in *William Whewell: An Account of His Writings* by I. Todhunter, 2 vols. (London: Macmillan 1876), 2:433–434.
96. Letter to Darwin by Ernst Haeckel (July 9, 1864), in Correspondence XII (2001), pp. 482–483.
97. *Criticisms on the "The Origin of Species"* by T.H. Huxley (1864), in *Natural History Review*, 4(16): 566–580.
98. *The Origin and Antiquity of Man: Darwin, Huxley and Lyell*, by Henry A. DuBois (*Anon.*) (1865) in *American Quarterly Church Review, and Ecclesiastical Register:* 17(2):169–198.
99. *Darwin and Domestication* by S.T. Frost (1867), in *Harper's New Monthly Magazine* 36(211):58–63.
100. *The Origin of Species* by F. Jenkin (*Anon.*) (1867), in *North British Review*, 46: 277–318.
101. *Neue Zusätze zu Charles Darwin's Schöpfungsgeschichte der organischen Welt* by Oscar Perschel *(Anon.)* (1867), in *Das Ausland*, 40: 73–80.
102. *Les Précurseurs Francais de Darwin* by A. de Quatrefages (1868), in *Revue des Deux Mondes*, 78: 832–860.
103. *Mr. Darwin's Hypotheses* by G. H. Lewes, in *Fortnightly Review* (1868), 3: 353–373, 3:611–628, 4:61–80, 4:492–509.

104. *Carlo Darwin e il suo ultimo libro* by P. Mantegazza (1868), in *Nuova Antologia*, 8:70–98.
105. *La Théorie de Darwin* by A. de Quatrefages (1869), in *Revue des Deux Mondes*, 79: 208–240.
106. *Discussion des Théories Transformistes* by A. de Quatrefages (1869), in *Revue des Deux Mondes*, 80: 64–95.
107. *Darwin et les Théories Transformistes, L'Espèce et la Race* by A. de Quatrefages (1869), in *Revue des Deux Mondes*, 80: 397–452.
108. *Théories de la Transformation progressive et de la Transformation brusque, Origine Simienne de l'homme*, by A. de Quatrefages (1869), in *Revue des Deux Mondes*, 80: 638–672.
109. *Difficulties of the Theory of Natural Selection* by St. George Jackson Mivart (*Anon.*) (1869), in *Month* 11: 35–53, 134–153, 274–289.
110. *Beitrage zur Lehre Darwin von des Entstehung der Arten* by Begouen (1870), in *Das Ausland*, 43:59–62.
111. *L'Instinct chez les Insectes* by George Pouchet (1870), in *Revue des Deux Mondes* (1870), 85: 691–703.
112. Darwin's Theory on the Origin of Species by J. Bascom (1871), in *The American Presbyterian Review* (July) 3:349–379.
113. *On the Genesis of Species*, by St. George Jackson Mivart (1871).
114. *The Origin of Species* by B. G. Wilder (*Anon.*) (1871), in *Nation*, 12:199–201.
115. *The Philosopher Among the Apes* by A. Weld, in *The Month* (1871), 4 (N.S.):71–101.
116. *Darwinism, or Evolution* by Rev. R.P. Stebbens, in *The Religious Magazine and Monthly Review*, 50:496–502.
117. *The Origin of Species, by Means of Natural Selection; or the Preservation of Favoured Races in the Struggle for Life* by A. W. Bennett (1872), in *Nature*, 5: 318–319.
118. *Mr Darwin and the Theory of Natural Selection* by L.T. Adams (1874), in *New Englander*, 33(129): 741–770.
119. *Coming of Age of the Origin of Species* by T.H. Huxley (1880), in *Nature*, 17: 337–344.

2.14.3 Reception, Part II: Notable Reviews and Comments

Despite Darwin's declaration to the botanist John Scott (1836–1880) in a Dec 3, 1862, letter that most of the reviews of the *Origin* were "not worth the paper they are printed on" (Darwin & Seward, 1903b, p. 311, Vol. II), Darwin was sensitive to many of them. While reading some of the more important reviews and comments below, it may be necessary to remember three important points:

- The opponents of Darwin's theory consisted not only of clergymen but also of very prominent scientists.

- There may be a certain temptation to consider those who opposed Darwin's theory as bigoted or foolish, but this is Whiggish[101] and historically false (as it is to label those who supported his theory as completely enlightened or heroic).
- Many of the criticisms were not directed against the secondary component of Darwin's theory, namely, descent with modification (or evolution), per se. Instead, they were aimed at its primary component, namely natural selection, which was viewed to be overly materialistic and devoid of any purpose (since it operated on *random* variations).

The first feedback that Darwin received on *Origin* was not in the form of a review. Rather, it was an enthusiastic letter (dated Nov 18, 1859) from the Christian socialist Charles Kingsley (1819–1875):

> **QUOT. 185** . . .All I have seen of it [the *Origin*] awes me; both with the heap of facts, & the prestige of your name, & also with the clear intuition, that if you be right, I must give up much that I have believed & written. (Burkhardt & Smith, 1991, p. 379)

Darwin was so pleased with the letter, especially since it came from a member of the clergy (most of whom were dead set against the ideas in the *Origin*), that he applauded the "celebrated author and divine" in all subsequent editions of the book.

Darwin's happiness was short-lived.

The very next day (5 days before the official release of the *Origin*!), he read the sarcastic review from an anonymous writer (who was later found to be the coalfield commissioner and man of letters John Leifchild) in the popular *Athenaeum* (Anonymous, 1859b). Leifchild scoffed at Darwin's concept of a struggle for existence:

> **QUOT. 186** How do our cherished poetical dreams mislead us, when we sing of the peacefulness and repose and harmlessness of animated nature, while the whole fields and forest are but one wide theatre of war! (*ibid.*, p. 659)

Concerning natural selection, Leifchild called it a "welcome fancy":

> **QUOT. 187** Yes, an unbroken, sure, though slow, living progress towards animal perfectibility is a delightful vision; natural and gradual optimism is a welcome fancy. What need of distinct creation? If a monkey has become a man—what may not a man become? (*ibidem*)

Leifchild's criticism was one of the most commonly made against Darwin's theory, namely that the latter degraded man's unique moral position by linking him to previous barbaric living forms as natural selection worked through its course. However, the charge that man descended from the monkey was truly a strawman argument since Darwin's theory implied no such thing. What it implied is that the current monkey and man must have shared some common ancestor in the past. However, even the latter fact would have been repugnant to Darwin's detractors.

---

[101] The term is due to Herbert Butterfield. In the book *The Whig Interpretation of History*, Butterfield (1931) indicated that historians, in light of the present, often interpret history as continued progress over reaction. Most historians have agreed on the dangers of Whig history, although Mayr (1990) has contended that an analysis of the past based on present knowledge is not only inevitable but also necessary.

Fig. 2.15 Adam Sedgwick (1785–1873). (From: https://commons.wikimedia.org/wiki/File:Adam_Sedgwick.jpg)

Darwin was understandably infuriated by the Athenaeum review[102] and, for years, wrongly begrudged the geologist Samuel Pickworth Woodward (1821–1865) for writing it.

On the other hand, Darwin received great support from his key lieutenants: Huxley, Hooker, Gray, and Wallace.[103] Huxley professed his conversion to the central idea in the *Origin* (Sect. 2.12.2), even acknowledging "[h]ow extremely stupid not to have thought of that!" (Huxley, 1900, p. 246, Vol. I). Hooker and Gray had already embraced transmutation and were now ready to defend it.

A scathing Nov 23, 1859 letter from Darwin's former teacher, the geologist and catastrophist Adam Sedgwick (Fig. 2.15), was a totally different affair:

> **QUOT. 188** I have read your book with more pain than pleasure. Parts of it I admired greatly, parts I laughed at till my sides were almost sore; other parts I read with absolute sorrow, because I think them utterly false and grievously mischievous. You have *deserted*—after a start in that tram-road of all solid physical truth—the true method of induction… (Darwin, 1888a, p. 43)

[102] See, for example, Darwin's Nov 22, 1959, letter to Hooker in Darwin (1888a, p. 24).

[103] As for Lyell, he was supportive of Darwin but had not accepted transmutation yet.

QUOT. 189 ...I humbly accept God's revelation of Himself both in his works and in His word, and do my best to act in conformity with that knowledge which He only can give me, and He only can sustain me in doing. If you and I do all this we shall meet in heaven. (*ibid.*, p. 45)

Sedgwick later wrote an equally negative review in the *Spectator* (Anonymous, 1860c).

A letter from Robert Fitzroy, expressing a similar sentiment, soon followed Sedgwick's letter. Fitzroy's review next came in the *Times*, written under the pen name *Senex* (1859).

To placate the clergy, Darwin made an important change in the second edition (published on Jan 7, 1860) of the *Origin*: the first edition's "...from some one primordial form, into which life was first breathed" (Darwin, 1859, p. 484) was changed to "...from some one primordial form, into which life was first breathed by the Creator" in the second edition (Darwin, 1860b, p. 484).

The year 1859 ended with a highly positive review by Huxley in *The Times*:

QUOT. 190 Mr. Darwin abhors mere speculation as nature abhors a vacuum. He is as greedy of cases and precedents as any constitutional lawyer, and all the principles he lays down are capable of being brought to the test of observation and experiment. The path he bids us follow professes to be, not a mere airy track, fabricated of ideal cobwebs, but a solid and broad bridge of facts. (Huxley, 1859, p. 9)

However, in his third review of the *Origin*, Huxley (1860) seemed to retract his full endorsement of Darwin's theory on two fronts: (i) he cast some doubt on the validity of natural selection, and (ii) he disagreed with Darwin's adoption of *Natura non facit saltum*,[104] which implied continuous variation. Despite these misgivings, Huxley still remained one of the most ferocious defendants of Darwin and his theory.

Two of the men (Sect. 2.10.2) who had major philosophical influences on Darwin reacted negatively to the *Origin*. First, John Herschel allegedly called natural selection the law of higgledy-piggledy, to which Darwin replied in dismay to Lyell: "What exactly this means I do not know, but it is evidently very contemptuous. If true, this is a great blow and discouragement" (Burkhardt & Smith, 1991, p. 423).[105] Then, William Whewell allegedly barred the *Origin* from the library of Trinity College, where he was Master (Darwin, 1868b, p. 261 footnote, Vol. II).

Of all the reviews ever written on the *Origin*, Darwin thought (many would argue, quite justifiably) that Richard Owen's[106] (Fig. 2.16) was one of the most vicious and dishonest. Owen had already been indisposed against the Darwinian circle, especially his arch-rival Huxley, for trying to block his proposed construction of a natural history museum. Writing anonymously in the April 1860 issue of the *Edinburgh*

[104] See QUOT. 176.

[105] However, Darwin later noted in a June 5, 1861, letter to Asa Gray that Herschel had written a favorable note on the *Origin* in the new edition of his book entitled *Physical Geography,* although Herschel still denied natural selection any substantive role.

[106] It was the same Owen who would later prep Wilberforce for the famous Oxford debate with Huxley (see Sect. 2.12.2).

Fig. 2.16 Richard Owen (1804–1892). (From: https://commons.wikimedia.org/wiki/File:Lithograph;_portrait_of_Sir_R._Owen,_Wellcome_L0014079.jpg)

Review (Anonymous, 1860b). Owen first seemed to laud several observations made by Darwin during his voyage. However, this was just a ploy to give the appearance of sympathy toward Darwin. For, in the very next lines, Owen launched a relentless attack on the *Origin*, even questioning Darwin's basic competency in the English language:

> **QUOT. 191** Mr. Darwin must be aware of what is commonly understood by an "uninhabited island;" he may, however, mean by the inhabitants of South America, not the human kind only, whether aboriginal or otherwise, but all the lower animals. Yet again, why are the fresh-water polypes or sponges to be called "inhabitants" more than the plants? (*ibid.*, p 496)

Owen thought that the *Origin* contained a fatal flaw:

> **QUOT. 192** …it seems to us, the fundamental mistake is committed, of confounding the questions, of species being the result of a secondary cause or law, and of the nature of that creative law. (*ibidem*)

That is, according to Owen, Darwin's gross mistake was not to describe a (secondary) law operating in nature but to assert knowledge of the exact nature of such a law!

As for evolution, Owen maintained that "the most exact reasoners in biology" (like himself, presumably) had been aware that it proceeded by continuous creation. In fact:

QUOT. 193 [I]n his last published work Professor Owen does not hesitate to state "that perhaps the most important and significant result" of palaeontological research has been the establishment of the "axiom of *the continuous operation of the ordained*" *becoming of living things'* (*ibidem*)

Thus, while writing anonymously, Owen claimed that he had held evolutionary views. He resented that these were ignored by Darwin in his *Origin of Species* (Darwin, 1859, p. 310) when he associated Owen's name with other avowed anti-evolutionists such as Cuvier, Agassiz, and Sedgwick.

Indeed, although most of the time Owen embraced fixist views, from the mid-1840s onwards, Owen seemed to have entertained some transmutationist ideas. During that time, he developed a theory of archetypes (Owen, 1846, 1848), according to which species developed from a series of Platonic ideal forms (see Sect. 1.1.2), which he called archetypes. The latter were created by God, while the transformation of the archetypes to species occurred through some kind of evolution.

In Owen's diatribe, even Hooker was not spared when the former called him "short-sighted" for not distinguishing between the operation of a secondary law and its nature.

Darwin was taken aback by the vitriol in Owen's review. In an April 10, 1860, letter to Lyell, he confessed:

QUOT. 194 It [Owen's review] is extremely malignant, clever, and I fear will be very damaging. (Darwin, 1888a, p. 94)

In his April 18, 1860, letter to Hooker, Darwin very uncharacteristically hurled an epithet on Owen, especially regarding the latter's several laudatory self-references while writing under the guise of anonymity:

QUOT. 195 [W]hat a base dog [Owen] is. Some of my relations say it cannot *possibly* be Owen's article, because the Reviewer speaks so very highly of Prof. Owen. Poor dear simple folk! (Burkhardt et al., 1993, p. 162)

All bridges between Darwin and Owen were thereafter burnt.

In 1866, Owen went even further and claimed that he had discovered Darwin's theory of natural selection as early as 1850 (Owen, 1866). He was here referring to statements such as the following he had made in 1850:

QUOT. 196 The actual presence, therefore, of small species of animals in countries where larger species of the same natural families formerly existed, is not the consequence of any gradual diminution of the size of such species, but is the result of circumstances, which may be illustrated by the fable of the "oak and the reed"; the smaller and feebler animals have bent and accommodated themselves to changes which have destroyed the larger species. (Owen, 1851, p. 15)

In response to Owen's claim, Darwin stated the following in the fourth edition of the *Origin*:

QUOT. 197 This belief in Professor Owen that he then gave to the world the theory of natural selection will surprise all those who are acquainted with the several passages in his works, reviews, and lectures, published since the "Origin," in which he strenuously opposes the theory; and it will please all those who are interested on this side of the question, as it may be presumed that his opposition will now cease. It should, however, be stated that the

Fig. 2.17 Samuel Wilberforce (1870–1873). (From: https://commons.wikimedia.org/wiki/File:Samuel_Wilberforce_by_Mowbray_of_Oxford.jpg]

> passage above referred to in the "Zoological Transactions," as I find on consulting it, applies exclusively to the extermination and preservation of animals, and in no way to their gradual modification, origination, or natural selection. (Darwin, 1866, p. xviii)

Coming back to Owen's 1860 scathing review, this was not to be outdone by that of his friend and advisee on such matters, Samuel Wilberforce (Fig. 2.17), who was the Bishop of Oxford (Wilberforce, 1860). Wilberforce tried to diminish Darwin's stature in comparison with Owen, whom he claimed was "a far greater philosopher" (*ibid.*, p. 262). He then rebuked Darwin for indulging in "the jungle of fanciful assumption":

QUOT. 198 That Mr. Darwin should have wandered from this broad highway of nature's works into the jungle of fanciful assumption is no small evil. (*ibid*, p. 263)

Darwin's grandfather, Erasmus Darwin, was also not spared when Wilberforce alleged that the "philosophical arguments" of the former had been exposed by George Canning and John Frere in their satirical *Poetry of the Anti-Jacobin* (Canning & Frere, 1800).

Wilberforce would later (in June 1860) have a famous debate with Huxley on evolution (Sect. 2.12.2).

On the other side of the Atlantic, Asa Gray launched a series of favorable reviews of the *Origin*. The first one was anonymously in the *American Journal of Science and Arts* (Anonymous, 1860a). At the outset, Gray announced that the *Origin* was "exciting much attention" and that its success against competing theories would be judged by the "survival of the fittest":

QUOT. 199 A spirited conflict among opinions of every grade must ensue, which,—to borrow an illustration from the doctrine of the book before us—may be likened to the conflict in nature among races in the struggle for life, which Mr. Darwin describes; through which the views most favored by facts will be developed and tested by "Natural Selection," the weaker ones be destroyed in the process, and the strongest in the long run alone survive. (*ibid.*, p. 154)

Gray then proceeded to contrast Darwin's theory of natural selection against Agassiz's theory of theist creation (Sect. 1.10). Gray's tone clearly aimed at impartiality, but his preference could hardly be concealed when, for example, he noted that Agassiz's theory "may be said to be theistic in excess" while Darwin's theory was "strictly scientific" (*ibid.*, p. 156). Soon thereafter, the review took a different turn when referring to the two theories, Gray asked:

QUOT. 200 But does the one really exclude the other? Does the investigation of physical causes stand opposed to the theological view and the study of the harmonies between mind and Nature? (*ibid.*, p. 160)

Gray attempted to reconcile the two viewpoints (natural versus theist) as follows:

QUOT. 201 But the intellectual connection being undoubted, Mr. Agassiz properly refers the whole to "the agency of Intellect as its first cause." In doing so, however, he is not supposed to be offering a scientific explanation of the phenomena. Evidently he is considering only the ultimate why, not the proximate why or how.

Now the latter is just what Mr. Darwin is considering. He conceives of a physical connection between allied species: but we suppose he does not deny their intellectual connection, as related to a Supreme Intelligence. (*ibid.*, p. 161)

Gray's description above was accurate since Darwin did ascribe secondary causes to natural selection while not removing the Creator as being responsible for first (or final) causes.[107]

However, the similarity between Agassiz (Fig. 2.18) and Darwin was small compared to the chasm that separated Agassiz's special creation and Darwin's

[107] See QUOT. 183.

Fig. 2.18 Louis Agassiz (1807–1873). (From: https://commons.wikimedia.org/wiki/File:Agassiz_Louis_1807-1873.png)

descent with modification. This was made evident in both Agassiz's spoken and written reaction to the *Origin*. In a debate at the Boston Society of Natural History, held on Feb 15, 1860, Agassiz's views were thus reported:

> **QUOT. 202** Prof. Agassiz made a verbal communication in opposition to the theory of Mr. Darwin, recently put forth in his work on the origin of species. Mr. Darwin he acknowledged to be one of the best naturalists of England, a laborious and successful writer; his works on the coral reefs, on the cirripeds, and his narrative of the voyage of the Beagle, show him to be a skilful and well-prepared naturalist; but this great knowledge and experience had, in the present instance, been brought to the support, in his opinion, of an ingenious but fanciful theory… (Boston Society, 1861, p. 231, Vol. VII)

Agassiz's review of the *Origin* finally appeared in the July 1860 issue of the *American Journal of Science*. It may be summarized as follows:

> **QUOT. 203** …the arguments presented by Darwin in favor of a universal derivation from one primary form, of all the peculiarities existing now among living beings have not made the slightest impression on my mind, nor modified in any way the views I have already propounded. (Agassiz, 1860, p. 143)

Agassiz's main contention stemmed from his essentialist (Sect. 2.9) belief that species did not exist. Therefore, Darwin's concept of transmutation between species was incongruent:

QUOT. 204 . . .while individuals alone have a material existence, species, genera, families, orders, classes, and branches of the animal kingdom exist only as categories of thought in the Supreme Intelligence. (*ibid.*, p. 151)

QUOT. 205 If species do not exist at all. . .how can they vary? and if individuals alone exist, how can the differences which may be observed among them prove the variability of species? (*ibid.*, p. 143)

Darwin dismissed the charge on the grounds that he had never asserted that species did not exist temporarily:

QUOT. 206 I am surprised that Agassiz did not succeed in writing something better. How absurd that logical quibble—"if species do not exist, how can they vary?" As if anyone doubted their temporary existence. (Darwin, 1888a, p. 124)

Nevertheless, given the stature and influence of Agassiz, Gray took it upon himself to defend the *Origin* in the United States, especially against charges of atheism and materialism. Thus, the next year, in 1861, Gray published the pamphlet *Natural Selection Not Inconsistent with Natural Theology* (Gray, 1861). Half of the publishing cost was paid by Darwin.

2.14.4 Reception, Part III: Fleeming Jenkin and the Review (1867) that Was "of More Real Value Than Any Other Essay or Review"

2.14.4.1 Introduction

As Francis Darwin, Charles Darwin's son, correctly pointed out, it is quite remarkable that the most valuable critique of the *Origin*, "should have come, not from a professed naturalist but from a Professor of Engineering" (Darwin, 1888a, p. 289). The engineer in question was the Scottish Fleeming Jenkin (Fig. 2.19), who, in addition to being competent in his main vocation, was also an economist, lecturer, linguist, and critic. Although Jenkin's reputation stands on its own, his name is also associated with two major personalities: William Thomson (1824–1907) (later Lord Kelvin), his mentor and colleague, and Robert Louis Stevenson (1850–1894), his student and friend.

Jenkin's review (Jenkin, 1867) was particularly valuable because it eschewed the usual religious arguments against the *Origin* and instead competently focused on its scientific merits. The review was published anonymously in the *North British Review* and concerned the fourth edition of the *Origin*. It raised five major questions:

1. "Can natural selection choose special qualities, and so breed special varieties as man does?" (*ibid.*, p. 278). Here, Jenkin asked whether natural selection can really amplify (heritable) variation, as is possible in selective breeding.
2. "Does it appear that man has the power to indefinitely magnify the peculiarities which distinguish his breeds from the original stock?" (*ibid.*, pp. 278–279). That

Fig. 2.19 Fleeming Jenkin (1833–1885). (From: https://upload.wikimedia.org/wikipedia/commons/7/75/Fleeming_Jenkin.jpg)

is, is it possible for artificial selection to produce varieties beyond a certain limit? If it could be argued that this was not possible, then there would be no reason to suppose that natural selection could lead to the transmutation of species.

3. "Is there no other evidence than that of geology as to the age of the habitable earth?" (*ibid.*, p. 279) Jenkin here wished to question the uniformitarian principle that Darwin had adopted and that allowed extensive amounts of time for natural selection to operate.
4. "How far, in the absence of other knowledge, does the mere difficulty in classifying organised beings justify us in expecting that they have had a common ancestor?" (*ibidem*). That is, is it reasonable to accept a theory of common ancestry when the boundaries between species and varieties are so blurred?
5. "What value is to be attached to certain minor facts supposed to corroborate the new theory?" (*ibidem*). This was a common criticism of the *Origin* in that it proposed a theory that explained certain observations but was not scientifically proven.

2.14.4.2 Limits to Variation

Jenkin took each of these questions in turn, starting with (2) in Sect. 2.14.4.1. He here argued that just because artificial selection could produce varieties within a stock, there was no reason to infer that natural selection would produce new species, even after an extended time. Jenkin's reasoning was based on some type of law of diminishing returns:

QUOT. 207 We all believe that a breeder, starting business with a considerable stock of average horses, could, by selection, in a very few generations, obtain horses able to run much faster than any of their sires or dams; in time perhaps he would obtain descendants running twice as fast as their ancestors, and possibly equal to our racehorses. But would not the difference in speed between each successive generation be less and less? (*ibid.*, p. 281)

QUOT. 208 ...whatever new point in the variable beast, bird, or flower, be chosen as desirable by a fancier, this point can be rapidly approached at first, but that the rate of approach quickly diminishes, tending to a limit never to be attained. (*ibid.*, p. 282)

Jenkin concluded that a species was always constrained within a "sphere of variation":

QUOT. 209 A given animal or plant appears to be contained, as it were, within a sphere of variation; one individual lies near one portion of the surface, another individual, of the same species, near another part of the surface; the average animal at the centre. Any individual may produce descendants varying in any direction, but is more likely to produce descendants varying towards the centre of the sphere, and the variations in that direction will be greater in amount than the variations towards the surface. (*ibidem*)

Thus, Jenkin's fixist argument here is essentially the same as Lyell's during the pre-*Origin* period (QUOT. 79). Furthermore, Jenkin's concept of variation toward the center is a rudimentary version of Galton's later concept of regression towards the mean.

Darwin did not have much difficulty with Jenkin's argument on the limits to variation. He addressed it in the fifth edition of the *Origin* as follows:

QUOT. 210 Some authors have maintained that the amount of variation in our domestic productions is soon reached, and can never afterwards be exceeded. It would be somewhat rash to assert that the limit has been attained in any one case; for almost all our animals and plants have been greatly improved in many ways within a recent period; and this implies variation. It would be equally rash to assert that characters now increased to their utmost limit, could not, after remaining fixed for many centuries, again vary under new conditions of life. (Darwin, 1869, pp. 45–46)

In the above, Darwin conceded that a species would remain fixed near the inside edge of the sphere. However, given enough time to accumulate the necessary variation, the species could pass across the edge and give rise to a new species.

2.14.4.3 Propagation of a Single Variation (or the Swamping Argument)[108]

Question (1) in Sect. 2.14.4.1 is the one that has received the most attention. Jenkin considered the case of common individual differences and rare sports separately. It will be remembered[109] that Darwin considered the former much more important than the latter in his theory of evolution by natural selection. Concerning individual differences, Jenkin admitted that natural selection was able to act on them to produce varieties. However, as he had argued in (1), there was no chance of a species going beyond its sphere of variation.

Next, turning his attention to sports, Jenkin asked whether natural selection could produce a new species. He then put forward an elaborate mathematical argument for why the answer to such a question should be no, even if the sport was highly favorable.

In the first part of his argument, Jenkin contended that the survival advantage of a sport would be completely outweighed by its numerical inferiority. He considered the case of one million organisms, of which 10,000 survive to produce offspring. Suppose one of the 10^6 organisms is a sport with twice the survival probability of the others. The probability that the sport is one of the offspring is only $2 \times 10^4/10^6 = 1/50$, which means that the odds are stacked against a sport being preserved, even if it is favorable.

In the second part of his argument, Jenkin considered the consequence if the sport in the example above was preserved. Then:

> **QUOT. 211** Let us consider what will be its influence on the main stock when preserved. It will breed and have a progeny of say 100; now this progeny will, on the whole, be intermediate between the average individual and the sport. The odds in favour of one of this generation of the new breed will be, say 1½ to 1, as compared with the average individual; the odds in their favour will therefore be less than that of their parent; but owing to their greater number, the chances are that about 1½ of them would survive. Unless these breed together, a most improbable event, their progeny would again approach the average individual; there would be 150 of them, and their superiority would be say in the ratio of 1¼ to 1; the probability would now be that nearly two of them would survive, and have 200 children, with an eighth superiority. Rather more than two of these would survive; but the superiority would again dwindle, until after a few generations it would no longer be observed, and would count for no more in the struggle for life, than any of the hundred trifling advantages which occur in the ordinary organs. (Jenkin, 1867, p. 289)

In the above, Jenkin used the previous model according to which, out of every 100 normal adults, only one survived on average. He next assumed that each surviving adult mated to give 100 offspring. Now, if the surviving adult was a

[108] Jenkin's swamping argument is also discussed in Morris (1994), Gayon (1998, pp. 92–97), Vorzimmer (1970, pp. 121–122), Bulmer (2004), Hardin (1959, pp. 113–117), Mayr (1982, pp. 512–514), Bowler (1996, pp. 133–134), Eiseley (1958, pp. 209–211), Hull (1974, pp. 55–57), Gould (1985), Mukherjee (2016, pp. 44–45), Endersby (2007, pp. 76–78), Browne (1996, pp. 282–283), Dobzhansky (1965, pp. 205–206), and Nei (2013, p. 3).

[109] See QUOT. 169.

sport (with twice the survival probability of a normal adult) who then mated with a normal adult, the 100 offspring would each be half-sports (i.e., with 1½ times the survival probability of a normal). Of these, $1 \times 1.5 = 1.5$ on average would survive.

Thus, Jenkin had thus assumed the same model that Darwin had used in his work, namely, a blending mode of inheritance (Sects. 2.10.1 and 2.14.1). It would be Whiggish to claim that, now that we know there is no such thing as blending, Jenkin's ensuing argument was completely invalid. However, blending theory was viable at the time, based on countless observations where offspring characteristics did appear to be intermediate between those of the parents.[110]

Continuing with Jenkin's example, an average of 1.5 surviving half-sports would, by mating with normal adults, produce 150 offspring who would each be quarter-sports (i.e., with 1¼ times the survival probability of a normal). Of these, $1.5 \times 1.25 = 1.875$ on average would survive, and so on.

Therefore, the total "sport-value" (or "person-sport") in the population decreases according to $1, 1.5 \times 0.5, 1.875 \times 0.25, \ldots = 1, 0.75, 0.47, \ldots$, eventually converging to zero. Jenkin's point was that any sport would soon be *swamped*, thus rendering natural selection ineffective. Swamping emerges as a necessary consequence of blending inheritance.

Jenkin went on to give a compelling graphic example to illustrate his point:

> **QUOT. 212** Suppose a white man to have been wrecked on an island inhabited by negroes, and to have established himself in friendly relations with a powerful tribe, whose customs he has learnt. Suppose him to possess the physical strength, energy, and ability of a dominant white race, and let the food and climate of the island suit his constitution; grant him every advantage which we can conceive a white to possess over the native; concede that in the struggle for existence his chance of a long life will be much superior to that of the native chiefs; yet from all these admissions, there does not follow the conclusion that, after a limited or unlimited number of generations, the inhabitants of the island will be white. (*ibid.*, pp. 289–290)

For all its persuasiveness, Jenkin's argument contained a serious error that almost invalidated its main conclusion. The error was in supposing that each adult pair mated to produce 100 individuals, of which *only one* survived on average. This leads to a rapidly decreasing population size. On the other hand, if each adult pair mated to produce 100 individuals, of which *two* (one male and one female) on average survived, then a constant population size would be maintained. Unfortunately, Jenkin's error has been ignored by many historians and was first noticed by Davis (1871), who corrected (in square brackets) Jenkin's numbers[111] as follows:

> **QUOT. 213** Let us consider what will be its influence on the main stock when preserved. It will breed and have a progeny of say 100; now this progeny will, on the whole, be intermediate between the average individual and the sport. The odds in favour of one of this generation of the new breed will be, say 1½ to 1, as compared with the average individual; the odds in their favour will therefore be less than that of their parent; but owing to their greater number, the chances are that about 1½ of them would survive

[110] Except for occasional observations of dominance and reversion.

[111] See QUOT. 211.

[*about 3 of them*, for without any advantage two would on average survive.] Unless these breed together, a most improbable event, their progeny would again approach the average individual; there would be 150 [300] of them, and their superiority would be say in the ratio of 1¼ to 1; the probability would now be that nearly two [6 x 5/4, or nearly 8] of them would survive, and have 200 [750] children, with an eighth superiority. Rather more than 2 [15] of these would survive. (*ibid.*, p. 161)

What are the implications of Davis' correction regarding Jenkin's main conclusion that the effect of a sport would soon be swamped and thus completely disappear? It is better to hear it from Davis himself:

QUOT. 214 The writer [Jenkin] thus concludes that the advantage derived by inheritance from the sport will ultimately die out. The true conclusion is, that the advantage never dies out, but only becomes distributed through the whole race; and, moreover, that the sum of the advantages of all the favoured individuals, when added together, is greater than the original advantage, and becomes greater and greater every successive generation, though it tends to a limit at which it never actually arrives. Thus, representing the original advantage by unity, the advantage in the next generation is 1½, in the next 1⅞, and so on. (*ibidem*)

Thus, under the correct model of two survivors out of 100, the sport is not swamped out completely and does not vanish but is evenly distributed in the population, albeit in very diluted form. The sporting value *does not decrease* according to 1, 0.75, 0.47,. . ., but instead *increases* according to 1, 1.5, 1.88,. . . . The white man wrecked on an island inhabited by blacks will surely not turn the population white over time (Darwin), but neither will the population be as dark-colored as before (Jenkin); instead, it will be slightly less dark (Davis).

However, as intimated before, Jenkin's error did not invalidate his argument. This is because, although natural selection is not rendered completely ineffective by swamping, its force is greatly reduced. Jenkin's argument was still valid because it showed the error in Darwin's belief that the slightest advantage would be propagated across generations.

Continuing with his argument, Jenkin made several additional points:

- Swamping could only be prevented if the variation occurred in many individuals and in the same direction. However, this would contradict Darwin's assumption of fortuitousness in variations (Jenkin, 1867, p. 290).
- Swamping would occur not only for sports but also for "slight and imperceptible variations" (*ibid.*, p. 291), i.e., Darwin's individual differences.
- If sports were non-blending and were transmitted to descendants, then a new species would indeed be formed. However, this would simply be a case of "successive creations" (*ibid.*, p. 294), which was contrary to Darwin's theory of gradual accumulation of differences.

All indications are that Darwin took Jenkin's review quite seriously, especially concerning the latter's argument on swamping. In a January 16, 1869, letter to Hooker, he confessed that "Fleeming Jenkin has given me much trouble, but has been of more real use to me than any other essay or review" (Darwin & Seward, 1903a, p. 379). Darwin realized that Jenkin had poked a hole in his argument that the slightest advantage would be propagated across generations. For example, in the *Origin*, Darwin had said:

QUOT. 215 Now, if any slight innate change of habit or of structure benefited an individual wolf, it would have the best chance of surviving and of leaving offspring. Some of its young would probably inherit the same habits or structure, and by the repetition of this process, a new variety might be formed. (Darwin, 1859, p. 91)

The quotation above, which can be found in the first four editions of the *Origin*, was dropped by Darwin in the fifth edition after Darwin had read Jenkin's review. The paragraph from which the quotation was taken now read as follows:

QUOT. 216 It should be observed that. . .I speak of the slimmest individual wolves, and not of any single strongly marked variation having been preserved. In former editions of this work I sometimes spoke as if this latter alternative had frequently occurred. I saw the great importance of individual differences, and this led me fully to discuss the results of unconscious selection by man, which depends on the preservation of the better adapted or more valuable individuals, and on the destruction of the worst. I saw, also, that the preservation in a state of nature of any occasional deviation of structure, such as a monstrosity, would be a rare event; and that, if preserved, it would generally be lost by subsequent intercrossing with ordinary individuals. Nevertheless, until reading an able and valuable article in the "North British Review" (1867), I did not appreciate how rarely single variations, whether slight or strongly marked, could be perpetuated. (Darwin, 1869, pp. 103–104)

In the above, Darwin claimed that he had been aware that sports were rare and would be lost if allowed to intercross with normal individuals. This claim seems to be borne by the facts, for in the 1844 Essay Darwin had indeed said that "[e]ven in the rare instances of sports, with the hereditary tendency very strongly implanted, crossing must be prevented with other breeds, or if not prevented the best characterized of the half-bred offspring must be carefully selected" (Darwin, 1909, pp. 63–64).

In QUOT. 216, taken from the fifth edition, Darwin also admitted that the article in the North British Review had convinced him that the effect of sports would be even more minimal than he had thought. Thus, Darwin henceforth put even more emphasis on the importance of individual differences (rather than sports) as the major type of variation on which natural selection acted.[112]

[112]This sentiment is also expressed in Darwin's Jan 22, 1869, letter to Wallace: "I have been interrupted in my regular work in preparing a new [5th] edition of the 'Origin,' which has cost me much labor, and which I hope I have considerably improved in two or three important points. I always thought individual differences more important than single variations, but now I have come to the conclusion that they are of paramount importance, and in this I believe I agree with you. Fleeming Jenkin's arguments have convinced me" (Darwin, 1888a, p. 288). Wallace misunderstood Darwin's sentence and thought the latter was now saying that sports were more important. Darwin replied as follows: "I must have expressed myself atrociously; I meant to say exactly the reverse of what you have understood. F. Jenkin argued in the 'North British Review' against single variations ever being perpetuated and has convinced me, though not in quite so broad a manner as here put. I always thought individual differences more important; but I was blind and thought that single variations might be preserved much oftener than I now see is possible or probable. I mentioned this in my former note merely because I believed that you had come to a similar conclusion, and I like much to be in accord with you. I believe I was mainly deceived by single variations offering such simple illustrations, as when man selects" (Darwin, 1887, pp. 288–289, Vol. I).

Some authors[113] have claimed that Jenkin's review forced Darwin to develop the (Lamarckian) "provisional hypothesis of pangenesis" (Sect. 3.7.2) in an attempt to show that variation could be constantly generated, which would thus overcome the effect of swamping. It is true that the provisional hypothesis of pangenesis first appeared in print in Darwin's *Variation of Animals and Plants under Domestication* (Darwin, 1868b) after Jenkin's review had come out. However, Darwin had developed this hypothesis before the review and had sent a manuscript on the subject to Huxley as early as 1865 (Olby, 1966, p. 1; Vorzimmer, 1970, pp. 120–121).[114] Thus, the claim that pangenesis was developed as a result of Jenkin's review is incorrect. Furthermore, Jenkin's criticism was mainly targeted against sports, but Darwin had always believed that individual differences (rather than sports) were more important.

On the other hand, Vorzimmer has adopted a diametrically opposite view to the aforementioned writers by claiming that Jenkin did not affect Darwin's thinking (Vorzimmer, 1963, p. 373, 1970, p. 99). This also seems to be exaggerated because, in the wake of Jenkin's review, Darwin was forced to point out that his (already developed) hypothesis of pangenesis would be able to counteract the effect of swamping.

2.14.4.4 Age of the Earth

Question (3) in Sect. 2.14.4.1 relates to Jenkin's questioning of the assumption of an extremely old Earth and also caused Darwin some trouble. As explained in Sect. 1.14, Lyell's uniformitarianism was essential to Darwin's theory since it allowed for extensive lapses of time during which natural selection could act and lead to the transmutation of species. In the first edition of the *Origin* (Darwin, 1859, p. 287), Darwin estimated the age of the Weald, a sedimentary rock formation in southeast England, to be approximately 300 million years. This number was noticed by William Thomson[115] (who was Jenkin's mentor), and he explicitly questioned its validity (Thomson, 1862, p. 375). In a famous follow-up paper (Thomson, 1863), Thomson attacked the uniformitarian principle. Basing his calculations on the cooling of the Earth to its present state, he hypothesized that the Earth's crust was only about 98 million years old[116] (*ibid.*, p. 160). Later the number decreased to approximately 24 million years (e.g., Burchfield, 2009, p. 43).

---

[113]For example, Willis (1940, pp. 5, 165), Eiseley (1959), Darlington (1959, pp. 38–39), and Hardin (1959, pp. 115–116).

[114]In fact, Darwin wrote to Lyell on August 22, 1867, that pangenesis was 26 or 27 years old (Darwin, 1888a, p. 72, Vol. II), which brings the date of the theory to about 1840. In March 1868, Darwin thanked the English doctor William Ogle (1827–1912) for pointing out to him that pangenesis had been used thousands of years before by Hippocrates.

[115]Later, Lord Kelvin (1824–1907).

[116]The current estimate is 4.5 billion years, which makes Thompson's estimate way off. This is because radioactivity, which is an additional source of heat, had not been discovered at the time of Thomson's calculations.

In his review, Jenkin took up Thomson's arguments and questioned Darwin's assumption of "countless ages":

> **QUOT. 217** Geology lends her aid to convince him that countless ages have elapsed, each bearing countless generations of beings, and each differing in its physical conditions very little from the age we are personally acquainted with. This view of past time is, we believe, wholly erroneous. (Jenkin, 1867, p. 295)

He thanked Thomson ("Professor Sir W. Thomson of Glasgow") for having exposed the fallacy of the geologists. Jenkin also dismissed Darwin's estimate of 300 million years for the age of the Weald as a "guess at the half and multiply by two" (*ibidem*).

Darwin took the Thomson–Jenkin criticism quite seriously[117] but could not find a compelling argument to counter it. In the third and later editions of the *Origin*, he had already deleted his estimate for the age of Weald because of wide criticism both in the popular and scientific press (Anonymous, 1859a). However, this was not in response directly to either Thomson or Jenkin, since the latter had criticized the *Origin* only after the third and fourth editions, respectively, had come out. In the fifth edition of the *Origin* (Darwin, 1869, p. 352), in the wake of Jenkin's review, Darwin cited James Croll's (1821–1890) assertion that it was difficult to estimate extended periods in years. Moreover, even supporters of Thomson were coming up with widely different estimates of the age of the Earth, so Darwin was able to emphasize the physicists' own uncertainties. In a Feb 28, 1866, letter to Hooker, he said:

> **QUOT. 218** I cannot think how you can attach so much weight to the physicists, seeing how Hopkins, Hennessey, Haughton, & Thompson [sic] have enormously disagreed about the rate of cooling of the crust. (Burkhardt et al., 2004, p. 84)

Finally, Darwin's emphasis on the Lamarckian effect of the environment in the generation of variation (through the hypothesis of pangenesis, see Sect. 2.14.4.3) was partly meant to speed up the rate of evolution and thus counter the argument that there might not have been enough time for natural selection to act.

2.14.4.5 Difficulties of Classification

Regarding Question (4) in Sect. 2.14.4.1, Jenkin argued that the lines of demarcation between species and varieties were blurred so that there was an element of arbitrariness in the way naturalists classified them. Sometimes, an organism was classified into a particular species group based on a certain organ, while on other occasions, it was based on another organ. Therefore, how valid was a theory of transmutation between species when the basis of classification itself was questionable? In Jenkin's own words:

> **QUOT. 219** [the] lines of demarcation, as drawn by different naturalists, vary much, being sometimes made to depend on this, sometimes on that organ, rather arbitrarily. This

[117] On April 14, 1869, Darwin acknowledged to Wallace that: "Thomson's views of the recent age of the world have been for some time one of my sorest troubles" (Darwin, 1888a, p. 296, Vol. II).

difficulty chiefly seems to have led men to devise theories of transmutation of species... (Jenkin, 1867, p. 305)

Darwin does not seem to have taken Jenkin's criticism of classification seriously, and this is probably Jenkin's weakest argument. Darwin had already acknowledged the difficulty of classifying closely related organisms, but this fact alone could not deny the possibility of transmutation between them.

2.14.4.6 Lack of a Definite Proof for Natural Selection

The last question (5) in Sect. 2.14.4.1 raised an issue that was not new to Darwin. The criticism was that Darwin's theory of descent with modification through natural selection was, in fact, a hypothesis that was founded entirely on empirical evidence and was not experimentally proved. Jenkin explained this criticism in the following way:

QUOT. 220 The chief arguments used to establish the theory rest on conjecture. Beasts may have varied; variations may have accumulated; they may have become permanent; continents may have arisen or sunk, and seas and winds been so arranged as to dispose of animals just as we find them... (*ibid.*, p. 313)

Although Darwin did not explicitly respond to the criticism, he often defended his scientific method. For example, in his autobiography, he stated:

QUOT. 221 I worked on true Baconian principles, and without any theory collected facts on a wholesale scale. (Barlow, 1958, p. 119)

As explained in Sect. 2.10.2, Darwin had also learned from Herschel and Whewell's writings that a theory would be more scientifically plausible if it was able to explain more (and often new) phenomena. This is indeed what happened to Darwin's theory, as an increasing number of his contemporaries soon came to realize the explanatory powers of natural selection. However, the road to acceptance was certainly not one without its vicissitudes. As will soon be seen, one of the thorniest issues was the inadequacy of the heredity component in Darwin's theory.

2.14.5 Reception, Part IV: St. George Mivart and the "Most Cutting Review" (1871) of Darwin's Theory

If there was one person whose criticism of Darwin's theory surpassed Owen's virulence, that person would be the English jurist and zoologist St George Jackson Mivart (Fig. 2.20). Mivart entered the communion of the Roman Catholic Church in 1844 when he was still 16. In his early years, he was influenced by Richard Owen, whom he communicated frequently with and learned much from. In January 1858, Mivart first attended Huxley's lectures and was profoundly impressed. He soon

Fig. 2.20 St George Jackson Mivart (1827–1900). (From: https://commons.wikimedia.org/wiki/File:St._George_Jackson_Mivart._Photograph_by_Barraud_%26_Jerrard._Wellcome_V0026864.jpg)

became a disciple of the latter and, like his mentor, embraced the *Origin*. However, as Mivart later recalled:

> **QUOT. 222** It was in 1868 that difficulties as to the theory of Natural Selection began to take shape in my mind, and they were strongly reinforced by the arguments of one who became, and remains, as gratitude demands, a highly valued friend, whose acquaintance I made at Professor Huxley's lectures, at which love of science had also made him a regular attendant. This was the Rev. W. W. Roberts. . . . The arguments he again and again urged upon me were the difficulties, or rather the impossibilities, on the Darwinian system, of accounting for the origin of the human intellect, and above all for its moral intuitions—not its moral *sentiments*,[118] but its ethical *judgments*. (Mivart, 1897, p. 994)

[118] Italics are Mivart's.

Thus, Mivart's main concern was that man's intellect was unique and could not be accounted for by Darwinian evolution. He became increasingly disturbed when Huxley published his *On Man's Place in Nature* (Huxley, 1863), where an all-pervasive theory of human evolution, acting on both man's physical and mental traits, was embraced. Mivart voiced his concerns to Huxley in July 1869. The latter was visibly surprised and saddened but still able to maintain a friendly relationship with Mivart.

Mivart first publicly voiced his misgivings about Darwinian evolution in a 3-part anonymous article entitled "Difficulties of the theory of natural selection," published in the Catholic journal *The Month* (Mivart, 1869). The article did not attract much attention, but when Mivart later expanded it into a book, *On the Genesis of Species* (Mivart, 1871c), a different effect was produced.

Mivart's *Genesis* was most aptly described by Darwin himself as a book that "collected all the objections which have ever been advanced" on the theory of natural selection, and which its author had written with "admirable art and force," so that "thus marshalled, they make a formidable array" (Darwin, 1872, p. 176). At the very beginning of his book, Mivart pointed out that he admitted both evolution and natural selection, but with some caveats:

> **QUOT. 223** The general theory of evolution has indeed for some time past steadily gained ground, and it may be safely predicted that the number of facts which can be brought forward in its support will, in a few years, be vastly augmented. But the prevalence of this theory need alarm no one, for it is, without any doubt, perfectly consistent with strictest and most orthodox Christian theology. Moreover, it is not altogether without obscurities, and cannot yet be considered as fully demonstrated. (Mivart, 1871c, p. 16)
>
> ...
>
> It is its [the book's] object to maintain the position that "Natural Selection" acts, and indeed must act, but that still, in order that we may be able to account for the production of known kinds of animals and plants, it requires to be supplemented by the action of some other natural law or laws as yet undiscovered. (*ibid.*, p. 17)

At the end of the first chapter, Mivart listed his major difficulties with Darwinian evolution. In the subsequent chapters, he went through each of these difficulties chapter by chapter with a tenacity that highlighted his lawyerly skills. Mivart's criticisms of Darwin's theory were as follows:

(i) *Initial stages of evolution.* Even if a completely developed structure or variation was advantageous, the initial stages of development (i.e., "the minute and rudimentary beginnings") would have no such advantage, and thus, natural selection could not account for them.

(ii) *Closely related structures from distant species.* While it is very likely that similar functional structures in closely related species could have evolved by natural selection, similar structures from distant species could not be accounted for by natural selection. This is because "the number of possible variations is indefinitely great, and it is therefore an indefinitely great number to one against a similar series of variations occurring and being similarly preserved in any two independent instances" (*ibid.*, p. 80).

(iii) *Sudden* vs. *gradualist modes of evolution.* The occurrence of "sudden" variations could not be ignored. As Galton pointed out (Sect. 4.3.4), only certain positions in the evolutionary path were stable. Huxley, too, had admitted as much. Instances given by Rev. R. Everett, Charles Lyell, M.C Naudin, and many others, all pointed to discontinuous evolution.
(iv) *Limits to the variability of species.* Several examples pointed to limits to the variability of species, with "occasional sudden and considerable modifications" (*ibid.*, p. 141). This view "had been well expressed in an able article" in the *North British Review.*[119]
(v) *Species and time.* First, several transitional forms, which would have been expected under Darwin's theory, remained absent. Second, as pointed out by William Thomson (Sect. 2.14.4.4), the Earth was not old enough for natural selection, acting on minute differences, to have produced any substantive changes.
(vi) *Species and space.* According to Darwin's theory, similar species should be expected to exist in geographically close regions.[120] However, there were several instances where this was not the case.
(vii) *Existence of homologies.* Homologies[121] could not be explained by the action of natural selection. Their existence pointed to the action of some "internal power, by the action of which these homological manifestations are evolved" (*ibid.*, p. 187).

Since Mivart admitted both evolution and natural selection, what then was his conception of these two forces, and how could such a conception avoid the difficulties above? Mivart first clarified that his views were independently derived but shared many similarities with Richard Owen's theory (called the "derivation" of species). According to Mivart, there was an "internal force" without which the action of natural selection was rendered useless. In his own words:

> **QUOT. 224** . . .an internal law presides over the actions of every part of every individual, and of every organism as a unit, and of the entire organic world as a whole. It is believed that this conception of an internal innate force will ever remain necessary, however much its subordinate processes and actions may become explicable:
>
> That by such a force, from time to time, new species are manifested by ordinary generation. . .these new forms not being monstrosities but harmonious self-consistent wholes. That thus, as specific distinctness is manifested by obscure sexual conditions, so in obscure sexual modifications specific distinctions arise.
>
> That these "jumps" are considerable in comparison with the minute variations of "Natural Selection"—are in fact sensible steps, such as discriminate species from species.

[119] By the engineer Fleeming Jenkin (see Sect. 2.14.4.2).

[120] This was more explicitly stated by Wallace in his Sarawak Law paper (Wallace, 1855) (see Sect. 2.13.2).

[121] That is the existence of two or more parts of similar nature, such as (i) the two hands of a human, (ii) the teeth in the two jaws, and (iii) the arm of a human, the foreleg of a horse, and the paddle of a whale.

> That the latent tendency which exists to these sudden evolutions is determined to action by the stimulus of external conditions.
>
> That "Natural Selection" rigorously destroys monstrosities, and abortive and feeble attempts at the performance of the evolutionary process.
>
> That "Natural Selection" removes the antecedent species rapidly when the new one evolved is more in harmony with surrounding conditions.
>
> That "Natural Selection" favors and develops useful variations, though it is impotent to originate them or to erect the physiological barrier which seems to exist between species. (*ibid.*, pp. 255–256)

Moreover:

> **QUOT. 225** ...we find a perfect harmony in the double nature of man, his rationality making use of and subsuming his animality; his soul arising from direct and immediate creation, and his body being formed at first (as now in each separate individual) by derivative or secondary creation, through natural laws ... [which,] for the most part as yet unknown but controlled by "Natural Selection". (*ibid.*, p. 305)

It is to be noted that each of the components above in Mivart's theory was in direct opposition to that of Darwin.

Mivart's *Genesis* was published immediately after Darwin's *Descent of Man* (Darwin, 1871b) and was well-received, especially by the clergy. Darwin seemed to be initially pleased, even declaring to Wallace on Jan 30, 1871, that:

> **QUOT. 226** ...I have just read (but not with sufficient care) Mivart's book, and I feel *absolutely certain*[122] that he meant to be fair (but he was stimulated by theological fervor)... I have since its publication received some quite charming letters from him. (Darwin, 1888a, p. 135)

Mivart's continued "charming" letters to Darwin, while at the same time bashing his work either anonymously or otherwise, would be a source of increasing irritation for Darwin. Although the latter did not initially seem displeased with Mivart's *Genesis*, that feeling changed when the American philosopher and mathematician Chauncey Wright (1830–1875) wrote to Darwin on June 21, 1871. Wright informed him that he had rebutted Mivart's arguments in an article to be published in the *North American Review* (Wright, 1871), and attached proofs of said article.

While acknowledging that *Genesis* was "the most effective general criticism of the theory of Natural Selection which has yet appeared" (*ibid.*, p. 66), Wright alleged that "the work falls far short of what we might have expected from an author of Mr. Mivart's attainments as a naturalist..." (*ibidem*). In particular:

(a) Mivart had completely misunderstood Darwin's work since he lacked "the genius that pursues successfully the researches for unknown causes by the skilful use of hypothesis and experiment" (*ibid.*, p. 78).
(b) Mivart had purposely quoted only parts of Darwin's sentences to distort their meanings. (*ibid.*, pp. 82–84).
(c) Mivart's insistence on the importance of final causes was not Christian in spirit, as he would have wanted it to be, but in fact Platonic (*ibid.*, pp. 96–97).

---

[122] Italics are Darwin's.

Wright's review apparently opened Darwin's eyes to the bias in Mivart's book. Taking up Wright's criticism (b), Darwin's July 9, 1871, letter to Wallace now read:

QUOT. 227 You will think me a bigot when I say, after studying Mivart, I was never before in my life so convinced of the general (i.e. not in detail) truth of the views in the "Origin". I grieve to see the omission of the words by Mivart, detected by Wright. I complained to Mivart that in two cases he quotes only the commencement of sentences by me, and thus modifies my meaning; but I never supposed he would have omitted words. There are other cases of what I consider unfair treatment. I conclude with sorrow that though he means to be honourable, he is so bigoted that he cannot act fairly... (Darwin, 1888a, pp. 144–145)

As Francis Darwin later pointed out (*ibid.*, p. 144, footnote), Mivart's omitted portions were not quotations from Darwin. Wallace too tried to reassure Darwin that the omitted words did not distort the meaning of the latter's sentences.

In the meantime, Mivart's devastating anonymous review of Darwin's *Descent of Man* appeared in the *Quarterly Review* (Mivart, 1871a). In his book, Darwin argued for the action of natural selection even in the evolution of man's mind. Mivart would have none of it:

QUOT. 228 Must we acknowledge that, "man with all his noble qualities, with sympathy which feels for the most debased, with benevolence which extends not only to other men but to the humblest living creature, with his god-like intellect which has penetrated into the movements and constitution of the solar system," must we acknowledge that man "with all these exalted powers" is descended from an Ascidian? Is this a scientific truth resting on scientific evidence, or is it to be classed with the speculations of a bygone age? (*ibid.*, pp. 52–53)

Ad hominem attacks toward Darwin were not spared:

QUOT. 229 Mr. Darwin's power of reasoning seems to be in an inverse ratio to his power of observation. (*ibid.*, p. 87)

Darwin was taken aback by the virulence of the review. Three days after his July 9 letter to Wallace, he wrote to the latter again. In a postscript to the letter, Darwin complained:

QUOT. 230 There is a most cutting review of me in the Quarterly. I have only read a few pages. The skill and style make me think of Mivart. I shall soon be viewed as the most despicable of men. (Darwin, 1868b, p. 146)

Darwin devoted a large portion of Chapter VII of the sixth edition of *The Origin* to address each of Mivart's criticisms. While he tried to maintain a neutral tone toward Mivart in the book, Darwin was annoyed.

What especially irked Darwin was Mivart's apparent duplicity: in his public reviews, he was bashing Darwin, but in his personal letters to him he seemed subservient. For example, when Mivart had received from Darwin a copy of the first volume of *Descent of Man*, he had said on Jan 22, 1871:

QUOT. 231 I do not see how anyone can reasonably feel anything but sympathy and esteem for a writer zealously & industriously labouring for the promotion of what he conscientiously believes to be truth.[123]

Huxley saw in Mivart's scathing *Quarterly* review someone who had been "insolent to Darwin," and he took it upon himself to "pin him [Mivart] out" (Huxley, 1918, p. 128, Vol. II). Huxley's rebuttal of Mivart's *Genesis* and *Quarterly* review was published in *The Contemporary Review* (Huxley, 1871). It mocked Mivart's scientific and theological arguments:

QUOT. 232 . . .Mr. Mivart does not hesitate to push his attempt to harmonize science with Catholic orthodoxy to its utmost limit; and, while assuming that the soul of man "arises from immediate and direct creation," he supposes that his body was "formed at first (as now in each separate individual) by derivative, or secondary creation, through natural laws". (*ibid.*, p. 331)

QUOT. 233 This means, I presume, that an animal, having the corporeal form and bodily powers of man, may have been developed out of some lower form of life by a process of evolution; and that, after this anthropoid animal had existed for a longer or shorter time, God made a soul by direct creation, and put it into the manlike body, which, heretofore, had been devoid of that *anima rationalis*, which is supposed to be man's distinctive character. (*ibid.*, p. 454)

Mivart tried to reply to Huxley in the forthcoming issue of *The Contemporary Review* (Mivart, 1871b), but he was no match to Huxley's rhetoric. Darwin was delighted.

Thereupon, a lull in the fight ensued. However, the pause was relatively brief. In July 1874, Mivart (1874) wrote a review of a recent book on primitive man by Tylor and Lubbock. In his review, Mivart hurled sharp attacks at Darwin. The latter had been used to those, but then Mivart decided to take George Darwin (Charles Darwin's fifth child) to task. George had previously published an article (Darwin, 1873) arguing that divorce should be made easier in cases of cruelty, abuse, or mental disorders. Mivart exaggerated George's stance and used phrases such as "unrestrained licentiousness" and "hideous sexual criminality of Pagan days" to describe the latter's views. Darwin was horrified and, in a letter dated Jan 12, 1875, informed Mivart he had decided to "refuse to hold for the future any communication" (Burkhardt et al., 2015, p. 92, Vol. XXIII) with him. Darwin's letter was the last correspondence between Mivart and himself.

Mivart attacked Darwin again in 1876 in his *Lessons from Nature* (Mivart, 1876). This time, Wallace took it upon himself to rebuke Mivart in the *Academy* (Wallace, 1876).

[123] Darwin Correspondence Project, "Letter no. 7451," accessed on 16 June 2020, https://www.darwinproject.ac.uk/letter/DCP-LETT-7451.xml.

2.14.6 Three "Precursors" of Natural Selection: Patrick Matthew (1831), William Charles Wells (1813), and Edward Blyth (1835)

2.14.6.1 Patrick Matthew (1790–1874)

On April 7, 1860, Patrick Matthew (Fig. 2.21), a Scottish horticulturist, wrote a letter to the *Gardener's Chronicle*. Referring to Huxley's review of the *Origin* in the *Times* (Huxley, 1859), Matthew stated that:

> **QUOT. 234** . . .[the] discovery recently published as "the results of 20 years' investigation and reflection" by Mr Darwin turns out to be what I published very fully and brought to apply practically to forestry in my work "[On] Naval Timber and Arboriculture," published as far back as January 1, 1831. . . (Matthew, 1860, p. 342)

The book by Matthew (mentioned in QUOT. 234) was concerned with the cultivation of trees for shipbuilding (!). In his letter, Matthew then offered several extracts from his book *On Naval Timber and Arboriculture* (Matthew, 1831) where he

Fig. 2.21 Patrick Matthew (1790–1874). (From: https://commons.wikimedia.org/wiki/File:Patrick_Matthew_1790.jpg)

adduced the principle of natural selection. A clear statement was provided in Note B of the Appendix:

> **QUOT. 235** THERE is a law universal in nature, tending to render every reproductive being the best possibly suited to its condition that its kind, or that organized matter, is susceptible of, which appears intended to model the physical and mental or instinctive powers, to their highest perfection, and to continue them so. This law sustains the lion in his strength, the hare in her swiftness, and the fox in his wiles. As Nature, in all her modifications of life, has a power of increase far beyond what is needed to supply the place of what falls by Time's decay, those individuals who possess not the requisite strength, swiftness, hardihood, or cunning, fall prematurely without reproducing—either a prey to their natural devourers, or sinking under disease, generally induced by want of nourishment, their place being occupied by the more perfect of their own kind, who are pressing on the means of subsistence. (*ibid.*, pp. 364–365)

As has been noted by Weale (2015), Matthew's book included all the essential elements of a viable theory of evolution by natural selection, namely variation within species (pp. 106, 385), heritability of traits (p. 385), and competition (pp. 364, 384).

In one interesting section of the book, Matthew considered the possibility of either repeated miraculous creation or evolution and favored the latter:

> **QUOT. 236** We are therefore led to admit, either of a repeated miraculous creation; or of a power of change, under a change of circumstances, to belong to living organized matter, or rather to the congeries of inferior life, which appears to form superior. The derangements and changes in organized existence, induced by a change of circumstance from the interference of man, affording us proof of the plastic quality of superior life, and the likelihood that circumstances have been very different in the different epochs, though steady in each, tend strongly to heighten the probability of the latter theory. (Matthew, 1831, pp. 381–382)

Darwin took immediate note of Matthew's letter and wrote to Lyell on April 10, 1860, essentially conceding Matthew's claim:

> **QUOT. 237** In last Saturday Gardeners' Chronicle, a Mr Patrick Matthews [sic] publishes long extract from his work. . .in which he briefly but completely anticipates the theory of Nat. Selection.—I have ordered the Book, as some few passages are rather obscure but it, is certainly, I think, a complete but not developed anticipation! Erasmus always said that surely this would be shown to be the case someday. Anyhow one may be excused in not having discovered the fact in a work on "Naval Timber". (Burkhardt et al., 1993, p. 154)

In the quotation above, Darwin raised two points that are in his favor:

(a) Matthew stated the principle of natural selection but did not develop it sufficiently in his book, certainly not to the extent Darwin did.
(b) Darwin may indeed be excused for not having read a book on naval timber, published at a time when he was voyaging, and having an explanation of natural selection in its appendix.

On April 21, 1860, Darwin acknowledged Matthew's claim in the *Gardener's Chronicle* (Darwin, 1860a) and promised to add a note in a subsequent edition of the *Origin*. The note was added in the *Historical Sketch* of the third edition of the *Origin* (Darwin, 1861, p. xiv).

Although Matthew wrote back to the *Gardener's Chronicle* on May 12, 1860, stating that "Mr Darwin here seems to have more merit. . .in the discovery than I have had" (Matthew, 1860), he went on to print "Discoverer of the Principle of Natural Selection" on his calling cards!

2.14.6.2 William Charles Wells (1757–1817)

In a strange twist, Darwin was informed in 1865 of yet another "precursor" of natural selection in the name of the Scottish-American physician William Charles Wells (1757–1817).[124] Wells wrote a paper that was first read at a meeting of the Royal Society in 1813. The paper dealt with observations he had made on a woman named Hannah West, who had been admitted to St Thomas' Hospital. The woman was of white race but had parts of her skin "similar to that of a negro." Wells included his paper, entitled "An account of a female of the white race of mankind, part of whose skin resembles that of a negro; with some observations on the causes of the differences in colour and form between the white and negro races of men," in the appendix of the 1818 edition of his book (Wells, 1818). In his paper, Wells stated:

> **QUOT. 238** ... those who attend to the improvement of domestic animals, when they find individuals possessing, in a greater degree than common, the qualities they desire, couple a male and a female of these together, then take the best of their offspring as a new stock, and in this way proceed, till they approach as near the point in view, as the nature of things will permit. But, what is here done by art, seems to be done, with equal efficacy, though more slowly, by nature, in the formation of varieties of mankind, fitted for the country which they inhabit. (*ibid.*, p. 435)

Wells' natural selection involved only the adaptation of humans. In particular, he did not refer to any transmutation between species.

On hearing of Wells' paper, Darwin made the following tongue-in-cheek comment to Hooker in an October 28, 1865, letter:

> **QUOT. 239** So poor old Patrick Matthew, is not the first, & he cannot or ought not any longer put his Title pages "Discoverer of the principle of Natural Selection"! (Darwin, 1888a, p. 225)

In the fourth edition of the Origin, Darwin added a note in the Historical Sketch: "In this [1813] paper, he [Wells] distinctly recognizes the principle of natural selection...but he applies it only to the races of man, and to certain characters alone" (Darwin, 1866, p. xiv).

2.14.6.3 Edward Blyth (1810–1873)

Edward Blyth (Fig. 2.22) was an English zoologist who, in 1841, traveled to India to become the curator of the museum of the Royal Asiatic Society of Bengal. He stayed

[124] William Charles Wells' biographer, Nicholas Wade, has noted that: "It is a sad reflection on both the life and the legacy of Wells that a portrait of him has not been unearthed" (Wade, 2003, p. viii).

Fig. 2.22 Edward Blyth (1810–1873). (From: https://upload.wikimedia.org/wikipedia/commons/a/a2/EdwardBlyth.jpg)

in that position until 1862, when he returned to England. Even while in India, Blyth corresponded regularly with Darwin.

Some years before his trip to India, in 1835, Blyth wrote an article entitled "An attempt to classify the "varieties" of animals, with observations on the marked seasonal and other changes which naturally take place in various British species, and which do not constitute varieties" in the *Magazine of Natural History* (Blyth, 1835) where he essentially stated the principle of natural selection:

> **QUOT. 240** It is a general law of nature for all creatures to propagate the like of themselves...and thus, among ourselves, we see a family likeness transmitted from generation to generation. When two animals are matched together, each remarkable for a certain given peculiarity, no matter how trivial, there is also a decided tendency in nature for that peculiarity to *increase*;[125] and if the produce of these animals be set apart, and only those in which the same peculiarity is most apparent, be selected to breed from, the next generation will possess it in a still *more* remarkable degree: and so on, till at length the variety I designate a *breed*, is formed, which may be very unlike the original type. (*ibid.*, pp. 45–46)

[125] All italics in this quotation are Blyth's.

Reference was also made to sexual selection, as was to the struggle for existence (*ibid.*, p. 46). The 1835 paper was followed by a lengthier paper in 1837 (Blyth, 1837).

In QUOT. 240, Blyth alluded to artificial selection (breeding). However, when he considered selection in nature, he thought it acted in an anti-evolutionary way:

> **QUOT. 241** The same law, therefore, which was intended by Providence to keep up the typical qualities of a species, can be easily converted by man into a means of raising different varieties; but it is also clear that, if man did not keep up those breeds by regulating the sexual intercourse, they would all naturally soon revert to the original type. (*ibid.*, p. 46)

Thus, according to Blyth, natural selection acted to prevent changes in animals in the wild. Species were fixed and created by "a wise provision":

> **QUOT. 242** . . .it may not be impertinent to remark here, that, as in the brute creation, by a wise provision, the typical characters of a species are, in a state of nature, preserved. (*ibid.*, p. 48)

Blyth's enunciation of the principle of natural selection was first pointed out by Geldart (1879). In 1911, Vickers provided an accurate summary of Blyth's viewpoint:

> **QUOT. 243** Though Blyth seems to have recognized the principle of natural selection, he fails in its true application in that he regards his "principle" as operating for the conservation rather than the progression of the type. (Vickers, 1911, p. 511)

Geldart and Vickers' papers did not gain much traction, but the idea was later picked up with full vigor by the anthropologist Loren Eiseley (1907–1977) (Eiseley, 1959). The latter went further than his predecessors in claiming that, by giving credit only to Malthus, Darwin concealed the fact that he had been inspired by Blyth. Eiseley pointed to the fact that Darwin was acquainted with the *Magazine of Natural History*, which to Eiseley meant that Darwin must have read Blyth's articles before. Eiseley also drew attention to Darwin's use of the word "inosculate"[126] in his 1836 notebook. This word had been used by Blyth in his 1836 and 1837 papers (Blyth, 1836, 1837). Eiseley later wrote a whole book, entitled *Darwin and the Mysterious Mr X* (Eiseley, 1979), on the whole matter.

Despite the respect that Eiseley enjoyed in the scientific community, few scholars have found his arguments convincing.

[126]Meaning "to pass into each other."

Part II
The Problem of Heredity

Chapter 3
Hybridization from the Earliest Times till Darwin

Abstract Questions regarding the transmission of characteristics from parents to offspring have been asked since very early times. Around 420 BC, Hippocrates proposed the idea of pangenesis, whereby body parts contributed their characteristics to the gametes. Camerarius first demonstrated sexuality in plants in 1694, while Kölreuter later obtained the first hybrids experimentally, and Knight was the first to observe segregation. In 1862, Naudin proposed that a hybrid consisted of two specific essences that were separated in the pollen and ovule. Unaware of Mendel's work, Darwin proposed the provisional hypothesis of pangenesis in 1868.

Keywords Sexuality in plants · Hybrids · Segregation · Theory of mosaic hybrids · Specific essence · Provisional hypothesis of pangenesis · Gemmules

3.1 Early Notions of Heredity

Although Darwin recognized the importance of heredity (i.e., the transmission of characteristics from parents to offspring)[1] in his theory of evolution, he was not able to propose a viable theory that would explain the hereditary process. This made his evolutionary theory more vulnerable to criticism. His 1868 provisional hypothesis of pangenesis, which was meant in part to explain heredity, failed to gain the approval of most of his contemporaries and was dealt a serious blow by Galton.

Of course, Darwin was not the first to contemplate the exact mechanism by which characters were passed across generations. Ideas about heredity have existed since ancient times. For example, the physician Hippocrates of Cos (460–377 BC) believed that an offspring tended to resemble one parent more than the other if the former's "sperm was stronger" than the latter's:

> **QUOT. 244** Sperm is a product which comes from the whole body of each parent, weak sperm coming from the weak parts, and strong sperm from the strong parts. The child must

[1] Heredity was first used in a biological sense by J. de Maistre in 1821 (de Maistre, 1821, p. 79, T. I). Its first use in English was due to Spencer in 1864 (Spencer, 1864, p. 80, Vol. I). Its first use in a juridical sense dates back to c. 1050.

P. Gorroochurn, *The Development of Evolutionary Genetics*,
https://doi.org/10.1007/978-3-031-69374-8_3

> necessarily correspond. If for any part of the father's body a greater quantity of sperm is derived than from the corresponding part of the mother's body the child will, in that part, bear a closer resemblance to its father; and vice versa. (Lonie, 1981, p. 4)

In the above, Hippocrates' assertion that sperm came from the body parts was later called *pangenesis*. Hippocrates was the first to describe it. The basic premise of pangenesis, namely that the body parts of the parents contributed particles to their sex cells and thus affected the characteristics of the offspring, was later embraced by many scholars, of whom Darwin was perhaps the most prominent. The latter was the first to use the term "pangenesis" (Darwin, 1868b, Vol. II, Chap. XXVII).

Approximately a century after Hippocrates, Aristotle (384 B.C.–322 B.C.) voiced skepticism against pangenesis because it was unable to account for the recurrence of features from ancestors to progeny. Aristotle further believed that the two sexes played different roles in procreation. The male sperm provided the "form" (*causa formalis*), which contributed to the development of the offspring. On the other hand, the female menstrual blood provided the "matter" (*causa matarialis*), which contributed to the formation of the fetus.

Later scholars such as Leonardo da Vinci (1452–1519), Rene Descartes (1596-1650), and Pierre Louis Moreau de Maupertuis (1698–1759) all believed in equal maternal and paternal contributions. Maupertuis, in particular, held views close to Hippocrates' pangenesis:

> **QUOT. 245** Elements suitable to form the fetus swim in the seeds of the male & female animal; but each one extracted from the similar part to the one it has to form. . .(Maupertuis, 1751, p. 158)

While the existence of sexuality in animals was never really questioned, the same cannot be said of plants. For thousands of years B.C., merchants sold male and female date palms, and pollination of these plants was practiced. However, Aristotle refused to acknowledge sexuality in plants for theoretical reasons. He argued that plants occupied an intermediate position between the animal and mineral worlds and should thus be incapable of reproduction. Aristotle's position was countered by his disciple Theophrastus (c.371–c.287 B.C.) but backed by later illustrious biologists such as Andrea Cesalpino (1519–1603) and Marcello Malpighi (1628–1694).

3.2 Rudolf J. Camerarius (1694): First Experimental Demonstration of Plant Sexuality

The experimental demonstration of sexuality in plants was first performed by Rudolf J. Camerarius (Fig. 3.1), a Professor of Natural Philosophy at the University of Tübingen, Germany. He removed the anthers from the castor oil plant *Ricinus* and observed that the seeds from the plant did not develop. Likewise, when he removed the stigmas, the seeds did not develop. In 1694, he published a memoir in the form of an 80-page letter to his colleague Michael Bernard Valentin (of the University of

Fig. 3.1 Rudolf J. Camerarius (1665–1721). (From: https://commons.wikimedia.org/wiki/File:Rudolf_Jacob_Camerarius._Line_engraving_by_J._C._Dehne._Wellcome_V0000975.jpg)

Giessen). In the letter entitled "De Sexu Plantarum Epistola" (Camerarius, 1694), Camerarius stated:

> **QUOT. 246** They [the two parent plants] behave indeed to each other as male and female, and are otherwise not different from one another. They are thus distinguished with respect to sex, and this is not to be understood as it is ordinarily done, as a sort of comparison, analogy, or figure of speech, but it is to be taken actually and literally as such. (Camerarius, 1694; Ostwald, 1899, p. 28)[2]

Camerarius described the pollen as "male" and the ovary as "female." In the letter, he also suggestively asked whether cross-pollination could give rise to hybrids:

> **QUOT. 247** The difficult question, which is also a new one, is whether a female plant can be fertilized by a male of another kind, the female hemp by the male hops; the castor bean from which one has removed the staminate flowers, through pollination with the pollen of Turkish wheat (maize); and whether, and in what degree altered, a seedling will arise therefrom. (Camerarius, 1694; Ostwald, 1899, p. 29)[3]

---

[2]English translation from Roberts (1929, p. 14).

[3]English translation from Roberts (1929, pp. 15–16).

Although Camerarius' ideas were not universally accepted, he set the stage for a host of plant hybridization experiments. His main conclusion about sexuality in plants was supported by Carl Linnaeus (1707–1778).[4] The latter used the sexual organs of plants as a basis for plant classification with great success. In 1751, he published *Plantae Hybridae* (Linnaeus, 1751), which was concerned with the formation of new species by hybridization.[5] Linnaeus' views were very influential, and some of the subsequent plant hybridization experiments by others were meant to address the question of whether plant hybridization leads to novel species, as Linnaeus had asserted. Many others had the more practical aim of producing better varieties, while very few, if any, were concerned with understanding heredity. However, it is through hybridization that the mechanism of heredity was eventually elucidated, as will be shown in what follows.

Some of the more important works on hybridization will next be examined, starting with Kölreuter and ending with Darwin. In Chap. 4, Galton's theory of heredity will be discussed. This will hopefully offer a better perspective on Mendel's breakthrough , which will be described in Chap. 6.

3.3 Joseph Gottlieb Kölreuter (1761): "First Botanical Mule"

Joseph Kölreuter (Fig. 3.2) was a German botanist and the first to perform systematic studies on plant hybridization. He is also credited with the discovery of insect pollination. Between 1761 and 1766, Kölreuter published a series of reports describing his experiments and their interpretation. The first report was entitled "Vorlaufige Nachricht von einigen das Geschlecht der Pflanzen betreffenden Versuchen und Beobachtunge"[6] (Kölreuter, 1761) and was succeeded by three follow-ups (Kölreuter, 1763, 1764, 1766). These works were reprinted in *Ostwald's Klassiker der exakten Wissenschaften, No. 41* (Ostwald, 1893).

Kölreuter had taken cognizance of Camerarius' work, especially the latter's ideas on the sexuality of plants and on the potential of pollination to produce hybrids. He was also aware of Linnaeus' belief that hybridization invariably led to new species. Equipped with a microscope, Kölreuter performed systematic studies on pollination and reported some important results. The first hybrids he obtained were from a cross between two species of *Nicotiana* (tobacco): he used the pollen of *N. paniculata* to fertilize castrated flowers of *N. rustica.* The first-generation (F_1) hybrids were all essentially intermediate between the two parents, and Kölreuter observed that they seemed perfectly healthy. Regarding the hybrids he obtained, Kölreuter declared:

---

[4]See Sect. 1.4.

[5]It is now known that new species may be produced by the hybridization of different species, but not exclusively by the latter process.

[6]Preliminary Report of Some Experiments and Observations Concerning Sex in Plants.

Fig. 3.2 Joseph Gottlieb Kölreuter (1733–1806). (From: https://commons.wikimedia.org/wiki/File:Josef_Gottlieb_Koelreuter_(original_upload).jpg)

QUOT. 248 This plant is thus in the real sense a true, and, so far as it is known to me, the first botanical mule which has been produced by art.[7] (*ibid.*, p. 31)

However, he later observed that the hybrids were also sterile. This observation led him to conclude that, in general, hybridization was not favored by nature, i.e., the parents "were not destined for each other by the wise Creator" (*ibid.*, p. 44).[8] Kölreuter found a way to rationalize this conclusion: the sterility of hybrids was a means of maintaining harmony in nature, for:

QUOT. 249 ... [w]hat an astonishing confusion would not the peculiar and unchanged hybrid characters, and the continually retained fertility of such plants give rise to in Nature. (*ibid.*, p. 45)

Kölreuter's observation that hybridization did not, in general, lead to new species was contrary to Linnaeus' belief.

[7]English translation from Roberts (1929).

[8]English translation from Roberts (1929, p. 45).

Kölreuter also performed reciprocal crosses (i.e., using the pollen of *N. rustica* to fertilize flowers of *N. paniculata*) and observed that the hybrids were essentially the same as those of the *N. paniculata* × *N. rustica* crosses. This implied the equality of contributions from male and female parents.

Although Kölreuter's initial crosses yielded sterile hybrids, he was more successful when he tried backcrosses (i.e., crosses between the hybrids and one of the parents). For example, a backcross of the pollens of *N. rustica* on the sterile *N. rustica* × *N. paniculata* hybrids produced new hybrids that had a greater degree of fertility than before. However, the new hybrids also showed reversion to the maternal parent through repeated backcrossing. Kölreuter interpreted this reversion again in terms of the inability of hybridization to produce new (and stable) species.

Crosses between pinks, carnations, and sweet Williams also produced a higher degree of fertility in the hybrids. Kölreuter found that crosses between related plants were more likely to produce fertile hybrids than crosses between distant plants.

Although the F_1 hybrids were mostly uniform and intermediate in characteristics between the two parents, Kölreuter also observed increased variability in hybrids beyond F_1. This would later be correctly explained once Mendel's work was understood.

In the cross between *Dianthus chinensis* (normal flowers) × *D. hortensis* (double flowers), Kölreuter also observed the phenomenon of *dominance*,[9] which would later first be delineated most clearly by Mendel. This is how Kölreuter described it:

> **QUOT. 250** Its flowers were all reduplicate, and consisted commonly of 15-20 quite carmine-red leaves; from which one plainly sees, that the pollen of doubled flowers possesses the character of reduplicating simple ones which are pollinated with it.[10] (*ibid.*, p. 152)

Kölreuter made an interesting observation that hybrids, although often sterile, grew faster, had a prolonged flowering time, and were generally vigorous. This phenomenon is now known as *heterosis* or *hybrid vigor*. Kölreuter explained it by noting that:

> **QUOT. 251** ...these plants [hybrids] cannot, like the natural ones, be exhausted and impoverished through the development of the seed.[11] (*ibid.*, p. 194)

The fact that hybrids, whether sterile or fertile, were formed prompted Kölreuter to propose a theory of heredity. According to this theory, the pollen and stigma secreted fluids, which were the actual sexual agents, such that:

> **QUOT. 252** ...(f)rom the union and commingling of these two materials, which occurs in the most intimate and orderly manner, according to a definite relationship, there originates

[9]According to Roberts (1929, p. 121), the term was first used by the French botanist Augustin Sageret in 1826 (Sageret, 1826). However, Martini (1961) has pointed out that the Italian botanist Gallesio had used it earlier in 1816 (Gallesio, 1816).

[10]English translation from Roberts (1929).

[11]English translation from Roberts (1929, p. 55).

another, which is of an intermediate sort, and which consequently also possesses an intermediate, compounded force, sprung from those two simple forces.[12] (*ibid.*, p. 42)

Kölreuter later rejected that the fluids were the actual sexual agents but that they only carried the sexual elements. Kölreuter's theory was incorrect since fertilization proceeds by the fusion of the pollen itself with the female gamete, as first established by Pringsheim (1856). In addition to this error, Kölreuter's definition considers the process of fertilization as the mixing of two "essences," which is an example of typological (or essentialist) thinking (Sect. 2.9). This tradition was continued by Gärtner (Sect. 3.5), Naudin (Sect. 3.6) and others, but a break took place with Mendel. The latter was interested in single-character differences rather than species essences. It was with the first type of approach that understanding the mechanism of heredity was possible.

3.4 Thomas Knight (1799): First Observation of Segregation

The botanist Thomas Andrew Knight (Fig. 3.3) was one of the pioneers of hybridization in Britain. In 1799, he published a paper entitled "An account of some experiments of the fecundation of vegetables" (Knight, 1799). Although Knight had worked with various fruits, in the 1799 paper, he expressed a predilection for the garden pea:

QUOT. 253 ...as I foresaw that several years must elapse, before the success or failure of this process could possibly be ascertained, I wished, in the interval, to see what would be its effects on annual plants. Amongst these, none appeared so well calculated to answer my purpose as the common pea; not only because I could obtain many varieties of this plant, of different forms, sizes, and colours; but also, because the structure of its blossom, by preventing the ingress of insects and adventitious farina, has rendered its varieties remarkably permanent. (*ibid.*, p. 196)

Mendel later experimented with the garden pea for similar reasons. In one of the first experiments described in the paper, Knight crossed gray peas with white peas. Regarding the F_1 hybrids thus obtained, he observed that they had:

QUOT. 254 ...excessive luxuriance, and the colour of their leaves and stems clearly indicated, that they had all exchanged their whiteness for the colour of the male parent: the seeds produced in autumn were dark gray. (*ibid.*, p. 197)

The gray color dominated the white one. Roberts has reported this to be the "first recorded instance of color-dominance in peas" (Roberts, 1929, p. 89). Knight then pollinated the gray hybrids with pollen from a white variety and found that the white color was discharged in the resulting hybrids. When gray seeds were crossed with the latter white hybrids, gray seeds were produced. However, the process was reversed (i.e., white hybrids × gray seeds):

[12] English translation from Roberts (1929, p. 44).

Fig. 3.3 Thomas Andrew Knight (1759–1838). (From: https://commons.wikimedia.org/wiki/File:Mezzotint_of_Thomas_Andrew_Knight,_1838_Wellcome_L0006140.jpg)

> **QUOT. 255** . . .a part only of them afforded plants with white blossoms; this part sometimes occupying one end of the pod, and being at other times irregularly intermixed with those which, when sown, retained their colour (*ibidem*)

Knight's observation was the first instance of *segregation*[13] (separation of traits in the progeny into parental traits) in a hybridization experiment. However, Knight did not count the relative proportion of each color, nor did he think in terms of segregation; he instead attributed the difference in the two crosses to the different quantities of pollen employed.

Almost a quarter of a century later, similar observations of segregation were made by John Goss (1824, p. 235) and Alexander Seton (1823, p. 237), again in garden peas, but this time by self-fertilization of F_1 hybrids obtained from white × blue crosses.

---

[13] Segregation contradicts blending inheritance. Despite several instances of segregation, blending inheritance continued to be believed in, based on observations that seemed to indicate the latter phenomenon. It was only later that what had seemed to be blending inheritance was understood to be actual segregation occurring at multiple loci.

Fig. 3.4 Karl Friedrich von Gärtner (1849). (From: https://commons.wikimedia.org/wiki/File:Karl_Friedrich_von_Gärtner_von_Emil_Orth.jpg)

3.5 Karl Friedrich von Gärtner (1849): Extension of Kölreuter's Work

Kölreuter's work was extended by the German botanist Karl Friedrich von Gärtner[14] (Fig. 3.4). Gärtner first published his studies on plant sexuality and hybridization in 1826 (Gärtner, 1826). In 1837, he won a prize competition by the Dutch Academy of Sciences in Haarlem. The topic of the competition was related to the production of new species and varieties through artificial fertilization. Gärtner's essay for the competition was published in Dutch in 1838. A greatly expanded German edition was published in 1849 (Gärtner, 1849) and used as the standard text on the subject for several generations to come.[15] The book listed more than 10,000 artificial fertilizations in 700 plant species. Gärtner's results confirmed many of Kölreuter's conclusions, namely:

---

[14]Karl Friedrich von Gärtner's father, Joseph Gärtner (1732–1791), was himself a botanist and close friend of Kölreuter.

[15]The book was thus cited by Darwin in *The Origin of Species* and more extensively referenced in *The Variation of Animals and Plants Under Domestication.*

- Increased vigor in hybrids.
- Decreased fertility in hybrids.
- Similarity of hybrids from reciprocal crosses.
- Increased variability in hybrids after F_1.
- No new species arise from natural hybridization.

3.6 Charles Naudin (1862)

3.6.1 Charles Naudin (1862): Theory of Mosaic Hybrids, "Mendel Precursor"?

Charles Naudin (Fig. 3.5) was a French botanist and naturalist who had the distinction of being the one who most nearly anticipated Mendel's results.[16] Naudin started plant hybridization experiments in 1854, which continued for the next two decades. His main concern was the formation of stable species through hybridization, as had been claimed by Linnaeus earlier (Sect. 3.2). A prize competition on the topic was held in 1860 by the Académie des Sciences. Naudin won the prize[17] with an essay entitled "Nouvelles recherches sur l'hybridité dans les végétaux". The second part of this essay, containing Naudin's conclusions, was published in 1863 (Naudin, 1863), followed by the full essay in 1865 (Naudin, 1865). It is in this essay that Naudin expressed views on heredity that were close to those of Mendel.

In the preface of the essay, Naudin highlighted an aspect of his experiments that differentiated him from his predecessors (and that was later critical for Mendel to correctly interpret his results):

> **QUOT. 256** My manner of proceeding has been a bit different from that followed in this type of research. I have not been content, like most of my predecessors, to observe a limited number of hybrid plants of each species; on the contrary, I have thought that, in order to obtain convincing results, plants of the same origin had to be multiplied so as to have the chance to find all the modifications that the hybrids are capable of. (*ibid.*, p. 26)

Naudin thus wanted to observe a large number of hybrids, not because he could observe statistical relationships between the different hybrid types (like Mendel did), but because he could observe more varieties within the hybrids.

In the essay, Naudin described hybridization experiments involving 60 different species. One of the experiments continued until the fifth generation, and three of them continued until the third generation.

Several of the findings of Naudin were related to observations made previously by Kölreuter and others, namely:

---

[16] See Sect. 6.2.

[17] The only other participant in the competition was Dominique Alexandre Godron (1807–1880), the dean of the Faculty of Sciences at the University of Nancy.

Fig. 3.5 Charles Victor Naudin (1815–1899). (From: https://commons.wikimedia.org/wiki/File:Portrait_Charles_Victor_Naudin.png)

(a) Uniformity of traits in F_1 hybrids (*ibid.*, p. 146)
(b) Similarity of reciprocal crosses (*ibid.*, p. 147)
(c) Intermediate characters of F_1 hybrids, compared to the two parents (*ibid.*, p. 148)

Another result that had been previously observed by others was:

(d) More variability in F_2 hybrids (*ibid.*, p. 149)

But here, Naudin gave more details:

> **QUOT. 257** Very often, the perfect uniformity of the first generation is followed by an extreme diversity of forms ("bigarrure de formes"), some approaching the specific type of the father, others that of the mother, some others suddenly and completely reverting to one or the other. (*ibidem*)

Naudin indicated that the above occurred in every hybrid he observed in *Datura Stramonium*, *Datura Meteloido-Metel, Nicotiana angustifolia* × *Nicotiona macrophylla*, and many others.

Regarding these reversions that Naudin observed in F_2 and later, he offered the following explanation:

QUOT. 258 All these facts are naturally explained through the disjunction of two specific essences in the pollen and in the ovules of the hybrids.[18]

A hybrid plant is an individual in which two specific essences are united, each having its own mode of development and final direction, which are mutually opposed to each other and which are constantly in a struggle to separate from each other. (*ibid.*, pp. 150–151)

In the above, Naudin seems to indicate some kind of segregation of two units within a hybrid. This may appear to be analogous to what would later be called Mendel's law of segregation. Naudin went even further and stated that the hybrid was a "living mosaic" whose pollen, especially, and ovules "are precisely the parts of the plant where the specific disjunction takes place with the most energy". Moreover:

QUOT. 259 Let us assume that, in the hybrids of the first generation, the disjunction takes place both in the anther and inside the ovary; that some of the pollen grains belong completely to the species of the father, others completely to the species of the mother; that in other grains, disjunction has not occurred or has just begun: let us also assume that the ovules are, to the same degree, disjointed towards the father and towards the mother... If the tube from a grain of pollen tending to revert to the species of the male parent encounters an ovule disjointed in the same direction, *a perfectly legitimate*[19] fertilization will take place, as a result of which a plant will be *completely reverted to the paternal species*; if the same combination takes place between a pollen grain and an ovule both disjointed in the direction of the female parent of the hybrid, the product will return in the same way to the species of the latter; if, on the contrary, the combination takes place between an ovule and a pollen grain disjointed in opposite directions to each other, there will result a true *cross-fertilization*, like that which gave birth to the hybrid itself, and there will result again an intermediate form between the two specific types. (*ibid.*, p. 153)

Note that, again, Naudin explicitly refers to disjunction occurring only in *hybrids.*[20] Nevertheless, the above looks very much like a Mendelian mechanism for the formation of homozygotes and heterozygotes. Indeed, Naudin's theory of heredity appears so close to that of Mendel that a few have argued that the law of segregation (Mendel's first law) should be attributed to the former. A notable scientist who made such an argument was the French botanist Louis Blaringhem (1878–1958), who supported both Lamarck's inheritance of acquired characters and de Vries' mutation theory. Writing on the occasion of the *IVe Conférence Internationale de Génétique*, Paris, 1911, Blaringhem titled his paper "Sur l'hérédité en mosaïque" (Blaringhem, 1913). The title obviously referred to Naudin's theory and, in the paper, Blaringhem stated:

QUOT. 260 I would like to also to take the opportunity on this occasion to pay tribute to the French scientist Charles Naudin, a contemporary of Mendel, who arrived at the theory of segregation of characters in the sex cells some years before Mendel (1860). (*ibid.*, p. 101)

As another example, in his book *Darwin's Place in History* (Darlington, 1959), the English geneticist Cyril Darlington (1903–1981) stated:

[18] Italics are Naudin's.

[19] All italics are Naudin's.

[20] Similarly, Mendel thought of segregation in hybrids only; see Sect. 6.2.7.

QUOT. 261 . . .Charles Naudin, who had expounded evolution by natural selection in 1852, announced the results of his experiments on the hybridization of varieties and species in 1861. They lay three years in the archives of the museum before they appeared in print. They then revealed, a year before Mendel, the same principles that Mendel also discovered. (*ibid.*, pp. 51–55)

Despite the statements of both Blaringhem and Darlington, the closeness of Naudin's theory to that of Mendel's in appearance only, not in substance. There is a fundamental aspect of Naudin's theory that disqualifies him from being the true discoverer of the law of segregation, namely that his thinking was typological (Sect. 2.9). According to Naudin, what was segregated in the hybrids was the whole "specific essence," rather than separate factors for each trait (as in Mendel's population thinking). Naudin would have been right only if there was a single factor responsible for all traits. Indeed, Mendel was the first to investigate separate traits for his hybridization experiments rather than the organism as a whole.

Although Naudin cannot be given priority for the discovery of Mendelism,[21] he did recognize the following:

(i) Segregation of two parental types in the germ cells of an individual
(ii) The formation of three possible combinations after segregation
(iii) The actual combination being formed through chance

Each of the above would soon be an integral part of Mendel's theory.

3.6.2 *Naudin, Mendel, and Darwin: Would Darwin Have Understood Mendel?*

The current section on Naudin and the closeness of his work to that of Mendel allows an important question to be addressed: given that Darwin never read Mendel's 1866 paper,[22] would Darwin have understood it if he had read it? Recall that a coherent theory of heredity was a badly missing component of Darwin's theory of evolution by natural selection. As early as 1902, Bateson made the following claim in his influential book *Mendel's Principles of Heredity: A Defence*:

[21] Similarly, Naudin cannot be given priority for the discovery of natural selection despite his important paper of 1852 (Naudin, 1852). In that paper, he stated: "We do not believe that nature has proceeded, so as to produce its species, in any other manner that we ourselves proceed to produce our varieties" (*ibid.*, p. 104). Thus, he pointed out the important analogy between natural and artificial selection. However, there is no mention of natural selection, nor is there a description of how this analogous process actually operates. Instead, Naudin appealed to a *principle of finality* that brings about changes.

[22] A few scholars have argued that Mendel sent Darwin a copy of his paper but that the latter did not read it. For a discussion of the issue, see Galton (2017, pp. 183–184), Porter & Graham (2015, p. 133), and Henig (2000, p. 143).

QUOT. 262 Had Mendel's work come into the hands of Darwin, it is not too much to say that the history of the development of evolutionary philosophy would have been very different from that which we have witnessed. (Bateson, 1902b, p. 39)

Similar claims were later repeated by others (e.g., Rose, 1998, p. 33).

However, there is little evidence that Darwin would have understood Mendel's paper. In fact, there is at least one strong piece of evidence that Darwin would not have grasped its significance and would have discarded it. This conclusion is based on Darwin's reaction when he read Naudin's paper. When Naudin sent Darwin the second part of his paper (Naudin, 1863) (which was first published) at the end of 1863, the latter discussed it with George Bentham (1800–1884) and sent Naudin a detailed criticism. This letter is not extant, but the gist of it may be inferred through the following comment made by Darwin next to Naudin's description of segregation:

QUOT. 263 This view will not account for distant reversion. [Darwin's annotation to Naudin (1863) on p. 193, Cambridge Reprint No. 300][23]

Thus, Darwin's objection to Naudin's theory explaining the reversion to *parental* types was that it did not explain reversion to *distant ancestral* types. This objection is made more explicit in Darwin's statement to Hooker about Naudin's work on September 13, 1864:

QUOT. 264 I have read with interest Bentham's address on hybridism. I am glad that he is cautious about Naudin's view, for I cannot think it will hold. The tendency of hybrids to revert to either parent is part of a wider law (which I am fully convinced that I can show experimentally), namely, that crossing races as well as species tends to bring back characters which existed in progenitors hundreds and thousands of generations ago. Why this should be so, God knows. But Naudin's view throws no light, that I can see, on this reversion of long-lost characters. (Darwin and Seward, 1903a, pp. 339–340)

A similar sentiment was expressed in Darwin's *Variation of Animals and Plants Under Domestication* (Darwin, 1868b, p. 49, Vol. II).

Now, since Mendel's theory also does not explicitly address reversion to distant ancestral types, Darwin would likely *not* have grasped the significance of Mendel's paper. In his 1868 book, Darwin invoked his provisional hypothesis of pangenesis (Sect. 3.7.2 (c)) to explain reversion to both parental and distant ancestral types. He stated that, instead of a mosaic of two parental specific essences, it was more likely that a hybrid was made up of two types of gemmules, namely "blended together" (i.e., hybridized) and "completely pure" (i.e., pure) (*ibidem*).

[23] Darwin's copy of Naudin's essay is preserved at Cambridge.

3.7 Charles Darwin

3.7.1 Hybridization (1839–) and Mendelian Ratios (1859, 1877)

Darwin's first known crossing experiments were recorded in the "Questions & Experiments" (Q&E) notebook (Table 2.1), which started in mid-1939. In 1856, when Mendel started hybridization experiments in peas, Darwin started the same in legumes although he had already done several such experiments before. Overall, Darwin did more hybridization than Mendel.

Darwin's first formal treatment of hybridization occurred in the epochal *Origin of Species* (Darwin, 1859). As mentioned in Sect. 2.14.1,[24] Darwin devoted a whole chapter (Chap. VIII) to the subject. Hybridization was very important for Darwin because, by showing that different species could produce fertile hybrids, he was able to undermine the notion that species were reproductively isolated entities and thus lay the ground for transmutation between them. He cited examples from the works of Kölreuter and Gärtner, which showed that crossings between different species did produce some fertile offspring. Darwin also wished to show that varieties and species were different only in degree so that domesticated animals were not fundamentally different from wild animals. Darwin summarized his viewpoint as follows:

> **QUOT. 265**...neither sterility nor fertility affords any clear distinction between species and varieties; but that the evidence from this source graduates away, and is doubtful in the same degree as is the evidence derived from other constitutional and structural differences" (*ibid.*, p. 248).

Following Kölreuter, Knight, and others, Darwin noted (*ibid.*, pp. 96, 267) hybrid vigor when distinct individuals of the same species were crossed. However, he also recognized the deleterious effects of inbreeding (now known as *inbreeding depression*):

> **QUOT. 266** ... *close* interbreeding diminishes vigour and fertility; that these facts alone incline me to believe that it is a general law of nature...that no organic being self-fertilises itself for an eternity of generations; but that a cross with another individual is occasionally-perhaps at very long intervals-indispensable. (*ibid.*, pp. 96–97)

The above statement that "no organic being self-fertilises itself for an eternity of generations" is known as the Knight–Darwin law in honor of the two men who articulated it. However, Francis Darwin has alleged that Thomas Knight[25] did not actually make such a statement but only went so far as to say that cross-fertilization was necessary (Darwin, 1898).

Like most of his predecessors, Darwin also observed the intermediate characteristics of F_1 hybrids and the tendency to revert to parental traits in later generations (Darwin, 1868b, p. 48). Well before Mendel's landmark paper, Darwin was aware of

---

[24] See 3rd paragraph after QUOT. 176.

[25] See Sect. 3.4.

the concepts of dominance and recessiveness, which he called "prepotency" and "latency," respectively. For example, in the *Origin of Species* (Darwin, 1869), one can read:

QUOT. 267 When two species are crossed, one has sometimes a prepotent power of impressing its likeness on the hybrid. (*ibid.*, p. 274)

In some of his hybridization experiments, Darwin investigated traits that would later be called Mendelian (i.e., which are independently assorted). For example, in the second volume of his *Variation of Animals and Plants Under Domestication* (Darwin, 1868b, Vol. II), Darwin reported the results of crossings between zygomorphous (common) and peloric *Antirrhinum majus* as follows:

QUOT. 268 ...I raised a large bed of plants from the peloric Antirrhinum, artificially fertilised by its own pollen, and sixteen plants, which alone survived the winter, were all as perfectly peloric as the parent-plant...The crossed plants, which perfectly resembled the common snapdragon, were allowed to sow themselves, and, out of a hundred and twenty-seven seedlings, eighty-eight proved to be common snapdragons, two were in an intermediate condition between the peloric and normal state, and thirty-seven were perfectly peloric, having reverted to the structure of their one grandparent. (*ibid.*, p. 70)

In the experiment above, the common snapdragon was dominant over the peloric, and the frequencies of the two in the F_2 generation were 88 and 37, respectively. Commenting on these numbers, Stubbe claimed that Darwin "had therefore found a regular case of a 3:1 segregation ratio" (Stubbe, 1972, p. 171). However, the segregation ratio (which was not actually calculated by Darwin) is 88/37 = 2.4. David Galton also stated that the "ratio of 1:2.4 [peloric:normal]...is not statistically different from a Mendelian ratio of 1:3" (Galton, 2017, p. 58). The suggestion seems to be that Darwin should have been intrigued by the segregation ratio of 3:1. However, it is difficult to understand *how* and *why* Darwin would have inferred his numbers implied such a ratio.

Similarly, in his book *The Different Forms of Flowers on Plants of the Same Species* (Darwin, 1877), Darwin gave frequencies of dominant and recessive types for a series of crosses (Fig. 3.6). In the case of self-fertilized short-styled *Primula auricula*, 75 short-styled and 25 long-styled offspring were observed. This is consistent with a Mendelian model with short-styled dominant over long-styled *Primula,* and the segregation ratio is exactly 3:1. However, it is again difficult to see how Darwin could have been struck by these frequencies, given the randomness of the proportions of offspring for the other crossings.

In the absence of any notion of segregation, Darwin found that his concepts of prepotency and latency were inadequate to explain the frequencies he observed in his crosses. In his *Antirrhinum* experiments, he could not use these concepts to explain the reappearance of the recessive peloric in F_2:

QUOT. 269 This case seems at first sight to offer an exception to the rule formerly given, namely, that a character which is present in one form and latent in the other is generally transmitted with prepotent force when the two forms are crossed. (Darwin, 1868b, p. 71, Vol. II)

CHAP. VI. ON HETEROSTYLED PLANTS. 269

TABLE 36.

Nature of the Offspring from Illegitimately fertilised Dimorphic Plants.

—		Number of Long-styled Offspring.	Number of Short-styled Offspring.
Primula veris . .	Long-styled form, fertilised by own-form pollen during five successive generations, produced	156	6
" " . .	Short-styled form, fertilised by own-form pollen, produced .	5	9
Primula vulgaris .	Long-styled form, fertilised by own-form pollen during two successive generations, produced	69	0
Primula auricula .	Short-styled form, fertilised by own-form pollen, is said to produce during successive generations offspring in about the following proportions .	25	75
Primula Sinensis . .	Long-styled form, fertilised by own-form pollen during two successive generations, produced	52	0
" " . .	Long-styled form, fertilised by own-form pollen (Hildebrand), produced	14	3
" " . .	Short-styled form, fertilised by own-form pollen, produced .	1	24
Pulmonaria officinalis	Long-styled form, fertilised by own-form pollen, produced .	11	0
Polygonum fagopyrum	Long-styled form, fertilised by own-form pollen, produced .	45	4
" "	Short-styled form, fertilised by own-form pollen, produced .	13	20

Fig. 3.6 Table taken from Darwin's *The Different Forms of Flowers on Plants of the Same Species* (Darwin, 1877, p. 269) showing a segregation ratio of 3:1 by self-fertilizing short-styled *Primula auricula* (the short-styled is dominant and the long-styled is recessive)

Since he had discarded Naudin's theory of mosaic hybrids (Sect. 3.6.2), Darwin preferred to use his provisional hypothesis of pangenesis to explain the reversions he had observed. This theory will be explained next.

3.7.2 *The Provisional Hypothesis of Pangenesis*[26] *(1868)*

Darwin's provisional hypothesis of pangenesis essentially holds that body cells eject particles (called gemmules) that are transported to sex cells to influence the course of heredity. Both the environment and the use/disuse of organs affect the characteristics of body cells. These characteristics are represented in gemmules and are, therefore, transmitted to offspring. Darwin described pangenesis, with proverbial modesty, as follows: "a provisional hypothesis or speculation; but until a better one be advanced, it may be serviceable by bringing together a multitude of facts which are at present left disconnected by any efficient cause" (Darwin, 1868b, p. 357). Indeed, Darwin proposed this hypothesis to explain not only prepotency and reversion but also a host of other phenomena linked to heredity and variation, as will be indicated later.

Some authors[27] have claimed that Darwin was forced to develop pangenesis as a result of the swamping argument skillfully put forward by Fleeming Jenkin (Jenkin, 1867) in his review of the 4th edition of the *Origin of Species.* These authors have stated that, through pangenesis, Darwin developed a mechanism which showed that variation could be constantly generated, which would thus overcome the effect of swamping.

While it is true that pangenesis came out in 1868 (Darwin, 1868b, Chap. XXVII) in the wake of Jenkin's review, Darwin had sent a manuscript on the subject to Huxley as early as 1865 (Olby, 1966, p. 1; Vorzimmer, 1970, pp. 120–121). Furthermore, Darwin wrote to Lyell on August 22, 1867, stating that pangenesis was 26 or 27 years old (Darwin, 1888a, p. 72), which brings the date of the hypothesis to about 1840. Thus, although Darwin might have published pangenesis as a result of Jenkin's review, his "speculation" on the matter had started well before.

When Darwin wrote on pangenesis in 1868, he was unaware that the phenomenon had been described by Hippocrates (c. 460–c. 370 BC) thousands of years ago:

> **QUOT. 270** For the seed comes from all parts of the body, healthy seed from the healthy parts, diseased seed from the diseased parts. If, therefore, bald parents for the most part beget bald children, grey-eyed parents grey-eyed children, squinting parents squinting children, and so on with other physical peculiarities, what prevents a long-headed parent having a long-headed child? (Jones, 1923, p. 111).[28]

---

[26] Pangenesis is also discussed in Holterhoff (2014), Ghiselin (1975), Liu (2018a, Part A, 2018b, Part B), Olby (1963), Hodge (2010), Geison (1969), and Zirkle (1946, Chap. II).

[27] For example, see Willis (1940, pp. 5, 165), Eiseley (1958, pp. 210–211), Darlington (1959, pp. 38–39), and Hardin (1959, pp. 115–116).

[28] See Zirkle (1946) for an early history of pangenesis.

In March 1868, Darwin thanked the English doctor William Ogle (1827–1912) for pointing this fact out to him.

The provisional hypothesis of pangenesis was proposed by Darwin as follows:

> **QUOT. 271** . . .It is almost universally admitted that cells, or the units of the body, propagate themselves by self-division or proliferation, retaining the same nature, and ultimately becoming converted into the various tissues and substances of the body. But besides this means of increase I assume that cells, before their conversion into completely passive or "formed material," throw off minute granules or atoms, which circulate freely throughout the system, and when supplied with proper nutriment multiply by self-division, subsequently becoming developed into cells like those from which they were derived. These granules for the sake of distinctness may be called cell-gemmules, or, as the cellular theory is not fully established, simply gemmules. They are supposed to be transmitted from the parents to the offspring, and are generally developed in the generation which immediately succeeds, but are often transmitted in a dormant state during many generations and are then developed. Their development is supposed to depend on their union with other partially developed cells or gemmules which precede them in the regular course of growth. . .Gemmules are supposed to be thrown off by every cell or unit, not only during the adult state, but during all the stages of development. Lastly, I assume that the gemmules in their dormant state have a mutual affinity for each other, leading to their aggregation either into buds or into the sexual elements. Hence, speaking strictly, it is not the reproductive elements, nor the buds, which generate new organisms, but the cells themselves throughout the body. These assumptions constitute the provisional hypothesis which I have called Pangenesis. (Darwin, 1868b, p. 374, Vol. II)

Thus, according to Darwin's provisional hypothesis of pangenesis, every part of the body throws off "minute atoms" or "gemmules" that are specific to body parts. These gemmules collect in the sperm or egg (or in the bud in plants) in a dormant state. During fertilization, dormant gemmules for a given body part from one parent combine through a mutual affinity (or chemical attraction) with the gemmules for that body part from the other parent. Some of the dormant gemmules develop into cells, while others are transmitted in a dormant state for several generations before development occurs. The growth of the fetus takes place by the division of the combined gemmules and their development into cells.

Darwin's pangenesis was thus *particulate*.

It is noteworthy that the production of gemmules is affected by the use and disuse of body parts, as well as by environmental factors. This adds a Lamarckian component to Darwin's theory of heredity and also diminishes the role of natural selection.

Although Darwin was unaware of Hippocrates when he wrote about pangenesis in 1868, he did acknowledge that Buffon, Bonnet, Spencer, and others used similar ideas. In particular, in 1864, Herbert Spencer proposed the concept of "physiological units," which were similar to Darwin's gemmules:

> **QUOT. 272** . . .physiological units will, because of their special polarities, build themselves into an organism of a special structure; so, on the other hand, if the structure of this organism is modified by modified function, it will impress some corresponding modification on the structures and polarities of its units. The units and the aggregate must act and re-act on each other." (Spencer, 1864, p. 256, Vol. I)

However, only in Spencer's theory were all physiological units made up of molecules that were the same for all species and were arranged in different combinations (units) for different species. Furthermore, unlike Darwin, he assumed that the units had specific *polarities* that endowed the corresponding tissues with their particular structure.[29]

It was mentioned at the end of Sect. 3.6.2 that Darwin believed that both reversion and prepotency (dominance) could more readily be explained by pangenesis than by Naudin's theory of mosaic hybrids. At the beginning of Sect. 3.7.1, it was also claimed that pangenesis explained a variety of phenomena having to do with variability and heredity. How this was done by Darwin will now be explained:

(a) *Origin of variability.* Darwin attributed the origin of variability to the action of disturbances produced by the environment: "...supposing that at any stage of development, certain cells or aggregates of cells had been slightly modified by the action of some disturbing cause, the castoff gemmules or atoms of the cell-contents could hardly fail to be similarly affected, and consequently would reproduce the same modification. This process might be repeated until the structure of the part at this particular stage of development became greatly changed, but this would not necessarily affect other parts whether previously or subsequently developed. In this manner we can understand the remarkable independence of structure in the successive metamorphoses, and especially in the successive metageneses of many animals." (*ibid.*, p. 389)
(b) *Prepotency (dominance).* When two individuals are crossed, those traits that are associated with gemmules that are more numerous, that are not dormant, or that have more affinity for each other will be prepotent: "Hybrids...occasionally...closely resemble one parent in one part and the other parent in another part, or even in their whole structure: nor is this difficult to understand on the admission that the gemmules in the fertilised germ are superabundant in number, and that those derived from one parent have some advantage in number, affinity, or vigour over those derived from the other parent." (Darwin, 1868b, pp. 385–386, Vol. II). This mechanism also explains variability from already existing gemmules.
(c) *Reversion.* Darwin hypothesized that gemmules were often transmitted across generations in a dormant state and subsequently developed. Each hybrid consisted of hybridized gemmules and pure gemmules that had previously been dormant. The interaction between the hybridized and pure gemmules determined whether reversion occurred: "Each organic unit in a hybrid must throw off, according to the doctrine of pangenesis, an abundance of hybridised gemmules...; but by the same hypothesis there will likewise be present dormant gemmules derived from both pure parent-forms; and as these latter retain their normal condition, they would, it is probable, be enabled to multiply largely during the lifetime of each hybrid. Consequently the sexual elements of a hybrid will include both pure and hybridised gemmules; and when two hybrids pair, the

---

[29]To Spencer, the polarities were simply "the intrinsic aptitude to aggregate into the form of that species: just as in the atoms of a salt, there dwells the intrinsic aptitude to crystallize in a particular way" (Spencer, 1864, p. 181, Vol. I).

combination of pure gemmules derived from the one hybrid with the pure gemmules of the same parts derived from the other would necessarily lead to complete reversion of character. . .Pure gemmules in combination with hybridised gemmules would lead to partial reversion. And lastly, hybridised gemmules derived from both parent-hybrids would simply reproduce the original hybrid form. All these cases and degrees of reversion incessantly occur." (*ibid.*, pp. 400–401)

(d) *Sterility and bud formation*. Darwin had previously noted that crosses between different species often (but not always) resulted in sterile hybrids owing to their reproductive organs being affected. He posited that the cells of the hybrids threw off hybridized gemmules that aggregated into buds (a term Darwin used for mutations) instead of being aggregated in the reproductive organs: "the cells of the hybrids throw off hybridised cell-gemmules, which become aggregated into buds, but fail to become aggregated within the reproductive organs, so as to form the sexual elements. In a similar manner, many plants, when placed under unnatural conditions, fail to produce seed, but can readily be propagated by buds" (*ibid.*, p. 387)

(e) *Sex-limited inheritance*. Darwin postulated that certain characters appeared in only one sex because gemmules for those characters were in a dormant state in the other sex: "Most, or perhaps all, the secondary characters, which appertain to one sex, lie dormant in the other sex; that is, gemmules capable of development into the secondary male sexual characters are included within the female; and conversely female characters in the male." (*ibid.*, 399)

Chapter 4
Galton's Theory of Heredity

Abstract Galton was Darwin's cousin, and his work on heredity was inspired by his belief in eugenics. His investigation of exceptional qualities in parents and offspring lead him to the concept of regression to the mean. Galton believed in both hard heredity and discontinuous evolution, both of which were at odds with Darwin's theory. As a reaction to Darwin's pangenesis, Galton proposed a hard theory of heredity based on the notion of stirp in 1876. Galton's stirp comprised gemmules that were found in the fertilized eggs and were thus different from Darwin's gemmules. At the same time, Galton also proposed a statistical theory (the ancestral law), which purported to account for an individual's received proportions of ancestral contributions.

Keywords Eugenics · Hard heredity · Reversion · Regression to the mean · Stability of types · Discontinuous evolution · Stirp theory · Law of ancestral heredity

4.1 Francis Galton (1822–1911)

Francis Galton[1] (Fig. 4.1) was a polymath genius and a cousin of Charles Darwin, both were grandsons of Dr Erasmus Darwin.[2] When Galton was young, his parents were keen on him joining medical school. Galton's mother, in particular, wished her son to follow her illustrious father's footsteps. At 16, Galton was sent to Birmingham to study medicine. After his practical training, he joined the medical school at King's College. Galton soon interrupted his medical studies to study mathematics at Trinity College, Cambridge. However, his performance in mathematics was not satisfactory, and he obtained only a pass degree. Disappointed, Galton returned to his medical studies. When his father died in 1844, Galton was left with a

[1] Galton's life and work are also discussed in Pearson (1914, 1924, 1930), Gillham (2002), Bulmer (2003), Cowan (1972a, 1972b, 1977), Keynes (1993), Froggatt & Nevin (1971a), Forrest (1974), Edwards (1993), Platt (1959), Sweeney (2001), Stigler (2010), Goldstein (2012), Crow (1993), and Gorroochurn (2016b).

[2] See Sect. 1.7.

P. Gorroochurn, *The Development of Evolutionary Genetics*,
https://doi.org/10.1007/978-3-031-69374-8_4

Fig. 4.1 Francis Galton (1822–1911). (From: https://commons.wikimedia.org/wiki/File:Francis_Galton,_1912.jpg)

considerable fortune and abandoned medicine altogether. He spent the next few years in amusement and travel.

After the initial setback with his studies, Galton started making a scientific comeback[3] from 1850 onward. During 1850–1852, he financed and led a reconnaissance expedition in South-West Africa (now Namibia). On his return, he published a book (Galton, 1853) and a memoir for the Royal Geographical Society (Galton, 1852). The latter awarded Galton a gold medal in 1854 for his work. Two years later, Galton was made a Fellow of the Royal Society.

Galton soon displayed his predilection for quantitative work. In between various scientific achievements (such as developing European weather charts and discovering the anticyclone), in the early 1860s, he familiarized himself with the normal curve through his friend, mathematician, and physicist William Spottiswoode (1825–1883).[4] He was also impressed by the social statistician Adolphe Quetelet (1796–1874), who had applied the normal distribution to human measurements. However, Galton differed from Quetelet (and the latter's predecessors) in a major conceptual way in the application of the normal curve. Previously, mathematicians had viewed individual deviations from the mean as "errors," which had to be minimized. Quetelet, in particular, focused on the average and popularized the typological concept of the "average man" ("*l'homme moyen*"). This approach

[3] Or, as more aptly described by Pearson, a "reawakening" (Pearson, 1914, Chap. VII).

[4] "My first serious interest in the Gaussian Law of Error was due to the inspiration of William Spottiswoode, who had used it long ago in a Geographical memoir for discussing the probability of the elevations of certain mountain chains being due to a common cause. He explained to me the far-reaching application of that extraordinarily beautiful law, which I fully apprehended." (Galton, 1908, p. 204)

presupposes that the mean represents the "ideal type" and is an example of typological thinking (Sect. 2.9). Galton's conceptual innovation was to use (rather than "eliminate") deviation from the mean to gain information about the individual members of the population[5] and is an example of population thinking (Sect. 2.9). In Galton's own words:

> **QUOT. 273** The primary objects of the Gaussian Law of Error were exactly opposed, in one sense, to those to which I applied them. They were to get rid of, or to provide a just allowance for errors. But these errors or deviations were the very things I wanted to preserve and to know about. (Galton, 1908, p. 305)

As was the case with many scientists of the time, a key event in Galton's life was the publication of Darwin's *Origin of Species*. In a letter dated December 9, 1859 (approximately two weeks after the release of the book), Galton thanked his cousin for having "initiated [him] into an entirely new province of knowledge which, nevertheless, connects itself with many other things in a thousand ways." (Darwin and Seward, 1903a, p. 129). His letter to Darwin 10 years later, however, revealed a much more profound personal impact:

> **QUOT. 274** I used to be wretched under the weight of the old fashioned 'arguments from design', of which I felt though I was unable to prove to myself, the worthlessness. Consequently the appearance of your 'Origin of Species' formed a real crisis in my life; your book drove away the constraint of my old superstition as if it had been a nightmare and was the first to give me freedom of thought. (Pearson, 1914, Dec 24, 1869, letter from Galton to Darwin, Plate II b/w pp. 6 and 7)

Motivated by Darwin's idea of natural selection, in 1865, Galton came to believe that human intelligence and other desirable qualities could be inherited and that, through careful artificial selection, the human stock could thus be improved. This was, in essence, his doctrine of "eugenics,"[6] a term that he would introduce only in 1883 (Galton, 1883, p. 24). The underlying reasons for Galton's belief in eugenics seem to be both personal and political. Darwin's monumental achievement had probably convinced Galton that his cousin had inherited Erasmus Darwin's genius. If that were true, then the prospects for Galton's own future intellectual achievement and stature would be very promising. Moreover, pressure was mounting in mid-Victorian Britain to extend the franchise to the lower classes. Galton's eugenics may be viewed as a statement of the futility of such political reform. Intelligence and high achievement were allegedly the exclusive province of the middle and upper classes, transmitted from and inherited by its members throughout generations. Galton's insistence on the heritability of intelligence may thus be interpreted as a call for preserving the middle class to which he himself belonged.

---

[5]This approach was explicit in Galton's later investigations of regression and correlation.

[6]From the Greek *eugenes*, which means "good in stock." Eugenics was initially a very active field. It was often used as a positive method of selecting the best varieties for breeding but was quickly subverted by the Nazis as a means of "cleansing of undesirables," resulting in some of the most atrocious episodes in the history of mankind. Currently, eugenics has been more or less thoroughly discredited.

Since eugenics dealt with the transmission of desirable human characteristics, it was natural for the problem of heredity to be central to Galton's doctrine. As will be shown, Galton was less successful than Mendel in developing a viable theory of heredity. Nevertheless, he struck success in at least four other respects:

1. As Cowan (1969, pp. 200–201) has argued, Galton changed the definition of the word "heredity" from its previous vague juridical meaning, namely, those "possessions" that one acquired from one's ancestors, to a new precise biological meaning, namely, those qualities that one acquired by virtue of one's ancestry (nature) as opposed to environmental factors (nurture).[7]
2. Galton's theory of heredity (which used the concept of 'stirps') rejected any environmental influence on the process of heredity transmission (i.e., was based on *hard* inheritance) and thus anticipated Weismann's later influential theory of the continuity of the germ plasm (Weismann, 1885b, 1893).[8]
3. By applying statistics to biological problems using a population approach (Sect. 2.9), Galton, in effect, created the field of biometry. The mantle of biometry was later taken up by Karl Pearson, Raphael Weldon, and others who promulgated a continuous model of evolution (as Darwin had originally intended). The biometricians met with strong opposition from the saltationists (William Bateson, Hugo de Vries, and others) who believed that evolution occurred in jumps. Ironically, Galton did not strictly belong to the biometrical group since he rejected continuous evolution. Instead, he was more aligned with saltationism, which later emerged as the dominant doctrine.
4. In his investigation of a physiological theory of heredity, Galton invented the concepts of regression and correlation. Correlation methods soon became a powerful mathematical tool in the hands of the biometricians in their dispute with the saltationists.

Galton's path toward establishing the foregoing ideas will now be examined in detail.

[7] Galton's first use of "heredity" (in its biological sense) occurred in his 1869 *Hereditary Genius*: "I was desirous of obtaining facts bearing on heredity from China" (Galton, 1869a, p. 334). It had previously been first used in a biological sense by J. de Maistre in 1821 (Maistre, 1821, p. 79, T. I). Its first use in English is due to Spencer in 1863 (Spencer, 1864, p. 80, Vol. I). The use of "heredity" in a juridical sense dates back to c. 1050.

[8] Although an even earlier statement of continuity of the germ plasm can be found in Owen's *On Pangenesis*: "Not all the progeny of the primary impregnated germ-cell are required for the formation of the body in all animals: certain of the derivative germ-cells may remain unchanged and become included in that body which has been composed of their metamorphosed and diversely combined or confluent brethren." (Owen, 1849, p. 5)

4.2 Hereditary Talent and Character (1865): Eugenics Manifesto and First Paper on Heredity

4.2.1 Introduction

Galton's first paper on heredity was entitled "Hereditary Talent and Character" (Galton, 1865) and consisted of two parts, "Part I" and "Second Paper." Galton's paper was also his eugenics manifesto, albeit an imperfect one. The very first sentences of Part I clearly stated what would remain Galton's guiding principle throughout his life:

> **QUOT. 275** The power of man over animal life, in producing whatever varieties of form he pleases, is enormously great. It would seem as though the physical structure of future generations was almost as plastic as clay, under the control of the breeder's will. It is my desire to show, more pointedly than – so far as I am aware – has been attempted before, that mental qualities are equally under control. (*ibid.*, p. 157)

It will soon be seen that almost all of Galton's subsequent heredity tenets were aligned with his primary belief in the truth of eugenics.

However, at this stage, what proof did Galton offer for eugenics? Unfortunately, none. What Galton did offer was highly selective data from various sources, which, on their own, constituted weak evidence. For example, he first consulted Sir Thomas Phillips' *The Million of Facts* and found that:

> **QUOT. 276** ...605 notabilities lived between the years 1453 and 1853. And among these are no less than 102 relationships, or 1 in 6, ... (*ibid.*, p. 159)

Galton thus concluded that there was one relationship on average among every six talented men, and this probability was supposedly much greater than that among ordinary men. Next, Galton considered the brief dictionary by a certain Mr. C Hone in an attempt to have "a biographical list of much greater extension." The dictionary contained approximately 19,000 names, and Galton decided to consider only those (1141 in all) that started with an M. He found that 103 were related, giving a ratio of approximately 1 in 11.

The remainder of Part I consisted of similar documentation, which purportedly showed a high degree of relatedness among people (mostly men) with high achievement. Galton's statistical method suffered from two major drawbacks:

(a) It is very questionable that the data sources consulted by Galton were representative of the population of people of high achievement.
(b) Galton offered no reason why the relatedness among people of high achievement should be ascribed to heredity (nature) rather than the environment (nurture).

The second part of Galton's paper ("Second Paper") was more instructive than the first in that it shed light, albeit in elementary form, into all (except one[9]) of his major heredity tenets, as will be discussed next.

---

[9]Namely, discontinuous evolution.

4.2.2 Regression to the Mean

Galton later (1877) formalized the observation that the offspring of exceptional individuals tended to be, on average, less exceptional than their parents. He first explained it through reversion (or atavism). In 1885, when he realized that the phenomenon was bidirectional (i.e., exceptional offspring also had fewer exceptional parents), he used the term "regression" to describe it (Galton, 1885a).

However, even in 1865, the germ of the idea of regression could be found in Galton's paper:

> **QUOT. 277** . . .though the talent and character of both of the parents might, in any particular case, be of a remarkably noble order, and thoroughly congenial, yet they would necessarily have such mongrel antecedents that it would be absurd to expect their children to invariably equal them in their natural endowments The law of atavism prevents it.[10] (Galton, 1865, p. 319)

Galton's regression may be reconciled with his eugenics principle because the former stresses the importance of the ancestral (or racial) average. Even if the lower classes were educated and became exceptional, regression would ensure that their offspring would regress to the unexceptional, ancestral mean. On the other hand, even if the higher classes could sometimes give rise to unexceptional offspring, regression would ensure that the next generation would regain the exceptional character of their ancestors.

Regression would also later be one[11] of the major reasons for Galton's belief that discontinuous variation was much more important than continuous variation in evolution.

4.2.3 Hard Heredity

Galton's belief in hard heredity in 1865 meant that there was no room for environmental (or Lamarckian) influences in the process of heredity transmission. In light of his eugenic beliefs, it is not difficult to understand why Galton stuck to this tenet. By rejecting any Lamarckian influence on the process of heredity transmission, in effect, Galton reinforced the eugenics principle that desirable human qualities could only be inherited from "superior" ancestors.[12] There was thus no hope for the lower classes to acquire such qualities through education and other environmental factors and tot pass them to their offspring.

---

[10] A similar statement can also be found on p. 158 of Part I.

[11] The other reason was his belief in the stability of types. See Sect. 4.3.4.

[12] A diametrically opposite approach, again for political (this time, Marxist) reasons, would later be adopted by Trofim Lysenko (1898–1976), Stalin's favorite biologist. Lysenko advocated for the complete inheritance of acquired characteristics at the expense of Mendelian genetics. See Sect. 16.5.1.

Thus, in the 1865 paper, Galton asked the rhetorical questions:

QUOT. 278 Will our children be born with more virtuous dispositions, if we ourselves have acquired virtuous habits? Or are we no more than passive transmitters of a nature we have received, and which We have no power to modify? (*ibid.*, p. 322)

Galton's answers to the two questions were "no" and "yes," respectively:

QUOT. 279 ...a list of life-long habits in the parents might be adduced which leave no perceptible trace on their descendants. (*ibidem*)

Galton's belief in hard heredity was made even more emphatic through the following startling statement, which anticipated Weismann's later theory of the germ plasm:[13]

QUOT. 280 We shall therefore take an approximately correct view of the origin of our life, if we consider our own embryos have sprung immediately from those embryos whence our parents were developed, and these from the embryos of their parents, and so on for ever. We should in this way look on the nature of mankind, and perhaps on that of the whole animated creation, as one continuous system, ever pushing out new branches in all directions, that variously interlace, and that bud into separate lives at every point of interlacement. (*ibid.*, p. 322)

4.2.4 Law of Ancestral Heredity

Galton's Law of Ancestral Heredity[14] describes the relative ancestral contributions of heritable characteristics to an individual. The law was first formally stated in 1885 (Galton, 1885b), but Galton's 1865 paper already contained an elementary, albeit inaccurate, version:

QUOT. 281 The share that a man retains in the constitution of his remote descendants is inconceivably small. The father transmits, on an average one-half of his nature, the grandfather one-fourth, the great-grandfather one-eighth; the share decreasing step by step, in a geometrical ratio, with great rapidity. (*ibid.*, p. 326)

The above initial enunciation of the Law of Ancestral Heredity was offered without any proof or justification. Galton later corrected it when he realized he had meant to say that each individual actually inherited 1/4 of their heritable characteristics from any one parent, 1/16 from any one grandparent, 1/64 from any one great-grandfather, and so on.[15]

As with his previous two hereditary tenets (Sects. 4.2.2 and 4.2.3), Galton's adherence to the ancestral law was very much aligned with his eugenics. This is

[13]Weismann (1885a, 1893). See Sects. 5.2.3 and 5.2.4 of the current book.

[14]Named as such by Karl Pearson in 1898 (Pearson, 1898).

[15]Or the numbers in QUOT. 281 could stay, but "father" should be changed to "mid-parent," "grandfather" to "mid-grandparents," and so on. The correction and statement of the law in terms of mid-ancestors occurs in Galton (1885b, p. 261). See also QUOT. 328.

because the law stressed the importance of ancestors in the acquisition of hereditary traits and demoted the relevance of environmental factors.

Galton later further refined the Law of Ancestral Heredity (Sect. 4.6). The law later also proved to be a powerful tool in the hands of the biometricians in their dispute with the saltationists (see Part III, in particular, Sect. 7.1).

4.3 Hereditary Genius (1869)

4.3.1 Introduction

Galton's 1865 paper (Galton, 1865) was followed by the paper "Hereditary Genius: The Judges of England between 1660 and 1865" (Galton, 1869b). This was a continuation of Galton's argument in Part I of the 1865 paper (Sect. 4.2.1), this time applied to judges. Of the 286 judges he examined, he found "no less than 133, or nearly one-half, to have one or more kinsmen of little or no less eminence than themselves" (*ibid.*, p. 426). However, as before, the argument here was not convincing.

The material in the 1865 and 1869 papers was expanded in Galton's 1869 book, *Hereditary Genius* (Galton, 1869a). The latter used the same type of argument to make a case for eugenics, and it is not surprising that the book did not gain much traction. However, it is remarkable for four important concepts that were included, and one key concept that was excluded.

4.3.2 Hard Heredity Out, Pangenesis In

Notably absent in *Hereditary Genius* is any mention of hard heredity. As mentioned in Sect. 4.2.3, hard heredity was central to Galtonian eugenics. However, this omission is not hard to explain. Darwin's provisional hypothesis of pangenesis,[16] which advocated soft inheritance, was recently proposed in *The Variation of Animals and Plants Under Domestication* (Darwin, 1868b, Vol. II, Chap. XVII). Moreover, with the publication of the *Origin of Species,* Darwin had acquired the status of one of the foremost naturalists of his times. Being no expert in natural history, Galton did not wish to get into an open controversy with his cousin and thus preferred to withhold any mention of his hard heredity views.

Thus, instead of hard heredity, Galton chose to include Darwin's pangenesis in *Hereditary Genius*:

QUOT. 282 Mr. Darwin, in his remarkable theory of Pangenesis,... supposes, first, that each cell, having of course its individual peculiarities, breeds nearly true to its kind, by

[16]See Sect. 3.7.2.

propagating innumerable germs, or to use his expression, "gemmules," which circulate in the blood and multiply there; remaining in that inchoate form until they are able to fix themselves upon other more or less perfect tissue, and then they become developed into regular cells. (Galton, 1869a, p. 363)

Galton's positive attitude toward pangenesis stemmed in part from his predilection for quantitative reasoning:

QUOT. 283 It becomes an interesting inquiry to determine how much of a person's constitution is due, on an average, to the unchanged gifts of a remote ancestry, and how much to the accumulation of individual variations. The doctrine of Pangenesis gives excellent materials for mathematical formulae. (*ibid.*, p. 370)

At the same time, Galton acknowledged that pangenesis was only "'provisional,' and avowedly based, in some degree, on pure hypothesis and very largely on analogy" (*ibid.*, p. 364).

However, QUOT. 282 shows that Galton had assumed that the gemmules circulated in the *blood*, a word that Darwin had never actually used in his exposition of pangenesis. The issue of blood circulation soon became a source of discord between Galton and Darwin.[17] Galton eventually became dissatisfied with Darwin's pangenesis and openly challenged it.

4.3.3 Applicability of the Normal Distribution to Mental Ability

In *Hereditary Genius,* Galton stressed Quetelet's use of the normal distribution in modeling human physical characteristics. However, Galton also moved away from Quetelet's typological thinking (Sect. 2.9) by recognizing the importance of deviations from the average. In the preface of his book, Galton stated:

QUOT. 284 ...I may claim to be the first to treat the subject [of heredity] in a statistical manner, to arrive at numerical results, and to introduce the "law of deviation from an average" [normal distribution] into discussions on heredity. (*ibid.*, p. vi)

In Galton's population thinking,[18] variations about a hypothetical mean were essential to the theory. In Quetelet's typological thinking, variations were errors that needed to be eliminated, and his concept of the average man ("*homme moyen*") took the status of reality.

Galton described two consequences of the addition of a large number of variables: (i) the constancy of the average and (ii) the central limit theorem:

[17] See Sect. 4.4.

[18] Although, as Bulmer (2003, p. 181) has pointed out, Galton still shared Quetelet's idea that the mean represented a racial type which was maintained from generation to generation. In this sense, Galton was not a complete "populationist."

QUOT. 285 Wherever there is a large number of similar events, each due to the resultant influences of the same variable conditions, two effects will follow. First, the average value of those events will be constant; and, secondly, the deviations of the several events from the average, will be governed by this [normal] law. (*ibid.*, p. 29)

By extrapolating on Quetelet's work, Galton argued that mental ability should also be normally distributed:

QUOT. 286 Now, if this [normality in distribution] be the case with stature, then it will be true as regards every other physical feature - as circumference of head, size of brain, weight of grey matter, number Now, if this be the case with stature, then it will be true as regards every other physical feature – as circumference of head, size of brain, weight of grey matter, number of brain fibres, &c.; and thence, by a step on which no physiologist will hesitate, as regards mental capacity. (*ibid.*, pp. 32–33)

As debatable as Galton's argument was, it remained influential. IQ tests were later designed so that they included more items for which people generally performed well so that the distribution of IQ would then fit the normal distribution. In their highly controversial book *The Bell Curve: Intelligence and Class Structure in American Life*, Herrnstein and Murray (1994) argued that: (i) intelligence was genetically determined, (ii) Asian Americans had higher intelligence than white Americans, who in turn had higher intelligence than African Americans, and (iii) life outcomes were related to races. Their conclusion thus was that life outcomes were genetically determined.

4.3.4 Discontinuity of Evolution and Stability of Types

Perhaps the most significant innovation in *Hereditary Genius* (Galton, 1869a) was Galton's argument for the importance of sports in evolution and his advocacy of evolution in jumps. Discontinuous evolution was antithetical to Darwin's theory, but Galton adduced it within the framework of Darwin's pangenesis so that it did not appear to contradict Darwin completely.

In his theory of pangenesis, Darwin had referred to *dormant gemmules* to account for reversion (Sect. 3.7.2). Building on this, Galton distinguished between *latent* (Darwin's *dormant*) and *patent* (Darwin's *developed*) gemmules. Unlike the patent gemmules, the latent ones did not develop into cells, although they could still be inherited. On the other hand, the patent gemmules were the ones that formed the character of the living form. Galton posited that the latent gemmules were vastly more numerous than the patent ones, but it was possible for this inequality to be reversed. This would occur when several latent gemmules become patent,[19] giving rise to a *sport*.

[19] Galton (1872, p. 396) later indicated that the reverse process, that of patent gemmules becoming latent, was doubtful (it could only occur in embryonic stages), or else distant progeny would resemble parents more than recent progeny. See Sect. 4.5.1.

Galton next crafted a skillful metaphor in an attempt to show how, within Darwin's theory of continuous evolution, only certain positions in the evolutionary path were actually realizable, making evolution in effect discontinuous. In his own words:

> **QUOT. 287** It is shown by Mr. Darwin, in his great theory of "The Origin of Species," that all forms of organic life are in some sense convertible into one another,[20] for all have, according to his views, sprung from common ancestry, and therefore A and B having both descended from C, the lines of descent might be remounted from A to C, and redescended from C to B. Yet the changes are not by insensible gradations; there are many, but not an infinite number of intermediate links; how is the law of continuity to be satisfied by a series of changes in jerks? The mechanical conception would be that of a rough stone, having, in consequence of its roughness, a vast number of natural facets, on anyone of which it might rest in "stable" equilibrium. That is to say, when pushed it would somewhat yield, when pushed much harder it would again yield, but in a less degree; in either case, on the pressure being withdrawn, it would fall back into its first position. But, if by a powerful effort the stone is compelled to overpass the limits of the facet on which it has hitherto found rest, it will tumble over into a new position of stability, whence just the same proceedings must be gone through as before, before it can be dislodged and rolled another step onwards. The various positions of stable equilibrium may be looked upon as so many typical attitudes of the stone, the type being more durable as the limits of its stability are wider. We also see clearly that there is no violation of the law of continuity in the movements of the stone, though it can only repose in certain widely separated positions. (*ibid.*, pp. 368–369)

Galton denoted the existence of 'various positions of stable equilibrium' in the above by a special name: *the stability of types* (*ibid.*, p. 368). The latter was the first reason for his saltationist viewpoint.

Galton would soon propose a second, powerful reason to buttress the saltationist viewpoint, namely, regression.

4.4 Investigation of Pangenesis in 1869–1871: Controversy with Darwin

As mentioned in Sect. 4.3.2, Galton withheld any mention of his hard heredity views in *Hereditary Genius* and instead accepted Darwin's pangenesis. However, his approval of pangenesis was explicitly tentative. Moreover, Galton's understanding of pangenesis was that the gemmules traveled in the *bloodstream* to the sperm and egg, after which they were transmitted to the offspring.

In December 1869, Galton started a correspondence with Darwin in an attempt to seek the latter's advice regarding experiments he wished to conduct to test pangenesis. The specific experiment Galton had in mind involved a pure-breeding rabbit *A* that would be transfused with blood from a donor rabbit *B* that had a different characteristic (such as color). Rabbit *A* would then be crossed with another rabbit

[20]This statement is not accurate with respect to Darwin's theory but does not invalidate Galton's argument that follows.

that may or may not have also been transfused. The presence of the characteristic of *B* in the offspring would then support pangenesis, as Galton had understood it.

Karl Pearson (1924, pp. 157–162) was able to find 12 letters from Galton to Darwin concerning the pangenesis experiments during the period 1869–1870, but none from Darwin.[21] In the first letter, dated December 11, 1869, Galton informed his cousin that he wanted "to procure a few couples of rabbits of marked and assured breeds." The letter gave no details about the experiment except to say that it was "really a curious one" (*ibid.*, p. 157).

Darwin must have known where Galton was heading with the rabbits because he had undoubtedly read Galton's reference to gemmules circulating in the blood in *Hereditary Genius* (Sect. 4.3.2). But even if Darwin was still ignorant of this detail, Galton's second letter of March 15, 1870, must have dispelled any such ignorance, for it explicitly referred to a doe being "transfused from yellow and the buck also from yellow" (*ibidem*). Yet, Darwin never warned Galton that it was wrong to assume that the gemmules were contained in the blood. In fact, Darwin continued to encourage his cousin in his rabbit experiments,[22] which made his later reaction to Galton's experiments all the more disconcerting.

In his letter of March 17, 1870, Galton gave bad news regarding the transfused silver-grays (rabbit *A*) he had used:

> **QUOT. 288** ... we opened 2 boxes and examined the litters. The first contained four dead young ones all true silver greys. One, however, has a largish light-coloured patch on its nose, but Bartlett tells me that this is not unusual with silver greys as the very tips of their noses are often white. (*ibid.*, p. 158)

Thus, the offspring of silver-grays (rabbit *A*) that had been transfused with blood from different-colored rabbits (rabbit *B*) did *not* show the color of the latter.

Galton's letters show that he really wanted the pangenesis experiments to succeed. When he noticed that some of the transfused bucks were sterile, he suspected that the defibrinated blood[23] he had used for the transfusions might be the reason and that the gemmules might, in fact, be in the fibrin. To preserve the fibrin in the transfused blood, Galton decided to perform direct cross-circulations between the carotids from rabbit *B* to rabbit *A*. Darwin was informed of this situation in a July 25, 1870, letter from Galton. However, the new experiments also failed to support pangenesis: of the 88 rabbits bred in 13 litters, none were "mongrels."

At this stage, Galton decided to publish the results of his experiments. However, while Galton had hitherto kept Darwin abreast of all the details of his pangenesis experiments, he did not consult his cousin before reporting his results to the Royal

[21] The only extant letters from Darwin are those written after the publication of Galton's paper "Experiments on Pangenesis" (Galton, 1871a).

[22] For example in Letter (6), dated April 8, 1870, Galton thanked Darwin as follows: "You are very kind in giving me so much valuable advice and so much encouragement." (Pearson, 1924, p. 160)

[23] The blood was defibrinated to prevent it from clotting.

Society on March 30, 1871.[24] The reason is clear enough: Galton was about to deliver a damming verdict on pangenesis.

In his paper, entitled "Experiments in Pangenesis" (Galton, 1871a), Galton started by pointing out that Darwin's provisional hypothesis of pangenesis had several implications and that "it seemed of pressing importance that these postulates should be tested" (*ibid.*, p. 394). He reiterated the main points of the theory, one of which he wrote as:

QUOT. 289. . .before it [the body] developed, and in all stages of its development, it throws "gemmules" into the circulation, which. . .swarm in the blood, in large numbers of each variety, and circulate freely with it. (*ibidem*)

As noted before, "swarm in the *blood*" was Galton's, rather than Darwin's, words, although the latter had not corrected Galton on this supposition during the pangenesis experiments.

Based on the postulates of Darwin's pangenesis, Galton's inference was as follows:

QUOT. 290. . .the gemmules in each individual must therefore be looked upon as entozoa[25] of his blood, and, so far as the problems of heredity are concerned, the body need be looked upon as little more than a case which encloses them. (*ibidem*)

Thereupon, he described the transfusions and cross-circulations he had performed on the silver-grays. His conclusions based on these experiments were unequivocal:

QUOT. 291 . . .the doctrine of Pangenesis, pure and simple, as I have interpreted it, is incorrect. (*ibid.*, p. 404)

Having pronounced his verdict on pangenesis, Galton gave two alternatives to the theory, the first of which later proved to be correct in principle, and the second of which showed that he still left an off-ramp for pangenesis:

(a) *The reproductive elements were, in fact, in the reproductive glands rather than in the blood*: "It seems *à priori* that, if the reproductive elements do not depend on the body and blood together, they must reside either in the solid structure of the gland, whence they are set free by an ordinary process of growth, the blood merely affording nutriment to that growth, or else that they reside in the blood itself. My experiments show that they are not independent residents in the blood, in the way that Pangenesis asserts. . ." (*ibid.*, p. 405)
(b) *The reproductive elements resided in the blood only temporarily:* "My experiments show that they [the gemmules] are not independent residents of the blood, in the way that Pangenesis asserts; but they prove nothing against the possibility of their being temporary inhabitants of it, giving off by existing cells, either in a

[24]The only hint Galton gave was to inform George Darwin (Charles Darwin's eldest son) one week after the paper was read that it would appear in the *Proceedings of the Royal Society* the following week and that he would send Charles a copy.

[25]Entozoa = animals living inside other animals (used especially in reference to tapeworms) (*our note*).

fully developed state or else in one so rudimentary that we could only ascertain their existence by inference. In this latter case, the transfused gemmules would have perished, just like the blood-corpuscles, long before the period had elapsed when the animals had recovered from the operations." (*ibidem*)

Darwin was uncharacteristically upset with Galton's paper, most likely because of the rush with which Galton had announced the demise of pangenesis. His letter is missing, but its tenor may be inferred from Galton's apologetic response on April 25, 1871:

QUOT. 292 MY DEAR DARWIN, I am grieved beyond measure to learn that I have misrepresented your doctrine, and the only consolation I can feel is that your letter to 'Nature' may place that doctrine in a clearer light and attract more attention to it. I write hurriedly, as time is important to save the morning's post, in order to point out two passages which, I hope, in your letter to 'Nature' you will explain at length, so as to remove the false impression of Pangenesis under which I and probably others labour. In "Domestication of Animals etc." p. 374 "...throw off minute granules or atoms, which circulate freely throughout the system " And p. 379 "...the granules must be thoroughly diffused; nor does this seem improbable considering...the steady circulation of fluids throughout the body." (Is there not also a passage in which the words "circulating fluid" are used? I cannot hurriedly lay my hand on it, but believe it to exist.) Believe me—necessarily in great haste—Very sincerely yours, FRANCIS GALTON. (Pearson, 1924, p. 162)

However, the reader cannot escape noticing something else in addition to Galton's apology in the letter above. It is that Galton was standing by his paper and asking his older cousin to please explain why he had used words such as "circulate freely" and "steady circulation of fluids in the body" in *The Variation of Animals and Plants Under Domestication* (Darwin, 1868b, Vol. II) if he had meant something other than blood. Galton hoped Darwin would explain these discrepancies in the latter's forthcoming response in *Nature*.

Darwin's short response in *Nature* was published in the April 27, 1871, issue (Darwin, 1871a). That Darwin chose to respond in print to criticism was quite unusual. He hardly, if ever, responded in print to the several highly critical reviews of the *Origin*. It was usually left to his lieutenants (Huxley, Hooker, and Gray) to do so. However, the situation was different this time. Pangenesis did not have the elaborately built structure that natural selection had, a fact that Darwin was all too aware of. This fact perhaps made him touchier to criticism. In addition, Darwin might have felt annoyed that Galton had consulted him throughout the pangenesis experiments, except when it came to publishing.

In his response, as to be expected, Darwin took exception with Galton's working assumption that the gemmules resided in the blood:

QUOT. 293 Now, in the chapter on Pangenesis in my "Variation of Animals and Plants under Domestication," I have not said one word about the blood, or about any fluid proper to any circulating system. (*ibid*., p. 502)

Darwin noted that, in his book, he had explicitly used Protozoa and plants (both of which do not possess blood) to illustrate pangenesis. Therefore, he contended, it should have been obvious that the presence of gemmules in the blood could have no place in his hypothesis. To answer Galton's query as to why he had used words such as "the steady circulation of fluids throughout the body", Darwin offered the following explanation:

QUOT. 294 . . .I used these latter words and other similar ones, I presume that I was thinking of the diffusion of the gemmules through the tissues, or from cell to cell, independently of the presence of vessels. . . (*ibidem*)

He further admitted that he had not sufficiently reflected on the subject when he had first heard of Galton's experiments and that he "saw not the difficulty of believing in the presence of gemmules in the blood." However, although Galton deserved "the highest credit for his ingenuity and perseverance," the fact remained that:

QUOT. 295 . . .it does not appear to me that Pangenesis has, as yet, received its death blow. . . (*ibid*., p. 503)

Galton's letter soon followed in the May 4, 1871, issue of *Nature* (Galton, 1871b). This was a lengthier version of his April 25 letter to Darwin. Galton expanded on several instances in *The Variation of Animals and Plants Under Domestication*, where Darwin had used phrases that could easily have been interpreted to imply that gemmules resided in the blood. One would also have expected Galton to raise the issue of Darwin having been kept abreast of all the details of the experiments, but this was not included in the letter. Instead, Galton then empathized with Darwin, admitting "how difficult it is to put thoughts into accurate speech", and ending his letter with "Vive Pangenesis!"

Galton's tact in dealing with Darwin ensured that he maintained a good relationship with the latter after this episode. Pearson has reported the intriguing fact that Galton and Darwin collaborated on further rabbit experiments for at least three more years! (Pearson, 1924, p. 165).

None of the experiments were positive.

4.5 Galton's Theories of Heredity

4.5.1 "On Blood Relationships" (1872): First Construction of a Physiological Theory of Hard Heredity

Although Galton, in consultation with Darwin, continued the pangenesis experiments beyond 1972, he already had the elements of a physiological theory of hard heredity by that time. In a May 28, 1872, letter, he informed Darwin that his paper "On blood-relationship" (Galton, 1872) would soon be read at the Royal Society and that he would soon send his cousin a copy. Galton described the paper as attempting "to define what the kinship really is, between parents and their offspring" (Pearson, 1924, p. 169) or (as he stated in the paper) "to set forth the doctrines of heredity in a more orderly and explicit manner than is otherwise practicable" (Galton, 1872, p. 394).

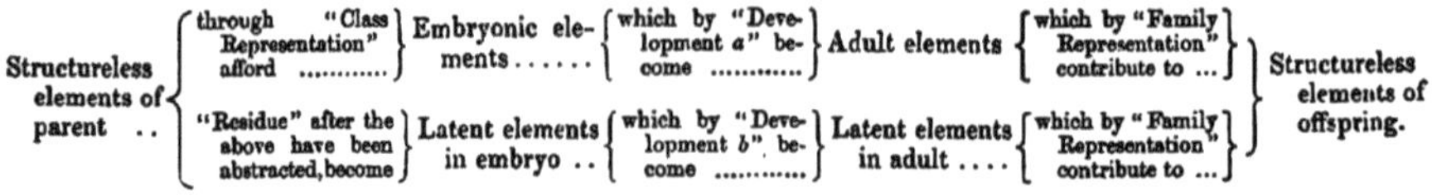

Fig. 4.2 Galton's first enunciation of a theory of heredity (Galton, 1872, p. 398)

Neither Darwin nor pangenesis were mentioned in the paper, but Galton's theory was, in fact, an offshoot of Darwin's provisional hypothesis of pangenesis. Just as Galton had previously (Sect. 4.3.4) adapted pangenesis to argue for the importance of discontinuous variation (sports) in evolution, he now used elements of pangenesis to argue for a theory of hard heredity. His belief in hard heredity, which he had held as early as 1865, was undoubtedly reinforced through the pangenesis experiments.

Galton's starting point was the dormant and developed gemmules that Darwin had previously introduced in his *Variation of Animals and Plants Under Domestication* (Darwin, 1868b), which Galton had subsequently used as *latent* and *patent* gemmules in his *Hereditary Genius* (Galton, 1869a):[26]

QUOT. 296 ...each individual may properly be conceived as consisting of two parts, one which is latent and only known to us by its effects on his posterity, while the other is patent, and constitutes the person manifest to our senses. (Galton, 1872, p. 394)

The distinction between patent and latent gemmules was thus meant to explain, respectively, how offspring inherited characteristics sometimes possessed by parents and sometimes absent in parents. The latter process, known as *reversion,* could thus be accounted for by latent gemmules. It was also because of reversion that Galton assumed that the latent gemmules were more numerous than the patent ones.

In Galton's particulate theory of heredity, primary *structureless elements* (or undifferentiated gemmules) do not travel from body organs (as in Darwin's pangenesis). Instead, they are initially found in a fertilized ovum (Fig. 4.2). A small portion of the gemmules is selected to form patent embryonic elements, which are then differentiated and developed into adult elements (*Development a*).

A large residue of the structureless elements remains latent in the embryo and then develops into latent adult elements (*Development b*). The two developmental paths (*a* and *b*) converge when a subset of patent and latent elements in the adult is selected to form structureless elements in the offspring. Galton noted that:

QUOT. 297 ...the span of each of the links in the general chain of heredity extends from one structureless stage to another, and not from person to person:–

[26]See Sect. 4.3.4.

structureless elements in the Father	...........Adult Father........... ....Latent in Father........	structureless elements in the Father

(*ibid.*, p. 394)

The above is one of Galton's clearest statements of the continuity of the germ line.

Now, before the two developmental paths (*a* and *b* in Fig. 4.2) converge, a portion of the latent elements become patent mainly in the embryo and, to some extent, in the adult. Galton introduced this idea "because ancestral qualities indicated in early life frequently disappear and yield place to others" (*ibid.*, p. 396). The latter are the sports whose occurrence Galton had previously (Sect. 4.3.4) explained in much the same way. Galton warned that the reverse process (that of patent elements becoming latent) was "doubtful" because this would mean that "the later children of a union would resemble their parents more nearly than the earlier ones" (*ibidem*). Galton's argument here was essentially one of hard heredity, because allowing for patent elements to become latent meant that personal qualities that were environmentally acquired would be inherited. On the other hand, the latter possibility was an integral part of Darwin's pangenesis where environmental influences were carried by the gemmules to the germ cells.

Galton's belief in hard heredity also meant that when the two developmental paths (*a* and *b* in Fig. 4.2) converge, *very few* of the adult patent elements are transmitted in combination with some adult latent elements as structureless elements in the offspring.

The above describes the hereditary process for a single parent. When both parents are considered, the process is shown in Fig. 4.3.

One important aspect of Galton's physiological theory of heredity involves the rules of selection from structureless elements in the fertilized egg to patent elements in the embryo (Fig. 4.2). Galton's answer was that:

> **QUOT. 298**...it must have been on some principle whose effects may be described as those of "Class Representation" ...I give as broad a meaning to the expression as a politician would give to the kindred one, a "representative assembly." By this he means to say that the assembly consists of representatives from various constituencies, which is a distinct piece of information so far as it goes, and is a useful one, although it deals with no matter of detail. (*ibid.*, p. 395)

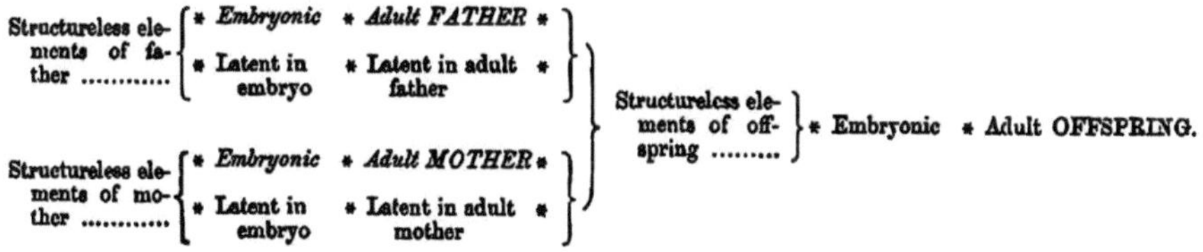

Fig. 4.3 Galton's theory of heredity in 1872, with both parents included (Galton, 1872, p. 398)

Galton was thus unaware of the principle behind the selection, but its effects were those of "class representation." This was meant to be a political metaphor to indicate that patent elements were selected to represent the various characteristics of the individual. Similarly, Galton was unaware of the exact principles behind the selection of patent elements in the adult to structureless elements in the offspring but was able to describe its effects as those of "family representation."

One frequently sees a certain fondness by Galton for political metaphors, which were often somewhat fanciful, in trying to understand purely biological phenomena. In contrast, Darwin's writing was rich in zoological and botanical details. These two styles are reflective of the different backgrounds of the two men, with Darwin being extremely well versed in natural history but Galton being much less so (and more effective in quantitative methods).

4.5.2 "A Theory Of Heredity" (1876): Elaboration of 1872 Theory, Stirp Introduced, Comparison with Darwin's Pangenesis

Galton's article "A theory of heredity" (Galton, 1876)[27] was an elaboration of his previous article (Galton, 1872) and an opportunity to introduce some key terminology by which his physiological theory would later be known. While the 1872 article did not mention Darwin's pangenesis, Galton now also put his theory into proper perspective by relating it to pangenesis.

It is precisely with this perspective, which also served as a clarification, that Galton started his paper. He stated that a complete theory of heredity should be able to account for two classes of phenomena, namely those relating to characters that are (i) hereditary, and (ii) environmentally acquired. However:

> **QUOT. 299** The first of these two groups is of predominant importance, in respect to the number of well-ascertained facts that it contains, many of which it is possible to explain... The second group includes much of which the evidence is questionable or difficult of verification... [Darwin's] Pangenesis appears more especially framed to account for the cases which fall in the second of the above-mentioned groups, which are of a less striking and assured character than those in the first group, and it will be seen that I accept the theory of Pangenesis with considerable modification, as a supplementary and subordinate part of a complete theory of heredity, but by no means for the primary and more important part. (Galton, 1876, pp. 329–330)

Thus, Galton's view was that pangenesis was based on evidence that was "questionable or difficult" and, even with considerable modification, could only be a "subordinate" part of a complete theory of heredity.

---

[27] A slightly shorter version of the paper, with the same title, was published in the *Contemporary Review* (Galton, 1875).

Galton next introduced the term by which his physiological theory of heredity would later be identified, namely, "stirp":

> **QUOT. 300** . . .I beg permission to use, in a special sense, the short word "stirp," derived from the Latin stirpes, a root, to express the sum-total of the germs, gemmules, or whatever they may be called, which are to be found, according to every theory of organic units, in the newly fertilised ovum – that is, in the earliest pre-embryonic stage – from which time it receives nothing further from its parents, not even from its mother, than mere nutriment. (*ibid.*, p. 330)

Thus, Galton's stirp was the collection of "germs," "gemmules," or "elements" and was found in the fertilized egg. It was *not* the same as Darwin's gemmules which were carried from specific body parts to the ovum.

Galton next (*ibid.*, p. 331) listed the four postulates on which his stirp theory was based (these postulates would also hold for pangenesis since the latter was a "subordinate" part of his complete theory):

1. Each independent part of the body originates from a particular germ.
2. The number and variety of germs in the stirp are much greater than the organic units that are derived from them, so only a few of the potential units actually develop.
3. The structureless elements that become latent in the parent contribute to the stirp of the offspring.
4. The germs in the stirp possess specific forces of attraction and repulsion that operate in the orderly development of a structure.

In retrospect, if the germs in the stirp are compared to the genes in the genome, then Galton's four postulates are remarkably close to current thinking. They may be interpreted as follows:

1'. One gene determines one protein (or, more accurately, one polypeptide).
2'. There are many more genes than there are proteins (since not all genes code for proteins)
3'. Genes (and not proteins) are transmitted from generation to generation.
4'. Gene expression is regulated during development.

Equally accurate are Galton's observations regarding (i) bisexual vs. unisexual organisms and (ii) the transmission of gemmules from parents to offspring.

Regarding (i), Galton noted that, at least in the higher forms of life, bisexual organisms were much more common than unisexual organisms. This is because, in the case of unisexual organisms, "a deterioration, which we may reasonably ascribe to a deficiency of some of the structural elements, is always observed to set in and gradually to increase, the race ultimately perishing from that cause" (*ibid.*, p. 332). The following passage, which is an even starker statement of the advantage of double parentage, foreshadows the concept of dominance: "when there are two parents, and therefore a double supply of material, the chance deficiency in the contribution from either of them, of any particular species of germ, tends to be supplied by the other" (*ibid.*, p. 334).

Regarding (ii), Galton postulated a process reminiscent of what would later be called meiosis: "as the stirp whence the child sprang, can be only half the size of the combined stirps of his two parents, it follows that one half of his possible heritage must have been suppressed" (*ibidem*). Since half of the stirp from each parent could be transmitted from parent to offspring, Galton concluded there was "a sharp struggle for place among the competing germs." This broadened the concept of "class representation" in his 1872 paper (Galton, 1872).[28]

Toward the end of his paper, Galton summarized the key differences between Darwin's pangenesis and his stirp theory:

(a) Darwin assumed that the gemmules originally existed in the cells, whereas Galton maintained that they resided in the stirp, which was found in the fertilized egg.
(b) Darwin's major assumption was that the body cells threw off gemmules. On the other hand, this process was of very minor importance in Galton's theory.
(c) Darwin earlier suggested that gemmules developed into cells[29] but later seemed to distance himself from this suggestion.[30] As for Galton, he regarded the conversion of gemmules into cells as the primary process of heredity.
(d) Darwin viewed the development of gemmules as a process similar to fertilization, when gemmules united with very young cells that preceded them in the regular order of growth. On the other hand, Galton viewed the development of gemmules as "the successive segmentations of the host of gemmules that are contained in the newly fertilized ovum," and "the development of the dominant or representative gemmules in each segmentation."
(e) Darwin believed that the sexual elements are formed by the aggregation of gemmules that traveled freely in the body, while Galton held that the sexual elements were formed from the residue of germs in the stirp and that the gemmules did *not* travel freely.

Galton's 1875 paper was quite influential. The German biologist August Weismann (1834–1914) later acknowledged that it was the one that contained the idea of continuity of the germ plasm before his own papers. However, as shown in Sect. 4.5.1, the idea had already been expressed in Galton's 1872 paper, but admittedly, the latter was hard to read. In a February 23, 1889, letter addressed to Galton, Weismann said:

[28] See end of Sect. 4.5.1.

[29] For example, ". . .I assume that cells, before their conversion into completely passive or "formed material," throw off minute granules or atoms, which circulate freely throughout the system, and when supplied with proper nutriment multiply by self-division, subsequently becoming developed into cells like those from which they were derived" (Darwin, 1868a, p. 373, Vol. I).

[30] For example, Darwin to Miles Joseph Berkeley on Sep 7, 1869: "I have never supposed that they [gemmules] were developed into free cells, but they penetrated other nascent cells and modified their subsequent development" (Darwin Correspondence Project online, letter DCT-LETT-6353).

QUOT. 301 …I am indeed very glad to have now all your memoirs at once at hand for consulting them. Till now I did not know all of them, but some ones, for instance "A Theory of Heredity" from 1875. It was Mr Herdman of Liverpool who—some years ago—directed my attention to this paper of yours, after having read my own papers on heredity, on continuity of germ-plasm and others. I regret not to have known it before, as you have exposed in your paper an idea which is in one essential point nearly allied to the main idea contained in my theory of the continuity of germ-plasm. (Pearson, 1930, pp. 340–341, Vol. IIIA)

4.5.3 *Galton's "Mendelian Ratios" of 1875*

Weismann might have later praised Galton's 1875 paper, but Darwin was far from impressed. On November 7, 1875, he wrote to Galton to tell him that he found his paper quite opaque:

QUOT. 302 My dear Galton, I have read your essay with much curiosity and interest, but you probably have no idea how excessively difficult it is to understand. I cannot fully grasp, only here and there conjecture, what are the points on which we differ—I daresay this is chiefly due to muddle-headiness on my part, but I do not think wholly so. Your many terms, not defined "developed germs"—"fertile" and "sterile" germs (the word 'germ' itself from association misleading to me), "stirp,"—"sept," "residue" etc. etc., quite confounded me. If I ask myself how you derive and where you place the innumerable gemmules contained within the spermatozoa formed by a male animal during its whole life I cannot answer myself. Unless you can make several parts clearer, I believe (though I hope I am altogether wrong) that very few will endeavour or succeed in fathoming your meaning. (Pearson, 1924, p. 187)

The quotation above shows that Darwin was unable to understand the several differences between his "provisional" theory of pangenesis and Galton's stirp theory. On December 18, 1875, he wrote to Galton, informing him that his son George had been able to shed light on the differences. Darwin conceded Galton's argument that gemmules were multiplied in the reproductive organs but he could not reject that the organs also threw off gemmules. He gave Galton an example and asked the latter how his theory could explain it:

QUOT. 303 …If 2 plants are crossed, it often or rather generally happens that every part of stem, leaf—even to the hairs—and flowers of the hybrid are intermediate in character; and this hybrid will produce by buds millions on millions of other buds all exactly reproducing the intermediate character. I cannot doubt that every unit of the hybrid is hybridised and sends forth hybridised gemmules. Here we have nothing to do with the reproductive organs. (*ibid.*, p. 189)

Thus, suppose two varieties of plants (e.g., white and black) were crossed. Then, the hybrid would often be intermediate in character (gray), and this gray plant would produce millions of buds, all resulting in gray plants. For Darwin, such an occurrence could only be explained by pangenesis, i.e., by the cells in the gray plant throwing off *gray gemmules* to form gray buds.

(1) (2)

Fig. 4.4 Galton's depiction of two types of hybrids in his Dec 19, 1875, letter to Darwin (Pearson, 1924, p. 189)

Galton replied on December 19, 1875, explaining that Darwin's example could be accounted for by his stirp theory as well. He reasoned that the hybrid cells could be of two types, as shown in Fig. 4.4.

In Fig. 4.4(1), it takes only one gemmule to form a cell, i.e., the structural unit (cell) is also the organic unit (gemmule). The hybrid consists of an equal number of white and black cells, which could thus be distinguished under a microscope but appear gray to the normal eye. In Fig. 4.4(2), it takes more than one gemmule to form a cell, i.e., the structural unit (cell) is an organic molecule (two or more gemmules). Moreover, in this case, Galton explained that:

> **QUOT. 304** The larger the number of gemmules in each organic molecule, the more uniform will the tint of greyish be in the different units of structure. . .If there were 2 gemmules only [per cell], each of which might be either white or black, then in a large number of cases one-quarter would always be quite white, one-quarter quite black, and one-half would be grey. If there were 3 molecules [per cell], we should have 4 grades of colour (1 quite white, 3 light grey, 3 dark grey, 1 quite black and so on. . . (*ibid.*, p. 190)

Now, Galton's predicted ratios of 1:2:1 and 1:3:3:1 are precisely those that Gregor Mendel had already predicted in 1866 (Mendel, 1866) (but which remained unknown to the scientific community until 1900). This has prompted some scholars to declare that Galton had all but independently discovered Mendel's work. For example, upon reviewing Galton's December 19, 1875, letter to Darwin, the historian Robert Olby announced:

> **QUOT. 305** Here we find all the elements of the Mendelian explanation save the independent segregation of different pairs of characters. (Olby, 1965, p. 637, 1966, p. 72)

However, although Galton's letter shows that he would probably have appreciated the significance of Mendel's 1865 paper, he had *not* actually derived Mendelian ratios. This is because, as pointed out by Lints and Delcour (1968), Galton's ratios of 1:2:1 and 1:3:3:1 were related to the constitution of the somatic cells *in one member of the* F_1 *(i.e., hybrid) generation*. He did not infer this ratio *among the members of the* F_2 *(i.e., crossing of hybrids) generation,* which is what Mendel had done.

In the second edition of his book, Olby (1985, p. 57) admitted that his interpretation of 1966 had "rightly been criticized."

4.5.4 "Typical Laws of Heredity" (1877): A Statistical Theory of Heredity, The Birth of Biometry

Having proposed a hypothetical physiological theory of heredity acting within individuals, Galton set out to investigate heredity at the population level. To do this, he had to introduce probabilistic and statistical techniques into the life sciences, thus creating the field of biometry. This was first explicitly done in Galton's paper, "Typical law of heredity" (Galton, 1877) (Fig. 4.5).

Galton started his paper with a problem that had preoccupied him for a while, indeed since his *Hereditary Genius* (Galton, 1869a): why do the characteristics of a hereditary attribute (such as height) from an isolated human population remain constant from generation to generation? In his own words:

> **QUOT. 306** Why is it, when we compare two groups of persons selected at random from the same race, but belonging to different generations of it, we find them to be closely alike? Such statistical differences as there may be, are always to be ascribed to differences in the general conditions of their lives; with these I am not concerned at present; but so far as regards the processes of heredity alone, the resemblance of consecutive generations is a fact common to all forms of life. (Galton, 1877, p. 492)

The main tool in Galton's statistical armory was the normal distribution, which at this stage he called "the law of deviation" and whose use in humans he credited to Quetelet. Galton announced the aim of his paper as one involving a statistical problem in heredity:

> **QUOT. 307** The outline of my problem of this evening is, that since the characteristics of all plants and animals tend to conform to the law of deviation, let us suppose a typical case, in which the conformity shall be exact, and which shall admit of discussion as a mathematical problem, and find what the laws of heredity must then be to enable *successive* generations to maintain statistical identity. (*ibid.*, p. 493)

Galton resorted to experiments with sweet peas to answer his questions. He sorted a large number of sweet pea seeds into seven equally spaced size (weight) classes and sent each of his friends seven packets, each containing ten seeds of a given class size. The seeds from the offspring were then collected and sent back to Galton. Before discussing his results, Galton made the following clarification:

> **QUOT. 308** The only processes concerned in simple descent that can affect the characteristics of a sample of a population are those of Family Variability and Reversion. It is well to define these words clearly. By family variability is meant the departure of the children of the same or similarly descended families, from the ideal mean type of all of them. Reversion is the tendency of that ideal mean filial type to depart from the parent type, "reverting" towards what may be roughly and perhaps fairly described as the average ancestral type. If family variability had been the only process in simple descent that affected the characteristics of sample, the dispersion of the race from its mean ideal type would indefinitely increase with the number of the generations; but reversion checks this increase, and brings it to a standstill, under conditions which will now be explained (*ibid.*, p. 513).

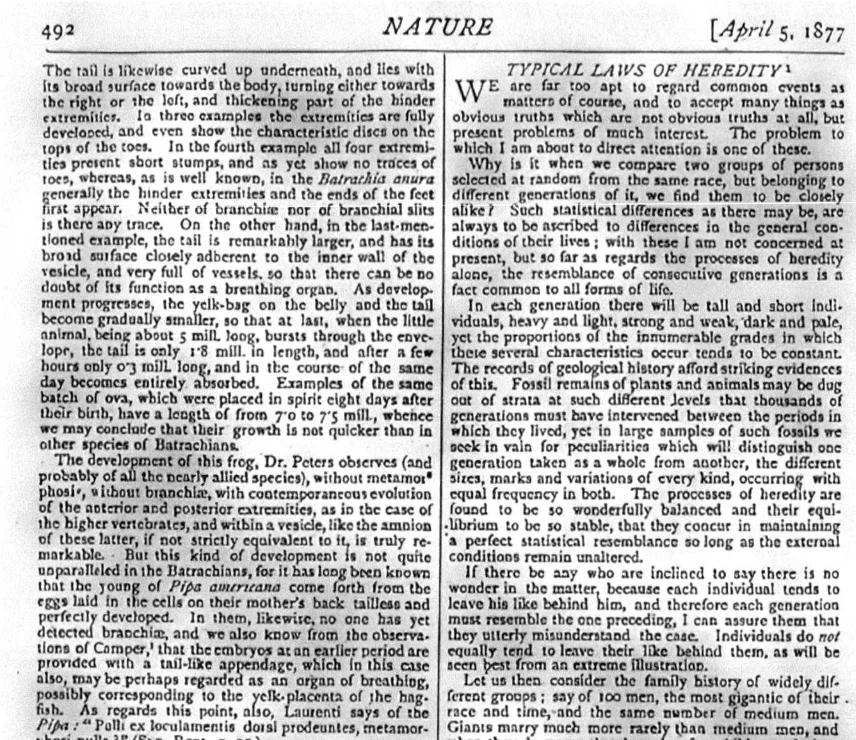

The tail is likewise curved up underneath, and lies with its broad surface towards the body, turning either towards the right or the left, and thickening part of the hinder extremities. In three examples the extremities are fully developed, and even show the characteristic discs on the tops of the toes. In the fourth example all four extremities present short stumps, and as yet show no traces of toes, whereas, as is well known, in the *Batrachia anura* generally the hinder extremities and the ends of the feet first appear. Neither of branchiæ nor of branchial slits is there any trace. On the other hand, in the last-mentioned example, the tail is remarkably larger, and has its broad surface closely adherent to the inner wall of the vesicle, and very full of vessels, so that there can be no doubt of its function as a breathing organ. As development progresses, the yelk-bag on the belly and the tail become gradually smaller, so that at last, when the little animal, being about 5 mill. long, bursts through the envelope, the tail is only 1·8 mill. in length, and after a few hours only 0·3 mill. long, and in the course of the same day becomes entirely absorbed. Examples of the same batch of ova, which were placed in spirit eight days after their birth, have a length of from 7·0 to 7·5 mill., whence we may conclude that their growth is not quicker than in other species of Batrachians.

The development of this frog, Dr. Peters observes (and probably of all the nearly allied species), without metamorphosis, without branchiæ, with contemporaneous evolution of the anterior and posterior extremities, as in the case of the higher vertebrates, and within a vesicle, like the amnion of these latter, if not strictly equivalent to it, is truly remarkable. But this kind of development is not quite unparalleled in the Batrachians, for it has long been known that the young of *Pipa americana* come forth from the eggs laid in the cells on their mother's back tailless and perfectly developed. In them, likewise, no one has yet detected branchiæ, and we also know from the observations of Camper,[1] that the embryos at an earlier period are provided with a tail-like appendage, which in this case also, may be perhaps regarded as an organ of breathing, possibly corresponding to the yelk-placenta of the hag-fish. As regards this point, also, Laurenti says of the *Pipa*: "Pulli ex loculamentis dorsi prodeuntes, metamorphosi nulla?" (Syn. Rept., p. 25.)

It would be of the highest interest, Dr. Peters adds, to follow exactly this remarkable development on the spot. The development of the embryo of these Batrachians in a way very like that of the scaled Reptilia makes one suspect that an examination of the temporary embryonic structures of *Hylodes* and *Pipa* would result in showing remarkable differences from those of other Batrachians. The general conclusions which might be drawn from this discovery are so obvious, says Dr. Peters, in conclusion, that it would be superfluous to put them forward.

A subsequent communication of Dr. Peters to the Academy informs us that it had escaped his notice that M. Bavay, of Guadaloupe, had already published some observations on the development of *Hylodes martinicensis*.[2] According to his observations, on each side of the heart there is a branchia consisting of one simple gill-arch, which on the seventh day is no longer discernible. On the ninth day there is no longer a trace of a tail, and on the tenth day the little animal emerges from the egg. M. Bavay also observed the contemporaneous development of the four extremities, and hints at the function of the tail as an organ of breathing.

The observations of Dr. Gundlach, therefore, says Dr. Peters, differ in some respects from those of M. Bavay. It would be specially desirable, however, to ascertain whether the arched vessel on each side of the heart is really to be regarded as a gill-arch, or only as the incipient bend of the aorta.

[1] Comm. Soc. Reg. Gotting. Cl. phys. ix. p. 133 (1788).
[2] Ann. Sc. Nat. ser. 5, xvii, art. No. 16 (1873).

TYPICAL LAWS OF HEREDITY[1]

WE are far too apt to regard common events as matters of course, and to accept many things as obvious truths which are not obvious truths at all, but present problems of much interest. The problem to which I am about to direct attention is one of these.

Why is it when we compare two groups of persons selected at random from the same race, but belonging to different generations of it, we find them to be closely alike? Such statistical differences as there may be, are always to be ascribed to differences in the general conditions of their lives; with these I am not concerned at present, but so far as regards the processes of heredity alone, the resemblance of consecutive generations is a fact common to all forms of life.

In each generation there will be tall and short individuals, heavy and light, strong and weak, dark and pale, yet the proportions of the innumerable grades in which these several characteristics occur tends to be constant. The records of geological history afford striking evidences of this. Fossil remains of plants and animals may be dug out of strata at such different levels that thousands of generations must have intervened between the periods in which they lived, yet in large samples of such fossils we seek in vain for peculiarities which will distinguish one generation taken as a whole from another, the different sizes, marks and variations of every kind, occurring with equal frequency in both. The processes of heredity are found to be so wonderfully balanced and their equilibrium to be so stable, that they concur in maintaining a perfect statistical resemblance so long as the external conditions remain unaltered.

If there be any who are inclined to say there is no wonder in the matter, because each individual tends to leave his like behind him, and therefore each generation must resemble the one preceding, I can assure them that they utterly misunderstand the case. Individuals do *not* equally tend to leave their like behind them, as will be seen best from an extreme illustration.

Let us then consider the family history of widely different groups; say of 100 men, the most gigantic of their race and time, and the same number of medium men. Giants marry much more rarely than medium men, and when they do marry they have but few children. It is a matter of history that the more remarkable giants have left no issue at all. Consequently the offspring of the 100 giants would be much fewer in number than those of the medium men. Again these few would, on the average, be of lower stature than their fathers for two reasons. First, their breed is almost sure to be diluted by marriage. Secondly, the progeny of all exceptional individuals tends to "revert" towards mediocrity. Consequently the children of the giant group would not only be very few but they would also be comparatively short. Even of these the taller ones would be the least likely to live. It is by no means the tallest men who best survive hardships, their circulation is apt to be languid and their constitution consumptive.

It is obvious from this that the 100 giants will not leave behind them their quota in the next generation. The 100 medium men, on the other hand, being more fertile, breeding more truly to their like, being better fitted to survive hardships, &c., will leave more than their proportionate share of progeny. This being so, it might be expected that there would be fewer giants and more medium-sized men in the second generation than in the first. Yet, as a matter of fact, the giants and medium-sized men will, in the second generation, be found in the same proportions as before. The question, then, is this:—How is it that although each individual does *not* as a rule leave his like behind him, yet successive generations resemble each other with great exactitude in all their general features?

[1] Lecture delivered at the Royal Institution, Friday evening, February 9, by Francis Galton, F.R.S.

Fig. 4.5 First page of Galton's "Typical Laws of Heredity" (Galton, 1877)

In the above, Galton asserted that the filial generation is the product of two opposing forces, namely, family variability and reversion. Family variability is kept in check (i.e., remains constant) due to reversion.

Galton next considered his sweet pea results and made the following two key conclusions:

> **QUOT. 309** On weighing and sorting large samples of the produce of each of the seven different classes of the peas, I found in every case the law of deviation to prevail, and in every case the value of 1° of deviation to be the same…The next great fact was that reversion followed the simplest possible law; the proportion being constant between the deviation of the mean weight of the produce generally and the deviation of the parent seed, reckoning in every case from one standard point. (Galton, 1877, p. 513)

The last two observations indicate the following:

(a) For a given parental class size, the size of the filial seeds was normally distributed, with the same probable error e_p within each class (i.e., the family variability within each class was the same).
(b) Reversion was a simple linear function of the deviation from the grand mean (M). Thus, if the parent size in a given class has mean $M + ke_p$ ($k = 0, \pm 1, \pm 2, \pm 3, \ldots$) then the corresponding filial size will have mean $M + k\rho e_p$, where $0 < \rho < 1$.

In an Appendix to the 1877 paper, Galton provided the algebraic conditions necessary for stability in population variability in terms of the modulus c of a distribution.[31] Galton first considered the case of two parents each of whom could be productive, but later turned to the case of simple descent. The latter case will be described here. First, Galton wrote the distribution of the "amount of deviation" x in a present population as

$$y = \frac{1}{c\sqrt{\pi}} \exp\left(-\frac{x^2}{c^2}\right).$$

Second, "reversion is expressed by a simple fractional coefficient of the deviation," which will be denoted by ρ[32] ($0 < \rho < 1$). Then, in the "reverted parentages,"

$$y = \frac{1}{\rho c\sqrt{\pi}} \exp\left(-\frac{x^2}{\rho^2 c^2}\right).$$

Galton then denoted the modulus of the present population by c_1 and that of the "reverted parentages" by c_2, so that

$$c_2 = \rho c_1. \tag{4.1}$$

[31] The modulus of a normal distribution was historically used to represent $\sigma\sqrt{2}$.

[32] Galton himself used the symbol r.

The next step for Galton was to consider the variation in the number of progeny for a given parental class, i.e., the family variability:

> **QUOT. 310** Family variability was shown by experiment to follow the law of deviation, its modulus, which we will write v, being the same for all classes. Therefore the amount of deviation of any one of the offspring from the mean of his race is due to the combination of two influences – the deviation of his "reverted" parentage and his own family variability; both of which follow the law of deviation. This is obviously an instance of the well-known law of the "sum of two fallible measures"[33] (*ibid.*, p. 533).

Denoting the modulus of the family variability by v, Galton wrote the law above as

$$c_4^2 = c_2^2 + \nu^2, \tag{4.2}$$

where c_4^2 is the overall modulus of the progeny. Combining Eqs. (4.1) and (4.2),

$$c_4^2 = \rho c_1^2 + \nu^2.$$

Now, for variability to remain constant across generations, we need $c_4 = c_1$ which implies

$$c_1^2 = \frac{\nu^2}{1-\rho^2} \tag{4.3}$$

(*ibidem*). The statistical law of heredity above is Galton's answer to the problem he initially posed. It expresses the fact that the variability across generations remains constant due to the balancing forces of family variability (which tends to increase spread) and reversion (which tends to reduce spread).

A modern version of Galton's analysis can be written as follows. Let X and Y be the present and parental population class sizes, respectively, such that (X, Y) is bivariate normal. Then each of X and Y is normally distributed. Let their variances be σ_X^2, σ_Y^2, respectively. The variance of the present population generation for a given parental class is $\sigma_{X|Y}^2 = \text{var}(Y|X)$, where

$$\sigma_{X|Y}^2 = \sigma_X^2(1-\rho^2)$$

and $\rho = \text{corr}(X, Y)$. Stability in population variance means that $\sigma_X^2 = \sigma_Y^2$, so that we finally obtain

$$\sigma_{X|Y}^2 = \sigma_Y^2(1-\rho^2).$$

Galton's 1877 paper was groundbreaking but still not entirely satisfactory. It did not justify why reversion should be proportional to the deviation from the grand

[33] Airy, "Theory of Errors," $ 43 (*Galton's footnote*).

mean. More importantly, Galton's use of reversion to signify "the tendency of that ideal mean filial type to depart from the parent type, reverting toward what may be roughly and perhaps fairly described as the average ancestral type" showed that, at this stage, he probably thought of the process as being both *unidirectional* and *particular to the hereditary process*. To his amazement, the contrary turned out to be true, and this was to be shown in his 1885 presidential address to the *Anthropology Section of the British Association* (Galton, 1885a).

The 1885 presidential address is also where the term "regression" was first used by Galton.

4.5.5 Galton's 1885 Presidential Lecture: Reversion Changed to Regression

In 1885, Galton made another breakthrough following his 1877 paper, where he had proposed a statistical law of heredity. In his 1885 Presidential Lecture to the Anthropology Section of the British Association (Galton, 1885a), Galton recalled his initial observation but now he used the term "regression," instead of "reversion," to describe it:

> **QUOT. 311** ...It appeared from these experiments that the offspring did *not* tend to resemble their parent seeds in size, but to be always more mediocre than they were – to be smaller than the parents, if the parents were large; to be larger than the parents, if the parents were small...
>
> The experiments showed further that the mean filial regression towards mediocrity was directly proportional to the parental deviation from it. (*ibid.*, p. 507).

Galton then admitted that:

> **QUOT. 312** ...I was then [in 1877] blind to what I now perceive to be the simple explanation of the phenomenon... (*ibidem*)

The presidential lecture was a breakthrough because Galton was able not only to show that regression (his previous "reversion") was symmetric in nature but also to provide the correct mechanism for it.[34]

Galton explained that he had taken pains to collect the heights of parents and adult children from 205 families. He then multiplied the height of each female parent by 1.08 to make the male and female parents comparable and "no objection grounded on the sexual difference of stature need be raised." He then took the average for each pair of parents to obtain a "mid-parent" height. The data are shown in Fig. 4.6, which appeared in the accompanying paper "Regression towards mediocrity in hereditary stature" (Galton, 1885b). From these data, Galton calculated the median for both

---

[34]Galton was correct in explaining regression from one generation to the other. However, Galton also made an argument for perpetual regression which turned out to be erroneous. For more details, see Bulmer (2003, p. 285).

TABLE I.

NUMBER OF ADULT CHILDREN OF VARIOUS STATURES BORN OF 205 MID-PARENTS OF VARIOUS STATURES.
(All Female heights have been multiplied by 1·08).

Heights of the Mid-parents in inches.	Heights of the Adult Children.														Total Number of		Medians.
	Below	62·2	63·2	64·2	65·2	66·2	67·2	68·2	69·2	70·2	71·2	72·2	73·2	Above	Adult Children.	Mid-parents.	
Above ..	..	..	..	..	..	..	..	..	..	..	..	1	3	..	4	5	..
72·5	..	..	..	..	..	..	..	1	2	1	2	7	2	4	19	6	72·2
71·5	..	..	..	..	1	3	4	3	5	10	4	9	2	2	43	11	69·9
70·5	1	..	1	..	1	1	3	12	18	14	7	4	3	3	68	22	69·5
69·5	..	..	1	16	4	17	27	20	33	25	20	11	4	5	183	41	68·9
68·5	1	..	7	11	16	25	31	34	48	21	18	4	3	..	219	49	68·2
67·5	..	3	5	14	15	36	38	28	38	19	11	4	..	..	211	33	67·6
66·5	..	3	3	5	2	17	17	14	13	4	..	..	..	..	78	20	67·2
65·5	1	..	9	5	7	11	11	7	7	5	2	1	..	..	66	12	66·7
64·5	1	1	4	4	1	5	5	..	2	..	..	..	..	..	23	5	65·8
Below ..	1	..	2	4	1	2	2	1	1	..	..	..	..	..	14	1	..
Totals ..	5	7	32	59	48	117	138	120	167	99	64	41	17	14	928	205	..
Medians ..	..	..	66·3	67·8	67·9	67·7	67·9	68·3	68·5	69·0	69·0	70·0	..	..	..	..	..

NOTE.—In calculating the Medians, the entries have been taken as referring to the middle of the squares in which they stand. The reason why the headings run 62·2, 63·2, &c., instead of 62·5, 63·5, &c., is that the observations are unequally distributed between 62 and 63, 63 and 64, &c., there being a strong bias in favour of integral inches. After careful consideration, I concluded that the headings, as adopted, best satisfied the conditions. This inequality was not apparent in the case of the Mid-parents.

Fig. 4.6 Galton's table on stature (Galton, 1885b, p. 248)

mid-parents (X) and offspring (Y) as 68.25 in. and the probable deviations as $e_{p,\ X} = 1.2$ and $e_{p,\ Y} = 1.7$, respectively. Next, for each row in Fig. 4.6, Galton calculated the median height of the adult children and plotted the latter median heights against the mid-parental heights. The line is fitted by inspection and Galton estimated a slope of 2/3.[35] Galton was now ready to define his law of regression for these data:

QUOT. 313 It is that the height-deviate of the offspring is, on the average, two-thirds of the height-deviate of its mid-parentage. (Galton, 1885b, p. 508)

In modern notation, using M to denote the mean of both X and Y, the above can be written as

$$\mathrm{E}(Y|X) - M = \frac{2}{3}(X - M) \quad \Rightarrow \quad \mathrm{E}(Y|X) = \frac{2}{3}X + \frac{1}{3}M, \tag{4.4}$$

Galton now provided the correct explanation for the regression effect:

QUOT. 314 The explanation of it is as follows. The child inherits partly from his parents, partly from his ancestry. Speaking generally, the further his genealogy goes back, the more numerous and varied will his ancestry become, until they cease to differ from any equally numerous sample taken at haphazard from the race at large. Their mean stature will then be

[35] Galton later noted that his initial estimate of the slope was 0.6 but that "the fraction of two-thirds was preferable as being the more simple expression" (Galton, 1889, p. 98). The initial estimate of 0.6 would later be commented on by Pearson in his investigation of Galton's Law of Ancestral Heredity. See Sect. 4.6.5.

> the same as that of the race; in other words, it will be mediocre. Or, to put the same fact into another form, the most probable value of the mid-ancestral deviates in any remote generation is zero.
>
> For the moment let us confine our attention to the remote ancestry and to the mid-parentages, and ignore the intermediate generations. The combination of the zero of the ancestry with the deviate of the mid-parentage, is that of nothing with something, and the result resembles that of pouring a uniform proportion of pure water into a vessel of wine. It dilutes the wine to a constant fraction of its original alcoholic strength, whatever that strength may have been. (*ibidem*)

Indeed, the explanation above is one of the most intuitive ways of understanding the regression effect: In general, suppose a first measurement X is made on a given subject, followed by a second measurement Y. Assume X is exceptionally high. As long as X and Y are imperfectly correlated, X can be thought to comprise two components:

(a) A first component that is unusually extreme and is expected to remain extreme
(b) A second component that is not extreme and is expected to remain near the center of the distribution

The first measurement, X, is extreme because both of the components are high. However, for the second measurement, Y, the first component is expected to remain high but the second component is expected to be near the center. Hence, the average value of Y will be less extreme than X and closer to the center of the distribution.

Notice that the above explanation does not require any hereditary force to be present for regression to occur. The regression effect (or regression to the mean) is a purely mathematical artifact arising from the imperfect correlation[36] between X and Y. As such, it is also symmetric, in the sense that the same reasoning as above can be made to argue that, for a given extreme Y, the value of X is expected to be less extreme and closer to the center.

Thus, what started as an investigation of heredity ended up as a discovery of a purely statistical phenomenon of wide generality, namely regression.

The regression of X on Y was the problem that Galton next addressed in his 1885 lecture and was probably the reason for his realization of the correct mechanism of regression. Having plotted offspring median heights (Y) against mid-parental heights (X), Galton also plotted the median mid-parental heights against the offspring heights and obtained a slope of 1/3 (cf. Fig. 4.7).

In modern notation,

$$\mathrm{E}(X|Y) - M = \frac{1}{3}(Y - M) \quad \Rightarrow \quad \mathrm{E}(X|Y) = \frac{1}{3}Y + \frac{2}{3}M \tag{4.5}$$

Galton correctly warned:

[36]Note that the concept of correlation was not known to Galton when he first discovered the regression effect.

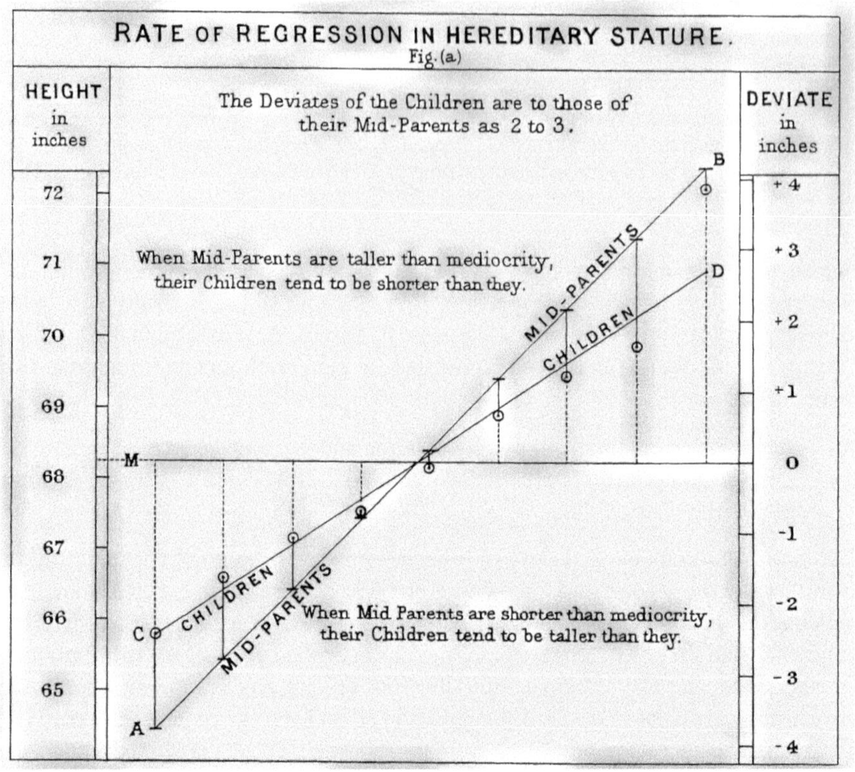

Fig. 4.7 Galton's plot of offspring median heights (Y) against mid-parental heights (X) (Galton, 1885b, Plate IX)

> **QUOT. 315** Because the most probable deviate of the son is only two-thirds that of his mid-parentage, it does not in the least follow that the most probable deviate of the mid-parentage is $\frac{3}{2}$, or $1\frac{1}{2}$. (*ibid.*, p. 509)

As will now be shown, if X and Y had the same variance, the two most probable deviates (or regression coefficients) would have been the same, namely the correlation coefficient between X and Y. In general, using modern notation, the regression coefficient of Y on X is

$$\beta_{Y|X} = \rho \frac{\sigma_Y}{\sigma_X}, \tag{4.6}$$

where $\sigma_X \equiv \sqrt{\text{var}X}$, $\sigma_Y \equiv \sqrt{\text{var}Y}$, and ρ is the correlation between X and Y. Similarly, the regression coefficient of X on Y is

$$\beta_{X|Y} = \rho \frac{\sigma_X}{\sigma_Y}. \tag{4.7}$$

If $\sigma_X = \sigma_Y$, then $\beta_{Y|X} = \beta_{X|Y} = \rho$.

On the other hand, in Galton's case, it is *not* true that $\sigma_X = \sigma_Y$ because X is the average of the male parent height (X_1) and adjusted female height (X_2), i.e., $X = (X_1 + X_2)/2$. Thus, if Y, X_1, X_2 are assumed to have the same variance σ^2, then

$$\text{var}Y = \sigma^2$$
$$\text{var}X = \text{var}\left(\frac{X_1 + X_2}{2}\right) = \frac{1}{4}\left(\sigma^2 + \sigma^2\right) = \frac{\sigma^2}{2},$$

so that $\beta_{Y|X} = \rho\sqrt{2}$, $\beta_{X|Y} = \rho/\sqrt{2}$, and $\beta_{Y|X}\beta_{X|Y} = 2$.[37]

However, back in 1885, the fact that the two regression coefficients were not the same troubled Galton because "some suspicion may remain of a paradox lurking in these strongly contrasted results." Therefore, Galton set out to graph the joint distribution of X and Y and, in the process, discovered the concept of the bivariate normal distribution.[38]

It now remained for Galton to discover the concept of correlation. Correlation, or rather "co-relation," first appeared in Galton's 1888 "Co-relations and their Measurements." This is how Galton defined it:

QUOT. 316 Two variable organs are said to be co-related when the variation of the one is accompanied on the average by more or less variation on the other, and in the same direction. (Galton, 1888, p. 135)[39]

In a later 1890 paper, Galton explained his reason for the 1888 paper:

QUOT. 317 In a book of mine called "Natural Inheritance," published about a year ago, I showed that the problems of family likeness fell entirely within the scope of the higher laws of chance...

These two problems – namely, that of estimating the stature of an unknown man from the length of one of his bones, and that of the relation between the various bodily dimensions of the same person – are clearly identical...

Reflection soon made it clear to me that not only were the two new problems identical in principle with the old one of kinship which I had already solved, but that all three of them were no more than special cases of a much more general problem – namely, that of Correlation.

Fearing that this idea, which had become so evident to myself, would strike many others as soon as "Natural Inheritance" was published, and that I should be justly reproached for

---

[37] The variances above are also in line with the probable errors reported by Galton on p. 19: $e_{p,\,X} = 1.2$, $e_{p,\,Y} = 1.7$ so that $e_{p,X}/e_{p,Y} = .71 = \sqrt{\text{var}X/\text{var}Y}$. Finally, note that, from Eqs. (4.6) and (4.7), $\rho^2 = \beta_{Y|X}\beta_{X|Y}$, which gives a correlation between offspring and mid-parent of $\rho = \sqrt{2}/3$, or a correlation between offspring and a single parent of 1/3.

[38] Galton's discovery of the bivariate normal surface was explained in more detail in the paper "Regression towards mediocrity in hereditary stature" (Galton, 1885b). In a subsequent paper (Galton, 1886), he enlisted the help of the mathematician Hamilton Dickson to provide a mathematical basis for his empirical findings. All these results were included in Galton's important book *Natural Inheritance* (Galton, 1889). More details can be found on pp. 202–206 of Gorroochurn (2016a).

[39] Galton was, of course, referring to *positive* correlation.

having overlooked it, I made all haste to prepare a paper for the Royal Society with the title of "Correlation". (*ibid.*, pp. 81–82)

Galton had thus realized that his previous investigations of regression were not limited to heredity problems but could be applied to many other measurements. Indeed, all these problems were the manifestations of one underlying phenomenon, namely correlation.

Continuing with the 1888 paper, Galton provided a fundamental insight through an example:

QUOT. 318 The relation between the cubit[40] and the stature will be shown to be such that for every inch, centimetre, or other unit of absolute length that the cubit deviates from the mean length of cubits, the stature will on the average deviate from the mean length of statures to the amount of 2.5 units, and. in the same direction. Conversely, for each unit of deviation of stature, the average deviation of the cubit will be 0.26 unit. These relations are not numerically reciprocal, but the exactness of the co-relation becomes established when we have transmuted the inches or other measurement of the cubit and of the stature into units dependent on their respective scales of variability... After this has been done, we shall find the deviation of the cubit as compared to the mean of the corresponding deviations of the stature, to be as 1 to 0.8. Conversely, the deviation of the stature as compared to the mean of the corresponding deviations of the cubit will also be as 1 to 0.8. Thus the existence of the co-relation is established, and its measure is found to be 0.8. (Galton, 1888, p. 136)

Here, Galton made the important observation that the regression coefficients for the two regressions, that of Y on X and that of X on Y, are generally different. However, when both X and Y are measured in terms of their respective variabilities (σ_X and σ_Y), the two regression coefficients become exactly the same, namely the correlation. In modern notation, Galton had just stated the relationships in Eqs. (4.6) and (4.7), which can now be written as

$$\beta_{Y|X}\frac{\sigma_X}{\sigma_Y}=\beta_{X|Y}\frac{\sigma_Y}{\sigma_X}=\rho. \tag{4.8}$$

Galton also mentioned that the number ρ[41] measures "the closeness of co-relation in any particular case."

4.5.6 *Two Limiting Cases of Galton's Particulate Theory of Heredity: Blending Inheritance and Alternative (Exclusive) Inheritance (1889)*

As explained in Sects. 4.5.1 and 4.5.2, Galton's theory of heredity, which was based on the notion of gemmules (that formed the individual's stirp), was *particulate*. This is how Galton described particulate inheritance in his *Natural Inheritance*:

[40] Derived from the Latin *cubitus*, the lower arm.

[41] Galton used the symbol *r*.

QUOT. 319 We seem to inherit bit by bit, this element from one progenitor that from another, under conditions that will be more clearly expressed as we proceed, while the several bits are themselves liable to some small change during the process of transmission. Inheritance may therefore be described as largely if not wholly "particulate," and as such it will be treated in these pages. (Galton, 1889, p. 7)

According to Galton's theory of particulate inheritance, every individual was thus a *mosaic* of its two parents: part of any given trait was inherited from one parent, and the remaining part was inherited from the other parent. Galton further distinguished two limiting cases of particulate inheritance:

(a) *Blending inheritance*. Using the same ideas that appeared in his 1875 correspondence with Darwin (see Sect. 4.5.3, in particular Fig. 4.4), Galton reasoned that if the parental gemmules were extremely minute and numerous in each hybrid cell, the hybrid trait of the hybrid would appear as a continuous mixture (or blend) of the parental traits:

QUOT. 320 The children of the white and the negro are of a blended tint; they are neither wholly white nor wholly black, neither are they piebald, but of a fairly uniform mulatto brown. . .Skin-colour is therefore a good example of what I call blended inheritance. It need be none the less "particulate" in its origin, but the result may be regarded as a fine mosaic too minute for its elements to be distinguished in a general view. (*ibid.*, p. 12)

(b) *Alternative (exclusive) inheritance.* A second limiting case of particulate inheritance occurred when an offspring inherited the whole of a parental trait (instead of a fraction) and no part from the other parent:

QUOT. 321 Next as regards heritages that come altogether from one progenitor to the exclusion of the rest. Eye-colour is a fairly good illustration of this, the children of a light-eyed and of a dark-eyed parent being much more apt to take their eye-colours after the one or the other than to have intermediate and blended tints. (*ibidem*)

4.5.7 Galton's Use of Regression to Argue Against the Importance of Continuous Variation in Evolution (1897)

In Sect. 4.3.4, it was shown how Galton used the notion of stability of types to highlight the importance of sports and advocate for discontinuity in evolution. Continuous variation (or individual differences), which was central to Darwin's theory, was thus downplayed. This was first done in Galton's *Hereditary Genius* (Galton, 1869a, pp. 368–369).

With the discovery of regression, Galton found another powerful argument to buttress his case against continuous variation. In 1885, he expressed the view that the regression coefficient (i.e., slope) was a measure of instability of a type:

QUOT. 322 The stability of a Type, which I should define as "an ideal form towards which the children of those who deviate from it tend to regress," would, I presume, be measured by the strength of its tendency to regress; thus a mean regression from 1 in the mid- parents to 2/3 in the offspring would indicate only half as much stability as if it had been to 1/3. (Galton, 1885b, p. 258)

In the preface of the second edition of his *Hereditary Genius* (Galton, 1891), Galton explained the effect of regression on continuous variation (referred to below by Galton simply as "variations"):

> **QUOT. 323** The filial centre falls back further towards mediocrity in a constant proportion to the distance to which the parental centre has deviated from it, whether the direction of the deviation be in excess or in deficiency. All true variations are (as I maintain) of this kind, and it is in consequence impossible that the natural qualities of a race may be permanently changed through the action of selection upon mere variations. The selection of the most serviceable variations cannot even produce any great degree of artificial and temporary improvement, because an equilibrium between deviation and regression will soon be reached, whereby the best of the offspring will cease to be better than their own sires and dams. (*ibid.*, p. xviii)

While natural selection acting on individual differences would be helpless in producing substantive permanent change, such was not the case when natural selection acted on sports:

> **QUOT. 324** The case is quite different in respect to what are technically known as "sports." In these, a new character suddenly makes its appearance in a particular individual, causing him to differ distinctly from his parents and from others of his race. Such new characters are also found to be transmitted to descendants. Here there has been a change of typical centre, a new point of departure has somehow come into existence, towards which regression has henceforth to be measured, and consequently a real step forward has been made in the course of evolution. When natural selection favours a particular sport, it works effectively towards the formation of a new species, but the favour that it simultaneously shows to mere variations seems to be thrown away, so far as that end is concerned. (*ibid.*, pp. xviii–xix)

Similarly, in his 1892 book *Finger Prints*, Galton noted that continuous variation was of "no permanent value to evolution, because there is a constant tendency in the offspring to 'regress' towards the parental type" (Galton, 1892, p. 20).

In 1897, Galton set out to quantitatively illustrate his argument on the inefficacy of natural selection acting on continuous variation (Galton, 1897a). He considered the stature of a hypothetical breed of British men, assumed to be normally distributed with a racial mean of 68.5 inches. Galton assumed a quartile of 1.5 inches and that individuals who were selected as parents were those in the 90th percentile of the distribution, i.e., with a height of 71.73 inches.[42] Now, given that the regression of offspring on mid-parents is 2/3 (i.e., the deviation of offspring is 2/3 that of mid-parents),[43] Galton calculated the deviation of an offspring in the first generation (from parents selected in the 90th percentile) as $2/3 \times (71.73 - 68.5) = 2/3 \times 3.23 = 2.15$ inches. Therefore, the first generation has a mean of $68.5 + 2.15 = 70.65$ inches and a quartile, as before, of 1.5 inches, with a 90th percentile at 73.5 inches. Individuals with the latter height will be selected as parents

---

[42] In modern notation, these numbers may be interpreted as follows. Let the height of a parent be denoted by X_0, where $X_0 \sim N(68.5, \sigma^2)$ and σ is the standard deviation of X. Now $\Pr\{X_0 > 68.5 + 1.5\} = .25$, so that $\Pr\{Z > 1.5/\sigma\} = .25$ and $\sigma = 2.224$ from tables. The value on the 90th percentile is t_0 where $\Pr\{X_0 > t_0\} = .10$. From tables, $t_0 = 71.35$, which is close to Galton's approximate value of 71.73.

[43] See Eq. (4.5).

(for offspring in the second generation) and are $73.5 - 68.5 = 5$ inches above the racial mean. Regression means that the second generation will be $2/3 \times 5 = 3.33$ inches above the racial mean.

Thus, the deviations from the racial mean of the first and second generations are 2.15 and 3.33 inches, respectively. The deviations of subsequent generations can be similarly calculated.

In general, suppose that the height of a member of the nth generation is X_n and its deviation from the racial mean is d_n. The mean of the nth generation is $68.5 + d_n$ and its 90th percentile is t_n, where $\Pr\{X_n > t_n\} = .10$ Using the calculated standard deviation of $\sigma = 2.224$ (last footnote),

$$\Pr\left\{Z > \frac{t_n - d_n - 68.5}{2.224}\right\} = .10$$

Using standard normal tables,

$$t_n = d_n + 68.5 + (1.28)(2.224).$$

The value of t_n gives the height of individuals who are selected as parents in the nth generation. In the next generation, regression will give rise to a deviation among the offspring of

$$d_{n+1} = \frac{2}{3}(t_n - 68.5) = \frac{2}{3}(d_n + 2.85) \qquad (4.9)$$

Using $d_1 = 2.15$ and $n = 1, 2, 3, \ldots$, the deviations from the racial mean (=68.5) for the 1st, 2nd, 3rd, 4th, ... generations are:

$$2.15, 3.33, 4.12, 4.65, \ldots$$

The limiting value of the above sequence may be found by setting $d_{n+1} = d_n = d$ in Eq. (4.9) so that $d = 5.7$ is the limiting deviation.

Galton's analysis showed that regression curbed the effect of selection so that starting with a mean height of 68.5 inches and selecting at the 90th percentile of a normal distribution with a quartile of 1.5 inches, the increase in height did not exceed a mere 5.7 inches. Furthermore, he warned that (with $w = 2/3$ here):

> **QUOT. 325** ...as soon as selection ceases it will regress to the level of the rest of the population through stages in which the deviation at starting, sinks successively to w, w^2, w^3, ... of its value. (*ibid.*, p. 606)

In his massive biography of Galton, Pearson (1930, p. 32, Vol. IIIA) called Galton's argument in the analysis above "perpetual regression." Many years before, in 1898, Pearson had already produced an effective counter analysis using Galton's own extension of regression, namely, his Law of Ancestral Heredity (Pearson, 1898). This will be described in Sect. 4.6.5.2.

4.6 Culmination of Galton's Statistical Theory of Heredity: The Law of Ancestral Heredity

4.6.1 Galton's First Enunciation of the Law of Ancestral Heredity in 1865

As mentioned in Sect. 4.2.4, Galton's Law of Ancestral Heredity describes the relative ancestral contributions of heritable characteristics to an individual. It was first stated in incorrect form in Galton's 1865 two-part paper "Hereditary talent and character", as follows:

> **QUOT. 326** The share that a man retains in the constitution of his remote descendants is inconceivably small. The father transmits, on an average, one-half of his nature, the grandfather one-fourth, the great-grandfather one-eighth; the share decreasing step by step, in a geometrical ratio, with great rapidity. (Galton, 1865, p. 326)

Three aspects of the above statement merit attention:

(a) Galton later corrected the Law of Ancestral Heredity so that each individual actually inherited, on average, 1/4 of its heritable characteristics from any one parent, 1/16 from any one grandparent, 1/64 from any one great-grandparent, and so on. Alternatively, the proportions he gave would hold if "father" is replaced by "mid-parent," "grandfather" by "mid-grandparent," and so on.
(b) The ancestral law is concerned with the individual's *received proportions* of hereditary characteristics, not the *transmitted proportions* by its ancestors.
(c) Although most scientists (especially Karl Pearson) correctly interpreted Galton's law to be statistical in nature, Galton also attempted to give it a physiological justification.

Most writers (e.g., Pearson, 1924, p. 84; Swinburne, 1965, p. 22; Froggatt and Nevin, 1971a, p. 5, 1971b, p. 5; Cowan, 1977, p. 145) have accepted the interpretation above that Galton initially misstated the Law of Ancestral Heredity. One exception is the population geneticist Michael Bulmer (1998, p. 508; 2003, p. 104), who has argued that Galton's numbers were *not* incorrect in QUOT. 326, since he *did not* have the ancestral law in mind but rather was merely stating the well-known law of halving. One of the earliest explicit occurrences of this law may be found in the 1725 book by William Wollaston:

> **QUOT. 327** ...Let C be the son of A and B, D the son of C, E of D, F of E: and let the *relation*[44] of C to A and B be as 1: then the *relation* D to A and B will be but ½; because C is but one of the parents of D, and so the *relation* of D to A and B is but the half of that which C bears to them. By proceeding after the same manner it will be found that the *relation* of E to A and B is ¼ (or half of the half), of F 1/8: and so on. (Wollaston, 1724, p. 166)

[44] Italics are Wollaston's.

Essentially, the law of halving correctly states that an individual inherits, on average, half of its hereditary material separately from each of its two parents. However, since one parent itself inherits half of its hereditary material from each of its own two parents, the individual *indirectly* inherits one-quarter of its hereditary material from each of its four grandparents, and so on. On the other hand, the ancestral law does not apportion the individual's inheritance into the separate average contributions of its two parents only (as the law of halving does) but rather into the *separate* average contributions of all its ancestors: one-quarter separate contribution from each of its two parents, 1/16 separate contributions from each of its four grandparents, and so on.

A reasonable compromise between Bulmer's position and that of most other scholars is that, in 1865, Galton did have the ancestral law in mind and that his formulation of it was inspired by the law of halving. That Galton had the ancestral law in mind is consistent with his later declaration that the 1865 paper did "contain the germs of many of my subsequent memoirs ... notwithstanding some crudeness here and there." (Galton, 1908, p. 289)

A final question relates to whether, in 1865, Galton's Law of Ancestral Heredity revealed his allegiance to a particular kind of inheritance. At first glance, how the specific inheritances are added would suggest compatibility with blending inheritance. However, as Cowan (1977, p. 145) has argued, this does not seem to be the case since Galton most likely did not know about blending inheritance at the time. Galton first publicly announced his particulate theory of inheritance in 1872 (Galton, 1872).[45] As discussed in Sect. 4.5.6, Galton also considered two limiting cases of particulate inheritance, namely blending inheritance and alternative (exclusive) inheritance. In 1885, when he first formally stated the ancestral law, Galton (1885b) illustrated the law by using a continuous character such as stature, which was believed to blend. Galton's ancestral law was then used as a law of *blending inheritance.* On the other hand, in 1889, Galton used a non-blending character, namely, blue eye color (Galton, 1889, Chap. 8). In this case, the ancestral law becomes: one-quarter of the offspring inherited their eye color from one parent, one-sixteenth of the offspring from one grandparent, and so on.[46] When used in this way, Galton's ancestral law becomes a law of *alternative inheritance*.[47]

---

[45] See Sect. 4.5.1.

[46] However, as will be shown in the next section, the way in which Galton and others interpreted the ancestral law for alternative inheritance was as follows: the total probability that a particular offspring inherits an ancestral trait is equal to ½ of the fraction of its parents who have the trait, ¼ of the fraction of its grandparents who have the trait, and so on.

[47] See Sect. 4.5.6.

4.6.2 *Galton's Formal Proof of the Law of Ancestral Heredity in 1885*

Based on his work on regression in his 1885 Presidential lecture (Sect. 4.5.5) and especially Eqs. (4.4) and (4.5),[48] Galton set out to present a theoretical justification for the ancestral law in the paper "Regression towards mediocrity" (Galton, 1885b).

From Eq. (4.4), Galton stated that, given that a mid-parental deviation (from the mean) is a, the offspring deviation is $2a/3$. On the other hand, from Eq. (4.5), if the offspring deviation is a, then the mid-parental deviation is $a/3$. Applying Eq. (4.5) again, if the mid-parental deviation is a, then the mid-grandparental deviation is $a/3$, and the mid-great grandparental deviation is assumed to be $a/3^2$, and so on. Suppose a constant fraction f_1 $(0 < f_1 < 1)$[49] of each mid-ancestral deviation is contributed to the offspring deviation ($=2a/3$). Then

$$\begin{aligned} af_1 + \frac{a}{3}f_1 + \frac{a}{3^2}f_1 + \cdots &= \frac{2a}{3} \\ f_1\left(a + \frac{a}{3} + \frac{a}{3^2} + \cdots\right) &= \frac{2a}{3} \\ f_1 \frac{a}{2/3} &= \frac{2a}{3} \\ \therefore f_1 &= \frac{4}{9} \end{aligned}$$

On the other hand, suppose the fraction contributed by the mid-parents is f_2 $(0 < f_2 < 1)$, that contributed by the mid-grandparent is f_2^2, that contributed by the mid-great grandparent is f_2^3, and so on. Then,

$$\begin{aligned} af_2 + \frac{a}{3}f_2^2 + \frac{a}{3^2}f_2^3 + \cdots &= \frac{2a}{3} \\ a\left(\frac{f_2}{3} + \frac{f_2^2}{3^2} + \frac{f_2^3}{3^3} + \cdots\right) &= \frac{2a}{9} \\ \frac{f_2/3}{1 - f_2/3} &= \frac{2}{9} \\ \therefore f_2 &= \frac{6}{11} \end{aligned}$$

Therefore:

> **QUOT. 328** The results of our two valid limiting suppositions are therefore (1) that the mid-parental deviate, pure and simple, influences the offspring to 4/9 of its amount; (2) that it influences it to 6/11 of its amount. These values differ but slightly from 1/2, and their mean is closely 1/2, so we may fairly accept that result. (*ibid.*, p. 261)

[48] Recall that, in both of these equations, Galton had *estimated* the slopes by inspection.

[49] Galton called $1 - f_1$ "the succession tax" (*ibid.*, p. 261).

One sees much hand-waving in Galton's reasoning:

(a) A regression coefficient of 1/3 for the regression of mid-grandparents on mid-parents is casually (and wrongly) assumed to imply a regression coefficient of $1/3^2$ for the regression of mid-great grandparents on mid-parents, and so on.[50]
(b) A constant ratio is assumed here and a geometrical ratio there, which are then averaged without asking what the meaning of such an average is, but just so that the predetermined number of 1/2 can be obtained.

Such was how the law of ancestral heredity was first "proved". Galton then stated the law as follows:

> **QUOT. 329** Hence the influence, pure and simple, of the mid-parent may be taken as ½, of the mid-grandparent ¼, of the mid-great-grandparent 1/8, and so on. That of the individual parent would therefore be ¼, of the individual grandparent 1/16, of an individual in the next generation 1/64, and so on. (*ibidem*)

However, a careful examination of Galton's derivation and the above statement of the ancestral law reveals an even greater problem:

(c) Galton's proof is concerned with the separate contributions from the *deviations of the ancestors* (a_1, a_2, a_3... for the deviations of the mid-parent, mid-grandparent, mid-great grandparent...) to the offspring deviation (a_0), as follows:

$$a_0 = \frac{1}{2}a_1 + \frac{1}{2^2}a_2 + \frac{1}{2^3}a_3 + \cdots \tag{4.10}$$

On the other hand, Galton's Law of Ancestral Heredity (as he had stated it) is concerned with the separate proportions of the *offspring's own deviation* (a_0) which have been contributed by ancestors, as follows:

$$a_0 = \underbrace{\frac{1}{2}a_0}_{\substack{\text{from} \\ \text{mid-parent}}} + \underbrace{\frac{1}{2^2}a_0}_{\substack{\text{from} \\ \text{mid-grandparent}}} + \underbrace{\frac{1}{2^3}a_0}_{\substack{\text{from} \\ \text{mid-great grandparent}}} + \cdots \tag{4.11}$$

Thus, in addition to the mathematical weaknesses in Galton's proof, it tried to prove something other than what was intended.

[50] For example, denote the mid-parent trait by Z, the mid-grandparent trait by Y, and the mid-great grandparent trait by X. Then the regression of mid-grandparent on mid-parent has coefficient $\beta_{YZ} = \text{cov}(Y, Z)/\text{var } Z$, the regression of the mid-great grandparent on the mid-grandparent has coefficient $\beta_{XY} = \text{cov}(X, Y)/\text{var } Y$, and the regression of the mid-great grandparent on the mid-parent has coefficient $\beta_{XZ} = \text{cov}(X, Z)/\text{var } Z$. Therefore, $\beta_{XZ} \neq \beta_{XY}\beta_{YZ}$.

Unfortunately, although Galton stated his ancestral law in the form (4.11), he most often used it in the form (4.10),[51] the latter being an extension of simple linear regression to multiple linear regression.

Galton's proof was reproduced in his major book *Natural Inheritance* (Galton, 1889, pp. 134–136).

4.6.3 Galton's Law of Ancestral Heredity According to Karl Pearson in 1896

Karl Pearson (1857–1936)[52] first learned of the ancestral law shortly after the release of Galton's *Natural Inheritance* (Galton, 1889), during a talk based on that book to the Men and Women's Club. Consistent with his positivist outlook, Pearson regarded the ancestral law as a purely statistical hypothesis whose physiological underpinnings were irrelevant:[53]

> **QUOT. 330** It [The ancestral law] is not a biological hypothesis, but the mathematical expression of statistical variates, which obey, as many measurable characters in man, certain forms of frequency distribution, these being maintained in successive generations. (Pearson, 1930, p. 21, Vol. IIIA)

In 1895, Pearson wrote a landmark paper titled "Mathematical contributions to the theory of evolution.-III. Regression, heredity, and panmixia" (Pearson, Lee and Bramley-Moore, 1895), which was the third in a series of mathematical papers on Darwin's theory of evolution. Pearson's 1895 paper was the one where the well-known product moment formula for the sample correlation coefficient was first derived (*ibid*., p. 265), and although the derivation contained an error (Gorroochurn, 2016a, p. 223) the paper also contained the first explicit derivation of the multivariate normal distribution and the first explicit multivariate regression model.

In one section of his paper (Pearson, Lee and Bramley-Moore, 1895, p. 302), Pearson derived the multivariate normal distribution explicitly for the first time, to represent the correlation surface of p-correlated organs in an organism:[54]

$$z = \frac{\mu n}{(2\pi)^{p/2}\sigma_1\sigma_2\sigma_3\ldots\sqrt{R}} \exp\left\{-\frac{1}{2R}\left(R_{11}\frac{x_1^2}{\sigma_1^2} + R_{22}\frac{x_2^2}{\sigma_2^2} + \cdots + 2R_{12}\frac{x_2 x_2}{\sigma_1\sigma_2} + \cdots\right)\right\} \tag{4.12}$$

---

[51] See Sect. 4.6.4.2.

[52] For details about Pearson's life, see Sect. 7.2.2.

[53] This was in contrast to Galton's view that the ancestral law had to be both statistically and physiologically consistent. See Sect. 4.6.4.

[54] For the equation below, the n in the numerator of z implies that the latter here integrates to n (and not to unity).

In the above, x_i ($i = 1, 2, \ldots, p$) is the deviation from the mean of the ith organ, n is the sample size ("the total number of sets of p organs"), μ is a "numerical factor denoting the number of $(p + 1)$th organs corresponding to each set...which is assumed to be practically constant...over the selected portion of it [i.e., of the correlation surface]," σ_i is the standard deviation of x_i, and R_{ij} is the (i,j)th minor of the correlation matrix $\{\rho_{ij}\}$ with determinant R. Pearson called Eq. (4.12) *Edgeworth's Theorem* on account of Edgeworth's prior investigation (Edgeworth, 1892) of the multivariate normal distribution.

By taking the distribution of $x_1, x_2, \ldots, x_p$ and dividing by that of $x_2, x_3, \ldots, x_p$, Pearson was able to obtain the conditional distribution of x_1 given $x_2, x_3, \ldots, x_p$. Since this conditional distribution is normal, the conditional mean and conditional variance of x_1 can be written down. The conditional mean of x_1 given $x_2, x_3, \ldots, x_p$ is the multiple regression of x_1 on $x_2, x_3, \ldots, x_p$.

Equipped with these mathematical tools, Pearson investigated Galton's Law of Ancestral Heredity. Following Galton, he assumed that the correlation of an offspring with a single parent was r, with a single grandparent was r^2, with a single great-grandparent was r^3, and so on. In general, the correlation between two individuals i generations apart was assumed to be r^i. Pearson then calculated the mean of the offspring conditional on *all* its ancestors as[55]

$$\zeta = r\frac{\sigma_1}{\sigma_2}h_2 + r\frac{\sigma_1}{\sigma_3}h_3 \tag{4.13}$$

(*ibid.*, p. 306), where h_2, h_3 are the deviations of each of the two parents, and σ_1, σ_2, σ_3 are the standard deviations of the offspring and each of the two parents, respectively. The above equation is one of the earliest multiple regression equations.[56]

What is surprising in Pearson's Eq. (4.13) is that, given the parents' deviations, the contributions of the grandparents and all earlier ancestors vanish! This is a direct contradiction of Galton's law of ancestral heredity. Pearson declared:

> **QUOT. 331** ...a knowledge of the ancestry beyond the parents in no way alters our judgment as to the size of organ or degree of characteristic probable in the offspring, nor its variability. *An exceptional father is as likely to have exceptional children if he comes of a mediocre stock as if he comes of an exceptional stock.*[57]
>
> ...
>
> This result seems to me somewhat surprising, but I cannot see how it is to be escaped... (*ibidem*)

Suddenly, it appeared as if Galton's ancestral law might be incorrect after all. However, within the next year, Galton came up with two *a priori* arguments and data on basset hounds, all of which supported his ancestral law.

Pearson would soon be forced to reassess his 1896 position.

---

[55] Assuming $\sigma_2 = \sigma_3$, the equation below can also be written as $\zeta = r_{\mathrm{mid}}(\sigma_1/\sigma_{\mathrm{mid}})h_{\mathrm{mid}}$, where r_{mid} is the offspring-mid-parent correlation, σ_1 is the offspring standard deviation, σ_{mid} is the mid-parent standard deviation, and h_{mid} is the mid-parent deviation.

[56] The first one occurred on p. 287 of Pearson's paper.

[57] Italics are Pearson's.

4.6.4 *Galton's Two* A Priori *Arguments for the Law of Ancestral Heredity in 1897: The Basset Hounds Data*

4.6.4.1 Two *A Priori* Arguments

Galton returned to the topic of the ancestral law in the paper "The average contribution of each several ancestors to the total heritage of the offspring," (Galton, 1897b) and provided two *a priori* (i.e., purely deductive) arguments in support of it. However, it will be shown that both arguments were as deficient as his 1885 demonstration (Sect. 4.6.2).

In his paper, Galton confidently announced that "there are strong grounds for believing it [the ancestral law] to be a general law of heredity" (*ibid.*, p. 401). Then, he gave his first *a priori* argument:

> **QUOT. 332** It should [be] noted that nothing in this statistical law contradicts the generally accepted view that the chief, if not the sole, line of descent runs from germ to germ and not from person to person. The person may be accepted on the whole as a fair representative of the germ, and, being so, the statistical laws which apply to the persons would apply to the germs also, though with less precision in individual cases... Now this law is strictly consonant with the observed binary subdivisions of the germ cells, and the concomitant extrusion and loss of one-half of the several contributions from each of the two parents to the germ-cell of the offspring. The apparent artificiality of the law ceases on these grounds to afford cause for doubt; its close agreement with physiological phenomena ought to give a prejudice in *favour*[58] of its truth rather than the contrary. (*ibid.*, p. 403)

In the above, Galton was referring to the recently observed process of the meiotic halving of germ cells. However, this is the law of halving which, as described in Sect. 4.6.1, is different from Galton's Law of Ancestral Heredity. Thus, Galton's first *a priori* argument, just like his demonstration of 1885 (Sect. 4.6.2), proved something other than his ancestral law.

QUOT. 332 also reveals Galton's insistence that his law was "strictly consonant with the observed binary divisions of the germs cell," i.e., he wanted his statistical law to be also *physiologically justified*. Immediately after giving his first *a priori* argument, Galton proceeded to give a second *a priori* argument:

> **QUOT. 333** ...it is reasonable to believe that the contributions of parents to children are in the same proportion as those of the grandparents to the parents, of the great-grandparents to the grandparents, and so on; in short, that their total amount is to be expressed by the sum of the terms in an infinite geometric series diminishing to zero. Lastly, it is an essential condition that their total amount should be equal to 1, in order to account for the whole of the heritage. All these conditions are fulfilled by the series of $1/2 + (1/2)^2 + (1/2)^3 +$ &c., and by no other. (*ibidem*)

Galton's argument above may be understood as follows: assume that an individual inherits a fraction f of its inherited characteristics from its parents, a fraction f^2 of inherited characteristics from its grandparents, a fraction f^3 of inherited

[58] Italics are Galton's.

characteristics from its great-grandparents, and so on. Then $f+f^2+f^3+\cdots=1$. But $f+f^2+f^3+\cdots=f/(1-f)$ for $|f|<1$. Solving $f/(1-f)=1$ yields $f=1/2$.

Galton's second argument is also weak because he does not justify why a geometric progression should hold. Although, as the law of halving predicts, while an individual *indirectly* receives its hereditary material from each ancestor according to a geometric series, it does not follow that the *separate* contributions also follow such a series.[59]

It may be conjectured that Galton's second *a priori* argument was the one he had used when he first announced the ancestral law in 1865 (see Sect. 4.6.1). Such a conjecture seems reasonable given Galton's keen interest in the geometrical series right from the start of his investigations.

4.6.4.2 The Basset Hounds Data

Notwithstanding the weaknesses in his *a priori* arguments, Galton put his ancestral law to the test by using a non-blending character, namely coat color in basset hounds. In this case, he used the ancestral law as follows:[60] the total probability that a particular offspring inherits an ancestral trait is equal to ½ of the fraction of its parents who have the trait, ¼ of the fraction of its grandparents who have the trait, and so on.

In Galton's example, two colors were present: tricolor (white, yellow, and black) and non-tricolor (white and yellow). Galton used Eq. (4.10) with all the a_i's now denoting proportions and showed that the numbers of tricolor offspring expected under his formula were in excellent agreement with the observed numbers. The results are shown in Table 4.1.

For example, consider the (1, 1)th data entry, $\left\{\begin{matrix}119\\0.91\end{matrix}\right\}$ in Table 4.1. There were 119 offspring for which both parents were tricolors and all four grandparents were tricolors. Therefore, $a_1=2/2=1$ and $a_2=4/4=1$ in Eq. (4.10), so that

$$a_0=\left(\frac{1}{2}\right)(1)+\left(\frac{1}{4}\right)(1)+4(.0408)=.91$$

is the total probability that an offspring is a tricolor. In the above, Galton added a correction factor (to account for earlier ancestors) of .0408 for each tricolor grandparent (similarly, a correction factor of .0243 was added for each non-tricolor grandparent) (*ibid.*, p. 406). Using the value of $a_0=.91$ and the total number of

[59] For more on *indirect* and *separate* ancestral contributions, see the paragraph after QUOT. 327.

[60] See last paragraph of Sect. 4.6.1, especially footnote 46 of this chapter.

Table 4.1 Galton's basset hounds example (Galton, 1897b, p. 411, Table V)

		Number of tricolours in grand-parents				Total tricolor offspring	
		4	3	2	1	Calculated	Observed
		a	*bcei*	*dfgjkm*	hlno		
2	All cases....	111	119	28	11		
	Coefficient....	0.91	0.83	0.76	0.68		
	Tricolor calc'd	*108*	*99*	*21*	*8*	*236*	
	" observed	106	101	24	8		
1	All cases.....	37	150	60	6		
	Coefficient....	0.66	0.58	0.51	0.43		
	Tricolor calc'd	*24*	*92*	*30*	*3*	*149*	
	" observed	20	79	36	4		139
0	All cases.....	..	..	18	6		
	Coefficient....	..	..	0.26	0.18		
	Tricolor calc'd	..	..	*5*	*1*	*6*	
	" observed	..	..	7	2		9
					Grand totals	*391*	387

offspring of 119, the expected number of tricolor offspring is $119 \times .91 \approx 108$, which agrees well with the observed number 106 (the (2, 1)th data entry, $\left\{\begin{matrix}108\\106\end{matrix}\right\}$).

Similarly, for the (3, 2)th data entry $\left\{\begin{matrix}158\\0.58\end{matrix}\right\}$, there were 158 offspring for which one of the two parents was a tricolor and three of the four grandparents were tricolors. Therefore, $a_1 = 1/2$ and $a_2 = 3/4$ in Eq. (4.10), so that

$$a_0 = \left(\frac{1}{2}\right)\left(\frac{1}{2}\right) + \left(\frac{1}{4}\right)\left(\frac{3}{4}\right) + 3(.0408) + 1(.0243) = .58$$

is the total probability that an offspring is a tricolor. Using the value of $a_0 = .58$. and the total number of offspring of 158, the expected number of tricolor offspring is $158 \times .58 \approx 92$, which is not too far off the observed number 79 (the (4, 2)th data entry, $\left\{\begin{matrix}92\\79\end{matrix}\right\}$).

Table 4.1 shows the excellent agreement between the observed and expected numbers of tricolor offspring. Many of Galton's contemporaries inferred from the basset hounds experiment that his ancestral law must be correct. This was even though Galton had used his ancestral law in the form (4.10), which is different from the form (4.11) in which he had stated it. Pearson himself seemed to have been more swayed by Galton's *a priori* arguments and reconsidered the conclusions of his 1896 paper (Pearson et al., 1895) (Sect. 4.6.3)

4.6.5 Pearson's Generalized Law of Ancestral Heredity in 1898: Refutation of Galton's Argument Against Continuous Variation

4.6.5.1 Pearson's Generalized Law of Ancestral Heredity in 1898

The publication of Galton's 1897 paper (Galton, 1897b) forced Pearson to reconsider his 1896 assertion that "knowledge of the ancestry beyond the parents in no way alters our judgment as to the size of organ or degree of characteristic probable in the offspring" (Pearson et al., 1895, p. 306).[61] In 1898, Pearson added a new paper (Pearson, 1898) to the "mathematical contributions to the theory of evolution" series. The paper was dedicated as "a New Year's greeting to Francis Galton."

However, while it would have been expected that Pearson had been swayed by how well the observed data on basset hounds fitted the ancestral law in Galton's paper, it was the latter's *a priori* argument that had persuaded him:

> **QUOT. 334** The confidence I put in the truth of the law is not measured by Mr. Gallon's researches on stature or on colour in Basset hounds, however strong evidence these may provide, but rather on the fact – that the theory gives a priori the correlation between parents and offspring, and that this correlation is practically identical with the value I have myself determined from these and other observations. (*ibid.*, p. 387)

If any inference can be made from the above, it is again that Pearson regarded the ancestral law as a purely statistical, predictive relationship whose biological basis was immaterial.

The 1898 paper also introduced the term "Galton's Law of Ancestral Heredity". It enunciated a generalized form of the law as follows:

$$k_0 = \frac{1}{2}\frac{\sigma_0}{\sigma_1}k_1 + \frac{1}{4}\frac{\sigma_0}{\sigma_2}k_2 + \frac{1}{8}\frac{\sigma_0}{\sigma_3}k_3 + \frac{1}{16}\frac{\sigma_0}{\sigma_4}k_4 + \cdots \tag{4.14}$$

(*ibidem*), where k_s is the deviation (from the mean) of the sth mid-parent in the sth ancestral generation, and k_0 is the deviation of the offspring, σ_s is the standard deviation of the sth mid-parental generation, and σ_0 is the standard deviation of the offspring.

[61] See Sect. 4.6.3.

It now remained for Pearson to reconcile the ancestral law with his 1896 result in Eq. (4.13). How was it that Galton assumed a geometric series for correlations between offspring and mid-parents, and obtained Eq. (4.10), while Pearson apparently made the same assumption and obtained Eq. (4.13) instead? The resolution of the paradox lies in the fact that, as Pearson correctly realized in 1898, Galton had phrased his correlations as *bivariate (or total) correlations* but Eq. (4.10) shows that these correlations had been entered as *partial correlations*. On the other hand, Pearson's assumed correlations were entered as *bivariate* correlations.

Thereupon, Pearson set out to determine what bivariate correlations between offspring and mid-ancestors would lead to Galton's ancestral law in Eq. (4.10). He assumed a correlation of r_i between an offspring and an ancestor i generations ago. This is in contrast to the pure geometric series he had used in 1896 (see Eq. (4.13)). Pearson now found that the modified geometric series

$$r_i = .6\left(\frac{1}{2}\right)^i \quad \text{or} \quad .3, .15, .075, \ldots \tag{4.15}$$

(*ibid.*, p. 395) leads exactly to Galton's Eq. (4.10). That is to say, the partial correlations that are implied in Galton's Eq. (4.10) are equivalent to the bivariate (total) correlations in Eq. (4.15).

Equation (4.15) thus gives the correlations between offspring and parent, grandparent, great-grandparent, and so on, as opposed to Galton's corresponding correlations of[62]

$$1/3, (1/3)^2, (1/3)^3, \ldots \quad \text{or} \quad .33, .111, .0370, \ldots \tag{4.16}$$

Pearson also noted that Eq. (4.15) implied that the slope of Galton's regression of offspring on mid-parent should have been $2r_1 = .6$, Galton's first estimate,[63] instead of 2/3. Indeed, the slope of the (simple) regression of offspring on *any* mid-ancestor is also .6. Pearson also observed that neither Galton's correlations of $(1/3)^i$, nor its modification to $.6(1/2)^i$ adequately fit the data: his own calculations on Galton's data showed a greater parental correlation (~.4).

Pearson finally turned to the issue of the uniqueness of the partial regression coefficients (½, ¼, ⅛, . . .) in Eq. (4.10). Galton's second *a priori* reasoning of 1887 (Sect. 4.6.4) had argued that assuming that an offspring obtained a fraction f^i $(0 < f < 1, i = 1, 2, \ldots)$ of its inherited characteristics from its ancestors i generations ago, then $f + f^2 + f^3 + \ldots = 1$ so that $f = 1/2$. Pearson (*ibid.*, p. 403) now argued that Galton's assumed geometric series $\sum_i f^i$ was a special case of a more general series

[62] See Sect. 4.6.2.

[63] See footnote 35 of this chapter.

$\sum_i \gamma \widetilde{\beta}^i$, where $\gamma = \left(1 - \widetilde{\beta}\right)/\widetilde{\beta}$[64] (since the series must add to unity) is a measure of the strength of inheritance. Thus, Galton's $\frac{1}{2} + \frac{1}{2^2} + \frac{1}{2^3} + \cdots$ could, in fact, be generalized to $\left(1 - \widetilde{\beta}\right) + \left(1 - \widetilde{\beta}\right)\widetilde{\beta} + \left(1 - \widetilde{\beta}\right)\widetilde{\beta}^2 + \cdots$, and his version of the law of ancestral heredity in Eq. (4.10) could be generalized to

$$a_0 = \left(1 - \widetilde{\beta}\right)a_1 + \left(1 - \widetilde{\beta}\right)\widetilde{\beta}a_2 + \left(1 - \widetilde{\beta}\right)\widetilde{\beta}^2 a_3 + \cdots \tag{4.17}$$

For example, suppose $\gamma = .9$ (i.e., the "succession tax" is .1). Then $\widetilde{\beta} = 1/(1 + \gamma) = .5263$ and the equation above becomes:

$$a_0 = .4737a_1 + .2493a_2 + .1312a_3 + \cdots$$

Regarding his generalization of the ancestral law, Pearson claimed that:

QUOT. 335 It indeed enables us to predict a priori the values of all the correlation coefficients of heredity, and forms, I venture to think, the fundamental principle of heredity from which all the numerical data of inheritance can in future be deduced, at any rate, to a first approximation. (*ibid.*, 386)

At the end of his paper, Pearson even compared the ancestral law to Newton's law of gravitation:

QUOT. 336 If Darwinian evolution be natural selection combined with heredity, then the single statement which embraces the whole field of heredity [i.e., the Ancestral Law] must prove almost as epoch-making to the biologist as the law of gravitation to the astronomer. (*ibid.*, p. 412)

Pearson's claim must have seemed over-the-top even to his contemporaries, but Galton was elated by the paper. In a Jan 4, 1898, letter to Pearson, he stated:

QUOT. 337 You have indeed sent me a most cherished New Year greeting. It delights me beyond measure to find that you are harmonizing what seemed disjointed, and cutting out and replacing the rotten planks of my propositions. We shall make something out of heredity at last. (Pearson, 1930, p. 504, Vol. IIIB)

However, despite his excitement, it could hardly have passed unnoticed to Galton that, in QUOT. 336, Pearson considered Galton's ancestral law to be completely consistent with Darwin's natural selection.[65] This was despite Galton's and Darwin's views on evolution being diametrically opposite (discontinuous vs. continuous). But then, Pearson also believed that Galton's main argument for discontinuous evolution was flawed, as will be described in the next section.

---

[64] Pearson's paper (Pearson, 1898) erroneously gives the normalizing condition as $\gamma\widetilde{\beta} = 1/2$. Note also that setting $\gamma = 1$ leads to $\widetilde{\beta} = 1/2$ and Galton's version of the ancestral law.

[65] In fact, Pearson and the biometricians later (Sect. 7.1.2) used Galton's ancestral law to validate Darwinian selection in their fight against the Mendelians.

4.6.5.2 Refutation of Galton's Argument Against Continuous Variation in 1898

It will be recalled from Sect. 4.5.7 that, in 1897, Galton (1897a) used the idea of perpetual regression to argue against the ability of natural selection to act on continuous variation to produce substantive permanent change. Using stature as an example, he had shown that, if parents who were in the 90th percentile of a normal distribution (with a mean of 68.5 inches and a quartile of 1.6 inches) were selected, then the deviation in subsequent generations exceeded no more than 5.7 inches under selection. However, as soon as selection was relaxed, Galton argued that regression would cause the gain of 5.7 inches to eventually vanish. In his 1898 paper, Pearson produced a relatively short, yet powerful, proof that refuted Galton's claim (Pearson, 1898).

Pearson considered a current population such that the deviation of the mid-parents in the jth generation afterward is k_j ($j = 1, 2, \ldots$). He now assumed that mid-parents who had a deviation of K in each of the next n generations were selected to produce offspring. Then, using the version of the ancestral law in Eq. (4.10), the deviation of the offspring in the $(n + 1)$th generation is

$$\begin{aligned} k_{n+1} &= \frac{1}{2}k_n + \frac{1}{2^2}k_{n-1} + \cdots + \frac{1}{2^n}k_1 \\ &= \left(\frac{1}{2} + \frac{1}{2^2} + \cdots + \frac{1}{2^n}\right)K \\ &= \frac{\frac{1}{2}\left\{1 - \left(\frac{1}{2}\right)^n\right\}}{1 - \frac{1}{2}}K \\ &= \left(1 - \frac{1}{2^n}\right)K \end{aligned}$$

(*ibid.*, p. 400). Pearson next assumed that the offspring in the $(n + 1)$th generation were now allowed to mate randomly (i.e., the selection was relaxed). Then, the deviation of the offspring in the $(n + 2)$th generation was

$$\begin{aligned} k_{n+2} &= \frac{1}{2}k_{n+1} + \frac{1}{2^2}k_n + \cdots + \frac{1}{2^{n+1}}k_1 \\ &= \frac{1}{2}\left(1 - \frac{1}{2^n}\right)K + \left(\frac{1}{2^2} + \cdots + \frac{1}{2^{n+1}}\right)K \\ &= \frac{1}{2}\left(1 - \frac{1}{2^n}\right)K + \frac{1}{2}\left(\frac{1}{2} + \cdots + \frac{1}{2^n}\right)K \\ &= \frac{1}{2}\left(1 - \frac{1}{2^n}\right)K + \frac{1}{2}\left(1 - \frac{1}{2^n}\right)K \\ &= \left(1 - \frac{1}{2^n}\right)K \end{aligned}$$

(*ibid*., p. 401). Similarly, the deviation of the offspring in the $(n + 3)$th generation was

$$
\begin{aligned}
k_{n+3} &= \frac{1}{2}k_{n+2} + \frac{1}{2^2}k_{n+1} + \cdots + \frac{1}{2^{n+2}}k_1 \\
&= \frac{1}{2}\left(1 - \frac{1}{2^n}\right)K + \frac{1}{2^2}\left(1 - \frac{1}{2^n}\right)K + \left(\frac{1}{2^3} + \frac{1}{2^4} + \cdots + \frac{1}{2^{n+2}}\right)K \\
&= \frac{3}{4}\left(1 - \frac{1}{2^n}\right)K + \frac{1}{2^2}\left(\frac{1}{2} + \frac{1}{2^2} + \cdots + \frac{1}{2^n}\right)K \\
&= \frac{3}{4}\left(1 - \frac{1}{2^n}\right)K + \frac{1}{4}\left(1 - \frac{1}{2^n}\right)K \\
&= \left(1 - \frac{1}{2^n}\right)K
\end{aligned}
$$

(*ibidem*). Thus, once selection was relaxed in the $(n + 1)$th generation and random mating was allowed, the deviations in the $(n + 1)$th, $(n + 2)$th, ... generations remained constant at $(1 - 2^{-n})K$. For example, if parents with a deviation of K had been selected for 10 generations, after which selection was relaxed, the offspring in all subsequent generations would carry 99.9% of the deviation K. This repudiated Galton's claim that any deviation would vanish geometrically once selection was relaxed.

Pearson had, in fact, shown that for each generation, Galton had considered the effect of only the previous generation and had thus used simple linear regression. By doing so, Galton had also ignored the contributions of previous generations and thus failed to use his ancestral law.

Chapter 5
The Theories of Heredity of Nägeli, Weismann, and de Vries

Abstract Unaware of Mendel's 1866 paper, Nägeli, Weismann, and de Vries each proposed elaborate theories of heredity in the latter parts of the nineteenth century. Nägeli proposed a theory based on micellae, which were crystalline groups of invisible molecules. Regularly aggregated micellae formed the idioplasm. During fertilization, maternal and paternal idioplasms fused and were transmitted to offspring. Weismann proposed a clear distinction between somatic and reproductive cells, with only the latter containing the hereditary material. He denoted the reproductive substance in the germ cells by the term "germ plasm" and proposed that the germ plasm was continuous and could affect the soma plasm, but not vice versa. De Vries' theory was based on Darwin's gemmules (which he called pangens) but excluded any transport of these gemmules from the body to reproductive cells. The pangens resided inside the nucleus of every cell.

Keywords Micellae · Idioplasm · Biophor · Germ plasm · Soma plasm · Weismann barrier · Pangens · Transportation hypothesis

5.1 Carl Nägeli: Micellae—Idioplasm[1]

5.1.1 Introduction

Carl Nägeli (Fig. 5.1) was a Swiss-German botanist who studied under the French-Swiss botanist Alphonse de Candolle (1806–1893). Nägeli was a first-class scientist in his own right, but his fame (notoriety?) also lies in the somewhat uneventful correspondence he had with Gregor Mendel during the period 1866–1874.[2]

After earning his doctorate in 1840, Nägeli studied Hegel's philosophy and was influenced by the latter's search for universal laws. This viewpoint was also shared

[1]Nägeli's theories of heredity and evolution are also discussed in Wilkie (1960a, b, c, 1961), Hall (1969, pp. 334–338), Churchill (2015, Chap. 10), Robinson (1979, Chap. 6), and Tort (1996, pp. 3147–3150, Vol. 2: F:N).

[2]See Sect. 6.3.

P. Gorroochurn, *The Development of Evolutionary Genetics*,
https://doi.org/10.1007/978-3-031-69374-8_5

Fig. 5.1 Carl Wilhelm von Nägeli (1817–1891). (From: https://commons.wikimedia.org/wiki/File:Carl_Wilhelm_N%C3%A4geli_(1817-1891).jpg)

by Franz Unger (1800–1870) and especially Mathias Schleiden[3] (1804–1881), under whom Nägeli worked in 1842. In that same year, based on his cytological research (Nägeli, 1842), Nägeli admitted the possibility of Schleiden and Schwann's (erroneous) theory of free-cell formation and showed the greater generality of cell formation by the division of preexisting cells.

5.1.2 Micellae Theory (1858, 1863, 1877)

In 1858, Nägeli (1858) introduced the first elements of a speculative theory, later to be called the *micellae*[4] *theory*, which was meant to account for the swelling in water and the birefringence of starch. The micellae theory posited that the cell walls and starch grains were made of crystalline groups of invisible molecules called *micellae*, which grew through intussusception. In 1863, Nägeli provided the following

[3] See Sect. 6.1.2.

[4] A term first used by Nägeli in 1877 (Nägeli and Schwendener, 1877).

summary of his theory, while drawing attention to a similar conclusion he had already reached in 1858:

> **QUOT. 338** Organized materials [Substanzen] consist of crystalline, birefringent molecules (which themselves consist of atoms) which lie freely, but with a definite regular arrangement, alongside one another. As a result of a preponderating attraction, each is surrounded, in the moistened state, by an envelope of water; in the dry state, they touch one another. Hence there is a duplex cohesion in organized material: one unites the atoms to form crystals; the other unites the molecules.[5] (Nägeli, 1863; Wilkie, 1960a, p. 29)

5.1.3 *Idioplasm Theory of Heredity (1879, 1884)*

In 1879, Nägeli further developed the micellae theory (Nägeli, 1879; Wilkie, 1960a, pp. 39–40) by assuming that the micellae in solution tended to aggregate either regularly (in parallel orientations) or irregularly. In the latter state, they served as a nutritive plasm for the regularly aggregated micellae, which were more tightly packed and had a definite structure. These regularly aggregated micellae were called *idioplasm* in Nägeli's major work on heredity and evolution, entitled *Mechanische-physiologische Theorie der Abstammungslehre*[6] (Nägeli, 1884). In the 1898 English summary of this work, it is explained that the idioplasm:

> **QUOT. 339** . . .has the tendency by nature to assume a reticular arrangement. The strands of this network consist, in conformity with their origin, of parallel rows of micellae extending lengthwise. These rows of micellae are combined into more or less complex arrangements, so that the cross section of the strand represents the configuration of the idioplasm.
>
> ... In the cell divisions, by which the organism grows, the idioplasm divides into as many parts as there are single cells, while it continually increases in quantity in a corresponding degree. The ontogenetic increase of the idioplasm takes place by length growth of the strands – that is, by intercalation of micellae in each row of cells of the strands, which thereby grow in length without changing the configuration of the cross section. (Nägeli, 1898, pp. 11–12)

Each cluster of micellae of the idioplasm was the determinant or *anlage* of a particular characteristic:

> **QUOT. 340** Since a particular cluster or group of micellae of the idioplasm produces a particular phenomenon in the organism, the former is designated as the determinant (*Anlage*, see p. 49) of the latter. Thus the organism must contain at least as many determinants in its idioplasm as there are different phenomena in its inheritable ontogeny. (*ibid.*, p. 10)

On the other hand, the sets of micellae with no regular structure in the idioplasm had a deactivated *anlage*.

---

[5]Italics are Nägeli's.

[6]A 53-page English summary by M.A. Clark and F.A. Waugh of Nägeli's book was published in 1898 under the title *Mechanico-Physiological Theory of Organic Evolution* (Nägeli, 1898).

In Nägeli's theory, the maternal and paternal idioplasms in sex cells fused[7] during fertilization and were thus transmitted to the offspring. During the fusion, slight changes in the regular structure of the micellae occurred. However, it was only when such changes had accumulated that the characteristics being transmitted could be altered or new characteristics could arise.[8] Since a previously deactivated *anlage* could later be reactivated, Nägeli's theory implied that offspring in later generations should show reversion to ancestral characters. The theory was thus incompatible with the possibility of constant (both dominating and recessive) traits in F_2 of Mendel's *Pisum* experiments (Sect. 6.2.5).

5.1.4 Nägeli's Theory of Evolution

As early as 1856 (i.e., before Darwin's *Origin of Species*), Nägeli had expressed belief in *directed evolution*[9] (Nägeli, 1856). In 1865, he stated his views on evolution more fully and doubted that natural selection played a central role in the process of evolution (Nägeli, 1865). In his 1884 work, Nägeli (1884, 1898) explained his orthogenetic views through the "automatic progressive or perfecting transformation of the idioplasm" (*ibid.*, p. 16). That is, the effect of the nutritive protoplasm on the idioplasm meant that different groups of micellae were activated in a specific order, thus leading to directed evolution. Moreover, after each generation, there was an increase in complexity:

> **QUOT. 341** Then, as the idioplasmic body becomes continually more complex in the further course of phylogeny, and consists of a greater number of groups of micellea differing from each other, the organisms become more complex and differentiate into a greater number of parts, because each group of micellae of the idioplasm produces its specific effect with regard to inner structure, outer form, and function... (Nägeli, 1898, p. 10)

What about the role of the environment and adaptation? Nägeli allowed the Lamarckian influence of the environment on the idioplasm and the consequent shaping of heritable characters:

> **QUOT. 342** ...And since idioplasmic bodies are formed under the influence of external stimuli, their plastic products always appear with a definite character of adaptation to environment. (*ibid.*, pp. 9–10)

However, for visible changes to appear within an organism, the environment must act in the same direction for an extended period.

---

[7]Nägeli's theory of heredity was thus a *blending* theory.

[8]Nägeli's theory of evolution was thus *discontinuous*.

[9]This later came to be known as *orthogenesis*, which is the intrinsic tendency of an organism to develop from one species to another without the necessity of adaptive constraints.

It is thus seen that Nägeli's theory gives prime importance to the micellae of the idioplasm as the driving force of evolution, and not to natural selection:

> **QUOT. 343**According to Nägeli, the organic world would have become much what it is, if natural selection and adaptation had performed no part in the operations of nature. He aptly says, that natural selection prunes the phylogenetic tree, but does not cause new branches to grow. He allows that the principle of selection is well suited to explain the adaptation of organisms to their environment and the suitableness and physiological peculiarities of their structure, but he asserts that in the definiteness of variation of plants and in their progressive differentiation there is evidence of a higher and controlling perfecting principle. (*ibid.*, p. 18, translator's footnote)

5.2 August Weismann: Biophors—Germ Plasm[10]

5.2.1 Introduction

Born in Frankfurt, Germany, August Weismann (Fig. 5.2) was a Professor of Zoology and is regarded as one of the most important evolutionary biologists of the nineteenth century. The latter part of that century witnessed considerable progress in cytology. Based on these advances, Weismann was able to propose a highly influential theoretical model[11] of heredity, many elements of which (but not all) have stood the test of time. Weismann also adopted Darwin's theory of natural selection. However, in light of his theory of heredity, Weismann's version of Darwinism became more extreme than Darwin's own. In particular, while Darwin allowed the environment to produce heritable characters (most notably through his provisional hypothesis of pangenesis[12]), Weismann admitted no such Lamarckian effects. Weismann's extreme version of the Darwinian theory is often described as "Neo-Darwinian",[13] a term that is usually ascribed to the Darwin stalwart George

[10]Weismann's theories of heredity and evolution are also discussed in Churchill (1968, 2015), Winther (2001), Robinson (1979, Chap. 7), Fothergill (1952, pp. 152–160), Mayr (1985), Hall (1969, pp. 340–345, Vol. II), Stubbe (1972, pp. 256–259), and Tort (1996, pp. 4613–4617, Vol. 3: O–Z).

[11]Weismann turned to theoretical evolutionary biology because of his poor eyesight (which started around 1864) and the resulting difficulty in using the microscope. His vision improved between 1874 and 1884, only to worsen soon afterward, making him nearly blind.

[12]See Sect. 3.7.2.

[13]Neo-Darwinism would later (in the 1930s) also be used to describe the synthesis of Mendel's theory of heredity and Darwin's theory of evolution by natural selection.

Fig. 5.2 August Weismann (1834–1914). (From: https://commons.wikimedia.org/wiki/File:August_Weismann.jpg)

Romanes[14] but seems to have first been used by Samuel Butler in his *Unconscious Memory* (Butler, 1880).[15] These issues will be described in more detail next.

[14]Romanes is often reported to have used "Neo-Darwinian" in his 1895 book (Romanes, 1895, p. 12, Vol. II), but, in fact, he had first used it in a short 1888 letter to *Nature* (Romanes, 1888, p. 413): "The school of Weismann may properly be called Neo-Darwinian: pure Darwinian it certainly is not." In the 1895 book, Romanes also used "ultra-Darwinian" to describe Weismann's (and Wallace's) version of the Darwinian theory.

[15]Samuel Butler (1835–1902) is usually remembered as a British novelist, but he was also an early adherent to Darwin's theory. However, when he read Mivart (Sect. 2.14.5), he turned against Darwin. In his famous book *Evolution, Old and New* (Butler, 1879), he accused Darwin of stealing the idea of evolution from Buffon and Lamarck and claimed Lamarck's theory was superior. In *Unconscious Memory*, while referring to Darwin's provisional hypothesis of pangenesis, Butler said: "I had not as yet seen that the principle I was contending for was Darwinian, not Neo-Darwinian" (Butler, 1880, p. 33). Butler's use of 'Neo-Darwinian' thus mirrors that of Romanes.

5.2.2 Acceptance of Lamarckism Before 1881

Together with his colleague Ernst Haeckel, Weisman read the first edition of Darwin's *Origin of Species* (Darwin, 1859) within a year of its appearance in German, and both became converts to Darwinism. They welcomed Darwin's mechanistic explanation of evolution and subsequently rejected any vitalistic, teleological, or other such causes.

Haeckel was a Lamarckian and his influence seems to have rubbed off on Weismann's early (pre-1882) views on evolution. Thus, in *Über den Saison-Dimorphismus der Schmetterlinge* (Weismann, 1875, 1882a), one can read:

> **QUOT. 344** By the application of even the greatest warmth nobody would be able to change the winter form of *A. Levana* into the summer form; nevertheless, the summer warmth, acting regularly on the second and third generations of the year, has, in the course of a lengthened period, stamped these two generations with a new form without the first generation being thereby changed. (*ibid.*, p. 87)

In the above, one sees Weismann's belief in the inheritance of acquired characters. In the preface of the English edition of *Studien zur Descendenztheorie* (in which the above-mentioned work also appears), Weismann was even more explicit:

> **QUOT. 345** Nor can the transforming influence of direct action, as upheld by Lamarck, be called in question, although its extent cannot as yet be estimated with any certainty. (Weismann, 1882a, p. xvii)

Owing to Weismann's later criticism of Lamarckism, his early adherence to this doctrine is relatively unknown and may be surprising to many.

5.2.3 Distinction Between Reproductive Cells and Somatic Cells in 1881

It was during the presentation of the paper "Uber die dauer des lebens" (Weismann, 1882b, 1889) at the Association of German Naturalists at Salzburg, on September 21, 1881, that Weismann first distinguished between somatic and reproductive cells.[16] While discussing the life span of various species, Weismann drew attention to the unlimited life span of unicellular organisms such as amoebae. To explain this phenomenon, he hypothesized that the two types of cells (somatic and reproductive) belonged to two very distinct groups:

> **QUOT. 346** As the complexity of the Metazoan body increased, the two groups of cells became more sharply separated from each other. Very soon the somatic cells surpassed the reproductive in number, and during this increase they became more and more broken up by the principle of the division of labour into sharply separated systems of tissues. As these changes took place, the power of reproducing large parts of the organism was lost, while the power of reproducing the whole individual became concentrated in the reproductive cells alone. (*ibid.*, p. 28)

[16] Weismann's famous "germ plasm" (*Keimplasma*) was not used yet.

Thus, according to Weismann, the reproductive cells contain the hereditary material but the somatic cells do not.[17] In addition:

> **QUOT. 347** . . .The reproductive cells cannot lose the capacity for unlimited reproduction, or the species to which they belong would suffer extinction. But the somatic cells have lost this power to a gradually increasing extent, so that at length they became restricted to a fixed, though perhaps very large number of cell-generations. (*ibidem*)

That is, the reproductive cells are immortalized through successive production across generations, while the somatic cells have a limited life and do not participate in heredity.

Weismann was thus able to explain the unlimited life of unicellular organisms and the limited life of multicellular organisms:

> **QUOT. 348** Normal death could not take place among unicellular organisms, because the individual and the reproductive cell are one and the same: on the other hand, normal death is possible, and as we see, has made its appearance, among multicellular organisms in which the somatic and reproductive cells are distinct. (*ibidem*)

Although this is not explicitly spelled out in the current paper, the logical conclusion from Weismann's separation of the somatic and reproductive tissues is the absence of any effect from the somatic to the reproductive cells, leading to the impossibility of the inheritance of acquired characters.

It should be noted that Weismann was not the first to propose the continuity of the germ line. As Weismann (1892, 1893, pp. 198–201) later acknowledged, the first to have proposed such an idea was Owen in 1849 (Owen, 1849). He was followed by Galton[18] (1872, 1875), Jäger (1877), Rauber (1880), and Nussbaum (1880).

5.2.4 *Explicit Rejection of Lamarckism in 1883: Introduction of the Concept of Germ Plasm (*Keimplasma*)*

Two years after the 1881 paper, Weismann delivered an inaugural public lecture, entitled "Uber die Vererbung" (Weismann, 1882c, 1883b) as Pro-Rector of the University of Freiberg. This 1883 lecture broke ground on two fronts:

1. *An explicit rejection of the inheritance of acquired characteristics.* Commenting on Darwin's use of the provisional hypothesis of pangenesis to explain the process of heredity, Weismann said:

   > **QUOT. 349** . . .an explanation [of heredity] can in this case be reached by an appeal to known forces, if we suppose that characters acquired (in the true sense of the term) by the parent cannot appear in the course of the development of the offspring, but that all the characters exhibited by the latter are due to primary changes in the germ. (*ibid.*, p. 78)

---

[17] This is now known to be untrue.

[18] See QUOT. 280 and Sect. 4.5.1.

Weismann's 1883 lecture was the first of a continued rejection of any Lamarckism in Darwin's theory of evolution.[19] It thus marked the birth of Neo-Darwinism (in Weismann's sense).

2. *The introduction of the concept of the germ plasm as the reproductive substance in the germ cells.* In the 1881 paper, Weismann had previously focused on the reproductive or germ cells and contrasted them with the somatic cells. Weismann now introduced the *Keimplasma* (germ plasm) as the reproductive substance (*ibid.*, p. 75). He hypothesized that the reproductive cells contained the germ plasm, i.e., the germ plasm was the reproductive substance of the germ cells:

   QUOT. 350 We have an obvious means by which the inheritance of all transmitted peculiarities takes place, *in the continuity of the substance of the germ-cells, or germ-plasm.* If, as I believe, the substance of the germ-cells, the germ-plasm, has remained in perpetual continuity from the first origin of life, and if the germ-plasm and the substance of the body, the somatoplasm, have always occupied different spheres, and if changes in the latter only arise when they have been preceded by corresponding changes in the former, then we can, up to a certain point, understand the principle of heredity. (*ibid.*, p. 104)

Thus, in the above, Weismann also indicated that the relationship between the soma plasm and germ plasm was unidirectional, with the germ plasm being able to influence the soma plasm but not vice versa. This unidirectional relationship is now known as the *Weismann barrier*. The inadmissibility of Lamarckism follows logically from it.

Later in the paper, Weismann gave the following precision about his concept of the germ plasm:

QUOT. 351 ...the germ-plasm was that part of a germ-cell of which the chemical and physical properties—including the molecular structure—enable the cell to become, under appropriate conditions, a new individual of the same species. (*ibid.*, p. 174)

In the same paragraph, he likened his theory of the germ plasm to Nägeli's idioplasm.

Weismann's 1883 lecture additionally introduced the important concept of "panmixia," a term he used to denote the deteriorating effects on a population if natural selection was paused:[20]

QUOT. 352 This suspension of the preserving influence of natural selection may be termed 'Panmixia' for all individuals can reproduce themselves and thus stamp their characters upon the species, and not only those which are in all respects, or in respect to some single organ, the fittest. (*ibid.*, p. 90)

In 1883 still, Weismann published a two-volume work (Weismann, 1883a) summarizing his investigations (which started in 1881) in hydroids. He later claimed that

---

[19] As stated in the previous section, the impossibility of Lamarckian influences was already implied in 1881 (Weismann, 1882b, 1889).

[20] Panmixia is now used more generally to denote random mating.

this work was the basis of his theory of the continuity of the germline.[21] However, after closely examining the 1883 work, Berrill and Liu (1948) have poured cold water on this claim, stating that:

QUOT. 353 ...ideas which Weismann arrived at intuitively or by induction from various sources, blinded him in his studies of hydroids and caused him to see imaginary migrations of visible and invisible germ cells, and that whatever the intrinsic merit of his ideas, they are not based upon the study to which they are credited. (*ibid.*, p. 131)

5.2.5 Formal Exposition of Germ Plasm Theory in 1885: Continuity of the Germ Plasm, Acknowledgment of the Cell Nucleus as a Repository of Hereditary Material

Weismann's theory of the continuity of the germ plasm as the basis of a theory of heredity was put forward more fully in the 1885 paper "Die kontinuitat des keimplasmas" (Weismann, 1885a, b):

QUOT. 354 ...the theory of 'The Continuity of the Germ-plasm,' ... is founded upon the idea that heredity is brought about by the transference from one generation to another, of a substance with a definite chemical, and above all, molecular constitution. I have called this substance 'germ-plasm,' and have assumed that it possesses a highly complex structure, conferring upon it the power of developing into a complex organism. (*ibid.*, p. 168)

Weismann also believed that, in each generation, only a specific portion of the germ-cells in an organism was used in the formation of body cells of its offspring, the rest being conserved for building the germ-cells in the following generation:

QUOT. 355 ...I have attempted to explain heredity by supposing that in each ontogeny, a part of the specific germ-plasm contained in the parent egg-cell is not used up in the construction of the body of the offspring, but is reserved unchanged for the formation of the germ-cells of the following generation. (*ibidem*)

Weismann's 1885 paper was written in the wake of major discoveries in cytology and in anticipation of still greater breakthroughs. As far back as 1866, Ernst Haeckel (1834–1919) had proposed (without experimental evidence) that the nucleus was the reproductive component of the cell (Haeckel, 1866, pp. 287–288). The experimental evidence came later in 1875 when Oscar Hertwig[22] (1875) and Edouard van Beneden (1875) independently showed that fertilization involved the fusion of not

---

[21] For example, in his 1885 "Die kontinuitat des keimplasmas," Weismann stated: "As a result of my investigations on Hydroids [*Die Entstehung der Sexualzellen bei den Hydromedusen*], I concluded that the germ plasm is present in a very finely divided and therefore invisible state in certain somatic cells, from the very beginning of embryonic development, and that it is then transmitted through innumerable cell generations, to those remote individuals of the colony in which sexual products are formed" (Weismann, 1885a, b, p. 207).

[22] Hertwig had studied under Haeckel at Jena.

only cells but also nuclei. Although Eduard Strasburger (1844–1912) had incorrectly denied the continuity of the nucleus in 1875 (Strasburger, 1875, pp. 308–309), he had abandoned this position in 1880 (Strasburger, 1880, p. 321). In 1884, he provided a definite confirmation that the nucleus alone was involved in heredity when he found that it was the pollen tube nucleus that entered the egg and fertilized the egg nucleus and that the nuclear segments remained preserved in the resulting nucleus (Strasburger, 1884, p. 86).

Drawing upon his contemporaries, Weismann concluded that:

> **QUOT. 356** . . .The nature of heredity is based upon the transmission of nuclear substance with a specific molecular constitution. This substance is the specific nucleoplasm of the germcell, to which I have given the name of germ-plasm. (Weismann, 1885a, b, p. 180)

5.2.6 Disagreement with Nägeli's Idioplasm Theory in 1886: All-Sufficiency of Natural Selection and Origin of Variations

As indicated in Sects. 5.1.3 and 5.1.4, Nägeli's major work, *Mechanische-physiologische Theorie der Abstammungslehre* (Nägeli, 1884), came out in 1884. In it, Nägeli had not only diminished the role of natural selection in evolution but also explicitly adopted Lamarckism. In the 1886 paper[23] "Die Bedeutung der Sexuellen Fortpfl anzung für die Selections-Theorie" (Weismann, 1886), Weismann strongly opposed both of these viewpoints. While expressing his admiration for the scientific accomplishments of the older botanist, Weismann questioned the very basis of Nägeli's demotion of natural selection, namely, the latter's theory of the idioplasm. Weismann's greatest objection to that theory was that it did not explain the process of adaptation:

> **QUOT. 357** . . .That species are, from time to time, transformed into new ones might perhaps be understood by means of an internal transforming force, but that they are so changed as to become better adapted to the new conditions under which they have to live, is left entirely unintelligible by this theory. For we certainly cannot accept as an explanation Nägeli's statement that organisms possess the power of being transformed in an adaptive manner simply by the action of an external stimulus. (*ibid.*, pp. 257–258)

Furthermore, Weismann contended that Nägeli's theory was entirely hypothetical: there was no proof of either the existence of the idioplasm or its supposed mode of action.

Regarding Nägeli's adoption of Lamarckism, Weismann repeated arguments that he had already made before, namely, that the separation of the germ plasm and the soma plasm prevented acquired characteristics from being transmitted. However, the rejection of Lamarckism led to a problem that Weismann quickly saw, since it concerned the validity of his theory of the germ plasm:

[23] This paper was based on a shorter lecture he gave the previous year.

QUOT. 358 Individual variability forms the most important foundation of the theory of natural selection: without it the latter could not exist, for this alone can furnish the minute differences by the accumulation of which new forms are said to arise in the course of generations. But how can such hereditary individual characters exist if the changes wrought by the action of external influences, during the life of an individual, cannot be transmitted? (*ibid.*, pp. 267–268)

That is, if environmental effects did not produce heritable changes, what, then, was the nature of the individual differences on which natural selection could act? To answer this question, Weismann proposed that individual differences were created through new combinations of germ plasm during sexual reproduction:

QUOT. 359 . . .in amphigonic [sexual] reproduction two groups of hereditary tendencies are as it were combined. . . I regard this combination as the cause of hereditary individual characters, and I believe that the production of such characters is the true significance of amphigonic reproduction. (*ibid.*, p. 272).

QUOT. 360 . . .When we remember that, in the tenth generation, a single germ contains 1024 different germ-plasms, with their inherent hereditary tendencies, it is quite clear that continued sexual reproduction can never lead to the re-appearance of exactly the same combination, but that new ones must always arise. (*ibid.*, p. 276)

However, Weismann's individual differences were simply the different combinations of *existing* variations. What was the *origin* of such variations? Weismann pointed to two possible sources:

(1) *Variations originating from unicellular organisms.* Weismann pointed to unicellular organisms where the germ plasm and soma plasm are the same:

QUOT. 361 . . .The origin of hereditary individual variability cannot indeed be found in the higher organisms—the Metazoa and Metaphyta; but it is to be sought for in the lowest—the unicellular organisms. In these latter the distinction between body-cell and germ-cell does not exist. Such organisms are reproduced by division, and if therefore any one of them becomes changed in the course of its life by some external influence, and thus receives an individual character, the method of reproduction ensures that the acquired peculiarity will be transmitted to its descendants. (*ibid.*, pp. 277–278)

Since all life forms are ultimately derived from unicellular organisms, the origin of variations is thus explained.

(2) *Variations originating from changes in nutrition to germ cells.* Although Weismann dismissed the possibility that changes in nutrition in the germ cells could produce individual differences in 1886 (Weismann, 1886, p. 272), he later changed his opinion. In 1892, he said:

QUOT. 362 The assumption that. . .very minute fluctuations occur, naturally follows from the impossibility of a complete uniformity as regards nutrition existing during growth. (Weismann, 1885a, b, p. 417)

The occurrence of rare, random genetic mutations was as yet unknown to Weismann (and his contemporaries).

Weismann's argument for the all-sufficiency of natural selection was attacked by Herbert Spencer (1893), a Lamarckian, and a controversy ensued between the two.[24] It was during this controversy that Weismann proposed the concept of *germinal selection* (last paragraph of Sect. 5.2.8)

5.2.7 Reduction Division in 1887

Whereas the process of mitosis (or indirect cell division) had been described by Walther Flemming (1843–1905) in 1879,[25] the same could not be said of meiosis in 1887. In 1883, Wilhelm Roux (1850–1924) suggested that the mechanism of germ-cell formation fundamentally differed from that of indirect (i.e., mitotic) nuclear division (Roux, 1883). An understanding of the former process took place through the study of *polar bodies*.

As early as 1824, Carl Gustav Carus (1789–1869) had observed that little bodies were expelled by the maturing ovum of the water-snail *Linnaeus*. These bodies did not function as eggs and were called *polar bodies*[26] by Charles-Philippe Robin (1821–1885) in 1862. To investigate the significance of polar bodies in fertilization, van Beneden chose the egg of *Ascaris megalocephala*. The latter is presently known to have a small number of relatively large chromosomes (2 or 4), which makes the process of nuclear division easy to follow with a microscope. Following a suggestion by Charles Minot (1852–1914) in 1877, van Beneden postulated in 1883 that the cell was, in fact, hermaphroditic and that the maturation of the egg proceeded by throwing off the male element as a polar body (van Beneden, 1883). He also believed that the female half must also be thrown off from a maturing sperm. *Since both the egg and sperm each had half of the hereditary material, fertilization ensured that the full hereditary material was restored.* This was a groundbreaking discovery, although van Beneden got some of the details wrong. In particular, he believed that the throwing away of half of the genetic material by each nucleus meant that two *half-nuclei* participated in the process of fertilization.

In 1885, Weismann had reported that parthenogenetic eggs (which developed without fertilization) expelled one polar body (Weismann, 1885b, p. 249). This was in contrast to the eggs in sexually reproducing organisms where two polar bodies were expelled. It was with the intent of investigating the significance of the second polar body that Weismann wrote the paper of 1887 (Weismann, 1887), which Wilson later described as "one of those brilliant essays on heredity which contributed in so important a way to the enlargement of our views concerning cytological research" (Wilson, 1925, p. 500).

---

[24] See Chapter 17 of Churchill (2015).

[25] The name was first given by Flemming (1882, p. 376).

[26] Only oogenesis (but not spermatogenesis) is associated with polar bodies.

In his paper, although Weismann incorrectly associated the halving of genetic material to the second (meiotic) division (i.e., the second polar body), he provided the correct rationale for reduction division as follows:

QUOT. 363 If then the germ-cells of the offspring contain the united germ-plasms of both parents, it follows that such cells can only contain half as much paternal germ-plasm as was contained in the germ-cells of the father, and half as much maternal germ-plasm as was contained in the germ-cells of the mother. (Weismann, 1885b, p. 356)

Moreover:

QUOT. 364 In each succeeding generation the number of the ancestral germ-plasms is doubled, while their quantities are reduced by one half. (*ibidem*)

The above is simply a statement of Galton's law of ancestral heredity (Sect. 4.6). Weismann would soon use the term "ids" to denote the different ancestral germ plasms.

Even more remarkable was Weismann's discussion, on purely theoretical grounds, of the process of meiosis and how it contrasted with mitosis:

QUOT. 365 The vast number together with the minute quantity of the ancestral germ-plasms permit us to conclude that they are, upon the whole, arranged in a linear manner in the thin thread-like loops: in fact the longitudinal splitting of these loops appears to me to be almost a proof of the existence of such an arrangement, for without this supposition the process would cease to have any meaning.

This is the only kind of karyokinesis which has been observed until recently; but if the supposed nuclear division leading to a reduction in the number of ancestral germ-plasms has any real existence, there must be yet another kind of karyokinesis, in which the primary equatorial loops are not split longitudinally, but are separated without division into two groups, each of which forms one of the two daughter-nuclei. In such a case the required reduction in the number of ancestral germ-plasms would take place, for each daughter-nucleus would receive only half the number which was contained in the mother-nucleus. (*ibid.*, pp. 359–360)

The first paragraph in the citation above refers to mitosis, while the second refers to meiosis. Weismann's "loops" would soon be called "chromosomes" by Wilhelm Waldeyer (1888, p. 27). A more complete description of the two cell divisions in meiosis was later provided by Theodor Boveri (1887) and Oscar Hertwig (1890). In 1887, Flemming called the first cell division (which differed from the typical mitotic division) *heterotype mitosis*, and the second cell division (which was similar to the typical mitotic division) *homotype mitosis.* In 1905, Farmer and Moore used *maiosis* to describe the whole process (1905, p. 489). This word was soon replaced by *meiosis*.

5.2.8 Elaboration of Germ Plasm Theory in 1892 in Light of de Vries' Intracellular Pangenesis*: Biophors, Determinants, Ids, and Idants*

In 1892, Weismann made a major overhaul to his germ plasm theory in the book *Das Keimplasma. Eine Theorie der Vererbung* (Weismann, 1892). This was in the wake of de Vries' *Intracellulare Pangenesis* (de Vries, 1889), where a theory of heredity based on Darwin's provisional hypothesis of pangenesis[27] had been proposed. The details of this theory will be explained in Sect. 5.3 below, but in essence it admitted Darwin's free hereditary particles (which de Vries called *pangens*) but denied their transport to the sexual organs (i.e., denied Darwin's *transportation hypothesis*). One sees in the latter de Vries' influence (and acceptance) of Weismann's rejection of Lamarckian influences. However, what was of concern to Weismann regarding de Vries' theory was the latter's insistence that the pangens were free (ungrouped) particles each containing the whole hereditary material inside the nucleus. Such a characterization obviated the necessity of both Weismann's ancestral plasm and its reduction division before fertilization.

Weismann's 1892 monograph was written to consolidate his germ plasm theory while at the same time admitting the existence of de Vries' pangens. To achieve this, he had to build a hierarchical heredity model of high sophistication. At the lowest level of this model were the *biophors:*

> **QUOT. 366** ...protoplasm invariably consists of *groups*[28] of molecules, each of which is composed of *different kinds* of chemical molecules. I shall call these units [i.e. groups of molecules] the '*bearers of vitality'* ('Lebensträger') or '*biophors,*' because they are the smallest units which exhibit the primary vital forces, viz., *assimilation and metabolism, growth, and multiplication by fission.* (Weismann, 1892, 1893, pp. 39–40)

In accordance with his own (and others') research on the importance of the nucleus as the repository of the hereditary material, Weismann's invisible biophors resided inside the nucleus in a dormant state and were activated once "transported" to the cell body. This was consistent with de Vries' views of the pangens (*ibid*., pp. 46-47).

Weismann's biophors were similar to Spencer's *physiological units*, although they differed in their function. They were not the same as Nägeli's *micellae* both in terms of their properties and construction. Finally, they were exactly the same as de Vries' *pangens*, except that they were part of a higher structure in Weismann's model.

At the next level of Weismann's hierarchical model were the *determinants*, which were groups of biophors responsible for a given character expressed by the cell:

> **QUOT. 367** I shall designate the cells or groups of cells which are independently variable from the germ onwards as the '*hereditary parts'* or '*determinates*,' and the particles of the germ-plasm corresponding to and determining them, as the 'determining parts' or '*determinants.*' (*ibid*., p. 57)

For example, the millions of red blood cells in the human body were controlled in the germ plasm by a single determinant. The determinant was 'not a disconnected mass of different biophors, but *a vital unit of a higher order than the biophor, possessed of special characters*' (*ibid*., p. 60).

[27] See Sect. 3.7.2.

[28] All italics in the quotations of this Section are Weismann's.

At a still higher level in Weismann's hierarchical model were the *ids*,[29] which were groups of determinants:

QUOT. 368 ...we are led to the assumption of groups of determinants, each of which represents a separate vital unit of the third degree, since it is composed of determinants, which in their tum are made up of biophors. These are the units which I formulated on different lines long ago, and to which the name of *ancestral germ-plasms* was then given. I shall now speak of them as *'ids,'* a term which recalls the 'idioplasm' of Nägeli. (*ibid.*, p. 62)

Furthermore:

QUOT. 369 ...*each idioplasm is composed of several or many ids, which are capable of growth and multiplication by division.* (*ibid.*, p. 63)

That is, the germ plasm was composed of different ancestral plasms (ids).

Weismann claimed that the ids, unlike the biophors and determinants, were visible under the microscope. They were arranged along the chromosomes[30] (which he called *idants*) and could be seen as chromosomal granules (Fig. 5.3):

QUOT. 370 ...I have elsewhere expressed the opinion that those rod-like, looplike, or granular masses of chromatin in the nucleus, – the chromosomes, - are to be considered equivalent, not to single ids, but to series or aggregations of ids. I have therefore proposed to call the chromosomes *idants*, in order to keep up a certain uniformity in the nomenclature. (*ibid.*, p. 67)

An important component of Weismann's theory was the occurrence of both *similar* and *dissimilar* nuclear divisions:

1. *Similar divisions* ("erbgleich"): the usual mitotic divisions, "the ids produced by the division naturally contain precisely similar determinants" (*ibid.*, p. 68)
2. *Dissimilar (differential) division* ("erbungleich"): divisions occurring during development "which ensure that the two daughter nuclei contain combinations of determinants which are usually entirely different from one another" (*ibidem*). As these divisions occurred, an increasing number of irrelevant determinants disintegrated until ultimately, the cell contained one specific determinant. Weismann's dissimilar cell divisions thus accounted for cell differentiation.

Dissimilar cell divisions also enabled Weismann in 1894 (Weismann, 1894) to explain the degeneration of useless organs through the process of *germinal selection*, whereby determinants competed for nutrients. Those determinants that were not competitive enough led to the disuse and eventual disappearance of the affected organs. Germinal selection was meant to obviate the Lamarckian influences of the environment. However, Weismann was somewhat correctly accused of resorting to Lamarckism by admitting the inherited effects of use and disuse (albeit not brought about by the external environment).

[29]The term *id* is now used to denote one of Sigmund Freud's three components of the human mind: the *id*, *ego*, and *superego*.

[30]Chromosomes had already been named by Wilhelm Waldeyer in 1888 (Waldeyer, 1888, p. 27). Genes were named later in 1909 by Johannsen (1909, p. 124).

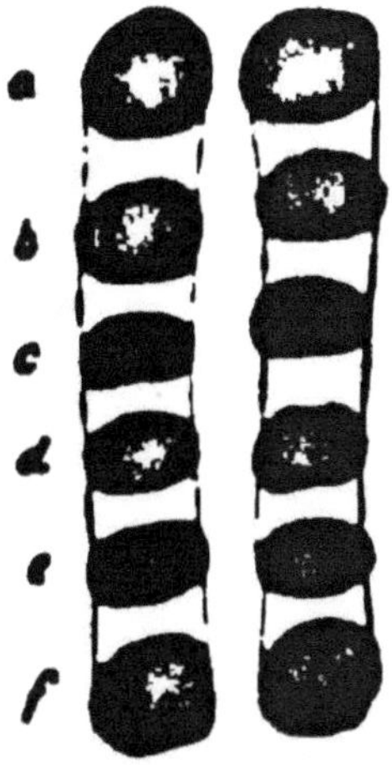

FIG. 2.
Two Idants with their contained Ids of *Ascaris megalocephala*. (After Boveri.)

Fig. 5.3 Weismann's diagram in 1892 (Weismann, 1892, 1893 p. 67) depicting five ids (**a**–**f**) alongside two idants (chromosomes)

5.3 Hugo de Vries: Pangens—Mutation Theory[31]

5.3.1 Introduction

Hugo de Vries (Fig. 5.4) was a Dutch botanist and one of the most influential evolutionary biologists of the late nineteenth century and early twentieth century. When he matriculated at the University of Leiden in 1866, he was already an expert on botany. De Vries was greatly influenced by the *Origin of Species* when he first read it (in German) in 1867 and soon became an adherent to the Darwinian theory. He met Charles Darwin in 1877 during a trip to England. De Vries had been conducting research on plant physiology, but, around 1880, he turned his attention to heredity and variation. In an Oct 15, 1881, letter to Darwin, he wrote:

> **QUOT. 371** For some time I have been studying the causes of the variations of animals and plants, as described in your treatise on the variations of animals and plants under domestication, and have endeavoured to collect some more facts on this theme. . .(van der Pas, 1970, p. 200)

[31] De Vries' theories of heredity and evolution are also discussed in Stamhuis et al. (1999), Stamhuis (2003), Allen (1969), Robinson (1979, Chap. 8), Lenay (2000), Stubbe (1972, pp. 226–232), Hall (1969, pp. 346–348, Vol. 2), and Gaudant & Gaudant (1971, Chap. 15).

Fig. 5.4 Hugo de Vries (1848–1935). (From: https://upload.wikimedia.org/wikipedia/commons/e/e5/Hugo_de_Vries.jpg)

QUOT. 372 I have always been especially interested in your hypothesis of Pangenesis and have collected a series of facts in favour of it... (*ibid.*, p. 201)

The results of de Vries' research on heredity and variation were published in *Intracellular Pangenesis* (de Vries, 1889, 1910a). It was inspired by Darwin's provisional hypothesis of pangenesis[32] but was not completely in favor of it. The book met with little success but was nevertheless a landmark in de Vries' scientific output.

[32] See Sect. 3.7.2.

5.3.2 *Pangens (***Intracellulare Pangenesis,** *1889)*

The primary aim of *Intracellulare Pangenesis* was to improve Darwin's provisional hypothesis of pangenesis,[33] but its influence on Weismann's latest researches, especially his germ plasm theory, cannot be ignored. At the very start, de Vries stated that Darwin's

QUOT. 373 . . .gemmules threw an unexpected light on a series of facts which had hitherto been in absolute darkness. (*ibid*., p. 4)

Thus, de Vries was convinced of the utility of Darwin's pangenesis, which he correctly summarized as consisting of two main assumptions:

QUOT. 374 1. In every germ-cell (egg-cell, pollen-grain, bud, etc.) the individual hereditary qualities of the whole organism are represented by definite material particles. These multiply by division and are transmitted during cell-division from the mother-cell to the daughter-cells.
2. In addition, all the cells of the body, at different stages of their development, throw off such particles; these flow into the germ-cells, and transmit to them the qualities of the organism, which they are possibly lacking. (Transportation-hypothesis). (*ibid*., p. 5)

However, de Vries was also aware of the unimpeachable arguments Weismann had made,[34] which "have shattered the generally accepted doctrine of the heredity of acquired characters" (*ibid*., p. 6). Therefore, the task was to build a theory of heredity and variability based on assumption 1 only (i.e., by excluding the transportation hypothesis). De Vries considered this task to be quite feasible because:

QUOT. 375 . . .if one reads attentively Darwin's discussion, he sees more and more clearly that the transmission of gemmules by cell-division, from the mother-cell to the daughter-cell, suffices to explain large groups of phenomena. Only isolated groups of facts demand in addition the hypothesis of transportation. (*ibid*., p. 4)

Like Weismann, de Vries was cognizant of the latest cytological research, in particular of the nucleus as the repository of the hereditary material. His theory of pangenesis was thus one of *intracellular pangenesis.* Now:

QUOT. 376 To the smallest particles, of which each represents one hereditary characteristic, I shall give a new name and call them *pangens*,[35] because with the designation "gemmule" (*Keimchen*) is associated the idea of a transportation through the whole organism. (*ibid*., p. 7)

[33] See Sect. 3.7.2.

[34] See Sect. 5.2.4.

[35] All italics in this quotation are de Vries'.

Thus, while Darwin's gemmules traveled freely inside the body, de Vries' pangens were hereditary characters[36] restricted to the cell nucleus. Moreover:

> **QUOT. 377** The pangens are not chemical molecules, but morphological structures, each built up of numerous molecules. They are the life-units, the characters of which can be explained in an historical way only. (*ibid.*, p. 79)

In de Vries' theory, each character was independently represented by several pangens of the same kind, such that:

> **QUOT. 378** ...[i]n the protoplasm, or at least in the nuclei, of the egg- and sperm-cells, as well as in that of all buds, all the pangens of the respective species are represented; every kind of pangen in a definite number. Predominating characters correspond to numerous pangens, slightly developed attributes to less numerous ones. (*ibid.*, p. 72)

That is, each nucleus consisted of all the pangens of a given species. Moreover, dominance occurred through the numerical superiority of pangens for a given character.

However, if each nucleus consisted of all the pangens, how then did differentiation and development take place? De Vries answer was that:

> **QUOT. 379** ...[t]he differentiation of the organs must be due to the fact that individual pangens or groups of them develop more vigorously than others. The more a certain group predominates, the more pronounced becomes the character of the respective cell. (*ibidem*)

Although de Vries eschewed Darwin's assumption of the transportation of gemmules from body cells to sex organs, his theory still admitted *intracellular* transport during development: the pangens were dormant inside the nucleus but differentially activated when transported to the cytoplasm. One sees in this aspect of de Vries' theory of pangenesis a similar approach adopted by Weismann in his theory of biophors.[37] In fact, Weismann's biophors were the same as de Vries' pangens, except that:

(a) Weismann's biophors were part of a higher structure while de Vries' pangens were ungrouped. The absence of grouping in the pangens also meant that the concepts of a germ plasm and its reduction division were redundant in de Vries' theory.

Two further important differences between the theories of de Vries and Weismann were the following:

[36]In this respect, de Vries' pangens were more similar to Nägeli's *micellae* (Sect. 5.1.2) than to Darwin's gemmules.

[37]See Sect. 5.2.8.

(b) In de Vries' theory, the pangens resided in every cell,[38] whereas, in Weismann's theory, the biophors were restricted to the germ plasm in the germ cells.
(c) In de Vries' theory, each cell nucleus had *all* the pangens of a given species, whereas, in Weismann's theory each cell nucleus ended up with a specific determinant.[39]

Weismann's postulate that heredity material was restricted to the germ plasm (as pointed out in (a) above) was an easy criticism of his theory, as de Vries was quick to point out through an example:

QUOT. 380 On continuing, under favorable conditions, the cultivation of pieces cut off from these plants, one can grow a new plant from every part that is not too small. (*ibid.*, p. 96)

A final point regards de Vries' assumption of two kinds of variability in his theory. He called the first type 'fluctuating variability', which was the same as Darwin's *individual differences*:[40]

QUOT. 381 . . .the pangens present may vary in their relative number, some may increase, others may decrease or disappear almost entirely some that have long been inactive may resume activity, and finally the grouping of the individual pangens may possibly change. All of these processes will amply explain a strongly fluctuating variability. (*ibid.*, pp. 73–74)

Thus, fluctuating variability was due to the varying number of pangens (of the same kind) for a given character. Regarding the second type of variability, de Vries postulated a process in which the first part was very similar to what would later be called a genetic mutation:

QUOT. 382 In the second place some pangens may change their nature more or less in their successive divisions or, in other words, new kinds of pangens may develop from those already existing. And when the new pangens, perhaps in the course of several generations, gradually increase to such an extent that they can become active, new characters must manifest themselves in the organism. (*ibid.*, p. 74)

De Vries' second type of variability was similar to Darwin's sports and may be called "species-forming variability" (*ibid.*, p. 214) or "specific variability" (de Vries, 1895, 1918, Opera V, p. 495).

In the years following *Intracellulare Pangenesis*, de Vries built an experimental program to study both types of variabilities, especially species-forming variability. He also conducted hybridization experiments, which led to his encounter with the now-famous Mendelian ratios.

---

[38] According to de Vries, the pangens were active in the somatic cells but latent in the germ cells.

[39] Concerning these three differences, Weismann's reduction division in (a) has been retained by modern biologists. However, in both of the differences in (b) and (c), today's beliefs are closer to de Vries than to Weismann. Instead of de Vries' several pangens determining one character, it is now believed that each character is determined by one pair of alleles of a given gene (or several pairs of alleles of different genes).

[40] See QUOT. 169.

Whereas Darwin had downplayed the second type of variability (sports) and stressed the importance of the first type (individual differences) on which natural selection acted, the reverse was true with de Vries.

5.3.3 *Mutation Theory (*Die Mutationstheorie, *1901–1903): Weldon's Reaction (1902)*

5.3.3.1 Die Mutationstheorie (1901–1903)

As early as 1886, de Vries observed a large number of specimens of *Oenothera lamarckiana*[41] in a field not far from Amsterdam. He found several varieties growing side by side, and this observation was striking enough that de Vries started to experiment with the species. Together, he found a total of seven stable varieties. To de Vries, these distinct, stable varieties were proof of the evolutionary importance of *discontinuous variations*.

As described in the last section, in *Intracellulare Pangenesis* (de Vries, 1889), de Vries proposed his theory of heredity based on pangens. He also introduced two kinds of variability: (i) fluctuating variability, due to the varying number of similar pangens for a given character, and (ii) species-forming variability, due to new pangens which led to the formation of specific characters.

During the 1890s, de Vries also became interested in Galton's statistical research on heredity. The latter had been an early proponent of discontinuous evolution.[42] In 1894, de Vries published a paper that found half-normal versions[43] of the fully normal curves Galton had previously used[44] (de Vries, 1894). De Vries argued that these "half-Galton curves" were superpositions of two normal curves, one (i.e., the progeny) shifted from the other (i.e., the parent) by the action of natural selection. To de Vries, the half-Galton curves were clear indications of discontinuous variations, which arose abruptly and spontaneously.

De Vries was also greatly influenced by William Bateson's *Materials for the Study of Variation* (Bateson, 1894).[45] Bateson had publicly declared his discontinuous view of evolution as early as 1891 (Bateson, 1891). The evolutionary views of the two men were extremely close, especially before Mendel's laws were rediscovered.

[41] This plant was first described by Lamarck (hence the name) and is commonly known as the evening primrose. It was introduced in Europe from America. Its flowers are usually yellow in color, with white and purple varieties also present.

[42] See Sect. 4.3.4.

[43] That is, either the right or left section of a normal curve.

[44] See Sect. 4.3.3.

[45] See Sect. 8.1.

De Vries' theory of discontinuous evolution was fully exposed in the two-volume opus *Die Mutationstheorie* (de Vries, 1901, 1903). This was right in the wake of the rediscovery of Mendelism, one of the rediscoverers being de Vries himself.[46] In *Mutationstheorie*, de Vries introduced the term "mutability" for the species-forming variability he had used previously in *Intracellulare Pangenesis*:

> **QUOT. 383** The name I propose to give to this "species-forming" variability is Mutability...the best-known examples of such Mutations are the so-called spontaneous variations (the "single variations" of Darwin) by which new and distinct variations arise. They are also termed, fitly enough, sports. (de Vries, 1901, 1909, pp. 4–5)

De Vries' mutability was the same discontinuous variability he had observed in *Oenothera lamarckiana* in 1886. Mutability was due to mutations, also known as *saltations, sports*, or *monstrosities*,[47] and was caused by a change in the nature of the pangens as described in QUOT. 382

De Vries' fundamental thesis in *Mutationstheorie* was that new species were formed by mutations:

> **QUOT. 384** The object of the present book is to show that species arise by saltations... (*ibid.*, p. viii)

Furthermore, regarding natural selection:

> **QUOT. 385** The Mutation theory is opposed to that conception of the theory of selection which is now prevalent. According to the latter view the material for the origin of new species is afforded by ordinary or so-called individual variation. According to the Mutation theory individual variation has nothing to do with the origin of species. This form of variation, as I hope to show, cannot even by the most rigid and sustained selection lead to a genuine overstepping of the limits of the species and still less to the origin of new and constant characters. (*ibid.*, p. 4)

That is, according to de Vries' mutation theory, natural selection acting on individual variations could not cross the species barrier. Only mutations could do that.

In his theory, de Vries identified three kinds of mutations:

(a) *Progressive mutations*, leading to the formation of new characters: "Progressive development is due to the formation of new characters, to increasing differentiation" (de Vries, 1903; 1910a, pp. 65–66).
(b) *Retrogressive mutations,* associated with the loss of characters: "...the absence of characters which are otherwise proper to the group to which the species belongs...[c]ases such as these are spoken of instances of retrogressive metamorphosis. And it is probably not too much to say that there are possibly more

[46]*Mutationetheorie* was started in November 1899 and finished by April 1903. De Vries wrote about his rediscovery of Mendelism in March 1900.

[47]De Vries' mutations are thus different from the way they are used today, namely, a change in the DNA sequence of a gene. Moreover, it turns that the different varieties of *Oenothera lamarckiana* de Vries had observed were not different mutants but varieties that had been created by genetic recombination.

species on the face of the earth at present that have arisen in retrogressive than on progressive lines" (*ibid.*, p. 66).

(c) *Degressive mutations,* leading to the reappearance of latent (i.e., atavistic) characters: "Degressive formation of species is therefore due to the activation of long established characters" (*ibid.*, p. 71).

It is important to note that although de Vries defined the three types of mutations as species-forming variability, he did not strictly adhere to this rule in his book: although progressive mutations were usually associated with new species, retrogressive and degressive mutations were often associated with varieties within species.

De Vries cited the absence of transitional forms in the characteristics of organisms as a major reason for his belief in discontinuous evolution (de Vries, 1901, 1909, p. 3). This was a familiar argument and had been used before by Huxley (to also argue for discontinuous evolution[48]), Mivart (to attack natural selection[49]), and several others.

De Vries was careful to note that his conclusions were based on his work in botany. He was confident that they would also hold in zoology as soon as they were confirmed in the first case. Based on his work on *Oenothera lamarckiana*, he stated seven key principles behind his mutation theory:

QUOT. 386 i. New elementary species arise suddenly, without transitional forms.
ii. New elementary species are, as a rule, absolutely constant from the moment that they arise.
iii. Most of the new forms that have appeared are elementary species, and not in the strict sense of the term.
iv. New elementary species appear in large numbers at the same time or at any rate during the same period.
v. The new characters have nothing to do with individual variability.
vi. The mutations, to which the origin of new elementary species is due, appear to be indefinite, that is to say, the changes may affect all organs and seem to take place in almost every conceivable direction.
vii. Mutability appears periodically. (*ibid.*, pp. 248–258)

De Vries also acknowledged Galton's pioneering work when he used the latter's stability of types metaphor (Sect. 4.3.4) to illustrate how discontinuous evolution could take place (*ibid.*, pp. 53–56).

If, according to de Vries, new species were formed by mutations rather than by natural selection, what role did the latter have in de Vries' theory? De Vries' answer was that:

QUOT. 387 The struggle for existence. . .may refer to two entirely different things. On the one hand the struggle takes place between the individuals of one and the same elementary species, on the other between the various species themselves. The former is a struggle between fluctuations, the latter between mutations. In the former case those that survive are the individuals which find conditions favorable to them. . .It is by this process that local races arise. . .If the new conditions of life are relaxed, the adapted race reverts to the form from which it sprang.

---

[48] See Sect. 2.12.2.

[49] See Sect. 2.14.5.

> The natural selection of newly arisen elementary species in the struggle for existence is an entirely different matter. They arise suddenly and without any obvious cause; they increase and multiply because the new characters are inherited. When this increase leads to a struggle for existence the weaker succumb and are eliminated....
>
> In a word, from the standpoint of the theory of mutation it is clear that the role played by natural selection in the origin of species is a destructive, and not a constructive one. (*ibid.*, pp. 211–212)

De Vries thus pointed to two roles for natural selection in his theory:

1. *Within a species*: it favored a new race that was most adapted to a new environment but then led to reversion to the previous race once the old environment was brought back.
2. *Between species:* it led to the elimination of less adapted species.

That is, insofar as the *origin of species* was concerned, the sole role of natural selection was one of an *eliminating force.*[50]

A final remark concerns perhaps the most important question regarding de Vries' mutation theory: was it Mendelian? To answer this question, one first needs to differentiate between de Vries' *unisexual* and *bisexual* crosses. Unisexual crosses were between two organisms that differed in a *specific* character; on the other hand, bisexual crosses were between two organisms that differed in a *varietal* character.

De Vries' thesis was that bisexual crosses were Mendelian whereas unisexual crosses were not:

> **QUOT. 388** The former [bi-sexual] conform to the laws of segregation, they lead to various combinations of elementary characters, and thus can lead to the origin of as many new races as the number of possible combinations indicates. These races are constant; the hybrids, however, always exhibit segregation in the formation of their sexual cells and sometimes even in the formation of buds. The hybrids of uni-sexual crosses on the other hand are constant; so far as my experience goes, they do not segregate. (de Vries, 1903, 1910b, p. 576)

Since new species often (but not always) corresponded to progressive mutations and new varieties often (but not always) corresponded to retrogressive and regressive mutations, de Vries' thesis also implied that Mendel's laws held for retrogressive and regressive mutations, but not for progressive mutations (*ibid.*, p. 577).

Overall, de Vries viewed Mendel's laws as a subordinate component of his mutation theory and his *Mutationstheorie* was *not* Mendelian.

5.3.3.2 Weldon's Reaction (1902)

As mentioned in Sect. 5.3.3.1, during the 1890s, de Vries became acquainted with Galton's use of statistics, in particular, the latter's use of regression to argue against the importance of continuous variation in evolution.[51] Since de Vries was on the

---

[50] This particular view of de Vries is not adhered to by modern biologists: natural selectin can in fact be both constructive and destructive.

[51] See Sect. 4.5.7.

same side of the issue as Galton, it is not surprising that he used similar regression arguments in *Mutationstheorie*.

In particular, de Vries carried out repeated crossings in plants, each time choosing the most extreme offspring for further crossing, so as to produce a new center of regression. Then when selection was relaxed, regression to the original parental mean was interpreted by de Vries as illustrating the ineffectiveness of natural selection. This is how he explained it:

> **QUOT. 389** If we regard GALTON'S [regression] formula as generally true the mean of the offspring deviates from the mean of the type in such a way that it retains only a third of the deviation of the parent. So that to produce a given advance in the whole family we would have to sow seed from a plant which had advanced three times so far. (1901, 1908, p. 83)

Now, Weldon pounced on de Vries' use of the concept of regression, claiming that it was incorrect and thus invalidated his argument. Weldon's criticism was written amid the biometry-saltationism war and appeared in the third issue of the newly created *Biometrika* (Weldon, 1902c). He correctly stated that individuals in a given generation regress to their own mean, not to the mean of their parents, as de Vries had implied in QUOT. 389. Citing de Vries' example of a race of maize in which the mean number of rows per head initially was 12–14 and then increased to 20 by subsequent breeding of the more extreme offspring, Weldon said:

> **QUOT. 390** A further proof that regression to the original race-mean does not occur is given by the subsequent history of this maize. In 1897 an attempt was made to reverse the process of selection, and for this purpose individuals were chosen out of the generation of that year which had only 16 rows of seeds per head, the mean of the generation being 20. Now 16 is a greater number than the mean number of rows [12–14] in the original race; and if regression to the original race-mean occurred, the number of rows of seeds per head among the offspring of individuals with 16 rows should have been at all events less than 16. As a matter of fact it was 20! (*ibid.*, p. 369)

Weldon then declared that de Vries' whole argument for the instability of continuous change was thus invalid:

> **QUOT. 391** So that this experiment, taken as a whole, forms a fairly conclusive proof that the statements concerning the focus of regression on which the whole theory of the instability of varieties depends, are erroneous, and a main part of the argument fails. (*ibidem*)

Weldon next took on de Vries' argument that mutations produced a new, stable center of regression. He referred to the latter's experiment with *Trifolium pratense* (red clover), where the number of leaflets on normal leaves was three. In 1886, de Vries had cultivated some plants with leaves bearing four to five leaflets in addition to leaves with three leaflets. In successive generations, the seeds from such plants were sown (each time discarding plants with mostly normal leaves). In August–September 1890, de Vries was able to obtain four seed plants which had a total of 69 leaves with 4 leaflets and 44 leaves with 5 leaflets. In his review, Weldon disagreed that de Vries had shown that a new center of regression had been created by mutations. His reasoning was that, if the normal plants had not been thrown away in each generation, the average number of leaflets would still be close to three rather than four to five. Therefore:

QUOT. 392 This postulated regression failed to occur in spite of the fact that the offspring were reared in a garden, under conditions shown to favour an excessive production of abnormal leaves. (*ibid.*, p. 371)

In this argument too, Weldon had a point, since de Vries' example did not show the existence of constant-breeding varieties that regressed to a new mean without the need for selection (throwing out the normal plants) in each generation.

However, de Vries did indeed also show the existence of constant-breeding mutants in several cases (seven examples in Volume 1, Part II, Chapter 1 of *Mutationstheorie*). For example, regarding *O. albida*, de Vries wrote:

QUOT. 393 In 1897 however I succeeded in getting five biennial plants to flower; and in obtaining seed from them after artificial fertilization. From these seeds I raised a second generation in 1898 and a third in 1899, both of annual individuals. Both generations were absolutely constant and exhibited no signs of reversion. (de Vries, 1901, 1909, pp. 229–230)

In his review, Weldon did not pay attention to such examples, contenting himself instead with the conclusion that regression would finally be understood when

QUOT. 394 . . .such naturalists as Professor de Vries and Mr Bateson will abandon their attempts to distinguish between "variations" and "mutations," or between "normal" and "differentiant" variations. (Weldon, 1902c, p. 374)

De Vries' did not respond to Weldon's review but was glad that his name was associated with that of Bateson. In a May 12, 1902, letter to Bateson, he wrote:

QUOT. 395 Weldon names at the end such biologists as Bateson, and de Vries, and I was glad, when reading this, to take leave from him in such good company. If you will defend me against him I will be much indebted to you. (Provine, 2001, p. 68)

However, it would be wrong to infer from the above that de Vries and Bateson were completely aligned in their views on heredity, which will be described next.

5.3.4 *De Vries and Bateson*

Shortly before the first (July 1899) International Conference on Hybridization, held in London, de Vries traveled to England and met William Bateson[52] for the first time in Cambridge. This was before de Vries' rediscovery paper on Mendel's work. De Vries and Bateson quickly became friends, largely due to their similar, discontinuous views on evolution. Writing on July 8 to his wife, Bateson said:

QUOT. 396 De Vries is a really nice person, very simple and rather rough in style. It is delightful to have a person with whom one can change a word. He is an enthusiastic discontinuitarian and holds the new mathematical school in contempt. So we hit if [*sic*] off to admiration. (Cock and Forsdyke, 2008, p. 186)

[52]For more on Bateson, see Sect. 7.2.3.

Later, on May 8, 1900, while Bateson was traveling by train to the second International Conference on Hybridization,[53] he was excited to read Mendel's work from de Vries' *Comptes Rendus* article (de Vries, 1900c) (although Mendel was not mentioned in that article). Referring to de Vries' own experiments on hybridization, Bateson declared:

QUOT. 397 In the present year Professor de Vries published a brief account of experiments which he has for several years been carrying on, giving results of the highest value. (Bateson, 1901, 1928, p. 175)

Similarly, de Vries' *Mutationstheorie* (1901, 1903) made several sympathetic references to Bateson, especially to the latter's *Materials for the Study of Variation* (Bateson, 1894). On October 18, 1900, de Vries sent an advance copy of the first volume of his book to Bateson. In an accompanying letter, the former confided:

QUOT. 398 I have now the pleasure of offering you my work on the origin of species, as discontinuous as you could hope it. (Provine, 2001, p. 67)

As QUOT. 394 shows, Bateson and de Vries' evolutionary views were initially so aligned that they were often lumped together when the biometricians attacked the saltationists.

However, de Vries' *Mutationstheorie* marked a gradual separation between de Vries' and Bateson's views. This is because de Vries now started to deny the universality of Mendel's laws and treat it as a special case of his mutation theory.[54] On the other hand, Bateson's support of Mendelism and his advocacy of the latter's primacy had not flinched.

Seeing the growing distance between Bateson and himself, on October 20, 1901, de Vries implored:

QUOT. 399 I prayed you last time, please don't stop at Mendel. I am now writing the second part of my book which treats of crossing, and it becomes more and more clear to me that Mendelism is an exception to the general rule of crossing. It is in no way the rule! It seems to hold good only in derivative cases, such as real variety-characters. (*ibid.*, p. 68)

However, the differences between the two men soon became irreconcilable. When Charles Druery (1843–1917), who had previously translated Mendel's paper with the assistance of Bateson,[55] asked for the latter's support in finding a publisher for an English translation of *Mutationstheorie*, Bateson declined. In an October 27, 1906, letter to Druery, he said:

QUOT. 400 If the matter proceeds you must kindly take it that my part is limited to a friendly desire to see the book in English and that I cannot accept any responsibility direct or indirect. (Cock and Forsdyke, 2008, p. 70)

Similarly, in de Vries' retrospective paper entitled "The origin of the mutation theory" (de Vries, 1917), Bateson was not even mentioned.

---

[53] See Sect. 9.2.2.

[54] See end of Sect. 5.3.3.1.

[55] See start of Sect. 9.2.3.

Chapter 6
Mendel's Theory of Heredity

Abstract In 1866, Mendel, the father of genetics, proposed a theory of heredity based on the results of Pisum experiments he had conducted since 1856. Mendel's experiments were meticulously planned, and he used a large number of plants for easier interpretation of results. He also focused on specific (binary) traits rather than examining the plants in toto. When one form of a trait was crossed with the other form, all F_1 hybrids were of one of the two forms. This form was called the dominating form, while the other form was called the recessive form. When the F_1 hybrids self-pollinated, dominants and recessives were obtained in the ratio 3:1 in F_2. Mendel was able to further resolve the dominants in F_2 into one part true-breeding (homozygote) and two parts hybrid (heterozygote). To explain his observations, Mendel postulated the existence of a factor inside each of a germinal and pollen cell, such that the two factors united during fertilization. Although Mendel's work was referred by a few, it was largely forgotten until its independent rediscovery in 1900 by de Vries, Correns, and Tschermak.

Keywords Cyril Napp · Franz Unger · Carl Nägeli · Pisum sativum · Dominant · Recessive · Law of segregation · Law of independent assortment · Anlage · Hieracium experiments · Mendel-Fisher controversy · Rediscovery papers

6.1 Mendel Before the 1856–1863 Experiments

6.1.1 Childhood and Education

The man (Fig. 6.1) who would later be acknowledged as the father of genetics was born on July 22, 1822, in the small village of Heinzendorf (now Hynčice), Moravia, then part of the Austrian Empire (now the Czech Republic). Mendel's parents were peasant farmers and christened their only son Johann—the latter took the name

P. Gorroochurn, *The Development of Evolutionary Genetics*,
https://doi.org/10.1007/978-3-031-69374-8_6

Fig. 6.1 Gregor Johann Mendel (1822–1884) (https://commons.wikimedia.org/wiki/File:Portrait_of_Gregor_Johann_Mendel,_Garrison._Wellcome_M0002351.jpg)

Gregor only in 1843, when he became a monk. Mendel's[1] primary education was greatly influenced by the progressive Reverend Johann Schreiber (1769–1850), who taught him natural science. Schreiber was a founding member of Brno's Pomological Association and a corresponding member of the Brno Agricultural Society. He kept a small garden beside the school in Heinzendorf, and it is most likely that Johann learned the basics of pollination and seed germination from him.

Because of his good performance, Mendel was recommended by both Schreiber and the schoolmaster Thomas Makitta to be sent to the school in Leipnik (Lipnik). The next year, he was admitted to the *Gymnasium* in Troppau (Opava) where he was

[1] Mendel's life and work are also discussed in Orel (1996), Iltis (1932), Mawer (2006), Corcos and Monaghan (1993), Monaghan and Corcos (1990, 1993), Falk and Sarkar (1991), Henig (2000), Klein (2000), Klein and Klein (2013), Sosna (1966), Olby (1985), Tudge (2002), Keynes et al. (2004), Liang et al. (1998), Blanc (1984), Di Trocchio (1991), Hartl and Orel (1992), Weiling (1991), Sapp (1991), Zirkle (1951), Adams (1969, Chap. 14), and Zhang et al. (2017).

expected to spend the next 6 years. However, when Mendel had done four years in his studies, his father was seriously injured at work in 1838. As Mendel himself later recalled, writing in the third person on his application for the State Certificate:

QUOT. 401 ...his [Mendel's] parents found themselves unable to meet the expenses incurred by the continuation of his studies and it therefore happened that the respectfully undersigned, then only sixteen years old, found himself in the sad position of having to fend for himself. (Crew, 1966, p. 18)

The physical and mental strains soon took a toll on Mendel's health, and he spent several months in the fifth year of his studies at home. However, Mendel soon returned to Troppau and started studying for a private tutor's certificate at the District High School so that he could earn a living. In 1840, having graduated from the *Gymnasium*, Mendel enrolled in a philosophy course at the Philosophical Institute in advance of higher studies at the University of Olmütz (Olomouc). However:

QUOT. 402 He [Johann] made repeated attempts in Olmütz to earn a living as a private tutor, but all his efforts were fruitless for the reason that he had no influential friends to recommend him. He became so distressed by continued disappointment and so anxious about his future that he fell ill, and was obliged to spend a year in his parents' home convalescing. (*ibidem*)

Mendel's inability to find "influential friends to recommend him" no doubt had to do with his poor Czech, the spoken language in Olomouc. Fortunately, there was a glimmer of hope while Johann was convalescing at home. His younger sister, Theresia, gave him a portion of her dowry so that Mendel was able to return to Olomouc and finish his philosophy course, which lasted two years. During that time, he was able to learn logic, physics, and mathematics, including, as Kruta and Orel (1974, p. 277, Vol. IX) have pointed out, combinatorial analysis. The latter would be an important component of Mendel's path-breaking 1866 paper.

Having completed his philosophy course, Mendel would normally have entered university. However, because of his still precarious financial situation, he did not. At that point, Mendel's physics professor, Friedrich Franz (1763–1860), wrote a letter of recommendation for his admission as a novice in the monastery of Brünn (Brno). Interestingly, Mendel excelled in physics, for Franz wrote:

QUOT. 403 ...I have made known to my pupils the Right Reverend Prelate's decision to accept satisfactory candidates at your monastery. Up to now, two candidates have given me their names, but I can only recommend one of them. This is Johann Mendel, born at Heinzendorf in Silesia. During the two-year course in philosophy he has had, almost invariably, the most unexceptionable reports, and is a young man of very solid character. In my own branch he is almost the best. (Iltis, 1932, p. 42)

Equally noteworthy is the fact that Mendel did not enter the monastery out of some deep religious conviction but because (as he later recalled):

QUOT. 404 He felt himself compelled to seek some position that would free him from the bitter struggle for existence. His circumstances determined his choice of a vocation. (Crew, 1966, p. 18)

Fig. 6.2 Abbot Cyril F. Napp (1792–1876) (https://commons.wikimedia.org/wiki/File:Cyril_Frantisek_Napp_1867_Maixner.png)

Abbot Cyril F. Napp (Fig. 6.2), who was to have a significant influence on Mendel, accepted the latter in the monastery without any interviews on September 7, 1843. On October 9, 1843, Mendel started as a novice in the convent of Augustinians under the name Gregor.

The monastery of Brno was a center of learning and scientific research, with many of its members being teachers at the Gymnasium and Philosophical Institute. Gregor Mendel welcomed his new position since he was now financially secure and saw the possibility of pursuing his education. In addition, in Abbot Napp, he found not only a highly learned polymath who was interested in agricultural research but also a very supportive figure.

Having joined the monastery, Mendel did his theological studies from 1844 to 1848. During that time, he also attended lectures in agriculture at the Philosophical Institute under Franz Diebl (1770–1859). The latter had published a five-volume textbook on agriculture; the second volume, in particular, described how more productive varieties could be obtained through artificial pollination.

At the end of his theological studies, Mendel started parish work which required visiting the sick at St. Anne's Hospital. However, Mendel's temperament was so sensitive that he could not bear to see the suffering of the sick and he went into depression. Napp intervened and sent Mendel in October 1849 as a substitute teacher in the local *Gymnasium* of the nearby town of Znaim (Znojmo).

Mendel found his true vocation in teaching. Both students and colleagues were very impressed. At the end of the first year, the headmaster of the *Gymnasium* in Znojmo recommended that Mendel take the university examination so that he could work as a full-time teacher. Mendel managed to do well in physics and meteorology, but he failed in geology and zoology. He was called for a further examination in Vienna in August 1850. Again, Mendel passed in physics but failed the natural history paper. Mendel's failures were no doubt due to his lack of a university education.

The plan in 1851 was now for Mendel, under the patronage of Napp, to attend the University of Vienna.

6.1.2 University Years and Influence of Franz Unger[2]

From 1852 onward, Mendel's university years played a decisive role in his intellectual development. Since he saw himself primarily as a physics teacher, it was natural for Mendel to take physics as one of his courses. He learned experimental physics from none other than the famous Christian Doppler (1803–1853). When the latter passed away, he was replaced by Andreas von Ettingshausen (1796–1878) who had previously written a book on combinatorial analysis.[3]

In addition to physics, Mendel also took courses in biology. His teacher was the renowned Austrian botanist Franz Unger (Fig. 6.3). There are at least three things that Mendel most likely learned from Unger and that influenced his later research:

(a) *Knowledge of Nägeli's work.* Unger had a deep admiration for the Swiss-German botanist Carl Nägeli (1817–1891), especially concerning the latter's significant work on cell division in 1844 (Nägeli, 1844).[4] Mendel later chose to

---

[2]Unger's influence on Mendel has also been discussed in Orel (1996, pp. 69–73) and Dröscher (2015).

[3]The reader will recall that Mendel had previously encountered combinatorics while doing his philosophy course at Olomouc, and it was then mentioned that combinatorics would be an important part of his groundbreaking 1866 paper.

[4]Nägeli was among the first to infer from observations that cells were formed by division of other cells. This was in accordance with Unger's own views (Unger, 1841), but contrary to the (incorrect) views held by the two fathers of cell theory, namely, Mathias Schleiden (1804–1881) and Theodor Schwann (1810–1882), according to whom the cell was formed from sap by coagulation (Schleiden, 1838; Schwann, 1839). Unger's praise of Nägeli may be found in his *Botanical Letters:* "... we cannot too highly esteem the endeavour to establish the architecture of the plant upon a true scientific basis, and to obtain, in the knowledge of its elementary composition, a key to all that which is now a mystery to us. The man who has hitherto penetrated the furthest into this obscure region of research, who has attempted to examine stone for stone in their production and application, who has given us both ground-plans and elevations of some plant structures in which each element is marked with the number its architect intended for it, this man is Karl Nägeli. It is undoubtedly from him that we may look for the unveiling, not only of the most simple, but likewise of the most complicated structures of the plant organism, which we have hitherto merely wondered at but not understood" (Unger, 1854, p. 16).

Fig. 6.3 Franz Unger (1800–1870). (From: https://commons.wikimedia.org/wiki/File:Franz_Unger.jpg)

communicate his results to Nägeli, and it is through Unger that he most likely learned about Nägeli's work.[5]

(b) *Importance of mathematics and the search for laws of nature*. Unger was also influenced by Schleiden's *Grundzuege der wissenschaftlichen Botanik nebst einer Methodologischen Einleitung als Anleitung zum Studium der Pflanze*[6] (Schleiden, 1842, 1849), which intended to place Botany on par with the sciences of Physics and Chemistry through the rigorous investigation of "the laws and forms of the Vegetable Kingdom" (*ibid.*, p. 1). In the book, Schleiden also stressed the importance of "mathematical laws confirmed with perfect exactness by experience" (*ibid.*, p. 264). In his groundbreaking 1866 paper, Mendel similarly stressed the importance of *rigor* in experiments, spoke in terms of *laws* about the formation of hybrids, and, in contrast to the qualitative

[5]Mendel also read Gärtner's 1849 book (Sect. 3.5) that described over 10,000 artificial fertilizations in 700 plant species. He marked and annotated some pages on pea hybrids.

[6]Translated as *Principles of Scientific Botany or Botany as an Inductive Science* in 1849.

approaches of Naudin[7] and other contemporaries, adopted a *quantitative* approach based on his observations.

(c) *Formation of new species by hybridization.* Unger had espoused the transmutation of species in his *Botanical Letters* (Unger, 1854)[8] and had *opposed* both Kölreuter's (Sect. 3.3) and Gärtner's (Sect. 3.5) views that new species could not be formed by hybridization. For example, in his book, Unger had said: "...who can deny that new combinations of the elements arise out of this permutation of vegetation ever reducible to a certain law—combinations which emancipate themselves from the preceding characteristics of the species, and appear as new species?" (*ibid.*, p. 92). Undoubtedly, this was the question that Mendel alluded to in 1866 when he mentioned "a question whose significance for the evolutionary history of organic forms must not be underestimated" (Mendel, 1866, 1966a, p. 2).

During the last year of his university studies, Mendel joined the Vienna Zoological and Botanical Society and read a short paper on a pyramid moth that attacked radishes. This was followed by another paper on the pea pest *Bruchus pisi.* That paper was presented in Vienna in 1854 by Dr. V. Kollar, as Mendel had finished his studies and returned to Brno at the end of July 1853.

Back in Brno, Mendel was appointed as a substitute teacher for physics and natural history at the *Realschule*, which focused on science with industrial applications. He most likely also experimented on peas and *Bruchus pisi* (which formed the basis of the paper presented by Kollar). On May 2, 1855, Mendel went to Vienna to take the first part of the teacher's exam again, in an attempt to become a full-time teacher. He reported for the second part of the exam on May 5, 1856, but had a breakdown and returned to Brno.[9]

6.2 The 1856–1863 Pisum Experiments and the 1866 Paper

6.2.1 Introduction

Leaving behind the obvious disappointment of a second failure at the teacher's exam, Mendel started his landmark experiments on peas in 1856, shortly after coming back from Vienna. Aside from the painstaking nature of these numerous experiments, conducting them while attending to his other duties was not especially problematic for Mendel. This is because, as mentioned, Abbot Napp was deeply interested in agricultural research and had made the monastery garden available for such activities.

---

[7] See Sect. 3.6.

[8] This was, of course, before Darwin's *Origin of Species* (Darwin, 1859) and cost Unger violent attacks by the Catholic press.

[9] Mendel remained a substitute teacher for the rest of his life (he stopped teaching after 1868).

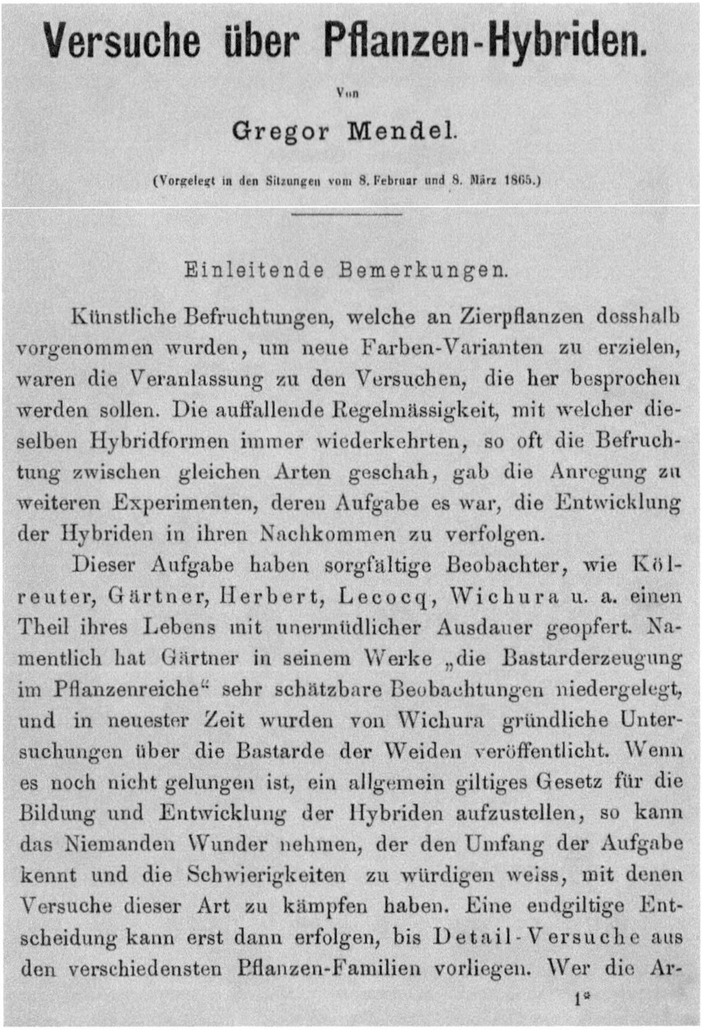

Versuche über Pflanzen-Hybriden.

Von

Gregor Mendel.

(Vorgelegt in den Sitzungen vom 8. Februar und 8. März 1865.)

Einleitende Bemerkungen.

Künstliche Befruchtungen, welche an Zierpflanzen desshalb vorgenommen wurden, um neue Farben-Varianten zu erzielen, waren die Veranlassung zu den Versuchen, die her besprochen werden sollen. Die auffallende Regelmässigkeit, mit welcher dieselben Hybridformen immer wiederkehrten, so oft die Befruchtung zwischen gleichen Arten geschah, gab die Anregung zu weiteren Experimenten, deren Aufgabe es war, die Entwicklung der Hybriden in ihren Nachkommen zu verfolgen.

Dieser Aufgabe haben sorgfältige Beobachter, wie Kölreuter, Gärtner, Herbert, Lecocq, Wichura u. a. einen Theil ihres Lebens mit unermüdlicher Ausdauer geopfert. Namentlich hat Gärtner in seinem Werke „die Bastarderzeugung im Pflanzenreiche" sehr schätzbare Beobachtungen niedergelegt, und in neuester Zeit wurden von Wichura gründliche Untersuchungen über die Bastarde der Weiden veröffentlicht. Wenn es noch nicht gelungen ist, ein allgemein giltiges Gesetz für die Bildung und Entwicklung der Hybriden aufzustellen, so kann das Niemanden Wunder nehmen, der den Umfang der Aufgabe kennt und die Schwierigkeiten zu würdigen weiss, mit denen Versuche dieser Art zu kämpfen haben. Eine endgiltige Entscheidung kann erst dann erfolgen, bis Detail-Versuche aus den verschiedensten Pflanzen-Familien vorliegen. Wer die Ar-

1*

Fig. 6.4 First page of Mendel's "Versuche über Pflanzen-Hybriden" (Mendel, 1866)

As stated in (c) of Sect. 6.1.2, Mendel's university lecturer, Franz Unger, had claimed, contrary to common belief, that hybridization produced new species. In (b), Unger's influence by Schleiden's advocacy for the search for mathematical laws in botany was also mentioned. *Mendel's aims were both to test the hypothesis of the formation of new species by hybridization and to uncover the quantitative laws determining the formation of hybrids.*

The sections that follow deal with Mendel's experiments during the period 1856–1863, as described in his groundbreaking 1866 paper "Versuche über Pflanzen-Hybriden"[10] (Mendel, 1866, 1966a) (Fig. 6.4).

6.2.2 Preparatory Work

From his initial work in 1854, Mendel had cultivated garden peas (*Pisum sativum*) with 34 distinct traits. These were grown for two years and tested for purity. Both Kölreuter and Gärtner had previously experimented with the *Pisum*. Knight had noted that he was able to obtain "many varieties of this plant, of different forms, sizes, and colors; but also, because the structure of its blossom, by preventing the ingress of insects and adventitious farina, has rendered its varieties remarkably permanent."[11] This characteristic, namely, the existence of several traits, each of which was pure breeding, was a special desideratum for Mendel. He called this property "constant differing traits," which are now called homozygous (or pure-breeding) discontinuous traits. Mendel finally selected seven traits for his experiments in 1856. An added advantage was that each of the seven traits existed only in *two* forms (i.e., was dichotomous). The seven chosen traits were related to the following:

QUOT. 405

1. To the difference in the shape of the ripe seeds [round or nearly round vs. irregularly angular and deeply wrinkled]
2. To the difference in coloration of seed albumen (endosperm) [pale yellow, bright yellow and orange vs. intense green]
3. To the difference in coloration of the seed coat [white vs. grey, grey-brown, leather-brown with or without violet spotting]
4. To the difference in shape of the ripe pod [smoothly arched and never constricted anywhere vs. deeply constricted between the seeds and more or less wrinkled]
5. To the difference in color of the unripe pod [colored light to dark green vs. vivid yellow]
6. To the difference in position of flowers [axillary (i.e. distributed along the main stem) vs. terminal (i.e. bunched at the end of the stem)]
7. To the difference in stem length [varied greatly in individual varieties but for a given condition, stem length was long (about 6–7 ft) vs. short (about 9–18 in)]

(*ibid.*, pp. 6–7)

Mendel was very meticulous in the planning and execution of his experiments. Since *Pisum sativum* is self-pollinating, self-fertilization was prevented by opening the immature flower buds and removing the anthers before they were ripe. Pollen from another form was then transferred to the stigma of the emasculated flower. Reciprocal crossings were also performed by transferring the pollen of the first form to the stigma of the emasculated second. Mendel was especially careful about the possibility of foreign pollination but noted that this cannot easily occur since the

[10] "Experiments in plant hybridization."

[11] See Sect. 3.4.

fertilization organs are closely surrounded by the keel. To gauge the importance of foreign pollination through insects, Mendel grew some peas in a greenhouse. This served as a control: by comparing the quality of the hybrids thus obtained against those of the main experiment in the open, Mendel concluded that foreign interferences occurred in very few out of the more than 10,000 plants examined.

6.2.3 The F_1 Hybrids (1857): Dominance and Recessiveness

Mendel's project in 1856 started with a series of seven separate experiments involving the crossing of the two forms in each of the traits in (1)–(7) above. In particular, for the:

QUOT. 406

1st	experiment	60	fertilisations	on	15	plants	were	undertaken
2nd	..	58	..	..	10	..	..	..
3rd	..	35	..	..	10	..	..	..
4th	..	40	..	..	10	..	..	..
5th	..	23	..	..	5	..	..	..
6th	..	34	..	..	10	..	..	..
7th	..	37	..	..	10	..	..	..

(*ibid.*, p. 7)

The seeds from the first crossings were observed[12] and then planted. The first-generation (F_1) hybrids were obtained in 1857. Mendel observed that although hybrids were usually intermediate in appearance between the parents, this was not always the case. This was especially not the case with the *Pisum* hybrids:

QUOT. 407 Each of the seven hybrid traits either resembles so closely one of the two parental traits that the other escapes detection, or is so similar to it that no certain distinction can be made. (*ibid.*, p. 9)

This prompted Mendel to introduce the following two fundamental definitions:

QUOT. 408 ...those traits that pass into hybrid association entirely or almost entirely unchanged, thus themselves representing the traits of the hybrid, are termed *dominating*,[13] and those that become latent in the association, *recessive*. (*ibidem*)

---

[12]These would be the seed portion of the first-generation hybrids.

[13]Instead of italics, Mendel's original manuscript in German used letter-spacing (or "sperratz") for emphasis. Mendel's own German words were: "...werden jene Merkmale, welche ganz oder fast unverändert in die Hybride-Verbindung übergehen, somit selbst die Hybriden-Merkmale repräsentiren, als d o m i n i r e n de, und jene, welche in der Verbindung latent werden, als r e c e s s i v e bezeichnet" (Mendel, 1866, p. 11). Mendel's "dominirende" is closer to Stern and Sherwood's "dominating," which is used here, than to Bateson's (1902b, p. 49) "the dominant." However, "dominant" has since been the commonly used word in the literature.

Table 6.1 Number of dominants and recessives obtained among the F_2 hybrids in Mendel's experiments

Trait	Total	Dominant	Recessive	Ratio
Seed shape	7324	5474	1850	2.96:1
Albumen coloration	8023	6022	2001	3.01:1
Color of seed coat	929	705	224	3.15:1
Shape of pods	1181	882	299	2.95:1
Coloration of unripe pods	580	428	152	2.82:1
Position of flowers	858	651	207	3.14:1
Length of stem	1064	787	277	2.84:1

Thus, if two parents, each with a different form, produce an offspring having the form of one of the parents, then that particular form is said to be *dominant*, and the other form is said to be *recessive.*[14]

While noting the equal contributions of the male and female organs,[15] Mendel listed the following dominant forms for the seven traits he had chosen: (1) round seed, (2) yellow coloration of seed albumen, (3) gray seed coat, (4) smooth ripe pod, (5) green unripe pod, (6) axillary flowers, and (7) long stem.

6.2.4 *The F_2 Hybrids (1858): Two Types of Dominants*

The F_1 hybrids were now allowed to self-pollinate, after which their seeds were examined[16] and grown. When the peas had grown, Mendel noted:

> **QUOT. 409** In this generation, *along with the dominating traits,*[17] the *recessive* ones also reappear, their individuality fully revealed, and they do so in the decisively expressed average proportion of 3:1, so that among each four plants of this generation three receive the dominating and one the recessive characteristic. This is true, without exception, of all traits included in the experiments. (*ibid.*, p. 10)

Thus, while the F_1 hybrids all showed the dominant form, the F_2[18] hybrids showed both the dominant and recessive forms, in the ratio 3:1. Thereupon, Mendel provided the composition of the F_2 hybrids for each of the seven traits (Table 6.1). The phenomenon that Mendel had just observed, namely, reversion (going back to

[14]As indicated in footnote 9 of Chap. 3, the phenomenon of dominance had been observed before Mendel's times, and the word "dominance" had been used too.

[15]This was not at all a settled question at the time.

[16]These would be the seed portion of the F_2 hybrids.

[17]Unfortunately, Mendel uses the same word "trait" for both the different types of characters and the different forms of a given character. For the first one, we have stuck to "trait"; for the second one, we have used "form." In this quotation, Mendel is thus referring to the dominating *forms.*

[18]Mendel's language in describing successive generations is different from the one currently used. Our current F_2 is Mendel's *first* generation from hybrids, and our current F_3 is Mendel's *second* generation from hybrids, and so on.

ancestral characters), had also been observed by several scientists before him, but he was the first to look for a numerical relationship between the number of different forms within the hybrids and the first to realize that large numbers were needed to find such a relationship if it existed.

Of special significance is Mendel's observation that there were two kinds of dominants, the first being a parental characteristic, and the second being a hybrid form. Moreover:

> **QUOT. 410** In which of the two meanings it appears in each individual case only the following generation can decide. As parental trait it would pass unchanged to all of the offspring; as hybrid trait, on the other hand, it would exhibit the same behavior as it did in the first generation. (*ibid.*, p. 13)

In hindsight, Mendel's first type of dominant is simply *homozygous* (or true-breeding) dominants, and his second type of dominant is simply *heterozygous* dominants.

6.2.5 The F_3 Hybrids (1859): Resolution of 3:1 Ratio in F_2 into 2:1:1 and the "Law of Segregation"

The F_3 hybrids were obtained similarly to the F_2 hybrids, by collecting the seeds from the latter, examining them,[19] and allowing them to self-pollinate. This time, Mendel observed that:

> **QUOT. 411** Those forms that receive the recessive character in the first generation do not vary further in the second [in F_3] with respect to this trait; they remain constant in their progeny.
>
> The situation is different for those possessing the dominating trait in the first generation [in F_2]. Of these, two parts yield offspring that carry the dominating and the recessive trait in the proportion of 3:1, thus showing exactly the same behavior as the hybrid forms; only one part remains constant for the dominating trait. (*ibid.*, p. 14)

Thus, all recessives from F_2 gave rise to only recessives in F_3. On the other hand, one-third of the F_2 dominants produced only dominants in F_3, while two-thirds of the F_2 dominants gave dominants and recessives in the ratio 3:1.

Now, the F_2 hybrids consisted of three parts dominant to one part recessive, and 1/3 of the dominants produced only dominants in F_3. It was thus easy for Mendel to infer that, of the three parts dominant in F_2, one part was a parental character (now homozygous dominant), and two parts were hybrid traits (now heterozygous dominant). Therefore:

> **QUOT. 412** The ratio of 3:1 in which the distribution of the dominating and recessive traits takes place in the first generation [in F_2] therefore resolves itself into the ratio of 2:1:1 in all experiments if one differentiates between the meaning of the dominating trait as a hybrid trait and as a parental character. (*ibid.*, p. 15)

---

[19] These would be the seed portion of the F_3 hybrids.

Continuing with his train of thought, Mendel next made a statement that is close to what later became known as *Mendel's Law of Segregation* (or *Mendel's first law*):

> **QUOT. 413** *...the seeds formed by the hybrids with one pair of differing traits, one half again develop the hybrid form while the other half yield plants that remain constant and receive the dominating and the recessive character in equal shares.*[20] *(ibidem)*

However, the reader may at once remark that, in Mendel's statement above, there are no notions of genes (or factors or elements) yet and segregation seems to occur only in the heterozygous hybrids.[21]

6.2.6 *Mendel's Notation*

6.2.6.1 One Trait

To explain the ratio of 1:2:1 for homozygous dominants to hybrids to recessives, Mendel introduced a symbolic notation that attempted to delineate the hereditary process:

> **QUOT. 414** If A denotes one of the two constant traits, for example, the dominating one, a the recessive, and Aa the hybrid form in which both are united, then the expression
>
> $$A + 2Aa + a$$
>
> gives the series for the progeny of plants hybrid in a pair of differing traits. (*ibid.*, p. 16)

Mendel thus denoted the homozygous dominant *form* by A and called it the dominating constant trait; the recessive *form* was denoted by a and was called the recessive constant trait; and the heterozygous dominant *form* was denoted by Aa and was called the hybrid form in which both constant traits are united.[22]

At this stage, Mendel's series $A + 2Aa + a$ was simply a means of representing the ratios of different forms of a given trait. However, imperfect Mendel's notation is by today's standards, it became useful when he next considered further generations and

[20] Italics are Mendel's. The modern version of Mendel's first law is: "During gamete formation, each gamete randomly receives one of the two homologous alleles that control for a trait." The phrase "law of segregation" is due to de Vries' first two rediscovery papers (de Vries, 1900a, c) and was first used in 1902 by Weldon (1902a, p. 229). See Sect. 6.6.1 for the Mendelian rediscovery.

[21] Recall that Naudin had made a similar assumption (Sect. 3.6.1). This has prompted some scholars to ask if Mendel was really a Mendelian (Olby, 1979; Serre, 1984; Blanc, 1985).

[22] In today's language, Mendel is thus referring only to *phenotypes* and not to *genotypes*. A clear-cut notion of these two concepts would emerge only in 1909 with Johannsen's work (1909, pp. 123, 130). We now write the genotype AA for the homozygous (or pure-breeding) dominant form, the genotype Aa for the heterozygous dominant form, and the genotype aa for the recessive form. The allele A is thus dominant for the trait and the allele a recessive. The parents have genotypes AA and aa, and when they are crossed we have $(A + A)(a + a) = 4Aa$, i.e., the F_1 genotypes are all Aa. When F_1 offspring self-fertilize, we have $(A + a)(A + a) = AA + 2Aa + aa$, i.e., the F_2 genotypes are AA, Aa, aa in the ratio 1:2:1.

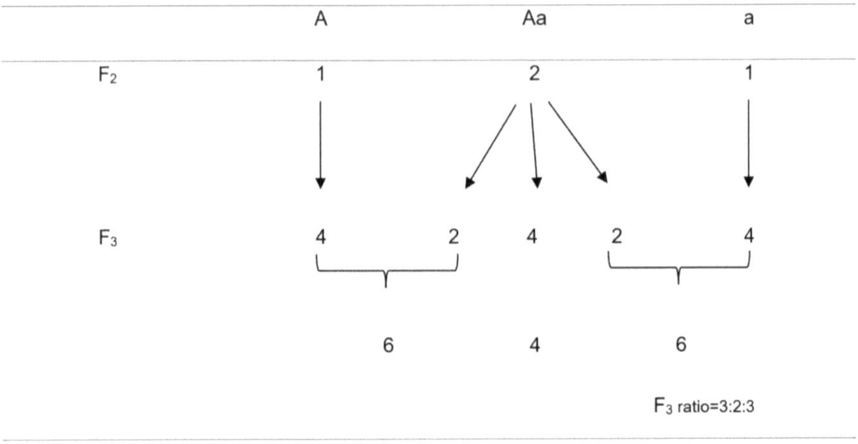

Fig. 6.5 Composition of F_3 from that of F_2

two or more traits at the same time. In addition, it enabled him to come close to the realization that there were two "factors" or "elements" that segregated in the sex cells and united randomly during fertilization.

Equipped with his notation, Mendel was able to predict the composition of offspring in any given generation by self-fertilization from the previous generation. For example, the composition of offspring in F_3[23] may be obtained from that in F_2 as shown in Fig. 6.5 (assuming that each plant produces, say four seeds[24]):

For F_{n+1} (generation n in Mendel's language), Mendel gave (A: Aa: a) as $2^n - 1 : 2 : 2^n - 1$. This formula may be proven by induction as shown in Fig. 6.6

It has thus been proven that if the formula is assumed to be true for generation n, it must also be true for generation $n + 1$. Now, it is obviously true for $n = 1$, and hence, it is true for all natural numbers. In today's language, it shows that homozygosity increases with self-fertilization, which is an extreme form of inbreeding.

6.2.6.2 Two or More Traits: Combination Series and the "Law of Independent Assortment"

Mendel's notation also enabled him to investigate the composition of offspring with respect to two or more traits and to establish a law for the formation and development of hybrids. He started by investigating whether the 1:2:1 ratio held for each trait independently of the other. Based on the several experiments he had conducted, Mendel stated that in general:

[23] Generation 2 in Mendel's language.

[24] The same final ratios are obtained irrespective of the number of seeds assumed.

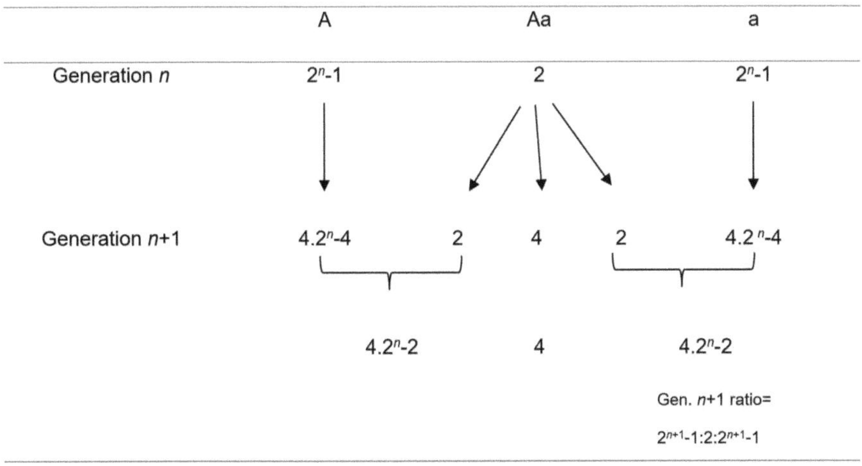

Fig. 6.6 Composition of F_{n+2} derived from that of F_{n+1}

> **QUOT. 415** The experiments demonstrated throughout that in such a case the hybrids always resemble more closely that one of the two parental plants which possesses the greater number of dominating traits. (*ibid.*, p. 17)

He denoted the homozygous dominants ("dominating constant traits") by *A*, *B*, *C*. . ., the recessives ("recessive constant trait") by *a*, *b*, *c*, . . . and the heterozygous dominants ("hybrid form in which both constants traits are united") by *Aa*, *Bb*, *Cc*, . . .

In his first experiment with multiple traits, Mendel crossed pure-breeding peas that were round and had yellow albumen (i.e., *AB*) with pure pure-breeding peas that were angular and had a green albumen (i.e., *ab*). The fertilized seeds (F_1) were round and yellow. When the latter seeds were self-fertilized, the resulting 556 seeds (F_2) were of the following type:

A. 315 round and yellow
B. 101 angular and yellow
C. 108 round and green
D. 32 angular and green

Using Mendel's notation, if the first and second dominant[25] traits are written as A' and B', respectively, then the composition of the F_2 offspring is approximately given by the following series:[26]

$$9A'B' + 3aB' + 3A'b + ab.$$

[25]The dominants here include both the homozygous and heterozygous dominants.

[26]This type of series, regrouping two or more traits, will soon be called a "combination series" by Mendel.

The dominant-to-recessive ratio for both the first and second traits is (9 + 3): (3 + 1) = 3:1, suggesting that the (segregation) ratio for the first trait does not affect that of the second trait (and vice versa).

However, Mendel was not interested in the (now classic) ratio of 9:3:3:1. Just as in the single-trait case[27] where he had to self-fertilize the F_2 offspring to ascertain which were constant types (i.e., true-breeding or homozygous) and which were hybrids (i.e., heterozygous), here too Mendel self-fertilized each of the four F_2 types above. He obtained the following results:

I. Of the 315 round and yellow seeds in F_2, 311 bore fruit and:

(a) 38 produced round yellow seeds and were therefore of type *AB*.
(b) 65 produced round yellow and green seeds and were therefore of type *ABb*.
(c) 60 produced round yellow and angular yellow seeds and were therefore of type *AaB*.
(d) 138 produced round yellow and green and angular yellow and green seeds and were therefore of type *AaBb*.

II. Of the 101 angular and yellow seeds in F_2, 96 bore fruit and:

(a) 28 produced angular yellow seeds and were therefore of type *aB*.
(b) 68 produced round yellow and green seeds and were therefore of type *ABb*.

III. Of the 108 round green seeds in F_2, 102 bore fruit and:

(a) 35 produced round green seeds and were therefore of type *Ab*.
(b) 67 produced round and angular green seeds and were therefore of type *Aab*.

IV. Of the 32 angular green seeds in F_2, 30 bore fruit and:

(a) All 30 produced angular green seeds and were therefore of type *ab*.

From the numbers above and in analogy with the single-trait case (Sect. 6.2.6.1), Mendel was able to give an expression for the composition of F_2 for the two-trait case, as follows:

$$AB + Ab + aB + ab + 2ABb + 2aBb + 2AaB + 2Aab + 4AaBb$$

(*ibid.*, p. 20). Mendel called this a "combination series," since its terms were obtained through a combination of the expressions $A + 2Aa + a$ and $B + 2Bb + b$, i.e.,

$$\begin{aligned} &AB + Ab + aB + ab + 2ABb + 2aBb + 2AaB + 2Aab + 4AaBb \\ &= (A + 2Aa + a)(B + 2Bb + b) \end{aligned}$$

[27] See Sect. 6.2.5.

Mendel had thus discovered a mathematical formula for the prediction of the relative number of offspring for each of the nine types when two traits were considered. He verified his formula for further cases of two traits. He also performed three-trait experiments and was able to verify that the relative numbers produced in F_2 were according to the expansion of

$$(A + 2Aa + a)(B + 2Bb + b)(C + 2Cc + c).$$

Thus, Mendel was able to declare that:

QUOT. 416 *The progeny of hybrids in which several essentially different traits are united represent the terms of a combination series in which the series for each pair of differing traits are combined.*[28] (*ibid.*, p. 22)

This was no doubt the "generally applicable law of the formation and development of hybrids" that he had sought and had referred to at the start of his 1866 paper[29] (*ibid.*, p. 2). Mendel further noted that, if the number of different traits between two parental plants is n, then the number of terms in the combination series is 3^n, the number of individuals belonging to the series is 4^n, and the number of combinations that remain constant is 2^n.

Equally important was Mendel's next statement of a preliminary version of what later came to be known as *Mendel's Law of Independent Assortment* (or *Mendel's second law*):

QUOT. 417 ... *the behavior of each pair of differing traits in a hybrid association is independent of all other differences in the two parental plants.*[30] (*ibid.*, p. 22)

Two points need to be made regarding this law:

(a) Mendel could have deduced it earlier by simply considering the numbers of the four types in F_2 in A-D above, as explained there.
(b) Mendel was fortunate to have considered traits (at least for his pea experiments) that are due to "factors" (genes) that are on different chromosomes (i.e., are not *linked*). Otherwise, the second law does not hold.

---

[28] Italics are Mendel's. Mendel had presumably assumed that, for n traits, and with A_i and a_i representing the dominating and recessive constant forms, respectively, for the ith trait, then the relative numbers of the various forms in F_2 would be given by the expansion of $(A_1 + a_1)(A_2 + a_2) \ldots (A_n + a_n)$.

[29] See also Sect. 6.2.1.

[30] Ditto. The modern version of Mendel's second law is: "During gamete formation, each pair of homologous alleles segregates independently of any other pair."

6.2.7 Mendel's "Anlage," "Factoren," and "Elementen": Toward the Conceptions of Particulate Inheritance and of the Gene, the Backcrosses

On the basis of his results on the observed traits of the crossed peas, Mendel was now ready to propose a mechanism of heredity involving the sex cells themselves. Given the types and compositions of the offspring in the various generations, was it possible to say something about the types and compositions of the egg and pollen cells? Using his knowledge of cell theory, he postulated that the germinal cells and fertilizing pollen were equipped with some kind of "potential" (*anlage*) for producing identical individuals. In addition:

> **QUOT. 418** ...we must consider it inevitable that in a hybrid plant also identical factors (*Factoren*) are acting together in the production of constant forms. Since the different constant forms are produced in a single plant, even in just a single flower, it seems logical to conclude that in the ovaries of hybrids as many kinds of germinal cells (germinal vesicles), and in the anthers as many kinds of pollen cells are formed as there are possibilities for constant combination forms and that these germinal and pollen cells correspond in their internal make-up to the individual forms. (*ibid.*, p. 24)

Mendel's "identical factors" are currently known as alleles (different forms of a gene). They are particulate (or discrete) units that reside in the sex cells of plants and combine in the production of hybrids. Mendel's hypothesis above was that there should be as many different types of factors inside the cells of pollens and eggs as there were constant forms. For example, in Sect. 6.2.6.2 it was shown that the *AaBb* dihybrids, when self-fertilized, produced the four constant types *AB*, *Ab*, *aB*, and *ab* in the ratio 1:1:1:1 (as well as other nonconstant types). According to Mendel's hypothesis, this implies that each plant (indeed each flower) produced pollen or egg cells containing factors for each of the four types in equal proportions.

To test the hypothesis above, Mendel performed a new type of cross, namely, a backcross (or test-cross). He crossed some pure-breeding round yellow (*AB*) seeds and some pure-breeding angular green (*ab*) pollen, producing round yellow (*AaBb*) hybrids. These hybrids were then crossed back (hence the name "backcross") with the parents as follows:

1. The	hybrid	with	pollen	from	*AB*.
2. The	hybrid	"	"	"	*ab*.
3. *AB*	"	"	"	the	hybrid.
4. *ab*	"	"	"	the	hybrid.

If Mendel's hypothesis was correct, then the *AaBb* hybrids should form germinal and pollen cells of the forms *AB*, *Ab*, *aB*, and *ab* in equal proportions. In terms of pollen and germinal cells, the backcrosses in (1)–(4) above are then represented as:

1′.	Germinal	cells	*AB, Ab, aB, ab*	with	pollen	cells	*AB.*
2′.	"	"	*AB, Ab, aB, ab*	"	"	"	*ab.*
3′.	"	"	*AB*	"	"	"	*AB, Ab, aB, ab*
4′.	"	"	*ab*	"	"	"	*AB, Ab, aB, ab*

Under Mendel's hypothesis, the following offspring types should be obtained in equal proportions:

1″. AB, ABb, AaB, AaBb.
2″. AaBb, Aab, aBb, ab.
3″. AB, ABb, AaB, AaBb.
4″. AaBb, Aab, aBb, ab.

Mendel's hypothesis was indeed confirmed. All the identical crossings (1) and (3) produced round yellow seeds. However, more importantly, the identical crossings (2) and (4) produced each of the four types in an approximate 1:1:1:1 ratio:

2‴. 31 round yellow (*AaBb*), 26 round green (*Aab*), 27 angular yellow (*aBb*), 26 angular green seeds (*ab*);
4‴. 24 round yellow (*AaBb*), 25 round green (*Aab*), 22 angular yellow (*aBb*), 27 angular green seeds (*ab*).

Mendel performed further experiments which still confirmed his hypothesis.

By considering the "factors" inside the pollen and egg cells, Mendel seems to have moved quite close to the concept of a gene. Again, in the concluding section of Mendel's paper, one can read:

QUOT. 419 …propagation in phanerogams is initiated by the union of one germinal and one pollen cell into one single cell, which is able to develop into an independent organism through incorporation of matter and the formation of new cells. This development proceeds in accord with a constant law based on the material composition and arrangement of the elements (*Elementen*) that attained a viable union in the cell. When the reproductive cells are of the same kind and like the primordial cell of the mother, development of the new individual is governed by the same law that is valid for the mother plant. When a germinal cell is successfully combined with a dissimilar pollen cell we have to assume that some compromise takes place between those elements of both cells that cause their differences. The resulting mediating cell becomes the basis of the hybrid organism whose development must necessarily proceed in accord with a law different from that for each of the two parental types. (*ibid.*, pp. 41–42)

The gist of Mendel's theory is clearly indicated in the following paragraph, which involves the crossing of two *Aa* hybrids in the production of four offspring (two per hybrid):

QUOT. 420 In their production, pollen and germinal cells of form *A* and *a* participate, on the average, equally in fertilization; therefore each form manifests itself twice, since four individuals are produced. Participating in fertilization are thus:

$$\text{Pollen cells } A + A + a + a$$

$$\text{Germinal cells } A + A + a + a$$

It is entirely a matter of chance which of the two kinds of pollen combines with each single germinal cell. However, according to the laws of probability, in an average of many cases, it will always happen that every pollen form *A* and *a* will unite equally often with every germinal cell form *A* and *a*; therefore, in fertilization, one of the two pollen cells *A* will meet a germinal cell *A*, the other a germinal cell *a*, and, equally, one pollen cell *a* will become associated with a germinal cell *A*, the other with *a*.

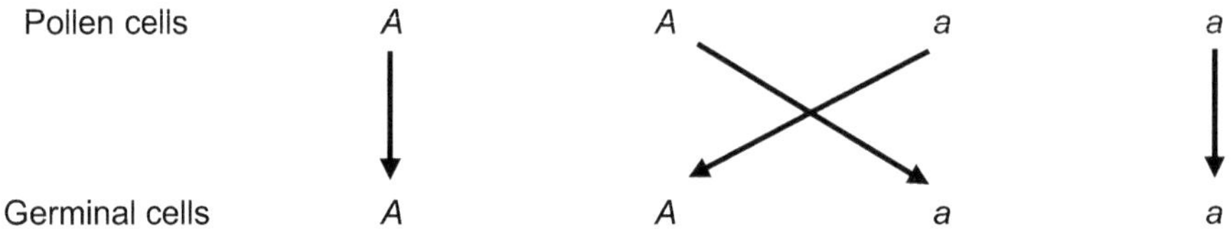

The result of fertilization can be visualized by writing the designations for associated germinal and pollen cells in the form of fractions, pollen cells above the line, germinal cells below. In the case under discussion one obtains:

$$\frac{A}{A}+\frac{A}{a}+\frac{a}{A}+\frac{a}{a}$$

In the first and fourth terms, germinal and pollen cells are alike; therefore, the products of their association must be constant, namely, *A* and *a*; in the second and third, however, a union of the two differing parental traits takes place again; therefore, the forms arising from such fertilizations are absolutely identical with the hybrid from which they derive. *Thus, repeated hybridization takes place.*[31] The striking phenomenon that hybrids are able to produce, in addition to the two parental types, progeny that resemble themselves is thus explained: $\frac{A}{a}$ and $\frac{a}{A}$ both give the same association, *Aa*, since, as mentioned earlier, it makes no difference to the consequence of fertilization which of the two traits belongs to the pollen and which to the germinal cell. Therefore,

$$\frac{A}{A}+\frac{A}{a}+\frac{a}{A}+\frac{a}{a}=A+2Aa+a.$$

(*ibid.*, pp. 29–30)

The above is almost how the *Aa* × *Aa* cross is currently visualized. The only difference is that Mendel wrote $\frac{A}{A}$ and $\frac{a}{a}$ as *A* and *a*, instead of *AA* and *aa*, respectively. However, as mentioned in Sect. 6.2.6.1, Mendel's writing of $\frac{A}{A}$ and $\frac{a}{a}$ as *A* and *a* is an indication that his concepts of two factors per trait and of segregation of these factors were limited to hybrids only.

[31] Italics are Mendel's.

6.2.8 *Mendel's Explanation of Continuous Variation*

It is perhaps little known that, in the *Pisum* paper, Mendel had already outlined the explanation that the biometricians and saltationists would spend years looking for in their debate over whether evolution proceeded continuously or in steps. To explain the continuous variation in color in the flowers and seeds of *Phaseolus multiflorus*, Mendel stated:

QUOT. 421 …these puzzling phenomena, too, could probably be explained by the law valid for *Pisum* if one might assume that in *Ph. multiflorus* the color of flowers and seeds is composed of two or more totally independent colors that behave individually exactly like any other constant trait in the plant. (*ibid.*, p. 35)

Thus, the continuity of color could be explained by the independent action of several colors each acting as a trait. If each trait is due to a factor, then Mendel's explanation of continuous variation leads directly to the multifactorial hypothesis.

When Mendel's laws were later rediscovered, Bateson and Saunders made a similar point in the first Report to the Evolution Committee of the Royal Society (Sect. 9.2.4(c)).

However, biologists were at the time too consumed by the battle over the validity of the ancestral law as opposed to Mendel's laws to pay serious attention to Mendel's early recommendations.

6.2.9 *Formation of New Species by Hybridization*

6.2.9.1 Constant versus Variable Hybrids: Formation of New Species Through Constant Hybrids

In the concluding remarks of his paper, Mendel returned to one of the questions that motivated his paper, namely, can new species be formed by hybridization? As mentioned in (c) of Sect. 6.1.2, Mendel's lecturer, Franz Unger, believed that hybridization resulted in the production of new species, but Kölreuter and Gärtner had opposed this view.

Before Mendel's answer to the species question is examined, it is important to appreciate how the term "species" was then understood. For Mendel and many of his contemporaries,[32] there was no clear-cut distinction between species and varieties (i.e., different forms of the same species): morphologically different varieties were often viewed as different species. Mendel phrased it thus:

QUOT. 422 Just as it is impossible to draw a sharp line between species and varieties, it has been equally impossible so far to establish a fundamental difference between the hybrids of species and those of varieties. (Mendel, 1866, 1966a, p. 5)

---

[32] Such as Darwin himself, see Sect. 2.14.1.

Having mentioned the case of *Pisum* and other "species" whose F_1 hybrids gave rise to different types of offspring, Mendel observed that a completely different situation existed whereby F_1 hybrids bred true. He called the latter *constant hybrids*, that is, "[F_1] hybrids that remain constant in their progeny and propagate like pure strains" (*ibid*., p. 41). Mendel noted that the distinction between constant versus variable hybrids was of:

QUOT. 423. . .particular importance to the evolutionary history of plants, because constant hybrids attain the status of *new species.*[33] (*ibidem*)

Although he had not obtained any constant hybrids in his research thus far, Mendel mentioned that Gärtner had done so with *Aquilegia atropurpurea-canadensis*, *Lavatera pseudolbia-thuringiaca*, *Geum urbano-rivale*, and some *Dianthus* hybrids.[34]

Mendel explained the occurrence of constant hybrids through a hypothetical mechanism involving *permanently associated mediated cells*:

QUOT. 424 When a germinal cell is successfully combined with a dissimilar pollen cell we have to assume that some compromise takes place between those elements of both cells that cause their differences. The resulting mediating cell becomes the basis of the hybrid organism whose development must necessarily proceed in accord with a law different from that for each of the two parental types. If the compromise be considered complete, in the sense that the hybrid embryo is made up of cells of like kind in which the differences are *entirely and permanently mediated*,[35] then a further consequence would be that the hybrid would remain as constant in its progeny as any other stable plant variety. (*ibid*., p. 42)

In variable hybrids, mediating cells are also formed through a compromise between germinal and pollen cells, but this compromise is disrupted as soon as the hybrids produce reproductive cells. At this point, the cells are completely free. The mediating cells are therefore only in *temporary association*.

The difference between constant and variable hybrids would be a major issue in Mendel's later *Hieracium* paper.

6.2.9.2 Species Transformation Through Variable Hybrids

Thus far, Mendel had explained the formation of new species in the context of constant hybrids. In the case of variable hybrids (e.g., in *Pisum*), Mendel understood species transformation as follows:

QUOT. 425 When species A was to be transformed into B, the two were combined by fertilization and the resulting hybrids once more fertilized with pollen from B; from among their various descendants those closest to species B were then chosen and repeatedly

---

[33] Italics are Mendel's.

[34] It is now known that most hybrids are variable: the hybrids *Aa* when self-fertilized must produce *AA*, *aa*, and *Aa*. In rare cases, such as apomixis (a term first applied by Winkler (1906)), offspring develop as a material clone without the need for male fertilization, and the hybrids are therefore constant.

[35] Italics are Mendel's.

> fertilized by pollen from B, and so on, until finally a form that was like B and remained constant in its progeny was obtained. Thus species A was transformed into the other species, B. (*ibid.*, p. 44)

Based on the results of his experiments thus far, Mendel in the above explained how a "species" could be transformed into another as follows. Consider a plant with three traits. Suppose it is intended to transform the "species" *ABC* into the "species" *abc* by repeated fertilization (backcrossing) with pollen from the latter. Then, the hybrid has the following eight types of germinal cells:

$$ABC, ABc, AbC, aBC, Abc, aBc, abC, abc.$$

When the hybrid is again fertilized with the *abc* pollen cells, the progeny obtained is according to the combination series

$$AaBbCc + AaBbc + AabCc + aBbCc + Aabc + aBbc + abCc + abc.$$

Therefore:

> **QUOT. 426**. . .[s]ince the form *abc* occurs once in the series of 8 terms there is little likelihood that it would be missing among the experimental plants, even if only a fairly small number were raised, and transformation would thus be complete after two fertilizations. (*ibid.*, p. 45)

Mendel rightly noted that fertilization would have to be continued if the number of experimental plants was small or if the number of differing traits was large (i.e., A and B were widely divergent species).

Mendel thus explained the process of "species transformation," both through constant and variable hybrids. In retrospect, Mendel was neither dealing with different species nor had he described a transformation in the evolutionary sense. What he did achieve, and where his predecessors had failed, was the transformation of a variety to another stable one through hybridization and the development of a theory behind it. Again, his success was due to three key ingredients in his experiments: the utilization of a large number of plants, the adoption of a mathematical approach, and the focus on specific traits (rather than the organism in toto).

6.2.10 Two Lectures in 1865 and Publication of Mendel's Groundbreaking Paper in 1866

Mendel finished his experiments in 1863.[36] Two years later, in the February 8, 1865, meeting of the Natural Sciences Society of Brno, Mendel presented some of the results in a first lecture. This was followed by a second lecture on March 8, 1865. In a

[36] One year earlier, in 1862, the German translation of the first edition of Darwin's *Origin of Species* had come out and reached the Brno naturalists. In 1863, Mendel bought the German translation of the second edition which had also come out in that same year.

December 31, 1866, letter to Nägeli, Mendel mentioned that his 1866 paper was "the unchanged reprint of the draft of the lecture[s] mentioned" (Correns, 1905, 1966b, p. 61). However, there are no sources regarding the specific sections of his paper that were covered in each lecture, except for the articles by an anonymous reporter of the Brno daily newspaper, *Neuigkeiten.*[37]

The first newspaper article (dated February 9, 1865) reported that, in his long (first) lecture, Mendel pointed out the fertility of hybrids and their tendency to revert to their parental types. This reversion was accelerated by repeated backcrossing with the parent form. Mendel performed experiments with many species of the genus *Pisum*, namely, *Pisum* sativum, *P. saccharatum*, and *P. quadratum*. The article also explained the different traits with which Mendel had worked and noted that the numerical relationships of the different types of hybrids were of particular interest. The audience had listened with "enthusiastic interest." The first lecture thus probably contained the material in Sects. 6.2.2, 6.2.3, 6.2.4, 6.2.5, and 6.2.6 above.

The second newspaper article (dated March 10, 1865) reported that, in his second lecture, Mendel continued from his first lecture. He spoke about the production of germ cells, fertilization, and the formation of seeds in general. The article also mentioned that Mendel had performed successful experiments with other closely related species of plants during the past years and that he intended to pursue his experiments in the coming summer. The second lecture received "general approval." This lecture thus probably contained the material in Sects. 6.2.7 and 6.2.9.2 above.

The contents of Mendel's two lectures were published the following year, in 1866,[38] in the *Proceedings* of the Natural Science Society of Brno (Mendel, 1866, 1966a). The *Proceedings* were distributed to 134 scientific institutions in various countries. Mendel requested 40 reprints of his paper, but the exact names of those to whom he sent a copy of his paper are not known.[39] Mendel sent a reprint together with an introductory letter to Carl Nägeli (1817–1891) on December 31, 1866, and another to Anton Kerner (1831–1898) on January 1, 1867. The latter had been Mendel's lecturer at Vienna University, and Nägeli was a leading botanist of his time. As mentioned before (Sect. 6.1.2), Mendel most likely learned about Nägeli and his work through his university lecturer, Franz Unger. A reprint might also have been sent to the latter,[40] who greatly influenced Mendel during his university years. It is also noteworthy that, as noted by Stomps (1954), a reprint was sent by the microbiologist and botanist M.W. Beijerinck (1851–1931) in 1900 to Hugo de Vries when he learned that the latter was conducting hybridization experiments.

[37] The contents of Mendel's two lectures have been analyzed by de Beer (1966), Olby and Gautrey (1968), and Zhang et al. (2017).

[38] This is the same year that Prussia defeated Austria in a 7-week war and assumed control of Germany.

[39] A few have argued that Mendel sent a reprint to Darwin, who did not read it. See footnote 22 in Chap. 3.

[40] See Sect. 6.1.2.

6.2.11 The Aftermath

Although Mendel's 1866 paper is currently viewed as a paradigm of clarity, it was all but forgotten between the time of its publication and its rediscovery in 1900. This has prompted some scholars to blame Mendel's contemporaries for failing to appreciate the real significance of his work. Thus, the prominent British zoologist Arthur Darbishire (1879–1915) stated that:

> **QUOT. 427** The publication of Mendel's paper in 1865 [*sic*] was the throwing of pearls before swine. (Darbishire, 1911, p. 189)

Darbishire was surely right in likening Mendel's paper to a pearl in the history of science, but it is probably Whiggish to castigate such blame. There are valid reasons that may explain the neglect of Mendel's work:

(a) *Lack of name recognition.* Mendel's difficult personal circumstances make comparisons with Darwin inevitable. Before the theory of evolution by natural selection was proposed, Darwin was relatively well-known through his *Beagle* expedition around the world. Not only did Darwin's expedition increase his stature among his contemporaries, but it also enabled him to publish. Mendel made no such worldwide travels nor did he have the means to afford such an endeavor (unlike Darwin, who was well-to-do). When Mendel published his 1866 paper, he was a virtual nobody in the field. While Darwin had Huxley (his "bulldog"), Gray (his "retriever"), and Wallace (his "knight") to fiercely defend and popularize his ideas, Mendel did not have anyone who even understood his work. The only person with whom he actively communicated, Carl Nägeli (Sect. 6.3 below), had other ideas about hybridization and heredity and therefore did not popularize Mendel's work.

(b) *Perceived lack of relationship between Darwin's theory of evolution and Mendel's hybridization experiments.* The world of biology was still grappling with Darwin's ideas on evolution in 1866. Mendel's paper on hybridization appeared to many as isolated observations specifically on *Pisum* and a few other plant species and thus seemed to be unrelated to Darwin's powerful theory. This perception was further accentuated since Mendel's experiments had shown variations were *discrete* while Darwin's theory had postulated that natural selection acted on *continuous* ones.

(c) *Peculiarities of Mendel's paper.* Various characteristics of Mendel's paper may also have contributed to its neglect. First, if the paper is again compared to Darwin's *Origin of Species*, one finds in the latter an explicit statement and delineation of natural selection. On the other hand, although Mendel's paper was clearly written, it did not explicitly address the issue of heredity. Nor there is any indication that Mendel appreciated that his theory complemented that of Darwin. Second, the quantitative nature of Mendel's results may have discouraged many biologists. A third point relates to the fact that the understanding of heredity in

Mendel's times usually proceeded from characteristics to sex cells,[41] whereas Mendel's paper proceeded in the reverse direction.

Researchers[42] have found that, from the time of publication in 1866 till the rediscovery in 1900, Mendel's work managed to garner citations from only approximately 11 sources (only the last of which showed some understanding of his work):

1. *The journal Flora* (1866): The German botanical journal *Flora* lists the botanic literature for the year 1866 and mentions Mendel's paper at no. 354. (p. 26 of pp. 17–55) Flora c.1 v.49 (1866, p. 17).
2. *Hoffmann* (1869): Hermann Hoffmann (1819–1891) was a German Professor of Botany and an opponent of Darwin's theory. In his book, he cited Mendel five times, but only two of these referred to his *Pisum* experiments (pp. 52, 136). These citations mentioned nothing more than some of the details of the experiments and the fact of reversion among offspring of the F_1 hybrids. Darwin referred to Hoffmann's book in his *The Effects of Cross and Self Fertilisation in the Vegetable Kingdom* (Darwin, 1876, pp. 151–152). He must therefore have read the book, but it is doubtful that he ever read Mendel's paper. It was argued in Sect. 3.6.2 that, even if he did, it is unlikely that Darwin would have appreciated the significance of Mendel's work.
3. *Besnard* (1872): This paper was published in the journal *Flora* by Dr. Anton Besnard in Munchen. It merely lists the names of authors who had performed research on the genus *Hieracium*, and Mendel's 1866 and 1869 papers are both mentioned on p. 392.
4. *Blomberg* (1872): Albert Blomberg was a student at Uppsala University in Sweden. In his 1873 thesis, entitled "On Hybrid Formation," he discussed Mendel's *Pisum* experiments and compared his work to that of Naudin.[43]
5. *Schmalhausen* (1874): Of all those who cited Mendel before 1900, the Russian botanist Johannes Schmalhausen (1849–1849)[44] has the distinction of being the only one who accurately reported the major results of Mendel's experiments. Schmalhausen became aware of Mendel's 1866 paper only after he had submitted his dissertation for a master's degree in 1874. Thereupon, he added a footnote explaining in detail the numerical results obtained by Mendel. He also pointed to the similarity between the works of Mendel and Naudin.
6. *Focke* (1881): Wilhelm Focke (1834–1922) was a Bremen physician and botanist who published a major book on plant hybrids in 1881, entitled *Die Pflanzen-Mischlinge, Ein Beitrag zur Biologie dfer Gewächse*[45] (Focke, 1881).

[41] A prime example is Darwin's provisional hypothesis of pangenesis (Sect. 3.7.2).

[42] See Olby and Gautrey (1968), Stubbe (1972, pp. 161–165), Blanc (1984), and Gallais (2018, pp. 51–52).

[43] See Sect. 3.6.

[44] Frequently confused by many authors with his son, the later Russian evolutionary biologist Ivan Schmalhausen (1884–1963).

[45] The Plant Hybrids, A Contribution to the Biology of Plants.

Mendel's name appears 15 times in the book, and his work on *Pisum* is treated on pp. 108–110. On p. 110, Focke stated that "Mendel's numerous crossing gave results similar to Knight's, but Mendel believed he had found constant numerical ratios between the different types of hybrids." Two additional observations may be made regarding Focke's book. First, it was from this book that both Correns and Tschermak found the reference to Mendel's paper in 1900. Second, a copy of the book was sent from Darwin to Romanes, as will be explained next.

7. *Romanes* (1881): George Romanes (1848–1894) was a Canadian-Scots evolutionary biologist and a disciple of Darwin. Romanes wanted a list of hybridists for an article entitled "Hybridism" that he was writing for the *Encyclopedia Britannica*. He asked Darwin for help and the latter sent him Focke's book. Romanes found the reference very useful and drew up a list of hybridists that included Mendel (p. 426 of Vol. 12).
8. *Jackson* (1881): In 1881, Focke's book (1881) was used again to reference Mendel's 1866 paper, this time by the British botanist Benjamin Daydon Jackson (1846–1927). The guide merely lists the works of various botanists and Mendel's paper appears on p. 100.
9. *Nägeli and Peter* (1885): It has often been stated that, despite his correspondence with Mendel, Nägeli never referenced the latter's works. This is not true because Mendel's 1866 paper was indeed referenced by Nägeli in a joint paper (Nägeli & Peter, 1885) with the German botanist Albert Peter (1853–1937). On p. 65, the authors stated that: "Artificial hybrids have been raised within the section Piloselloiden by F. Schultz and Mendel, also we have recently made many attempts to produce them, but with less success. It is evident that many of these artificial hybrids are either identical or very like hybrids observed growing in the wild."[46] Mendel's name was further mentioned three other times.[47]
10. *Bailey* (1892): Yet another reference to Mendel's 1866 paper was picked from Focke's book by the American horticulturist and botanist Liberty Hyde Bailey (1858–1954). In an 1892 article (Bailey, 1892), Bailey merely listed Mendel's 1866 paper on p. 32 (and his 1869 paper on p. 34). Bailey had not read Mendel's work and included it again in the second edition of his book *Plant Breeding* (Bailey, 1902).[48] Later, de Vries informed Bailey that he had become aware of Mendel's work from his 1892 paper.[49]

[46] This translation is taken from Olby and Gautrey (1968, p. 15).

[47] Nägeli referenced Mendel a second time in the foreword he wrote for E. Widmer's monograph of 1891 (Widmer, 1891, p. 4). This reference was only to Mendel's *Hieracium* experiments, though.

[48] The first edition of Bailey's book (Bailey, 1895) did not contain a bibliography.

[49] However, de Vries also informed H. F. Roberts (while the latter was writing his major book *Plant Hybridization Before Mendel* (Roberts, 1929)) that he had learnt of Mendel's work from Bailey's (1895) book. But this seems to be a faulty recollection, given that the 1895 edition contained no mention of Mendel.

11. *Correns* (1899): Correns' paper on xenia[50] in maize was read on December 29, 1899, and has the distinction of being "the first printed indication that anyone had understood any part of Mendel's work" (Sturtevant, 1965, p. 28).[51] In the last paragraph of the paper, Correns criticized Trevor Clarke for attributing to xenia the color change in the seeds of *Matthiola* hybrids. He then stated that the color change was in fact due to the color in the cytoledons, adding that "it is the same state of affairs as in the hybridisation of yellow and green seeded races of peas, which Darwin and Mendel have already pointed out correctly."

After the *Pisum* paper, Mendel published only one more work on hybridization, more specifically a report on *Hieracium* in 1869 (next section). He became an abbot in 1868, and his monastery duties prevented him from pursuing research with the same intensity as before. He passed away in 1884.

6.3 Correspondence with Carl Nägeli (1866–1874) and the Hieracium Experiments (1870)

6.3.1 Introduction

Of all those to whom Mendel sent a reprint of his 1866 paper, Carl Nägeli (1817–1891)[52] seems to have been the only one to reply. In addition, he was certainly one of the most qualified people to critique his work.

Mendel sent a reprint together with an introductory letter on December 31, 1866. He received a response from Nägeli on February 27, 1866. Thereupon a correspondence between the two men ensued. Mendel's letters to Nägeli were found in 1904 and given by the latter's family to Correns, who had been Nägeli's student. Correns published the letters in 1905 (Correns, 1905). Unfortunately, Nägeli's letters are no longer extant, except for fragments from three letters. A portion of Nägeli's first letter appears in Iltis (1932, pp. 191–192). Other fragments may be found in Hoppe (1971).

The dates of Mendel's letters to Nägeli are December 31, 1866 (1st), April 18, 1867 (2nd), November 6, 1867 (3rd), February 9, 1868 (4th), May 4, 1868 (5th), June 12, 1868 (6th), April 15, 1869 (7th), July 3, 1870 (8th), September 27, 1870 (9th), November 18, 1874 (10th).

The start of Mendel's first letter reads as follows:

[50] Xenia is a process of double-fertilization whereby characters of the pollen-donating plant start appearing in the seeds and fruit of the fertilized plant already, rather than in the offspring. The term was coined in 1881 by Wilhelm Focke (1881, p. 510). Note that, as Correns correctly pointed out, xenia causes changes in the endosperm rather than the embryo.

[51] However, see Sect. 6.6.3.2.

[52] See Sect. 5.1 for Nägeli's views on heredity.

QUOT. 428

HIGHLY ESTEEMED SIR:

The acknowledged pre-eminence your honor enjoys in the detection and classification of wild-growing plant hybrids makes it my agreeable duty to submit for your kind consideration the description of some experiments in artificial fertilization.

The experiments, which were made with different varieties of Pisum, resulted in the offspring of the hybrids forming curious series, the members of which in equal measure resembled the two original types. The presence of nonvariant intermediate forms, which occurred in each experiment, seems to deserve special attention. In the series for two and three differing traits, discussed in the monograph (pp. 20–21), the notations for the constant forms have been placed first, since the terms are arranged according to their coefficients; but they receive a more correct position when the terms are placed according to their natural relationship to the two parental types, whereby that term which represents a hybrid in all traits and has, at the same time... (Correns, 1905, 1966b, p. 57)

The dates of the three fragments of Nägeli's letters are: February 25, 1867 (1st), May 11, 1868, and April 27, 1870. The start of the extant portion of Nägeli's first letter to Mendel reads as follows:

QUOT. 429

HONOURED COLLEAGUE:

It seems to me that the experiments with Pisum, far from being finished, are only beginning. The mistake made by all the more recent experimenters is that they have shown so much less perseverance than Koelreuter and Gartner. I note with pleasure that you are not making this mistake, and that you are treading in the footsteps of your famous predecessors. You should, however, try to excel them, and in my view this will only be possible (and thus alone can any advance be made in the theory of hybridisation) if experiments of an exhaustive character are made upon one single object in every conceivable direction. No such complete series of experiments, providing irrefutable proofs for the most momentous conclusions, has ever yet been made...I refrain from comment upon other points in your communication, for, without detailed knowledge of the experiments upon which your monograph is based, what I can say would be mere surmise. Your design to experiment on plants of other kinds is excellent, and I am convinced that with these different forms you will get notably different results (in respect of the inherited characters). It would seem to me especially valuable if you were able to effect hybrid fertilisations in Hieracium, for this will soon be the species about whose intermediate forms we shall have the most precise knowledge... (Iltis, 1932, pp. 191–192)

From Mendel's second letter, it can be gathered that Nägeli's letter above also contained the following statement: "You [Mendel] should regard the numerical expressions as being only empirical, because they cannot be proved rational" (Correns, 1905, 1966b, p. 63).

After his 1866 paper, Mendel published only one other paper on hybridization in 1870 entitled "Uber einige aus künstlicher Befruchtung gewonnenen Hieracium-Bastarde"[53] (Mendel, 1869, 1966b, pp. 49–55) (Fig. 6.7). The latter was, in fact, a short report on his initial findings on *Hieracium.*[54] The Mendel-Nägeli correspondence provides important insights into Mendel's thinking during these periods. At

[53]"On Hieracium-Hybrids Obtained by Artificial Fertilisation."

[54]Also known by the common name "hawkweed."

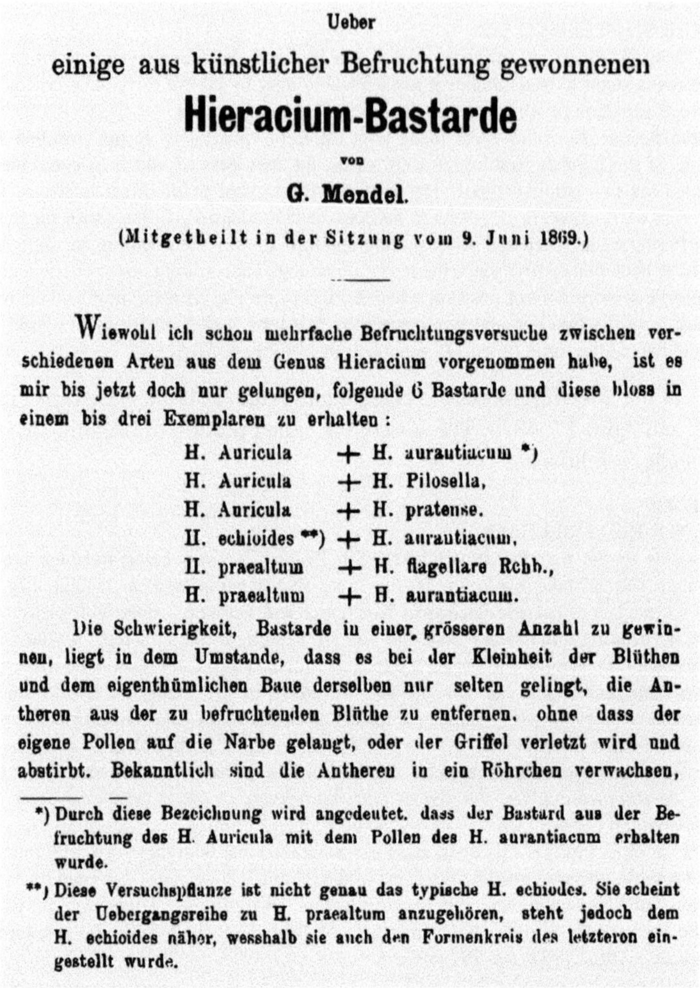

Ueber

einige aus künstlicher Befruchtung gewonnenen

Hieracium-Bastarde

von

G. Mendel.

(Mitgetheilt in der Sitzung vom 9. Juni 1869.)

Wiewohl ich schon mehrfache Befruchtungsversuche zwischen verschiedenen Arten aus dem Genus Hieracium vorgenommen habe, ist es mir bis jetzt doch nur gelungen, folgende 6 Bastarde und diese bloss in einem bis drei Exemplaren zu erhalten:

H. Auricula + H. aurantiacum *)
H. Auricula + H. Pilosella,
H. Auricula + H. pratense.
H. echioides **) + H. aurantiacum,
H. praealtum + H. flagellare Rchb.,
H. praealtum + H. aurantiacum.

Die Schwierigkeit, Bastarde in einer grösseren Anzahl zu gewinnen, liegt in dem Umstande, dass es bei der Kleinheit der Blüthen und dem eigenthümlichen Baue derselben nur selten gelingt, die Antheren aus der zu befruchtenden Blüthe zu entfernen, ohne dass der eigene Pollen auf die Narbe gelangt, oder der Griffel verletzt wird und abstirbt. Bekanntlich sind die Antheren in ein Röhrchen verwachsen,

*) Durch diese Bezeichnung wird angedeutet, dass der Bastard aus der Befruchtung des H. Auricula mit dem Pollen des H. aurantiacum erhalten wurde.

**) Diese Versuchspflanze ist nicht genau das typische H. echioides. Sie scheint der Uebergangsreihe zu H. praealtum anzugehören, steht jedoch dem H. echioides näher, wesshalb sie auch den Formenkreis des letzteren eingestellt wurde.

Fig. 6.7 First page of Mendel's "Uber einige aus künstlicher Befruchtung gewonnenen Hieracium-Bastarde" (Mendel, 1869, 1966b, pp. 49–55)

the same time, because of a general lack of documents, some misunderstandings have arisen regarding Mendel's views and motivations. Proper reading and interpretation of the correspondence help dispel many of these fallacies.

6.3.2 *Why Mendel Chose* Hieracium

A widely held mistaken belief is that, after his *Pisum* experiments, Mendel turned to *Hieracium* at Nägeli's suggestion. It is now known that this genus develops through apomixis, whereby an offspring is formed as a material clone without the need for male fertilization. Nägeli is therefore often blamed for Mendel's failure to replicate his 1866 findings and for the subsequent neglect of Mendel's work. For example, Moore wrote:

> **QUOT. 430** Nägeli suggested that Mendel try another plant – *Hieracium*, the hawkweed. Mendel did and failed to find consistent rules for inheritance. (Moore, 1999, p. 298)

Everdell has been even more critical of Nägeli:

> **QUOT. 431**. . .only one of the botanists wrote back. He was Karl Nägeli. . .Why not study a different plant, Nägeli advised – hawkweed, for example? So Mendel had abandoned his peas and studied hawkweeds (genus Hieracium) for the next five years, trying and failing to find true breeding traits in it, while Nägeli, like Laban, gently encouraged his Jacob from afar. (Everdell, 1997, p. 160)

However, a careful reading of Mendel's first letter to Nägeli reveals a quite different state of affairs. In his letter, as soon as Mendel had provided a few details about his *Pisum* experiments, he mentioned that he intended to experiment with *Hieracium, Circium*, and *Geum.* He stated that he had already performed some preliminary tests with *Hieracium* and found artificial fertilization extremely difficult because of the small size and structure of the flower:

> **QUOT. 432** *Hieracium*, *Cirsium*, and *Geum* I have selected for further experiments. In the first two, manipulation in artificial pollination is very difficult and unreliable because of the small size and peculiar structure of the flowers. Last summer I tried to combine *H. Pilosella* with *pratense*, *praealtum*, and *Auricula*; and *H. murorum* with *umbellatum* and *pretense*, and I did obtain viable seeds; however. I fear that in spite of all precautions, self-fertilization did occur. (Correns, 1905, 1966b, p. 58)

Thus, Mendel's decision to experiment with *Hieracium* was *not* made upon Nägeli's suggestion. Rather, Mendel intended to do so before Nägeli wrote back to him. At the same time, being an expert in *Hieracium* himself, Nägeli certainly did not dissuade him from experimenting with this genus.

Why then did Mendel turn to *Hieracium* for his next set of experiments? There may be two reasons for this:

1. The first reason was alluded to by Mendel himself in his 1870 paper and has to do with the large number of possible varieties of *Hieracium*: "This [*Hieracium*] genus possesses such an extraordinary profusion of distinct forms that no other genus of plants can compare with it" (Mendel, 1869, p. 50).
2. A second reason is that Mendel knew that Nägeli was an expert on *Hieracium* and that he could obtain assistance with the design and interpretation of his experiments from the renowned professor. In fact, at the end of his first letter to Nägeli, Mendel stated: "I am afraid that in the course of my experiments, especially with Hieracium, I shall encounter many difficulties, and therefore I am turning confidently to your honor with the request that you not deny me your esteemed interest when I need your advice" (Correns, 1905, 1966b, p. 59). This second reason would make Nägeli indeed responsible, albeit in an indirect manner, for Mendel's choice of *Hieracium* for further experiments.

6.3.3 *Mendel's Intentions in the* Hieracium *Paper*

Some confusion also surrounds the issue of what exactly Mendel wished to do in his *Hieracium* paper of 1869. Many authors (e.g., Iltis, 1932, p. 174; Hartl & Orel, 1992, p. 253) have claimed that Mendel wanted to verify whether the same segregation ratios he had already obtained for *Pisum* also held for *Hieracium.* After all, if that was the case, then it would lend more support to his "generally applicable law" (Sect. 6.2.6.2). These authors base their claim on Mendel's own words in his first letter to Nägeli:

QUOT. 433 In order to determine the agreement, if any, with Pisum, a study of those forms which occur in the first generation should be sufficient. (Correns, 1905, 1966b, p. 57)

Other authors (e.g., Callender, 1988, p. 59; van Dijk & Ellis, 2016), however, have indicated that the *Hieracium* paper had an altogether different aim from that of the *Pisum* paper. These authors have argued that Mendel was not so much interested in segregation ratios as he was in the formation of *constant hybrids* (i.e., F_1 hybrids that bred true),[55] which would have thus proven that new species could be formed by hybridization. This argument is based on the fact that, in both his 1869 paper and his communication with Nägeli, Mendel's focus was on constant versus variable hybrids. The following excerpt from the *Hieracium* paper is noteworthy because it links constant hybrids with both Darwin (who is mentioned for the first time by Mendel) and Nägeli (whose name is implied):

QUOT. 434 The question of the origin of the numerous and constant intermediate forms has recently acquired no small interest since a famous *Hieracium* specialist has, in the spirit of the Darwinian teaching, defended the view that these forms are to be regarded as [arising] from the transmutation of lost or still-existing species. (Mendel, 1869, 1966b, p. 51)

Authors questioning Mendel's interest in segregation ratios also point out that the issues of constant versus variable hybrids and the formation of new species by hybridization had been in Mendel's mind right from the start (e.g., in his 1866 paper, see Sect. 6.2.1).

There is no doubt an element of truth in these claims, but it defies logic that Mendel would not wish to test for segregation ratios in *Hieracium* also since he had always been interested in generally applicable laws. Therefore, it is reasonable to infer that, for his *Hieracium* experiments, Mendel was interested *both* in the numerical ratios he had previously obtained and in constant hybrids. This would also be consistent with the stated aims in his 1866 paper (Sect. 6.2.1). Mendel obtained constant hybrids with *Hieracium* and, in his 8th letter to Nägeli, he stated:

QUOT. 435 ...I can not resist remarking how striking it is that the hybrids of *Hieracium* show a behavior exactly opposite to those of *Pisum*. Evidently we are here dealing only with individual phenomena, that are the manifestation of a higher, more fundamental law. (*ibid.*, p. 90)

---

[55] See Sect. 6.2.9.1.

6.3.4 Nägeli's Assessment of Mendel's Work

A final question regards Nägeli's views on Mendel's experiments. There is almost no disagreement among scholars that Nägeli did not believe that Mendel's results were correct. In his second letter, Mendel stated:

QUOT. 436 I am not surprised to hear your honor [Nägeli] speak of my experiments with mistrustful caution. (*ibid.*, p. 61)

There are two main grounds for Nägeli's disagreement with Mendel, namely, a philosophical and a technical one.

On the philosophical level, in his first letter to Mendel, Nägeli stated that the latter's numerical expressions were "only empirical, because they can not be proved rational." What Nägeli probably implied here is that Mendel's numerical expressions had not been derived *deductively* from a theory of hybrids that would hold universally.[56] Instead, Mendel had made observations on a limited number of species and therefore the theory he had built from then would probably be limited. However, Nägeli was wrong to believe that theories can only be built deductively. He had disregarded that Mendel had, in fact, used the same *hypothetico-deductive procedure* that both Newton and Darwin had previously put to such efficient use. The hypothetico-deductive method (or scientific method) that Mendel used may be condensed as follows:

(i) First, Mendel built a *hypothesis* based on observations (the hypothesis is a tentative theory).
(ii) He then made *deductions* (or logical conclusions) based on his hypothesis (this is the rational part).
(iii) The deductions are then tested against further observations (agreement lends support to, but does not necessarily prove the hypothesis; disagreement falsifies it).

This is exactly the argument that Mendel used in his second letter to explain to Nägeli that his method was a valid one:

QUOT. 437 My experiments with single traits all lead to the same result: that from the seeds of hybrids, plants are obtained half of which in turn carry the hybrid trait (*Aa*), the other half, however, receive the parental traits A and a in equal amounts. Thus, on the average, among four plants two have the hybrid trait Aa, one the parental trait *A*, and the other the parental trait *a*. Therefore $2Aa + A + a$ or $A + 2Aa + a$ is the empirical simple series for two differing traits. Likewise it was shown in an empirical manner that, if two or three differing traits are combined in the hybrid, the series is a combination of two or three simple series. Up to this point I don't believe I can be accused of having left the realm of experimentation. If then I extend this combination of simple series to any number of differences between the two parental plants, I have indeed entered the rational domain ...These experiments should be repeated and verified. (*ibid.*, p. 63)

[56] An example of such a rational theory is Nägeli's *micellae* and *idioplasm* theory (Sect. 5.1).

On the technical level, Nägeli disagreed with Mendel because his theory (Sect. 8.1) did not allow for constant forms and instead believed in inevitable reversal to parental types. Thus, Nägeli questioned "whether one may conclude that constancy of type has been obtained if the hybrid *Aa* produces a plant *A*, and this plant in turn produces only *A*" (*ibid.*, p. 61). Further evidence of Nägeli's disbelief in constant types is provided by a personal note he made when writing his first letter to Mendel:

QUOT. 438 The constant forms require to be tested further (A, a, AB, Ab, aB, ab). I expect that (when inbred) they would sooner or later be found to vary once more. (Iltis, 1932, p. 193)

This personal note may provide yet another reason why Nägeli disagreed with Mendel's result: it is that Nägeli did not understand his theory. This charge on Nägeli is based on what he wrote next in his note:

QUOT. 439 A, for instance, has half a in its body [being bred out of Aa], and when inbred cannot lose that element. (*ibidem*)

The above shows that Nägeli did not understand Mendel's concept of complete separation (or segregation) in the hybrid *Aa* and hence misinterpreted his notation.

6.4 Mendel's Attitude Toward Evolution and Darwin's Theory

As mentioned in Sect. 6.1.2, Mendel was influenced by the transformationist views of his lecturer, Franz Unger (1800–1870), during his university years in 1852–1853. Unger believed in the transmutation of species and the formation of new species by hybridization. While explaining one of the motivations of his experiments, Mendel stated that:

QUOT. 440 ...this seems to be the one correct way of finally reaching the solution to a question whose significance for the evolutionary history of organic forms must not be underestimated. (Mendel, 1866, 1966a, p. 2)

Thus, in addition to investigating the quantitative laws of hybrid formation, Mendel was also interested in the evolutionary question of how new species could be formed. As explained in Sect. 6.2.9, Mendel used the theory he built from his *Pisum* results to explain how "species" could be transformed by repeated hybridization. Thus, despite some claims to the contrary (e.g., Callender, 1988; Endersby, 2007, p. 125), Mendel was not against the idea of species transformation.

Regarding Mendel's attitude toward Darwin's theory of evolution by *natural selection*, it should be noted that Mendel probably first learned of Darwin's theory from the geologist Carl Schwippel. At a meeting of the Natural Science Society of Brno in September 1861, the latter reviewed Chap. X ("On the Geological Succession of Organic Beings") of Darwin's *Origin of Species*. In 1863, Mendel bought the

German translation of the second edition of the *Origin*.[57] He was thus acquainted with Darwin's views before he wrote his *Pisum* paper (Mendel, 1866), but not when he started his experiments.

A strong statement was first made by Bateson to the effect that Mendel, in fact, began his *Pisum* experiments because he *opposed* Darwin's theory:

> **QUOT. 441** With the views of Darwin which at that time were coming into prominence Mendel did not find himself in full agreement, and he embarked on his experiments with peas, which as we know he continued for eight years. (Bateson, 1909, p. 311)

The statement above can be given hardly any credence since Mendel started his experiments in 1856, at which time, except for a closed circle of friends, nobody knew of Darwin's theory. Given Bateson's fight against Darwin's theory of natural selection acting on *continuous* variation, his claim is thus tendentious at best. Similar statements were made by others later (e.g., Orel, 1933, p. 175; Hercik & Novak, 1952) but seem more political than scientific.

Mendel's first reference to Darwin occurs in the *Hieracium* paper:

> **QUOT. 442** The question of the origin of the numerous and constant intermediate forms has recently acquired no small interest since a famous *Hieracium* specialist has, in the spirit of the Darwinian teaching, defended the view that these forms are to be regarded as [arising] from the transmutation of lost or still-existing species. (Mendel, 1869, 1966b, p. 51)

Not much can be inferred from the above regarding Mendel's views on Darwin's theory, but a near endorsement seems to happen in his 10th letter to Nägeli: in describing the effect of environmental disturbances on fertilization with the pollen of *H. Pilosella incanum*, Mendel explained that if those disturbances increased, then the species involved would eventually disappear; however, species that were more adapted to the changed conditions might take up the struggle for existence and successfully continue for a long stretch of time. In Mendel's own words:

> **QUOT. 443** ...the naturally-occurring hybridizations in *Hieracium* should be ascribed to temporary disturbances, which, if they were repeated often or became permanent, would finally result in the disappearances of the species involved, while one or another of the more happily organized progeny, better adapted to the prevailing telluric and cosmic conditions, might take up the struggle for existence successfully and continue it for a long stretch of time, until finally the same fate overtook it. (*ibid.*, p. 102)

The above is nothing but a classic Darwinian explanation of species survival and reproductive success.

Although Mendel does not seem to have been opposed to Darwin's theory, he did have some reservations on certain aspects of the theory, and he did believe that new species could also be formed by hybridization.

Mendel's reservations about Darwin's theory seem to mainly concern the latter's provisional hypothesis of pangenesis.[58] On p. 497 of the German edition of the second volume of Darwin's *Variations of Animals and Plants Under Domestication*

---

[57] See footnote 36.

[58] See Sect. 3.7.2.

Darwin had said: "As each unit, or group of similar units throughout the body, casts off its gemmules, and as all are contained within the smallest egg or seed."[59] Referring to this sentence, Mendel wrote the following footnote: "To indulge in an impression without reflection [!]"

Although Mendel read all of Darwin's books, there is no indication that the latter read Mendel's papers (especially his 1866 paper). Some authors[60] have contended that Mendel sent Darwin a copy of his *Pisum* paper but the latter did not read it. This does not seem to be the case.

On July 24, 1862, Mendel also traveled via Vienna and Paris to London, but there is no indication that he met Darwin.

6.5 Were Mendel's Data Falsified? The Mendel-Fisher Controversy

6.5.1 Weldon (1902)

In 1900, the battle was still raging between the biometricians and the saltationists when Karl Pearson (1857–1936) published his groundbreaking paper on the χ^2 goodness-of-fit (Pearson, 1900b). Shortly after, the biometrician Raphael Weldon (1860–1906) alerted Pearson to Mendel's 1866 paper and pointed out that he intended to subject Mendel's data to a χ^2-test.

Weldon's paper was published in 1902 (Weldon, 1902a), where he compared the observed (Table 6.1) and expected dominant counts in F_2 plants under Mendel's theory (i.e., by multiplying the total number of F_2 hybrids per trait by 3/4). Weldon then noted:

> **QUOT. 444** ...Only one determination has a deviation from the hypothetical frequency greater than the probable error of the determination, and one has a deviation sensibly equal to the probable error; so that a discrepancy between the hypothesis and the observations which is equal to or greater than the probable error[61] occurs twice out of seven times, and deviations much greater than the probable error do not occur at all. These results then accord so remarkably with Mendel's summary of them that if they were repeated a second time, under similar conditions and on a similar scale, the chance that the agreement between observation and hypothesis would be worse than that actually obtained is about 16 to 1.[62] (*ibid.*, p. 233)

---

[59] This sentence appears on p. 378 of the English edition (Darwin, 1868b).

[60] See footnote 22 in Chap. 3.

[61] The probable error (*PE*) of a random variable *X* is its semi-interquartile range. For a symmetric distribution with mean μ, *PE* satisfies $\Pr\{|X-\mu|<PE\}=.5$ (*our note*).

[62] This number seems to be erroneous. Let *X* be the number of experiments which give a discrepancy equal to or greater than the probable error. Then, under the hypothesis that the segregation ratios are true, *X*-binomial (7, .5) and $\Pr\{X\geq 2\}=\sum_{i=2}^{7}\binom{7}{i}\frac{1}{2}=\frac{15}{16}$. Therefore, the probability of obtaining the same number or more discrepancies than Mendel is 15/16 and the corresponding odds are 15 to 1.

He next calculated the value of Pearson's χ^2 for each trait and combined them, obtaining a final p-value of .95.[63]

Although Weldon never questioned Mendel's integrity, he did wonder if the results were too good in his letters to Pearson.

6.5.2 *Fisher's Two Arguments for Why Mendel's Data Were Tampered With (1911, 1936)*

6.5.2.1 The "Too-Good-to-be-True" Argument

Almost a decade later, Weldon's statement about the "chance of 16 to one" was probably picked up by Fisher when, as a starting third-year undergraduate student at Gonville and Caius College, he said in a talk to the Cambridge University Eugenics Society in 1911 that:

> **QUOT. 445** It is interesting that Mendel's original results all fall within the limits of probable error; if his experiments were repeated the odds against getting such good results is about 16 to one. It may have been just luck; or it may be that the worthy German abbot, in his ignorance of probable error, unconsciously placed doubtful plants on the side which favoured his hypothesis. (Norton & Pearson, 1976, p. 160; Bennett, 1983, p. 57)

During the ensuing years, little attention was given to Mendel's results until Fisher returned to the issue in 1936. In a Jan 2, 1936, letter to E.B. Ford, he summarized his conclusions as follows:

> **QUOT. 446** . . .I have had the shocking experience lately of coming to the conclusion that the data given in Mendel's paper must be practically all faked. I cannot conceive that Mendel himself had any hand in it, and quite independently, and this is what I was really studying his paper for, I have come to the conclusion that his experiments were planned and set out exactly as he records. I mean, for example, that his primary crosses really were unifactorial, and that he had carefully selected them to be so. So, if the data were faked, I presume it was by some assistant who knew too well what was expected. (*ibid.*, p. 19)

In that same year, Fisher published a full-length paper (Fisher, 1936) in which he attempted to substantiate his conclusions above through a detailed analysis of Mendel's results. Fisher's findings are summarized in Table 6.2, and they all show extremely good fit (especially the later results). The combined p-value for all of Mendel's experiments is .99997 (which is mistakenly given as .99993 in Fisher's table). This implies that, under the hypothesis that all segregation ratios really hold, there is a probability of .00003 of obtaining results as good as or better than Mendel's.

---

[63]The combined value of $\chi 2$ may be shown to be 2.1389 on 7 d.f. A p-value of .95 means that there is only a 1/20 probability of obtaining results as good as or better than those obtained by Mendel, assuming the segregation ratios are true.

Table 6.2 Summary of Fisher's findings regarding Mendel's results (Fisher, 1936, p. 131, Table V)

	Expectation		χ^2		p-value
3 : 1 ratios — Seed characters	2		.2279		
3 : 1 ratios — Plant characters	5		1.8610		
	—	7	—	2.1389	.95
2 : 1 ratios — Seed characters	2		.5983		
2 : 1 ratios — Plant characters	6		4.5750		
	—	8	—	5.1733	.74
Bifactorial experiment	—	8		2.8110	.94
Gametic ratios		15		3.6730	.9987
Trifactorial experiment		26		15.3224	.95
Total		64		29.1186	.99987
Illustrations of plant variation		20		12.4870	.90
Total		84		41.6056	.99997[a]

[a]This p-value was mistakenly reported as .99993 in Fisher's paper

The 3:1 ratios in the table above refer to the dominant–recessive relationships among the F_2 hybrids in Mendel's single-trait experiments (Table 6.1). These are the same ratios that Weldon had analyzed in his paper (Weldon, 1902a), with both Weldon and Fisher reporting a combined p-value of .95 for those experiments.

6.5.2.2 The "Approach-to-the-Incorrect-Ratio" Argument

The 2:1 ratios in Table 6.2 demand some explanation. The reader will recall from Sect. 6.2.5 that Mendel had hypothesized that the F_2 dominants consisted of two parts hybrid (Aa) to one part true-breeding (A).[64] To test this hypothesis, he selected the F_2 dominants and allowed them to self-fertilize. For the case of two seed traits, Mendel then inferred the type of a dominant in F_2 simply by examining the seeds of that dominant when self-fertilized. However, when it came to the five plant traits, Mendel had to allow the F_3 plants to grow. Ideally, he would have needed a large number of F_3 plants for each F_2 dominant to score the type of that dominant. Since this was not feasible due to lack of space, Mendel instead proceeded as follows: for each of the five plant traits, Mendel selected 100 F_2 dominant plants, and from each of the latter, he grew 10 F_3 descendants; now, if all 10 descendants had the dominant trait, then Mendel classified the F_2 dominant as true-breeding (AA); otherwise, he classified it as hybrid (Aa).

As the reader may have realized, Mendel's procedure of inferring the type of an F_2 dominant based on the traits of 10 descendants is liable to a misclassification error, since 10 F_3 descendants can have the dominant trait and yet the F_2 dominant

[64]Mendel's true-breeding A is today's homozygote AA.

parent is a hybrid (*Aa*). This error in turn distorts the expected segregation ratio of 2: 1. In his paper (Fisher, 1936, p. 125), Fisher saw this source of error and calculated the new expected segregating ratio when allowance was made for it. Equipped with this new ratio and given Mendel's results, he was able to mount a powerful attack against the trustworthiness of Mendel's data. In his attack, Fisher made use of an argument that was different from the one he had made elsewhere in his paper, namely, that the *p*-values associated with Mendel's results were simply too large (or that the calculated values of χ^2 were simply too small). How Fisher made this new argument will now be described.

Fisher first noted that a heterozygote *Aa* has a probability 3/4 of producing a dominant (i.e., either *AA* or *Aa*) offspring by selfing. Thus, 10 F_3 descendants that are all dominant may still be produced by a heterozygote parent, and such an event has probability

$$\left(\frac{3}{4}\right)^{10} \approx .0563. \tag{6.1}$$

Therefore, each F_2 dominant that was a hybrid (*Aa*) was misclassified with a probability of .0563 by Mendel to be true-breeding (*AA*). Hence, instead of the expected 2*X* hybrids and *X* true-breeding plants (where *X* is a natural number), there should have been $2X - 2X(.0563) = 1.8874X$ expected hybrids and $X + 2X(.0563) = 1.1126X$ expected true-breeding plants. As Fisher had seen, the expected segregating ratio should have been

$$1.8884 : 1.1126 \text{ (or } 1.6964 : 1),$$

instead of 2:1. He then reasoned that if Mendel's data had been obtained properly, the observed counts should have been near the ratio 1.6964:1, not the ratio 2:1. However, for the 600 F_2 dominants that Mendel had classified,[65] 399 were hybrids (*Aa*), and 201 were true-breeding (*AA*), thus giving a ratio 1.985:1. Fisher thus concluded:

> **QUOT. 447** ...A serious and almost inexplicable discrepancy has, however, appeared, in that in one series of results the numbers observed agree excellently with the two to one ratio, which Mendel himself expected, but differ significantly from what should have been expected had his theory been corrected to allow for the small size of his test progenies. To suppose that Mendel recognized this theoretical complication, and adjusted the frequencies supposedly observed to allow for it, would be to contravene the weight of the evidence supplied in detail by his paper as a whole. Although no explanation can be expected to be satisfactory, it remains a possibility among others that Mendel was deceived by some assistant who knew too well what was expected. This possibility is supported by independent evidence that the data of most, if not all, of the experiments have been falsified so as to agree closely with Mendel's expectations. (*ibid.*, p. 132)

[65] Since Mendel had selected 100 F_2 dominants for each of the 5 plant traits, the total should have been 500. However, because the 100 F_2 dominants for pod color (Experiment 5) yielded results which were too far off the predicted ratio, Mendel analyzed 100 additional F_2 dominants.

It is surprising that Fisher's 1936 paper, despite containing strong arguments against the veracity of Mendel's results, did not attract much attention. It is necessary to go down the years until 1964 when the botanist Conway Zirkle (1895–1972) first drew attention to the issue, without, however, mentioning Fisher (Zirkle, 1964, p. 66). In the same year, de Beer (1964) published the first article that mentioned Fisher's 1936 work on Mendel's results. From then on, the pace quickly picked up, and there has been an enormous literature on the "Mendel-Fisher" controversy.[66]

6.5.3 *Attempts to Resolve the Mendel-Fisher Controversy*

As mentioned in the last section, Fisher's explanation for why Mendel's data had such a good fit was that "Mendel was deceived by some assistant who knew too well what was expected" (Fisher, 1936, p. 132). Subsequently, several other explanations have been provided, all intending to resolve the Mendel-Fisher controversy. The main ones are given below.

6.5.3.1 The Claim That Fisher's Binomial Model Over-estimates the True Variance in Segregations

Weiling (1989) has argued that the binomial model implicitly assumed by Fisher in his too-good-to-be-true argument of Sect. 6.5.2.1 is erroneous. Consider one of the experiments carried out by Mendel, involving a total number of N seeds from a parent plant, of which M showed the dominant trait. Suppose that, out of the N seeds, n were sampled (without replacement) by Mendel. Then, the number of dominants (X) in the sample n has a hypergeometric distribution. Fisher implicitly assumed that $n < < N$ so that X is approximately binomial with

$$\mathrm{E}\,X = np = n\left(\frac{M}{N}\right), \quad \mathrm{var}\,X = np(1-p) = n\left(\frac{M}{N}\right)\left(\frac{N-M}{N}\right).$$

Weiling (*ibid.*, p. 138) observed that Mendel had consistently pointed out that some plants were always lost during the experiment (e.g., Mendel, 1866, 1966a, p. 18), which means that n should be only slightly smaller than N, and the full hypergeometric model should be used:

[66]See, for example, Franklin et al. (2008), Orel (1996, pp. 199–207), Weiling (1966, 1986, 1989), Olby (1985, pp. 209–211), Root-Bernstein (1983), Radick (2022, 2023), Pilgrim (1984, 1986), Edwards (1986a, b), Monaghan and Corcos (1985), Piegorsch (1986), Pilpel (2007), van der Waerden (1968), Schwartz (2008, pp. 95–96), Campbell (1976), de Beer (1964, 1966), Fairbanks and Rytting (2001), Novitski (1995, 2004a, b), Sturtevant (1965, pp. 12–16), and Magnello (2004a, b).

$$\mathrm{E}\,X = np = n\left(\frac{M}{N}\right), \quad \operatorname{var} X = np(1-p) = n\left(\frac{M}{N}\right)\left(\frac{N-M}{N}\right)\underbrace{\left(\frac{N-n}{N-1}\right)}_{c}.$$

Thus, Weiling suggested that the true variance should be less than that assumed by Fisher by a factor $c = (N - n)/(N - 1)$ so that Fisher's calculated χ^2 (which is inversely proportional to the variance) should be corrected to χ^2/c.[67] To estimate c, Weiling proposed the relation

$$c = \frac{\chi^2}{\operatorname{median}\left(\chi_f^2\right)}, \tag{6.2}$$

where median (χ_f^2) is the median of the chi-squared distribution with f degrees of freedom (corresponding to the degrees of freedom associated with χ^2).

The problem with Weiling's estimate in (6.2) is that it cannot be substituted back in χ^2/c to obtain the corrected test statistic (since the latter would then always be $\chi^2/\left\{\chi^2/\operatorname{median}\left(\chi_f^2\right)\right\} = \operatorname{median}\left(\chi_f^2\right)$).[68] Moreover, even if there was no reduction in the variance, approximately half of the observed χ^2 values would lie below the theoretical median. It does not seem appropriate, therefore, to conclude that, based on a *single* aggregated χ^2 value, there is indeed a reduction in variance. However, this is exactly what Weiling seems to have done in his various estimates of c ($.6 < c < .9$ in Weiling (1986), $.2 < c < .9$ in Weiling (1989)). Thus, based on the 20 illustrations of plant variation performed by Mendel (1866, 1966a, p. 12), the aggregated value of the χ^2 statistic is 12.487.[69] Weiling then calculated $c = 12.487/\operatorname{median}(\chi_{20}^2) = 12.487/19.3 = .647$. However, when the distribution of the 20 individual χ^2 values is examined, 12 are above and 8 are below the median of the χ_1^2 distribution. The extent to which the variance of segregations is reduced seems to remain an open question.[70]

---

[67] Elsewhere, Weiling (1966, 1980, p. 18) has also used a different argument to argue why Fisher's χ^2 should be corrected to χ^2/c, namely, that the pollen sack of a heterozygous *Aa* plant has half *A*-pollen and half *a*-pollen which are not randomly mixed, but only "semi-randomly" (because of their tetrad arrangement). This is turn leads to a reduction in the variance.

[68] A better estimate of c, one which could be substituted back in χ^2/c to provide the corrected test statistic, is $c = \operatorname{median}\left\{\chi_{(1)}^2, \chi_{(2)}^2, \ldots, \chi_{(k)}^2\right\}$, where $\chi_{(1)}^2, \chi_{(2)}^2, \ldots, \chi_{(k)}^2$ are k chi-squared test statistics, each with the same degrees of freedom as χ^2, obtained from similar experiments.

[69] See the penultimate row in Table 6.2.

[70] Edwards has gone even further and stated that: "The diminished-variance hypothesis is not supported by other more extensive data, and. . .simply reducing the variance is in any case not enough to explain the peculiarities of Mendel's data" (Edwards, 1986a, p. 301).

6.5.3.2 The Claim That Fisher's Calculation of the Probability of Misclassification Is Incorrect

Weiling (1986) also questioned Fisher's "approach-to-the-incorrect-ratio" argument of Sect. 6.5.2.2. Again reasoning as in Sect. 6.5.3.1 that Mendel's data were sampled from a slightly larger set, Weiling argued that Fisher's calculation of the misclassification probability in (6.1) ignored this aspect of Mendel's data and was therefore erroneous. Weiling assumed that each plant in Mendel's experiments yielded 30 seeds on average of which $3/4 \times 30 \approx 23$ were expected to have the dominant trait. Mendel's 10 F_3 descendants that were examined were actually sampled from the expected 30, and the probability that all 10 F_3 descendants are dominant must be calculated from a hypergeometric distribution as

$$\frac{\binom{23}{10}\binom{7}{0}}{\binom{30}{10}} \approx .0381.$$

Weiling's misclassification probability of .0381 is 67.7% of Fisher's .0563 and results in a segregation ratio of 2–2(.0381): 1–2(.0381) = 1.9238:1.0762 ≈ 1.893: 1. The latter is closer to the segregation ratio of 2:1 than to Fisher's 1.6964:1.

However, Weiling's analysis is correct only when each of Mendel's plants produced 30 F_3 seeds *actually*, not when each produced 30 F_3 seeds *on average*. An analysis that overcomes this shortcoming of Weiling's analysis proceeds as follows. Let A_i be the event that there are i dominants out of a total of n seeds produced by a plant, and let B be the event that, of the sampled ten seeds, all ten are dominant. Then

$$\Pr\{A_i\} = \binom{n}{i}\left(\frac{3}{4}\right)^i\left(\frac{1}{4}\right)^{n-i}, \qquad i = 0, 1, \ldots, n$$

and

$$\Pr\{B|A_i\} = \frac{\binom{i}{10}\binom{n-i}{0}}{\binom{n}{10}} = \frac{\binom{i}{10}}{\binom{n}{10}}, \qquad i = 10, 11, \ldots, n.$$

Therefore,

$$
\begin{aligned}
\Pr\{B\} &= \sum_{i=10}^{n} \frac{\binom{i}{10}}{\binom{n}{10}} \binom{n}{i} \left(\frac{3}{4}\right)^{i} \left(\frac{1}{4}\right)^{n-i} \\
&= \sum_{i=10}^{n} \frac{(n-10)!}{(i-10)!(n-i)!} \left(\frac{3}{4}\right)^{i} \left(\frac{1}{4}\right)^{n-i} \\
&= \sum_{i=0}^{n-10} \frac{(n-10)!}{i!(n-10-i)!} \left(\frac{3}{4}\right)^{i+10} \left(\frac{1}{4}\right)^{n-i-10} \\
&= \left(\frac{3}{4}\right)^{10} \approx .0563
\end{aligned}
$$

Note that the last result holds irrespective of the value of n. Thus, when Weiling's argument is corrected, the probability of misclassification becomes .0563, as Fisher had found.

6.5.3.3 The Claim That Mendel Had Used Optimal Stopping

In his 1965 book, Dunn offered a possible explanation for Mendel's results that the latter may have stopped counting when the data fit his theory (Dunn, 1965a, p. 12). One year later, Olby also found an extremely good fit in Tschermak's results and stated that:

> **QUOT. 448** Both Tschermak and Mendel stopped scoring their results when the totals gave a striking confirmation of a simple ratio. (Olby, 1966, p. 183)

In 1968, van der Waerden (1968) went even further than the previous two authors: he showed how a sequential approach to the experiment results in low values of χ^2 so that Mendel's results should not be surprising.

Van der Waerden's analysis considered the F_2 dominants that were produced from the F_1 hybrids (Sect. 6.2.5). Mendel had hypothesized that the F_2 dominants consisted of two parts hybrid (*Aa*) to one part true-breeding (*A*)[71] and tested this hypothesis by allowing the F_2 dominants to self-fertilize. Table 6.3 shows the number of (true-breeding, hybrid) he obtained for the seven experiments (one for each trait) he did. As mentioned in footnote 65, Experiment 5 at first yielded numbers (40, 60) that were too far off and had to be repeated, yielding (35, 65).

Van der Waerden modeled the number X of true-breeding plants in each of Experiments 3–7[72] as a random variable from a binomial (n, p) distribution, with $n = 100$ and $p = 1/3$. Then the standard deviation of X is

[71] Mendel's true-breeding *A* is today's homozygote *AA*.

[72] Since the χ^2 values for Experiments 1 and 2 were not extremely small, they did not have too-good-to-be-true issues and were not analyzed by van der Waerden.

Table 6.3 Mendel's results for the number of true-breeding and hybrids among F_2 dominants (Mendel, 1866, 1966a)

Experiment no.	True-breeding	hybrid
1	193	372
2	166	353
3	36	64
4	29	71
5	40(35)	60(65)
6	33	67
7	28	72

The bracketed experiments were analyzed by van der Waerden (1968)

$\sigma = \sqrt{100(1/3)(2/3)} = 4.7$. All of Experiments 3–7 have an absolute deviation $|X - np| \le 1.3\sigma$, except for Experiment 5. Since the latter experiment was repeated, van der Waerden imagined a series of experiments similar to 3–7, except that each one was run *sequentially*. In particular, each experiment would be repeated if $|X - np| > 1.3\sigma$; otherwise it would simply stop. Now, let

$$\begin{aligned} \chi^2 &= \text{chi-squared test statistic for a sequential experiment,} \\ P &= \text{probability of stopping after the first step,} \\ Q &= 1 - P, \\ \mathrm{E}_1\chi^2 &= \mathrm{E}(\chi^2|\text{sampling is stopped after first step}), \\ \mathrm{E}_2\chi^2 &= \mathrm{E}(\chi^2|\text{sampling continues after first step}). \end{aligned}$$

Then,

$$\mathrm{E}\chi^2 = P\,\mathrm{E}_1\chi^2 + Q\,\mathrm{E}_2\chi^2 \tag{6.3}$$

(*ibid.*, p. 284). Now, even if sampling continues after the first step, it will eventually stop. When that happens, it will be because $|X - np| \le 1.3\sigma$ for the total value of X and p in that experiment, i.e., $\chi^2 \le 1.3^2 = 1.69$. Hence,

$$\mathrm{E}_2\,\chi^2 \le 1.69. \tag{6.4}$$

Moreover,

$$\begin{aligned} P &= \Pr\{\chi_1^2 \le 1.69\} = .8064, \\ Q &= 1 - .8064 = .1936. \end{aligned} \tag{6.5}$$

Additionally, $E_1\chi^2$ is the expectation of a χ_1^2-distribution truncated from above at 1.69. Since $\chi_1^2 = Z^2$, where $Z \sim N(0, 1)$,

$$E_1\chi^2 = E_1 Z^2 = \frac{1}{P}\int_{-1.3}^{1.3} \frac{1}{\sqrt{2\pi}} e^{-z^2/2} dz$$

so that

$$P\,E_1\chi^2 = \int_{-1.3}^{1.3} \frac{1}{\sqrt{2\pi}} e^{-z^2/2} dz = .3608. \tag{6.6}$$

Substituting the values in (6.4)–(6.6) into Eq. (6.3):

$$E\chi^2 \le .3608 + (.1936)(1.69) = .688$$

(*ibid*., p. 285). Hence, while for an unconditional approach using a fixed sample size, the expected value of χ^2 is unity, for the sequential approach described above, the expected value of χ^2 is substantially less. Therefore, Mendel's extremely low χ^2 should not be surprising. In fact, as van der Waerden said:

> **QUOT. 449** . . .by a suitable choice of the stopping limit for χ^2, the expectation of χ^2 may be made as small as we please. (*ibidem*)

Using the idea of sequential experiments, van der Waerden next explained why Fisher's probability of misclassification of .0563 (Eq. 6.1) was an overestimate in Mendel's case. Van der Waerden hypothesized that for each of the 100 F_2 dominants, Mendel had planted 12–15 seeds instead of only 10. This seriously reduced any chance of misclassification.

Van der Waerden used the sequential argument to also explain Fisher's other extremely low χ^2 values (viz., for the 3:1 ratio for the 5 plant characters, see Table 6.2). For some of Mendel's experiments (viz., the bifactorial experiments), the sample size was fixed, but the χ^2 value was not too low. The only exception occurred for Mendel's experiments in Sect. 9 of his paper, involving backcrosses.[73] Here, a fixed sample was used, and Fisher's calculated χ^2 was 3.6730, with a p-value of .9987.[74] Commenting on these results, van der Waerden admitted:

[73] See Sect. 6.2.7 of current book.

[74] See "Gametic ratios" row of Table 6.2.

QUOT. 450 . . .I cannot but agree with Fisher's conclusion that the data have probably been biased. By the time when the last series of five experiments was performed, Mendel and his gardening assistants knew too well what they had to expect. (*ibid.*, p. 287)

Thus, while van der Waerden's sequential approach explains *some* of Mendel's "too-good" results, it does not explain *all* of them. Furthermore, Mendel's statement does not support the sequential hypothesis:

QUOT. 451 To discover the relationships of hybrid forms to each other and to their parental types it seems necessary to observe *without exception all*[75] members of the series of offspring in each generation. (Mendel, 1866, 1966a, p. 4)

6.5.3.4 The Claim That Mendel May Have Been Subject to Unconscious Bias

In addition to the reason given in Sect. 6.5.3.3, Dunn also conjectured that Mendel's data had such a good fit because "[o]pportunity for some unintentional selection may have existed" (Dunn, 1965a, p. 12). In other words, when the trait of an offspring was not sharply defined, there was a tendency to classify it in the direction of expectation.

A similar explanation was provided by Sewall Wright when Curt Stern elicited his opinion during the publication of *The Origin of Genetics: A Mendel Source Book* (Stern & Sherwood, 1966). In a short letter (pp. 173–175), Wright remarked:

QUOT. 452 I do not think that Fisher allows enough for the cumulative effect on χ^2 of a slight subconscious tendency to favor the expected result in making tallies. (*ibid.*, p. 173)

He further added:

QUOT. 453 . . .Anyone who doubts the difficulty in making repeatable counts should read chapter 5 in Pearl's *Introduction to Medical Biometry*. He reports an experiment in which 15 trained observers obtained extraordinary differences in sorting and counting the same 532 kernels of corn. Checking of counts that one does not like, but not of others, can lead to systematic bias toward agreement. I doubt whether there are many geneticists even now whose data, if extensive, would stand up wholly satisfactorily under the χ^2 test. (*ibidem*)

However, Fairbanks and Rytting (2001) have countered this explanation because the five plant traits that Mendel considered had very distinct phenotypes. Moreover, even if there was more potential for misclassification of the two seed traits, the misclassification was probably negligible. The authors therefore concluded:

QUOT. 454 From a botanical point of view, this explanation [of unconscious bias] can account for only a negligible degree of bias. (*ibid.*, p. 742)

Fairbanks and Rutting have also provided an explanation that seems to be the most plausible for why Mendel's data have such a good fit, namely, that Mendel had performed many experiments and that in his paper, he presented those that best described his theory:

---

[75] Italics are Mendel's.

QUOT. 455 We believe that the most likely explanation of the bias in Mendel's data is also the simplest. If Mendel selected for presentation a subset of his experiments that best represented his theories, χ^2 analyses of those experiments should display a bias. His paper contains multiple references to experiments for which he did not report numerical data, particularly di- and trihybrid experiments. (*ibid.*, p. 743)

6.6 The Rediscovery of Mendelism[76] in 1900: de Vries, Correns, and Tschermak[77]

6.6.1 Introduction

In 1900, 16 years after Mendel's death, three biologists (Hugo de Vries,[78] Carl Correns, and Erich von Tschermak) (Fig. 6.8) each claimed that they had independently arrived at the same conclusions as Mendel 34 years ago (Mendel, 1866). Several papers were published in quick succession by these biologists to document how they had rediscovered Mendel's laws before actually seeing the late abbot's work. The papers, in order of their date of publication, are as follows:

1. **V1**: de Vries, Hugo (1900c): Sur la loi de disjonction des hybrides.[79] *Comptes Rendus de l'Academie des Sciences* (Paris) 130: 345–347 (submitted on March 19, 1900, published on March 26)
2. **V2**: de Vries, Hugo (1900a): Das spaltungsgesetz der bastarde.[80] *Berichte der Deutschen Botanischen Gesellschaft* 18: 83–90 (submitted on March 14, published on April 25).
3. **C1**: Correns, Carl (1900a): G. Mendel's regel über das verhalten der nachkommenschaft der bastarde.[81] *Berichte der Deutschen Botanischen Gesellschaft* 18: 158–168 (submitted on April 22, received on April 24, published on May 23)

---

[76]"Mendelism" was first used in 1903 by L.H. Bailey as follows: "The breeder wants to preserve the desirable characters or traits and eliminate the undesirable ones, but under the strict interpretation of Mendelism this is difficult" (Bailey, 1903, p. 451). "Mendelian" was first used earlier, in 1901, by Bateson and Saunders as follows: "Cases which follow Mendelian principles" (Bateson & Saunders, 1901, p. 15).

[77]The rediscovery of Mendelism has also been described in Orel (1996, pp. 282–291), Stubbe (1972, pp. 273–285), Olby (1985, Chap. 6), Bowler (1989, pp. 113–116), Shan (2020, p. Chap. 3), Galton (2017, pp. 130–132), Müller-Wille and Rheinberger (2012, pp. 137–145), Stern and Sherwood (1966, pp. viii–xii), and Williams (2019, pp. 56–58).

[78]For de Vries' theory of heredity, see Sect. 5.3.

[79]Published in English as "Concerning the law of segregation of hybrids" in de Vries (1950).

[80]Published in English as "Law of segregation of hybrids" in de Vries (1966).

[81]Published in English as "G. Mendel's law concerning the behavior of progeny of varietal hybrids" in Correns (1966a).

Fig. 6.8 From left to right: Hugo de Vries (1848–1935), Erich von Tschermak (1871–1962), and Carl Correns (1864–1933). (From: https://commons.wikimedia.org/wiki/File:PSM_V67_D338_Hugo_de_Vries.png, https://commons.wikimedia.org/wiki/File:Acta_Horti_berg._-_1905_-_tafl._124._-_Erich_Tschermak.jpg, https://upload.wikimedia.org/wikipedia/commons/8/8e/Carl_Correns_1910s.jpg)

4. **V3**: de Vries, Hugo (1900d): Sur les unités des caractères spécifiques et leur application à l'étude des hybrides. *Revue Générale de Botanique*, 12, 257–71 (submitted on March 19, published on July 15)
5. **T1**: Tschermak, Erich von (1900a): Ueber künstliche Kreuzung bei *Pisum sativum*. *Berichte der Deutschen Botanischen Gesellschaft*, 18, 232–239 (received on June 2, published on July 24)
6. **T2**: Tschermak, Erich von (1900b): Ueber künstliche Kreuzung bei *Pisum sativum*. *Zeitschrift für das landwirtschaftliche Versuchswesen in Österreich*, 3, 465–555 (initially submitted on Jan 17, published soon after **T1**)
7. **C2**: Correns, Carl (1900b): Gregor Mendel's 'Versuche über Pflanzen-Hybriden' und die Bestätigung ihrer Ergebnisse durch die neuesten Untersuchungen," *Botanische Zeitung*, 58, 229–235 (published on August 1)
8. **C3**: Correns, Carl (1900c): Über Levkojenbastarde. Zur Kenntniss der Grenzen der Mendelschen Regeln. *Botanisches Centralblatt*, 84, 97–113 (published on October 17, 1900)

Commenting on Mendel's 1866 paper (Mendel, 1866), de Vries stated in his German manuscript (**V2**) that:

> **QUOT. 456** This important [i.e. Mendel's] treatise is so seldom cited that I first learned of its existence after I had completed the majority of my experiments and had deduced from them the statements communicated in the text. (de Vries, 1900a, 1966, p. 10, footnote 6)

In his rediscovery manuscript (**C1**), Correns similarly said:

> **QUOT. 457** I thought that I had found something new. But then I convinced myself that the Abbot Gregor Mendel in Brunn, had, during the sixties, not only obtained the same result through extensive experiments with peas, which lasted for many years, as did de Vries and I, but had also given exactly the same explanation, as far as that was possible in 1866. (Correns, 1900a, 1966a, p. 120)

Finally, Tschermak (**T1**) also claimed that he had discovered Mendel's laws before seeing the original paper:

> **QUOT. 458** The simultaneous "discovery" of MENDEL by CORRENS, DE VRIES and myself appears to me especially gratifying. Even in the second year of experimentation, I too still believed that I had found something new. (Tschermak, 1900a; Brooklyn Botanical Garden, 1950, p. 47)

In the late 1920s, when Herbert Fuller Roberts was writing his book on the history of plant hybridization, he contacted de Vries, Correns, and von Tschermak separately and obtained similar accounts to the three quotations above (Fig. 6.8). This prompted Roberts to say:

> **QUOT. 459** This concludes the account, rather extended in detail, of the three remarkable, practically simultaneous, discoveries of Mendel's celebrated papers published in the year 1900. Without regard to the actual precedence in respect to dates of publication, the entirely independent and practically equal merit of the contributions of the three separate investigators leaves little room for contention from the standpoint of the contributions themselves. (Roberts, 1929, p. 357)

However, there were cracks in the nice story that scholars were at first led to believe. These cracks relate to an important question, namely, did the three biologists understand Mendelism *before* reading the 1866 paper, as each of their rediscovery papers claimed?

The next three sections will describe the "rediscoveries" of each of the three biologists in turn while at the same time attempting to answer the question above.

6.6.2 Hugo de Vries (March 1900)

6.6.2.1 Introduction

In a Dec 18, 1924, letter to Roberts, de Vries described the circumstances under which he came across Mendelian segregation ratios:

> **QUOT. 460** When preparing my book on the Mutation Theory, I worked on the basis of Darwin's Hypothesis of pangenesis, and of the version of it proposed in my Intracellular Pangenesis. The main principle of Pangenesis is the conception of unit characters. This led on the one side to the theory of the origin of species by means of mutations, and on the other to the description of the phenomena of hybridization as recombinations of these units. In 1893, I crossed *Oenothera lamarkiana* with *O. lam. brevistylis*, and found their progeny to be uniform, and true to the specific parent in 1894, but splitting in the second generation 1895, giving 17–26 individuals with the recessive character (Mut. The. II, p. 157). Many other species were tried with the same result, and dihybrid crosses showed the laws of chance to be valid for them also. After finishing most of these experiments, I happened to read L. H. Bailey's 'Plant Breeding' of 1895. In the list of literature of this book, I found the first mention of Mendel's now celebrated paper, and accordingly looked it up and studied it. Thereupon I published in March 1900 the results of my own investigations in the Comptes Rendus de l'Academie des Sciences, T. CXXX, p. 845, under the title of 'Sur la loi de disjonction des hybrides,' and shortly afterwards, in the same year, in the Berichte der

deutschen botanischen Gesellschaft, T. XVIII, p. 83, (March 14, 1900). A full account of my experiments was given in the second volume of the German edition of my Mutation Theory, 1903. (Roberts, 1929, p. 323)

In the above, de Vries provided the motivation for his hybridization experiments, namely, the validation of his theory of pangens, as described in his 1889 *Intracellular Pangensis*[82] (de Vries, 1889). It will be recalled that according to this theory, each character of species was represented by a set of similar pangens. Therefore, if de Vries was successful in transferring a character from one species to the other through hybridization, the support for his theory would increase.

Several hybridizations performed by de Vries in the pre-1900 period were documented in Vol. 2 of his *Mutationstheorie* (de Vries, 1903, Vol. II). The ones that were added to his second rediscovery paper (**V2**) (de Vries, 1900a) are shown in Table 6.4.

One of the earliest artificial crossings not mentioned in Table 6.4 was performed in 1892 and involved *Lychnis diurna* (hairy) × *L. vespertina glabra* (hairless). The following year, all F_1 hybrids were hairy. De Vries reported the F_2 offspring obtained as a 2:1 ratio (for hairy to hairless) (de Vries, 1897, 1920, p. 25, Vol. VI) but as a 3:1 ratio at the International Conference on Hybridization in 1899 (de Vries, 1899a, p. 74).

Some troubling questions about de Vries' rediscovery will be examined next.

6.6.2.2 When Did de Vries First Read and Understand Mendel's Paper?

In his December 18, 1924, letter to Roberts, de Vries stated that he had already found 17–26% recessives in F_2 before he discovered the first mention of Mendel's paper in the literature list of L.H. Bailey's book *Plant Breeding* (Bailey, 1895). However, the 1895 (i.e., first) edition of Bailey's book contained no literature list and no mention of Mendel. In fact, in his book, Roberts pointed out[83] that de Vries had previously thanked Bailey for the latter's 1892 article (Bailey, 1892) through which de Vries had learned about Mendel's article. Assuming that de Vries had observed 3:1 ratios before reading Mendel's paper in 1892, it is unlikely that he understood its full implications, as will now be explained.

In all the hybridization experiments mentioned in this section or elsewhere, nowhere in de Vries' writing before 1900 can a clear Mendelian explanation be found, i.e., an explanation in terms of two factors (or elements or anlagens) that are responsible for a given trait and of which one from each parent is randomly passed over to the offspring.

---

[82] See Sect. 5.3.2.

[83] On p. 323, Roberts pointed to the footnote that Bailey had introduced in the 1908 (4th) edition of his book on p. 155 where de Vries had thanked him.

Table 6.4 De Vries' segregation ratios in the *Beritche* paper (de Vries, 1900a, 1966, p. 113)

Dominating	Recessive	% recessive	Year of crossing
(a) Artificial crossings			
Lychnis diuma	*L. vespert.* (white)	26	1892
Lychnis vespertina (hairy)	*glabra*	28	1892
Papaver somnif. Mephisto	*Danebrog*	28	1893
Papaver somnif. Nanum (simple)	filled	24	1894
Agrostemma Githago	*nicaeensis*	24	1898
Chelidonium maius	*laciniatum*	26	1898
Hyoscyamus niger	*pallidus*	26	1898
Oenothera Lamarckiana	*brevistylis*	22	1898
Zea Mays (starchy)	*saccharata*	25	1898
(b) Free crossings			
Coreopsis tinctoria	*brunnea*	25	1894
Solanum nigrum	*chlorocarpum*	24	1894
Veronica longifolia	*alba*	22	1894
Viola comuta	*alba*	23	1894
Chrysanthemum Roxburghi (yellow)	*album*	23	1896
Aster tripolium	*album*	27	1897

The closest de Vries came to a Mendelian interpretation can be found in some notes he wrote on August 10, 1896. These notes were discovered in the Hugo de Vries Archive and appear in Stamhuis et al. (1999). Commenting on purple offspring obtained from white flowers, on p. 10 de Vries stated:

> **QUOT. 461** Discussion. According to the law of pangene hybridization (p. 187), the purple specimens from white mother must have purple fathers and be central hybrids. They show therefore that the white specimens at Huizen (preferably almost entirely, partially? [inserted: 95%]) have been fertilized by purple ones. Just as my Trifol. pat. alb. have been fertilized by my 7-leaved race at 16b VI.
>
> Therefore gain seed and sow it. If there are no white flowers this year and therefore all specimens are central hybrids, the seed must give 75% purples and 25% whites. This to be investigated [*sic*].
>
> At the same time this is a new principle in the transfer of varieties from the wild into the garden. If this happens after fertilization in the field, then all specimens from the seed can be look like old-types; then the variant will still emerge from their seed (namely in 25% of the specimens).
>
> (*ibid.*, pp. 249–250)

In another note written on Aug 18, 1896, de Vries commented on Veronica longifolia as follows:

> **QUOT. 462** According to the 1.2.1 law the old-types, no matter how they are pollinated, always give have to give [*sic*] 100% blue ones, while the central bastards in the case of free pollination (with the exception of whites), therefore by central bastards and by old-types, would have to give between 0 and 25% whites. (*ibid.*, p. 250)

Although the 3:1 and 1:2:1 ratios are reminiscent of Mendelian segregation, in these 1896 notes, de Vries was, in fact, thinking in terms of pangens (with several pangens for a trait) rather than the Mendelian two-factor per character model.

There is evidence that even as late as December 1899, de Vries was not thinking in terms of Mendelian segregation. In his first paper on xenia[84] in *Zea mays* (maize) submitted on Dec 4, 1899, de Vries obtained a clear 3:1 ratio for starchy vs. sugary seeds in F_2 (de Vries, 1899b, 1920, p. 198, Vol. VI), but nowhere in the paper did he invoke the law of segregation, let alone the name of Mendel. In a second paper on xenia in maize, submitted on April 15, 1900, de Vries incorrectly ascribed to xenia the yellow color of the cotyledons (which are inside the embryo and not part of the endosperm) in F_1 when yellow and dark seeds were crossed (de Vries, 1900b, 1920, pp. 273–274, Vol. VI). On the other hand, in a similar study, submitted on Dec 22, 1899, Correns correctly attributed the yellow color to a Mendelian crossing rather than to xenia (Correns, 1899, p. 417).

A more likely period when de Vries fully understood Mendel's paper might be inferred from two additional pieces of information, namely, (i) Correns' mention of Mendel in his xenia paper of 1899[85] (Correns, 1899), and (ii) the microbiologist and botanist M.W. Beijerinck dispatch of a copy of Mendel's paper[86] in early 1900 to Hugo de Vries when he learned that the latter was conducting hybridization experiments. Even if de Vries had heard of Mendel's paper before, Correns' mention of it might have further aroused his interest, and when Beijerinck sent him a copy, he might well have read and understood it fully. Such a chain of events would explain why he hurriedly sent three papers in quick succession to be published. However, it also implies that he deduced Mendel's laws only *after* reading (or rereading, if he had already done so in 1892) the latter's paper.

It is in the *Comptes Rendus* (**V1**) article that a first delineation of Mendelian segregation (Mendel's first law) is to be found in de Vries' writings:

> **QUOT. 463** In the hybrid the simple differential character from one of the parents is accordingly visible or *dominant* while the antagonistic character is in the latent condition or *recessive*.
>
> The antagonistic characters ordinarily remain combined during all of the vegetative life, one dominant, the other latent. But in the generative period, they are segregated. Each grain of pollen and each oosphere receives only one of the two. (de Vries, 1900c; Brooklyn Botanical Garden, 1950, p. 30)

The law of independent assortment (Mendel's second law) was stated as follows:

> **QUOT. 464** ...And the same proposition [as above] can be supported for the others (di- and polyhydrids), when one considers only a single character at a time. (*ibidem*)

De Vries also improved on Mendel's expression $A + 2Aa + a$ for the F_2 hybrids when F_1 hybrids Aa self-pollinated (Sect. 6.2.6.1). In Mendel's notation, A and

[84] See footnote 50.

[85] See (a) above and also Sect. 2.10, (11).

[86] This was pointed out by Correns' student, Th. J. Stomps, in a 1954 article (Stomps, 1954).

a represented only the dominant and recessive *traits*, respectively. In de Vries' notation, however, *A* and *a* were used to represent the *pollen* (*or ovule*) with the dominant and recessive characters, respectively:

QUOT. 465 If one calls D the grains of pollen or the ovules having a dominant character and R those which have the recessive character, one can represent the number and the nature of the hybrids by the following representative formula in which the numbers of D and R are equal:

$$(D + R)(D + R) = D^2 + 2DR + R^2$$

(*ibid.*, p. 32)

De Vries' notation, unlike Mendel's, expressed the actual process of fertilization of segregated factors.

His rediscovery papers notwithstanding, de Vries did not believe in the universality of Mendel's laws. He regarded his theory of mutation based on pangens as supreme, with Mendel's laws being only a subordinate part of it.[87] Moreover, despite Mendel having made no such claim, de Vries insisted that one of the two characters that made up the F_1 hybrid ought to be dominant:

QUOT. 466 The hybrid always shows the character of one of the two parents and that in all of its force.[88] (*ibid.*, p. 30)

The above statement was later singled out by Correns who correctly pointed out that dominance was not necessary in Mendel's theory. It was also later used in the form of the *Law of Dominance* by Weldon to undermine Mendelism.

6.6.2.3 Did de Vries Wish to Claim Priority Over Mendel's Results?

Surprisingly, de Vries omitted any reference to Mendel in the rediscovery paper that was first *published*, namely, the *Comptes Rendus* paper (**V1**) (de Vries, 1900c). In a postscript dated May 16, 1900, which was added to the proofs of his rediscovery paper (**C1**), Correns complained:

QUOT. 467 In the meantime de Vries has published in these Proceedings [in *Berichte der Deutschen*...] (No.3 of this year) some more details concerning his experiments. There he refers to Mendel's investigations, which *were not even mentioned in the "Comptes rendus."*[89] (Correns, 1900a, 1966a, p. 132)

[87] See Sect. 5.3.3 for more details.

[88] A similar statement was made in de Vries' *Berichte* article: "*Of the two antagonistic characteristics, the hybrid carries only one*, and that in complete development. Thus in this respect the hybrid is indistinguishable from one of the two parents. There are no transitional forms" (de Vries, 1900a, 1966, p. 110).

[89] Italics are Correns'.

Referring to dominating and recessive traits, Correns explained why he thought de Vries must have known of Mendel's paper, since "Mendel named them in this way, and, by a strange coincidence, de Vries now does likewise."

Correns' choice to use Mendel's name in the title of his rediscovery paper was undoubtedly meant to foil any attempt by de Vries to claim priority over Mendel's results, as Correns had seen it.

Similarly, de Vries' *Revue Générale* paper (**V3**) does not seem to have initially contained any reference to Mendel. A note was probably later added to the last page of the paper, stating that:

> **QUOT. 468** This law [of segregation] is not new. It was announced, for a particular case (peas), more than 30 years ago. It is Gregor Mendel who proposed it... (de Vries, 1900d, 1920, p. 269)

A footnote was also inserted pointing to the rediscovery papers of Correns and Tschermak.[90]

However, de Vries' rediscovery paper that was first *submitted*, namely, the *Berichte* paper (**V2**) (de Vries, 1900a), did contain several references to Mendel. Sturtevant (1965, p. 27) later noted that de Vries extensively altered the proofs of the *Berichte* paper. Perhaps de Vries had seen a reprint of Correns' accusatory paper and then decided to quickly insert references to Mendel?

In a 1979 paper, van der Pas poured cold water on such a scenario (van der Pas, 1976, pp. 6–7). He discovered a postcard written to de Vries by the editor (Carl Müller) of *Berichte* who apologized for *not* having been able to include de Vries' alterations in the paper. In addition, the corrected galley proofs show that references to Mendel had already been included in the *original* text, not after having seen a preprint of Correns' letter. Based on this evidence, de Vries cannot be accused of attempting to claim priority on Mendel's results.

6.6.2.4 Conclusion

De Vries was one of the greatest biologists of his time. Of all three rediscoverers, he was the only one who had already built a theory of variation and heredity. His contributions to genetics are outstanding and cannot be diminished. At the same time, his account of the rediscovery of Mendel's laws has several inconsistencies. Contrary to his claim, de Vries probably did not understand Mendel's law on his own before 1895. He probably read Mendel's paper in 1892 but did not understand it then. He might have noticed 3:1 ratios before reading Mendel's paper, but he did not interpret them in terms of Mendelian segregation. He most certainly did so after having read (or reread) Mendel's paper carefully in early 1900. However, it does not seem that he wished to usurp the credit for Mendel's work. In any case, de Vries

[90] The note and footnote are both dated March 19, 1900. This is actually the date of submission of the original paper; the note and footnote were added much later since the latter refers to both Correns and Tschermak.

thought Mendel's laws were not universally applicable and were only a subordinate part of his theory of intracellular pangenesis. In 1908, he refused to sign Tschermak's petition to organize a commission for the creation of a Mendel memorial in Brünn. De Vries also refused to attend the 1922 celebration of Mendel's work. Writing to his friend Friedrich Went in September 1922, he said:

QUOT. 469 To my regret I cannot accede to your request. I just do not understand why the academy would be so interested in the Mendel's celebrations. The honoring of Mendel is a matter of fashion which everyone, also those without much understanding, can share; this fashion is bound to disappear. The celebration in Brünn is nationalistic and anti-English, directed especially against Darwin and thus unsympathetic to my mind, but therefore also very popular. (Galton, 2017, p. 132)

6.6.3 Carl Correns (May 1900)

6.6.3.1 Introduction

In a Jan 23, 1925, letter to Roberts, Correns (who had been a student of Nägeli) explained his rediscovery of Mendelism as follows:

QUOT. 470 ...The operation of a principle was soon found in the case of peas and maize. I was able accordingly soon to proceed systematically in the experiments, as the two genealogies for peas in my first contribution show. I did not come at first upon the explanation of the regular relationship [Gesetzmassigkeit], but I likewise, however, did not seek intensively after it. For I wished, for various reasons, to first finish an extensive book upon the sexual propagation of the foliose mosses, upon which I had worked for years. I then wished first to push intensively the elaboration of the investigations on xenias and hybrids, which had been carried on at the same time since 1894. The printing of the book lasted until in August 1899; then I was able to devote myself earnestly to the genetics researches. The date of the day upon which, in the autumn (October) of 1899, I found the explanation. I no longer know; I do not make note of such matters. I only know that it came to me at once 'like a flash,' as I lay toward morning awake in bed, and let the results again run through my head. Even as little do I know now the date upon which I read Mendel's memoir for the first time: it was at all events a few weeks later. I then first made ready for the press the contribution on xenias in maize. In it, it is already pointed out, that in crosses between maize races I had found very interesting but very complicated relationships. That other investigators also worked in the same direction I naturally did not know, otherwise I would have hastened more with the preparation of the publication.

On the morning of the 21st of April, 1900, I received a separate 'Sur la loi de disjonction des hybrides,' of de Vries, and by the evening of the 22nd of April, my contribution, 'G. Mendels Regel über das Verhalten der Nachkommenschaft der Bastarde,' was ready. I sent it to the German Botanical Society in Berlin, where it was received April 24, and was reported in the session of April 27. The issue in question of the 'Berichte' appeared at the end of May, about the 25th...

(Roberts, 1929, pp. 335 & 337)

In a later Jan 30, 1925, letter, Correns added:

QUOT. 471 I did not discover the constant relationship [Gesetzmiissigkeit] in *Pisum* alone but in *Zea* and *Pisum* simultaneously. In the publication, I placed *Pisum* in the foreground on Mendel's account, and out of didactic considerations.

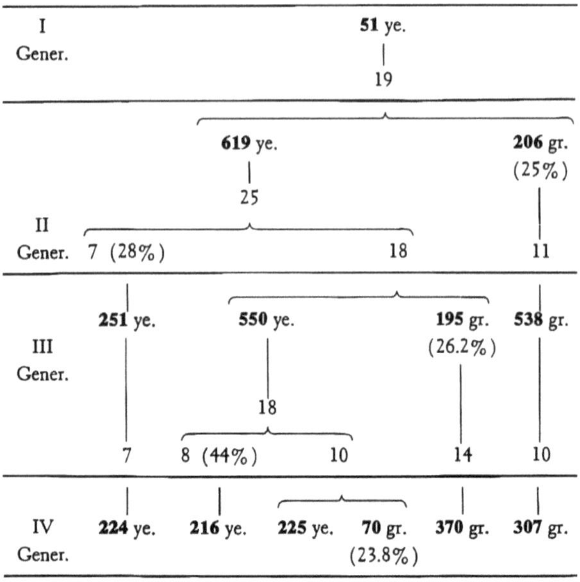

Fig. 6.9 Hybridization experiments performed by Correns (1900a) involving the color of the embryo within the seeds of *Pisum*. The two colors were yellow (dominant) and green (recessive)

> . . .
>
> Besides through Focke's book, I had been made cognizant of Mendel's investigations through my teacher Nägeli. And I believe also to remember that he told me of Mendel, but certainly only of the Hieracium investigations, in which alone he was permanently interested...
>
> (*ibid.*, pp. 337–338)

In the extracts above, Correns indicated that he rediscovered Mendelism through his research on xenia[91] and hybridization and dated his first reading of Mendel's paper to a few weeks after October 1899.

The hybridization experiments that were performed by Correns in his first rediscovery paper (**C1**) (Correns, 1900a) are shown in Fig. 6.9. They relate to the two colors (yellow, dominant vs. green, recessive) of the embryo within the seeds of *Pisum*. No date is mentioned in Correns' paper, but the first crosses seem to have been made on June 19, 1896 (Rheinberger, 2003, p. 228)

Some questions about Correns' rediscovery will be examined next.

6.6.3.2 When Did Correns First Read and Understand Mendel's Paper?

Despite Correns' indicated time frame of a few weeks after October 1899 for his first reading of Mendel's paper, Hans-Jörg Rheinberger has produced evidence of a different date. In 1995, he recovered a folder from the protocols of Correns'

[91] See footnote 50.

experiments in the Archive for the History of the Max-Planck Society in Berlin. In one of the notes in the folder, one can read the following text on *Pisum* written by Correns on April 16, 1896:

QUOT. 472

16. IV. 96.
Mendel (66) distinguishes:
dominant and recessive characters. For our cases is

	dominant:	recessive:
– form of seed	round	wrinkled
– seed coat: ("albumen")	grey to brown	*white*
– cotyledons:	yellow	pale-yellow, *green*
– pod:	inflated	constricted
–	green (unripe)	yellow (ripe)

The dominant and recessive characters are expressed already in the first generation in such a way that the former are present in 3, the latter in 1 individual, respectively.

| The hybrid form of seed shape and cotyledons develops immediately and directly through fertilization |

Cot. Yellow ♀ + green ♂ = yellow ♂ + green ♀ = 3/4 yellow + 1/4 green
Form. round ♀ + wrinkled ♂ = wrinkled ♀ + round ♂ = 3/4 round + 1/4 wrinkled
The seed coat, the form, and the color are *not changed*
But later Mendel notes, e.g., that A (seed round, cot. (p. 19) yellow) pollinated wrinkled cot. green), *exclusively* yielded *yellow* seeds which were *round.*
(Rheinberger, 1995, pp. 224–225)

The extract above, especially Correns' reference to p. 19 of Mendel's *Versuche* (Mendel, 1866), leaves no doubt that Correns must have read Mendel's paper on or before April 16, 1896. Therefore, Correns' statement in the Jan 23, 1925, letter to Roberts that he first read Mendel a few weeks after October 1899 seems to be the result of misrecollection. There are at least two reasons that can be inferred from Correns' *Pisum* notes of April 16, 1896, which would explain why Correns might have dismissed Mendel's paper at the time to the point of forgetting about it later:

(a) In these experiments, Correns was more interested in the phenomenon of xenia (which produces immediate effects on the seeds of the fertilized plant) than in numerical ratios, as the emphasized statement (between the two vertical bars) in his notes indicates: "The hybrid form of seed shape and cotyledons develops immediately and directly through fertilization."
(b) Correns may have thought there were inconsistencies in Mendel's paper. In the notes, when Correns spoke of 3:1 ratios *in Mendel's first generation*, he probably did not realize that the latter was, in fact, F_2 (since Mendel counted the offspring generations from the hybrids). When later Correns saw that all the seeds were

yellow and round in the first generation (Mendel's hybrids), he saw an inconsistency (wrongly, of course) and may therefore have dismissed Mendel's paper.

Thus, Correns most likely did not understand Mendel's paper when he first read it on or before April 16, 1896.

The experiments with *Pisum* that were carried out in June 1896 (last section) also yielded Mendelian ratios. These must have attracted Correns' attention, but he made no mention of Mendel this time.

As expected, Correns' search for xenia effects in *Pisum* did not succeed. However, he also experimented with *Zea Mays*, sometimes in parallel with the *Pisum* experiments.

Indeed, a note written on Jan 2, 1898, concerning crossings in *Zea Mays* shows that Correns correctly understood Mendelian segregation at that time (Rheinberger, 2003). Concerning the self-fertilization of "caesia impure" (which is the hybrid from the crossing between the varieties *caesia* (blue, dominant) and *alba* (white, recessive), he said:

QUOT. 473 . . .if we assume that one half of the pollen grains of caesia (impure) be disposed for white, one half of them for blue (and that blue alters *always*!), then, because the "white" pollen grains never just will come to the 'white' stigmas, the blue pollen grains to the 'blue' stigmas, and because 'caesia' ovaries are not directly influenced by alba pollen and the crosses a ♀ + a ♂ give white, whereas a ♀ + c ♂ , c ♀ + a ♂ and c ♀ + c ♂ therefore give blue grains, the blue grains must amount to approximately 3/4 instead of 1/2. (*ibid.*, p. 241)

Having found segregation in maize, Correns returned in the winter of 1899/1900 to revisit his previous *Pisum* experiments.

Correns was thus quite prepared when he read de Vries' papers on xenia in maize (Sect. 6.6.2.2). In the second paper, submitted on April 15, 1900, de Vries incorrectly ascribed to xenia the yellow color of the cotyledons in F_1 when yellow and dark seeds were crossed (de Vries, 1900b, 1920, pp. 273–274, Vol. VI). On the other hand, in a similar study, submitted on Dec 22, 1899, Correns correctly attributed the yellow color to a Mendelian crossing rather than to xenia, arguing that the latter cannot produce changes inside the embryo (Correns, 1899, p. 417).

Correns was equally prepared when he received de Vries' *Comptes Rendus* paper (de Vries, 1900c) on April 21, 1990. He was able to quickly put together his *Pisum* results, and as he stated in his January 1925 letter to Roberts, he submitted his rediscovery paper (**C1**) (Correns, 1900a) the next day, on April 22. As explained in Sect. 6.6.2.3, Correns had suspected that de Vries wished to claim priority on Mendel's results since the *Comptes Rendus* article used Mendel's language but made no mention of him.

Correns' enunciation of Mendel's law of segregation (Mendel's first law) in his rediscovery paper was almost the same as de Vries':

QUOT. 474 ...following fusion of the reproductive nuclei the "anlage" for one trait, the "recessive" one ...is suppressed by the other trait, the "dominating" one...The anlage, however, although "latent" is preserved, and prior to *the definitive formation of the reproductive nuclei a complete separation of the two anlagen occurs, so that one half of the* *reproductive nuclei receive the anlage for the recessive trait... the other half the anlage for the dominating trait...*[92] (*ibid.*, pp. 124–125)

A noticeable difference from de Vries' version (Sect. 6.6.2.2) is that, following the term used by his teacher Nägeli, Correns referred to the segregation of the *anlagen* (potentials) whereas de Vries' referred to the segregation of the *characters*. In this respect, Correns' statement was more accurate than de Vries'.

Correns also took issue with de Vries' insistence that dominance of one trait in the F_1 hybrids was a necessary condition:

QUOT. 475 I can not understand why de Vries assumes that in *all* pairs of traits which differentiate two strains, one member must always dominate.[93] Even in peas, where some traits completely conform to this rule, other trait pairs are also known in which neither trait dominates, as for instance the color of the seed coat being either reddish-orange or greenish-hyaline.[94] In this case the hybrid may show all transitions, (this is true especially for the seed coat of peas), or may show either more of one or of the other trait (for example in hybrids of stocks; here a certain hybrid may be *just barely* distinguished from one parental form by its hardly noticeable, slighter covering of hairs, although with some care separation is always possible, while it is *highly distinct* from the other, i.e., the glabrous parental type). (*ibid.*, pp. 121–122)

Similarly, Correns' statement of Mendel's second law (law of independent assortment) was more explicit than de Vries':

QUOT. 476 ...*the behavior of each pair of differing traits in a hybrid association is independent of all other differences in the two parental plants.*[95] (*ibid.*, p. 130)

However, like de Vries, Correns also believed that Mendel's first law was not universally applicable[96]:

QUOT. 477 At present, however, this law is applicable only to a certain number of cases, i.e., those where one member of a pair of traits dominates, and probably only to hybrids between varieties. It seems impossible that all pairs of traits of all hybrids should behave according to this law. Some hybrids of peas bear this out. (*ibid.*, p. 131)

Concerning Mendel's second law, a very important exception was later added by Correns in the proofs of his paper as a footnote:

QUOT. 478 There are again exceptions to this rule; strains with linked traits exist. (*ibid.*, p. 130f)

---

[92] Italics are Correns'.

[93] For instance "D'autre part, l'étude des caractères simples des hybrides peut fournir la preuve la plus directe du principe énoncé. *L'hybride montre toujours Ie caractère d'un des deux parents, et cela dans toute sa force*; jamais Ie caractère d'un parent, manquant à l'autre, ne se trouve réduit de moitié" (loq. cit. paragraph 3, italics mine). (*Correns' note*).

[94] The color of the hilum on the other hand (whether black, brownish, etc.) represents a dominating character. (*Correns' note*).

[95] Italics are Correns'.

[96] It is now known that Mendel's first law is almost universally true, with only rare exceptions (e.g., in meiotic drive, first proposed by Sandler and Novitski (1957), there is a distortion in the 1:1 segregation ratio).

Correns' reservation regarding Mendel's second law was the first explicit recognition of the phenomenon of *linkage*.[97]

6.6.3.3 Conclusion

There is evidence to suggest that Correns observed 3:1 ratios before reading Mendel's paper on or before 1896 but did not understand the full implications of these ratios, even after reading Mendel's paper at the time. Therefore, strictly speaking, his claim that he understood Mendelism before reading Mendel's paper cannot be agreed with. However, because his initial research interest was on xenia, he most probably forgot about his 1896 reading of Mendel's paper. More light on Mendelian segregation was shed in later years as Correns continued his experiments with *Zea Mays* and *Pisum*. His notes show that he had understood Mendelian segregation in January 1898 on his own, i.e., before rereading Mendel's paper. However, the influence on his rediscovery of his initial reading of Mendel's paper in or before 1896 cannot be discarded.

6.6.4 *Erich von Tschermak (July 1900)*

6.6.4.1 Introduction

On January 7, 1924, Erich Tschermak[98] wrote a letter to Roberts explaining his rediscovery of Mendel's laws:

> **QUOT. 479** . . .Following the suggestion of a pupil of my father, the mineralogist, Professor Renard, at the university of Ghent, to extend my practical horticultural knowledge among some well-known horticultural enterprises of Belgium, Holland and France, I betook myself in the spring of 1898 to Ghent...I selected peas, impelled by the reading of Darwin on the effects of cross- and self-fertilization. I was also urged by the incompleteness of the observation-material in the case of this plant with Darwin (only four pairs of plants were measured and compared to complete these experiments). The yield-results from my experiments I took with me to Vienna...Since the prospective assistant's position was still not yet

[97] In QUOT. 478 above, Correns was referring to experiments he had been conducting on *Matthiola* (Correns, 1900c). He distinguished between two causes for the correlation of characters: (i) a single *anlagen* contributing to two or more traits (now known as *pleiotropy*), (ii) the physical organization of the *anlagens* (now known as *linkage*). Correns indicated that his results on *Matthiola* conformed more with (ii). See Rheinberger (2004) for more details. As for the term "linkage," it was first used by Morgan and Lynch (1912) in 1912 (see Sect. 13.4.4.3). Prior to this, the term "coupling" was in common use.

[98] Erich Tschermak's father, Gustav Tschermak, had been a famous mineralogist and university professor and was elevated in 1906 by Emperor Franz Joseph I (1830–1916) into hereditary nobility with the title Gustav Tschermak, Edler von Seysenegg. From then on, Erich Tschermak was also known as Erich von Tschermak-Seysenegg.

open in 1899. I volunteered for a year on the estate of the Imperial family's foundation in Esslingen, near Vienna...

There, in the garden of a neighboring estate owner, my experiments with peas, begun in Ghent, were continued, and at the same time, new crosses instituted with grain and garden beans. The working up of the F.. of my peas hybrids, followed in the fall of 1899. In this I was struck especially by the different value of the characters of the individual races with respect to their structure, cotyledon-color and form (conclusion III of my first paper, in which I emphasized besides that, instead of 'dominieren' [dominate], one should say rather 'prävalieren' [predominate], at least in certain cases [see conclusion VIII]). In counting out the seed-characters, the ever-recurring number relationship of 3:1 could naturally not escape me, any more than the number-relation of 1:1 on back-crossing of green-seeded peas with hybrid pollen of the F_1 generation. The rules of inheritance, quite intentionally, I expressed at first purely descriptively or phenomenologically, in order not at once to anchor the newly-beginning experimental phase of the doctrine of heredity – as had happened inexpediently with Darwinism – to definite theoretical terms. From this standpoint, I designated the regularities found by Mendel and myself, as those of the regularly varying values of the characters for the inheritance, under which I comprehended not only the rule of dominance, but mass-value (dominance-recessiveness, or equivalence of value); quantities-value (the relation of segregation), and inheritance-value, or splitting *per se*...

In the autumn of 1899, I received from Prof. A. v. Liebenberg the permission to volunteer in his department, and to make use of the library. The first work I seized upon was the well-known book of Focke: 'Pflanzenmischlinge,' of 1881. There I found, in the chapter on 'Peas,' the familiar obscure expression of Focke's concerning Mendel's treatise, as well as the views on Mendel's experiments with beans and Hieraceae. Since Mendel's work was not on hand in the library of the Hochschule für Bodenkultur, I had on the same day of this 'discovery' the 'Transactions of the Natural History Society of Brunn,' hunted out of the university library, which now gave me the information, to my greatest surprise, that the regular relationships discovered by me, had already been discovered by Mendel much earlier...

(Roberts, 1929, pp. 343–346)

In the extract above, Tschermak explained that his experiments on peas started in 1898 and were motivated by Darwin's cross-fertilization and self-fertilization experiments.[99] He traced his first discovery of segregation ratios of 3:1 in F_2 hybrids and 1:1 in backcrosses to the fall of 1899. This discovery prompted him to consult Focke's book (Focke, 1881), where he found the reference to Mendel's work and immediately procured the paper.

In his first published paper (**T1**) (Tschermak, 1900a) which summarized his 90-page dissertation (**T2**) (Tschermak, 1900b), Tschermak stated that his crosses with peas had yielded F_2 ratios close to 3:1 for dominant to recessive traits (seed shape and color of seed coat) (Tschermak, 1900a; Brooklyn Botanical Garden, 1950, p. 45). Regarding the backcrosses of the F_1 hybrids with the recessive parent, Tschermak declared:

QUOT. 480 The influence of the character "yellow" in the seeds in the hybrid was in this case reduced by 57 per cent, while that of the character "green" was reduced by 43.5 percent. (*ibid.*, p. 46)

[99] As published in the latter's 1876 book (Darwin, 1876).

The above statement was made without providing any explanation, except that it was based on Tschermak's concept of the *hereditary potency* ('Werthigkeit') of a character.

Tschermak also indicated in the paper that he wished his name to be associated with those of de Vries and Correns as one of the three rediscovers of Mendel's laws. However, there are strong arguments that can be made as to why this should not be the case, as will be explained in the next section.

6.6.4.2 Did Tschermak Understand Mendel's Laws at the Time of his Rediscovery?

In both de Vries' and Correns' cases, it was necessary to ask whether each of them had rediscovered Mendelism before reading Mendel's paper. However, in Tschermak's case, it seems more appropriate to ask if he had understood Mendelism even *after* reading Mendel's paper. This is because, although Tschermak claimed to have rediscovered Mendelism, the interpretations that he gave in his rediscovery papers, especially the first (**T1**), were not along the lines of Mendelian thinking. This prompted the geneticist Curt Stern (1902–1981) to vigorously argue that Tschermak ought not be considered a rediscoverer of Mendelism:

> **QUOT. 481** . . .De Vries and Correns not only claimed that they had discovered the essential facts and developed their interpretation, before they had found Mendel's article but they demonstrated in their own publications that they fully understood the essential aspects of Mendel's theory: the assumption that pairs of determinants were present in hybrids, that they were segregated in germ cells, and that this segregation accounted for the ratios in various types of crosses. On the other hand, Tschermak's papers of 1900 not only lack fundamental analysis of his breeding results but clearly show that he had not developed an interpretation. (Stern & Sherwood, 1966, p. xi)

Two aspects of Tschermak's paper (**T1**) lend support to Stern's claim:

(a) *Misunderstanding of Mendel's Concept of Dominance*. In Mendel's theory, dominance is the property of one form of a trait (the dominating form) being able to represent the form of a hybrid when it enters in association with another form (the recessive form). On the other hand, in Tschermak's framework, dominance occurs only when the overwhelming majority of hybrids have the dominating form. Thus, when F_1 hybrids self-fertilize, to Mendel a ratio 3:1 means that the proportion of *dominants* is .75. On the other hand, to Tschermak a ratio of 3:1 means that the first form is only *prepotent* (in the ratio 3:1). When contrasting the F_1 and F_2 hybrids, Tschermak therefore stated:

> **QUOT. 482** . . .while in the artificial production of products of heteromorphic xenogamy, the first named characters are almost without exception dominant, and the latter "recessives" appear in pure form or as admixtures only in a very few cases, in the seeds of the first hybrid generation, the former characters are expressed in pure form in a majority of cases only, and the recessive character appears in pure form in a minority. In the first case an almost absolute dominance exists, in the second a mere prepotency (in a fixed proportion). (Tschermak, 1900a; Brooklyn Botanical Garden, 1950, p. 45)

Thus, in Mendel's theory, one form of a trait has a constant relationship with another form. However, in Tschermak's framework, the relationship is variable and depends on the numerical ratio between the forms. This was one of the major implications of Tschermak's concept of the *hereditary potency* of a trait.

Unfortunately, as late as 1924, when Mendelism had gained all but universal acceptance, Tschermak was still harking back to prepotency:

> **QUOT. 483** ...see conclusion III of my first paper, in which I emphasized besides that, instead of 'dominieren' [dominate], one should say rather 'prävalieren' [predominate], at least in certain cases [see conclusion VIII]... (Roberts, 1929, p. 345)[100]

(b) *Misinterpretation of Backcrosses.* In the previous section, it was mentioned that Tschermak also investigated backcrosses. In his first rediscovery paper (**C1**), regarding backcrosses of the F_1 hybrids with the recessive parent, Tschermak failed to realize that the backcrosses should yield a segregation ratio of 1:1. Curt Stern was especially critical of this omission and disqualified Tschermak as a rediscoverer of Mendelism:

> **QUOT. 484** ...When he [Tschermak] reports on the results of the back cross of F_1 hybrids to their recessive parents he notes that it "results in an increase of the number of bearers of the recessive trait as compared to the number of recessives obtained after selfing of the F_1." Specifically, he adds that "the influence of the trait 'yellow' in the seeds of the hybrid was reduced by 57 per cent, that of the trait 'smooth' by 43.5 per cent." No discussion of these percentages is offered. The quoted sentences reveal that Tschermak had not realized that back crosses should give a 1: 1 ratio. (Stern & Sherwood, 1966, p. xi)

In fact, not only is the 1:1 ratio omitted in Tschermak's first paper, but segregation (of two factors or elements per character) itself is never mentioned.[101] This is perhaps not surprising since Tschermak's theoretical framework involved inherited potency, not Mendelian segregation.

6.6.4.3 Conclusion

Tschermak published extensively in diverse biological fields and had as illustrious a career as both de Vries and Correns. He did a great service by popularizing Mendel's ideas after the rediscovery. However, there is no indication that he, unlike de Vries and Correns, had a correct understanding of Mendelism at or before the time of the rediscovery.

In a series of papers, Michal Simunek and his colleagues have also pointed to the important role that Tschermak's older brother, Armin von Tschermak (1870–1952), played in shaping his views during the crucial 1900–1901 period (Simunek et al., 2011, 2012, 2017).

[100] See QUOT. 479, §2.

[101] The ratio 1:1 is also absent in his 90-page thesis (Tschermak, 1900b), although segregation is mentioned. Mendelian-type segregations are referred to only in 1901.

Part III
The Biometry-Saltationism Debate

Chapter 7
Introduction to the Biometry-Saltationism Debate

Abstract The Mendelian rediscovery at first only increased the discord between the biometric (represented mainly by Weldon and Pearson) and saltationist (represented mainly by Bateson) viewpoints on evolution. The biometricians held that evolution occurred continuously, with natural selection acting on individual differences (as Darwin had postulated). On the other hand, the saltationists held that evolution occurred discontinuously by the action of mutations. To the biometricians, heredity could be understood through Galton's ancestral law. The saltationists disagreed and insisted that heredity was explained though Mendel's laws.

Keywords Biometry · Individual differences · Ancestrian · Ancestral law · Saltationism · Sport · Mutation

7.1 Two Opposite Views of Evolution

7.1.1 Introduction

To the modern eye, the expectation might be that, once Mendelian heredity was rediscovered in 1900, biologists should at once have recognized that it was the missing component in Darwin's theory and set about synthesizing the two. However, the path of scientific advancement is all, except linear: the history of science is *not* a history of unrelenting progress. Interpreting history as if this was the case is dangerously Whiggish because it ignores the fact that scientists operate within the confines imposed by the state of knowledge at the time. This needs to be remembered before castigating blame for the considerable delay in a potential synthesis of Darwin's and Mendel's theories, which in turn led to a delay in the birth of the powerful field of population genetics.

Indeed, instead of harmonizing existing theories, the Mendelian rediscovery only exacerbated a controversy that had already been brewing between two parties, namely, the so-called biometricians and saltationists. Essentially, the two parties had diametrically opposite views on evolution: the biometricians believed that evolution was *continuous* (i.e., acting on individual differences, as Darwin had proposed), and the saltationists believed it was *discontinuous* (i.e., proceeding in

P. Gorroochurn, *The Development of Evolutionary Genetics*,
https://doi.org/10.1007/978-3-031-69374-8_7

jumps, or by the action of *sports* or *mutations*). With the Mendelian rediscovery, the evolutionary debate turned into a hereditary one. Most saltationists welcomed Mendelism as *the* hereditary theory compatible with their evolutionary beliefs, while the biometricians argued that Galton's Law of Ancestral Heredity[1] best described the phenomenon of heredity. The discreteness of the Mendelian theory was thus assumed to be compatible with only a discrete mode of evolution, while the continuity of the ancestral law was assumed to be compatible with only a continuous mode of evolution. It was not until a clear distinction between genotype and phenotype was made, and Mendelian heredity was shown to be compatible with continuous evolution that a reconciliation between the two viewpoints was possible.

7.1.2 Biometry

Lamarck, who was the first to propose a comprehensive theory of evolution, believed that species transformation was essentially continuous.[2] However, the continuous model the biometricians had in mind was Darwin's (and Wallace's) continuous model, with natural selection acting on *individual differences*. The biometricians were represented mainly by Raphael Weldon (1860–1906) and Karl Pearson (1857–1936).

The word biometry itself has an interesting history. It was first used in 1831 by Whewell to describe demographic calculations as follows:

> **QUOT. 485** By the way there is a problem in Biometry (if you choose to call your calculations on lives by a Greek name) which may perhaps be included. (Todhunter, 1876, p. 135, Vol. II)

Its current use in the sense of statistical methods applied to biological problems, especially within the framework of population thinking,[3] was initiated in 1877 by Galton, the founder of the field (see Sect. 4.5.4). Galton's definition of the term first occurred in 1901, as follows:

> **QUOT. 486** The primary object of Biometry is to afford material that shall be exact enough for the discovery of incipient changes in evolution which are too small to be otherwise apparent. (Galton, 1901, p. 9)[4]

Since Galton believed that evolution was discontinuous, it is ironic that the adherents to a continuous mode of evolution were called biometricians. The latter wished to identify themselves with Galton's *statistical methods* (together with the associated population thinking), not with his *evolutionary beliefs*.

Some of the key beliefs of the biometricians were as follows:

---

[1] See Sect. 4.6.

[2] See Sect. 1.8.

[3] See Sect. 2.9.

[4] Even earlier, Meldola had used "biometric station" in 1896 (Pearson, 1930, p. 133, Vol. IIIA).

(a) Evolution proceeds continuously by the action of natural selection acting on individual differences.
(b) Heredity may be explained by statistical methods alone, without the need for physiological explanations.
(c) Inheritance in Darwin's theory is blending and is mathematically expressed by Galton's Law of Ancestral Heredity.[5]

The biometricians were thus often called *Darwinians*, *ancestrians*, or *gradualists*.

7.1.3 Saltationism

It is noteworthy that most of the very early proponents of evolutionary concepts espoused a discontinuous view of evolution. One of the earliest modern saltationists was Geoffroy Saint-Hilaire.[6] He was followed by Kolliker, Huxley,[7] Galton,[8] and others.

Saltation was first used in an evolutionary sense by Huxley in 1864 as follows:

> **QUOT. 487** We greatly suspect that she [Nature] does make considerable jumps in the way of variation now and then, and that these saltations give rise to some of the gaps which appear to exist in the series of known forms. (Huxley, 1864, p. 576)

A sport of nature in the sense of a marked deviation from the normal type (i.e., a "freak" of nature or *Lusus naturae*) was first used by George Hakewill in his *An Apologie of the Power and Providence of God* (Hakewill, 1635, p. 230).[9] Although the term mutation is usually traced to de Vries' *Die Mutationstheorie*[10] (de Vries, 1901, 1903), it was first used in the sense of "species-forming" variability by the Austrian paleontologist and geologist Wilhelm Heinrich Waagen (1841–1900) in 1869. Referring to various forms of the ammonites, he said:

> **QUOT. 488** The name Variety, which has been used for a long time, will suffice to denote them, but for the sake of brevity I would like to suggest a new term 'mutation'. (Waagen, 1869, p. 186)

The main proponent of the saltationist viewpoint was William Bateson (1861–1926). Some of the characteristic beliefs of the saltationists were as follows:

---

[5] See Sect. 4.6.

[6] See Sect. 1.11, especially QUOT. 67.

[7] See Sect. 2.12.2, especially QUOT. 125.

[8] See Sect. 4.3.4, especially QUOT. 287.

[9] However, the phenomenon of marked abnormalities was known since ancient times. *Lusus naturae* was introduced by Pliny the Elder in the first century A.D.

[10] See Sect. 5.3.3.

A. Evolution proceeds discontinuously by the occurrence of abrupt changes (or sports or mutations), with natural selection having only a secondary role, namely, that of *selecting* existing variants (rather than *creating* new ones).
B. Heredity needs to be understood physiologically and experimentally before the application of statistical methods.
C. Heredity follows Mendel's laws, not Galton's ancestral law (this distinction was made once Mendelism was rediscovered in 1900).

The saltationists were thus often also called *mutationists*, *Batesonians*, or *Mendelians.*

The biometry-saltationism debate involved not only a clash of scientific theories but also scientific personalities. Therefore, before the debate is examined, it will be worthwhile to describe the main protagonists involved.

7.2 The Main Protagonists: Weldon, Pearson, and Bateson

7.2.1 Raphael Weldon (1860–1906)

Raphael Weldon[11] (Fig. 7.1) joined University College, London, in 1876 to pursue a medical career. At University College, Weldon learned mathematics from Olaus Henrici and biology from the zoologist E. Ray Lankester. The following year, he was transferred to King's College, London, then to St. John's College, Cambridge, in 1878. At Cambridge, Weldon studied under the zoologist Francis Maitland Balfour and learned about Darwin's writings on morphology and embryology. Weldon decided to give up his medical career for zoology, and in 1881, he obtained a first-class degree in the Natural Sciences Tripos.

In 1881 still, Weldon left for Naples to work at the Zoological Station and published his first paper entitled "Note on the early development of *Lacerta muralis*" (Weldon, 1883). In September 1882, Weldon returned to Cambridge as a demonstrator for Adam Sedgwick, who had succeeded Balfour. Weldon wrote his second paper from the Sedgwick laboratory. In 1884, Weldon was elected to a Fellowship at St. John's College, and shortly after, he was appointed University Lecturer in Invertebrate Morphology. Five years later, Weldon succeeded E. Ray Lankester in the Jodrell Chair of Zoology at University College, London. By that time, Weldon's interests had changed from morphology[12] to the distribution of variation and the organic correlation between characters, with a view to understanding how natural selection affected such correlations.

[11]Weldon's life and work are also discussed Pearson (1906), Magnello (2004a, b), and Cowan (1981).

[12]Several other young biologists (notably Bateson and de Vries) also moved away from morphology to experimental methods. The historian Garland Allen has described this trend as a "revolt against morphology" (Allen, 1975, p. 8).

Fig. 7.1 Walter Frank Raphael Weldon (1860–1906). (From: https://commons.wikimedia.org/wiki/File:WFR_Weldon.jpg)

In 1889, when Weldon read Galton's *Natural Inheritance* (Galton, 1889), he was elated because he saw in Galton's book the appropriate methods to pursue his investigations. Weldon's first paper (Weldon, 1890) using Galton's methods was written in 1890. It investigated the variations in *Crangon vulgaris* (brown shrimp), and the results showed that:

> **QUOT. 489** ...the frequency, with which deviations from the average size of an organ occurred, to obey the law of error in all cases, whether the animals observed were under the action of natural selection or not. (*ibid.*, p. 446)

This observation was in accordance with Galton's prediction, but more importantly, it showed that Weldon was already thinking in terms of natural selection acting on individual differences (i.e., *continuously*), as Darwin had prescribed.

Weldon's 1890 paper was followed by a second one on the same species (Weldon, 1892), where he calculated correlation coefficients between various organs of brown shrimp. These were the first correlation coefficients calculated for a wild population.

Weldon's third biometry paper (Weldon, 1893) was published in 1893. Its aim was to verify whether the previous conclusion on *Crangon vulgaris* also applied to *Crangon moenas* (shore crab) and is noteworthy, especially for two reasons:

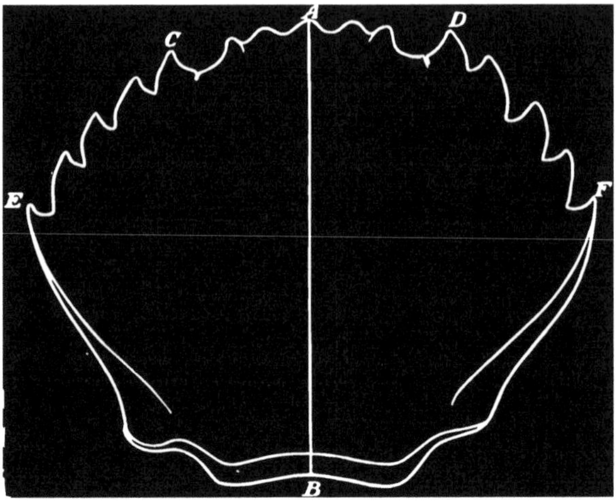

Fig. 7.2 Measurements on *Crangon moenas* (Weldon, 1893, p. 319). The frontal breath is the straight line joining C and D

(a) *Announcement of the Biometric Agenda.* At the end of the paper, Weldon spelled out what would be the agenda for the Darwinists, namely, a systematic investigation of evolutionary processes within a statistical framework:

QUOT. 490 It cannot be too strongly urged that, the problem of animal evolution is essentially a statistical problem: that before we can properly estimate the changes at present going on in a race or species we must, know accurately (a) the percentage of animals which exhibit a given amount of abnormality with regard to a particular character; (b) the degree of abnormality of other organs which accompanies a given abnormality of one; (c) the difference between the death rate per cent. in animals of different degrees of abnormality with respect to any organ; (d) the abnormality of offspring in terms of the abnormality of parents, and *vice versâ*. These are all questions of arithmetic; and when we know the numerical answers to these questions for a number of species we shall know the direction and the rate of change in these species at the present day – a knowledge which is the only legitimate basis for speculations as to their past history and future fate. (*ibid.*, p. 329)

(b) *Recruitment of Karl Pearson to the Biometric Program.* Out of the 11 sets of measurements taken by Weldon in the 1893 paper, all were normally distributed except for one, namely, the frontal breadth (distance *CD* between the two tips of the anterior lateral teeth in Fig. 7.2) of the carapace of the shore crab. The distribution of values in this case was left-skewed (Fig. 7.3). These values could not be analyzed by Galton's methods and Weldon suspected "the presence, in the sample measured, of two races of individuals, clustered symmetrically about separate mean magnitudes" (*ibid.*, p. 324). He therefore enlisted the help of his more mathematically sophisticated friend, Karl Pearson, whom he had met in 1891, to verify if that was the case. The latter showed that, indeed, "Carcinus moenas is slightly dimorphic in Naples with respect to its frontal breadth" (*ibidem*).

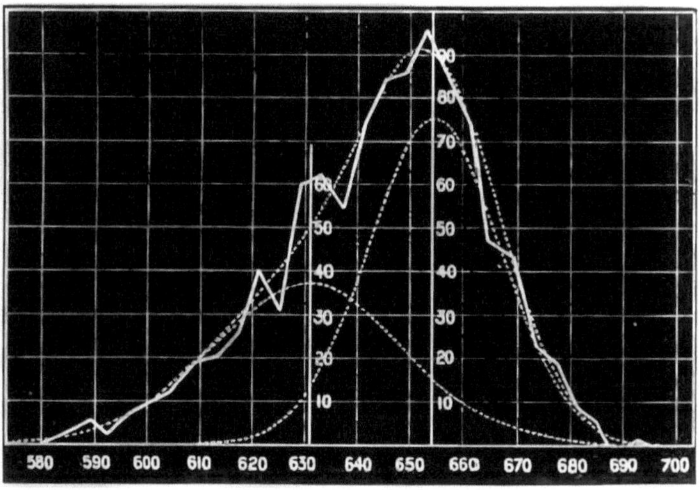

Fig. 7.3 The distribution of measurements of frontal breadth in *Crangon moenas* (top solid line) (Weldon, 1893, p. 324). Pearson showed that the original skewed distribution (top broken line) was the superposition of two normal distributions, represented by the two lower broken lines

Pearson published his intricate statistical analysis in the first of a series of 27 papers (Pearson, 1894)[13] with the main title "Contributions of the mathematical theory of evolution"[14] and this marked his entry into the biometric program, as Weldon has described it.

Although the results of Pearson's analysis strongly suggested the type of divergence Darwin had referred to previously,[15] Weldon was cautious and refrained from making any such evolutionary conclusions. On the other hand, Pearson went further and concluded that Weldon had caught (Darwinian) evolution in the act:

> **QUOT. 491** …it is shown that the Naples crabs are breaking up into two different sized families, owing to evolution in their foreheads. (Pearson, 1893a, p. 333)

In any case, with Pearson now in the biometric program, Weldon had a powerful ally. Many of the biological questions that Weldon asked were responsible for Pearson's major contributions to modern statistics. Pearson put it as follows:

> **QUOT. 492** …of this the writer [Pearson] feels sure, that his earliest contributions to biometry were the direct results of Weldon's suggestions and would never have been carried out without his inspiration and enthusiasm. (Pearson, 1906, p. 18)

---

[13]An abstract (Pearson, 1893a) of the article followed Weldon's 1893 paper and also appeared in the *Journal of the Royal Statistical Society* (Pearson, 1893b).

[14]Most of the times, the titles appeared as "Mathematical contributions to the theory of evolution." Pearson numbered only 19 of the 27 papers.

[15]See Sect. 2.13.5.

Fig. 7.4 Karl Pearson (1857–1936). (From: https://commons.wikimedia.org/wiki/File:Karl_Pearson.jpg)

Pearson has further remarked that Weldon's temperament was particularly fragile:

> **QUOT. 493** To see Weldon keen over a piece of work was to believe him robust and ready for any fray; but looking back on the past one can see what each piece of work cost him, and the strain on a highly nervous temperament began in even those early Cambridge days. (*ibid.*, p. 9)

Weldon remained close to Pearson until his untimely death from pneumonia in Oxford on April 13, 1906, at the age of 46.

7.2.2 Karl Pearson (1857–1936)

Karl Pearson[16] (Fig. 7.4) was educated at home until the age of nine when he was sent to University College School, London, for 7 years. During the final year, Pearson withdrew because of health reasons and spent the next year with a private tutor. In 1875, he obtained a scholarship at King's College, Cambridge, where he

[16]Karl Pearson was originally called Carl Pearson. While he was at the University of Heidelberg, Germany, his first name was misspelled. Haldane (1957a, p. 304) conjectured that Pearson decided to keep the name Karl owing to his love of Germany and possibly as a tribute to Karl Marx, whose writings Pearson admired. To his surroundings, Pearson was known simply as K.P. Pearson's life and work are also discussed in Porter (2004), Pearson (1938), Eisenhart (1974), and Magnello (2001, pp. 248–256).

was taught mathematics by James Clerk Maxwell, Arthur Caley, and William Burnside. Pearson obtained his B.A. in 1879 and soon after went to Germany, where, as he later recounted:

> **QUOT. 494** In Heidelberg I studied Physics under Quincke, but also Metaphysics under Kuno Fischer. In Berlin I studied Roman Law under Bruns and Mommsen, but attended the lectures of Du Bois Reymond on Darwinism. Pearson (1934, p. 20)

In November 1880, Pearson returned to England, took up rooms in the Inner Temple, and read law at Lincoln's Inn. He was called to the Bar the next year and practiced law for a very short time. During the early 1880s, Pearson also lectured on socialism, Marx, and Lassale.

Darwin's influence on Pearson's work showed quite early. In a lecture on socialism in February 1884, Pearson was already advocating gradual, continuous evolution:

> **QUOT. 495** ...*no great change ever occurs with a leap*[17]...It is the result of a gradual growth, a progressive change, what we term an *evolution*. This is as much a law of history as of nature. (Pearson, 1901, p. 347)

In 1884, Pearson was appointed Goldsmid Professor of Applied Mathematics and Mechanics at University College, London. From 1891 to 1894, he took up the position of Gresham Chair of Geometry while retaining his other position at University College. At Gresham College, Pearson delivered 12 lectures a year, free to the public. The first eight lectures formed the basis of his book *The Grammar of Science* (Pearson, 1892), which was subsequently translated into several languages.

Pearson was Galton's anointed successor to the biometric program. Before he met Galton in 1894, Pearson read the latter's *Natural Inheritance* (Galton, 1889) but does not seem to have been much impressed by it initially. In his 1889 talk based on the book to the Men and Women's Club,[18] Pearson declared:

> **QUOT. 496** Personally I ought to say that there is, in my own opinion, considerable danger in applying the methods of the exact sciences to problems in descriptive science, whether they be problems of heredity or of political economy: the grace and logical economy of the mathematical processes are apt to so fascinate the descriptive scientist that he seeks for sociological hypotheses which fit his mathematical reasoning and this without first ascertaining whether the basis of his hypothesis is as broad as that human life to which the theory is to be applied. I write therefore as a very partial sympathiser with Galton's methods.[19]

However, Pearson's interest in Galton's statistical methods increased when he met Raphael Weldon in 1891 at the Association for Promoting a Professional University for London. Weldon had come to realize that Darwin's theory was a working

[17] All italics here are Pearson's.

[18] A debating society founded by Pearson himself, to discuss social issues in relation to the two sexes. Incidentally, this is where Pearson met Maria Sharpe, his future wife. The society ran from 1885 to 1889.

[19] Taken from Pearson's talk, "The laws of inheritance according to Galton," which is kept in the Pearson archive.

hypothesis whose various parts remained to be proven. Many of the questions posed by Weldon required Galton's statistical methods. Pearson soon became deeply influenced by Galton's *Natural Inheritance* (Galton, 1889) and welcomed the latter's ideas on regression and correlation. These concepts were in line with his positivist view of science.[20] He later declared:

> **QUOT. 497** ... there was a category broader than causation, namely correlation, of which causation was only the limit, and that this new conception of correlation brought psychology, anthropology, medicine and sociology in large parts into the field of mathematical treatment. It was Galton who first freed me from the prejudice that sound mathematics could only be applied to natural phenomena under the category of causation. Here for the first time was a possibility, I will not say a certainty, of reaching knowledge – as valid as physical knowledge was then thought to be – in the field of living forms and above all in the field of human conduct. (Pearson, 1938, p. 19)[21]

Pearson's entry into the biometric program was occasioned when Weldon asked for his assistance in dealing with a skewed distribution that he had observed for the frontal breadth of the shore crab (Weldon, 1893).[22] Through a sophisticated statistical analysis (Pearson, 1894), Pearson decomposed the skewed distribution into two normal curves and observed evidence for Darwinian evolution. This was Pearson's first in a long series of mathematical contributions to the theory of evolution.

One of Pearson's most important contributions to the biometric cause was his mathematical development of Galton's Law of Ancestral Heredity. The nature of these developments has been described in Sects. 4.6.3 and 4.6.5 but what is especially important is Pearson's (and the other biometricians') belief that the ancestral law was the heredity component in Darwin's theory[23] and that any other law that did not account for distant ancestry was simply non-Darwinian.

Furthermore, there was no disagreement between Pearson and Weldon regarding the supreme status of statistical methods in the understanding of evolutionary processes (including heredity). For example, in the heat of the biometry-saltationism controversy, Pearson reiterated Weldon's view in QUOT. 490 that:

> **QUOT. 498** ...the whole problem of evolution is a problem in vital statistics – a problem of longevity, of fertility, of health, and of disease, and it is as impossible for the evolutionist to proceed without statistics, as it would be for the Registrar-General to discuss the national mortality without an enumeration of the population, a classification of deaths, and a knowledge of statistical theory. (Pearson, 1902, p. 320)

---

[20] The positivist school (as put forward by Comte, then Mach) holds that science should concern itself only with what is positively given and should refrain from speculating on matters that go beyond actual experience. Science does not *explain* natural phenomena, rather it tries to *describe* them.

[21] Similarly, in the third edition of Pearson's *Grammar of Science*, one can read: "No phenomena are causal; all phenomena are contingent, and the problem before us is to measure the degree of this contingency, which we have seen is between the zero of independence and the unity of causation. That, briefly, is the wider outlook we must now take of the universe as we experience it" (Pearson, 1911, p. 174).

[22] See Sect. 7.2.1.

[23] See QUOT. 336.

Such was Pearson's confidence in statistical methods that, as early as 1895, he proposed to represent the complex physiological process of heredity through a single quantity, namely, the correlation coefficient:

> **QUOT. 499** Given any organ in a parent and the same or any other organ in its offspring, the mathematical measure of heredity is the correlation of these organs for pairs of parent and offspring. (Pearson et al., 1895, p. 259)

The only excuse for such a lapse is that the distinction between phenotypes and genotypes had not been made in 1895. In any case, Pearson's writings in mathematics, statistics, and philosophy of science were widely read. Thus, Einstein himself both read and recommended[24] his *Grammar of Science* (Pearson, 1892).

In an important passage of *Grammar*, one can read:

> **QUOT. 500** In an age like our own, which is essentially an age of scientific inquiry, the prevalence of doubt and criticism ought not to be regarded with despair or as a sign of decadence. It is one of the safeguards of progress; *la critique est la vie de la science*, we must again repeat. One of the most fatal (and not so impossible) futures for science, would be the institution of a scientific hierarchy which would brand as heretical all doubt as to its conclusions, all criticism of its results. (*ibid.*, pp. 66–67)

Ironically, the man who so admirably preached the virtues of criticism was himself averse to it. Ad hominem attacks were often tools of predilection when Pearson was challenged, as evidenced by his dealings with Yule, Castle, Darbishire, Fisher, and many others.

7.2.3 William Bateson (1861–1926)

In 1879, Bateson[25] (Fig. 7.5) joined St. John's College, Cambridge, where his father was Master. From the start, Bateson had considerable difficulty with mathematics. As he later (in 1905) recalled:

> **QUOT. 501** Mathematics were my difficulty. Being destined for Cambridge I was specially coached in mathematics at school. Arrived here, I was again coached, but failed. Coached once more I passed, having wasted, not one, but several hundred hours on that study. Needless to say, my knowledge of mathematics is nil. (Bateson, 1928, p. 10)

Bateson's weakness in mathematics (and statistics) would later further alienate him from the biometricians whose evolutionary output almost always contained sophisticated mathematics. Bateson fared much better in the natural sciences.

Bateson's wife, Beatrice Bateson, has noted that, during his days at Cambridge, Bateson's most intimate friend was none other than Raphael Weldon (*ibid.*, p. 17). Under the latter's influence, Bateson was at first drawn to morphology and the Balfour Morphological Laboratory. When Bateson became interested in

[24]For example, see Robinson (2019, p. 190).

[25]Bateson's life and work are also discussed in Cock and Forsdyke (2008), Bateson (1928), Crowther (1935, pp. 248–310), and A Fellow of the International Education Board (1926).

Fig. 7.5 William Bateson (1861–1926). (From: https://commons.wikimedia.org/wiki/File:William_Bateson.jpg)

Balanoglossus (an ocean-dwelling acorn worm), Weldon helped him make contact with the renowned morphologist William Keith Brooks[26] (Fig. 7.6) at Johns Hopkins University. The acquaintance with Brooks in 1883 and 1884 proved to be an important milestone in Bateson's career. Brooks was finishing his book *The Law of Heredity* (Brooks, 1883), where he had clearly endorsed a saltationist view of evolution:

> **QUOT. 502** There is therefore no reason why natural selection should not often be presented with great and extended variations – the saltations which Mivart believes in – and the evolution of organisms may therefore be a much more rapid process than Darwin believes. (*ibid.*, pp. 297–298)

[26]Who, notably, also trained both E. B. Wilson and Thomas Morgan.

Fig. 7.6 William Keith Brooks (1848–1908). (From: https://upload.wikimedia.org/wikipedia/commons/7/73/PSM_V55_D302_William_Keith_Brooks.png)

The influence of Brooks on Bateson's evolutionary views cannot be underestimated. Beatrice Bateson later said:

> **QUOT. 503** He [Bateson] delighted in recalling the long hours of discussion (Brooks lying in his shirt-sleeves on his bed and Will sitting by), when problem and theory and practice passed in long review with ever-fresh interest. Professor Brooks, congenial and stimulating, was always remembered with affectionate respect. (Bateson, 1928, p. 18)

Bateson published four morphological papers on *Balanoglossus* (Bateson, 1884a, b, 1885, 1886), after which he was elected Fellow of his College in 1885. However, like many young biologists of the time,[27] Bateson lost interest in morphology. In the spring of 1886, he traveled to the salt lakes in the Steppe (in western Central Asia) to study the variation in aquatic species and the environmental conditions. More specifically, he wished to investigate any correlation between various degrees of salinity of the water and the modifications in *Cardium edule* (common cockle), *Dreissena polymorpha* (a small shellfish with characteristic stripes on its shell), and *Hydrobia ulvae* (mudsnail). Bateson found positive results only in the first case (Bateson, 1889; Punnett, 1928, p. 34, Vol. I). However, there was no clear-cut evidence for natural selection acting on individual differences or for Lamarckian effects.

[27] See footnote 12.

After his return, Bateson read the beginning chapters of Galton's *Natural Inheritance* with great satisfaction[28] since they were in line with the discontinuous views on evolution he had been receptive to since meeting William Brooks. From 1890 onward, Bateson started publishing on discontinuous variation. In the 1891 article "On the Variations in Floral Symmetry of Certain Plants Having Irregular Corollas" (Bateson, 1891; Punnett, 1928, Vol. I), he first explicitly endorsed the saltationist viewpoint:

QUOT. 504 It may be observed at this stage that in proportion as the process of Evolution shall be found to be discontinuous the necessity for supposing each structure to have been gradually modelled under the influence of Natural Selection is lessened, and a way is suggested by which it may be found possible to escape from one cardinal difficulty in the comprehension of Evolution by Natural Selection. (*ibid.*, p. 128)

The "one cardinal difficulty in the comprehension of Evolution by Natural Selection" in QUOT. 504 was this: in the continuous sequence of possible transitional forms, only particular ones could be useful and selected; therefore, it was

QUOT. 505 ...difficult to suppose both that the process of Variation has been a continuous one, and also that Natural Selection has been the chief agent in building up the mechanisms of living things... [thus] the evolution of a special contrivance of adaptation is not compatible with constant and perpetual usefulness. (*ibidem*)

Bateson thus used the same stability-of-type argument that Galton had first proposed in *Hereditary Genius* (Galton, 1869a)[29] and that Mivart had also proposed before.[30]

Bateson's saltationist views were further expanded in his 1894 *Materials for the Study of Variation* (Bateson, 1894), a big book entirely devoted to the study of variation. Weldon's review of *Materials* was also the first in a series of disputes between the saltationists and the biometricians. This will be described in Sect. 8.1.

One of Bateson's most enduring legacies is the bestowal of the name "genetics" to the scientific study of heredity and variation. The term had been used by Darwin in the 6th edition of *The Origin* in the sense of a common biological origin,[31] but its modern use originated with Bateson. In a letter written to Adam Sedgwick on April 18, 1905, he said:

QUOT. 506 If the Quick Fund were used for the foundation of a professorship relating to Heredity and Variation the best title would, I think, be "The Quick Professorship of the study of Heredity". No single word in common use quite gives this meaning. Such a word is badly

---

[28] Bateson's letter to Miss Biggs (July 7, 1909): "I often remember the thrill of pleasure with which I first read *Hereditary Genius* and the earlier chapters of *Natural Inheritance*, and every year when I read aloud pieces of those books to my class, as I always do, I can see what excitement they have the power to cause" (Pearson, 1930, p. 288, Vol. IIIA).

[29] See Sect. 4.3.4.

[30] See Sect. 2.14.5.

[31] On p. 101 of Darwin (1872): "It is incredible that the descendants of two organisms, which had originally differed in a marked manner, should ever afterwards converge so closely as to lead to a near approach to identity throughout their whole organisation. If this had occurred, we should meet with the same form, independently of genetic connection, recurring in widely separated geological formations; and the balance of evidence is opposed to any such an admission."

wanted, and if it were desirable to coin one, "Genetics" might do. Either expression clearly includes Variation and the cognate. (Bateson, 1928, p. 93)

This letter was followed by Bateson's review in *Nature* of a textbook (written by J. P. Lotsy) where he used the word again (Bateson, 1906, p. 306).

Bateson's enormous contribution to the field of genetics was also accompanied by his involvement in several controversies.[32] In this regard, in his excellent biographical book, Crowther has stated that Bateson's

QUOT. 507 . . .qualities appeared to some as the model of absolute honesty and fearlessness inspired by a burning passion for truth. To others, these same qualities seemed to show a lack of self-control and an inability to enter into other people's points of view, and cooperate with them. (Crowther, 1935, p. 306)

An example of Bateson's "lack of self-control" can be seen in the highly rhetorical nature of his book *Mendel's Principles of Heredity: A Defence* (Bateson, 1902b).

At the same time, Bateson's one redeeming quality concerning his controversies should be mentioned. Whenever his views were challenged, as a first reaction, Bateson almost always refrained (unlike several of his colleagues) from attacking his opponent either privately or in press. Instead, his attitude in general was to approach his opponents and attempt to convince them of the truthfulness of his views. This can be seen in his early interactions with Weldon, Pearson, Darbishire, and several others.

7.2.3.1 A Note on Bateson and the Chromosome Theory of Heredity

The pioneering nature of much of Bateson's scientific output makes his early denial of the fundamental importance of chromosomes to the theory of heredity an enigma. This seems all the more surprising given that the cytological evidence favored a *Mendelian* view of heredity, which Bateson was at the time fighting hard to support against the biometricians' assault. As early as 1902, shortly after the Mendelian rediscovery, the American biologist Walter Sutton (1877–1916) announced:

QUOT. 508 . . .the phenomena of germ-cell divisions and of heredity are seen to have the same essential features, viz. purity of units (chromosomes, characters) and the independent transmission of the same; while, as a corollary, it follows in each case that each of the two antagonistic units (chromosomes, characters) is contained by exactly half the gametes produced. (Sutton, 1903, p. 237)

A similar statement was made independently and simultaneously by the German biologist Theodor Boveri (1903). However, Bateson remained unconvinced: he did not deny the existence of chromosomes; rather, he doubted that they were the repositories of hereditary material. In the *Science* article "Facts limiting the theory of Heredity," Bateson justified his position mainly through the lack of correlation between chromosome number and organism complexity:

[32] These controversies are listed on p. 464 of Bateson (1928).

QUOT. 509 . . .I anticipate that the more immediate question, whether the chromosomes are or are not the bearers of hereditary characters, will be fully debated. Without presuming to a definite opinion on this question, I venture to state what seem to me formidable difficulties in the way of this expectation. If the chromosomes were directly responsible as chief agents in the production of the physical characteristics, surely we should expect to find some degree of correspondence between the differences distinguishing the types, and the visible differences of number or shape distinguishing the chromosomes. . . no correspondence between the chromosome numbers and complexity of structure has ever been asserted to exist. Low forms may have many; highly complex types may have few. (Bateson, 1907, pp. 651–652)

In the face of mounting cytological evidence in the 1910s, especially from T. H. Morgan and his *Drosophila* group, Bateson remained skeptical until early 1920. Finally, in an April 24, 1920, letter to Reginald Gates, he relented his position slightly:

QUOT. 510 I shall not be very surprised if some connexion is proved to exist between the number of linkages and the number of chromosomes. (Rushton, 2014, p. 157)

When Bateson was invited in late 1921 to speak at the American Association for the Advancement of Science meeting in Toronto, he visited the Morgan group in New York in December of that year. The visit proved to be a turning point:

QUOT. 511 I can see no escape from capitulation on the main point. The chromosomes must be in some way connected with our transferable characters. (Bateson, 1928, p. 143)

Bateson's conversion was made public in his lecture entitled "Evolutionary Faith and Modern Doubts," which was delivered in his Toronto Address:

QUOT. 512 We have turned still another bend in the track and behind the gametes we see the chromosomes, for the doubts – which I trust may be pardoned in one who had never seen the marvels of cytology, save as through a glass darkly – cannot, as regards the main thesis of the Drosophila workers, be any longer maintained. The arguments of Morgan and his colleagues, and especially the demonstrations of Bridges, must allay all scepticism as to the direct association of particular chromosomes with particular features of the zygote. The transferable characters borne by the gametes have been successfully referred to the visible details of nuclear configuration. (Bateson, 1922, p. 554)

Chapter 8
Biometry-Saltationism Controversies During the Pre-Mendelian Period

Abstract The opening salvo in the biometry-saltationism war was Weldon's critical review of Bateson's 1894 book on discontinuous evolution. In 1894 still, a Committee of the British Association, chaired by Galton, was set up to investigate evolutionary questions. The committee at first consisted mainly of biometricians. However, when Bateson was invited to join in 1897, the committee was soon overtaken by a saltationist agenda, much to the distress of the biometricians. The first public controversy between Weldon and Bateson occurred in 1895 over the nature of variations in Cineraria cruenta. The controversy lasted several years and involved several others. In 1901, Pearson published a controversial paper on the so-called principle of homotyposis, which claimed that fraternal correlations were equal to the mean correlation between undifferentiated like organs. Much to Pearson's distress, even before his paper was published, it was publicly criticized by Bateson at a Royal Society meeting. This unfortunate incident directly led to the creation of the journal *Biometrika* in 1901.

Keywords Materials of the Study of Variation · Carcinus moenas · Cineraria cruenta · Homotyposis · Undifferentiated organs · Intraclass correlation · Fraternal correlation · Biometrika

8.1 Round 1: Weldon's Review of Bateson's "Materials" (1894)

Following the 1891 paper (Bateson, 1891), where he first publicly announced his saltanonist views, Bateson amassed 886 cases of discontinuous evolution in the 1894 book *Materials for the Study of Variation, Treated with Especial Regard to Discontinuity in the Origin of Species* (Bateson, 1894). The book had an 80-page

P. Gorroochurn, *The Development of Evolutionary Genetics*,
https://doi.org/10.1007/978-3-031-69374-8_8

introduction[1] where Bateson delineated the motivations and scope of the book. In condensed form, the aim was to:

QUOT. 513 . . .solve the problem of the forms of living things is the aim with which the naturalist of to-day comes to his work. How have living things become what they are, and what are the laws which govern their forms? (*ibid.*, p. 1)

Bateson praised Darwin's trailblazing work but admitted that the theory of natural selection (as well as Lamarck's theory of evolution) suffered from a major difficulty:

QUOT. 514 According to both [Darwin's and Lamarck's] theories, specific diversity of form is consequent upon diversity of environment, and diversity of environment is thus the ultimate measure of diversity of specific form. Here then we meet the difficulty that diverse environments often shade into each other insensibly and form a continuous series, whereas the Specific Forms of life which are subject to them on the whole form a Discontinuous Series. (*ibid.*, p. 5)

It should be noted that the quotation above is in line with Bateson's opinions already expressed in the 1891 paper.[2] The reference to the continuity of the environmental conditions was undoubtedly influenced by Bateson's observations in the salt lakes of the Steppe.[3]

Thus, Bateson's understanding was that, according to Darwin's theory, the form of the organism should mirror that of the environment. Therefore, the existence of discontinuous forms amidst a continuously varying environment meant that the theory of natural selection was deficient. This led Bateson to diminish its role. Natural selection was not needed for evolution to take place since:

QUOT. 515 Variation, in fact, *is* Evolution. (*ibid.*, p. 6)

These are more or less the views that de Vries would later embrace in his *Die Mutationstheorie* (de Vries, 1901, 1903).[4]

If, according to Bateson, variation itself was evolution, what then was the role of natural selection? Very much like de Vries (see QUOT. 387), Bateson believed that *natural selection only selected existing variants, but did not create anything new*—creation fell under the purview of discontinuous variation (or mutation):

QUOT. 516 . . .the crude belief that living beings are plastic conglomerates of miscellaneous attributes, and that order of form or Symmetry have been impressed upon this medley by Selection alone; and that by Variation any of these attributes may be subtracted or any other attribute added in indefinite proportion, is a fancy which the Study of Variation does not support. (Bateson, 1894, p. 80)

[1] Below the heading of the Introduction was a verse from First Corinthians: "All flesh is not the same flesh: but there is one kind of flesh of men, another flesh of beasts, another of fishes, and another of birds" (1 Corinthians 15:39). In a letter written on July 22, 1893, Bateson expressed the opinion that this verse was the "best statement of the Problems of the Discontinuity of Species that I know." (Bateson, 1928, p. 52).

[2] See especially QUOT. 505.

[3] See Sect. 7.2.3.

[4] See Sect. 5.3.3.

Bateson was also aware of Galton's argument that natural selection was ineffective against continuous variation (see Sect. 4.5.7). In this regard, he repeated Galton in QUOT. 324 by saying that:

QUOT. 517 . . .characters which do not readily blend, are thus exempt from what Galton has called the Law of Regression. (*ibid.*, p. 40)

In the Introduction, Bateson also indicated (*ibid.*, p. 60) that a second volume would soon follow, with further evidence of discontinuity in substantive variation. However, the project had to be abandoned because, despite the efforts that went into the writing of *Materials*, it was far from a resounding success.

However, Galton was quite elated by the book since it endorsed his own saltatonist views. In a review written in *Mind*, he said:

QUOT. 518 It was, therefore, with the utmost pleasure that I read Mr. Bateson's work bearing the happy phrase in its title of 'discontinuous variation,' and rich with many original remarks and not a few trenchant expressions. (Galton, 1894, p. 369)

In his review, Galton also noted that "Mr. Bateson has thus far been more desirous to collect facts than to formulate theories" (*ibid.*, p. 371). The presentation of data without a theoretical framework may be one reason why Bateson's *Materials* was unsuccessful.[5] Another reason was the book's explicit departure from Darwinian principles, in particular its diminishment of the role of natural selection.

Weldon, who had hitherto been on friendly terms with Bateson, was not as enthusiastic about *Materials* as Galton had been. In a letter written to Bateson on February 15, 1894, he expressed his doubt about the latter's claim that discontinuous characters should be free from the regression effect (Provine, 2001, p. 44).[6] Weldon's tone was cordial but changed when he followed up with a review in *Nature*. After some congratulatory words, Weldon started an extensive criticism of Bateson's main thesis of discontinuous evolution:

QUOT. 519 The only contention which can here be noticed is that alluded to on the title-page, namely that variation frequently proceeds in such a way that changes in an organ occur only by steps of definite and measurable magnitude; and that discontinuous variation of this kind is necessary for the evolution of new species. (Weldon, 1894c, p. 24)

Weldon not only attacked Bateson's views on discontinuous variation but also his statement of a continuously varying environment. He also took exception to Bateson's sampling method in producing cases of discontinuity:

QUOT. 520 In the case of teeth, and in many other cases discussed, the method employed is not adapted to a determination of the least possible magnitude of variations. For in judging evidence based largely upon museum preparations and on printed records, it must always be remembered that there is a tendency among curators and others to regard a slight abnormality as not worth bottle, spirit, and a place on the shelf, or to think it too trivial for printed record.

---

[5]In contrast, de Vries' *Die Mutationstheorie* (de Vries, 1901, 1903) did both and was much more successful.

[6]See QUOT. 517.

> Anyone who has tried to obtain specimens for a museum knows that many persons will take pains to present a rare or striking specimen, who cannot be induced to send quite common things. (*ibid.*¸ p. 26)

After having second-guessed Bateson's methodology, Weldon recommended that the latter adopt a large-sample (biometric) approach for a better analysis.

Several other reviews also followed, and Weldon's was not the only negative review. In particular, Wallace wrote a report that criticized both Bateson's and Galton's saltationism (Wallace, 1895). However, given his close friendship with Weldon, there is no doubt that Bateson must have especially resented his "friend's" acidic review.

Weldon's criticism of *Materials* was thus the opening shot in the biometry-saltationism battle.

8.2 Round 2: Committee for Conducting Statistical Inquiries into the Measurable Characteristics of Plants and Animals (1894)

On February 3, 1891, Wallace wrote to Galton asking him to consider setting up a Committee of the British Association (probably financed through a British Association grant, a Royal Society Grant, and personal subscriptions) to look into what he deemed two fundamental but disputed points of organic evolution, namely:

> **QUOT. 521** (1) Whether individually acquired external characters are inherited, and thus form an important factor in the evolution of species,- or whether as you [Galton] or Weismann argue, and as many of us now believe, they are not so, & we are thus left to depend almost wholly on variation & natural selection.
>
> (2) What is the amount and character of the *sterility* that arises when closely allied but permanently distinct species are crossed, and then "hybrid" offspring bred together. Whether the amount of infertility differs between the hybrids of species that have presumably arisen in the *same area*, & those which seem to have arisen in very *distinct* or *distant areas* – as oceanic or other islands.
>
> (Pearson, 1930, p. 128, Vol. IIIA)

Fearing that there were not enough experiments to warrant such a project, Galton expressed his reservations to Wallace. The latter provided a list of possible experiments. However, after consulting with Herbert Spencer and Thiselton-Dyer and hearing their criticism, Galton could not see a way to start the project.

However, the project received a new impetus after the success of Weldon's biometry articles (Weldon, 1890, 1892, 1893).[7] A discussion took place between Raphael Meldola, Francis Galton, and Raphael Weldon on December 9, 1893, at the Savile Club.

[7] See Sect. 7.2.1.

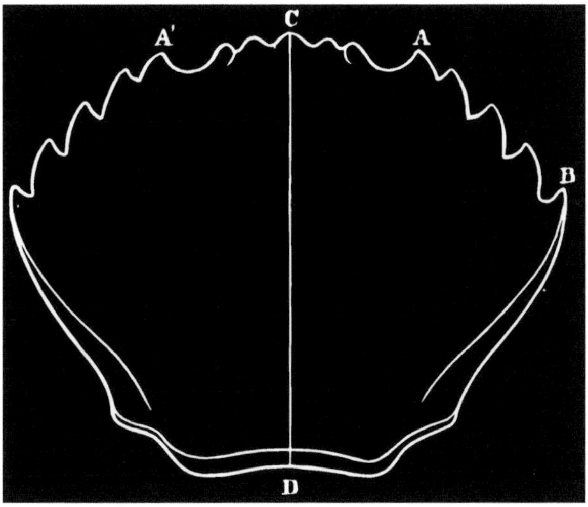

Fig. 8.1 Measurements on *Crangon moenas* (top solid line) (Weldon, 1894b, p. 361). Carapace length = *CD*, frontal breadth = *A′A*, right dentary margin = *AB*

Eventually, on January 18, 1894, a Committee was appointed by the Council of the Royal Society, consisting of Francis Galton (chairman), Raphael Weldon (secretary), Francis Darwin (son of Charles Darwin), Alex Macalister, Raphael Meldola, and Edward Poulton. The express purpose of the Committee was that of "Conducting Statistical Inquiries into the Measurable Characteristics of Plants and Animals." Thus, this was a biometric committee at heart. Its first report was received by the Royal Society on November 20, 1894, and consisted of two articles by Weldon, namely, a paper on *Carcinus moenas* (Weldon, 1894b) and a shorter paper entitled "Remarks on variation in animals and plants" (Weldon, 1894a).[8] The report was read at the Royal Society meeting on February 28, 1895.

Weldon's 1894 paper on *Carcinus moenas* was a landmark: it was the first study to almost[9] successfully demonstrate the action of natural selection in a wild population. In this paper, Weldon measured the carapace length, frontal breadth, and right dentary margin (Fig. 8.1) of 7000 young female shore crabs and 1000 adult females. The latter were sampled from the same locality as the young females and had already been studied in Weldon's third biometry paper (Weldon, 1893).[10] The young females were grouped into 7 categories according to carapace length, together with the 1000 adult females. The values of median carapace length (*C*) and mean quartile deviation (*Q*)[11] of frontal breadth obtained are shown in Table 8.1.

[8]The second paper was also published in *Nature* (Weldon, 1895).

[9]The adverb 'almost' is used because a truly successful demonstration would require a controlled experiment, which was not the case in Weldon's paper.

[10]See paragraphs above QUOT. 490.

[11]The mean quartile deviation was also known as the probable error. In today's language, it is known as the semi-interquartile range.

Table 8.1 Variation of mean carapace length (*C*) with mean quartile deviation of frontal breadth (*Q*) in *Carsinus moenas* (Weldon, 1894b, p. 367)

Carapace length (*C*)	Mean quantile deviation of frontal breadth (*Q*)
7.5	9.42
8.5	9.83
9.5	9.51
10.5	9.58
11.5	10.25
12.5	10.79
13.5	10.09
(Adult)	9.96

The first 7 rows of numbers refer to 7000 young female shore crabs. The last row refers to 1000 adult females

Assuming that carapace length also increases with age, Weldon concluded that there was an increase in the variability of frontal breadth with age, reaching a maximum that is greater than the variability in the adults, and then falling off again. He stated that his results, which indicated an initial increase in variability, were

QUOT. 522 ...simply confirmations of Darwin's statement, that many variations appear at a late period of development. (Weldon, 1894b, p. 367)

However, the most important conclusion of the paper was that the subsequent drop in variability was due to the effect of destructive selection:

QUOT. 523 ...the hypothesis may be provisionally adopted that the diminution in the frequency of individuals of given deviation is due to a selective destruction... (*ibid.*, p. 368)

Another important highlight of Weldon's paper was his ingenuous calculation of the selective death rate. Weldon's reasoning may be understood as follows. Consider a population of crabs at its maximum variability. Its distribution can be represented by the normal curve ADLE in Fig. 8.2, and it has a modulus of c_1[12] and a y-intercept of k_1 (= *BD*). In Weldon's notation, the total area under the curve is equal to the population size N_1, so he had implicitly assumed that the equation of ADLE is

$$y_1 = \frac{N_1}{c_1\sqrt{\pi}} e^{-x^2/c_1}.$$

Since the y-intercept is k_1,

$$k_1 = \frac{N_1}{c_1\sqrt{\pi}} \quad \Rightarrow \quad N_1 = k_1 c_1 \sqrt{\pi} \tag{8.1}$$

[12]The modulus was used before to denote variability and is equal $\sqrt{2}$ times the standard deviation.

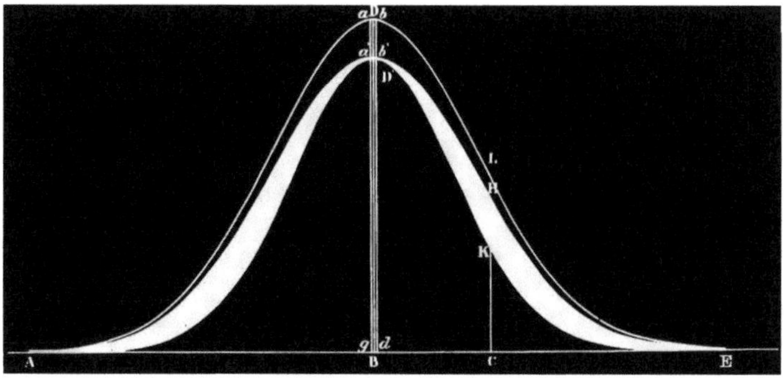

Fig. 8.2 Weldon's calculation of the effect of destructive selection (Weldon, 1894b, p. 368) (the rightmost point on the x-axis was wrongly denoted by D in the original paper and has been replaced by E here). The normal curve ADLE represents the distribution of crabs at maximum variability; AD'HE represents the distribution of crabs after the effect of nonselective destruction; AD'KE represents the distribution of crabs after the effect of selective destruction

Now, any destruction that is *nonselective* will destroy an equal percentage of crabs at any deviation. It will not change the modulus c_1 but will decrease the population size to N_2 ($<N_1$). The distribution showing the effect of nonselective destruction is denoted by AD'HE. The y-intercept is now k_2 ($= BD'$) and, in analogy with Eq. (8.1),

$$N_2 = k_2 c_1 \sqrt{\pi}$$

On the other hand, a destruction that is *selective* will more likely eliminate crabs with larger than smaller deviations. It will decrease both the modulus to c_2 ($<c_1$) and the population size to N_3 ($<N_1$). The distribution showing the effect of selective destruction is denoted by AD'KE. The y-intercept is k_2 ($= BD'$) and in analogy with Eq. (8.1),

$$N_3 = k_2 c_2 \sqrt{\pi}$$

Weldon was thus able to calculate the selective death rate as

$$\frac{N_2 - N_3}{N_2} = \frac{c_1 - c_2}{c_1},$$

which gives the proportion of crabs destroyed by the effect of destructive selection alone. The selective death rate may be estimated from the data in Table 8.1 as

$$\frac{Q_{\text{max}} - Q_{\text{adu}}}{Q_{\text{max}}} = \frac{10.79 - 9.96}{10.79} = .077,$$

where Q_{max} is the maximum quartile deviation and Q_{adu} is the quartile deviation for the adults. Weldon then declared:

QUOT. 524 . . .the hypothesis of selective destruction involves a death-rate of about 77 per thousand between the age corresponding to 12.5 mm. in carapace-length and maturity, as a consequence of deviation in frontal breadths, and in the group of structures, whatever these may be, which are directly correlated with it. (*ibid.*, p. 370)

Finally, Weldon also calculated the selective death rate $g(x)$ at any deviation x. Let the curves AD'HE and AD'KE in Fig. 8.2 be denoted by y_2 and y_3, where (remembering that the y-intercept in both cases is k_2)

$$y_2 = k_2 \exp\left(-\frac{x^2}{c_1}\right),$$

$$y_3 = k_2 \exp\left(-\frac{x^2}{c_2}\right).$$

Since the number of crabs that have deviations in $(x, x + dx)$ after nonselective destructions is y_2dx and after selective destructions is y_3dx, the selective death rate as a function of the deviation x is

$$\begin{aligned} g(x) &= \frac{y_2dx - y_3dx}{y_2dx} \\ &= \frac{y_2 - y_3}{y_2} \\ &= \frac{e^{-x^2/c_1^2} - e^{-x^2/c_2^2}}{e^{-x^2/c_1^2}} \\ &= 1 - e^{-x^2\left(1/c_2^2 - 1/c_1^2\right)} \\ &= 1 - e^{-hx^2}, \end{aligned}$$

where $h = 1/c_2^2 - 1/c_1^2$ (*ibid.*, p. 370). Substituting $c_1 = 10.79$ and $c_2 = 9.96$ from Table 8.1 gives $h = .0015$ so that $g(x) = 1 - e^{-.0015x^2}$ approximately. As expected, the selective death rate increases with x.

Weldon repeated the analysis with respect to the right dentary margins of the crabs. This time, he found that the mean quartile deviation of the young crabs was always less than that of the adults except for crabs with carapace lengths between 9 and 10 mm. Weldon inferred that:

QUOT. 525 . . .among female crabs in Plymouth Sound, during the period of life to which these observations refer, there is no indication of any destructive agency which acts selectively upon the dentary margin. (*ibid.*, p. 379)

He concluded his paper by triumphantly reiterating one of the key tenets of the biometric approach:

QUOT. 526 These are all the data which are necessary, in order to determine the direction and rate of evolution; and they may be obtained without introducing any theory of the physiological function of the organs investigate. (*ibidem*)

Furthermore, according to Weldon, the biometric approach was superior to any other approach that relied on physiology because the latter was nothing but "guess-work" at this stage:

QUOT. 527 The advantage of eliminating from the problem of evolution ideas which must often, from the nature of the case, rest chiefly upon guess-work, need hardly be insisted upon. (*ibidem*)

The second article (Weldon, 1894a) of the report was more of an appendix to the first paper and contained a further discussion of its results. It started with broadsides against those who were inclined to

QUOT. 528. . .regard change in specific character as an event which occurs, not slowly and continuously, but occasionally and by steps of considerable magnitude, as a consequence of the capricious appearance of "sports." (*ibid.*, p. 380)

Weldon further expanded on the virtues of the biometric approach: it was based on a method, namely, statistical, which was not only necessary but also sufficient. Thus:

QUOT. 529 The questions raised by the Darwinian hypothesis are purely statistical, and the statistical method is the only one at present obvious by which that hypothesis can be experimentally checked. (*ibid.*, p. 381)

So strong was Weldon's faith in the statistical method that he thought it obviated the need to demonstrate functional adaptation for natural selection to take place:

QUOT. 530 Knowing that a given deviation from the mean character is associated with a greater or less percentage death-rate in the animals possessing it, the importance of such a deviation can be estimated without the necessity of inquiring how that increase or decrease in the death-rate is brought about, so that all ideas of "functional adaptation" become unnecessary. (*ibidem*)

Not surprisingly, this statement did not go well with many Darwinians, especially Wallace.

Weldon's two articles, especially the first one, created quite a sensation and elicited public reactions from a wide range of scholars. These will be described in detail in Sect. 8.3. An important private reaction came from Bateson who sent "reams of criticism and numerous letters of protest" to Galton (Pearson, 1930, p. 126, Vol. IIIA). Bateson's main contention was that the 7000 young female crabs Weldon had used in his study were at different stages of molting. Since each molt altered the frontal breadth, Bateson implied that Weldon's study was invalid.

Bateson's attacks continued (but remained private) and Galton wrote in desperation to Weldon on November 17, 1896:

QUOT. 531 Herewith is another paper from Bateson, and I enclose with this his accompanying letter to myself. We must talk over what is the fairest course to adopt when we meet (as we probably shall) before the meeting of the R. Soc. on Thursday.

You see that he offers to print his four letters for circulation among members of the Committee. My greatest difficulty in thinking what should be done arises from the lengthiness of these papers. I wish the issue could be stated in much more condensed language.

> It would in many ways be helpful, if Bateson were made a member of our Committee, but I know you feel that in other ways it might not be advisable. The other members besides yourself hardly do enough.
>
> (*ibid.*, p. 127)

Bateson was thus invited to join the Committee on Jan 1, 1897, but he refused at first. Pearson had already joined in December 1896, and, with most other members being biometricians, Bateson did not feel comfortable with the direction the Committee was taking.

However, Galton was able to convince Bateson on January 6, 1897, and the latter attended his first meeting on January 14. During that and subsequent meetings, other members also joined, including Sir E. Clarke, F.D. Godman, W. Heape, E. Ray Lankester, E.J. Lowe, M.T. Masters, O. Salvin, and W.T. Thiselton-Dyer. During the second meeting of the Committee, held on February 11, 1897, it was decided to rename the Committee the "Evolution (Plants and Animals) Committee of the Royal Society."

It was already clear to the biometricians that Bateson and others unsympathetic to the biometric agenda would soon be taking over. Thus, on the day after the meeting, Weldon wrote to Galton:

> **QUOT. 532** I wanted to write a few words to you about yesterday's meeting. . .I felt sadly out of place in such a gathering of biologists, and little capable of expressing opinions, which would only have hurt their feelings and not have been productive of real good.
>
> . . .I venture to think that the Committee you have got together is entirely unsuited to direct such experiments. It is far too large, contains far too many of the old biological type, and is far too unconscious of the fact that the solutions to these problems are in the first place statistical, and in the second place statistical, and only in the third place biological.
>
> (*ibid.*, 127–128)

Indeed, the Evolution Committee was soon dominated by Bateson and a saltationist agenda. Weldon, Pearson, Galton, Thiselton-Dyer, and Meldola all simultaneously resigned from it on January 25, 1900.[13]

However, before that, an important battle was fought in the biometry-saltationism war.

8.3 Round 3: The Cineraria Affair—First Public Controversy Between Weldon and Bateson (1895)

Section 8.2 described Weldon's *Carninus moenas* article (Weldon, 1894b), which was the main paper in the first report of the Royal Society Committee. This paper was the first attempt to catch natural selection in action and elicited a wide range of commentaries from several scholars. It also brought about the first public

[13]The Committee continued and from 1900 to 1910 Bateson served as secretary and Godman as president.

confrontation between Weldon and Bateson. This was over the nature of the variations in *Cineraria cruenta*, which is a flowering plant in the sunflower family. In what follows, the back-and-forth between the two parties will be described through various articles, Letters to the Editor, correspondences, etc. (denoted by **Art.**)

8.3.1 Thiselton-Dyer (March 14, 1895, Art. 1)

The first published reaction to Weldon's paper came from a sympathetic William Thiselton-Dyer (Fig. 8.3). The latter was a botanist but had also trained in mathematics. He had married Joseph Hooker's daughter in 1877 and was director of the Royal Gardens at Kew from 1885 to 1905.

During the February 28, 1895, meeting of the Royal Society, Thiselton-Dyer was not able to make certain points when Weldon presented his work. He therefore wrote a Letter to the Editor in *Nature* (Thiselton-Dyer, 1895d). In this Letter, which

Fig. 8.3 William Thurner Thiselton-Dyer (1843–1923). (From: https://commons.wikimedia.org/wiki/File:ThiseltonDyer.jpg)

appeared on March 14, Thiselton-Dyer's main thesis was that there was a stability among species with discriminating characteristics and that this stability coincided with the mean of those characteristics. This led him to the concept of the *mean specific form*:

QUOT. 533 No two individual representatives of a species in nature are exactly alike. All differ in some respect. We may picture the aggregate, however, as grouped with respect to any discriminating character like the shots on the target. Our conception of the species to which they belong is an abstraction which we endeavour to represent in our museums by a specimen which would be placed as near as possible to the centre of greatest density. Such an abstraction we may call the *mean specific form*. (*ibid.*, p. 459)

Thiselton-Dyer's *mean specific form* was undoubtedly derived from Quetelet's *l'homme moyen*. However, the interesting connection with the *Carcinus moenas* paper was Thiselton-Dyer's claim that Weldon had actually proved the existence of the mean specific form: the destructive selection of the young crabs which Weldon had described in his paper was nothing but nature's way of restoring the mean specific form:

QUOT. 534 Prof. Weldon arrives at the conclusion "that the position of minimum destruction should be sensibly coincident with the mean of the whole system," is a condition which "may be expected to hold for a large number of species, which are sensibly in equilibrium with their present surroundings, so that their mean character is tensibly the best." The actual statistical demonstration of this fact, in my opinion, deserves to rank amongst the most remarkable achievements in connection with the theory of evolution. (*ibid.*, p. 460)

Thiselton-Dyer next decided to take the saltationists head-on:

QUOT. 535 I entirely agree with him [Weldon] in minimising the value of "sports" in evolution...

The value...of sports may be easily overrated. It appears to me that generally, so to speak, they attempt too much and overshoot the mark. The improbability of a casual sport being exactly what is wanted to bring the organism into an advantageous relation to the environment at any particular juncture seems to me very great. That such a thing may occur is not denied, but it can hardly be more than a "fluke."

(*ibid.*, p. 461)

Thiselton-Dyer's view was one of Darwinist (continuous) evolution:

QUOT. 536 As I conceive the process, it is one of continuous adjustment of "slight" variations on one side and the other. (*ibidem*)

However, the biometricians' Darwinism was different from how Thiselton-Dyer had interpreted it: he did not put an over-emphasis on statistical methods, and he instead believed in the utility of experimental methods:

QUOT. 537 I am not sure that I quite understand Prof. Weldon when he says that "the statistical method is the only one at present obvious by which [the Darwinian] hypothesis can be experimentally checked." In the first place, I should myself hardly call it experimental at all. In the next place, though I think it will throw important light on the stability problem, in the important cases where evolution is actually taking place, the mathematical analysis appears to me to be beset with very great difficulties. We must not, therefore, expect too much from it. (*ibidem*)

In this regard, Thiselton-Dyer may be called an "old-school" Darwinian.

At some point in his paper (p. 461), Thiselton-Dyer recalled the plants of the feral type and of a recently cultivated form of *Cineraria cruenta* he had presented during the Royal Society meeting. This was done quite innocently,[14] to illustrate the power of "the gradual accumulation of small variations" in transforming the wild form to the cultivated one. Little did Thiselton-Dyer realize at the time that his reference to Cineraria would soon become the launching pad for an all-out war between the biometricians and saltationists.

Before the confrontation began, the Lamarckian Joseph Cunningham (1859–1935) had an opportunity to reply to both Weldon's report and Thiselton-Dyer's complimentary review of it. Cunningham's Letter to the Editor (1895), which was published on March 28, was anything but complimentary. He stated that Weldon had only demonstrated the possibility of destructive selection in one of the dimensions of the crab. This in itself was not enough for Thiselton-Dyer to claim that the *Carcinus moenas* paper was a remarkable achievement. Cunningham took exception to Weldon's downplaying the role of functional adaptation in the Darwinian paradigm, claiming that this simply was not Darwinism. He stressed the need for controlled experiments and dismissed Weldon's claim that the questions raised by the Darwinian hypothesis were purely statistical.

8.3.2 Bateson (April 25, 1895, Art. 2)

Now it was Bateson's turn to intervene. He had just come back from abroad and got a chance to read Thiselton-Dyer's Letter. As mentioned before,[15] Bateson had refrained from making public his disapproval of Weldon's two articles and had instead privately communicated with the Committee. Thiselton-Dyer's dismissal of sports and his reference to Cineraria as an exemplar of continuous evolution caught his attention, and he decided to write to *Nature*. He did not wish to draw Weldon into the debate and took on only the arguments made by Thiselton-Dyer.

In a *Nature* Letter to the Editor,[16] published on April 25, Bateson (1895b) consulted a variety of historical sources. He found that James Drummond (1784–1863), who was curator of the botanic garden at Cork, had published an article on Cineraria in 1827 (Drummond, 1827), after which the latter was able to hybridize the flower. Bateson cited evidence from Loudon's *Ladies' Magazine of Gardening* (Loudon, 1842, p. 111) that such hybrids were indeed produced after 1827. Bateson thus inferred that the cultivated Cineraria was not directly

[14]Thiselton-Dyer would later describe it "as a merely incidental illustration." (Thiselton-Dyer, 1895b, p. 78).

[15]See paragraph above QUOT. 531.

[16]This and all subsequent *Letters to the Editor* in Nature had the same title "The Origin of the Cultivated Cineraria."

transformed from *C. cruenta*, as Thiselton-Dyer had claimed, but had been hybridized from at least four species: *C. cruenta, aurita, populifolia*, and *lanata*. Furthermore, he cited documented sources of sports occurring within these species. He concluded:

> **QUOT. 538** (1) that the modem Cinerarias arose as hybrids derived from several very distinct species; (2) that the hybrid seedlings were from the first highly variable; (3) that "sports" of an extreme kind appeared after hybridisation in the early years of the "improvement" of these plants; (4) that the subsequent perfection of the form, size and habit has proceeded by a slow process of selection. (Bateson, 1895b, p. 607)

Hence:

> **QUOT. 539** Mr. Dyer's statement that the modern Cinerarias have been evolved from the wild *C. cruenta* "by the gradual accumulation of small variations" is therefore, in my judgment, misleading, for this statement neglects two chief factors in the evolution of the Cineraria, namely, hybridisation and subsequent "sporting." (*ibidem*)

Therefore, the debate, which at heart was about continuous versus discontinuous evolution, was fought along the lines of *direct transformation through natural selection from a single species* versus *transformation through hybridization from multiple species (and possibly involving sports).*

8.3.3 Thiselton-Dyer (May 2, 1895, Art. 3)

Thiselton-Dyer (1895a) shot back in a *Nature* Letter published on May 2, challenging the technical details provided by Bateson. He stated that cultivated Cineraria and *C.cruenta* were both herbaceous while *C. aurita, populifolia*, and *lanata* were all shrubby species. Furthermore, the first pair had the same foliage, but the second pair had the foliage of the white poplar. Thiselton-Dyer implied that the cultivated Cineraria could not have arisen by hybridization from *C. cruenta, aurita, populifolia*, and *lanata*, as Bateson had claimed, but instead from *C. cruenta* directly. Thiselton-Dyer then attacked Bateson personally:

> **QUOT. 540** He [Bateson] will probably then regret, for the sake of his reputation as a naturalist, that he committed himself to print on a subject on which he evidently possesses little objective knowledge. (*ibid.*, p. 3)

The invective above was no doubt due to Bateson having previously called Thiselton-Dyer's statement "misleading," but it surely did not help the discussion.

8.3.4 Bateson (May 9, 1895, Art. 4)

Bateson (1895c) now wrote a second Letter to *Nature* to be published on May 9, 1895. He congratulated himself for having successfully made his point across for, in his last Letter to the Editor, Thiselton-Dyer no longer insisted that the cultivated Cineraria had been formed by continuous evolution. The latter had shifted the debate to the *origin* of the cultivated Cineraria but that was of "subordinate interest." Even on that "subordinate" debate, Bateson contrasted Thiselton-Dyer's (and Rolfe's[17]) views against those "of contemporaries and the almost universal opinion of practical gardeners."

8.3.5 Gardeners' Chronicle Editorial (May 11, 1895, Art. 5)

In the meantime, the editorial of *The Gardener's Chronicle* of May 11 took note of the debate taking place in *Nature*. Without taking sides, it stated:

> **QUOT. 541** An interesting discussion is being carried on in the columns of *Nature*, as to the origin of the garden Cineraria. On the one hand, Mr. DYER asserts that the varieties now so numerous originated by selection from Cineraria cruenta only; on the other, Mr. BATESON brings forward an imposing array of evidence from horticultural publications tending to show that more than one species has been concerned in the production of this favourite flower. (Anonymous, 1895f, p. 588)

8.3.6 Weldon (May 16, 1895, Art. 6)

The back-and-forth had hitherto involved only Thiselton-Dyer and Bateson but took a turn which was to have lasting consequences when Weldon decided to intervene on behalf of Thiselton-Dyer. In a Letter to the Editor published in *Nature* on May 16, 1895, Weldon declared:

> **QUOT. 542** ...I have consulted the principal authorities cited by Mr. Bateson in NATURE of April 25; I now wish to point out that Mr. Bateson has omitted from his account of these records some passages which materially weaken his case. (Weldon, 1895a, p. 54)

Weldon thus accused Bateson of deception. He then cited a passage from Loudon's *Ladies Magazine of Gardening* (Loudon, 1842, p. 112), which he alleged Bateson had willfully omitted and which purportedly showed that Cineraria species had *directly* descended from *C. cruenta* and not through *hybridization*. Furthermore, he stated that, upon examining one of the sports Bateson had mentioned (viz., *C. webberiana*), he failed to see any sport whatsoever. To finish off, Weldon accused Bateson of plain carelessness:

> **QUOT. 543** All I wish to show is that the documents relied upon by Mr. Bateson do not demonstrate the correctness of his views; and that his emphatic statements are simply evidence of want of care in consulting and quoting the authorities referred to. (*ibidem*)

---

[17]Robert Allen Rolfe (1855–1921) was an assistant to Thiselton-Dyer and had published a short article on the Cineraria in 1888 (Rolfe, 1888, p. 654). Rolfe's article showed that diverse forms developed when *C. cruenta* was cultivated. *C. cruenta* itself was so different from the garden variety that "we might almost doubt whether the one could have been directly derived from the other." Rolfe's view was thus one of direct Darwinian evolution from *C. cruenta* to the cultivated Cineraira.

8.3.7 Hemsley (May 16, 1895, Art. 7)

The botanist William Botting Hemsley (1843–1924) inserted his own Letter in the May 16 issue, stating that:

> **QUOT. 544** . . .Mr. Bateson has relied too implicitly on the literature of the subject. Many of the records of hybrid productions in the vegetable kingdom are based upon groundless assumptions; mere seminal variations having been mistaken for crosses. (Hemsley, 1895, p. 54)

Hemsley's main point was that hybrids were usually too unstable, while selected variations were stable and could "intercross most freely, and give birth to new varieties without end" (*ibidem*).

8.3.8 Bateson's Meeting with Weldon (May 21, 1895)

Bateson was offended, especially by Weldon's remarks. He sent a response to *Nature* on May 19, and on the next day, he sent a personal letter to the editor of the journal explaining that he needed to reply to both Weldon and Hemsley in a single Letter. Bateson further promised that, unless Thiselton-Dyer's forthcoming Letter contained new material, he would not respond again. On May 20, 1895, Bateson also addressed Weldon in a letter. Part of the letter is reproduced below; it reveals Bateson's hurt feelings, his gratitude toward Weldon, and a request for a meeting to sort things out:

> **QUOT. 545** There are facts which if I read then rightly put your charges in a very unpleasant light. On thinking it all over and after consulting my friends here I should like to give you an opportunity of hearing my defence.
>
> I am conscious that I owe a great deal in life to you and I am unwilling to treat you as I would Dyer and his like.
>
> Will you see me tomorrow?. . .[18]

The meeting between Bateson and Weldon took place on May 21. According to Bateson's contemporaneous notes of the meeting (Fig. 8.4):

> **QUOT. 546** He [Weldon] defended his letter and declined to withdraw, but said he regretted that it shd. [should] be thought he meant to suppose that I [Bateson] had suppressed evidence because damaging.. . .[Weldon] thinks 'hybrid' used loosely, and as equivalent of 'sports'. Says thinks Dyer is "bluffing." He told me of passage in "Cross-Fert.", p. 335 showing that Dyer had been in communication with Darwin and that the latter believed Cs. [Cinerarias] to be probably hybrids.[19]

---

[18]Letter retrieved online from Cambridge Digital Library, Bateson Archive at https://cudl.lib.cam.ac.uk/view/MS-ADD-08634-C/128

[19]Letter retrieved online from Cambridge Digital Library, Bateson Archive at https://cudl.lib.cam.ac.uk/view/MS-ADD-08634-C/130

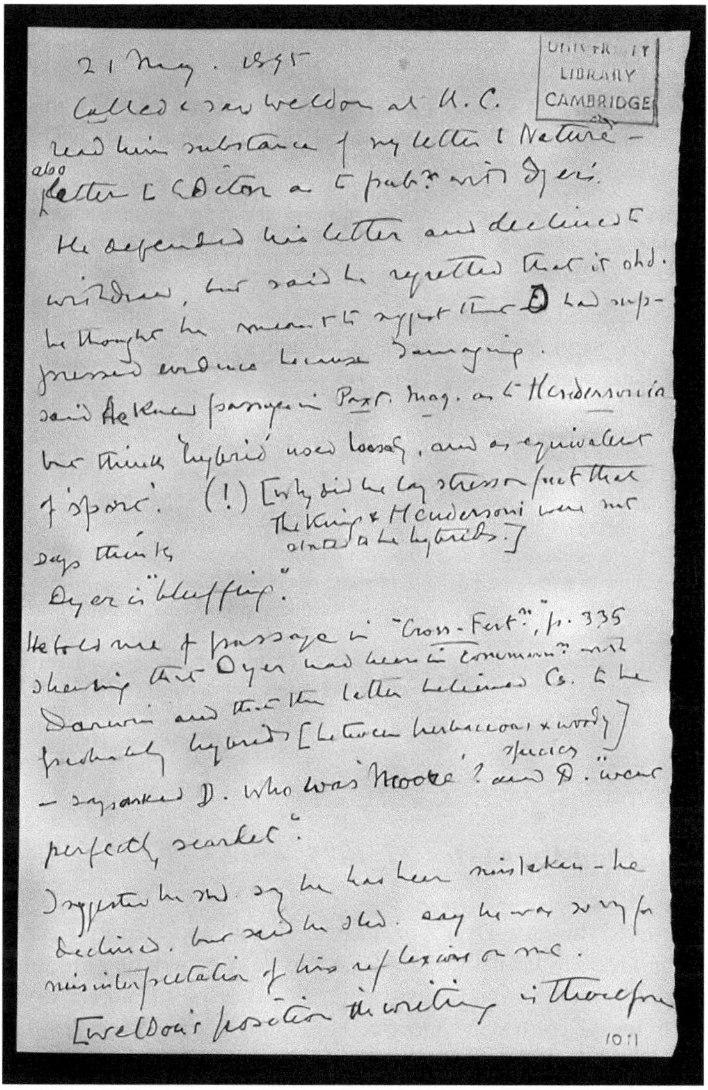
21 May. 1895
Called & saw Weldon at U.C.
read him substance of my letter to Nature —
also letter to Editor as to publ.n with Dyer's.
He defended his letter and declined to
withdraw, but said he regretted that it shd.
be thought he meant to suggest that D had sup-
pressed evidence because damaging.
said the passage in Pax. Mag. on the Hendersonii
but thinks 'hybrid' used loosely, and as equivalent
of 'sport'. (!) [why did he lay stress on fact that
The King & Hendersoni were not alluded to as hybrids.]
says thinks
Dyer is "bluffing".
He told me of passage in "Cross-Fert.n", p. 335
showing that Dyer had been in communication with
Darwin and that the letter believed Cs. to be
probably hybrids [between herbaceous & woody species]
— & asked D. who was Moore? and D. wrote
"perfectly scarlet".
I suggested he shd. say he has been mistaken — he
declined. but said he shd. say he was sorry for
misinterpretation of his reflexions on me.
[Weldon's position the writing is therefore

Fig. 8.4 Bateson's contemporaneous notes of his May 21, 1895, meeting with Weldon. (From Cambridge Digital Library, Bateson Archive at https://cudl.lib.cam.ac.uk/view/MS-ADD-08634-C/128)

On the face of it, this was a big revelation. The alleged p. 335 of Darwin's book (Darwin, 1876) indeed said:

> **QUOT. 547** *Senecio cruentus (greenhouse varieties, commonly called Cinerarias, probably derived from several fruticose or herbaceous species much intercrossed*)*

In the footnote to (*), Darwin acknowledged the help of Thiselton-Dyer (and one Mr. Moore) in providing him with the information on the varieties he experimented with!

In his next letter, Bateson would soon use this information and a flattering statement Thiselton-Dyer made about Darwin to deliver a powerful blow to his opponent.

Returning to Bateson's contemporaneous notes of the May 21 meeting with Weldon, Bateson's increasing distrust of Weldon is clearly shown in the two sentences below:

> **QUOT. 548** Weldon's position in writing is therefore that of the accomplice who creates a diversion to help a charlatan. I cannot at all understand his motives on how he can bring himself to play this part.[20]

Indeed, Bateson could not understand how Weldon declared Thiselton-Dyer to be a bluffer and then went on to support him. When he expressed this in a letter to Weldon, the latter seemed exasperated:

> **QUOT. 549**
>
> 24 May 1895
>
> Dear Bateson,
> I can do no more.
>
> First, you accuse me of attacking your personal character; and when I disclaim this, you charge me with a dishonest defense of some one else.
>
> I have throughout discussed only what appeared to me to be facts, relating to a question of scientific importance.
>
> If you insist upon regarding any opposition to your opinions concerning such matters as a personal attack upon yourself, I may regret your attitude, but I can do nothing to change it.
>
> (Provine, 2001, pp. 47–48)

8.3.9 *Thiselton-Dyer (May 23, 1895, Art. 8)*

In the meantime, Thiselton-Dyer's next Letter already appeared in *Nature* on May 23 (Thiselton-Dyer, 1895b). Thiselton-Dyer was quite patronizing towards Bateson and criticized him on several occasions. Regarding Bateson's historical evidence of a hybrid origin for the cultivated Cineraria, Thiselton-Dyer remarked that the nomenclature used by horticulturists, as opposed to expert botanists, was unreliable and that not much credence should be attached to the names they used. Thiselton-Dyer concluded that:

> **QUOT. 550** ...in the study of evolution we have had enough and to spare of facile theorising. I infinitely prefer the sober method of Prof. Weldon, even if it should run counter to my own prepossessions, to the barren dialectic of Mr. Bateson. (*ibid.*, p. 79)

However, before Thiselton-Dyer wrote the conclusion above, he had made an important concession that would soon be seized by Bateson:

---

[20]Letter retrieved online from Cambridge Digital Library, Bateson Archive at https://cudl.lib.cam.ac.uk/view/MS-ADD-08634-C/130 and at https://cudl.lib.cam.ac.uk/view/MS-ADD-08634-C/131

QUOT. 551 Mr. Darwin, as he has told us, spent the best part of his life in studying patiently and sifting critically a vast mass of observation and fact. Ultimately he permitted himself to draw certain conclusions. The result is that if you take any statement which Mr. Darwin has put forward, you may feel assured that behind it is a formidable body of carefully considered evidence not likely to be upset. (*ibid.*, p. 78)

8.3.10 Gardeners' Chronicle Editorial (May 25, 1895, Art. 9)

Meanwhile, an editorial note of the May 25 issue of *The Gardeners' Chronicle* quoted a passage from Focke's authoritative book *Die Pflanzen-Mischlinge* (Focke, 1881), supporting a hybrid origin of the cultivated Cineraria:

QUOT. 552 The crossing of Senecio cruentus with S. populifolius (non L.) gave as issue the so-called Cineraria of gardens. According to A. Orro, the first hybrid varieties were bicolor, coelestis, Formosa, Hendersoni, pulchellam and Waterhouseina. The Canary Islands are the home of the parent species. Afterwards several Canarian and Madeiran species were crossed with these mongrel varieties, viz., S. tussilaginis, Lees; S. Heritieri, D.C.; S. Maderensis, D.C. (=S. auritus Lowe), and S. Webbii, Schltz. Bip. All of these crosses have now united in gardens in one mongrel race of plants rich in variety of form and colour, A.S. Webbii, Schltz. X S. cruentus, D.C. was, according to BOUCHÉ (*Wittm. Monatsschr.* 22, p. 298), spontaneously produced in the Berlin botanical garden from two cultivated plants of these species standing in proximity to each other. (Anonymous, 1895a, p. 655)

8.3.11 Bateson (May 30, 1895, Art. 10)

Coming back to Bateson, as mentioned before, his initial response to Weldon and Hemsley had been sent to *Nature* on May 19, 1895. It was then modified to take Thiselton-Dyer's latest Letter into account. In the light of his conversation with Weldon on May 21, Bateson also toned down his Letter regarding the latter's accusation.

Bateson's Letter to *Nature* was finally published on May 30 (Bateson, 1895d). He first addressed Weldon's charge by reproducing the passage (from Loudon's *Ladies' Magazine of Gardening*) that Weldon alleged Bateson had intentionally omitted. The latter argued that the passage in question said that new varieties had arisen from the *seeds* of *C. cruenta* and that Weldon had interpreted this to mean that the varieties had arisen from *C. cruenta* directly without consideration of the *pollen.* Bateson then showed that the author of the passage (Mrs. Loudon) had believed in her writings elsewhere that other varieties had actually been hybridized, with *C. Cruenta* as one of the parents.

Bateson also addressed Hemsley's Letter by explaining the exact circumstances under which hybrid ancestry was inferred.

However, the brunt of Bateson's counterattack was reserved for Thiselton-Dyer. The latter had downplayed historical evidence on the basis that horticulturists were often incompetent with names. Bateson replied that:

QUOT. 553 . . .this is not a principle which Darwin, whom Mr. Dyer would claim as his master, would have endorsed. (*ibid.*, p. 103)

Bateson finally referred to Thiselton-Dyer's acknowledgment (see QUOT. 551) that Darwin's statements were always based on a vast body of carefully considered evidence. How then did Thiselton-Dyer explain Darwin's declaration (see QUOT. 547) that the cultivated Cineraria was indeed a hybrid from several varieties? Bateson thanked the "courtesy of an opponent" (Weldon) for passing this information.

8.3.12 Weldon (May 30, 1895, Art. 11)

Bateson's Letter was followed by a short note from Weldon (1895b) insisting that he never questioned Bateson's sincerity concerning the paragraph the latter had allegedly omitted.

8.3.13 Thiselton-Dyer (June 6, 1895, Art. 12)

As Bateson had promised to the editor, his May 30 Letter was the last to *Nature* in so far as the Cineraria affair was concerned. However the debate continued in *Nature* (by others) and especially in *The Gardener's Chronicle*.

In the June 6, 1895, issue of *Nature*, Thiselton-Dyer (1895c) repeated many of his previous arguments. In particular, he reiterated his distrust of historical evidence. His only response to the Darwin issue was that he simply passed the information from Thomas Moore to the latter. Thiselton-Dyer also continued to use highly charged language, for example:

QUOT. 554 I cannot but think that there is a growing danger nowadays of a pseudo-biology growing up for the especial use of evolutionists. This is not the first time by many that I have been so unlucky as to come into collision with it. (*ibid.*, p. 129)

8.3.14 Weldon (June 6, 1895, Art. 13)

In the June 6, 1895, issue of *Nature*, Weldon (1895c) stated that Bateson had misinterpreted the passage the latter had quoted in his earlier Letter and that several species of Cineraria had indeed been obtained directly from *C. cruenta.* Weldon concluded as follows:

QUOT. 555 . . .my interest in the matter ends, and I do not propose to speak further upon it. (*ibidem*)

8.3.15 Gardeners' Chronicle Editorial (June 15, 1895, Art. 14)

The June 15, 1895, editorial of *The Gardeners' Chronicle* editorial again alerted readers of the ongoing Cineraria controversy in *Nature* and informed them of a letter by Thiselton-Dyer requesting that the 1827 article by Drummond (1827) be published in *The Gardeners' Chronicle*. Thiselton-Dyer said that he had been alerted to the article by Mr. Nicholson, curator of the Royal Gardens, and assured that it was

QUOT. 556 ...pretty conclusive historical evidence that the garden Cineraria originated from C. cruenta, and from nothing else... (Anonymous, 1895c, p. 742)

8.3.16 Bateson (June 22, 1895, Art. 15)

In the June 22, 1895, issue of *The Gardeners' Chronicle*, Bateson (1895a) expressed surprise by Thiselton-Dyer's statement that the latter had been alerted to Drummond's article by Mr. Nicholson, when it was Bateson who drew attention to that article in his April 23, 1895, letter to *Nature* (Bateson, 1895b). He then summarized his historical research on the origins of the *Cineraria* and concluded that Cineraria:

QUOT. 557 ...was brought forward by Mr. Dyer, wrongly, as it appears to me, to illustrate what could be done by the accumulation of small variations without hybridization and without sudden variation. (Bateson, 1895a, p. 774)

8.3.17 Douglas (June 22, 1895, Art. 16)

In the same June 22, 1895, issue of *The Gardeners' Chronicle*, J. Douglas expressed his opinion on the origin of the cultivated Cineraria. He was clearly on Bateson's side in the debate:

QUOT. 558 ...I think it improbable that our present garden Cinerarias have been obtained from this species by ordinary generation, that is, by selection. (Douglas, 1895, p. 774)

Citing historical evidence, Douglas declared that the modern Cineraria was probably a hybrid between *Cineraria cruenta* and *C. lanata.*

8.3.18 Gardeners' Chronicle Editorial (July 6, 1895, Art. 17)

The July 6, 1895, editorial of *The Gardeners' Chronicle* (Anonymous, 1895d) informed the readers of a statement from Mr. Kelway, published in *The Gardener's*

Magazine, that he raised the cultivated *Cineraria* by selection from seedlings of *C. cruenta* in 1837, 10 years after Drummond's article (Drummond, 1827).

8.3.19 Harrow (July 6, 1895, Art. 18)

In the same July 6 issue of *The Gardeners' Chronicle*, Harrow (1895) stated what should have been obvious right from the start:

> **QUOT. 559** As authentic records are wanting upon the early history of the subject, a few experiments ought to be made with a view of ascertaining the value of the two conjectures. (*ibid.*, p. 17)

Following up on Douglas' statement,[21] Harrow suggested that the *Cineraria cruenta* should be crossed with *C. lanata* under controlled experimental conditions.

8.3.20 Gardeners' Chronicle Editorial (July 27, 1895, Art. 19)

The July 27, 1895, editorial of *The Gardeners' Chronicle* (Anonymous, 1895b) reiterated Mr. Kelway's claim (see Art. 17) that the cultivated Cineraria descended directly from *C. cruenta*, as per his observations in 1837. However, the editorial also noted that hybrid Cinerarias were definitely in existence around 1850. It next stated:

> **QUOT. 560** That the Cineraria, as we now know it, may have had a hybrid origin within comparatively recent times, is a hypothesis which the facts now to be recorded show, is not to be summarily dismissed as unworthy of credence. (*ibid.*, p. 89)

The editorial then gave two instances of hybrid Cinerarias obtained from crosses between *Cineraria lanata* (which is the same as *Senecio heritieri*) and the cultivated Cineraria. The first was by R. Irwin Lynch and the second was by J.T. Bennett Poe. The editorial acknowledged that these did not prove that the cultivated Cineraria were of hybrid origin, but they suggested that experiments could easily be carried out to settle the issue.

8.3.21 Gardeners' Chronicle Editorial (August 17, 1895, Art. 20)

The August 17, 1895, editorial of *The Gardeners' Chronicle* (Anonymous, 1895e) informed of a letter from Thiselton-Dyer, where the latter announced that he did not wish to continue the debate any further. However, before closing off, he commented on the July 27 editorial (see Art. 19). Thiselton-Dyer had probably sensed that the editorial was leaning toward a hybrid origin for the cultivated Cineraria. Therefore,

[21] See paragraph below QUOT. 558.

he stated that the formation of Cineraria hybrids proved nothing in the debate: what mattered to him was why, if the cultivated Cineraria was truly of hybrid origin, it retained only the characteristics of *C. cruenta*. He claimed that his explanation of direct parentage from *C. cruenta* was the simplest.

8.3.22 *Gardeners' Chronicle Note (February 22, 1896, Art. 21)*

A Kew Note of the February 22, 1896, issue of *The Gardeners' Chronicle* (Anonymous, 1896) claimed to make an interesting contribution to the debate on the garden Cineraria's origin. The note reported crossings between *Cineraria cruenta* and *C. lanata*, yielding hybrids very similar to the garden Cineraria, except for their taller stature and looser flowers. On the other hand, crosses between *C. lanata* and the garden Cineraria yielded hybrids that were very different from the latter. These observations could well have supported the hypothesis that the garden Cineraria originated from an initial crossing between *Cineraria cruenta* and *C. lanata* followed by natural selection, but the note was clearly on Thiselton-Dyer's and Weldon's side of the debate and made the following conclusion:

> **QUOT. 561** ...if C. Heritieri [*C. lanata*] was ever used by the early breeders of garden Cinerarias, it was soon bred out, and that the relationship between them and C. cruenta is so close that no breeder would hesitate to accept the latter as the sole progenitor of the garden Cineraria. (*ibid.*, p. 240)

8.3.23 *Thiselton-Dyer (May 14, 1896, Art. 22)*

Although he had previously declared that he wished to let the Cineraria matter rest (see Art. 20), Thiselton-Dyer (1896) wrote a further short note in the March 14, 1896, issue of *The Gardeners' Chronicle*. He referred to a passage of the July 27, 1895, editorial (Anonymous, 1895b, p. 90) (see Art. 19), where Thiselton-Dyer thought it had been stated as a matter of fact that the garden Cineraria was a cross between *Senecio cruentus* (*C. cruenta*) and *S. populifolius*. He wished to clarify that this was not a settled issue and that the statement was originally made *based on popular opinion*. In contrast, Thiselton-Dyer found no traces of *S. populifolius* in the garden Cineraria.

8.3.24 *Harrow (May 2, 1896, Art. 23)*

In his second article on the subject in *The Gardeners' Chronicle*, published on May 2, 1896, Harrow (1896) reported the results of experiments he had performed on a single plant of *Cineraria cruenta*. He confirmed that the plant was not pollinated by *C. lanata* and that 40 of its seedlings flowered. He stated that:

QUOT. 562 Judging from the varieties which we have raised from this single plant of C. cruenta, it would not be a work of many years duration before one could obtain by judicious selection and good cultivation a strain of Cinerarias from them, similar to those of the florists' varieties. (*ibid.*, p. 560)

Harrow concluded that the cultivated Cineraria originated solely from *C. cruenta*. However, this conclusion is not warranted since Harrow had not ruled out the possibility that the single plant of *C. cruenta* could have been pollinated by cultivated Cineraria.

8.3.25 Lynch (February 11, 1897, Art. 24)

R. Irwin Lynch[22] was the curator of the University Botanic Garden at Cambridge and had previously collaborated with Bateson. In the February 11, 1897 issue of *Nature*, Lynch (1897) appealed to the readers of the journal to send him seeds of all herbaceous species of *Senecio*. He reported that he had already performed crosses among four distinct types. In the biography of her husband, Beatrice Bateson revealed that Lynch's crosses were made at William Bateson's request (Bateson, 1928, p. 464). In his short article, Lynch reported the results of the first type of crosses, those between *S. multiflorus* (female) and several color forms of the cultivated Cineraria. The crosses showed the predominating influence of the latter species and the ease with which Cinerarias generally crossed.

8.3.26 Bateson (May 10, 1897, Art. 25)

In the May 10, 1897, issue of the *Proceedings of the Cambridge Philosophical Society*, Bateson (1897) reported the initial results of some of the crosses performed by R. Irwin Lynch and Miss Pertz. The crosses involved *S. cruentus* (*C. cruenta*), *S. multiflorus*, and *S. heritieri* (*C. lanata*). The *S. heritieri* x *S. cruentus* crosses showed considerable variety. To counter Thiselton-Dyer's argument that *S. heritieri* was substantially different from cultivated Cineraria and could not have been one of its parents, Bateson declared that the crosses proved:

QUOT. 563 ...how greatly the peculiar characters of *Heritieri* may be obscured in the offspring, even of the first cross. (*ibid.*, p. 308)

Considering the variability of the hybrids obtained, Bateson concluded that the experiments:

[22] See also Art. 19.

QUOT. 564 ...were entirely consistent with the view that the Cineraria was a hybrid between several species, *cruentus*, *Heritieri* and probably *multiflorus* being among them. The two first are named by most authorities as probable parents. (*ibid.*, p. 309)

8.3.27 Rolfe (Aug 6, 1898, Art. 26)

Rolfe was an assistant to Thiselton-Dyer and had published an article on the Cineraria in 1888 (Rolfe, 1888), where he had shown the great variety of hybrids obtained when *C. cruenta* was cultivated.[23] In the August 6, 1898, issue of *The Gardeners' Chronicle*, Rolfe (1898) reported the results of crosses performed by Mr. James, a plant grower. The latter's first set of crosses involved *Senecio cruentus* (*C. cruenta*) with the pollen of *S. Herritieri* (*C. lanata*). Since those on the other side of the debate had suggested that the garden Cineraria (which has no scrubbiness) was a cross between these two species, Rolfe was quick to point out that the hybrids obtained were

QUOT. 565 ...unmistakable hybrids, presenting the usual intermediate character. They are more shrubby than cruentus, and are plentifully covered with tomentum on the stems and undersides of the leaves. (*ibid.*, p. 101)

The hybrids obtained were intercrossed and showed "the most bewildering range of variation." Rolfe believed that the hybrids were, in fact, *Senecio hybridus* (*Cineraria hybrida*) and that there was "an enormous gap between it and the garden Cineraria of today." Against those who believed that the garden Cineraria was a cross between *Senecio cruentus* and *S. Herritieri* followed by a period of natural selection, Rolfe again made the counterstatement:

QUOT. 566 It seems to me highly probable that a race of plants with shrubby habit and improved flowers might be derived from these hybrids if selection were carried out in that direction, and this process would tend to eliminate the cruentus like character from the vegetative organs. (*ibid.*, p. 102)

8.3.28 Bateson (July 11, 1899, Art. 27)

At the International Conference on Hybridisation, held on July 11th and 12th 1899 and organized by the Royal Horticultural Society, Bateson gave a speech entitled "Hybridisation and Cross-Breeding as a Method of Scientific Investigation" (Bateson, 1899). Bateson praised hybridization as a method that was perhaps second to none "to bring forth results of first-rate importance." As examples, he mentioned several plants, especially the Cineraria:

---

[23] See footnote 17.

QUOT. 567 I mention the Cineraria, because I have personal knowledge of these hybrids, and because I notice that the view that our garden Cinerarias are not hybrids is being again repeated, in spite of the clear evidence, both of history and recent experiment, to the contrary. (*ibid.*, p. 61)

At the same time, Bateson warned that hybridization was not responsible for the origin of species. Its importance lay in that it was through hybridization of closely related species that information on the origin of species could be gleaned.

A similar paper by Bateson (1901, 1928, pp. 171–180) was read on May 8, 1900 at the next International Conference on Hybridisation. Both papers touched on the Cineraria issue and contained passages that showed the closeness of Bateson's views with Mendel's regarding the statistical investigation of the results of crosses and the need for investigations to be performed on a character-by-character basis.[24]

8.3.29 Lynch (July, 1899, Art. 28)

In the same Hybrid Conference Report of the meeting of July 1899, R. Irwin Lynch provided further results from his hybridization experiments[25] (Lynch, 1899). Lynch's article was more of a reply to Rolfe's (1898)[26] since its results were squarely opposite to those of the latter. The different species Lynch worked with were *Senecio cruentus, S. heritieri, S. multiflorus*, and *S. tussilaginis*. At the very start, Lynch noted that his specimens of *S. cruentus* were quite different from the usual ones. He further remarked that Cineraria was not very fertile but hybridism sometimes increased fertility. Rolfe had found that *Senecio cruentus* x *S. Herritieri* produced shrubby hybrids, which he probably thought contradicted Bateson's claim that the garden Cineraria (which has no scrubbiness) descended from this cross. In contrast, in his *Senecio Herritieri* x *S. cruentus* crosses, Lynch's finding was that:

QUOT. 568 ...the shrubby habit of *Heritieri* is lost by crossing a very few times... It [the scrubbiness] had practically disappeared in some of the hybrids named below, and its absence from the Cineraria is not, I believe, to be regarded as evidence that *Heritieri* is not in the parentage. There is however, very strong evidence, to my mind, of *Heritieri* parentage in the ringed arrangement of colour so familiar in the florists' Cineraria. (Lynch, 1899, p. 270)

However, because Lynch's *S. cruentus* was not typical, his *Senecio Herritieri* x *S. cruentus* crosses were not comparable to Rolfe's corresponding crosses.

Lynch reported that his *Cruentus* x *Tussilaginis* hybrids were very close to the garden Cineraria. He concluded that:

---

[24] See Sect. 9.1.

[25] See Arts. 24 and 25.

[26] See Art. 26.

QUOT. 569 The Cineraria I believe to be a hybrid, because it has the reduced fertility of a hybrid, because it has a mode of colouration foreign to the kind of plant from which no doubt it obtained its habit, because it has flower heads which to me suggest *Tussilaginis*, and I, may add now, because of the big jump to which I have referred. (*ibid.*, p. 273)

8.3.30 Barkley (1966, Art. 29)

In a 1966 review article, Barkley (1966) was understandably surprised that there was a paucity of factual historical evidence regarding the origin of the cultivated Cineraria. After a review of the literature, his conclusion was as follows:

QUOT. 570 The biological origin of the complex has been studied fitfully, but the available evidence suggests that the cineraria is of hybrid origin, involving *Senecio cruentus* and *S. heritieri* and possibly several other species. (*ibid.*, p. 394)

Barkley's statement is in line with current opinion. The garden Cineraria is now known as *Senecio* x *hybrida* (or *Pericallis* x *hybrida*) and is believed to have originated as a hybrid between *S. cruentus* and *S. heritieri* (*C. lanata*).

8.4 Round 4: The Homotyposis Affair—First Public Controversy Between Bateson and Pearson (1900–1903)

8.4.1 Pearson's Principle of Homotyposis

Whereas Pearson had not engaged in the Cineraria affair, his homotyposis paper (Pearson et al., 1901b) thrusted him into the heart of the war between the biometricians and saltationists. The paper's primary concern was not continuous versus discontinuous evolution, but it still dealt with an issue that was fundamental to both parties, namely, the place and amount of mathematics necessary in dealing with questions of heredity and evolution.

The Royal Society received Pearson's homotyposis paper on October 6, 1900, at which time the latter had not yet heard of Mendel's paper.[27] Pearson read a five-page abstract (Pearson et al., 1901a) on November 15, giving the broad outlines of his theory of homotyposis. First was the concept of *homotypes*:

QUOT. 571 ...an individual produces a number of like organs, which so far as we can ascertain are not differentiated... Such organs may be blood-corpuscles, hairs, scales, spermatozoa, ova, buds, leaves, flowers, seed-vessels, &c., &c. Such organs I term *homotypes* when there is no trace to be found between one and another of differentiation in function. (*ibid.*, p. 2)

[27]Pearson heard of Mendel for the first time from Weldon's Oct 16, 1900, letter to him: "About pleasanter things, I have heard of and read a paper by one, Mendel, on the results of crossing peas, which I think you would like to read. It is in *Abhandlungen des naturforschenden Vereines in Brünn* for 1865. I have the R.S. copy here, but I will send it to you if you want it." (Pearson, 1906, p. 36).

Pearson's primary questions concerning these undifferentiated "organs" were those of *variability* and *correlation*—"what is the degrees of diversity and likeliness among them?" He termed the principle of likeliness among undifferentiated organs *homotyposis*. To quantify the homotypic correlation, Pearson's compared the likeliness between undifferentiated organs belonging to the same organism with the likeliness between undifferentiated organs belonging to different organisms. He also calculated standard deviations and reported that:

> **QUOT. 572** The individual [homotypic] variation in the twenty-two series was measured and expressed as a percentage of the racial [belonging to different organisms] variation; the results range from 77 to 98 per cent., with a mean value of 87 per cent. (*ibid.*, p. 5)

Pearson further announced that, with the help of several assistants, he had found that the homotypic correlation was stable in both the animal and vegetable kingdoms, with a mean value (across 32 series of observations) of .4570.

Pearson's concept of correlation among homotypes and his ideas of "within" and "between" variations undoubtedly had an effect in the later development of the intraclass correlation and analysis of variance, respectively—in fact, his homotypic correlation *is* the well-known intraclass correlation coefficient.[28] This serves as a reminder that although their excessive use of mathematics was criticized at the time, the biometricians were responsible for several key theoretical and methodological developments in statistics. At the top of the list of such biometricians was unarguably Pearson.

Pearson next announced the major result of his upcoming paper, namely:

> **QUOT. 573** A theory of fraternal hereditary resemblance is given on the basis of the likeness of brothers being due to homotyposis in the characters of spermatozoa and ova put forth by the same two individuals and uniting for the zygotes whence the brothers arise. It is found that the mean value of fraternal correlation ought to be equal to the mean intensity of homotypic correlation. (*ibid.*, p. 4)

The above is indeed a most astounding proposition: the sperm and ova are akin to undifferentiated organs, and because the offspring are representative of these germ cells, the mean value of the homotypic correlation between undifferentiated organs of an individual must be equal to the mean value of the sib-sib (fraternal) correlation!

8.4.2 Pearson's Proof of "Fraternal Correlation=Homotypic Correlation"

The proof of the statement above was given in the full paper (Pearson et al., 1901b) and will now be explained with a few modifications in the original notation. Let the deviation of an individual character from its mean be z. Suppose the character depends on characters of the parental sperm having deviations $x_1, x_2, x_3, \ldots$ from

[28] See Sec. 2.3.8 (pp. 253–258) of Gorroochurn (2016a) for the origins and development of the intraclass correlation coefficient.

their respective means; similarly, let the character also depend on characters of the parental ovum having deviations y_1, y_2, y_3, ... from their respective means. Then, Pearson assumed

$$z = \alpha_1 x_1 + \alpha_2 x_2 + \alpha_3 x_3 + \ldots + \beta_1 y_1 + \beta_2 y_2 + \beta_3 y_3 + \ldots \tag{8.2}$$

where $\alpha_1, \alpha_2, \alpha_3, \ldots, \beta_1, \beta_2, \beta_3, \ldots$ are constants representing the "male and female inheritance." Squaring on both sides of the above and taking the mean,

$$\sigma^2 = \mathrm{E}z^2 = \sum_k \alpha_k^2 \mathrm{E}x_k^2 + 2\sum_{j,k} \alpha_j \alpha_k \mathrm{E}x_j x_k + \sum_k \beta_k^2 \mathrm{E}y_k^2 + 2\sum_{j,k} \beta_j \beta_k \mathrm{E}y_j y_k,$$

where $\sigma = \mathrm{SD}(z)$, and it has been assumed that x_j and y_k are independent. Now, with $\mathrm{SD}(x_k) = \sigma_k$, $\mathrm{SD}(y_k) = \sigma'_k$, $\mathrm{corr}(x_j, x_k) = r_{jk}$, $\mathrm{corr}(y_j, y_k) = r'_{jk}$, the above can be written as

$$\sigma^2 = \sum_k \alpha_k^2 \sigma_k^2 + \sum_k \beta_k^2 \sigma'^2_k + 2\sum_{j,k} \alpha_j \alpha_k \sigma_j \sigma_k r_{jk} + 2\sum_{j,k} \beta_j \beta_k \sigma'_j \sigma'_k r'_{jk}. \tag{8.3}$$

Pearson next considered two siblings who had, respectively, deviations z_1 and z_2 in their characters, deviations $x'_1, x'_2, \ldots$ and $x''_1, x''_2, \ldots$ in parental sperm characters, and deviations $y'_1, y'_2, \ldots$ and $y''_1, y''_2, \ldots$, in parental ovum characters. In analogy with Eq. (8.2), he wrote

$$z_1 = \sum_k \alpha_k x'_k + \sum_k \beta_k y'_k,$$
$$z_2 = \sum_k \alpha_k x''_k + \sum_k \beta_k y''_k.$$

By multiplying the two equations above, taking expectations and remembering that sperm and ovum characters are assumed to be independent,

$$\mathrm{cov}(z_1, z_2) = \sum_k \alpha_k^2 \mathrm{E}x'_k x''_k + \sum_{j,k} \alpha_j \alpha_k \mathrm{E}\left(x'_j x''_k + x''_j x'_k\right) + \sum_k \beta_k^2 \mathrm{E}y'_k y''_k$$
$$+ \sum_{j,k} \beta_j \beta_k \mathrm{E}\left(y'_j y''_k + y''_j y'_k\right)$$

Let R be the sib-sib correlation. Next, Pearson assumed that x'_k and x''_k are undifferentiated like organs from a single male and y'_k and y''_k are undifferentiated like organs from a single female, such that $\mathrm{corr}(x'_k, x''_k) = \rho_{kk}$ and $\mathrm{corr}(y'_k, y''_k) = \rho'_{kk}$. The equation above then becomes

$$\sigma^2 R = \sum_k \alpha_k^2 \sigma_k^2 \rho_{kk} + \sum_k \beta_k^2 \sigma'^2_k \rho'_{kk} + 2\sum_{j,k} \alpha_j \alpha_k \sigma_j \sigma_k \rho_{jk} + 2\sum_{j,k} \beta_j \beta_k \sigma'_j \sigma'_k \rho'_{jk}. \tag{8.4}$$

Pearson next made two further assumptions: (i) the homotypic correlations ρ_{kk} and ρ'_{kk} are stable across a range of characters (i.e., for different values of k) and may be replaced by their mean value ρ; (ii) since the correlations between two characters in the sperm and ovum, i.e., ρ_{jk} and ρ'_{jk}, vanish when either $\rho = 0$ or $r_{jk} = 0$ and are perfect when $\rho = r_{jk} = 1$, one may write $\rho_{jk} = \rho r_{jk}$ and $\rho'_{jk} = \rho r'_{jk}$. These two assumptions, written together, are

$$\begin{aligned} \rho_{kk} &= \rho'_{kk} = \rho \\ \rho_{jk} &= \rho r_{jk}, \ \rho'_{jk} = \rho r'_{jk} \end{aligned} \tag{8.5}$$

Substituting Eq. (8.5) into Eq. (8.4) and comparing with Eq. (8.3), Pearson finally obtained

$$R = \rho. \tag{8.6}$$

That is,

> **QUOT. 574** *The correlation between brothers will be equal to the mean correlation between the undifferentiated like organs put forth by an individual.*[29] (*ibid.*, p. 291)

Pearson's proof consists of many tenuous assumptions. The most important such assumption, one that invalidates both Eqs. (8.4) and (8.5) and his conclusion in QUOT. 574, is that germ cells from a given parent are undifferentiated like organs. As far back as 1866, Mendel had already correctly pointed out in his *Pisum* paper that:

> **QUOT. 575** It was proven experimentally that in *Pisum* hybrids form different kinds of germinal and pollen cells and that this is the reason for the variability of their offspring. (Mendel, 1866, 1966a, p. 41)

However, as mentioned at the start of Sect. 8.4.1, Pearson was not aware of Mendel's work when he wrote his homotyposis paper. Even if he had been aware of it, it is doubtful that he would have shared Mendel's beliefs. Thus, when Bateson sent him a copy of the *Pisum* paper on October 12, 1901, he expressed doubts about the generalizability of Mendel's results.

8.4.3 Pearson's Thesis: Heredity as a Special Case of Homotyposis

Pearson thought he had now discovered a new principle through the phenomenon of homotyposis. He had calculated 19 cases of fraternal correlation in the animal kingdom, yielding an average value of .4479, which was close to the homotypic

[29] Italics are Pearson's.

correlation in the vegetable kingdom. By relating the correlation between offspring arising by sexual reproduction to the correlation between homotypes, he claimed he had shown that heredity was a special case of the broader phenomenon of homotyposis:

> **QUOT. 576** It is, therefore, very probable that heredity is but a phase of homotyposis, and that the latter approximates to a certain value throughout living forms. (Pearson et al., 1901a, p. 4)

Pearson thus thought that he had *simplified* the process of inheritance by "throwing it back" to homotyposis. The question of heredity was not as important as the question of how resemblance between undifferentiated organs was modified: heredity had nothing to do with sexual reproduction but was a problem of *variability*:

> **QUOT. 577** ... if we can throw back the resemblance of offspring of the same parents upon the resemblance between the undifferentiated like organs of the individual, we shall have largely simplified the whole problem of inheritance. Inheritance will not be a peculiar feature of the reproductive cells. (*ibid.*, p. 4)
>
> ...
>
> if (v.) [Eq. (8.6)] represents an approximate or average truth of living nature, then the problem of inheritance is to a large extent the same as the problem of variability in the individual...[and] is part of a much more general problem having nothing to do with sexual reproduction. (*ibid.*, p. 7)

8.4.4 *Unfair Treatment of Pearson's Homotyposis Paper: Creation of* Biometrika *(October 1901)*

During the November 15, 1900, meeting, Bateson informed Pearson that he was one of the referees of his paper and that he had written an unfavorable report. Bateson had also been able to obtain the permission of the secretary of the Royal Society, Sir Michael Foster, to base his remarks during the meeting on Pearson's *full paper*, rather than the *abstract* which was read at the meeting and which was the version circulated to the audience![30] On December 13, 1900, a distressed Pearson wrote to Galton that:

---

[30]Bateson's letter of November 9, 1900, to Foster (retrieved online from Cambridge Digital Library, Bateson Archive, starting at https://cudl.lib.cam.ac.uk/view/MS-ADD-08634-C/169) reads as follows:

Dear Sir Michael,

I write to take part in the discussion of K. Pearson's paper on "Homotyposis" which is announced for reading at the R.S. on 15 Nov.

The programme is a long one and I doubt whether less than 10 minutes would be any use as the subject is complicated. Do you think you can manage to let me have so much time? Otherwise I shall not go up that day.

Secondly, as you may know, the main paper has been referred to me. If I shall speak on the subject it will be evident I have seen the MS. That I should refer for I cannot avoid references to vital points which are in the paper & not in the abstract. I myself have no objection to Pearson's knowing I have acted as a referee, but at least a breach of the etiquette of the societies would be involved. If you think this unreasonable I will stay away.

> **QUOT. 578** He [Bateson] came to the R.S. at the reading and said there was nothing in the paper – that it was a fundamental error to suppose that number had any real existence in living forms. That this criticism did not apply to this memoir only but to all my work, that variability was differentiation, etc., etc. (Pearson, 1930, p. 241, Vol. IIIA)

Pearson's letter seems to indicate that Bateson had generally denounced all work of a quantitative nature (and in particular Pearson's research). The unfair favor that had been bestowed to Bateson by Foster and the former's over-the-top statement in QUOT. 578 understandably upset Pearson, although Bateson's published response indicates that he might have questioned the existence of the average rather than that of the number.

Pearson's distress increased even more by the next turn of events. Foster invited Bateson to write down his response to Pearson's homotyposis paper in the form of a formal paper to be published in the *Proceedings*. Bateson obliged, and his response was received by the *Royal Society* on January 25 and read on February 14, 1901. However, this was before Pearson had been notified of this, and the paper that was being criticized was made available during the February reading! As Pearson later explained:

> **QUOT. 579** ...a detailed criticism of the [homotyposis] paper by one of the referees [Bateson] was actually printed by the Secretaries and issued to Fellows at a meeting, before the fate of the criticised paper itself had been notified, and before the paper itself was in the hands of those present. (Pearson, 1906, p. 35)

When Bateson sent proofs of his response to Pearson, the latter was understandably irked. Pearson complained to the Royal Society and returned Bateson's proofs "unexamined." The latter wrote an apologetic letter to Pearson on February 18, 1901 (Fig. 8.5), explaining that he had decided to withdraw his paper so that it would be published *after* Pearson's homotyposis paper. However, the harm had already been done: Weldon, too, was taken aback by Bateson's comment that numbers meant nothing and did not exist in Nature, and on the very next day of the November 15 meeting, he had already suggested to Pearson that perhaps they ought to start a new journal:

> **QUOT. 580** The contention 'that numbers mean nothing and do not exist in Nature' is a very serious thing, which will have to be fought. Most other people have got beyond it, but most biologists have not.
>
> Do you think it would be too hopelessly expensive to start a journal of some kind?...
>
> (*ibidem*)

Pearson's December 13 letter to Galton also indicated his belief that no biometric work would be published by the Royal Society as long as Bateson was a referee:

> **QUOT. 581** ...what is clear is that if the R.S. people send my papers to Bateson, one cannot hope to get them printed. It is a practical notice to quit. This notice applies not only to my work, but to most work on similar statistical lines. It seems needful that there should be some organ for publication of this sort of work. (Pearson, 1930, p. 241, Vol. IIIA)

The need for a separate outlet for the biometricians' scientific output was felt even more when Pearson complained to the Royal Society and, in return, received a request that in the future mathematics should be kept out of biological papers. As Pearson later recounted:

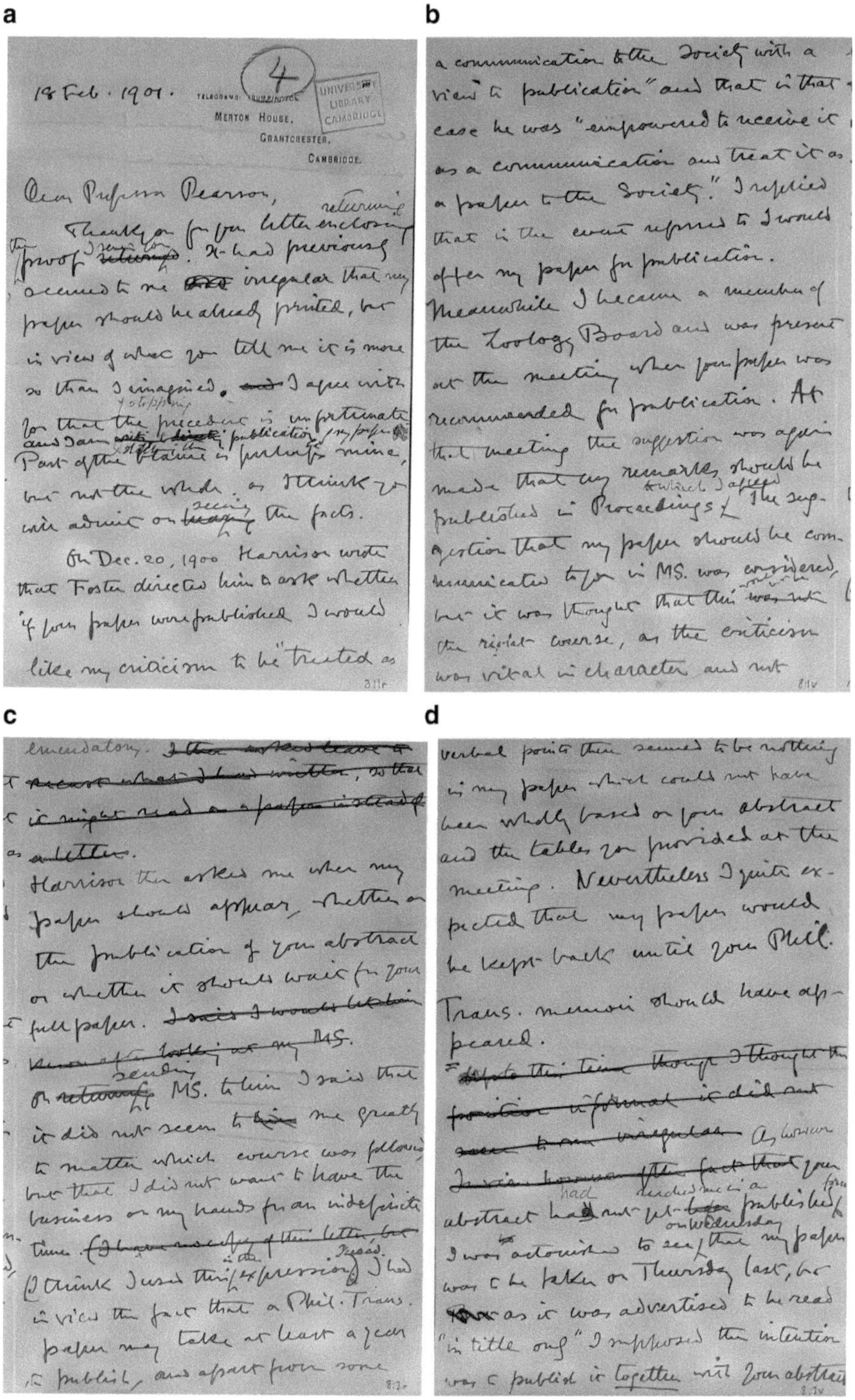

a

18 Feb. 1901.

MERTON HOUSE,
GRANTCHESTER,
CAMBRIDGE.

Dear Professor Pearson,

Thank you for your letter returning the proof I sent you. It had previously occurred to me irregular that my paper should be already printed, but in view of what you tell me it is more so than I imagined. I agree with you that the precedent is unfortunate and I am [illegible] publication of my paper. Part of the blame is perhaps mine, but not the whole, as I think you will admit on seeing the facts.

On Dec. 20, 1900 Harrison wrote that Foster directed him to ask whether if your paper were published I would like my criticism to be treated as

b

a communication to the Society with a view to publication" and that in that case he was "empowered to receive it as a communication and treat it as a paper to the Society." I replied that in the event supposed I would offer my paper for publication. Meanwhile I became a member of the Zoology Board and was present at the meeting when your paper was recommended for publication. At that meeting the suggestion was again made that my remarks should be published in Proceedings, to which I agreed. The suggestion that my paper should be communicated to you in MS. was considered, but it was thought that this would not be the right course, as the criticism was verbal in character and not

c

emendatory. ~~[illegible]~~

Harrison then asked me when my paper should appear, whether on the publication of your abstract or whether it should wait for your full paper. ~~[illegible]~~ On sending MS. to him I said that it did not seem to me greatly to matter which course was followed but that I did not want to have the business on my hands for an indefinite time. ~~[illegible]~~ I think I used this expression in view of the fact that a Phil. Trans. paper may take at least a year to publish, and apart from some

d

verbal points there seemed to be nothing in my paper which could not have been wholly based on your abstract and the tables you provided at the meeting. Nevertheless I quite expected that my paper would be kept back until your Phil. Trans. memoir should have appeared. ~~[illegible]~~ As however your abstract had not yet been published I was astonished to see that my paper was to be taken on Thursday last, but as it was advertised to be read "in title only" I supposed the intention was to publish it together with your abstract

Fig. 8.5 Bateson's Feb 18, 1901, letter (in four pages **a–d**) to Pearson. (From Cambridge Digital Library, Bateson Archive, starting at https://cudl.lib.cam.ac.uk/view/MS-ADD-08634-C/184)

QUOT. 582 Shortly afterwards a resolution of the Council was conveyed to me [Pearson], requesting that in future papers mathematics should be kept apart from biological applications. *βίος* was an admissible topic, *μέτον* also, but their combination was anathema, and that at a time when statistical theory had to be worked out step by step as the biological applications demanded. (*ibid.*, p. 100, footnote)

A draft of the first editorial (written by Weldon) of the new journal was circulated on November 29, 1900. By December 2, the name *Biometrika* had been coined for this new journal by Pearson (the *k* in *Biometrika* was after Pearson's first name, Karl). The subtitle "A Journal for the Statistical Study of Biological Problems" clearly indicated the scope of *Biometrika*.

Biometrika would thus be a journal created *by* the biometricians and (primarily) *for* the biometricians, at least in its early years.

Funding for several years was soon secured, and publication by Cambridge University Press had been arranged by June 1901. The first issue of *Biometrika* appeared in October 1901 (Fig. 8.6), with Galton as Consulting Editor, and Weldon, Pearson, and Davenport[31] as editors.

8.4.5 Bateson's Criticism of Pearson's Homotyposis Paper (April 1902)

As explained in the previous section, Bateson apologized to Pearson in a February 18, 1901, letter and indicated he would withdraw his response for the moment so that Pearson's paper could be published first. Bateson's response was received on January 25, 1901, read on February 14, 1901, and eventually published on April 4, 1902 (Bateson, 1902a), while Pearson's homotyposis paper came out earlier on September 1, 1901 (Pearson et al., 1901b).

In his response paper, Bateson started by congratulating himself for having first introduced the idea of merism, or repetition of parts, in his *Materials* (Bateson, 1894, p. 20). He thanked Pearson for pursuing his idea but then admitted that Pearson's mathematical reasoning was beyond him. This would generally be the case in Bateson's assessment of the biometricians' (in particular, Pearson's) work since he lacked mathematical training. Bateson's tone soon became critical. His main criticisms were as follows:

(a) *Pearson's failure to discriminate between regulated and unregulated variations.* Bateson argued that Pearson's paper provided no means to discriminate between variation occurring because of differentiation (i.e., regulated variation) and variation inherent in homotyposis (i.e., unregulated variation)

[31] Charles Davenport (1866–1944) was born in Stamford, Connecticut. From 1899 till 1901, he was Assistant Professor at the University of Chicago. He was Associate Professor from 1901 till 1904 and also curator of the Zoological Museum of the University of Chicago. Davenport was largely responsible for the introduction of biometrical methods in the United States. However, soon after the rediscovery of Mendelism, he became an adherent to the latter doctrine.

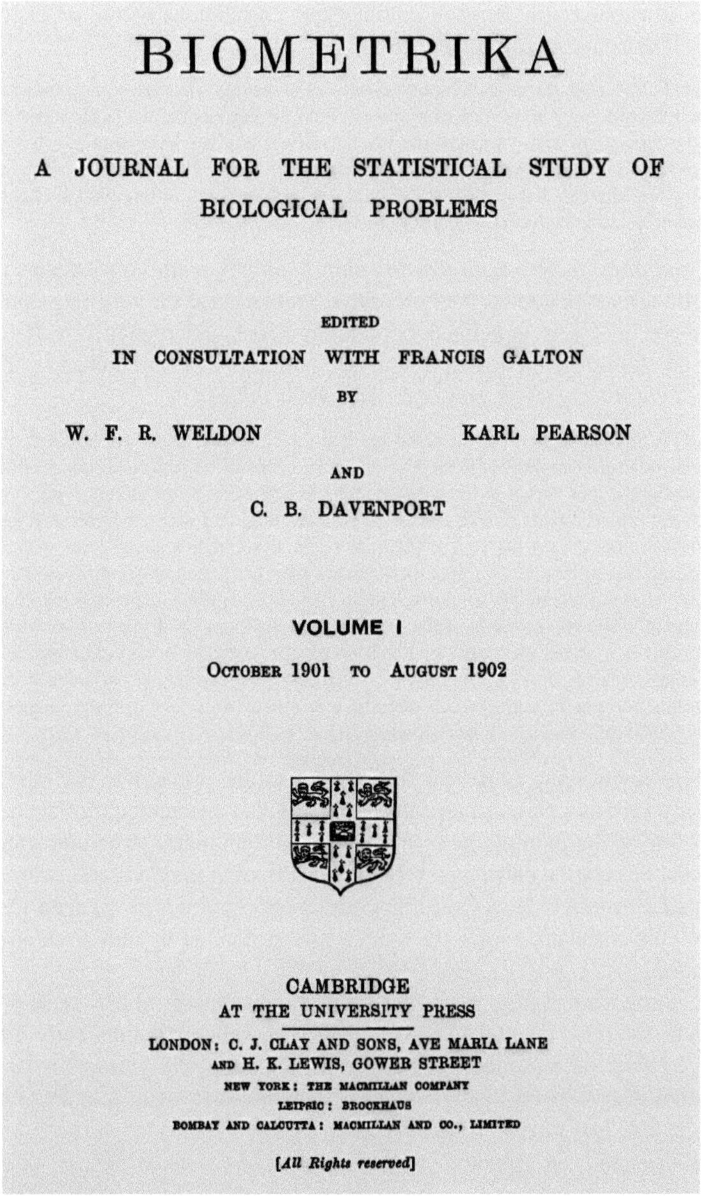

BIOMETRIKA

A JOURNAL FOR THE STATISTICAL STUDY OF BIOLOGICAL PROBLEMS

EDITED
IN CONSULTATION WITH FRANCIS GALTON
BY
W. F. R. WELDON KARL PEARSON
AND
C. B. DAVENPORT

VOLUME I

October 1901 to August 1902

CAMBRIDGE
AT THE UNIVERSITY PRESS

LONDON: C. J. CLAY AND SONS, AVE MARIA LANE
AND H. K. LEWIS, GOWER STREET
NEW YORK: THE MACMILLAN COMPANY
LEIPSIC: BROCKHAUS
BOMBAY AND CALCUTTA: MACMILLAN AND CO., LIMITED

[All Rights reserved]

Fig. 8.6 Title page of the first volume of *Biometrika*

QUOT. 583 In order, therefore, that the inquiry should have significance, it is necessary that differentiation occurring between members of a series of parts should have a clear meaning as distinct from variation occurring amongst them; and further, in order that the investigation should be carried through, we must be able to discriminate such differentiation from variation. (*ibid.*, p. 196)

He illustrated how Pearson's homotypic correlations would be distorted by the failure to make such a distinction:

> **QUOT. 584** Take the case of blood-corpuscles of a Frog. Measure some character of a hundred corpuscles in each of many Frogs. Find the correlation due to the individuality of the Frog. How can we determine whether in some of the individuals we have used there may not be differentiation such as was found in *Nigella*, so that the parts are not really "undifferentiated like parts"? Would not such irregular differentiation change the mean correlation between the corpuscles? (*ibid.*, pp. 197–198)

(b) *Sibs contained both regulated and unregulated variations.* Bateson attacked Pearson's formulation of the homotypic problem and the resulting conclusion. He stated that sibs contained both unregulated and regulated variations, so Pearson should have equated homotypic correlations of individuals with fraternal correlations *between non-differentiated brothers*:

> **QUOT. 585** Professor Pearson perceives that the correlation between "undifferentiated like parts" has an analogy with the resemblance or correlation between brothers. But does he recognise that variation between brothers is comparable not merely with variation between repeated parts, *but also with differentiation,*[32] and with predominantly orderly variation among such parts? The phenomena in a colonial or social form will clearly illustrate this principle. Ova and spermatozoa may be treated as "undifferentiated like parts" so long as their variations, judging by the resulting offspring, are sensibly irregular. Can we recognize differentiation among them as distinct from variation? Certainly we sometimes can. In determining the correlation of confraternities, the parentage enables us to distinguish the fraternal groups correctly, and consequently a fraternal correlation may be truly determined. *For to do so we are not compelled to distinguish differentiation from variation.* (*ibid.*, p. 200)

Bateson's reasoning, as indicated by the quotation above, was that sibs showed not only variation between repeated parts but also differentiation. The latter was the result of the parent's germ cells being differentiated. Therefore, the mean fraternal correlation calculated was affected by both *undifferentiated and differentiated variations*. However, this correlation could not be equated to a mean homotypic correlation since the latter was the affected by only *undifferentiated variations*.

It is noteworthy that, in his criticism above, Bateson did not use Mendel's explicit statement (see QUOT. 575) that germ cells were inherently differentiated. Instead, he reasoned from the traits of sibs to the germ cells. Bateson's criticism also showed that, if Bateson's mathematics was wanting, Pearson's biology also had room for improvement.

(c) *Regulated and unregulated variations may be impossible to distinguish.* Bateson's third criticism was perhaps his most formidable one. In his homotyposis paper, Pearson had spoken of regulated and unregulated variations on the assumption that they were two distinct types of variations. Whereas in the first criticism (a), Bateson indicated that Pearson had provided no means to

[32] All italics in this quotation are Bateson's.

differentiate the two, his criticism now was that *there might be no clear-cut* distinction between the two. Referring specifically to two siblings, he stated:

> **QUOT. 586** To find then a value comparable with the homotyposis of undifferentiated like parts we must find the fraternal correlation between undifferentiated like brothers. But differentiation has here again no meaning which can be determined with precision. It shades insensibly into variation. (*ibid.*, p. 201)

Based on his three criticisms, Bateson concluded that:

> **QUOT. 587** ...we could find homotypic coefficients of any required value. The average value of such coefficients taken at random has no significance in nature. (*ibid.* p. 197)

It is entirely possible that Bateson might have been referring to the average of arbitrarily obtained homotypic values (as shown in the quotation above) when Pearson and Weldon both complained about his rejection of the existence of numbers in real life.[33] Indeed, Pearson had calculated an average homotypic correlation of .4570,[34] which he then concluded was close to the average fraternal correlation of .4479.[35] However, to get the correlation of .4570, Pearson changed his initial range of correlations of .1733–.8607 by excluding some of the highest and lowest values, thereby obtaining a final range of correlations of .1733–.6311; hence, an average of .4570 was calculated. In today's language, Pearson's average probably had a large variance and lacked precision.

Although Bateson made several insightful remarks in his critique of Pearson's paper, some of his comments were uncalled for. For example, since Pearson's work on continuous evolution was largely based on the normal distribution, Bateson remarked that:

> **QUOT. 588** More fitly might this work be described as "Mathematical Contributions to a Theory of Normality." (*ibid.*, p. 203)

After reading his response on February 14, 1901, Bateson made private attempts at reconciliation with Pearson. He sent the latter a copy of Mendel's *Pisum* paper on October 12, 1901. In his response, Pearson cast doubt on the generalizability of Mendel's results. In the January 1902 issue of *Biometrika*, Weldon published a criticism of Mendelism. Bateson in vain wrote a flattering letter to Pearson on February 13, 1902, trying to win him over.

8.4.6 *Pearson's Response to Bateson (April 1902)*

Given his bad experience with the *Royal Society*, there was no way that Pearson would publish a detailed response to Bateson's criticism in that venue. Instead, he

[33] See QUOT. 578 and QUOT. 580.

[34] See paragraph below QUOT. 572.

[35] See first paragraph in Sect. 8.4.3.

instead sent a note to the *Proceedings* (received and read on February 20, 1902) stating that he fully intended to respond to Bateson but in a different journal. This was in effect Pearson's goodbye note to the *Royal Society* for, henceforth, *Biometrika* would be the outlet for his (and other biometricians') research.

Pearson's detailed response, entitled "On the fundamental conceptions of biology" (Pearson, 1902), was thus published in the April 1902 issue (No. 3) of *Biometrika*. As a result of the Royal Society embroilment, considerable resentment toward Bateson had been brewing inside of Pearson, and he let it all out in his response. The tone was set at the very start of the paper when Pearson attacked Bateson's competency to understand a biometric paper:

> **QUOT. 589** When one opponent has not even a preliminary training in biometry, and the other fails to attach any clear ideas to the terms used by his antagonist, used apparently as if they had universally accepted weight, it seems very hard to find a common ground for discussion. (*ibid.*, p. 322)

The clear implication was that if Bateson was not competent enough to understand a biometric paper, then his criticism of it could not have much value. In particular, Pearson criticized Bateson's understanding of the term 'variation'. For example, in *Materials* Bateson had written:

> **QUOT. 590** For though on the whole the offspring is like the parent or parents, its form is perhaps never identical with theirs, but generally differs from it perceptibly and sometimes materially. To this phenomenon, namely the occurrence of differences between the structure, the instincts or other elements which compose the mechanism of the offspring, and those which were proper to the parent, the name **Variation** has been given. (Bateson, 1894, p. 3)

Pearson correctly observed that, in biometry, variation had a clear-cut meaning: it measured the amount of deviation in a population from its mean, *without any reference to the parental population.* He also criticized Bateson's understanding of correlation and regression. The saltationist de Vries was also blamed for having a concept of regression that was "equally obscure."

Pearson next made sure there was no doubt as to where he stood in the biometry-saltationism debate, namely, opposite Bateson:

> **QUOT. 591** ...let me state once and for all that I consider Mr Bateson's peculiar theory of evolution by discontinuous variations untenable. It is, as he recognises, quite incompatible with much of my own work on evolution. (Pearson, 1902, p. 331)

Regarding Bateson's main criticism that a distinction needed to be made between regulated variation (differentiation) and unregulated variation (variation inherent in homotyposis), Pearson proposed a simple statistical test for means. That is, to test whether two segments were homotypes against the hypothesis they were differentiated:

> **QUOT. 592** Measure the characters of these two particular segments in a sufficiently large population and determine whether the differences of the means and of the standard deviations are or are not sensible in comparison with the probable errors of those differences. If they are not, then one set of segments may be looked upon as equivalent to a random sample of the other set and there is no class-differentiation. This method is so familiar to statisticians, who are using it every day to test whether a class is or is not differentiated from the

general population, that it appears somewhat surprising that Mr Bateson should believe we are in the habit of detecting differentiation solely by an inspection of modes in seriations. (*ibid.*, pp. 334–335)

However, notably absent in Pearson's rhetoric was any mention of Bateson's observation that the germ cells could be differentiated. In fact, in his response, Pearson repeated his early claim that "the production of gametes seems a process analogous to that of the production of any like organs by an individual" (*ibid.*, p. 338).

8.4.7 Bateson's Second and Final Response to Pearson (1903)

Pearson's response to Bateson, as described in the section above, was a *25-page* paper (Pearson, 1902) published in *Biometrika*, which had been newly created by the biometricians. When Bateson wrote down a *12-page* response to be published in *Biometrika*, he was told by the editors (which included both Pearson and Weldon, of course) that this was too long and had to be cut down to six pages instead. However, Bateson would not accede to such a request. Writing to Pearson on May 28, 1903, he stated:

QUOT. 593 I am obliged to the Editors of Biometrika for expression of their willingness to publish my article on Homotyposis if reduced to six pages, and I regret that I am unable to accept this offer.[36]

Having declined the offer by *Biometrika*, Bateson arranged for his response to be privately published by Cambridge University Press. The response was neither the original 12-page manuscript nor a 6-page trimmed version, but rather a 23-page rebuttal (Bateson, 1899) whose rhetoric matched that of Pearson's previous response.

In his response, Bateson defended his imprecise use of terms by pointing out that many of the concepts had not yet been fully developed. For instance, referring to the deliberate absence of a precise definition of 'discontinuity in variation' in *Materials*, Bateson explained:

QUOT. 594 Has not the reserve there shown found its fullest justification? Has not time brought exactly that "larger and more definite conception of Discontinuity" in Variation, which ten years ago it was impossible to form! Variation was then an empty name. Should I have pretended to define conceptions half-formed? (*ibid.*, p. 420)

[36]Letter retrieved online from Cambridge Digital Library, Bateson Archive at https://cudl.lib.cam.ac.uk/view/MS-ADD-08634-C/219

Regarding Pearson's proposal of a test of means to distinguish between regulated variation (differentiation) and unregulated variation (variation inherent in homotyposis), Bateson denounced such a test as "fallacious." This is because as he had already stated in his first response (Bateson, 1902a),[37] theoretically:

QUOT. 595 . . .differentiation is not in essence distinct from variation. (*ibid.*, p. 427)

A notable highlight of Bateson's response was the first mention in the public debate of Mendel and the "Mendelian revelation." However, as in his first response,[38] there was no mention of Mendel's statement that germ cells were differentiated.

[37] See also QUOT. 586.

[38] See Sect. 8.4.5, last paragraph of criticism (b).

Chapter 9
Biometry-Saltationism Controversies in the Immediate Post-Mendelian Period

Abstract With the Mendelian rediscovery, the biometry-saltationism controversy turned into a heredity debate involving Galton's ancestral law against Mendel's laws. When Bateson read Mendel's paper, he immediately embraced Mendelism since it was aligned with his saltationist viewpoint. The biometricians first attacked Mendelism in a 1902 paper by Weldon. This prompted Bateson to respond in a strongly worded book entitled Mendel's Principle of Heredity: A Defence, which also contained an English translation of Mendel's 1866 paper.

Keywords Gametic purity · Allelomorph · Homozygote · Heterozygote · Law of dominance · Lychnis experiments

9.1 Introduction

Before the Mendelian rediscovery, the biometry-saltationism debate centered on continuous versus discontinuous evolution. With the Mendelian rediscovery, the debate turned into one involving hereditary, in particular Mendelian versus ancestral heredity. Mendelian heredity was assumed to be compatible only with a discrete mode of evolution, while ancestral heredity was assumed to be consistent with a continuous mode.

9.2 Bateson Learns of Mendel's Laws (May 1900)

9.2.1 First (July 1899) International Conference on Hybridization

Before 1897, Bateson's work had largely been observational. However, the Cineraria controversy must have convinced him that experimentation would be important for further progress. After he joined the Evolution Committee in January 1897,

P. Gorroochurn, *The Development of Evolutionary Genetics*,
https://doi.org/10.1007/978-3-031-69374-8_9

Bateson started conducting hybridization experiments with Lychnis, poultry, and butterflies.

Bateson's views on how hybridization experiments needed to be conducted were close to those of Mendel 30 years ago. He recognized the need for a statistical treatment of the results of the crossings (meaning that a *large* number of crossings needed to be performed to discover any relationships) and the need to analyze them on a character-to-character basis (and not in toto). Thus, in his speech at the July 1899 International Conference on Hybridisation,[1] one can read:

> **QUOT. 596** At this time we need no more *general* ideas about evolution. We need *particular*[2] knowledge of the evolution of *particular* forms. What we first require is to know what happens when a variety is crossed with its *nearest allies*. If the result is to have a scientific value, it is almost absolutely necessary that the offspring of such crossing should then be examined *statistically*. It must be recorded how many of the offspring resembled each parent and how many shewed character intermediate between those of the parents. If the parents differ in several character, the offspring must be examined statistically, and marshalled, as it is called, in respect of each of those character separately . . . If it can only be noticed that the offspring came, say, halflike one parent and halflike the other, or that the whole shewed a mixture of parental character, a few brief notes of this kind may be a most useful guide to the student of evolution. . . .All that is really necessary is that some approximate numerical statement of the result should be kept. (Bateson, 1899, p. 63)

Considering the above, it is reasonable to speculate that even if Bateson had not later learned of Mendel and the Mendelian rediscovery, there is a chance that he could have discovered Mendel's laws on his own.

9.2.2 *Second (May 1900) International Conference on Hybridization*

On May 8, 1900, Bateson was en route to the second International Conference on Hybridisation (again organized by the Royal Horticultural Society) to read a paper entitled "Problems of heredity as a subject for horticultural investigation" (Bateson, 1901) and, as his wife Beatrice Bateson later recounted,

> **QUOT. 597** He had already prepared this paper, but in the train on his way to town to deliver it, he read Mendel's actual paper on peas for the first time. As a lecturer he was always cautious, suggesting rather than affirming his own convictions. So ready was he however for the simple Mendelian law that he at once incorporated it into his lecture. (Bateson, 1928, p. 73)

This version of events has been adopted by several scholars, who have repeated Beatrice Bateson's recollection that her husband read Mendel's paper on the train

[1] See Sect. 8.3.28.

[2] All italics in this quotation are Bateson's.

and incorporated it into his speech.[3] However, as will be explained below, William Bateson most likely read de Vries' *Comptes Rendus* article (de Vries, 1900c) on the train, not Mendel's original article. Furthermore, the offprint of the *Comptes Rendus* article had not been sent to Bateson by de Vries personally (whom the former first met shortly before the July 1899 International Conference on Hybridisation). In fact, de Vries had sent a copy to C.C. Hurst, from whom Bateson obtained it a few days before May 8.

That Bateson read de Vries' *Comptes Rendus* article (where no mention of Mendel was made) on his way to the conference would explain why the editorial of the May 12, 1900, issue of *The Gardeners' Chronicle* (Anonymous, 1900, p. 303) mentioned de Vries, but not Mendel, in its account of Bateson's speech. On the other hand, the printed version of the speech (Bateson, 1901, 1928, pp. 171–180), which appeared in 1901, contained references to de Vries, Correns, and Tschermak and a substantive description of Mendel's work. Thus, Bateson likely had not heard of Mendel during his trip to the May 8, 1900, conference: he had read only de Vries' *Comptes Rendus* paper. On his return, he also read the latter's *Berichte* paper (de Vries, 1900a), most likely after May 19, the date on which de Vries sent that paper to Hurst. Since the *Berichte* paper contained references to Mendel, Bateson must have then procured and read Mendel's paper and made substantial changes to his conference paper.

Beatrice Bateson has vividly described Bateson's excitement on reading Mendel's original paper:

> **QUOT. 598** His delight and pleasure on his first introduction to Mendel's work were greater than I can describe; as when with a very long line to hoe, one suddenly finds a great part of it already done by someone else and one is unexpectedly free to get on with other jobs. He was fortified with renewed faith in the largeness of his research; he found in it new interest, new possibilities, and drew from it new inspiration. (Bateson, 1928, p. 73)

Given his recent Cineraria fight[4] with Thiselton-Dyer and Weldon over the importance of hybridization, it is no wonder that Bateson probably felt vindicated through Mendel's hybridization paper. Moreover, Mendel's particulate theory was perfectly aligned with his saltationist views. From then on, Bateson became Mendel's most fervent defender and "bulldog" (much as Huxley was Darwin's).

Bateson's conference speech, as reported by *The Gardeners' Chronicle* (Anonymous, 1900, p. 303), acknowledged the importance of Galton's Law of Ancestral Heredity. The law seemed to hold for both blending and alternative inheritance. In the latter case, the law predicted that a particular *fraction* of all offspring inherited a trait depending on what fractions of the different ancestries possessed that trait.[5] Bateson noted, however, that it was also common for *all* offspring to inherit an ancestral trait. It was such cases of complete dominance (or prepotency), where

[3] See, for example, Gee (2004, p. 112), Griffiths et al. (2015, p. 5), Hamblin (2005, p. 21), Dronamraju (2018, p. 34), Hull (2010, p. 52), Haugen (2014, p. 18), and Birkhead (2014, p. 116).

[4] See Sect. 8.3.

[5] See Sect. 4.6.4.2.

Galton's ancestral law did not hold, that Bateson at first thought de Vries' paper addressed:

QUOT. 599 There were. . .the numerous classes of crosses in which the character of one parent wholly predominates the first generation of crossing. To these the simple [Galton] law could not be applied as originally stated. The recent work of Professor de Vries (*Comptes Rendus*, March, 1900), however; shows that in certain such cases subsequent breeding from the cross-bred plants leads to the reproduction of the parent species in such proportions that the facts can be expressed by a modification of Galton's law. (*ibidem*)

Thus, this is how Bateson at first viewed Mendelian heredity, namely, as a mere extension of Galton's ancestral law. A similar sentiment was conveyed in the printed version of Bateson's conference speech (Bateson, 1901, 1928, pp. 171–180), where Mendel was first mentioned by him. In describing Mendel's work, Bateson also declared:

QUOT. 600 That we are in the presence of a new principle of the highest importance is, I think, manifest. (*ibid.*, p. 179)

Except for Bateson and his allies, few at the time realized that this was indeed going to be the case.

9.2.3 First English Translation of Mendel's 1866 Paper (1901)

Shortly after the May 1900 International Conference on Hybridization, Rev. William Wilks (1843–1923), who was then secretary of the Royal Horticultural Society, requested that the botanist Charles Thomas Druery (1843–1917) translate Mendel's 1866 paper. This was the first English translation of Mendel's work and was done with the participation of Bateson, who provided an introductory note and footnotes. The translation appeared in a 1901 issue of the Journal of the Royal Horticultural Society (Mendel, 1901). That Bateson viewed Mendel's law to be fundamental is evident from the following paragraph in the introductory note:

QUOT. 601 In so far as Mendel's law applies, therefore, the conclusion is forced upon that a living organism is a complex of characters, of which some, at least, are dissociable and are capable of being replaced by others. We thus reach the conception of unit-characters, which may be rearranged in the formation of the reproductive cells. It is hardly too much to say that the experiments which led to this advance in knowledge are worthy to rank with those that laid the foundation of the Atomic laws of Chemistry. (*ibid.*, p. 1)

A copy of the English translation was sent by Bateson to Pearson on October 12, 1901. The latter expressed reservations as to the wider applicability of Mendel's results. As for Weldon, he had already obtained a copy of the translation by October 16, 1900, the date on which he sent a letter to Pearson alerting him of the paper.[6]

[6]See footnote 27 of chap. 8.

9.2.4 *First Report to the Evolution Committee of the Royal Society (1901)*

A first indication of the wider generality (i.e., beyond the vegetable kingdom) of Mendel's laws was provided in the first Report to the Evolution Committee of the Royal Society (cowritten by Bateson and E. R. Saunders[7]). This was presented on December 17, 1901 (Bateson & Saunders, 1901), and showed that Mendelian segregation also held in fowl. In addition, in an article written on March 12, 1902, the French biologist Lucien Cuénot (1866–1951) confirmed Mendel's results in mice (Cuénot, 1902a).

Bateson and Saunders also stated that previous breeding experiments could now be interpreted in Mendelian terms. For example, in the second volume of *The Variation of Animals and Plants Under Domestication* (Darwin, 1868b, p. 46, Vol. II), Darwin had crossed peloric *Antirrhinum* with the normal form and vice versa. The F_1 hybrids were all normal and, when self-fertilized, yielded 88 normals, 2 intermediates, and 37 peloric in F_2. Darwin had correctly reasoned that the normal trait was prepotent (dominant). However, regarding F_2, the rule was that "a character which is present in one form and latent in the other is generally transmitted with the prepotent force when the two forms are crossed" (*ibidem*). The reappearance of the peloric type in F_2 meant that Darwin had to resort to a complicated reasoning involving pangenesis.[8] Mendelism removed all such complications:

QUOT. 602 Now, of course, we can perceive that the zygomorphic form is dominant and the peloric recessive, and that the arguments based on other hypotheses [pangenesis] have no longer any significance. (Bateson and Saunders, 1901, p. 125)

Another important section of the first Report had to do with Galton's ancestral law. Whereas Bateson had hitherto considered Mendelian heredity as a simple extension of the ancestral law, by the time the first Report came out, he and Saunders had realized two fundamental differences between the two theories:

[7]Edith Rebecca (Becky) Saunders (1865–1945) obtained first-class honors in the Natural Sciences Tripos in 1888. The following year, she was appointed demonstrator in botany at Newnham College. She took over the direction of the Balfour laboratory in 1899. Her collaboration with Bateson was fruitful, and the latter made sure that her important contributions were fully acknowledged in papers. Similarly, Karl Pearson made sure Alice Lee was listed as co-author whenever she contributed to any joint work. However, women have often not been given proper credit in the past. In a revealing article, Dung et al. (2019) have explained how women often made important contributions to papers now considered as classics in population genetics, but their contributions were relegated to mere footnotes in those papers. A notorious example is Watterson's seminal paper "On the number of segregating sites in genetical models without recombination" (Watterson, 1975). In that paper, the population geneticist Margaret Wu made a key contribution to the development of the so-called Watterson's estimator, but her contribution was relegated to a mere "I thank Mrs. M. Wu for help with the numerical work, and in particular for computing table I." Such a practice would, of course, not be tolerated nowadays and would have her listed as a major co-author in Watterson's paper.

[8]See Sect. 3.7.2(c).

1. Mendelian heredity was based on the *purity* of gametes whereas Galton's ancestral law allowed for blending.
2. In Mendelian heredity, once the characters[9] of the parents were specified, the distribution of characters among the offspring was completely determined, whereas in Galton's ancestral law, the characters of all ancestors beyond the parents still mattered.

Bateson and Saunders summarized these two fundamental differences as follows:

> **QUOT. 603** We note at once that the Mendelian conception of heredity effected by *pure* gametes representing definite allelomorphs[10] is quite irreconcilable with Galton's conception in which *every* ancestor is brought to account in reckoning the probable constitution of every descendant. (*ibid.*, p. 157)

A third fundamental difference, not mentioned by Bateson and Saunders, is that:

3. Mendel's theory was *physiological*, whereas Galton's ancestral law was essentially *statistical* (even though Galton wished his law to also have a physiological meaning)

Bateson and Saunders' first Report of December 1901 (especially Part III of the Report, entitled "The Facts of Heredity in the Light of Mendel's Discovery," pp. 125–157) also broke new ground on several other fronts:

(a) *New terminology and notation: allelomorphs (D, R), homozygotes (DD, RR)* and *heterozygotes (DR)*. Having introduced the concept of gametic purity, Bateson and Saunders next put forward the logical consequence of this concept, namely, the existence of *allelomorphs, homozygotes,* and *heterozygotes*, defined as follows:

> **QUOT. 604** We thus reach the conception of unit-characters existing in antagonistic pairs. Such characters we propose to call *allelomorphs*, and the zygote formed by the union of a pair of opposite allelomorphic gametes, we shall call a *heterozygote*. Similarly, the zygote formed by the union of gametes having similar allelomorphs, may be spoken of as a *homozygote*.[11] (*ibid.*, p. 126)

Before enunciating these definitions, Bateson and Saunders had already introduced notations for the three entities:

> **QUOT. 605** The whole result, therefore, of self-fertilising the cross-breds is really 1DD:2DR:1RR, though apparently the proportion is 3D:1R because DR plants are not to the eye distinguishable from pure D. (*ibid.*, p. 8)

---

[9]In the modern interpretation of Mendelian heredity, once the *genotypes* of the parents are specified, the distribution of genotypes of the offspring is completely determined. If $\mathbf{X}_o, \mathbf{X}_p, \mathbf{X}_g, \mathbf{X}'_g, \ldots$ denote the genotypes of the offspring, parents, grandparents, great-grandparents, and so on, then Mendelian heredity implies $\Pr\left\{\mathbf{X}_o|\mathbf{X}_p, \mathbf{X}_g, \mathbf{X}'_g, \ldots\right\} = \Pr\left\{\mathbf{X}_o|\mathbf{X}_p\right\}$. This is also known as the *Markovian* property.

[10]See (a) below (*our note*).

[11]Both the terms "homozygote" and "heterozygote" have stayed. "Allelomorph" was later shortened to "allele," first by Jenkins in 1931.

Bateson and Saunders' *DD, DR,* and *RR* were marked improvements over both Mendel's and de Vries' notations. Recall that (see Sect. 6.2.6) Mendel had used *D* to denote the dominating constant trait, *DR* to denote the dominating hybrid trait, and *R* to denote the recessive constant trait. De Vries improved upon Mendel's notation by representing the fertilization of segregated factors through the following algebraic expression:

$$(D + R)(D + R) = D^2 + 2DR + R^2$$

(see QUOT. 465). With Bateson and Saunders' notation, the most logical development of these incomplete prior notations was achieved.

(b) *Distinction between heterozygotes and the traits they displayed.* Bateson and Saunders noted the important distinction between the heterozygotes and the characters they displayed:

QUOT. 606 We readily perceive that the heterozygotes may show either of the parental characters discontinuously, or various blends between them, while the gametes which composed the heterozygotes may still be pure in respect of the parental characters. The degree of blending in the heterozygote has nothing to do with the purity of the gametes. (*ibid.*, p. 152)

This would later be generalized to the *genotype-phenotype* distinction by Johannsen (1909, pp. 123 & 130).

(c) *Explanation of continuous variation through Mendelian heredity.* Having distinguished between zygotes and traits, Bateson and Saunders explained continuous traits through the possible action of more than one pair of allelomorphs:

QUOT. 607 It must be recognised that in, for example, the stature of a civilised race of man, a typically continuous character, there must certainly be on any hypothesis more than one pair of possible allelomorphs. There may be many such pairs, but we have no certainty that the number of such pairs and consequently of the different kinds of gametes are altogether unlimited even in regard to stature. If there were even so few as say, four or five pairs of possible allelomorphs, combinations might, on seriation, give so near an approach to a continuous curve, that the purity of the elements would be unsuspected, and their detection practically impossible. (*ibid.*, pp. 152–153)

If the ideas contained in the quotation above had been given more attention, the biometry-saltationism controversy could have been resolved sooner.

(d) *New notation: parental generation (P), first filial generation* (F_1), etc. Bateson and Saunders pointed to the confusion created by the different meanings of the *first generation* as used by various investigators.[12] Therefore, they proposed a uniform notation that they stated was inspired from Galton's *Hereditary Genius*:

[12] For example, Mendel used the hybrid generation as the reference generation and started counting offspring generations from that reference (see footnote 18 of chap. 6). On the other hand, Correns denoted the hybrid generation itself as the first generation and, as indicated in Sect. 6.6.3.2 (b), may have been confused at one point by Mendel's results.

QUOT. 608 We suggest as a convenient designation for the parental generation the letter P. In crossing, the P generation are the pure forms. The offspring of the first cross are the first filial generation F. Subsequent filial generations may be denoted by F_2, F_3, &c. (*ibid.*, p. 160)

Bateson and Saunders' recommendations have prevailed: the parental generation is now denoted by P, the hybrids are in the first filial generation (F or F_1), and subsequent generations are denoted by F_2, F_3, etc.

(e) *The phenomenon of linkage*. Like Correns before,[13] Bateson and Saunders found several instances of correlated traits, for example, between seed color and leaf character. They also referred to Correns' use of "*halb-identisch*" and "*conjugirte*" to describe perfect and imperfect correlations, respectively.

9.3 Round 5: The Biometricians' First Attack on Mendelism

9.3.1 Weldon's "Mendel's Laws of Alternative Inheritance in Peas" (1902)

Weldon must have read the Royal Horticultural Society translation of Mendel's paper just before October 16, 1900. On that very day, he sent a letter to Pearson alerting him to the paper.[14] Since it was providing a powerful boost to Bateson's saltationist views, Weldon decided to "nip Mendelism in the bud." He submitted a paper entitled "Mendel's laws of alternative inheritance in peas" (Weldon, 1902a) to *Biometrika*, the newly created journal that had him as one of the editors. The paper was received on December 9, 1901, and was published in the January 1902 issue (first volume, second issue) of *Biometrika*.

In the first part of the paper, Weldon provided an account of Mendel's experiments and the main conclusions that were drawn from them. The description was competent except that Weldon interpreted the appearance of the dominant trait in the F_1 hybrids as a necessary condition for Mendel's theory to be true. He called this the *Law of Dominance*. His insistence on dominance was probably inspired by similar utterances made by de Vries in the rediscovery papers,[15] although Correns had already warned that dominance was not necessary.[16]

It is also in Weldon's 1902 paper that the first observation was made that Mendel's observed values were perhaps too close to the theoretically expected ones.[17]

---

13 See footnote 97 of chap. 6.

14 See footnote 27 of chap 8.

15 See QUOT. 466.

16 See QUOT. 475.

17 See Sect. 6.5.1.

Having explained Mendel's laws, in the remaining two-thirds of the paper Weldon mounted three arguments against the validity of Mendel's conclusions:

1. *Failure of the Law of Dominance*. Weldon's first argument was that traits usually failed to show the type of dominance among the F_1 hybrids that Mendel had observed. For example, he cited studies performed by Tschermak whereby:

   QUOT. 609 [the] variety *Telephone* [which is quite wrinkled] especially, when crossed with smooth-seeded varieties, gave a large number of wrinkled seeds, or seeds of intermediate character. (*ibid.* p. 239)

   As another example, Weldon cited experiments where the second generation of the cross *Victoria* ♀ [smooth] x *Telephone* ♂ [wrinkled] gave both smooth and wrinkled seeds as per Mendelian expectations. However, the F_2 wrinkled seeds did not breed true and gave rise to both smooth and wrinkled seeds.
2. *Difficulty of classification due to continuous variability in traits*. Weldon's second argument was somewhat related to the first. It was that traits usually showed continuous variation, making their classification as dominant versus recessive arbitrary:

   QUOT. 610 The statement that smoothness of seed is dominant over the production of wrinkles means therefore that if a parent, belonging to a variable race which falls into the category "smooth-seeded," be crossed with a parent belonging to a variable race of the category "with wrinkled seeds," the offspring will themselves be variable, but will always belong to the category "smooth-seeded": and as it with shape, so it is with colour. (*ibid.*, p. 236)

3. *Ignorance of ancestry*. Weldon's third argument was that crosses often gave results that were inconsistent with Mendel's theory (as in the examples in (1) above) because the latter ignored any ancestry beyond the parents. Weldon declared that the ignorance of ancestry was a damning indictment of Mendel's whole work:

   QUOT. 611 The fundamental mistake which vitiates all work based upon Mendel's method is the neglect of ancestry, and the attempt to regard the whole effect upon offspring, produced by a particular parent, as due to the existence in the parent of particular structural characters; while the contradictory results obtained by those who have observed the offspring of parents apparently identical in certain characters show clearly enough that not only the parents themselves, but their race, that is their ancestry, must be taken into account before the result of pairing them can be predicted. (*ibid.*, p. 252)

Weldon's first argument had already been addressed by Correns, who had warned that dominance was not necessary in Mendel's theory.[18] His second and third arguments have elements of circularity: two theories are compared, one that assumes continuous evolution and incorporates ancestry, and the other that does not; the second one is then declared deficient because it does not make the assumptions of the first.

[18] See QUOT. 475.

9.3.2 Bateson's Defence *(1902)*

Bateson was taken aback by Weldon's 1902 *Biometrika* article (Weldon, 1902a), where the latter outright dismissed the Mendelian discovery. Before writing a detailed response to Weldon's criticism, Bateson approached the latter's ally, Pearson, in an attempt to win him over. In a letter written on February 13, 1902, he said:

> **QUOT. 612** I respect you as an honest man and perhaps the ablest and hardest worker I have met, and I am determined not to take up a quarrel with you if I can help it.
>
> I have thought for a long time that you are probably the only Englishman I know at the moment who first thought is to get that the truth in their problems...
>
> I write to you as I am doing because I know that we are standing at the beginning of a new epoch in our line of work. There has probably been no discovery made in theoretical biology that we can remember which approaches Mendel's in magnitude, and the consequences it leads to. This is not a matter of opinion but certain. You have worked well in the same field and if through any fault of mine you were to be permanently alienated from the work that is coming, I should always regret it.
>
> With Weldon it is different. He is a naturalist. He goes in with his eyes open. Besides, as between him and me it is too late. It was a bitter grief to me when he first made it clear to me that all partnership between us was at an end. At different times, as perhaps you know, we have each tried to renew our intercourse if not friendship, but it came to nothing and it is no use trying again. There are faults of temperament on both sides.
>
> In this matter he is now committed. How far he has mistaken not only Mendel's work but the gravity of the issue cannot be long unknown.[19]

Bateson's letter was written right amid his polemic on homotyposis with Pearson,[20] and the latter rejected the overture in a letter written 2 days later:

> **QUOT. 613** I think sometimes you cannot be aware that Weldon has been for many years past one of my closest and most valued friends; that I do not readily make friends, and that when I say a man is my friend I mean that I have tested the strength of his affection in the graver matters of life, and am prepared to do for him and to accept from him anything that one human being can or will do for another. I think, as I say, that you have not known this, or possibly your references to him, – only three or four, but my memory is very jealous in such matters – would have been more guarded. As to the scientific side of the present controversies, I am perfectly ready to hear both sides, and will willingly reserve space in Part III of Biometrika for your defence of Mendel, if you think our Journal a suitable locus for your paper. (Provine, 2001, pp. 63–64)

Bateson's rebuttal of Weldon's 1902 article was promptly written and published in the same year. However, Bateson did not respond in kind through an article; rather, he wrote a 200-plus page book entitled *Mendel's Principles of Heredity: A Defence* (Bateson, 1902b)! The latter was written in a style which, in the biography of her husband, Beatrice Bateson conceded was "fierce, and in some quarters this fierceness has been deplored and, probably, thought unreasonable" (Bateson, 1928, p. 74).

---

[19]Letter retrieved online from Cambridge Digital Library, Bateson Archive, starting at https://cudl.lib.cam.ac.uk/view/MS-ADD-08634-C/196

[20]See Sect. 8.4.

Right in the preface of *Defence*, Bateson expressed his utmost disapproval of Weldon's 1902 critique of Mendel's work: the latter's views needed to be sanctioned not simply because they were opposite to his but more so because they could sway the minds of the younger generation. In Bateson's own words:

> **QUOT. 614** In the world of knowledge we are accustomed to look for some strenuous effort to understand a new truth even in those who are indisposed to believe. It was therefore with a regret approaching to indignation that I read Professor Weldon's criticism [Biometrika, I., 1902, Pt. II.]. Were such a piece from the hand of a junior it might safely be neglected; but coming from Professor Weldon there was the danger—almost the certainty—that the small band of younger men who are thinking of research in this field would take it they had learnt the gist of Mendel, would imagine his teaching exposed by Professor Weldon, and look elsewhere for lines of work. (*ibid.*, p. vi)

Later, he compared Weldon's attitude toward Mendel's theory to the "writings of the orthodox which followed the appearance of the 'Origin of Species'" (*ibid.*, p. 106).

The first section of *Defence* dealt with the Mendelian concepts that had already been introduced in the first Report to the Evolution Committee of the Royal Society,[21] namely, gametic purity, allelomorphs, homozygotes, heterozygotes, and so on. The next section contained English translations of Mendel's *Pisum* paper[22] (Mendel, 1866) and *Hieracium* report (Mendel, 1869). The remaining section, which comprised slightly more than half of *Defence*, was devoted to a rebuttal of Weldon (1902a), point by point.

To Weldon's charge that "Mendel's Law of Dominance" was generally invalid,[23] Bateson replied that Mendel enunciated no such law and that, although dominance was important in Mendel's work, it was not necessary for his theory to be valid:

> **QUOT. 615** The whole question whether one or other character of the antagonistic pair is dominant though of great importance is logically a subordinate one. It depends on the specific nature of the varieties and individuals used, sometimes probably on the influence of external conditions and on other factors we cannot now discuss. There is as yet no universal law here perceived or declared. (Bateson, 1902b, pp. 117–118)

In fact, Bateson next accused Weldon's review of giving prominence to dominance at the expense of gametic purity. The latter, unlike dominance, was an essential component of Mendel's theory.

Regarding Weldon's argument that the existence of continuous variation made the classification of traits as either dominant or recessive arbitrary,[24] Bateson replied that *some* variability in hybrids was inevitable because of forces other than the crossing itself. These forces may be independent of the crossing, such as environmental conditions, or they may be dependent on it, such as sports:

---

[21] See Sect. 9.2.4. The Report had not yet appeared at the time of Bateson's writing of *Defence*.

[22] This was a slightly modified version of the Druery translation for the Royal Horticultural Society (see Sect. 9.2.3).

[23] See Sect. 9.3.1(1).

[24] See Sect. 9.3.1(2).

QUOT. 616 When Mendel asserts the purity of the germ-cells of cross-breds he cannot be understood to mean that they are more pure than those of the original parental races. These must have varied in the past. The wrinkled seed arose from the round, the green from the yellow (or vice versa, if preferred), and probably numerous intermediate forms from both. (*ibid.*, p. 127)

Bateson argued that the existence of these forces did not mean that Mendel's theory was invalid: after all, the theory dealt with the *transmission* of hereditary characters, not the *variability* between them.

Finally, regarding Weldon's criticism that hybridization results were often inconsistent with Mendelian expectations because of the neglect of ancestry,[25] Bateson attributed the so-called inconsistencies to Weldon's misinterpretation of the results. For example, Weldon had reported von Fischer's results that piebald rats crossed with albino varieties gave piebald young if only the father was piebald and white young if only the mother was piebald; on the other hand, Crampe had found that in either case, the offspring was a mixture of piebald forms and albinos. Weldon had then concluded that the traits of offspring depended not only on those of the parents but also on those of the parents' ancestors. Bateson regarded this as a misinterpretation because Weldon had not ascertained the gametes of the parents in the two cases:

QUOT. 617 Most if not all of such "conflicting" evidence can be reconciled by the steady application of the Mendelian principle that the progeny will be constant when—and only when*—similar gametes meet in fertilisation, apart from any question of the characters of the parent which produces those gametes. (*ibid.*, p. 177)

As mentioned before, Bateson's *Defence* was written in fiery language. In a tribute penned in the wake of Weldon's untimely death in 1906, Pearson revealed that:

QUOT. 618 ...the tone of Mr Bateson's defence deeply pained Weldon, and rendered it difficult for a finely strung temperament to maintain – as it did maintain to the end – the impersonal tone of scientific controversy. (Pearson, 1906, p. 42)

This is not surprising, since most of Bateson's fierceness and personal attacks in *Defence* were targeted at Weldon.

9.3.3 *Weldon's "On the Ambiguity of Mendel's Categories" (1902): The Lychnis Experiments*

9.3.3.1 Weldon's Statistical Criticism of Mendelian Categories

Given the highly contentious nature of Bateson's *Defence*, Weldon should have been commended for toning down the rhetoric in his reply, entitled "On the ambiguity of Mendel's categories" (Weldon, 1902b). However, the response contained little new

[25] See Sect. 9.3.1 (3).

information except for a statistical argument for why Bateson's classification of traits into distinct categories was unsatisfactory.

Weldon's basic statistical argument was that there was no legitimate way of classifying an individual into a particular category unless there was a measure of the amount of variation allowed in that category. This is the same as saying that it is not statistically possible to determine if a particular observation came from a population with a given mean unless there is some measure of variability of the population (or at least an estimate of the variability). Similarly, according to Weldon, the resemblance between two individuals could not be gauged without knowledge of the variability of the underlying population. In Weldon's own words:

QUOT. 619 The confusion between resemblance to a race and resemblance to an individual involved in Mr Bateson's treatment of Mendel's work is one of the many unfortunate results which follow when Mendel's system of dividing a set of variable characters into two categories, and of using these categories as statistical units, is carried too far. Unless the range of characters actually included in each category be constantly borne in mind, the degree of resemblance between two individuals, implied by placing them in the same category, cannot be estimated; and when, as constantly happens, the range of variation in one of the alternative categories differs widely from that included in the other, the Mendelian system becomes absolutely misleading without some explanation (nearly always withheld) of the real limits implied by the terms used. (*ibid.*, p. 46)

Weldon's point is quite apposite statistically but is largely unnecessary in Mendel's theory.

9.3.3.2 The Lychnis Experiments

To further illustrate his point, Weldon performed the experiments on hairy and glabrous (hairless) varieties of the flower *Lychnis* that had been reported first by de Vries (1897, p. 25) and later by Bateson and Saunders (1901). As mentioned at the end of Sect. 6.6.2.1, de Vries had performed *Lychnis diurna* (hairy) x *L. vespertina glabra* (glabrous) crossings as early as 1892. From these crossings, the F_1 hybrids were hairy, which Bateson and Saunders interpreted as a case of Mendelian inheritance (*ibid.*, p.15). However, Weldon accused Bateson and Saunders of failing to mention one of de Vries' experiments, namely, *Lychnis vespertina* x *Silene noctiflora*, where the hairy offspring had hairs belonging to *Silene noctiflora*, rather than *Lychnis vespertina*. Weldon interpreted this result as a clear case violating the Mendelian concept of gametic purity:

QUOT. 620 Surely we have here a clear proof that the "dominant" character, hairiness, may on the application of a suitable stimulus be manifested by the fertilised germ-cells of what is said to be a purely recessive plant; so that the theory of pure parental gametes, on which Mr Bateson lays such stress, is shown to be inadequate for this case, and a theory of inheritance, with reversion to particular ancestors, is indicated as likely to express the facts of "Mendelian" cases also. (Weldon, 1902b, p. 48)

Next, Weldon questioned de Vries' reported numbers. According to the description at the end of Sect. 6.6.2.1, de Vries gave F_2 results for crossing hybrids from

Lychnis diurna x *L. vespertina glabra* as a 2:1 ratio (for hairy to hairless) (de Vries, 1897, 1920, p. 25, Vol. VI) but then as a 3:1 ratio at the International Conference on Hybridization in 1899 (de Vries, 1899a, p. 74). In the *Beritche* rediscovery paper, de Vries (1900a, 1966, p. 113) gave the proportion of recessives (hairless) as 28%. Considering the various experiments de Vries had performed, it is very likely that de Vries' initial 2:1 ratio was the result of an innocent counting error. However, Weldon pounced on the mistake and sarcastically declared:

QUOT. 621 The odds against a deviation so largely exceeding Mendel's result with 500 individuals would be about 17 to 1. (Weldon, 1902b, p. 48)

Chapter 10
Round 6: Attempted Reconciliation Between Mendel's and Ancestral Laws: The Birth of the (Hardy-Weinberg) Equilibrium Law (Yule 1902, Castle 1903, Pearson 1904, Hardy 1908, Weinberg 1908)

Abstract Amid the bitter biometry-saltationism debate, Yule wrote a 1902 paper, which attempted to reconcile Galton's ancestral law and Mendel's laws. In his paper, Yule also derived, for equal gamete frequencies, the so-called Hardy-Weinberg (HW) law, although he showed some misunderstanding of Mendel's law of segregation. In a 1903 paper criticizing Yule's mistake, Castle derived the HW in its full generality. Like Yule earlier, Pearson also attempted a reconciliation in 1904 and derived the HW law for equal gamete frequencies. In response to a query from Punnett, the pure mathematician Hardy explicitly derived the HW law in 1908. Stern later indicated that the German physician Weinberg had independently derived a similar law a few months before Hardy's derivation. The HW law was subsequently extended to two loci by Weinberg, Jennings, Robbins, and Lewontin and Kojima.

Keywords Random mating · Multifactor model · Protozygote · Allozygote · Hardy-Weinberg law · Random union of gametes · Two-locus equilibrium · Linkage disequilibrium

10.1 Yule's Attempted Reconciliation of 1902

10.1.1 Introduction

George Udny Yule (Fig. 10.1) was born in Scotland on February 18, 1871. He received his early education at Winchester College and then at University College, where he studied civil engineering. After graduating, Yule spent one year in a small engineering works (George Wailes) and then another at Arc Works, Chelmsford. In 1892, he realized his interest was not in engineering and spent one year under Heinrich Hertz in Bonn. There, he wrote his first paper on the passage of electric waves through electrolytes. When Yule returned to London in 1893, he was offered a position by Karl Pearson as a demonstrator at University College. In 1896, Yule was promoted to Assistant Professor of Applied Mathematics. Because of financial reasons, he resigned from his position at University College and took an administrative post at the City and Guilds of London Institute. Because of his interest in

P. Gorroochurn, *The Development of Evolutionary Genetics*,
https://doi.org/10.1007/978-3-031-69374-8_10

Fig. 10.1 George Udny Yule (1871–1951). (From: https://en.wikipedia.org/wiki/Udny_Yule#/media/File:George_Udny_Yule.jpg)

statistics, he also held a position as Newmarch Lecturer on Statistics at University College during the period 1902–1909.

Having been a demonstrator to Pearson, Yule belonged to the biometric tradition and initially enjoyed a cordial relationship with the latter. However, a slight disagreement between them on the measurement of associations in fourfold tables started in 1900 (Yule, 1900), and took a turn for the worse in 1906–1907 when Yule (1906a, b) and Pearson (1907a) openly exchanged criticism. The debate reached a boiling point in 1911–1913 with several contentious exchanges between the two parties[1]

In the summer of 1901, when his relationship with Pearson was still cordial, Yule discussed his interest in biometry and Mendelian heredity. Since he did not have any training in biology, it took him a while to understand the two theories. However, the highly contentious character of Bateson's *Defence* in 1902 prompted him to submit a paper entitled "Mendel's laws and their probable relations to intra-racial heredity" (Yule, 1902) to *The New Phytologist*, a relatively neutral journal. One of Yule's major aims was to reconcile the viewpoints of the biometricians and Mendelians.

Right at the start of his paper, Yule expressed his strong disapproval of the belligerent tone of Bateson's *Defence*:

[1]See Sec. 2.3.7 of Gorroochurn (2016a) for a description of the Yule-Pearson controversy.

QUOT. 622 Mr. Bateson devotes many words to these questions [on heredity], but one cannot help feeling that his speculations would have had more value had he kept his emotions under better control; the style and method of the religious revivalist are ill-suited to scientific controversy. (*ibid.*, p. 194)

Having thus scolded Bateson, Yule went on to address several pertinent issues with an originality that earned his paper respect for future generations. Although Yule did not attract much attention at the time, this respect seems to be well deserved. The issues that Yule addressed in his paper will be described next.

10.1.2 Biologist's Heredity versus Statistician's Heredity

Yule's first point related to the difference in the interpretations of heredity by biologists and statisticians. For biologists, heredity usually meant "like begets like":

QUOT. 623 In this sense it denotes inter alia the phenomenon of the constancy of specific or racial types and of sexual characters; a character may be said to be *inherited* when it always, in one generation after another, is one of the characters of the species, of the race, or of the one sex of the race, as distinct from the other. The species, race or sex, so to speak, " begets its like" as a whole. (Yule, 1902, p. 196)

Thus, the biologist's understanding of heredity was in the constancy across generations of characters in *races* or *species*. On the other hand, for the statistician, heredity meant the constancy of characters in *individuals* within races (i.e., do individuals beget their like?):

QUOT. 624 . . .even if the type of the race is constant, do individual types within the race beget their like? In so far as any individual diverges in character from the mean of the race do his offspring tend to diverge in the same direction, or not? It is to this question that statisticians have confined themselves, and they speak of a character being "inherited" or not according as the answer to the question is yes or no - they deal solely with what we may term "individual heredity." (*ibidem*)

Yule believed that individual (or statistical) heredity could be identified simply by a positive slope in the regression equation:

$$Y = A + BX,$$

where X is the character of the parent and Y is the mean character of the offspring:

QUOT. 625 Quite generally, the statistician speaks of a character as *inherited* whenever the number or "constant" B is greater than zero; if it does not differ sensibly from zero the character is held to be non-heritable, quite apart from the question whether the mean is more or less constant from one generation to the next, a consideration which does not affect the conception of individual heredity. (*ibid.*, p. 197)

Using the concept of regression, Yule went on to illustrate a fundamental difference between race (or biological) and individual (or statistical) heredity:

QUOT. 626 . . .the maxim "like begets like" does not hold in the same rigid sense for the individual and the race. One generation of a race is (approximately speaking) the same as the preceding; apart from such changes as are hardly revealed except by measurement, the mean of the offspring is that of the parents. It would seem natural perhaps to assume that the same law holds for individual types within the race—why should an isolated group of individuals behave differently from the race as a whole? Yet the assumption would be false; the offspring of any abnormal individual, any individual differing from the mean of the race, *are always, on the average, more mediocre than himself.* (*ibid.*, p. 197)

Yule's argument in the above was that because of the phenomenon of regression, racial heredity did not always imply individual heredity.

Regarding continuous and discontinuous variation, Yule believed that the distinction was not fundamental because "successive discontinuities may be so slight as to be undiscoverable by the most careful and repeated measurements." What was more consequential for Yule was the fact that biometricians had investigated heredity *within races* while Mendelians had focused on hybridization, or crossing *between races.* He therefore admonished that laws of heredity should in no way be interpreted as laws of hybridization.

10.1.3 Broadening the Law of Ancestral Heredity

In his paper (Yule, 1902), Yule broadened the form and application of Galton's law of ancestral heredity. The reader will recall that Galton's form of the law in Eq. (4.10) (as well as Pearson's generalization in Eq. (4.17)) imposed a strict mathematical relationship between the coefficients of the parental and ancestral deviations (i.e., between the coefficients of a_1, a_2, . . .). Furthermore, following Galton, the law had been interpreted in terms of ancestral contributions and heritage. However, Yule declared that he did not believe:

QUOT. 627 . . .the facts indicate any fixity of formula even for intra-racial heredity. (*ibid.*, p 204)

Concerning ancestry "contributing" to "heritage," Yule found:

QUOT. 628 ...two conceptions of which I find it difficult to grasp the exact meaning. (*ibid.*, p. 205)

The two difficulties in QUOT. 627 and QUOT. 628 motivated Yule to write his version of the ancestral law as a *multiple regression equation*

$$Y = A + B_1X_1 + B_2X_2 + \dots \tag{10.1}$$

(*ibid.*, p. 202), where Y is the mean character of the offspring, X_1, X_2, . . . are the characters of successive ancestry, A is a constant, and B_1, B_2, . . .is a series of decreasing coefficients, which are otherwise free. Yule's version of the ancestral law in Eq. (10.1) was, therefore, a *predictive equation* such that the mean character

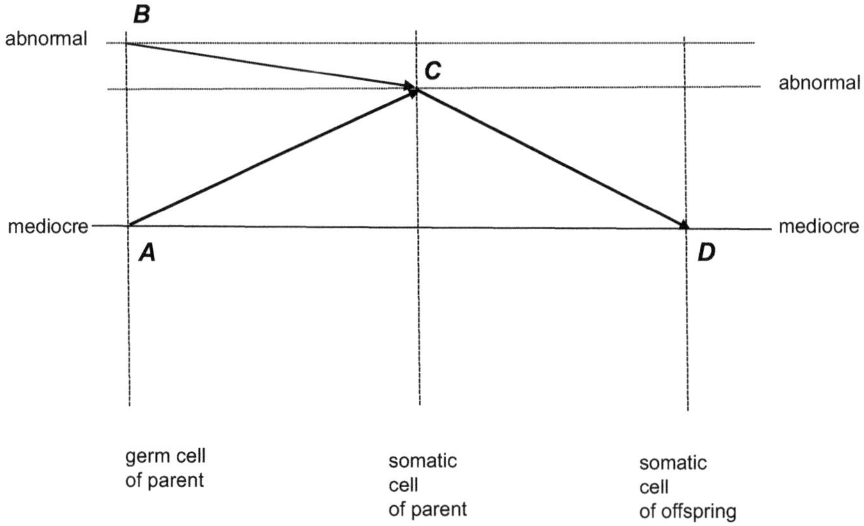

Fig. 10.2 Yule's justification for a regression interpretation of the ancestral law

of the offspring "*can be calculated with the more exactness, the more extensive our knowledge of the corresponding characters in the ancestry*".[2]

To justify the use of regression, Yule provided the following argument:

> **QUOT. 629** ...a series of absolutely identical germ cells will not produce a series of absolutely identical individuals, but a group or array of individuals differing more or less inter se. The somatic character of an individual is not therefore an absolute guide to the character of the ovum from which he sprang nor, *a fortiori*, to the mean character of the germ cells which he produces. If we conceive a number of individuals of the same somatic type, some will be abnormal developments of mediocre germ cells (i.e. cells producing on the average mediocre types); others, but these will be fewer, will have sprung from germ cells producing on the average, or usually, more abnormal types. The odds are, therefore, that a given abnormal somatic type is an abnormal development of a mediocre germ cell rather than a mediocre (or subnormal) development of an abnormal cell. But the somatic characters of offspring will follow the germ cell characters and not the somatic characters of the parent; therefore *regression* occurs. (*ibid.*, pp. 205–206)

Yule's reasoning may be understood with the help of Fig. 10.2. An abnormal somatic character (C) in a parent may be produced either by the abnormal development (AC) of mediocre germ cells or the mediocre development (BC) of abnormal germ cells. However, because most of the parent germ cells will be clustered around the average A (Yule presumably assumed a normal distribution here), many more individuals will follow the path AC than BC. Now, the somatic characters of the offspring will follow the germ cell characters, rather than the somatic characters, of the parent, so that the somatic characters will be clustered around D. Hence, a regression effect (CD) is observed when comparing the somatic characters of the offspring compared with those of the parent.

[2]Italics are Yule's.

10.1.4 Equilibrium Law for Equal Gamete Frequencies, Attempted Reconciliation Between Biometricians and Mendelians, Multifactor Hypothesis

10.1.4.1 Introduction

As mentioned in Sect. 10.1.1, one of the primary aims of Yule's paper (Yule, 1902) was a reconciliation between the biometricians and Mendelians. To this end, he put forward two important arguments, the first of which would later be generalized as the Hardy-Weinberg equilibrium, and the second of which would later serve as the basis of the modern synthesis.

10.1.4.2 First Reconciliation Argument: Mendel's Laws and Ancestral Law Not Incompatible

Yule modeled a random-mating population under the influence of Mendelian heredity. Yule was thus the first to operationalize the definition of random mating in a population, although his model of random mating was problematic. Moreover, if population genetics is understood as the study of populations subject to Mendelian inheritance in the presence or absence of evolutionary forces, then Yule's paper ought to be regarded as the first population genetics paper.

Yule considered a Mendelian human population that, for a given trait, consisted of a dominant gamete A and a recessive gamete a at equal frequencies for each sex. Next, he stated:

> **QUOT. 630** Now when A's and a's are first inter-crossed we get the series of *uniform* hybrids; when these are inter-bred we get the series of three dominant forms (two hybrids, one pure) to one recessive. If all these are again intercrossed at random the composition remains unaltered. (*ibid.*, p. 225)

In the above, AA is crossed with aa in the parental generation, giving Aa hybrids in F_1. The frequency of A and a in both males and females is now $^1/_2$. When the F_1 hybrids are crossed, an equilibrium for AA:Aa:aa is achieved in F_2, in the ratio 1:2:1. Yule's "unaltered" 1:2:1 ratio is a special case (for p = freq. $(A) = {}^1/_2$) of the equilibrium law, later called the *Hardy-Weinberg (HW) law*.[3] The HW law turns out

[3]The Hardy-Weinberg (HW) law for two alleles at a locus states that, in an infinitely large, random-mating population with no selection, mutation, or migration, there is a simple relationship between the allele frequencies and the genotype frequencies: if the allele frequencies of the two alleles among both male and female parents are p and q $(= 1 - p)$, then the genotype frequencies among the progeny are p^2, $2pq$, q^2 and these remain constant from generation to generation. For multiple alleles $A_1, A_2, \ldots, A_k$ with respective frequencies $p_1, p_2, \ldots, p_k$, the genotype frequencies can be obtained from the multinomial expansion of $(p_1A_1 + p_2A_2 + \ldots + p_kA_k)^2$. There are two components to the HW law: (i) a square-law component, relating genotype and allele frequencies; (ii) an equilibrium component, pertaining to the constancy of allele and genotype frequencies after one generation of random mating.

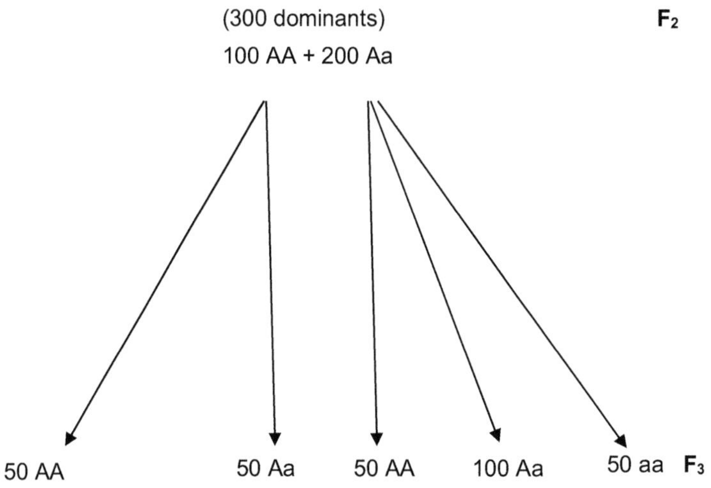

Fig. 10.3 Yule's example of a "randomly-mating" population

to be a foundational principle of population genetics. It represents the effect of random mating in a purely Mendelian population, i.e., when no evolutionary forces are acting.

Yule next followed the trajectory of 100 *AA* homozygotes and 200 *Aa* heterozygotes, both of unknown parentage, in an F_2 population of constant size across generations, and made up of an equal number of males and females. Still assuming a frequency of $^1/_2$ for the dominant gamete *A* and the recessive gamete *a*, he wrote:

> **QUOT. 631** The 100 pure individuals will give rise to dominant forms in the proportion of 50 pure to 50 hybrids; the 200 hybrids may, as segregation takes place, be considered as 100 pure dominants and 100 pure recessives, the former giving rise to 50 pure dominants and 50 hybrids, the latter to 50 hybrid dominants and 50 *pure recessives*. The 300 parent dominants, therefore, give rise to offspring in the proportion of 250 dominant forms to 50 recessive, i.e., the chance of a dominant parent producing a dominant form as offspring is 5/6. (*ibidem*)

Yule's reasoning may be understood with the help of Fig. 10.3.

It is difficult to make sense of Yule's model of random mating. With 100 *AA* and 200 *Aa*, the assumption of a frequency of 1/2 for each of *A* and *a* is untenable. Moreover, the statement "the 200 hybrids may, as segregation takes place, be considered as 100 pure dominants and 100 pure recessives" in QUOT. 631 seems to show a misunderstanding of Mendel's law of segregation.[4] These points were later seized by Castle (1903) in a major rebuttal of Yule's analysis and conclusions.

---

[4]That Yule's model of random mating is problematic can be seen from its incompatibility with his own result in QUOT. 630. For, according to Yule's model, if *AA* and *aa* randomly mated, one would expect *AA* to give ½ *AA* and ½*Aa*, and *aa* to give ½*Aa* and ½*aa*. Thus, *AA* × *aa* would give ½*AA*, *Aa*, ½aa, instead of 2*Aa*.

In any case, Yule's model yielded 250 dominants and 50 recessives in the F_3 offspring, so that

$$\Pr\left\{\text{dominant offspring}\,\middle|\,\begin{array}{l}\text{dominant parents of}\\ \text{unknown parentage}\end{array}\right\}=\frac{250}{300}=\frac{5}{6}. \tag{10.2}$$

Yule next followed the trajectory of the 250 dominants (100 *AA* + 150 *Aa*) among the F_3 offspring, all of whose parents had also been dominants. Assuming the same model as before, he reasoned that, upon mating, the 100 *AA* homozygotes were expected to yield 50 *AA* homozygotes and 50 *Aa* heterozygotes, and the 150 *Aa* heterozygotes were expected to yield 37.5 *AA* homozygotes, 75 *Aa* heterozygotes, and 37.5 *aa* homozygotes. The ratio of dominant to recessive F_4 offspring was therefore 212.5 to 37.5, so that

$$\Pr\left\{\text{dominant offspring}\,\middle|\,\begin{array}{l}\text{dominant parents of}\\ \text{dominant parentage}\end{array}\right\}=\frac{212.5}{250}=\frac{51}{60}. \tag{10.3}$$

From Eqs. (10.2) and (10.3), it is seen that Yule's Mendelian model resulted in an increased probability of dominance in offspring with dominant parents whose parents were also dominant, as opposed to unknown. Yule declared:

QUOT. 632 But this is precisely a case of the law of ancestral heredity! (*ibid.*, p. 226)

He mathematically formulated the above model as follows. Let

$$\begin{aligned} T_n &= \left\{\begin{array}{l}\text{total number of dominants in } n^{\text{th}} \text{ generation}\\ \text{all of whose ancestors are also dominant}\end{array}\right.\\ p_n &= \text{number of } AA \text{ homozygotes in } n^{\text{th}} \text{ generation}\\ i_n &= \text{number of } Aa \text{ heterozygotes in } n^{\text{th}} \text{ generation}\\ &= T_n - p_n \end{aligned} \tag{10.4}$$

Since half of the *AA* homozygotes and one-quarter of the hybrids in generation *n* gave rise to *AA* homozygotes in generation *n* + 1 in Yule's model (Fig. 10.3),

$$p_{n+1}=\frac{1}{2}p_n+\frac{1}{4}i_n \tag{10.5}$$

Similarly, since half of the *AA* homozygotes and half of the hybrids in generation *n* gave rise to hybrids in generation *n* + 1,

$$i_{n+1}=\frac{1}{2}p_n+\frac{1}{2}i_n=\frac{1}{2}T_n \tag{10.6}$$

Adding the last two equations,

$$T_{n+1} = \frac{1}{2}p_n + \frac{1}{4}i_n + \frac{1}{2}T_n = \left(\frac{1}{2}p_n + \frac{1}{2}i_n\right) + \frac{1}{2}T_n - \frac{1}{4}i_n = T_n - \frac{1}{4}i_4.$$

From Eq. (10.6), $i_{n+1} = T_n/2$, therefore

$$T_{n+1} = T_n - \frac{1}{8}T_{n-1}. \tag{10.7}$$

Defining $C_n \equiv T_{n+1}/T_n$ to be the probability that a parent in generation n with completely dominant ancestors produces a dominant offspring,

$$C_n = 1 - \frac{1}{8C_{n-1}}. \tag{10.8}$$

With $C_1 = 5/6 = 0.83333$, Yule then obtained $C_2 = 0.85000, C_3 = 0.85294, \ldots, C_\infty = 0.85355$. In fact, the limiting value C_∞ may be obtained by solving $C_\infty = 1 - 1/(8C_\infty)$ so that

$$C_\infty = \frac{1}{2} + \frac{1}{4}\sqrt{2}.$$

Having shown in his model that the C_i's increased to a limiting value, Yule stated that "the chance of an *A* producing an *A* is increased if the ancestry be also A's." Therefore:

QUOT. 633 Mendel's Laws, so far from being in any way inconsistent with the Law of Ancestral Heredity, lead then directly to a special case of that law, for the *dominant* attribute at least. (*ibid.*, pp. 226–227)

Yule also cautioned that the consistency did not hold for a recessive attribute since:

QUOT. 634 ...the chance of a recessive producing a recessive offspring is, on the above hypothesis of random mating, one-half, whatever the parentage may be. (*ibid.*, p. 227)

On the other hand, consistency still applied for partial dominance and co-dominance.

There are two reasons why Yule's Eqs. (10.2) and (10.3) do not, in fact, reconcile Galton's ancestral law with Mendel's law:

(i) As indicated before,[5] Yule's model was not truly Mendelian since it did not accord with the law of segregation.
(ii) Even if Yule's model had been *truly* Mendelian, the fact that pure recessives were discarded in each generation implies that the model was not *purely* Mendelian. Indeed, genotypes can be added or removed in each generation to make the probability in Eq. (10.3) any real number between 0 and 1.

[5] See footnote 4.

It is also noteworthy that Yule was on the biometric side of the debate, and as James Tabery (2004) has indicated, he used the ancestral law as the reference in his analysis: for non-recessive cases, *Mendel's law* is declared to be consistent with the ancestral law, and not vice versa.

10.1.4.3 Second Reconciliation Argument: Multifactor Hypothesis

Yule's second argument was based on the so-called multifactor hypothesis, which later became the founding principle for the modern synthesis. However, the hypothesis had already been referred to many years earlier by Mendel (see Sect. 6.2.8).

Yule considered a Mendelian character and took a length as an example. Suppose this Mendelian character assumed two possible values x_1 and x_2, one of which was dominant over the other. Similarly, another length was also Mendelian and could assume two values, x_1' and x_2' (one dominant, one recessive). Yule next considered a compound character that was the sum of the two Mendelian characters. He correctly noted that this compound character was no longer Mendelian and could give rise to four possible 'somatic forms'[6]:

$$x_1 + x_2, \quad x_1' + x_2', \quad x_1' + x_2, \quad x_1 + x_2'$$

This is in effect a two-factor hypothesis, whereby a trait is the result of the (additive) action of genes at two loci.

Yule noted that, for the additive effect of n Mendelian factors, 2^n possible somatic forms were possible. Then:

> **QUOT. 635** But how great is *n* likely to be in such a case as stature, assuming that it can be analysed into a set of Mendelian units? As Mr. Galton has remarked ((2) p, 83) "... human stature is not a simple element, but a sum of the accumulated lengths or thicknesses of more than a hundred bodily parts, each so distinct from the rest as to have earned a name by which it can be specified ...." Surely it would be a very moderate estimate that the number of units could not be less than 50? Yet this would suffice to give, on the simplest Mendelian assumption that each unit can only exhibit two types, not some mere ten thousand different values of stature, the run of which would be quite indistinguishable from strictly continuous variation, but *over a thousand-million million different types*[7]! Even then if the variations of "units" do take place by discrete steps *only* (which is unproven), discontinuous variation must merge insensibly into continuous variation simply owing to the compound nature of the majority of characters with which one deals. There does not seem any escape from this conclusion. (Yule, 1902, p. 235)

Yule's second reconciliation argument was thus that if a trait is acted upon by multiple Mendelian factors then, although Mendelian heredity is discrete, it is likely that the trait will vary continuously. Unfortunately, this compelling argument was all but ignored amid the bitter biometry-saltationism debate. It would later serve as the founding principle for the synthesis of Darwinism and Mendelism.

---

[6] Yule's somatic forms would later be called "phenotypes."

[7] Italics are Yule's.

10.1.5 Conclusion

Despite the several useful ideas it contained, Yule's paper failed to attract much attention. There are two major reasons for this:

(a) Yule was a statistician and had little to no training in evolutionary biology. This made his first paper on the subject less impactful.
(b) The biometry-saltationism debate had become extremely personal, with each side wanting to show the correctness of its own approach. Therefore, there was not much room for Yule's ideas of reconciliation.

However, Yule's paper did not go completely unnoticed. In 1903, Pearson published a paper entitled "The law of ancestral heredity" (Pearson, 1903b), where he first publicly commented on Mendel's theory. In Appendix 1 to that paper, Pearson also explicitly addressed Yule's 1902 paper. However, Pearson's work was preceded by an important paper by the American evolutionary biologist William Castle (1867–1862) (Castle, 1903), who threw a wrench into the works by questioning both Galton's ancestral law and its supposed reconciliation with Mendel's law by Yule. This will be described next.

10.2 William Castle's 1903 Response to Yule: Equilibrium Law in the General Case

10.2.1 Introduction

Born on a farm in Ohio, William Ernest Castle (1867–1962) was later recognized as the father of mammalian genetics. After undergraduate studies in biology, he worked as an assistant to one of his teachers, Charles Davenport, at Harvard Zoological Laboratory. Before 1900, he studied the effect of selection on the sex ratio in mice and guinea pigs. After the Mendelian rediscovery in 1900, Castle turned his attention away from morphology toward heredity.[8]

The exact circumstances around Castle's acquaintance with Mendel's work are relatively unknown. He may have read any one of the rediscovery papers by de Vries, Correns, or Tschermak. It is also possible that he learned of Mendel through Davenport's "Mendel's law of dichotomy in hybrids," published in the *Biological Bulletin* (Davenport, 1901). In any case, Mendel's work was a major revelation to Castle. In a 1959 letter written to his former student Leslie Dunn, he recalled:

> **QUOT. 636** The rediscovery of Mendel's work threw a flood of light on the situation, the full significance of which was first realized by Bateson. Here was where genetics began. (Dunn, 1965b, p. 39)

---

[8]This is yet another example of the "revolt against morphology," which involved many biologists around the turn of the nineteenth century (see footnote 12 of chap. 7).

The Mendelian rediscovery set the direction for Castle's future research, which consisted of investigating several questions of a hereditary nature. These were later listed by him as follows:

QUOT. 637 1. First came the problem, how extensive is the applicability of Mendel's law? Does it apply to all discontinuous variations? Does it apply to cases of intermediate or blending inheritance?

2. The assumed purity of the gametes produced by a hybrid after the association of contrasted characters in the same zygote for many cell generations. Is it true?

3. The assumption that a character segregating as a unit in hybridization cannot be modified by selection however long and persistently continued. Is it true?

4. The conclusion of Weismann that germ cells and body cells are distinct, germ cells alone being the vehicle of heredity, and consequently that acquired characters are not inherited. Is it true?

(Castle, 1951, p. 67)

The unique nature of Castle's work will be examined later[9] but for now, the focus will be on his fourth Mendelian paper (Castle, 1903). This paper was a critique of the biometric position and questioned Yule's attempted reconciliation between Mendel's law and the ancestral law (Yule, 1902). It is also in this paper that the equilibrium law (later the Hardy-Weinberg law) was first worked out for unequal allele frequencies. However, the paper also contained an unfortunate mistake in Castle's understanding of the law of ancestral heredity and its modification by Pearson, as will be explained next.

10.2.2 Castle's Confusion of Ancestral Contributions with Ancestral Correlations

In the first section of his paper, Castle explained Galton's ancestral law and its modification by Pearson. He referred to Galton's paper of 1897 (Galton, 1897b) and quoted the Law of Ancestral Heredity as follows:

QUOT. 638 The two parents contribute between them, on the average one-half, or (0.5) of the total heritage of the offspring; the four grandparents, one-quarter, or $(0.5)^2$; the eight great-grandparents, one-eighth, or $(0.5)^3$, and so on. Thus the sum of the ancestral contributions is expressed by the series $[(0.5) + (0.5)^2 + (0.5)^3$, etc.], which, being equal to 1, accounts for the whole heritage. (Castle, 1903, pp. 223–224)

It should be noted that the geometric series above, namely, $[(0.5) + (0.5)^2 + (0.5)^3 \ldots]$, gives the coefficients in Galton's multiple regression in Eq. (4.10). Since the series also refers to *ancestral contributions*, it necessarily adds up to unity. Furthermore, to obtain the geometric series above, Galton had assumed *ancestral correlations* of $(1/3)^i$ (cf. Sect. 4.6.2). These correlations are also in geometric progression but do not necessarily add up to unity.

---

[9] See Sect. 11.8.

Table 10.1 Castle's (1903, p. 224) table of ancestral correlations (Pearson's series I and II) taken from Pearson (1903b, p. 222) and of Galton's ancestral contributions (Galton's series)

	Pearson's series I	Pearson's series II	Galton's series
Parental influence	0.49	0.50	0.50
Grandparental influence	0.32	0.33	0.25
Great-grandparental influence	0.20	0.22	0.125
Great-great-grandparental influence	0.13	0.15	0.0625

Unfortunately, Castle completely confused the geometric series of ancestral contributions with that of ancestral correlations. Referring to Pearson (1903b), he stated:

QUOT. 639 He [Pearson] believes that "the contributions of the ancestry follow a geometrical series, although not that originally proposed by Mr. Galton." (Castle, 1903, p. 224)

Thereupon, Castle reproduced two series *of ancestral correlations*, which Pearson had given on p. 222 in his paper, side by side with Galton's series of *ancestral contributions*, for comparison!

Referring to Table 10.1, Castle stated:

QUOT. 640 Comparing Pearson's series with that of Galton, we see that parental influence is reckoned as substantially the same by both Galton and Pearson, but that Pearson assigns a much greater influence to the more remote ancestors than does Galton. (*ibidem*)

A cursory look at the sums of each of Pearson's series would have convinced Castle that these could not be ancestral contributions (since each sum exceeds unity). However, such a check was not made and Castle's observation in QUOT. 640 is completely erroneous.

Castle's mistake would later be seized by Pearson in a fiery reply (see Sect. 10.5).

10.2.3 *Criticism of Yule's Reconciliation and Equilibrium Law in the General Case*

Fortunately, the rest of Castle's paper was relatively free from the type of error that was explained before. In fact, it successfully criticized Yule's reconciliation (Yule, 1902) and contained a result of major importance.

First, Castle referred to data on color inheritance in mice. The data had been provided by von Guaita (1898) and were later analyzed by Davenport (1900). Castle observed:

QUOT. 641 In this series the inheritance of color follows closely neither the law of Galton nor the series suggested by Pearson. (Castle, 1903, pp. 224–225)

He then gave three generations for which the observed and expected values disagreed substantially (see Table 10.2).

Table 10.2 One of the tables given by Castle (1903) to show disagreement between observed and expected counts using Galton's series. The

Generation	II (%)	IV (%)	V (%)
Observed	100	58	48
Calculated	0	48	60

As mentioned, these percentages were obtained from Davenport's paper, where they had been calculated in exactly the same way that Galton had done for the basset hounds data (see Sect. 4.6.4.2).

Of course, Castle was only partly correct in QUOT. 641, since the series by Pearson referred to correlations, not ancestral contributions, as explained in (a).

Next, Castle explained Mendelian heredity. Using calculated Mendelian proportions, he correctly obtained expected counts for various generations of von Guaita's data. He then concluded:

QUOT. 642 The Mendelian predictions are very close to the observed numbers throughout the table, generation by generation and pair by pair. (*ibid.*, p. 232)

Having argued for the Mendelian case, Castle next took on Yule's 1902 paper (Yule, 1902). Recall that, in Yule's peculiar model of random mating,[10] 100 *AA* homozygotes and 200 *Aa* heterozygotes were chosen in F_2, both of unknown parentage, in a population of constant size across generations. These yielded 250 dominants and 50 recessives in F_3. The trajectory of only the dominants, subjected to random mating, was then followed across generations. Yule's model yielded $C_\infty = 0.85355$ for the limiting probability that a parent with completely dominant ancestors produces a dominant offspring. Castle took issue with Yule's value for C_∞:

QUOT. 643 Now, what, in plain unmathematical language, does this mean? It means that when a dominant form has once been crossed with a recessive (as a pigmented animal, for example, with an albino), the stock of the former is forever contaminated, and cannot be freed entirely from the albino character by mere elimination of white individuals, however long the process is continued. Ever afterward the cross-bred dominant stock will produce on the average at least fourteen or fifteen white individuals in every hundred born. This conclusion is absurd, as every breeder knows. (Castle, 1903, p. 233)

In the above, Castle likened the focus on only the dominants in each generation to the successive elimination of recessives by natural selection. Yule's model therefore implied the impossible result that natural selection, however long it acted, would be unable to eliminate a recessive lethal condition.

Castle reasoned that 100 *AA* and 200 *Aa* in F_2 did not, in fact, yield 250 dominants and 50 pure recessives in F_3 under Mendelian random mating, as Yule had calculated (Fig. 10.3). He used a different model from Yule's, which was also different from

[10] See Fig. 10.3.

Hardy's and Weinberg's.[11] Assuming equal numbers of males and females in the 300 (AA + Aa) individuals, Castle reasoned that each of the 300 individuals had 50 (=100/2) possible AA mates and 100 (=200/2) possible Aa mates. Therefore, the expected number of possible crossings and corresponding F_3 offspring are:

	AA	Aa	aa
$\left(\frac{1}{3}\times 300\right)\times 50 = 5,000\ AA\times AA$ crossings yielding offspring	5,000		
$2\left\{\left(\frac{1}{3}\times 300\right)\times 100\right\} = 20,000\ AA\times Aa$ crossings yielding offspring	10,000	10,000	
$\left(\frac{2}{3}\times 300\right)\times 100 = 20,000\ Aa\times Aa$ crossings yielding offspring	5,000	10,000	5,000

Therefore, in Castle's model, the expected ratio of AA:Aa:aa among the F_3 offspring was 20,000:20,000:5000 = 4:4:1 instead of Yule's 100:150:50 = 2:3:1. Yule's conditional probability of 5/6 in Eq. (10.2) for F_3 dominants therefore becomes, under Castle's scheme,

$$\Pr\left\{\text{dominant offspring}\left|\begin{array}{l}\text{dominant parents of}\\ \text{unknown parentage}\end{array}\right.\right\} = \frac{4+4}{4+4+1} = \frac{8}{9}. \tag{10.9}$$

To determine the conditional probability of F_4 dominants, the aa's are first removed, yielding an equal number of AA's and Aa's in F_3. Considering the mating of 1 AA and 1 aa male with an equal number of females, the expected number of possible crosses and corresponding F_4 offspring are:

	AA	Aa	aa
$\left(\frac{1}{2}\times 4\right)\times 1 = 2AA\times AA$ crosses yielding offspring	2		
$2\left\{\left(\frac{1}{2}\times 4\right)\times 1\right\} = 4\ AA\times Aa$ crosses yielding offspring	2	2	
$\left(\frac{1}{2}\times 4\right)\times 1 = 2\ Aa\times Aa$ crosses yielding offspring	$\frac{1}{2}$	1	$\frac{1}{2}$

The AA:Aa:aa ratio is thus $4\frac{1}{2} : 3 : \frac{1}{2}$ and Yule's conditional probability of 51/60 in Eq. (10.3) for F_4 dominants therefore becomes, under Castle's scheme,

$$\Pr\left\{\text{dominant offspring}\left|\begin{array}{l}\text{dominant parents of}\\ \text{dominant parentage}\end{array}\right.\right\} = \frac{7\frac{1}{2}}{8} = \frac{15}{16}. \tag{10.10}$$

Castle gave the successive proportions of dominants as the terms of the series

[11] More precisely, unlike Hardy and Weinberg later, Castle did not make use of the equivalency between random mating of individuals and the random union of their gametes.

$$\frac{1}{1}+\frac{2^2-1}{2^2}+\frac{3^2-1}{3^2}+\frac{4^2-1}{4^2}+\frac{5^2-1}{5^2}+\dots$$

Thus, the successive conditional probabilities in Castle's model are 1.00000 (F_1), 0.75000 (F_2), 0.88889 (F_3), 0.93750 (F_4), 0.96000 (F_5), ... converging to 1. This may be compared to Yule's 1.00000 (F_1), 0.75000 (F_2), 0.83333 (F_3), 0.85000 (F_4), 0.85294 (F_5), ... converging to 0.85355. The convergence to unity in Castle's model was reassuring to the latter since it implied that:

QUOT. 644 ...it is entirely possible for a breeder, under the conditions stated, practically to eliminate an undesirable recessive character, in a very few generations, *merely by not breeding from individuals which manifest that character*[12]. This accords with experience. (*ibid.*, p. 236)

The eventual disappearance of *all* recessives in Castle's Mendelian model also meant that the influence of ancestry disappeared after a few generations. Castle noted that this result was in direct contradiction with Galton's ancestral law, thus rejecting Yule's attempted reconciliation.

Although Castle was correct in this interpretation, he did not comment on his values in Eqs. (10.9) and (10.10), which were based on his Mendelian model. Do these values not show the influence of ancestral dominance on the probability of offspring dominance and thus also support Galton's ancestral law? The answer to this question is no because although Castle's model (unlike Yule's) was *truly* Mendelian, the model (like Yule's) was not *purely* Mendelian, since natural selection was explicitly allowed. When evolutionary forces are allowed in a purely Mendelian population, genotypes get added or subtracted to give arbitrary probability values in Eq. (10.10).

Next, Castle removed the effect of natural selection so that the effect of random mating in a purely Mendelian population could be considered:

QUOT. 645 If the breeder eliminates recessives but once, namely, in the second generation following the cross, the series will be as follows:-

1st generation.........100 % dominants
2d generation...........75 „
3d generation...........88.8 „
4th generation..........88.8 „

etc., ad infinitum. If he eliminates recessives twice only, namely, in the second and third generations, the race will thereafter continue to contain 93.7 per cent dominant individuals, as follows:-

1st generation...........100 % dominants
2d generation..............75 „
3d generation..............88.8 „
4th generation.............93.7 „
5th generation............. 93.7 „
etc., ad infinitum.
(*ibid.*, pp. 236–237)

[12] Italics are Castle's.

Thus, once selection is halted in any generation, the population becomes purely Mendelian, and random mating stabilizes the genotype frequencies achieved in that generation. In Castle's own words:

QUOT. 646 In general, *as soon as selection is arrested the race remains stable at the degree of purity then attained*[13], provided of course that one form is as fertile as the other, and subject to no greater mortality. (*ibid.*, p. 237)

The results in QUOT. 645 and QUOT. 646 represent the earliest version, for general gamete frequencies, of the equilibrium law (later the Hardy-Weinberg law). Note that the equilibrium component[14] of the law was explicitly stated, but the square-law component was not.

Unfortunately, Castle's fundamental result was all but ignored. In his paper, Castle called the result in QUOT. 646 "the law governing race improvement," but he does not seem to have promoted it elsewhere. Nor did he claim any priority when the result came into full prominence after Hardy's demonstration in 1908 or when it was later realized that Weinberg had preceded Hardy by a few months.[15] Writing in 1951 about the beginnings of Mendelism in America, Castle (1951, p. 68) mentioned his 1903 paper but did not deem it necessary to point out that he had preceded Hardy and Weinberg by 6 years!

One reason why Castle and others did not highlight the equilibrium law, especially the square-law component, might be because the result seemed so obvious as to demerit attention. Thus, Keeler reported that Sewall Wright wrote to C.C. Li a letter in 1967, in which the former admitted that:

QUOT. 647 I remember that I read Castle's 1903 paper when I became a graduate student of his in 1912. I was well aware of the law at the time (though not from Hardy or Weinberg's papers) taking it merely as an obvious consequence of the multiplication principle of probability applied to random mating, $[(1 - q)a + qA]^2$, and so was not specially impressed by Castle's demonstration. (Keeler, 1968, p. 111)

Keeler has further pointed out that H.D. Fish, who had been Castle's student, had already indicated in 1914 in a paper by P.W. Whiting (1914, p. 343) that Castle had essentially anticipated the equilibrium law in 1903. However, this indication was not discovered until 1954 when C. C. Li was revising his book, *Population Genetics* (Li, 1955). Even then, Li at first incompletely stated Castle's achievement:

QUOT. 648 W. E. Castle (1903) and Karl Pearson (1904) showed that if the F_2 individuals from an original cross *AA* x *aa* are allowed to mate at random, the composition of the next generation would be the same as in the F_2, viz., (¼, ½, ¼). This is simply a particular case of the Hardy-Weinberg proportions with $p = q = ½$. (*ibid.*, p. 9)

This omission was soon addressed, first by Curt Stern[16]:

---

[13] Italics are Castle's.

[14] See footnote 3.

[15] Keeler (1968, p. 112) has attributed this to Castle's modesty.

[16] Curt Stern (1943) had earlier discovered that Hardy's law had independently been discovered by Wilhelm Weinberg a few months before Hardy. Stern called the law the "Hardy-Weinberg law." If Castle's priority is taken into account, the law should really be called the "Castle-Weinberg-Hardy Law."

QUOT. 649 He [Castle] found that whatever ratio of the genotypes had been reached during the generations of selection this would remain constant in future generations once selection had been discontinued. . .Thus, he had discovered an equilibrium law not only for the 1: 2: 1 ratio but for the infinitely large class of all ratios resulting from selection against recessives. With this successful defense of Mendelism against Yule's position, Castle returned to his breeding experiments, little aware of the gold which he had found in the still undefined area of population genetics. (Stern, 1965, p. 220)

Stern was soon followed by C.C. Li who devoted a full paper (Li, 1967a) on Castle's prior discovery.

10.3 Pearson's 1903 Response to Yule

Pearson's paper (Pearson, 1903b) was not only an elaboration of the assumptions underlying the law of ancestral heredity,[17] but also the author's first remarks on Mendel's theory. Pearson commented on Yule's 1902 work (Yule, 1902) in an appendix, which indicated that he had become aware of the paper only after his own was sent to the press.

Although Pearson stated that he did not want his paper to be regarded as a response to Bateson's incisive *Defence* (Bateson, 1902b), at the very start, he took an unequivocal stance against Mendel's theory. His main argument was that his observations on horses, dogs, or men showed the complete absence of true-breeding varieties as predicted by Mendel's theory. For example, if two varieties with recessive traits were crossed, Mendel's theory predicted all recessives across future generations. With a familiar tinge of sarcasm, Pearson argued that this prediction was contradicted by the evidence:

QUOT. 650 If black or grey coat-colour in horses were "recessive," when two blacks were mated we should expect only black offspring, but black can disappear for a generation or even two and then reappear. Or, take a case like that of a grey horse *Viscount*, where grey remained dominant for three generations only to disappear before the chestnut of the mare *Blue Stocking* in the *Viscount* and *Blue-Stocking* filly *Miss Johanna*. (Pearson, 1903b, p. 214)

Therefore:

QUOT. 651 We have therefore to classify large portions of the animal kingdom as exceptions to Mendel's Laws. (*ibid.*, p. 215)

Pearson next explained Galton's original version of the ancestral law (see Eq. (4.10) and why it needed to be replaced by a more general geometric series (see Eq. (4.17)). Pearson found the latter series to fit the "data for horse and man remarkably well."

Having explained the ancestral law, Pearson concluded as follows:

[17] See Sect. 4.6.

QUOT. 652...no Mendelian theory can (a) replace the statistical treatment unless it shows that offspring are absolutely determined by their parents, for if the correlation be not perfect, a better prediction must be obtainable by using the positively correlated ancestry, or (b) be proved without careful quantitative analysis of the variations falling within the wide categories adopted. (*ibid.*, p. 228)

Finally, Pearson took on Yule's 1902 paper in Appendix I. Since Yule had argued from the biometric side of the debate, it is not surprising that:

QUOT. 653 With much of his paper, I agree, for example, with his insistence on the point that the laws of intra-racial heredity are not incompatible with Mendelian principles holding for hybridization. (*ibid.*, p. 228)

However, Pearson disagreed with Yule's interpretation of the ancestral law in the form (10.1), in which the regression coefficients B_1, B_2, ...formed a decreasing series but were otherwise free. Pearson's main disagreement was that, since the coefficients had to be determined individually, it implied the "impossibility of working it out for all the ancestry, and for all characters of all races."

Against Yule's complaint that previous versions of the ancestral law[18] restricted the coefficients B_1, B_2, ... to fixed constants, Pearson explained that his generalization of 1898 (Pearson, 1898)[19] had introduced flexibility in the law. Thus, Galton's fixed series 1 + 1/2 + 1/4 + ... could now be replaced by the more flexible one $\left(1-\widetilde{\beta}\right)+\left(1-\widetilde{\beta}\right)\widetilde{\beta}+\left(1-\widetilde{\beta}\right)\widetilde{\beta}^2+\ldots$. Moreover, once $\widetilde{\beta}$ was determined, the contributions of *all* ancestors could be obtained.

10.4 Pearson (1904b): Multifactor Model, Equilibrium Law for Equal Gamete Frequencies, Attempted Reconciliation Between Mendel's Law and the Ancestral Law

10.4.1 Introduction

Shortly after the 1903 paper,[20] where he had first (unfavorably) commented on Mendel's theory, Pearson wrote a theoretical paper (Pearson, 1904b) that was the twelfth in his "Mathematical Contributions to the Theory of Evolution" series. Egon Pearson (1938, p. 41) has noted that the paper was apparently a joint effort by Karl Pearson and Raphael Weldon, but the latter refused any authorship since he did not contribute to the more technical aspects of the work.

In the paper, since Mendelians had recently had some difficulty explaining the results of several crossings, Pearson remarked that:

[18] See. Eq. (4.10).

[19] See Eq. (4.17).

[20] See Sec. 10.3.

QUOT. 654 The pure and simple Mendelian theory seems to have been discarded in the light of recent experimental results by more than one Mendelian, both in this country and abroad. The original Mendelian theory has been replaced by what are termed "Mendelian Principles." In this aspect of investigation the fundamental principles propounded by MENDEL are given up, and for each individual case a pure gamete formula of one kind or another is suggested as describing the facts. This formula is then emphasized, modified or discarded, according as it fits well, badly, or not at all with the growing mass of experimental data. (Pearson, 1904b, p. 53)

Pearson's aim was to develop a *generalized theory of alternative inheritance* based on the concept of *gametic purity*. Because of the reasons mentioned in QUOT. 654, he refused to associate the name of Mendel with his theory.

However, it is difficult to see how Mendel's theory can be separated from gametic purity. As will be seen, Pearson's theory of the pure gamete was essentially Mendelian. Moreover, Pearson's use of *alternative inheritance* suggests that he had (wrongly) assumed that dominance was inherent in Mendel's theory. Nonetheless, Pearson made a major innovation, namely, the first mathematical construction of a multi-locus model, as will be described next.

10.4.2 Multifactor Model

Assuming that the parental zygotes are denoted by (AA') and (aa'), the fundamental equation underlying Pearson's theory of the pure gamete is

$$(AA') \times (aa') = (A + A') \times (a + a') = \begin{cases} Aa \\ Aa' \\ A'a \\ A'a' \end{cases}$$

(Pearson, 1904b, p. 54). Thus, Pearson's theory of the pure gamete is, in fact, based on Mendelian segregation. Instead of "dominant" and "recessive," Pearson used the term *protogene* to denote gamete A (the "A-element") and the term *allogene* to denote gamete a (the "a-element"). AA is then called a *protozygote*, aa an *allozygote*, and Aa a *heterozygote*. So far, there was nothing new in Pearson's theory, but next, he introduced the concept of a zygote whose composition was specified by *n pairs* (or "couplets"):

$$A_1A_1 + A_2A_2 + \ldots + A_nA_n. \tag{10.11}$$

The above denotes a full protozygous zygote, which is then crossed with a fully allozygous zygote

$$a_1a_1 + a_2a_2 + \ldots + a_na_n, \tag{10.12}$$

yielding a "perfect multiple heterozygote" (in F_1):

$$a_1A_1 + a_2A_2 + \ldots + a_nA_n. \tag{10.13}$$

Pearson's notation in Eqs. (10.11)–(10.13) anticipates the concept of *genotypes at multiple loci*, but there is no concept of *haplotype* yet, since a_jA_j is treated the same as A_ja_j.

Next, when F_1 heterozygotes in Eq. (10.13) are intercrossed, the *j*th couplet in the F_2 offspring can be any of the four types: a_ja_j, a_jA_j, A_ja_j, A_jA_j. Pearson denoted this by the expression

$$a_ja_j + 2a_jA_j + A_jA_j. \tag{10.14}$$

The same notation has thus been used to denote something completely different now: in Eqs. (10.11)–(10.13), $\sum_i \alpha_i\beta_i$ is used to denote the composition of *all n* couplets, but in Eq. (10.14) it is used to denote the different possible compositions for *one* couplet. With this new meaning, Pearson next proceeded by symbolic analysis, representing the proportions of *aa*, *aA*, and *AA* by *u*, *v*, and *w*, respectively. He first assumed that $u + 2v + w$ counted as four possibilities arising from the cross $(aa) \times (AA)$. Then, the total number of ways of obtaining $n - p - q$ allozygotes, p heterozygotes, and q protozygotes across the n couplets is the coefficient of $u^{n-p-q}v^pw^q$ in the expansion of the multinomial $(u + 2v + w)^n$. This coefficient is

$$2^p \binom{n}{n-p-q, p, q} = 2^p \frac{n!}{(n-p-q)!p!q!} \tag{10.15}$$

(*ibid.*, p. 57).

10.4.3 Equilibrium Law for Equal Gamete Frequencies

Pearson (1904b) next used his multifactor model to derive the Hardy-Weinberg law for equal gamete frequencies as follows. He considered a male and female with compositions *P* and *Q*, respectively, at the first $(n - 1)$ couplets. Suppose that the mating between the two produces an offspring with composition *J*. A new couplet is now added to both male and female. The composition of the male at all *n* couplets is now any one of the four possibilities:

$$P + a_na_n, \quad P + a_nA_n, \quad P + A_na_n, \quad P + A_nA_n, \tag{10.16}$$

and similarly for the female $(Q + a_na_n, \ldots)$. Therefore, the total number of *new* possibilities for the cross between the male and female is 4^2. Since there are four possible offspring for each cross, the total number of new possible offspring is 4^3, of which 16 are due to crossings of each of the four possibilities in Eq. (10.16). Pearson wrote the composition of all new possible offspring as follows:

$$\overbrace{8(u+v)}^{\substack{\text{from male parent with}\\ n^{\text{th}}\text{ couplet } AA}} + \overbrace{8(u+2v+w)}^{\substack{\text{from male parent with}\\ n^{\text{th}}\text{ couplet } Aa \text{ or } aA}} + \overbrace{8(v+w)}^{\substack{\text{from male parent with}\\ n^{\text{th}}\text{ couplet } aa}} = 16(u+2v+w) \tag{10.17}$$

(*ibid.*, p. 59). Therefore, the composition of the offspring of n-couplet parents is

$$4^3\left(\frac{1}{4}u+\frac{2}{4}v+\frac{1}{4}w\right)J, \tag{10.18}$$

where J is the composition of the offspring of $(n-1)$-couplet parents. Now, again assuming that each cross produces four possible offspring, for parents with $n = 1$ couplet and composition $aa + 2aA + AA$, the composition of the offspring is

$$(aa+2aA+AA)\times(aa+2aA+AA) = 16aa+32aA+16AA = 4^3\left(\frac{1}{4}u+\frac{2}{4}v+\frac{1}{4}w\right).$$

By using this result and the recursion in Eq. (10.18), Pearson finally obtained the composition of offspring of n-couplet parents as

$$4^{3n}\left(\frac{1}{4}u+\frac{2}{4}v+\frac{1}{4}w\right)^n \tag{10.19}$$

(*ibid.*, p. 60). Since the composition of the n-couplet parents was $(u + 2v + w)^n$, Pearson remarked:

> **QUOT. 655** This, except for the constant factor 4^n x 4^n, is absolutely identical with the distribution of the parental population, and accordingly if the next generation also mates at random, the mixed race will continue to reproduce itself without change. We therefore reach the following result:-
>
> *However many couplets we suppose the character under investigation to depend upon, the offspring of the hybrids - or the segregating generation—if they breed at random* inter se, *will not segregate further, but continue to reproduce themselves in the same proportions as a stable population. (ibidem)*

The expression in Eq. (10.19) and the last paragraph in QUOT. 655 are statements, respectively, of the square-law and equilibrium component[21] of what later came to be known as the Hardy-Weinberg Law. However, as was the case with Yule,[22] this was stated only for the case of equal gamete frequencies (since Pearson assumed that each F_1 individual was a heterozygote at every couplet).

---

[21] See footnote 3.

[22] See Sect. 10.1.4.

10.4.4 Attempted Reconciliation Between Mendel's Law and the Ancestral Law

From the expression in Eq. (10.15) for the number of fathers of type $u^{n-p-q}v^p w^q$ and the expression in Eq. (10.17) for the composition of the corresponding offspring, Pearson wrote the distribution of such offspring as

$$\begin{aligned}&\binom{n}{n-p-q,p,q}\{8(u+v)\}^{n-p-q}\{8(u+2v+w)\}^p\{8(v+w)\}^q\\&=4^{2n}\binom{n}{n-p-q,p,q}\left(\frac{1}{2}u+\frac{1}{2}v\right)^{n-p-q}\left(\frac{1}{2}u+v+\frac{1}{2}w\right)^p\left(\frac{1}{2}v+\frac{1}{2}w\right)^q\end{aligned}$$

(*ibid.*, p. 61). From the above, the distribution of offspring from fathers with s $(= n - p - q)$ allozygotes is

$$\begin{aligned}&4^{2n}\sum_{p,q:p+q=n-s}\frac{n!}{(n-p-q)!p!q!}\left(\frac{1}{2}u+\frac{1}{2}v\right)^{n-p-q}\left(\frac{1}{2}u+v+\frac{1}{2}w\right)^p\left(\frac{1}{2}v+\frac{1}{2}w\right)^q\\&\quad=4^{2n}\sum_{i=0}^{n-s}\binom{n}{s,i,n-s-i}\left(\frac{1}{2}u+\frac{1}{2}v\right)^{n-i}(v+w)^i\end{aligned} \tag{10.20}$$

(*ibid.*, p. 62). Pearson next listed each of the component binomials above (C_i), together with the distance (m_i) of the mean of the binomial from the $(n + 1)^{\text{th}}$ couplet, and the total frequency of the binomial (f_i)[23] (assuming a frequency of 2 for $v + w$):

$$C_1=\left(\frac{1}{2}u+\frac{1}{2}v\right)^n,\qquad m_1=1+\frac{n}{2},\qquad f_1=4^{2n}\binom{n}{s,0,n-s}. \tag{10.21}$$

$$C_2=\left(\frac{1}{2}u+\frac{1}{2}v\right)^{n-1},\quad m_2=2+\frac{n-1}{2},\quad f_2=4^{2n}\binom{n}{s,1,n-s-1}2^1=2f_1\frac{n-s}{1}.$$

$$\begin{aligned}C_3=\left(\frac{1}{2}u+\frac{1}{2}v\right)^{n-2},\quad m_3=3+\frac{n-2}{2},\quad f_3&=4^{2n}\binom{n}{s,2,n-s-2}2^2\\&=2^2f_1\frac{(n-s)(n-s-1)}{2!}\dots\end{aligned}$$

[23] In Pearson's theory, the binomial distribution is denoted by $\alpha(p+q)^n$, where $p+q=1$ and α is the total area under the distribution (i.e., the total frequency). Furthermore, the mean is at a distance np from the origin and a distance $i-np$ from the term in p^i.

$$C_{i+1} = \left(\frac{1}{2}u + \frac{1}{2}v\right)^{n-i}, \quad m_{i+1} = i + 1 + \frac{n-i}{2},$$
$$f_{i+1} = 2^i f_1 \frac{(n-s)(n-s-1)\dots(n-s-i+1)}{i!}.$$

Therefore, the total frequency of all the binomials in Eq. (10.20) is

$$f_1 + f_2 + f_3 + \dots = f_1 3^{n-s}$$

(*ibid.*, p. 63), where f_1 is given in Eq. (10.21). Now, if the distance from the $(n + 1)^{th}$ couplet to the mean of the distribution in Eq. (10.20) is denoted by m_s, then

$$\begin{aligned} f_1 3^{n-s} \times m_s &= f_1 \times \left(1 + \frac{n}{2}\right) + f_2 \times \left(2 + \frac{n-1}{2}\right) + \dots + f_{i+1} \times \left(i + 1 + \frac{n-i}{2}\right) + \dots \\ &= f_1 \left\{3^{n-s} + (n-s)3^{n-s-1} + \frac{n}{2}3^{n-s}\right\} \end{aligned}$$

so that

$$m_s = 1 + \frac{1}{2}n + \frac{1}{3}(n-s)$$

(*ibidem*). Therefore, the mean number of offspring from parents with s allogenic couplets is

$$(n+1) - m_s = \frac{1}{2}n - \frac{1}{3}(n-s). \tag{10.22}$$

Furthermore, considering only allozygotes versus non-allozygotes among the whole population of offspring, the distribution is given by the terms in the expansion of $(1/4 + 3/4)^n$ and has mean $n/4$. Therefore, the deviation of the offspring mean in Eq. (10.22) from the mean of the whole population is

$$\left\{\frac{1}{2}n - \frac{1}{3}(n-s)\right\} - \frac{n}{4} = \frac{1}{12}(4s-n).$$

On the other hand, the deviation of fathers' mean from the mean of the whole population is

$$s - \frac{n}{4} = \frac{1}{4}(4s-n).$$

Hence,

$$\frac{\text{deviation of offspring from the population mean}}{\text{deviation of fathers from the population mean}} = \frac{\frac{1}{12}(4s-n)}{\frac{1}{4}(4s-n)} = \frac{1}{3} \tag{10.23}$$

(*ibid.*, p. 64). However, this is exactly Galton's result in QUOT. 313,[24] derived from the biometric approach. Furthermore, later in his paper, Pearson extended Eq. (10.23) and showed a geometric halving of ancestral regression (and correlation) coefficients:

$$\frac{\text{deviation of offspring from the population mean}}{\text{deviation of } m^{\text{th}} \text{ grandparent from the population mean}} = \frac{1}{3}\left(\frac{1}{2^m}\right)$$

(*ibid.*, p. 73). This result is in line with the geometric decrease in ancestral correlations predicted by the ancestral law. Regarding (10.23), Pearson declared:

> **QUOT. 656** A very general theory of the pure gamete type leads to linearity of the regression curve, a result amply verified by observations on inheritance in populations. . .the value of the correlation reached is numerically identical with the value obtained by FRANCIS GALTON in his original investigations on the inheritance of stature! The generalized theory of the pure gamete is thus shown, whatever the number of couplets taken, to lead to precisely the chief results already obtained by those who have studied heredity statistically. (*ibid.*, p. 64)

Those who had believed that, by showing its compatibility with biometry, Pearson had thus secured a place for Mendelism in the theory of evolution, found their hopes quickly crushed. For, Pearson next argued that recent biometric observations had indicated three facts that were in contradiction with the Mendelian result in Eq. (10.23):

> **QUOT. 657** *First.* - That the parental correlation appears to be markedly greater than $^1/_3$, nearer to .45 to. 5
> *Secondly.* - That this correlation appears to vary slightly from character to character.
> *Thirdly.* - That it does not appear to be absolutely the same for all species.
> (*ibid.*, pp. 64–65)

Pearson was therefore categorical: the general theory of the pure gamete (which was essentially the same as Mendel's theory) was too limited to be of general validity. In his own words:

> **QUOT. 658** It is most unfortunate for this general theory of the pure gamete, that it throws the Mendelian back into the position of the biometrician of 1885. One might have hoped that the generality involved in n couplets would have led to the requisite elasticity, or, failing this, to a numerical value of parental correlation nearer the cluster point of existing measurements than $^1/_3$. We can only say, at present, that a generalized theory of the pure gamete leads to precisely the same general features of regression as have been observed by the biometricians, but it appears numerically too narrow to describe the observed facts. (*ibid.*, p. 65)

Pearson thus rejected Mendelism even though his so-called general theory of the pure gamete was itself limited: it assumed equal gamete frequencies and dominance. The latter point would later be seized by Yule in a 1906 reply to Pearson.[25]

---

[24]The slope in Eq. (4.4) is 2/3 since the mid-parent was used; it would have been 1/3 if a single parent had been used. In that case, since the parent and offspring are assumed to have the same standard deviation, the regression coefficient is also the correlation coefficient. See footnote 37 of chap. 4.

[25]See Sect. 10.7.

10.5 Pearson's 1904 Response to Castle: Criticism of Castle's Mistake

Pearson's earlier paper (Pearson, 1904b) did not mention Castle (1903), presumably because Pearson had not yet read the paper. This omission was soon addressed in a paper that, as Castle later recalled (Keeler, 1968, p. 111), showed that Pearson was "very unhappy" about the former's earlier publication.[26] "Very unhappy" might still be an understatement because Pearson was relentless in his rebuke of Castle, especially regarding the latter's error.[27]

The tone was set in the title of Pearson's paper, "A Mendelian's view of the law of Ancestral inheritance" (Pearson, 1904a). In the opening paragraph, Castle was named and shamed as follows:

QUOT. 659 It is difficult to understand at all the attitude with which a biologist like Professor Castle approaches a problem in inheritance, but it is summed up in the words that he has not the least idea that there is any distinction between a correlation, a multiple regression coefficient, and the number of individuals who may take after an individual ancestor. (*ibid.*, p. 110)

Concerning Castle's confusion of Pearson's two correlation series with that of ancestral contributions, the latter said:

QUOT. 660 Had Professor Castle merely added up the numbers he gives in this "Pearson's Series" he would have realized that they could not possibly represent what he states them to do; for the four grandparents alone would have had more offspring "like" themselves than the total number of offspring! (*ibid.*, p. 111)

As explained in Sect. 10.2.2, Castle had indeed made a mistake in his interpretation of Pearson's series. However, regarding Castle's Table 10.2, which had been worked out in exactly the same way as Galton had done for the basset hounds data (see Sect. 4.6.4.2), Pearson now objected:

QUOT. 661 The Law of Ancestral Heredity, as I have repeatedly stated, makes no assumption as to the mean character of any generation; it gives no statement whatever as to the number of offspring who are like any individual ancestor. What it does give us is this: the means of determining the probable deviation of any individual from the type of his own generation, when we know deviations of some or all his ancestry from the types of their respective generations. (*ibidem*)

Pearson also expressed reservations about Galton's use of the law for discrete traits. He then ended his paper in the same fiery manner that he had started it, by throwing a challenge to Castle to either admit he was ignorant of the details behind the ancestral law or to retract the statements he had made against that law.

---

[26] Pearson had undoubtedly also been displeased with Castle's pro-Mendelian intervention (Castle & Allen, 1903) in the Darbishire affair.

[27] See Sect. 10.2.2.

Absent in Pearson's response paper was any comment on Castle's powerful argument in QUOT. 644 that Mendelian inheritance combined with natural selection can lead to the disappearance of a given trait, a conclusion directly at odds with the ancestral law.

10.6 Castle's 1905 Response to Pearson: The Ancestral Law Not a Law of Heredity

Did Castle take up the challenge that was thrown down to him by Pearson? Castle included a response to Pearson's 1904 paper (Pearson, 1904a) in a 1905 memoir published by the Carnegie Institution of Washington (Castle, 1905a). In his memoir, Castle did not acknowledge that he had wrongly interpreted Pearson's series in Table 10.2; instead, he pointed out that he had not used any of the Pearson's series to test the validity of the ancestral law. Next, Castle counterattacked some of the statements that Pearson had made in his response. Against the latter's admission in QUOT. 661 that the ancestral law cannot be used in the case of von Guita's data, Castle remarked that, if anything, this was a damning implication of the ancestral law:

> **QUOT. 662** Yet in Von Guaita's material we have a full record of the coat color of every animal in seven successive generations, and we know that back of this for an indefinite number of generations all the ancestors on one side of the ancestry were albinos, on the other side spotted black-and-white. That is sufficient basis on which to make very reliable predictions as to the character of the offspring, under Mendel's law, as I have elsewhere shown. (*ibid.*, p. 22)

Castle then went even further and concluded by pointing out a fundamental difference between Mendel's law and the ancestral law. The latter was a population-based statistical law and, unlike the former, could *not* be used to make predictions in individual crosses. Castle thought this was enough to disqualify the ancestral law as a law of heredity:

> **QUOT. 663** ...from a genuine law of heredity, such as that of Mendel, one is enabled, as I have shown, to predict with great accuracy what color types will prevail among the offspring in successive generations. Since Pearson's law, as now interpreted by its author, though dealing with alternative color types in successive generations, gives no information whatever as to *what* color types will prevail in one as compared with another, nor as to what proportion of the young will be *devoid* of color, it would seem desirable, to avoid confusion, that the law be renamed as something other than a law of heredity. (*ibid.*, p. 22–23)

10.7 Yule's 1906 Response to Pearson: Theory of Pure Gamete More Flexible than Proposed by Pearson

Yule's second and final paper (Yule, 1906c) on the reconciliation between Mendel's law and the ancestral law was presented at the Third International Conference on Genetics in 1906. In fact, this short paper was a commentary on Pearson's theoretical paper of 1904 (Pearson, 1904b). It will be recalled that Pearson's so-called model of the pure gamete had yielded a parental correlation of $^1/_3$ (see Eq. (10.23)), whence the latter rejected Mendelism since the observed correlations were closer to $^1/_2$ than to $^1/_3$. Yule's main point was that, by insisting on a dominant model, Pearson had limited his general theory of the pure gamete, and that a more flexible model could yield correlations that approximated the observed ones.

Yule's more flexible model consisted of an individual's total "length" L (across multiple loci, or "couplets" using Pearson's language) defined by

$$L \equiv am_1 + bm_2 + cm_3, \tag{10.24}$$

where m_1, m_2, and m_3 are the number of protozygotes (or homozygote dominants), heterozygotes, and allozygotes (or homozygote recessives), respectively, and a, b, and c are the dominance coefficients such that:

- for a dominant model, $a = b$;
- for a recessive model, $b = c$;
- for a codominant model, $b = (a + c)/2$.

Yule did not give details as to how his model worked, but he pointed out that it involved "methods which are relatively much simpler than those employed by Professor Pearson," yielding a parental correlation of

$$R = \frac{(a-c)^2}{2(a-c)^2 + (a-2b+c)^2}$$

(Yule, 1906c, p. 141). When $a = b$ or $b = c$, then $R = {}^1/_3$. Moreover, when $b = (a + c)/2$, a parental correlation of $R = {}^1/_2$ is achieved, which is close to the observed correlations. Yule then concluded:

> **QUOT. 664** ...the theory of the pure gamete, as applied to compound characters, is much more flexible than would appear from Professor Pearson's work, and can hardly be summarily dismissed as inapplicable to cases in which the coefficients of correlation approximate to 0.5. (*ibid.*, p. 142)

10.8 Pearson's 1907 Response to Yule

Pearson seldom let any criticism of his work go unanswered, and Yule's short 1906 paper (Yule, 1906c) was no exception. In an equally short paper published the next year (Pearson, 1907b), Pearson appeared quite defensive, since Yule had successfully poked some holes in his general theory of the pure gamete.

By quoting sentences from his paper, Pearson denied that he had ever claimed his theory was a generalized Mendelian theory of heredity. Next, he corrected Yule's assertion that $a = b$ or $b = c$ in Eq. (10.24) reduced to Pearson's theory. In fact, Pearson claimed, he had assumed that the Mendelian character had been determined solely by protozygotes, so that $b = c = 0$. He further declared that this assumption was truly Mendelian, as opposed to Yule's assumption that all of protogenic, heterogenic, and allogenic pairs had an effect on the character. Finally, Pearson denied that he had summarily dismissed the theory of the pure gamete, although this was, in fact, exactly what he had done in his paper.

10.9 Hardy's 1908 (July 10) Derivation of the Equilibrium Law

On February 28, 1908, Reginald Punnett (1875–1967), who had been Bateson's associate since 1903,[28] delivered a lecture to the Royal Society of Medicine. The lecture was entitled "Mendelism in Relation to Disease" (Punnett, 1908) and was attended by several scholars, including Yule. As mentioned before,[29] Yule's views were much more aligned with those of the biometricians than with those of the Mendelians. During the discussion session, one of his comments on Punnett's lecture was reported as follows:

> **QUOT. 665** Assuming that brown or duplex eye-colour was dominant over blue, if matings of persons of different eye-colours were random (and that was very nearly true), it was to be expected that in the population there would be three persons with brown eyes to one with blue; but that was not so. There were more blues than browns. The same applied to the examples of brachydactyly. The author said that brachydactyly was dominant. In the course of time one would then expect, in the absence of counteracting factors, to get three brachydactylous persons to one normal, but that was not so. (*ibid*, p. 165)

Thus, Yule believed that a random mating population should yield a 3:1 ratio for dominants to recessives, as in hybridization experiments. However, he did not take into account that, in the latter experiments, all F_1 crosses were of the $Aa \times Aa$ type, so a 3:1 ratio was to be expected in F_2. On the other hand, in a random mating population, F_1 crosses can be of any type and 3:1 ratios cannot be expected in F_2.

Punnett's later recounting of the episode seems to indicate that he had misunderstood Yule's question to mean why the dominants did not eventually take over completely:

> **QUOT. 666** In the subsequent discussion I was asked why it was that, if brown eye were dominant to blue, the population was not becoming increasingly brown-eyed... I could only answer that the heterozygous browns also contributed their quota of blues, and that somehow this must lead to equilibrium. (Punnett, 1950, p. 9)

It is probably in the above form that, on his return to Cambridge, Punnett put the question to the pure mathematician Godfrey Harold Hardy (Fig. 10.4), with whom he used to play cricket and had previously acted as joint secretaries to the Committee for the retention of Greek in the Previous Examination. Hardy had no interest in

---

[28] Bateson's formal offer to Punnett in 1903 was that of a professional "partner," with a stipend of £80 a year. Punnett was so excited by the offer that he refused the stipend. See pp. 216–219 of Cock and Forsdyke (2008).

[29] See Sect. 10.1.3 and end of Sect. 10.1.4.2.

Fig. 10.4 Godfrey Harold Hardy (1877–1947). (From: https://upload.wikimedia.org/wikipedia/commons/6/6f/Godfrey_Harold_Hardy_1.jpg)

genetics but he was already one of the finest mathematicians England had ever known.[30] His simple solution to Punnett's question earned him a lasting legacy in the history of population genetics.

Hardy's solution was written on April 5, 1908, and was published as a short *Letter* in the July 10, 1908, issue of *Science*[31] (Hardy, 1908). The *Letter* started modestly as follows:

> **QUOT. 667** I am reluctant to intrude in a discussion concerning matters of which I have no expert knowledge, and I should have expected the very simple point which I wish to make to have been familiar to biologists. However, some remarks of Mr. Udny Yule, to which Mr. R. C. Punnett has called my attention, suggest that it may still be worth making. (*ibid.*, p. 49)

Referring to Yule's comment in QUOT. 665, Hardy declared that "such an expectation would be groundless." To prove this, he let the number of pure dominants (*AA*), heterozygotes (*Aa*), and pure recessives (*aa*) be in the ratio $p : 2q : r$. Then:

[30] In 1940, Hardy also wrote *A Mathematician's Apology* (Hardy, 1941) a must-read for any mathematician, where he expressed his distaste of applied mathematics and his love for pure mathematics, which he considered to be "creative art." His discovery of one of the most applied equations in population genetics is thus quite ironical. Hardy also helped the Indian mathematical prodigy Srinivasa A. Ramanujan (1887–1920) come to England in 1914 and collaborated with him until the latter's early demise in 1920.

[31] The *Letter* was not published in *Nature* because the latter's editorial policy was strongly anti-Mendelian at the time. See Edwards (2008, p. 1147).

QUOT. 668...suppose that the numbers are fairly large, so that the mating may be regarded as random, that the sexes are evenly distributed among the three varieties, and that all are equally fertile. A little mathematics of the multiplication-table type is enough to show that in the next generation the numbers will be as

$$(p+q)^2 : 2(p+q)(q+r) : (q+r)^2,$$

or as $p_1 : 2q_1 : r_1$, say. (*ibidem*)

Hardy's $(p + q)^2 : 2(p + q)(q + r) : (q + r)^2$ is simply the square-law component[32] of a fundamental principle in population genetics, later known as the Hardy-Weinberg law. If the frequencies of gametes A and a are denoted by P and Q $(= 1 - P)$, respectively, then the square-law component can be written as

$$P^2 : 2PQ : R^2.$$

Thus, when the square-law component is written in the form above, Hardy's "little mathematics of the multiplication-table type" is based on the equivalence between *random mating of individuals* and the *random union of their gametes*:

$$\text{freq}(\text{AA}) = P\text{x}P = P^2,$$
$$\text{freq}(\text{Aa}) = P\text{x}Q + Q\text{x}P = 2PQ,$$
$$\text{freq}(\text{aa}) = Q\text{x}Q = Q.^2$$

Concerning the equilibrium component of the law:

QUOT. 669 The interesting question is - in what circumstances will this distribution be the same as that in the generation before? It is easy to see that the condition for this is $q^2 = pr$. And since $q_1^2 = p_1 r_1$, whatever the values of p, q and r may be, the distribution will in any case continue unchanged after the second generation. (*ibidem*)

Hardy thus asked the question: under what condition will genotype frequencies be preserved across generations? If one puts $p = (p + q)^2$, then:

$$p = p^2 + 2pq + q^2$$
$$p(1-p) = 2pq + q^2$$
$$p(2q+r) = 2pq + q^2$$
$$\therefore\ q^2 = pr$$

Note that this condition is *not* satisfied in Hardy's initial generation of $p : 2q : r$. However, it *is* satisfied in the second generation since $\{(p + q)(q + r)\}^2 = (p + q)^2(q + r)^2$, i.e., $q_1^2 = p_1 r_1$. Hence, Hardy's statement in QUOT. 669 that "the distribution will in any case continue unchanged after the second generation."

[32] See footnote 3.

Hardy concluded his analysis as follows:

QUOT. 670 . . .there is not the slightest foundation for the idea that a dominant character should show a tendency to spread over a whole population, or that a recessive should tend to die out. (*ibidem*)

The above probably reflects the way in which Punnett had posed the question to him.

In a postscript to his *Letter*, Hardy mentioned that Pearson (1904b) had already shown stability of the 1:2:1 ratio.[33] However, as described in Sects. 10.1.4 and 10.2.3, even earlier Yule (1902) had obtained the law for the 1:2:1 case, followed by Castle (1903) in the general case.

Punnett later recalled (1950, p. 9) that he promised Hardy that the latter's result ought to be known as "Hardy's Law," a promise that Punnett declared he fulfilled in the next (third) edition of *Mendelism* (Punnett, 1911). Indeed, *Mendelism* does contain an interesting discussion of Hardy's result, although it is there called a "suggestive contribution" (*ibid.*, p. 136) by G.H. Hardy to the subject, rather than "Hardy's Law."

In any case, no less due to the efforts of Punnett, the importance of Hardy's law was quickly realized. In his important book *Genetics and the Origin of Species,* Dobzhansky summarized its implication as follows:

QUOT. 671 Hardy's formula is important because it shows that the hereditary variability once gained by a population is automatically maintained on a constant level, instead of being eroded and finally leveled off by crossing. As mentioned already, this is a corollary to the particulate as contrasted with the blending type of inheritance. The maintenance of the genetic equilibrium is evidently a conservative and not a progressive factor. Evolution is essentially a modification of this equilibrium. (Dobzhansky, 1937, p. 124)

The conservation of genetic variation was an important consequence of Hardy's formula. Recall that, under Darwin's (and later, the biometricians') model of blending inheritance, variability was soon lost[34]; hence, the latter needed to appeal to the provisional hypothesis of pangenesis.[35] On the other hand, the preservation of genetic variation in a Mendelian population, as predicted by Hardy's formula, forgoes the need for any such hypothesis.

Hardy's result was thus a welcome advantage for the Mendelians in their fight against the biometricians.

[33] See Sect. 10.4.3.

[34] The loss of variability was eloquently expressed in Fleeming Jenkin's swamping argument. See Sect. 2.14.4.3.

[35] See Sect. 3.7.2.

10.10 Weinberg's 1908 (Jan 13) Derivation and Use of the Equilibrium Law

In 1943, the German-born American geneticist Curt Stern wrote a note in *Science* (Stern, 1943), indicating that Hardy's 10 July 1908 result had been anticipated by a few months by the German physician and medical statistician Wilhelm Weinberg (Fig. 10.5).

Weinberg's interest in genetics started as a result of his obstetrical practice.[36] From 1866 until 1935, he wrote more than 160 papers in genetics and other fields. His important 84-page paper of 1901, "Beiträge zur Physiologie und Pathologie der Mehrlingsgeburten beim Menschen" (Weinberg, 1901), introduced the difference method for estimating the number of monozygotic and dizygotic twins. Weinberg's attention was turned toward Mendelian problems when, in 1905, he attended a lecture by Heinrich Ernst Ziegler (1858–1925), a former student of August Weismann. This resulted in the publication of the following four fundamental Mendelian papers:

- Über den Nachweis der Vererbung beim Menschen[37](Weinberg, 1908a)
- Über Vererbungsgesetze beim Menschen I and II[38] (Weinberg, 1908b, 1909), and
- Weitere Beiträge zur Theorie der Vererbung[39] (Weinberg, 1910)

In view of his several pioneering contributions, Weinberg deserves to be called one of the greatest (and most underrated) geneticists of the twentieth century.

In the first of the four publications above, Weinberg (1908a) developed the equilibrium law. His lecture was delivered to the "Vereins für vaterländische Naturkunde in Württemberg" on January 13, 1908. Weinberg's aim was to verify Mendelian (as opposed to blending) inheritance in twins and his reason for considering a panmictic population rather than a pedigree is explained as follows:

> **QUOT. 672** Following a suggestion of Professor Haecker's I investigated my previously collected material for evidence supporting Mendelian inheritance in twins. However, it would have been very difficult to assemble a sufficient number of cases in which children of mothers of twins had married one another and then determined the frequency of twin births among their children. I therefore attempted to construct a formula for the frequency of dominant and recessive traits among the mothers and siblings and daughters of carriers of traits of the same character under the assumption that absolute panmixia is present. (Weinberg, 1908a, 1977, p. 120)

Weinberg first reported the ratio of dominants to recessives under repeated self-fertilization. Recall that this result was first obtained by Mendel in his groundbreak-

[36] See Stern (1962) for an informative biography on Weinberg.

[37] On the proof of inheritance in Man.

[38] On the law of inheritance on Man I and II.

[39] Further contributions to the theory of inheritance.

Fig. 10.5 Wilhelm Weinberg (1862–1937). (From: https://www.sutori.com/en/item/wilhelm-weinberg-1862-1937-wilhelm-weinberg-a-german-physician-didn-t-collabor)

ing paper (Mendel, 1866, 1966a, p. 16). Assuming *A* to be dominant over *a*, the required ratio is $2^{n-1} + 1 : 2^{n-1} - 1$ for the *n*th generation.[40] However:

> **QUOT. 673** This situation appears much different when Mendelian inheritance is viewed under the influence of panmixia. (Weinberg, 1908a, p. 122)

Then:

> **QUOT. 674** I start by assuming that m male and female pure representatives of Type A and likewise n representatives of Type B are present. If these are randomly crossed then, by application of the symbolism of the binomial theorem, the composition of the daughter generation is obtained:

$$(\mathrm{mAA} + \mathrm{nBB})^2 = \frac{\mathrm{m}^2}{(\mathrm{m}+\mathrm{n})^2}\mathrm{AA} + \frac{2\mathrm{mn}}{(\mathrm{m}+\mathrm{n})^2}\mathrm{AB} + \frac{\mathrm{n}^2\mathrm{B}^2[sic]}{(\mathrm{m}+\mathrm{n})^2} \tag{10.25}$$

or, if m + n = 1

[40]In Mendel's notation, this is the ratio for the $(n - 1)$th generation. See Fig. 6.6 for a proof by induction of the result.

$$\mathrm{m^2AA + 2mnAB + n^2BB} \tag{10.26}$$

(*ibidem*)

The expression in (10.26) is the square-law component of the equilibrium law. Like Hardy later, Weinberg presumably equated random mating with the random union of gametes. This is evident when the binomial in Eq. (10.25) is more appropriately written as

$$\mathrm{(mA + nB)^2 = m^2AA + 2mnAB + n^2BB}$$

Regarding the equilibrium component of the law, Weinberg proceeded differently from Hardy, by modeling the actual mating of the different genotypes:

QUOT. 675 If the male and female members of the first generation are randomly crossed, then the following frequencies of different combinations of crosses are obtained: or

$$\mathrm{m^2m^2(AA\ x\ AA) = m^4AA}$$

$$\mathrm{4m^2mn\ (AA\ x\ AB) = 2m^3nAA + 2m^3nAB}$$

$$\mathrm{2m^2n^2(AA\ x\ BB) = 2m^2n^2AB}$$

$$\mathrm{4(mn)^2(AB\ x\ AB) = m^2n^2AA + 2m^2n^2AB + m^2n^2BB}$$

$$\mathrm{4mnn^2(AB\ x\ BB) = 2mn^3AB + 2mn^3BB}$$

$$\mathrm{n^2n^2(BB\ x\ BB) = n^4BB}$$

or the relative frequencies are:

$$\mathrm{AA : m^2(m+n)^2}$$

$$\mathrm{AB : 2m(m+n)^2n}$$

$$\mathrm{BB : (m+n)^2n^2}$$

and the composition of the second or daughter generation is again:

$$\mathrm{m^2AA + 2mnAB + n^2BB}$$

We thus obtain the same distribution of pure types and hybrids for each generation under the influence of panmixia and thereby the possibility of calculating for each generation how these types are represented in panmixia and Mendelian inheritance among the parents, sibling, and children of the various types and hybrids. (*ibid.*, pp. 122–123)

In fact, Weinberg's proof can also be written without using the equivalence between random mating and the random union of gametes, as follows. Consider a large random mating population subject to Mendelian segregation, with homozygotic and heterozygotic (AA, AB, BB) frequencies of M, P, and N, respectively, such that the frequencies of the gametes are:

$$m = \mathrm{freq(A)} = M + \frac{P}{2},$$

$$n = \mathrm{freq(B)} = 1 - \mathrm{freq(A)} = N + \frac{P}{2}.$$

Assuming no evolutionary forces, the possible parental crossings, the associated frequencies, and the frequencies and types of the resulting offspring are shown in Table 10.3.

Table 10.3 Modern proof of equilibrium law using Weinberg's Idea

Parental crossing	Frequency of parental crossing	Frequency in offspring population		
		AA	AB	BB
AA × AA	M^2	M^2	0	0
AA × AB	$2MP$	MP	MP	0
AA × BB	$2MN$	0	$2MN$	0
AB × AB	P^2	$P^2/4$	$P^2/2$	$P^2/4$
AB × BB	$2NP$	0	NP	NP
BB × BB	N^2	0	0	N^2

The frequencies of the various homozygotes and heterozygotes in the offspring population are:

$$\text{freq}(\text{AA}) = M^2 + MP + \frac{P^2}{4} = \left(M + \frac{P}{2}\right)^2 = m^2,$$

$$\begin{aligned}\text{freq}(\text{AB}) &= MP + 2MN + \frac{P^2}{2} + NP \\ &= 2M\left(\frac{P}{2} + N\right) + P\left(\frac{P}{2} + N\right) \\ &= 2\left(M + \frac{P}{2}\right)\left(N + \frac{P}{2}\right) \\ &= 2mn,\end{aligned}$$

$$\text{freq}(\text{BB}) = \frac{P^2}{4} + NP + N^2 = \left(N + \frac{P}{2}\right)^2 = n^2.$$

By crossing the F_1 offspring as in Table 10.3 with $M = m^2$, $P = 2mn$, and $N = n^2$, the frequencies in the F_2 offspring are again $(m^2, 2mn, n^2)$, i.e. equilibrium has been achieved from F_1 onward.

Having modeled Mendelian inheritance in a panmictic population, Weinberg returned to the motivation for actually developing the model, namely, to verify whether twin births followed Mendelian (as opposed to blending) inheritance. Using the results of his model, he was able to predict that:

- assuming A is dominant over B:
  - (i) Among the parents of a given dominant individual, dominants:recessives = $(1 + mn) : n^2$
  - (ii) Among the siblings of a given dominant individual, dominants:recessives = $\{4(1 + mn) + mn^2\} : n^2(3 + 4n)$
  - (iii) Among the children of a given dominant individual, dominants:recessives = $(1 + mn) : n^2$

- assuming A is recessive over B:
  - (iv) Among the parents of a given dominant individual, recessives:dominants $= m : n$

(v) Among the siblings of a given dominant individual, recessives:dominants = $(2 + m)^2 : (3 + 4n)$

(vi) Among the children of a given dominant individual, recessives:dominants $= m : n$

Although Weinberg did not show the calculations, any of the results in (i)–(vi) above may readily be obtained. For example, regarding (iii), the possible crossings of a *given* dominant parent (written first in each row), the children and their frequencies (in brackets) are:

$$\begin{aligned}
&\mathrm{AA} \times \mathrm{AA} = \mathrm{AA}(m^4)\\
&\mathrm{AA} \times \mathrm{AB} = \mathrm{AA}(m^3n) + \mathrm{AB}(m^3n)\\
&\mathrm{AA} \times \mathrm{BB} = \mathrm{AB}(m^2n^2)\\
&\mathrm{AB} \times \mathrm{AA} = \mathrm{AA}(m^3n) + \mathrm{AB}(m^3n)\\
&\mathrm{AB} \times \mathrm{AB} = \mathrm{AA}(m^2n^2) + \mathrm{AB}(2m^2n^2) + \mathrm{BB}(m^2n^2)\\
&\mathrm{AB} \times \mathrm{BB} = \mathrm{AB}(mn^3) + \mathrm{BB}(mn^3)
\end{aligned}$$

Therefore,

$$\begin{aligned}
\frac{\text{freq. of dominants amongst children}}{\text{freq. of recessives amongst children}} &= \frac{\text{freq}(\mathrm{AA}\text{ or }\mathrm{AB})}{\text{freq}(\mathrm{BB})}\\
&= \frac{m^4 + 4m^3n + m^2n^2 + mn^3}{m^2n^2 + mn^3}\\
&= \frac{m(m+n)(m^2 + 3mn + n^2)}{mn^2(m+n)}\\
&= \frac{(m+n)^2 - 2mn + 3mn}{n^2}\\
&= \frac{1 + mn}{n^2}.
\end{aligned}$$

Regarding his results in (i)–(vi) above, Weinberg concluded:

QUOT. 676 For characteristics which are measurable, as in our case, this must lead to different average values for parents and sibling. Assuming a non-Mendelian character, in which the hybrids are represented by several intermediate forms, the average representation of measurable characteristics would be the same in parents and siblings. Thus a real difference exists between Mendelian and non-Mendelian characters. I have found that the trait for triplets results in essentially different values for parents and sibling, which supports a Mendelian trait. Similar small differences occur with twin births. (*ibid.*, p. 123)

Based on his observations on triplets and twins, and his Mendelian values, Weinberg thus favored a Mendelian model of inheritance over a blending one.

Furthermore, in the last part of his paper, Weinberg provided a method that is still used today to infer the mode of inheritance. By assuming the latter to be (I) recessive, or (II) dominant, (III) equivalent to single births, he calculated the equilibrium frequencies for the following groups:

Table 10.4 Weinberg's observed and expected (assuming HWE) counts of multiple birth under different modes of inheritance (I: recessive, II: dominant, III: same as single births) for different groups [(a): mothers of mothers of twins, (b): daughters of mothers of twins, (c): sisters of mothers of twins, (d): mothers of mothers of triplets, (e): sisters of mothers of triplets] (Weinberg, 1908a, 1977, p. 124)

Group	Expected counts under each mode of inheritance			Observed count
	I	II	III	
(a)	26	27	30	27
(b)	28	31	33	24
(c)	21	22	23	27
(d)	51	91	91	45
(e)	45	57	57	36
	171	228	232	159

(a) among the mothers of mothers of twins
(b) among the daughters of mothers of twins
(c) among the sisters of mothers of twins
(d) among the mothers of mothers of triplets
(e) among the sisters of mothers of triplets

The observed and expected counts under (I), (II), and (III) for each of the groups (a)–(e) are displayed in Table 10.4.

By comparing the observed and expected counts, Weinberg concluded:

QUOT. 677 The assumption that the tendency for twin births is recessive gives expectable values which are much closer than those expected under any other assumption...

The results found in the inheritance of multiple births therefore are best explained by the assumption that the tendency to multiple births is inherited according to Mendel's law and is recessive. (*ibid.*, p. 125)

Weinberg later generalized his result to multiple alleles in a similar way (Weinberg, 1908b, p. 443). While neither Hardy nor others were at first aware of Weinberg's equilibrium result, the latter soon learned of Hardy's derivation and mentioned it. For example, in 1910, Weinberg said:

QUOT. 678 In this case panmixis leads to a constant composition of successive generations. This was shown first by Pearson[41] for a sequence of generations originating from a single cross, while Hardy and, myself, more or less at the same time and independently of each other, proved it for any composition of the generation in which panmixis begins. (Weinberg, 1910, p. 43)

In his paper, Stern (1943) suggested that the equilibrium law, hitherto known as Hardy's Law, be called "The Hardy-Weinberg Law." This suggestion was gradually adopted. Stern later wrote another paper (Stern, 1965) where he indicated that the

[41] As shown in Sect. 10.1.4, this was first done by Yule, not Pearson (*our note*).

Hardy-Weinberg law in its general form had been anticipated by Castle (1903),[42] although not as explicitly as Weinberg and Hardy had independently done later.

10.11 Equilibrium at Two Loci (Weinberg, 1909; Jennings, 1917; Robbins, 1918b; Lewontin & Kojima, 1960)

10.11.1 Weinberg (1909)

Weinberg (1909) was the first to mathematically investigate the attainment of equilibrium in the case of two loci. Weinberg's model implicitly assumed no linkage. He was able to show that, even with independent assortment, equilibrium in a two-locus model was achieved only asymptotically (i.e. after a very long time).

Weinberg considered two bi-allelic loci A and B[43] with alleles (A_1, A_2) and (B_1, B_2), respectively, in a random-mating population. Initially (G_0), suppose there were only two gametes A_1B_1 and A_2B_2, with respective frequencies m and n, respectively, where $m + n = 1$. Note that this implies that freq(A_1) = freq(B_1) = m and freq (A_2) = freq(B_2) = n. After one generation of random mating, the distribution of gametes in G_1 is as follows:

$$(mA_1B_1 + nA_2B_2)^2 = m^2A_1A_1B_1B_1 + 2mnA_1A_2B_1B_2 + n^2A_2A_2B_2B_2$$

(*ibid.*, p. 286). After one further generation of random mating, the distribution of gametes in G_2 is

$$\begin{aligned} m^2\frac{1}{4}(A_1 + A_1)(B_1 + B_1) &= m^2A_1B_1 \\ + 2mn\frac{1}{4}(A_1 + A_2)(B_1 + B_2) &= \frac{mn}{2}A_1B_1 + \frac{mn}{2}A_1B_2 + \frac{mn}{2}A_2B_1 + \frac{mn}{2}A_2B_2 \\ +n^2\frac{1}{4}(A_2 + A_2)(B_2 + B_2) &= n^2A_2B_2. \end{aligned}$$

Thus, G_2 can be represented by

$$\left(m^2 + \frac{mn}{2}\right)A_1B_1 + \frac{mn}{2}A_1B_2 + \frac{mn}{2}A_2B_1 + \left(n^2 + \frac{mn}{2}\right)A_2B_2.$$

By writing the gamete distributions as follows,

[42] See Sect. 10.2.3 (end of section mentions Castle's paper).

[43] Note the change in notation from Weinberg's 1908 one-locus equilibrium paper (e.g., see QUOT. 674), where A and B denoted two alleles at *one* locus.

$$G_1 : (mA_1B_1 + nA_2B_2) + (mA_1 + nA_2)(mB_1 + nB_2)$$
$$G_2 : \frac{1}{2}(mA_1B_1 + nA_2B_2) + \frac{1}{2}(mA_1 + nA_2)(mB_1 + nB_2)$$

Weinberg inferred the gametic composition in generation $x(x = 0, 1, 2, \ldots)$ as

$$G_x : \frac{1}{2^x}(mA_1B_1 + nA_2B_2) + \left(1 - \frac{1}{2^x}\right)(mA_1 + nA_2)(mB_1 + nB_2)$$

(*ibidem*). Therefore, as $x \to \infty$, the gametic composition becomes

$$m^2A_1B_1 + mnA_1B_2 + mnA_2B_1 + m^2A_2B_2,$$

which corresponds to two-locus equilibrium (since freq(A_1B_1) = freq(A_1) × freq (B_1) = m^2, freq(A_1B_2) = freq(A_1) × freq(B_2) = mn, and so on). Weinberg concluded that, in the two-locus case, equilibrium is attained "after an infinite number of generations".

10.11.2 Jennings (1917)

Herbert Jennings (Fig. 10.6) was an American zoologist and geneticist. He was born in Tonica, Illinois and received his PhD from Harvard University in 1896. He is especially well-known for his work in invertebrate psychology.

Unaware of Weinberg's pioneering work, Jennings (1917) published a paper that incorporated linkage in a two-locus model with random mating. In his paper, Jennings made two important assumptions:

(a) The linkage parameter was denoted by r, which stood for the number of gametes showing the same combination of factors as in the previous generation.[44] Thus, zygotes of the type *ABab* give gametes *AB*, *Ab*, *aB*, *ab* in the ratio $r : 1 : 1 : r$
(b) Each zygote produced $(2r + 2)$ gametes

Then, Jennings listed all possible zygote → gamete pairs (*ibid.*, p. 108):

(i) *ABAB* → $(2r + 2)AB$
(ii) *AbAb* → $(2r + 2)Ab$
(iii) *aBaB* → $(2r + 2)aB$
(iv) *abab* → $(2r + 2)ab$
(v) *ABab* → $rAB + Ab + aB + rab$
(vi) *AbaB* → $AB + rAb + raB + ab$
(vii) *ABAb* → $(r + 1)AB + (r + 1)Ab$
(viii) *ABaB* → $(r + 1)AB + (r + 1)aB$

[44] The modern usage is to define the linkage parameter R as the fraction of new types or recombinants, i.e., $R = 1/(1 + r)$. For the discovery of recombination, see Sect. 13.4.4.3.

Fig. 10.6 Herbert Spencer Jennings (1868–1947). (From: https://commons.wikimedia.org/wiki/File:Herbert_Spencer_Jennings.jpg)

(ix) *abAb* → (*r* + 1)*Ab* + (*r* + 1)*ab*
(x) *abaB* → (*r* + 1)*aB* + (*r* + 1)*ab*

Furthermore, let p_n, q_n, s_n and t_n denote the proportions of the gametes *AB*, *Ab*, *aB*, and *ab* respectively, in the *n*th generation. Then, in the (*n* + 1)th generation, an *AB* gamete is obtained if, in the *n*th generation, it is produced either from an *AB/AB* genotype, or from an *AB/Ab* genotype, or from an *AB/aB* genotype, or from an *AB/ab* genotype without recombination, or from *Ab/aB* with recombination, i.e.,

$$p_{n+1} = (r+1)p_n^2 + (r+1)p_n q_n + (r+1)p_n s_n + r p_n t_n + q_n s_n.$$

A similar reasoning can be applied to the other three gametes, so that for all four gametes:

$$\begin{aligned} p_{n+1} &= (r+1)p_n(p_n + q_n + s_n) + r p_n t_n + q_n s_n, \\ q_{n+1} &= (r+1)q_n(p_n + q_n + t_n) + r q_n s_n + p_n t_n, \\ s_{n+1} &= (r+1)s_n(p_n + s_n + t_n) + r q_n s_n + p_n t_n, \\ t_{n+1} &= (r+1)t_n(q_n + s_n + t_n) + r p_n t_n + q_n s_n \end{aligned} \tag{10.27}$$

(*ibid.*, p. 144). Note that Jennings' proportions did not add to unity, since

$$p_{n+1} + q_{n+1} + s_{n+1} + t_{n+1} = (r+1)(p_n + q_n + s_n + t_n)^2.$$

10.11.3 Robbins (1918b)

Rainard B. Robbins (1886–1951) was born in Clinton County, Indiana. He had a most unusual career. He obtained an A.B. degree from Indiana University in 1909 and a PhD from Harvard University in 1914. His doctorate dissertation was entitled *A Method in the Calculus of Variations.* Thereupon, he wrote some key papers in population genetics. However, in 1923, he resigned from the Mathematical Faculty of the University of Michigan to work as an assistant actuary in the New York State Insurance Department. He subsequently served the Teachers Insurance and Annuity Association (T.I.A.A.), successively as secretary, actuary, vice-president, and member of the board of trustees, until his death in 1951.

Robbins (1918b) used Jennings' equations in Eqs. (10.27) to derive important results in two-locus theory. He introduced the additional condition that Jennings' proportions of *AB*, *Ab*, *aB*, *ab* should be standardized to P_n, Q_n, S_n, T_n, respectively, such that:

$$P_n + Q_n + S_n + T_n = 1.$$

This means that $P_n = p_n/(1 + r)$, $Q_n = q_n/(1 + r)$, and so on. Jennings noted that Eqs. (10.27) were greatly simplified by introducing the quantity

$$\Delta_n = Q_n S_n - P_n T_n \tag{10.28}$$

(*ibid.*, p. 2), for then the first equation in Eqs. (10.27) becomes

$$p_{n+1} = (r+1)p_n(p_n + q_n + s_n + t_n) + q_n s_n - p_n t_n = (r+1)p_n + \Delta_n.$$

Similarly for the other proportions, so that, in terms of the standardized proportions:

$$\begin{aligned} P_n &= P_{n-1} + \Delta_{n-1}/(1+r), \\ Q_n &= Q_{n-1} - \Delta_{n-1}/(1+r), \\ S_n &= S_{n-1} - \Delta_{n-1}/(1+r), \\ T_n &= T_{n-1} + \Delta_{n-1}/(1+r) \end{aligned} \tag{10.29}$$

(*ibidem*). Moreover, Eq. (10.28) can be written as

$$\Delta_n = \{Q_{n-1} - \Delta_{n-1}/(1+r)\}\{S_{n-1} - \Delta_{n-1}/(1+r)\} - \{P_{n-1} + \Delta_{n-1}/(1+r)\} \times \{T_{n-1} + \Delta_{n-1}/(1+r)\},$$

yielding

$$\Delta_n = \frac{r}{1+r}\Delta_{n-1} \tag{10.30}$$

(*ibidem*). This implies that

$$\Delta_n = \left(\frac{r}{1+r}\right)^n \Delta_0, \quad n = 0, 1, 2, \ldots \tag{10.31}$$

Note that:

(i) For independent assortment (the case considered by Weinberg in Sect. 10.11.1), $r = 1$ so that $\Delta_n = (1/2)^n\Delta_0$ and $\Delta_n \to 0$ as $n \to \infty$. In view of Weinberg's finding that two-locus equilibrium is attained only after an "infinite number of generations," this would suggest that Δ_n is a measure of departure from two-locus equilibrium.
(ii) For complete linkage, $r \to \infty$ so that $\Delta_n = \Delta_0$ is constant (i.e., "*the case of complete linkage is in essence a one-factor problem*").

10.11.4 Lewontin and Kojima (1960)

Building on the works of Geiringer (1944) and Kimura (1956) (who had used D to denote Robbins' $-\Delta$), Lewontin and Kojima (1960) introduced the modern terminology in two-locus theory and gave the correct interpretation of Kimura's D with respect to the two-locus equilibrium. In their paper, the authors stated:

QUOT. 679 Essential to a discussion of the two- locus problem is the concept of *linkage equilibrium*. (*ibid.*, p. 458)

This was the first use of the term *linkage disequilibrium*, defined as

$$D = g_{AB}g_{ab} - g_{Ab}g_{aB}, \tag{10.32}$$

where $g_{AB}, g_{Ab}, g_{aB}, g_{ab}$ are, respectively, the frequencies of the gametes *AB*, *Ab*, *aB*, *ab* at two bi-allelic loci *A* and *B*. Now, let

$$\begin{aligned} p_A &= g_{AB} + g_{Ab} = \text{freq}(A), \\ 1 - p_A &= g_{aB} + g_{ab} = \text{freq}(a), \\ p_B &= g_{AB} + g_{aB} = \text{freq}(B), \\ 1 - p_B &= g_{Ab} + g_{ab} = \text{freq}(b). \end{aligned} \tag{10.33}$$

Lewontin and Kojima defined linkage equilibrium as the condition under which each gametic frequency is equal to the product of the corresponding allele frequencies, i.e., under linkage equilibrium, $g_{AB} = p_A p_B$, $g_{Ab} = p_A(1 - p_B)$, $g_{aB} = (1 - p_A)p_B$, $g_{ab} = (1 - p_A)(1 - p_B)$. By substituting one equation at a time from (10.33) into (10.32), it can be shown that:

$$D = g_{AB} - p_A p_B = p_A(1 - p_B) - g_{Ab} = (1 - p_A)p_B - g_{aB} = g_{ab} - (1 - p_A)(1 - p_B),$$

which explains more explicitly than (10.32) why D is a measure of linkage *disequilibrium.*

Finally, using R to denote the recombination frequency (which is the same as Jennings' and Robbins' $1/(1 + r)$), Lewontin and Kojima obtained the following expression for the linkage disequilibrium parameter after n generations of random mating:

$$D_n = (1 - R)^n D_0,$$

in analogy with Robbins' Eq. (10.31). The authors noted that, in the presence of evolutionary forces (most especially natural selection), the approach to equilibrium may be disturbed.

10.12 Random Mating in a Subdivided Population: The Wahlund Effect (1928)

If several random-mating demes are treated as a single population, then the latter will show an excess of homozygotes (and a deficiency of heterozygotes) relative to Hardy-Weinberg expectations. The excess homozygosity, which is usually observed under inbreeding or selection for homozygotes, turns out to be simply the effect of subdivision. This interesting fact, also known as the *Wahlund effect*, was first observed by Wahlund in 1928.

Sten G.W. Wahlund (Fig. 10.7) was born in Bollnäs, Sweden. He graduated from Uppsala University in 1924 and obtained his doctorate in statistics from the University of Stockholm in 1933. His dissertation (Wahlund, 1932) dealt with the statistical analysis of the Lapponic population of Northern Sweden. He served as Professor of Statistics at the University of Stockholm during the period 1938–1953. He also served as center-party member of the Swedish parliament in 1944–1970. His single paper in population genetics is the one that deals with the Wahlund effect (Wahlund, 1928, 1975).[45]

Wahlund's paper was generally concerned with deviations from Hardy-Weinberg equilibrium in real populations. After studying one- and multi-locus models, Wahlund turned his attention to the effect of isolation in a one-locus model:

[45] Wahlund's paper remains the most cited article of the journal *Hereditas* (Tunlid et al., 2020, p. 11).

Fig. 10.7 Sten Gösta William Wahlund (1901–1976). (From: https://commons.wikimedia.org/wiki/File:Sten_G%C3%B6sta_William_Wahlund_SPA.jpg)

QUOT. 680 In a previous section (p. 226), formulas were presented for the composition of a population in monohybrid heredity under the assumption of random crossing. This hypothesis at times produces results that deviate from reality. A cause of such deviations in a population, which has up to now not been of fundamental concern, is the occurrence of limits, which will be designated as isolation boundaries. In this and the following sections, we shall attempt to determine how important such boundaries are and to elucidate under which circumstances one is allowed to work with the formulas and results that are built upon the presupposition of complete panmixia. (*ibid.*, p. 241)

Thereupon, Wahlund considered two demes (isolates) G and H of relative size $g : h$ where $g + h = 1$. Let the frequencies of gamete (D) for the dominant trait in G and H be d_g and d_h, respectively, and those of gamete (R) for the recessive trait be r_g and r_h, respectively. Then $d_g + r_g = d_h + r_h = 1$. Assuming random mating in each of G and H, the genotypic frequencies in each deme are given in Table 10.5.

If the isolation (subdivision) is ignored, the observed proportion of D gametes is $\overline{p}_R = gr_g + hr_h$ and the expected proportion of RR homozygotes under HWE is

Table 10.5 Genotypic frequencies in Demes G and H under random mating

Genotype	Proportions in G	in H
DD	d_g^2	d_h^2
DR	$2d_g r_g$	$2d_h r_h$
RR	r_g^2	r_h^2

$$\overline{p}_R^2 = (gr_g + hr_h)^2 = g^2 r_g^2 + h^2 r_h^2 + 2ghr_g r_h.$$

However, the observed proportion of *RR* homozygotes is

$$\overline{p_R^2} = gr_g^2 + hr_h^2.$$

There is an apparent excess of observed *RR* homozygotes compared to HW expectations, by an amount

$$\overline{p_R^2} - \overline{p}_R^2 = gr_g^2 + hr_h^2 - g^2 r_g^2 - h^2 r_h^2 - 2ghr_g r_h.$$

Writing $V_R = \overline{p_R^2} - \overline{p}_R^2$ for the variance in the frequency of the *R* gamete, Wahlund obtained

$$V_R = gh(r_g - r_h)^2 \tag{10.34}$$

(*ibid.*, p. 242). Similarly, the apparent excess of observed *DD* homozygotes is $gh(r_g - r_h)^2 = V_R$ and the apparent deficiency of observed *DR* heterozygotes is $2gh(r_g - r_h)^2 = 2V_R$. According to Eq. (10.34), the greater the difference in allele frequency in the two demes, the greater the apparent excess of homozygotes, i.e., the greater the Wahlund effect. In particular, there is no Wahlund effect when the allele frequencies are the same in G and H.

The above reasoning can readily be extended to *m* isolates of relative size $i_1 : i_2 : \ldots : i_m$, where $\sum_k i_k = 1$, such that the frequencies of the recessive gamete *R* are, respectively, $r_1, r_2, \ldots, r_m$. Let $r = i_1 r_1 + i_2 r_2 + \ldots + i_m r_m$. Then, when the isolates are lumped together, the apparent excess of *RR* homozygotes compared to HW expectations is

$$\sigma^2 = \sum_{k=1}^{m} i_k (r_k - r)^2$$

(*ibid.*, p. 247). Wahlund noted that the apparent excess disappears when there is inter-deme migration.

Chapter 11
Round 7: The Darbishire Affair (1902–1905)

Abstract An important episode in the biometry-saltationism debate occurred when, in 1901, Weldon urged his student Darbishire to conduct crossing experiments with albino and Japanese waltzing mice. In his first report, Darbishire found variability in the coat color of F_1 hybrids and interpreted these as evidence against Mendelism. In his second report, Darbishire reported results on coat color that were compatible with the Mendelian hypothesis when F_1 mice were intercrossed and backcrossed with parental albinos. However, he rejected a Mendelian explanation for eye color. In his fourth report, Darbishire performed a key experiment involving crosses between two heterozygotes (in F_2) in three sets of experiments such that the parents (F_1) in each set had a different number of recessive alleles. He reported that as the number of recessive alleles in F_1 increased, the proportion of recessive trait (albinism) in F_3 increased, as predicted by the ancestral law. However, Castle pointed out a serious error in Darbishire's key experiment: the F_2 crosses did not all involve a pair of heterozygotes, as Darbishire had thought. When Darbishire's error is accounted for, his results are in accordance with Mendelian predictions. In his next paper, Darbishire acknowledged that his results did not contradict Mendelism.

Keywords Albino · Japanese waltzing · Backcross · Mosaic inheritance · Compound character · Extracted hybrids

11.1 Introduction

The Darbishire affair[1] was a major turning point in the biometry-saltationism war and signaled that the biometricians were likely fighting a losing war against the saltationists.

---

[1]The Darbishire affair is also discussed in Kim (1994, pp. 63–79), Ankeny (2000), Provine (2001, pp. 73–80), Gillham (2002, pp. 314–319), Radick (2023, pp. 208–217), Schwartz (2008, pp. 121–132), and Cock & Forsdyke (2008, pp. 228–229).

P. Gorroochurn, *The Development of Evolutionary Genetics*,
https://doi.org/10.1007/978-3-031-69374-8_11

Weldon was aware that experiments on mice had been conducted by von Guaita (1898) and later analyzed by Davenport (1900) to verify whether the data conformed with Galton's ancestral law.[2] The crosses involved the albino mouse (white throughout, with pink eyes and normal movement) and the piebald Japanese waltzing mouse (white except for patches of yellow in the neck and haunches, with pink eyes and a waltzing movement). The F_1[3] hybrids were all gray in color, had dark eyes and showed no waltzing behavior. Similar results had been obtained by Haacke (1895). Furthermore, when von Guaita's F_1 hybrids were intercrossed, the F_2 offspring exhibited a wide range of colors but could be broadly classified as gray (57%), albino (25%), and waltzing (18%).

In the *Biometrika* paper "Mendel's law of alternative inheritance" (Weldon, 1902a, p. 244), which was his first public commentary[4] on the newly rediscovered Mendelian theory, Weldon used the results of the crosses to attack that theory. His main argument was that the F_1 hybrids clearly showed that dominance was violated. This argument is, of course, fallacious since dominance is not a necessary component of Mendelism. Bateson's answer (Bateson, 1902b, p. 174) to Weldon was that both the albino and waltzing phenotypes were recessive, and that the gray color in F_1 was the result of heterozygotes "of one or more combinations".[5]

In 1901, Weldon urged his student, Arthur Darbishire (Fig. 11.1), to conduct mice experiments that would once and for all validate the ancestral law and discredit Mendel's law. Darbishire had worked under Weldon for the Honor School of Zoology at Oxford and, in October 1901, was appointed Demonstrator in Comparative Anatomy. As his sister later recounted in the preface of Darbishire's 1917 book:

> **QUOT. 681** He [Darbishire] had a profound admiration for Professor Weldon, and was, not unnaturally, influenced at this time by his hostile attitude towards the Mendelian School. (Darbishire, 1917, p. vii)

11.2 Lucien Cuénot's Experiments on Albino Mice (March 1902)

Unbeknownst to both Weldon and Darbishire, the Frenchman Lucien Cuénot (Fig. 11.2) also started some pioneering experiments involving crosses with the albino mouse. He was the first to embrace Mendelism in France and was thus the first geneticist of that country.

---

[2]See Sect. 10.2.3.

[3]F_2 in Davenport's paper.

[4]See Sect. 9.3.1.

[5]In today's language, they were heterozygotes from alleles either at one or several loci.

Fig. 11.1 Arthur Dukinfield Darbishire (1879–1915). (From: https://en.wikipedia.org/wiki/Arthur_Dukinfield_Darbishire#Life). Darbishire lived a short life. In July 1915 (at the age of 36), he was enrolled as a private in the 14th Argyll and Sutherland Highlanders. On Christmas of that year, he fell ill from cerebral meningitis and died the following morning

Cuénot[6] was born in Paris in 1866 and was interested in zoology at an early age. He became attracted to what later came to be known as the field of genetics when, in 1894, he read Emile Zola's *Le Doctor Pascal,* where the renowned French novelist discussed his theories on heredity.

In 1900, when Cuénot was Professor of Zoology, he urged a young researcher (Ms Barthelet) to cross albino and gray house mice to test the theory of telegony.[7] Cuénot realized how easy it was to cross mice and set out to test in *animals* the

[6]Lucien Cuénot's life and work are also discussed in Chomard-Lexa (2004), Tétry (1981, pp. 492–494), Courrier (1952), Darden (1991, Chaps. 5 & 8), Schmitt (2002), Tort (1996, pp. 736–743, Vol. 1: A–E), Goldschmidt (1951), Burian et al. (1988), and Buican (1984, Chap. 2).

[7]Telegony claims that a mother is able to pass characteristics to her offspring based on her *first* impregnation; children thus inherit features not only from their current father but also from a previous partner their mother might have had. This theory is now known to be false but was very widely believed at the times.

Fig. 11.2 Lucien Cuénot (1866–1951). (From: https://upload.wikimedia.org/wikipedia/commons/f/f0/Le_biologiste_Lucien_CUÉNOT_%281866-1951%29.JPG)

recently rediscovered Mendel's laws, which had hitherto been observed only in *plants*. In 1902, he became the first to observe that albinism was recessive to pigmentation in mice:

> **QUOT. 682** If a grey mouse (♂ ou ♀) is crossed with a white mouse (♀ ou ♂), one always obtains, without exception, grey mice. The *pigmented* character is therefore dominant with respect to *absence of pigment.* (Cuénot, 1902a, pp. XXVIII–XXIX)

All F_1 mice were thus gray, and when these were intercrossed, Cuénot found that in F_2:

> **QUOT. 683** Experiments fully accord with this [i.e. Mendelian] prediction: I have obtained 270 offspring, which consist of 198 grey and 72 albinos, that is 26.6 % of the latter. (*ibid.*, p. XXIX)

Furthermore, to prove that the F_2 gray mice contained both pure dominants (i.e., homozygotes) and hybrids (i.e., heterozygotes), he intercrossed a random sample of these mice. Half of the pairs produced only gray F_3 offspring, and the other half produced gray and albino in the ratio 3:1, suggesting that each F_2 parent had at least one dominant gamete in the first half and at least one recessive gamete in the second half.

At the same time, Cuénot was initially not convinced of the generality of Mendel's laws:

> **QUOT. 684** In my breeding experiments, I incidentally obtained yellow mice, grey tainted with white, and tainted black: I am now trying to discover the laws of heredity which govern these variations, laws which seem very different from Mendel's. (*ibid.*, p. XXX)

However, in an article that followed, Cuénot was more decidedly on the Mendelian side of the heredity debate:

> **QUOT. 685** I have no doubt that Mendel's law [of segregation], when it is better understood, will find interesting zootechnical applications; it has considerable theoretical importance and de Vries has rightly felt the support it lends to theories of heredity based on the hypothesis of particle representatives. (Cuénot, 1902b, p. 398)

Cuénot wrote to Bateson about his initial results and in a letter dated July 23, 1902, the latter thanked him.

In the following year, Cuénot (1903b) wrote a paper that was meant to explain the conflicting results some researchers had obtained when crossing albino mice with pigmented ones. Cuénot's explanation was that not all albinos were the same because of different ancestry:

> **QUOT. 686** This observation [i.e. ancestry dependence], which had not yet been made so far, hence the key to the contradictory results obtained by the authors who have made crosses between albinos and pigmented individuals (rabbits, mice, rats); without suspecting it, they operated with different albinos, and, consequently, the color of the offspring seemed to escape all rule. Certainly, the albino mice sold by the merchants have varied ancestral origins, therefore, hereditary influence variable, despite the identity of their external appearance. (*ibid.*, p. 301)

Like a similar one proposed at about the same time by Castle and Allen (1903, p. 614),[8] Cuénot's hypothesis in the above is ancestrian in nature and therefore non-Mendelian.

Cuénot was more successful in his *chromogenic* theory of coat color in mice (Cuénot, 1903a). According to this theory, the fur on the gray mouse was determined by two enzymes (each an oxidizing diastase) that can react on a substance called the *chromogene* (factor C). These three substances were located on three "mnenons" (genes) in the germ cells. On the other hand, the black mouse possessed only one enzyme and one chromogene on three mnenons, while the albino mouse possessed two enzymes but no chromogene.[9] Based on this theory, CG yielded a gray mouse, but AG yielded an albino. Furthermore, Cuénot showed that gray was dominant over black, and yellow was dominant over both gray and black.

In 1905, Cuénot also became the first to discover gene lethality (Cuénot, 1905) while experimenting with yellow mice.

---

[8] See QUOT. 704.

[9] See Chomard-Lexa (2004, pp. 89–90) for more details.

11.3 Darbishire's First Report (Nov 1902): Lack of Uniformity in F_1 Hybrids

Meanwhile, under Weldon's direction, Darbishire crossed albino mice with Japanese waltzing mice. The results of the first generation of hybrids were hurriedly published in the second volume of *Biometrika* under the simple title "Note on the results of crossing Japanese waltzing mice with European albino races" (Darbishire, 1902). Having first briefly described the previous works of Haacke and von Guaita in the field, and the comments made by Bateson, Darbishire stated that the objective of his experiments was to answer the following two questions:

> **QUOT. 687** (1) Does the first generation of cross-bred individuals, produced by pairing albino and Japanese waltzing mice, exhibit such a uniformity of colour as will justify us in considering that Mendel's Law of Dominance applies to it?
>
> (2) Does the Mendelian Law of Segregation apply to the offspring of hybrids resulting from such crosses?
>
> (*ibid.*, p. 101)

Darbishire did not yet have the results for the F_2 hybrids to be able to answer (2) above, but regarding (1), he noted that he had obtained 48 F_1 hybrids from nine crosses thus far. The F_1 hybrids were grayish but quite variable and could be broadly categorized into four groups (Fig. 11.3). Of the 48 F_1 hybrids, 18 were of type 2, 13 of type 3, 15 of type 4, and 2 of the fourth type. Darbishire interpreted the failure of the white color to disappear in the F_1 hybrids (especially those of types 2 and 3) to mean that his experiments, in fact, showed that albinism was not recessive.

In view of the high variability in the colors in the F_1 hybrids, Darbishire thought he had the answer to the first question in QUOT. 687:

> **QUOT. 688** The answer to the first question has been obtained, and it is in the negative. (*ibid.*, p. 102)

Concerning the influence of ancestry, Darbishire stated that:

> **QUOT. 689** . . .it looks as if it could be said that the more in-bred an albino is the less power it has of transmitting its whiteness. But many more crosses must be made before any definite statement can be made on this point. (*ibid.*, p. 104)

To the modern eye, the lack of uniformity of color in Darbishire's F_1 hybrids does not necessarily disprove the Mendelian hypothesis. This is because uniformity assumes that mouse color is due to the action of "factors" (now genes) at a single locus. However, if a multi-locus hypothesis is allowed, a wide variety of colors in F_1 is possible.[10]

---

[10]For example, Silvers has stated that: "In the house mouse (*Mus musculus* L.) more genes have been identified which affect coat color than any other trait. Moreover, this number has increased enormously over the past 25 years. During this period the number of coat-color determinants has risen from 32 to more than 130 and the number of loci involved has increased from approximately 20 to more than 50." (Silvers, 1979, p. 2).

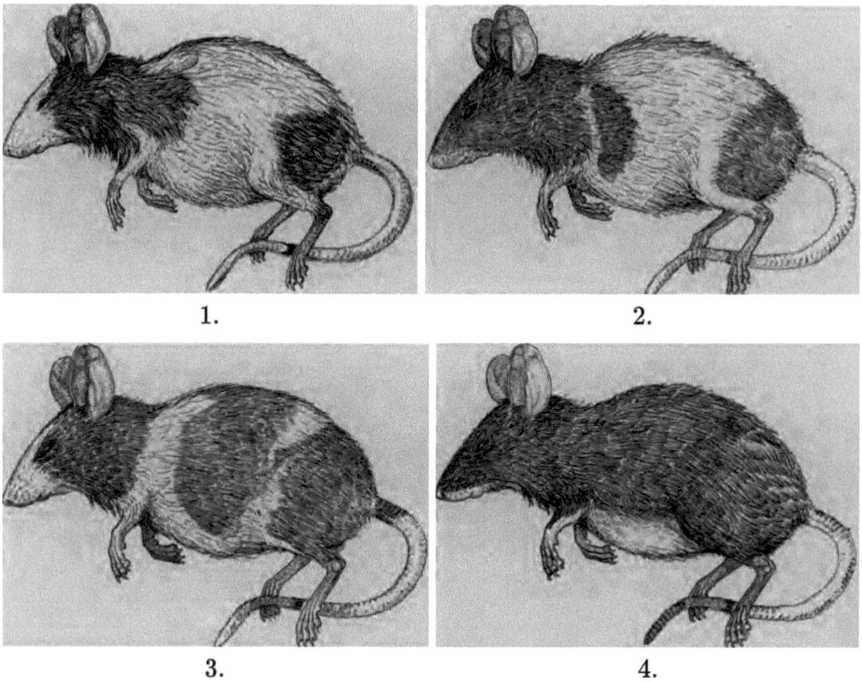

Fig. 11.3 The types of mice used in Darbishire's first experiment: 1. Pie-bald Japanese waltzing mouse (one of the two parental types, the other being the albino), 2. F_1 hybrid similar to type 1, 3. F_1 hybrid with much more gray, 4. F_1 hybrid which is all gray except for the belly and tail. A fourth category of F_1 hybrid had all gray except for the belly

11.4 Bateson's Reaction to Darbishire's First Report (Dec 31, 1902): Eye Color and Waltzing Behavior of F_1 Hybrids

Darbishire's (1902) dismissal of Mendelism and promotion of ancestral influence simply on the basis of F_1 hybrids elicited a short December 31, 1902, personal letter from Bateson (Fig. 11.4). The latter had two simple questions:

QUOT. 690 Can you kindly tell me
(1) the eye color of the cross-breds
(2) whether they "waltzed"?[11]

On January 3, 1903, Darbishire answered as follows:

QUOT. 691 Many thanks indeed for your note and the Report to the Evolution Committee which I have much pleasure in accepting.

[11] Letter retrieved online from Cambridge Digital Library, Bateson Archive at https://cudl.lib.cam.ac.uk/view/MS-ADD-08634-C/268

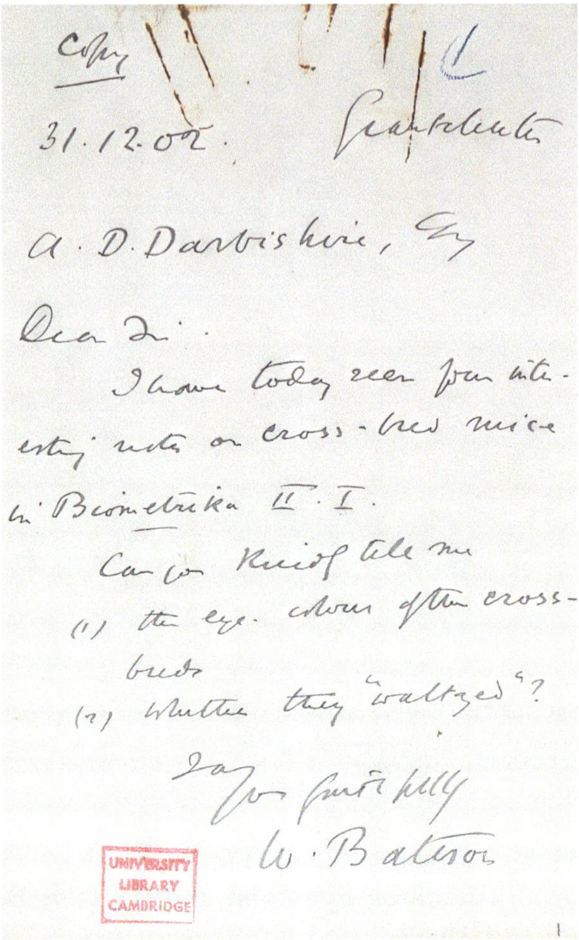

Copy

31.12.02. Grantchester

A. D. Darbishire, Esq

Dear Sir:

I have today seen your interesting note on cross-bred mice in Biometrika II. I.

Can you kindly tell me

(1) the eye-colour of the cross-breds

(2) whether they "waltzed"?

Yours faithfully

W. Bateson

Fig. 11.4 Bateson's first letter to Darbishire. (From Cambridge Digital Library, Bateson Archive, at https://cudl.lib.cam.ac.uk/view/MS-ADD-08634-C/268)

> In answer to your questions
> (1) The eye colour of the hybrids (of the first generation) is in all cases black.
> (2) As yet no sign of waltzing has been detected in the first generation. I should be very glad to answer any question you may want to ask me: I am absolutely unbiased about Mendel and am very keen to come to unprejudiced conclusion on it.
> (Kim, 1994, p. 65)

Darbishire's statement that he was "absolutely unbiased about Mendel" should be taken with a grain of salt given his later confession that this was not the case.[12] In any case, the two answers in QUOT. 691 are revealing because the absence of Mendelian

[12] Darbishire's sister also later explained that he was "not unnaturally, influenced at this time by his [Weldon's] hostile attitude towards the Mendelian School". See QUOT. 681.

dominance for coat color was mentioned in Darbishire's paper but the presence of Mendelian recessiveness for both pink eye and waltzing behavior was omitted. In a January 4, 1903 letter, Bateson said:

> **QUOT. 692** Please accept my thanks for the answer you have so kindly sent. The supplementary facts you have communicated are of such great significance that I am surprised they were not stated in your first note. From general acquaintance with such cases I naturally guessed that the cross-breds would have black eyes, but the casual reader hardly could...[13]

However, one point that was not yet addressed by Bateson was the following: if pink eye is recessive and both types of parents (albino and Japanese waltzing) had pink eyes, then how can the appearance of dark eyes in F_1 be explained? This would soon become a major point of contention and mark Weldon's entry into a hitherto cordial exchange.

11.5 Darbishire's Second Report (Feb 1903): Lack of Uniformity in F_1 Hybrids, Possible Mendelian Explanation for Coat Color, No Mendelian Explanation for Eye Color

By the time Darbishire's second report (Darbishire, 1903a) appeared in February 1903, the number of F_1 hybrids (from crossing albinos with Japanese waltzing mice) had increased from 48 to 154. The hybrids showed even greater variability than before and were classified into five groups according to the amount of whiteness. Because of the increased F_1 variability, Darbishire now claimed that, contrary to the Mendelian prediction of uniformity, *there was segregation of traits actually occurring in* F_1*:*

> **QUOT. 693** Now in these mice this segregation seems to take place in the first generation, for there appear in it besides the wild colour, yellow and black... (*ibid.*, p. 170)

In contrast, regarding ancestral influence:

> **QUOT. 694** ...the fact remains that the ancestry of a white mouse does make a difference in the character of its offspring when crossed with a waltzing mouse. (*ibid.*, p. 171)

In the second report, Darbishire also had results for crosses between F_1 hybrids and for backcrosses between F_1 hybrids and albinos. He noted that the offspring from the first set of crosses were much more heterogeneous than those from the second, and that this was consistent with the ancestral law since "the heterogeneous offspring of

[13] Letter retrieved online from Cambridge Digital Library, Bateson Archive at https://cudl.lib.cam.ac.uk/view/MS-ADD-08634-C/269

two hybrids have a more complex ancestry than do the less variable offspring of a hybrid and an albino" (*ibid.*, p. 272). Darbishire then added that:

QUOT. 695 [n]o theory of compound allelomorphs such as that put forward by Mendel will account for this striking difference between the variability of the two groups of offspring. (*ibidem*)

This is a surprising statement since, in fact, the Mendelian hypothesis generally predicts three types (AA, Aa, aa) for a cross between hybrids (Aa × Aa), as opposed to two types (AA and Aa) from a backcross (Aa × AA).[14]

Having hitherto discarded Mendelism, Darbishire finally declared that there were some observations that were "in possible accordance with some form of Mendelian hypothesis." These were:

QUOT. 696 (1) the apparently regular appearance of albinos when hybrids are crossed with albinos, although the evidence at present available does not suffice to shew whether these occur in Mendelian proportions (50% albino and 50% hybrid) or not, (2) the well-known fact that albinos ancestry when paired together produce albino young, exceptions to this rule being at least very rare, and (3) the appearance of waltzing and albino mice in the second hybrid generation. (*ibidem*)

In a postscript added on February 12, 1903, Darbishire noted that the number of F_1 hybrids had now increased to 203. All of them had dark eyes, none were albinos and none waltzed. By pairing the F_1 hybrids, 66 F_2 mice had been produced. Of these, 13 were pink-eyed albinos, 17 had pink eyes and more or less colored coats, and 36 were dark-eyed with some coat color. Moreover, by backcrossing F_1 hybrids with albinos, 205 offspring were produced, consisting of 111 albinos and 94 with dark eyes and some coat color. Thus, the two sets of numbers are approximately in accordance with the Mendelian 1:2:1 ratio for F_2 and 1:1 ratio for the backcrosses. While Darbishire admitted that the latter results were not in disagreement with Mendel's results, he also affirmed that:

QUOT. 697 The inheritance of eye-colour is not in accordance with Mendel's results. . .Pink eyes are however not "dominant," since the two pinkeyed parents of the first generation always produce dark-eyed young. For the same reason pink eyes are not "recessive". . . The behaviour of eye-colour is thus in every respect discordant with Mendel's results. (*ibid.*, p. 173)

As mentioned at the end of Sect. 11.4, the formation of black eyes in F_1 hybrids by crossing parents with pink eyes (a supposedly recessive trait) was enigmatic. In the above, Darbishire drew attention to the phenomenon in an attempt to discredit Mendelism. Bateson was now forced to intervene.

[14] If A is dominant over a, the Mendelian hypothesis still predicts more variety for the cross between hybrids than the backcross (2 varieties versus 1).

11.6 Bateson's March 19, 1903, Response to Darbishire's Second Report: Attempt at Resolving the Dark-Eye Enigma

Bateson's response to Darbishire's second report was written on March 11 and published in the March 19, 1903, issue of *Nature* (Bateson, 1903a). Since Darbishire had already conceded that the coat colors in the F_2 and backcrossed mice were in possible accordance with Mendel's results, Bateson focused on the results for eye color. Here, Darbishire had categorically denied the Mendelian hypothesis.

Bateson first alluded to the fact, not mentioned in Darbishire's first report, that all F_1 mice had dark eyes. He stated that this result:

QUOT. 698 ...though to some extent paralleled by certain plant cases, is probably as yet unique among animals, at least in degree. (*ibid.*, p. 462)

In contrast to Darbishire's statement, Bateson next declared that the latter's experiments, in fact, showed that the inheritance of eye color was strictly Mendelian. Recall that the parents were albino mice (white throughout, with pink eyes and normal movement) and the piebald Japanese waltzing mice (white except for patches of yellow in the neck and haunches, with pink eyes and a waltzing movement). Bateson denoted gametes of the first and second types by G and G′, respectively. When GG is crossed with G′G′, all F_1 hybrids are GG′ (or G′G). He then explained the gray-colored and dark-eyed G′G hybrids through atavism (or reversion to the ancestral type):

QUOT. 699 The case is closely comparable with that found by Miss Saunders in Malthiela (Rep. to Evol. Cttee. of Roy. Soc., 1902), when a white hoary form crossed with a white glabrous form gave purple hoary offspring; and with the production by Kölreuter (confirmed by Gärtner, Naudin and Godron) of purple flowered hybrids by the union of two while flowered Daturas, *D. ferox* and *D. laevis*. Why in these cases the heterozygotes are atavistic we do not know, but the problem need not be insoluble. (*ibid.*, p. 463)

Once this explanation was accepted, the rest of Darbishire's observations fitted extremely with Mendelian expectations:

QUOT. 700 Treating GG′ and G′G as identical, we thus expect the ratio

1 GG: 1 G′G′: 2 GG′.

Therefore the most probable distribution of the 66 mice is as follows:-

16.5 pink-eyed albinos: 16.5 pink-eyed with colour in coat: 33 dark-eyed with colour in coat,

and the experiments gave

13 pink-eyed albinos: 17 pink-eyed with colour in coat: 36 dark-eyed, (? all) with colour in coat,

Similarly, on crossing the hybrids with albinos, we expect equal numbers of GG and GG′. Therefore, the most probable distribution of the 205 mice so produced is

102.5 pink-eyed albinos: 102.5 dark-eyed with colour in coat,

and the experiment gave

111 pink-eyed albinos: 94 dark-eyed with colour in coat,

Experiment agrees well with expectation. In what respect are they discordant?

(*ibidem*)

In retrospect, although there is no question that Darbishire's results fit Mendelian expectations quite well, Bateson's statement in QUOT. 699 seems improvised and an incomplete explanation for the dark-eye phenomenon in F_1. Weldon certainly did not buy it, as evidenced in the short note he wrote to *Nature*.

11.7 Weldon's April 2, 1903, Response to Bateson: Bateson's "Mysterious" Heterozygotes

Recall that, in Darbishire's experiments, the parents were albino mice (white throughout, with pink eyes and normal movement) and the piebald Japanese waltzing mice (white except for patches of yellow in the neck and haunches, with pink eyes and waltzing movement). In his 1903 *Nature* paper, Bateson tried to explain the dark eyes in F_1 hybrids as follows: he denoted gametes of the albinos by G and those of the Japanese waltzing mice by G′, such that GG′ hybrids were gray-colored and had dark-eyes due to atavism.

In a short letter written on March 24 and published in the April 2, 1903, issue of *Nature*, Weldon (1903a) made his entry in the discussion by challenging Bateson's Mendelian interpretation of Darbishire's results. Regarding Bateson's explanation given in the paragraph above, he retorted:

> **QUOT. 701** . . .by reference to the mysterious properties of "heterozygote" any difficulties presented by their eye-colour are avoided. (*ibid.*, p. 512)

Weldon also referred to Darbishire's G′G′ offspring in F_2 when the GG′ hybrids in F_1 were intercrossed:

> **QUOT. 702** Now the mice G′G′ [in F_2] are of the same constitution in respect of all the characters concerned as their pure-bred grand-parent G′ [G′G′]. Mr. Darbishire has shown (Biometrika, vol. ii. part ii.) that they do not always resemble their grandparent, or either of their parents, in one of the characters (coat-colour) denoted by G′. They may show a new colour, "lilac," not present in any of their near ancestors. Six out of eighteen mice of this category, at present old enough for study, show lilac colour. (*ibidem*)

Weldon promised that a fuller discussion would be published in a forthcoming *Biometrika* article, but for the time being he challenged Bateson to explain: (i) the appearance in the G′G′ mice in F_2 of the lilac color not to be found in any of the ancestors, as described in QUOT. 702 above, (ii) the variation in coat color in the F_1 hybrids (G′G).

With his pointed remarks, suddenly it appeared as though Weldon had gained the upper hand on Bateson. However, the latter would soon receive strong support from a young biologist from the other side of the Atlantic, namely, the American William Castle.

11.8 Castle and Allen's April 1903 Response to Darbishire's First Report

11.8.1 Darbishire's Observations: A Clear Case of Mendelian Heredity

Castle[15] wrote his first paper (Castle & Allen, 1903) on albinism with Glover Allen,[16] shortly after the appearance of Darbishire's first report.[17] The paper was published in April 1903, at about the same time as the exchange between Bateson and Weldon was taking place.

At the start of their paper, Castle and Allen expressed their utter disagreement with Darbishire's conclusions. Recall that Darbishire's two parental types were albino and Japanese waltzing mice (or partial albino), and the F_1 offspring had varying shades of gray. Moreover, unlike the two parental types, which were pink-eyed, all F_1 mice had pigmented eyes. Whereas Darbishire had not mentioned any eye color and had interpreted the variation in F_1 coat color as non-Mendelian, in contrast Castle and Allen argued that coat and eye colors were both clear evidence of Mendelian heredity:

> **QUOT. 703** Not only were the offspring *not intermediate*[18] in pigmentation between the two parents, as we should expect on any hypothesis other than the Mendelian one, but *not a single mouse in the forty-eight contained as little pigment as was found in the pigmented parent.* In every case the pigmentation was greatly increased over what it had been in the pigmented parent, either in intensity or in extent, and usually in *both* respects. (*ibid.*, p. 608)

While Darbishire had taken the failure of white in the coat color of the F_1 hybrids to disappear as evidence against Mendelian recessiveness, Castle and Allen pointed out that not one of the F_1 hybrids was an albino. Therefore, it could not be denied that albino was, in fact, recessive.

Castle and Allen next proposed a theory to explain the several observations that had led Darbishire to reject Mendelism. However, while they wished to vindicate Mendelism, Castle and Allen in fact ended up undermining two major tenets of that theory, namely, segregation and gametic purity.

---

[15] See also Sects. 10.2 and 10.6 for descriptions of Castle's other works related to the biometry-saltationism debate.

[16] Glover Morrill Allen (1879–1942) graduated from Harvard College in 1901 and received his PhD in 1904. His doctoral thesis was entitled "The Heredity of Coat Color in Mice." He traveled widely and became an eminent zoologist, being elected a Fellow of the American Academy of Arts and Sciences in 1915.

[17] See Sect. 11.3.

[18] All italics in this quotation are Castle and Allen's.

11.8.2 Mosaic Inheritance: Violation of Mendelian Segregation and Gametic Purity

Having declared that Darbishire's results on the F_1 hybrids, in fact, supported the Mendelian hypothesis, Castle and Allen next provided a theoretical explanation for some of the peculiarities of those results (e.g., the occurrence of more extensive gray compared to the parents, the pigmented eyes, the substantial variation in coat-color).

Castle and Allen first considered the two parental types used by Darbishire, namely, the albino and the "spotted mouse" (previously referred to as the Japanese waltzing mouse or partial albino). Concerning the first type, contrary to Darbishire's assertion, the authors declared that albinism was recessive relative to the pigmented condition. Concerning the second type, and this is where their new theory starts, they declared that the spotted mouse was not a dominant, but a *mosaic*:

> **QUOT. 704** The spotted mouse is a peculiar kind of individual, in which are found both the dominant and the recessive characters, yet not in their usual relationship (one latent, the other alone visible), but *both* visible side by side, in distinct areas of the animal's body. Such an individual is called a *mosaic*. (*ibidem*)

Thus, the dominant–recessive relationship is suspended in the mosaic individual but that relationship is reactivated when the mosaic is crossed. Therefore:

> **QUOT. 705** . . .the crossing of a mosaic with a normal, or with a recessive individual, results usually in the production of normally pigmented individuals only. Thus when spotted rats or mice are crossed either with gray individuals (dominants) or with albinos (recessives), the offspring are commonly all gray or black in color, none spotted. Here we find that dominant and recessive characters have abandoned the balanced relationship which they had in the spotted parent, and returned to the ordinary relationship of a dominant to a recessive character. (*ibid.*, pp. 608–609)

Now, suppose that D is a gamete and R a recessive one. Let the three Mendelian types (later genotypes) DD, DR, and RR be denoted as D.D, D.R, and R.R, where the dot demarcates the parental contribution. Then, writing Castle and Allen's mosaic as D(R).D(R), a cross between a mosaic (i.e., spotted mouse) and a recessive (i.e., albino) may be written as

$$\mathrm{D(R).D(R)} \times \mathrm{R.R} = \mathrm{D(R).R} \tag{11.1}$$

When the F_1 hybrids are interbred:

$$\mathrm{D(R).R} \times \mathrm{D(R).R} = \underbrace{\mathrm{D(R).D(R)} + 2\mathrm{D(R).R} + \mathrm{R.R}}_{1:2:1} \tag{11.2}$$

When the F_1 hybrids are backcrossed with the recessives:

$$\mathrm{D(R).R} \times \mathrm{R.R} = \underbrace{\mathrm{D(R).R} + \mathrm{R.R}}_{1:1} \tag{11.3}$$

Thus, according to Castle and Allen's theory of mosaic inheritance, the gametes of mosaic individuals are also mosaic, i.e., "containing side by side the dominant and the recessive character." Therefore, crosses involving mosaics do not yield results consistent with Mendelian segregation. For example, the mosaic theory assumes that a spotted mouse contains both the D (pigmented) and R (albino) gametes; however, when the spotted mouse is crossed with an albino, not a single albino is produced in F_1. In Castle and Allen's own words, "the law of segregation is commonly suspended also" (*ibid.*, p. 609).

Thus, Castle and Allen explained the occurrence of the more extensive gray color in F_1 mice through the reactivation of the dominant-recessive relationship when the parental mosaics were crossed with the albinos (see QUOT. 705). Similarly, they explained the occurrence of pigmented eyes in F_1 mice through the release of the pigment in the mosaic. More specifically, they stated that the pigment in the mosaic was quite localized (hence the occurrence of only gray *patches* in F_1), but its release was more or less complete:

> **QUOT. 706** ...it is evident that in all his hybrids there occurred release, more or less complete, of the pigment-forming character from its localization in the original mosaic. In every case, apparently, the hybrid had pigmented eyes, though neither parent possessed this character. (*ibid.*, p. 610)

Castle and Allen further differentiated between *pure mosaics* and *hybrid mosaics*. Pure mosaics are mosaic types in which *both* gametes are mosaics, such as one of the parental mice (D(R).D(R)) used by Darbishire. On the other hand, hybrid mosaics are mosaic types in which a single gamete is mosaic, such as the F_1 mice ((D(R).R) in Darbishire's experiments.

When F_1 mice are intercrossed, the mosaic theory predicts that, in accordance with (11.2), 25% of the F_2 offspring should be pure mosaics (like one of the parental types, namely, the Japanese waltzing mouse), 50% should be hybrid mosaics (like the F_1 mice), and 25% should be recessives (like the other parental type, namely the albino). Having just received Darbishire's second report (Darbishire, 1903a), Castle and Allen triumphantly announced:

> **QUOT. 707** This prediction has been fulfilled sooner than we had expected. In a paper just received Darbishire (: 03) states that his hybrid mice bred inter se have produced 66 young. Of these 13 are albinos, and 17 are (in our terminology) pink-eyed mosaics, while the remaining 36 have dark eyes. The last named class consists clearly of hybrid mosaics like their parents. The Mendelian expectation is that half the offspring, or 33, will be of this sort; the observed number is 36. The other two classes (albinos and pure mosaics) should theoretically number 16.5 each, or together 33; they number 13 and 17 respectively, or together 30. (Castle & Allen, 1903, p. 612, footnote)

Finally, Castle and Allen addressed the substantial variation in gray coat-color in the F_1 hybrids of Darbishire's experiments. Recall that this variation was one of the major reasons Darbishire had rejected the Mendelian hypothesis. Castle and Allen's explained that Darbishire had used *two* kinds of albinos:

> **QUOT. 708** 1) "pure-bred" albinos purchased from breeders, and presumably descended from white parents, and (2) "cross-bred" albinos known to be descended from spotted [i.e. hybrid mosaic] parents. The latter, when crossed (four different pairs) with the fawn white dancing mice, produced only spotted gray-white offspring, nineteen in number. The former,

when crossed in the same manner (five different pairs), also produced gray-white mice, twelve in number, but produced in addition fifteen gray mice (with lighter bellies and tail) and two fawn-colored mice (likewise lighter below). (*ibid.*, pp. 613–614)

The first type was designated a *pure recessive*, and the second was designated an *impure recessive*. However, both recessives were the same Mendelian type R.R and by drawing a distinction between them, Castle and Allen were clearly treading along non-Mendelian paths. Thus, these astounding statements were made:

QUOT. 709 This result indicates that *not all albinos breed alike*[19] when crossed with the same pigmented stock, a conclusion which our own experiments fully substantiate.

. . . .

Darbishire is entirely right in concluding that the ancestry of white mice does "make a difference" in their breeding capacity.

(*ibid.*, p. 614)

The above is essentially the ancestrians' position and is in direct violation of Mendelian *gametic purity*. This violation was not a consequence unforeseen by Castle and Allen. They were fully aware of it and, in fact, fully endorsed it in yet another surprising statement:

QUOT. 710 . . .in the strictest sense, the doctrine of gametic purity is untenable. We cannot accept that interpretation of it which requires that the gametes formed by an individual be the *precise equivalents*[20] in all respects of the respective gametes which united to form that individual. (*ibid.*, p. 615)

Thus, the very observations that Castle and Allen had declared to be consistent with Mendelism were explained by a theory that was essentially non-Mendelian.

In fact, the denial of gametic purity would be one of the hallmarks of Castle's early research, culminating in the theory of "gametic contamination" in 1906 (Castle, 1906). He would relent on that position only in 1919, in the face of overwhelming evidence (Castle, 1919a, p. 385).[21] Earlier in 1909, he was able to provide a proper Mendelian interpretation of both coat color and eye color (Castle & Little, 1909). The D(R) mosaic was replaced by *an allele for spotting at a separate locus* in mice. Another locus contained the albino series (*C*), such that the *cc* mice were both white and pink-eyed. A third locus containing *dilute alleles* (D) was found to influence both coat and eye-color. Finally, a locus containing the *pink-eyed alleles* (P), in particular the *pink-eyed dilution* (p), also affected hair and eye color. Then, the occurrence of dark eyes by the crossing of two pink-eyed parents can readily be explained by a cross such as

$$\underbrace{\mathrm{cp'/cp'}}_{\text{pink-eyed}} \times \underbrace{\mathrm{c'p/c'p}}_{\text{pink-eyed}} = \underbrace{\mathrm{cp'/c'p}}_{\text{dark-eyed}}, \tag{11.4}$$

where both c and p are recessive.

[19] Italics are Castle and Allen's.

[20] Italics are Castle and Allen's.

[21] See also Sect. 13.2.3.

11.9 Anonymous April 9, 1903, Review in Nature: Darbishire's Observations Are "Prima Facie" Evidence of Ancestral Inheritance

A further indication of the growing interest in Darbishire's experiments was an anonymous note in the *Scientific Serial* section of the April 9, 1903, issue of *Nature* (Anonymous, 1903). Visibly sympathetic to the biometric program, the author was full of praise for Karl Pearson:

QUOT. 711 The excellence of Prof. Karl Pearson's elaborate studies in statistical theory is becoming widely recognised, and his comments and criticisms add much to the value of the work of other contributors. (*ibid.*, p. 550)

Although the difficulty in interpreting Darbishire's results was acknowledged, this did not prevent the author from concluding that:

QUOT. 712 ...it must be allowed that much of the evidence is prima facie in favour of ancestral inheritance. (*ibidem*)

11.10 Bateson's April 23, 1903 Response to Weldon: "Discussing the Perturbations of Uranus with a Philosopher Who Denies That the Planets Have Orbits"

While an often-fierce debate was taking place in public, Bateson and Darbishire remained engaged in a civil discussion. Responding to a suggestion by the latter, Bateson wrote on April 7, 1903, that:

QUOT. 713 It would be a pleasure to meet to discuss your results with you as you kindly propose and to show you some of my own...Perhaps however as I am already engaged in a public discussion of these mice it would be simpler to wait till this is over. A little later in the year also I should have more to show you.[22]

Regarding the public debate, it should be remembered that Bateson had hitherto referenced only Darbishire in his critique of the mice experiments. However, in the April 2, 1903 issue of *Nature*[23] (Weldon, 1903a), Weldon had challenged his interpretation. In particular, Weldon had asked Bateson two questions: how to explain (i) the appearance in some of the F_2 mice of a lilac color not to be found in any of the ancestors? (ii) the coat color of the F_1 mice?

[22] Letter retrieved online from Cambridge Digital Library, Bateson Archive starting at https://cudl.lib.cam.ac.uk/view/MS-ADD-08634-C/270

[23] See Sect. 11.7.

Bateson was now forced to reply to Weldon's queries. He wrote a long letter to *Nature* on April 7, 1903,[24] but the editor refused to publish it on account of its length. In a much-curtailed letter written on April 10 and published in the April 23 issue of *Nature*, Bateson expressed his downright frustration with Weldon:

QUOT. 714 To debate these finer points with one who doubts the Mendelian nature of the phenomena taken as a whole is like discussing the perturbations of Uranus with a philosopher who denies that the planets have orbits. (Bateson, 1903b, p. 585)

In retrospect, Bateson's frustration was more an indication of his inability to use Mendelian principles to comprehensively and decisively explain the peculiarities in Darbishire's results. However, he offered the following suggestions as possible explanations for the 'lilac mice' in F_2 and for the coat color in F_1:

1. *Regarding 'lilac mice' in* F_2: Bateson here suggested that the coat-color of the parental Japanese waltzing mice was a *compound trait*, a term he had previously used to denote the formation of new gamete types during cross-fertilization.[25] The 'lilac' color then occurred when the compound trait was crossed with an albino.
2. *Regarding the coat color in* F_1*:* Bateson argued that although both parental types were true-breeding with respect to color, the F_1 mice could still have a variety of colors if the coat color was determined by "factors (probably determinable) other than the visible colours of the parents, and having an independent distribution amongst their gametes." In other words, coat color could be due to the action of multiple gamete types (i.e. alleles at different loci).

In a postscript, Bateson also took aim at the anonymous author of the *Scientific Serial* section of the April 9, 1903 issue of *Nature* (Anonymous, 1903).[26] Recall that the anonymous author had claimed that much of the evidence from Darbishire's experiments thus far were "prima facie in favour of ancestral inheritance." Bateson replied that, in fact:

---

[24]A rough draft of the letter may be found at https://cudl.lib.cam.ac.uk/view/MS-ADD-08634-C/297 and spans six pages.

[25]In *Mendel's Principles of Heredity: A Defence,* a compound character is explained as follows: "The character brought in on fertilisation by one or other parent may be of such a nature that when the zygote, *AB,* forms its gametes, these are not individually bearers merely of *A* and *B*, but of a number of characters themselves again integral, which in, say *A*, behaved as one character so long as its gametes united in fertilisation with others like themselves, but on cross-fertilisation are resolved and redistributed among the gametes produced by the cross-bred zygote. In such a case we call the character *A* a compound allelomorph, and we can speak of the integral characters which constitute it as *hypallelomorphs*. We ought to write the heterozygote $(AA'A''\ldots)\,B$ and the gametes produced by it may be of the form $A, A', A'', A''', \ldots, B$. Or the resolution may be incomplete in various degrees, as we already suspect from certain instances; in which case we may have gametes $A, A'A'', A'''A''''$, $A'A''A^{v}, \ldots, B$, and so on. Each of these may meet a similar or a dissimilar gamete in fertilisation, forming either a homozygote, or a heterozygote with its distinct properties." (Bateson, 1902b, p. 29).

[26]See Sect. 11.9.

QUOT. 715 [i]t is scarcely too much to state that in each set of matings the distribution (1) of pink and dark-eyed, (2) of coloured and albino coats, (3) of "waltzers" and non-waltzers, is in punctilious agreement with Mendelian prediction. (Bateson, 1903b, p. 585)

11.11 Weldon's April 30, 1903, Response to Bateson: Bateson's Doctrine "Too Amazing for Brief Treatment"

Weldon's prompt letter (written on April 24) in the April 30, 1903, issue of *Nature* (Weldon, 1903b) was meant to show that, in his attempt to explain the various results of Darbishire's experiments, Bateson was becoming embroiled in self-contradictions. Weldon pointed to Bateson's March 19, 1903, letter to *Nature* (Bateson, 1903a),[27] where the latter had used two gametes (G and G′) such that each one influenced both coat- and eye-color. He next pointed out Bateson's April 23, 1903 letter (Bateson, 1903b)[28] where the previous hypothesis was abandoned and Bateson

QUOT. 716 ...now says (1) that more than two kinds of gametes take part in the first crosses, since the gametes of one or both pure races are heterogeneous; (2) that coat-colour is split into simpler elements when the hybrids form gametes. (Weldon, 1903b, p. 610)

In a tone that expressed more sarcasm than astonishment, Weldon declared that Bateson's assumption (1) in QUOT. 716 was "a doctrine too amazing for brief treatment," since both parental types bred true and Bateson had also previously conceded that the albino trait was recessive.

Regarding Bateson's assumption (2) in QUOT. 716, Weldon's counterargument was simple: if, as Bateson had said, there were multiple gametes for coat color and three main coat colors (albino, black, and yellow) to account for in F_2, then the expected proportion of (recessive) albinos in F_2 ought to be $(1/3)^2 = 1/9$. However, under the previous assumption of two gametes (albino and non-albino), the corresponding proportion was $(1/2)^2 = 1/4$, which agreed with observations. Using Bateson's own language, Weldon then asked sarcastically:

QUOT. 717 The two "Mendelian predictions" which Mr. Bateson has so far uttered ex post facto are mutually contradictory; with which of them is the inheritance of coat-colour "in punctilious agreement"? (*ibidem*)

[27] See Sect. 11.6.

[28] See Sect. 11.10.

11.12 Bateson's May 14, 1903, Response to Weldon: "No Question of One in Nine"

In a back-and-forth reminiscent of the Cineraria affair,[29] Bateson shot back in a letter (written on May 1) that appeared in the May 14, 1903, issue of *Nature* (Bateson, 1903c). He assured the reader that contrary to Weldon's allegation, there were no contradictions in his explanation.

If the aim was to distinguish only the albino with pink-eyes and the non-albino with pink eyes, the two types could be represented by gametes G and G′. Then, the F_1 hybrids were GG′ (dark-eyed and coat-colored), and the F_2 hybrids consisted of an equal number of GG (pink-eyed albinos) and G′G′ (pink-eyed with some coat color) and twice the number of GG′ (dark-eyed with some coat color). This was a reiteration of Bateson's position in the March 19, 1903, letter to *Nature.*[30]

On the other hand, if the aim was to explain the various shades of color in the GG′ hybrids of F_1 and F_2 and in the G′G′ hybrids of F_2, then Bateson's position was that of his April 23 letter,[31] namely, non-albino coat color needed to be treated as a *compound character.*[32] In other words, when Japanese waltzing mice (G′G′) were crossed with albinos (GG), the G′ gamete was resolved into various G′ gamete subtypes, which Bateson denoted by aG′, bG′, cG′, abcG′, and so on. Therefore, the F_1 hybrids (GG′) consisted of various subtypes such as G.aG′, G.bG′, G.cG′, and so on, and likewise the G′G′ hybrids in F_2 consisted of aG′.aG′, aG′.bG′, and so on. Consequently:

> **QUOT. 718** Each combination will have its appropriate colour and frequency, though (if the regularity be maintained) all *GG′* will have dark eyes and *some* colour, and all *G′G′* pink eyes and some colour. But as the hybrid produces *G* gametes equal in number to the various *G′* gametes collectively, *GG′* × *GG′* will give on an average one albino in four offspring (experiment gave nine in thirty-seven); and there is no question of one in nine. (*ibid.*, p. 34)

11.13 Weldon's May 14, 1903, Response to Bateson: Mendel's "Name Is Made to Shelter Almost Any Hypothesis, and Almost Any Experimental Test Is Evaded"

The back-and-forth between Weldon and Bateson had reached a point where each one was merely repeating his respective views in letter after letter, without adding anything new to the debate. Bateson kept reiterating that the coat-color of the

[29] See Sect. 8.3.

[30] See Sect. 11.6.

[31] See Sect. 11.10.

[32] See footnote 25.

Japanese waltzing mouse was a compound character, while Weldon kept insisting that the latter was simply improvising explanations and contradicting himself.

In the same issue of *Nature* where Bateson had last replied, in a letter written on May 6 Weldon used Bateson's own new notation in an attempt to prove him wrong (Weldon, 1903c). Writing Bateson's gamete subtypes[33] aG′, bG′, cG′, ... as G_1', G_2' and G_3', ... Weldon stated that, according to Bateson's explanation, the G′G′ homozygotes in F_2 thus consisted of both *homozygotes* such as $G_1'G_1'$, $G_2'G_2'$ and $G_3'G_3'$, ... and *heterozygotes* such $G_1'G_2'$, $G_1'G_2'$, $G_2'G_3'$, ... Since the G′G′ mice (Japanese waltzing) were known to breed true, not only would the homozygotes be similar, but also the heterozygotes. Weldon declared:

QUOT. 719 It is this consequence of heterogeneity in a pure-breeding race which seems to me amazing. (*ibid.*, p. 34)

Weldon further accused Bateson of abandoning Mendelian principles by refusing to assume that the various gamete subtypes had equal frequencies. He summarized his views as follows:

QUOT. 720 By modifying first one and then another of Mendel's statements, his name is made to shelter almost any hypothesis, and almost any experimental test is evaded. (*ibidem*)

Weldon further added that Darbishire would soon be publishing another report in *Biometrika* with "a series of new results, which have an important bearing on the application of Mendel's 'principles' to his mice." However, until then he did not wish to pursue the correspondence. An Editor's note after Weldon's letter simply added:

QUOT. 721 This correspondence must now cease. (*ibidem*)

11.14 Bateson's 1903 Review Paper: "The Present State of Knowledge of Colour-Heredity in Mice and Rats"

Although *Nature* refused to publish any more letters regarding Darbishire's experiments, the debate on the subject between the biometricians and saltationists was by no means over.

In a lengthy article published in the *Proceedings of the Zoological Society of London*, Bateson reviewed the state of knowledge on the heredity of color in mice and rats (Bateson, 1903d). He used the example of Cuénot's pioneering experiments on mice to argue that, contrary to Darbishire's and Weldon's suggestions, albinism in mice was indeed recessive (and therefore Mendelian). He noted that this was consistent with the frequent production of albinos when colored mice were crossed and with the absence of colored offspring when albinos were crossed. Bateson cited only one known exception to the latter prediction: in later editions of the book *Fancy Mice*, Dr Carter Blake (who provided criticisms and remarks on the crossings in the

[33] See Sect. 11.12.

book) gave an example where two albinos were crossed and produced both albino and colored offspring.[34] Lest his opponents jumped on this example to argue against Mendelian inheritance in mice, Bateson hurriedly warned that:

> **QUOT. 722** . . .we should be cautious in declaring the result impossible, for in Mendelian experiments the observer must be on the look out for the appearance of a character, elsewhere a definite dominant, *as the consequence of crossing two dissimilar recessives.*[35] (*ibid.*, p. 76)

Bateson did not explicitly show how two dissimilar recessive albinos could produce a colored mouse, but it is not hard to imagine a two-locus model with alleles (a, A) at locus L1 and alleles (b, B) at locus L2, where the alleles a and b are both recessive for albinism. Then, a cross between the two albinos aB/aB and Ab/Ab yields the double-heterozygote aB/Ab, which is colored. This mechanism is similar to that used in (11.4) to explain the production of dark-eyed mice by crossing (recessive) pink-eyed parents. Gillham (2002, p. 317) has thus wondered why Bateson did not use the concept of dissimilar recessives to also explain the dark-eyed phenomenon.

After reviewing previous crosses between Japanese waltzing mice and albinos, especially those of Haacke and von Guaita, Bateson next turned to Darbishire's experiments. To explain the variation in color in the F_1 hybrids, as before, he suggested the possibility that the coat color of waltzers was a compound character.[36] He also questioned the purity of the waltzers used by Darbishire. While acknowledging a few difficulties, Bateson's conclusion was that:

> Mendelian analysis provides a means of elucidating a large part of the phenomena. (Bateson, 1903d, p. 95)

11.15 Darbishire's Third Report (June 1903): Pure versus "Extracted" Recessives

Unphased by the ongoing debate, Darbishire released his third report in June 1903 (Darbishire, 1903b). This time, he decided to settle the issue of ancestral versus Mendelian heredity by showing the untenability of one of the key principles of

[34] In the third edition of *Fancy Mice,* for example, one can read on p. 16: "Two albino mice were mated. The birth produced nine. One of these, devoured by the mother almost at once, was dark brown and white; two were albinos, exactly like the parents; two were plum-coloured, with lightish tails; two were exactly similar to the common grey wild mice, but much smaller than it in size; and two were of a breed which, as it is almost unknown to me, I prefer to describe in detail. The individuals are large, very black, the tail being entirely black; the ears rounded and prominent, unlike those of all mice, and similar to those of the black rat (*Mus rattus*); the moustaches long, the hair on the head, snout, and back very long. The nose is longer and more acuminate than in most of the varieties of mice. I know not what name to give to this variety for the purpose of convenient distinction, but suggest that of 'uracoïd' mice, from the Greek for rat." (Old Fancier, 1893).

[35] Italics are Bateson's.

[36] See footnote 25.

Table 11.1 Darbishire's 2×2 table for the correlation of coat color in F_1 and F_3 hybrid mice (Darbishire, 1903b, p. 283)

	yellow	wild	totals
Yellow	8	4	12
Wild	1	67	68
Totals	9	71	80

Mendelism, namely, gametic purity. In particular, in addition to true-breeding pure recessives, he used "extracted" recessives, i.e., recessives obtained from at least one non-recessive parent, with the aim of verifying if:

> **QUOT. 723** . . .the extracted recessive is identical with the original pure recessive and is not affected by its unusual ancestry in any way. (*ibid.*, p. 282)

A first set of extracted recessive albinos was obtained by intercrossing F_1 hybrids obtained from pure-breeding waltzing mice and pure-breeding albinos. The extracted recessives were then crossed with pure-breeding waltzers. The resulting (F_3) hybrids were then compared with those in F_1. Darbishire noted that the two sets of hybrids "had invariably the same color."

A second set of extracted recessive albinos was obtained by backcrossing F_1 hybrids (resulting from the cross between pure-breeding waltzers and pure-breeding albinos) with pure-breeding albinos. The extracted recessives were then crossed with pure-breeding waltzers, resulting in F_3 hybrids. This time, Darbishire tabulated the correspondence between the coat colors of the F_1 and F_3 hybrids in Table 11.1.

Using the numbers in Table 11.1, Darbishire calculated Pearson's tetrachoric correlation (Pearson, 1900a, p. 6)[37] as $r = .951$ and interpreted the magnitude of this correlation as decisive evidence against gametic purity, hence Mendelism:

> **QUOT. 724** This correlation between grandparent and grandchild with regard to a "dominant character" transmitted through the recessive parent seems to dispose entirely of any theory which involves the gametic purity of the recessive parent. (*ibid.*, p. 283)

However, a high correlation between grandparent and grandchild, even for a dominant character transmitted through a recessive parent, cannot be taken as evidence against Mendelism. This is because the latter theory does not deny such a correlation. To understand why, recall an essential element of Mendelism, namely, that once the characters of the parents are specified, the distribution of characters among the offspring is completely determined.[38] This implies that, in Mendelian heredity, the correlation between grandchild and grandparent is zero only when the effect of the parent has been removed (partialed out). However, in the absence of any conditioning on the parents, correlations between offspring and ancestral traits exist. This point seems to have been poorly understood by early biometricians.

[37] See also p. 248 of Gorroochurn (2016a).

[38] See footnote 9 of chap. 9.

In view of the aim explicitly stated in QUOT. 724 by Darbishire to verify the effect of ancestry, a better and more decisive experiment would have been to perform sets of crosses, each involving the same parental types but different grandparental types. Differences in the offspring types across the sets could then have weighed against Mendelism. This is precisely the type of experiment that Darbishire later resorted to in his fourth report and that subsequently forced him to reconsider the Mendelian theory.

11.16 Weldon's June 1903 Paper: "Mr Bateson's Revisions of Mendel's Theory of Heredity"

In the same issue of *Biometrika* where Darbishire's third report came out, Weldon devoted a whole paper (Weldon, 1903) arguing that Bateson had distorted the original Mendelian doctrine in his attempt to explain Darbishire's results. Weldon's implication clearly was that Mendelism was inadequate as a theory and had to be modified each time to accommodate new observations.

In particular, Weldon accused Bateson of five "transgressions":

(a) *Nature of pure dominants and pure recessives*. Weldon recalled that, when Mendel intercrossed F_1 hybrids *Aa* (where gamete *A* is dominant over *a*), the F_2 offspring included pure dominants (*AA*) and pure recessives (*aa*).[39] Now, Mendel had stated that these remained constant for the dominating and recessive traits, respectively. Weldon explained that this meant that the *AA* and *aa* in F_2 were *similar* to their respective grandparents *within the limits of variation in a race*. He accused Bateson, however, of insisting that:

> **QUOT. 725** The pure dominant and pure recessive members of each generation are not merely like but identical with the pure parents.[40] (*ibid.*, p. 289)

Weldon complained that Bateson's prescription above was contradicted by actual experimental results, which Bateson still insisted were in accordance with Mendel's laws. Finally, Weldon claimed that the discrepancies between Mendelian expectations and the experimental results of others could, in fact, be accounted for by differences in ancestry.

(b) *Necessity of dominance*. Weldon next accused Bateson of abandoning what he believed to be one of the key tenets of Mendelism, namely, dominance:

> **QUOT. 726** The phenomenon of dominance, though always observed by Mendel, is not essential; and in asserting this proposition Mr Bateson does not hesitate to accuse his master of something very like a gross breach of good faith. (*ibid.*, p. 290)

[39] In addition to the heterozygotes *Aa*.

[40] This statement has been made in Bateson and Saunders (1901, p. 12).

The above is a reiteration of Weldon's position in his very first paper on Mendelism (cf. Sect. 9.3.1) and ignores Correns' previous admonishment that dominance was not necessary in Mendel's theory (cf. QUOT. 475).

(c) *Atavism in hybrids.* Weldon explained that, since dominance was *not* always observed in experiments, Bateson often resorted to *atavism* to explain the peculiar traits of hybrids. No doubt, Weldon here had a recent instance in mind, namely, Bateson's atavistic explanation for the appearance of dark eyes in F_1 hybrid mice when two pink-eyed parents (albinos and Japanese waltzing mice) were crossed.[41] In addition, Bateson's explanation in terms of atavism was also applied to hybrids that had arisen from the crossing of distinct races. Thus:

> **QUOT. 727** We have here a second addition to Mendel's statements—the manifestation of a spurious atavism as a property of hybrids, and the doctrine that several different gametic combinations may produce identical individuals. (*ibidem*)

Weldon complained that Bateson had judged it unnecessary to explain the two phenomena above through his theory of gametic purity.

(d) *Similarity of reciprocal crosses.* Weldon recalled that, in all his experiments, Mendel had affirmed the similarity of crosses, i.e., it did not matter whether the dominant character belonged to the seed or pollen. Weldon accused Bateson of departing from this principle in the horse, ass, and certain species of *Digitalis*:

> **QUOT. 728** He [Bateson] conceives a species in which the male gametes contain only one element *A* of a Mendelian pair, the female gametes containing only the complementary member *a* of the pair. When a male and female of such a species breed together, the fertilised zygote is of composition *Aa*; but the gametes produced by such a zygote will contain only *A* elements if the zygote body be a male, only a elements if it be a female. Each individual will therefore completely suppress half the heritage it receives from its parents during the course of its development. (*ibid.*, p. 291)

(e) *Compound characters.* Weldon took issue with Bateson's *compound allelomorphs* to explain the appearance of the lilac color on the coat of F_2 mice in Darbishire's experiments (see Sect. 11.10). Recall that[42] Bateson's allelomorphs behaved as a single factor (e.g., *A*) when united in fertilization but were then resolved into several factors A', A'', A'', ... when cross-fertilized. Weldon disagreed with this concept, alleging that it was antithetical to Mendel's independent factors for different traits and had been introduced simply to evade difficulties:

> **QUOT. 729** If a simple result, such as that obtained by Mendel when he crossed yellow and green peas, be obtained on crossing two animals or plants which differ in respect of any character, that character is simple, and there is no need to go further. Should the hybrids exhibit any unexpected character, a reference to the properties of "heterozygotes" will remove all difficulties; should the offspring of the hybrids be of more than two kinds, one or other of the foregoing hypotheses will generally be found to evade the difficulty. (*ibid.*, 294)

---

[41] See QUOT. 699. Weldon later mocked this explanation as a convenient escape when difficulties were presented by hydrids (see QUOT. 701).

[42] See footnote 25.

Weldon concluded his paper by taking exception to Bateson's treatment of sex as a Mendelian factor and denouncing his treatment of eye-color and albinism in Darbishire's mice experiments as "pseudo-Mendelian."

11.17 Darbishire's Fourth Report (Jan 1904)

At the end of Sect. 11.15, it was mentioned that, to test Mendelism against the ancestral law, an appropriate approach would be to perform sets of crosses, each involving the same parental types but different grandparental types. Differences in the distributions of offspring across the sets could then count as evidence against Mendelism. This is precisely the approach Darbishire adopted in one of the experiments in his fourth report (Darbishire, 1904b), which was published in January 1904 and was thus far the most comprehensive of all his reports.

Right at the start, Darbishire left no doubt as to what he intended to demonstrate in his big paper. Commenting on the variability of F_1 hybrids when albinos were crossed with Japanese waltzing mice, Darbishire declared:

QUOT. 730 How Mendelians will account for this variability. I do not know, but I have facts to show that it is due to the ancestry of the parent forms. (*ibid.*, p. 7)

Thereupon, Darbishire calculated correlations between the purity of albino parents and (i) amount of white coat in hybrids, (ii) color of coat in hybrids. Having shown that these correlations (−.21 and −.16, respectively) were "invariably negative," he added:

QUOT. 731 On any theory of gametic purity the correlations should of course be zero. (*ibid.*, p. 13)

Darbishire next detailed several experiments involving the offspring and grandchildren of hybrids, as described below:

(a) *Hybrids paired together*. Darbishire reported that the pairing of F_1 hybrid mice yielded 555 F_2 offspring with the distribution of eye and coat color shown in Table 11.2:

With respect to walking behavior, the distribution of the 555 offspring is shown in Table 11.3:

Table 11.2 Distribution of eye and coat colors of F_2 mice in Darbishire's fourth report (Darbishire, 1904b, p. 14)

Eye and coat color in F_2	Number
Albino	137
Colored/piebald with dark eyes	287[a]
Colored/piebald with pink eyes	131
Total	**555**

[a]Darbishire observed that he was unsure about three mice in this category: it was possible that their eyes were more pink than dark so that the 287 pink-eyed were really 284 and the 131 pink-eyed were really 134

Table 11.3 Distribution of waltzing behavior of F_2 mice in Darbishire's fourth report (Darbishire, 1904b, p. 14)

Walking behavior in F_2	Number
Waltzing	97
Normal	458
Total	**555**

Thus, for eye and coat colors (Table 11.2), the observed counts are approximately in the Mendelian ratio 1:2:1. Darbishire declared:

QUOT. 732 It is clear that the albinos occur in this by a purely Mendelian theory of albinism regarded as a simple unit character, allelomorphic to a single unit character in the waltzing race; that hypothesis leads us to expect that a quarter of the offspring of hybrids paired together will be albinos; and since 555/4 = 138.75, the observed result is in excellent accord with it; the difficulty in this, as in all attempts to apply Mendel's hypothesis, is to discover what are the unit characters concerned. (*ibid.*, pp. 14–15)

However, concerning waltzing behavior (Table 11.3):

QUOT. 733 *Waltzing* occurs in only 97 out of the 555 individuals resulting from the union of hybrids. When we compare this with the number of pink-eyed individuals (131–134) or of albinos (137) we see that the proportion of waltzing individuals cannot be regarded as a possible quarter. The probable error of the expectation that a quarter of the individuals will waltz is, on the Mendelian hypothesis, $0.6745\sqrt{\frac{1}{4}\times\frac{1}{4}\times 555} = 6.88$ only, and the observed deviation is $138.75 - 97 = 41.75$, the odds against so great a deviation being rather more than 50,000 to 1. As the result here obtained differs from Mendelian expectation in the same direction as that already obtained by von Guaita[43] and to an extent consistent with the agreement of both, the evidence that the waltzing character does not segregate in Mendelian proportions is very strong. (*ibid.*, p. 20)

Darbishire would change his position regarding waltzing just a few months later (see QUOT. 738), much to Pearson's ire. On the other hand, his statement at the end of QUOT. 732 concerning the proper identification of the relevant unit characters was quite apposite, especially with regard to the dark-eyes and variety of colors in the F_1 hybrids. Darbishire next addressed Bateson's explanation in the March 19, 1903 issue of *Nature* (Bateson, 1903a).[44] Let G and G' represent, respectively, pinkness of eye color with white coat (i.e., the trait of the albino) and pinkness of eye with some color in the coat (i.e., the trait of the Japanese waltzing mouse). Then, the crossing between the albino parents GG and $G'G'$ yielded F_1 hybrids of type GG'. When the latter were crossed, the distribution in F_2 was represented by $GG + 2GG' + G'G'$. Darbishire objected to this explanation since it did not explain the dark eyes of GG' and the variety of coat colors in F_1. Darbishire now considered Bateson's explanation of May 14, 1903 (Bateson, 1903c).[45] Using the concept of compound

[43] "Versuche mit Kreuzungen von verschiedenen Rassen der Hausmaus," Ber. D. naturforsch. Gesellsch. Freiburg, Bd. X. 1898, Bd. XI. 1900. (*Darbishire's footnote*).

[44] See Sect. 11.6.

[45] See Sect. 11.12.

characters,[46] Bateson had posited that the G' gamete was resolved into several subtypes $G'_1, G'_2, G'_3, \ldots$ upon fertilization. Darbishire assumed that the parental Japanese waltzing mice consisted of yellow pink-eyed forms (pure dominants $G'_1G'_1$), lilac pink-eyed forms (recessives $G'_2G'_2$), and fawn-yellow pink-eyed forms (heterozygotes $G'_1G'_2$). He noted that if the waltzing mice consisted of $G'_1G'_2$, the intercrossing would have resulted in the three colors for $G'_1G'_1$, $G'_1G'_2$, and $G'_2G'_2$ appearing in the offspring, but this was contradicted by observations. On the other hand, if the waltzing mice consisted of $G'_1G'_1$ and $G'_2G'_2$, then lilac color would have been present in them, and this also was contradicted by observations. Darbishire considered other similar possibilities and, despite his statement in QUOT. 732, concluded:

> **QUOT. 734** The behaviour of the waltzing mice when bred together is alone sufficient to destroy the validity of any hypothesis, involving numerous elements G', which does not depart from Mendel's doctrines. (Darbishire, 1904b, p. 16)

(b) *Hybrids paired with albinos.* Darbishire paired F_1 hybrid mice with albinos and obtained 746 offspring consisting of 368 albinos and 378 pie-bald or self-colored. The colored offspring all had dark eyes and showed no signs of waltzing. Darbishire made no mention that the offspring were approximately in the 1:1 Mendelian ratio but instead focused on the color of the colored offspring in an attempt to offer an explanation in terms of ancestry. For example, he emphasized that the gap between albinos and colored mice resulting from the backcross was wider than when hybrids were crossed. The implication clearly was that ancestry mattered. A similar suggestion was made when he next observed that:

> **QUOT. 735** It will be seen that the coloured offspring of hybrids and albinos, which have more white in their ancestry than the offspring of hybrids alone, are characterised first by having considerably less white fur, and secondly by the greater frequency of wild-colour and black, as compared with yellow and fawn-colour—that is, by the presence of black pigment in larger quantity. (*ibid.*, p. 21)

However, Darbishire acknowledged the difficulty in explaining why the correlation between the amount of whiteness in the ancestor and that in the offspring was negative.

(c) *Hybrids paired with extracted albinos.* Darbishire next crossed F_1 hybrids with *extracted* albinos (i.e., albinos having at least one non-recessive parent, see Sect. 11.15) to show a difference from (b). Such a difference would have corroborated his ancestral hypothesis. However, as in (b), he observed an offspring distribution (59 albinos out of 120), which was approximately in the 1:1 Mendelian ratio. Darbishire promptly discounted the agreement of his observations with Mendelian expectations on the grounds that: (i) some of his extracted albinos were derived from two hybrids, while others were from one recessive and one hybrid, (ii) the hybrids were in some cases the offspring of waltzers and extracted whites (instead of recessive albinos), and (iii) the attempt here to diminish the

[46] See footnote 25.

amount of albino ancestry by using extracted rather than pure albinos was "swamped" by the presence of albinos in the parents.

(d) *Grandchildren of hybrids.* Denoting the F_1 hybrids by *DR*, the F_2 children in (a) are of the types *DD* (pink-eyed with coat color), *DR* (dark-eyed with coat color), and *RR* (albino), and are approximately in the Mendelian ratio 1:2:1. Darbishire called the three types, extracted dominants, extracted hybrids, and extracted recessives, respectively. Extracted hybrids were also obtained by backcrossing F_1 *DR* with parental *RR*. The pairing of extracted types was the main concern of the rest of Darbishire's experiments.

Of all possible pairings of extracted types that he described in his report, the one that Darbishire recognized as providing the "most conclusive results" was that between extracted hybrids, because this enabled him to have the same extracted parental types while varying the grandparental types and investigating the effects on the grandchildren. The three types of crosses involving extracted hybrids are depicted in Fig. 11.5.

Thus, the grandchildren (i), (ii), and (iii) have, respectively, 0, 1, and 2 albinos (*RR*) as grandparents. Now, under Mendelian theory, Darbishire reasoned that the grandchildren in each case should contain 25% albinos. On the other hand, the law of ancestral inheritance implied that the grandchildren (i), (ii), and (iii) should contain proportionately more albinos. Darbishire triumphantly announced that the latter was "exactly what we find to be the case." As proof, he produced percentages as shown in Table 11.4.

Therefore, Darbishire's major conclusion in the fourth report was that:

QUOT. 736 The effect of varying ancestry is so great, in every case in which it has been examined, as to show that the phenomena observed cannot be adequately described except in terms of the ancestry of all the individuals used as parents. (*ibid.*, p. 27)

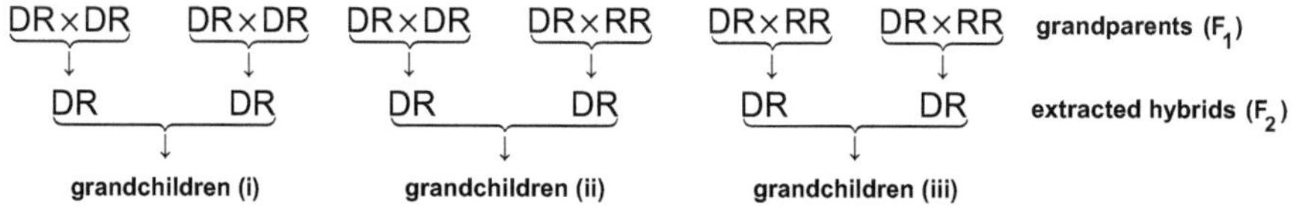

Fig. 11.5 Key experiment in Darbishire's fourth report (Darbishire, 1904b, pp. 24–25) involving the same parental types (extracted hybrids) but different grandparental types

Table 11.4 Percentage of Albinos among the grandchildren of hybrids in Darbishire's key experiment (Darbishire, 1904b, p. 25)

Grandchildren type	No. of grandchildren	No. of albinos	% age albinos
(i)	93	10	10.75
(ii)	107	20	18.69
(iii)	121	30	24.79

11.18 Darbishire's Summary Paper Delivered on March 15, 1904: First Public Admission of Both Mendelian and Ancestrian Theories

Darbishire wrote a paper (Darbishire, 1904a) just a few months later on March 15, 1904, which essentially summarized the findings of his fourth report.

Darbishire started his paper by briefly reviewing the difference between continuous and discontinuous variations and stating the ancestral law. He next gave a detailed explanation of Mendel's theory. Now, commenting on the relative validity of the two theories, Darbishire referred to the key experiments on the grandchildren of hybrids that he had described in his fourth report[47] and which validated the ancestral theory:

> **QUOT. 737** I refer to the phenomenon of segregation. We have seen what Mendel says (*see* Bateson :**02**, p. 57).[48] But this is flatly contradicted by the Galtonian generalization, according to which the greater number of generations a given hybrid is from the first hybrid (i.e., of course, also from the parents of the hybrid) the fewer pure recessive and dominant forms is it likely to produce when mated with another hybrid of its own generation (Darbishire :**04**, pp. 23 et seq.).[49]
>
> I refer to this point (which at first sight may appear insignificant, but in reality is not) because it seems to me to afford a means of deciding between the relative validity of the two theories, inasmuch as it is a matter about which the Mendelian and Galtonian predictions are totally at variance. (*ibid.*, pp. 9–10)

Instead of stopping there, Darbishire went on to also mention those results that validated the Mendelian theory. Referring to the crosses between hybrids obtained from albinos and Japanese waltzing mice, he said:

> **QUOT. 738** First, with regard to colour, we should expect 25% albinos which should breed true: this is in fact what we get. Secondly, with regard to their progression, we should expect to find 25% waltzing mice: this is very roughly what happens. . .we should expect one in every sixteen to be an *albino waltzer* ... What has happened is that we have taken the recessive character —albinism —from one parent of the hybrid, and the recessive character —'waltzing' —from the other. . .I have bred several examples of this new species. . . (*ibid.*, p. 13)

Note that the above statement is in direct contradiction with Darbishire's statement in the fourth report concerning observed and expected proportions (see QUOT. 733). This contradiction was later criticized by Pearson(1904c).[50]

The bigger issue here, of course, was that Darbishire had changed his position from being anti-Mendelian to now accommodating both the ancestral and Mendelian theories.

---

[47] See Sect. 11.17 (d).

[48] The reference here is to Bateson (1902b).

[49] The reference here is to Darbishire (1904b).

[50] See Sect. 11.21.

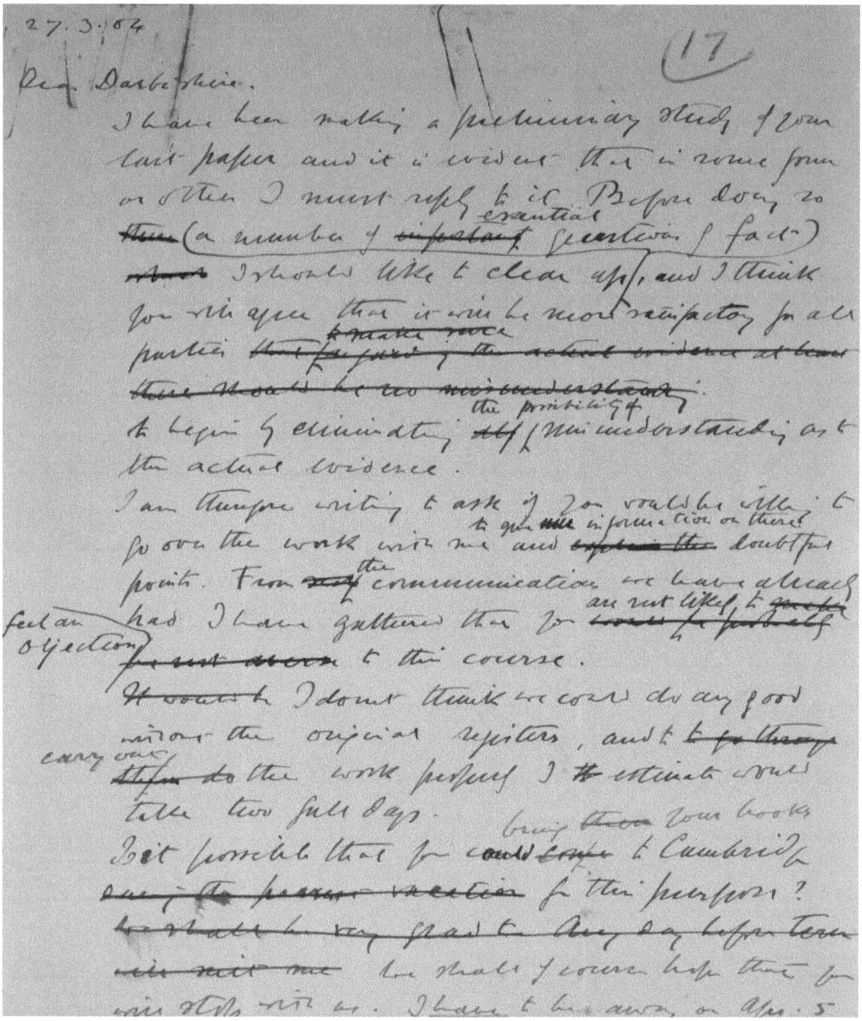
27.3.04

17

Dear Darbishire.

I have been making a preliminary study of your last paper and it is evident that in some form or other I must reply to it. Before doing so (a number of essential questions of fact) I should like to clear up, and I think you will agree that it will be more satisfactory for all parties to begin by eliminating the possibility of misunderstanding as to the actual evidence.

I am therefore writing to ask if you would be willing to go over the work with me and to give information on these doubtful points. From the communication we have already had I have gathered that you are not likely to feel an objection to this course.

I don't think we could do any good without the original registers, and to carry out the work properly I estimate would take two full days.

Is it possible that you could bring your books to Cambridge for this purpose? ... we shall of course hope that you will stay with us. I have to be away on Apr. 5

Fig. 11.6 Bateson's March 27 letter to Darbishire requesting a meeting to discuss the results in the latter's fourth report

11.19 Bateson's Reaction to Darbishire's Fourth Report

Bateson examined Darbishire's fourth report carefully and found discrepancies not only between it and Darbishire's other reports but also within the fourth report itself. In a letter dated March 27, 1904, he proposed a meeting to discuss Darbishire's experiments (Fig. 11.6):

> **QUOT. 739** I have been making a preliminary study of your last paper and it is evident that in some form or other I must reply to it. Before doing so, I should like to clear up a number of essential questions of fact, and I think you will agree that it will be more satisfactory for all parties to begin by eliminating the possibility of misunderstanding as to the actual evidence.

I am therefore writing to ask if you would be willing to go over the work with me and to give information on these doubtful points. From the communications we have already had I have gathered that you are not likely to feel an objection to this course.[51]

Darbishire declined any such meeting on the grounds that it would be preferable that all communication be put in writing:

QUOT. 740 Many thanks for your letter which has followed me here. I had hoped, and did my best, to keep clear of the discussion which is taking place; but since I have become involved and since either party may base conclusions on the statements, made by both parties, I think it is better that these statements should be written down; so that there can be no ground for future uncertainty. If therefore you will ask me what are the questions of fact about which you are uncertain I hope you can rely on me to give a complete answer. (Kim, 1994, p. 74)

Since Darbishire was not eager to meet, on April 3, 1904, Bateson sent him a list of inconsistencies he had found in his reports:

QUOT. 741 Preliminary questions suggested by Darbishire's fourth paper, Biometrika, III. I.

A comparison of paper IV with papers III and II shows several points of imperfect agreement. The system of classification is apparently (IV, pp 1, 2) that used in III and II. On the other hand, the changes cannot be due to revision of the individual classifications, for there is no statement that such revision has been made. Sometimes moreover the changes are such as not to be referrable to revision. If any of these differences are due to error, I should be glad to have the points decided before proceeding further.

Beginning with Paper II, Table I, p. 167, and comparing it with paper IV, Table A, p. 29, I find the following discrepancies in the account of the 29 crosses, which are common to both Tables.

	Paper II Offspring	Paper IV Offspring
Cross VI	2d	2c
___ XII	5a, 5a, 5d, 5d, 5d	5a, 5a, 5c, 5c, 5c
___ XXXV	2c, 2d, 3d, 3d, 4d, 4d	Omitted
___ XLV	2c, 2c, 2c, 2c	2d, 2d, 2d, 2d
___ LXXIII	3c, 3c, 5c, 5c, 5c, 5c	3c, 3c, 3c, 5c, 5c, 5c
___ LXXXV	3e, 3e, 4d, 5d, 5d, 5d, 5e	3e, 3e, 4d, 5d, 5d, 6d, 5e

...[52]

[51] Letter retrieved online from Cambridge Digital Library, Bateson Archive at https://cudl.lib.cam.ac.uk/view/MS-ADD-08634-C/278

[52] Portion of letter retrieved online from Cambridge Digital Library, Bateson Archive at https://cudl.lib.cam.ac.uk/view/MS-ADD-08634-C/283. The coding here was the one Darbishire had used in his reports: group 1 = mouse has more white and less extent of colored patches than the normal waltzing mouse, Group 2 = mouse has about as much white as a normal waltzing mouse, group 3 = mouse has less white (i.e., greater extent of colored patches) than a waltzer, group 4 = mouse has still less white, group 5 = mouse has no pure white, but the belly is whitish, not gray, group 6 = mouse whose belly is nearly the same colors as its backs; a = patches are yellow, b = patches are fawn yellow, c = patches are pale wild in color, d = patches are dark wild in color, e = patches are back, f = "lilac" (pale blue gray), g = chocolate; p after a mouse indicates it has a pink eye and W that it waltzes.

The above portion of the April 03 letter shows only 6 out of a total of 28 inconsistencies Bateson had found.

In the face of so many apparent discrepancies, Darbishire replied with a long explanatory letter on April 21, 1904. He started his letter as follows:

QUOT. 742 I got back yesterday; and hasten to reply to your letter, and to account for the errors, which you have discovered in my papers. I am returning you [a] list with the inconsistencies numbered, so as to able to refer to them more conveniently. (Kim, 1994, p. 75)

After a few more letters between Darbishire and Bateson, the latter expressed his growing frustration on May 22, 1904:

QUOT. 743 .... At the first glance, it was apparent that the account in paper IV differed frequently and sometimes emphatically from those of the previous papers, and that whenever it has been necessary to traverse the same ground twice, the several parts of paper IV differed similarly from each other. Notwithstanding these discrepancies, the results were presented in a series of 54 tables which though pretending to even mathematical precision are not always compatible with the statements in the text.

As a beginning I put certain preliminary questions designed to test whether this apparent confusion was superficial or inherent. The answers you have been good enough to prepare show that such latitude has been possible in the use of the various terms, and actual mistakes so frequent, and that to attempt to unravel the facts by correspondence would be an almost interminable task. Your replies in fact raise further difficulties. To take an illustration...[53]

Following this damming indictment of Darbishire's entire methodology, Bateson enumerated further inconsistencies. Next, in the same letter, Bateson referred to a possible public discussion of the issue:

QUOT. 744 Meanwhile, as regards any public reference to the subject, my action must depend on the measure of importance which is attributed to your paper. It seems to me likely that authors accustomed to the critical study of evidence will find in the publication report causes for suspicion sufficient to make any special warnings superfluous. (*ibidem*)

Even though Bateson was for now withholding any public discussion of the inconsistencies in Darbishire's reports, the latter rightly sensed that events were taking a particularly negative turn as far as his scientific competence and integrity were concerned. In a May 27, 1904 letter, he reversed his previous position and volunteered to meet Bateson:

QUOT. 745 I am in such a difficult position that ... even now I don't know that I am doing right. I should like this letter to be considered as a private one between man and man; and not as forming part of that correspondence which I suggested; your reply shall of course be considered in the same light; it shall be destroyed .... If I came to Cambridge, I should have to do so without anyone knowing. Do you think that would be fair or even possible? ... I don't mind your saying what you like about the interpretations and conclusions in the mouse paper; but to have my records discredited would be heart-breaking and render it useless and a waste of time for me. (Kim, 1994, p. 77)

[53] Letter retrieved online from Cambridge Digital Library, Bateson Archive at https://cudl.lib.cam.ac.uk/view/MS-ADD-08634-C/291

With growing impatience, Bateson replied on May 30, 1904:

QUOT. 746 In reply to your letter received yesterday, let me say that I appreciate the difficulties to which you allude. Nevertheless, though of course I shall give a careful consideration to any further materials you may send, it does not seem to me that I can add anything at this time to what I wrote you on May 22. It will, I think, be obvious to you on reflexion, that any communication between us which is to serve as a basis of discussion must be of a public nature.[54]

With the threat of possible public exposure, Darbishire was now in a precarious position.

11.20 Section D, Zoology Debate at the British Association Meeting in August 1904

The August 1904 meeting of the British Association was one of the biometricians' last battles in their losing war against the Mendelians. The meeting of Section D, Zoology started on August 18, 1904. Its proceedings were described both in the Sep 19, 1904 issue of *Nature* (Anonymous, 1904b) and the *Times* article of Aug 20, 1904 (Anonymous, 1904a). The latter, which provided some key details not to be found in the *Nature* article, described the general atmosphere of Section D as follows:

QUOT. 747 In anticipation of the encounter between the Mendelians, as represented by Mr. Bateson, and the biometricians, as represented by Professor Weldon and Professor Karl Pearson, the section room was very crowded all day; and if undue heat between the disputants was avoided, the discussion was not lacking in elements of the excitement, reflected by the audience in their demonstrations of partisanship so new to the serene abstraction of Section D. (Anonymous, 1904a, p. 8)

In his presidential address, delivered on August 18, Bateson presented his views on the new science he himself had largely contributed to. He did not yet have a definite name for that science, and tentatively wrote:

QUOT. 748 The breeding-pen is to us what the test-tube is to the chemist—an instrument whereby we examine the nature of our organisms and determine empirically what for brevity I may call their genetic properties. (Bateson, 1904, 1928, p. 243)

Bateson would use the formal name "genetics" first in a letter written to Adam Sedgwick[55] on April 18, 1905, and then again in a *Nature* review of a textbook written by J. P. Lotsy (Bateson, 1906, p. 306).

Coming back to the presidential address, Bateson deplored the fact that the acceptance of Mendelism had been so hard:

---

[54] Letter retrieved online from Cambridge Digital Library, Bateson Archive at https://cudl.lib.cam.ac.uk/view/MS-ADD-08634-C/296

[55] See QUOT. 506.

QUOT. 749 Convention requires that a president should declare all well in his science; but I cannot think it a symptom indicative of much health in our body that the task of assimilating the new knowledge has proved so difficult. (*ibid.*, p. 234)

Bateson was convinced that:

QUOT. 750 ...[h]ad a discovery comparable in magnitude with that of Mendel been announced in physics or in chemistry, it would at once have been repeated and extended in every great scientific school throughout the world. (*ibidem*)

He hailed the pioneering works of Kölreuter[56] and others as moving in the right direction but then noted that investigations of intra-specific variability almost came to a halt with Darwin's *Origin of Species*.

Next, Bateson took on the biometricians. Consistent with his previous writings, he admitted that natural selection acting on individual differences was a true phenomenon, but denied that it had any role in species formation:

QUOT. 751 The popular belief that Evolution, whether natural or artificial, is effected by mass-selection of impalpable differences arises from many errors which are all phases of one—imperfect analysis—though the source of the error differs with the circumstances of its exponent. (*ibid.*, p. 239)

Instead, he emphasized the role of discontinuous variation. Concerning the biometricians' statistical method (and its characteristic *en masse* approach), Bateson qualified it as "a misleading instrument." On the other hand, the Mendelians' method, which he called "descriptive zoology," had achieved successes which were:

QUOT. 752 ...so palpable and so attractive, that, not unnaturally, these which are the means of progress have been mistaken for the end. (*ibid.*, p. 234)

The following day was devoted to the presentation and discussion of papers. The first one was by a Mendelian, Miss E. R. Saunders, who presented "Heredity in Stocks." Saunders argued that the experimental evidence for the purity of germ cells was increasing rapidly. She explained that much of this evidence came from experiments similar to those of Mendel and that, in some cases, reversion and "gametic coupling of distinct characters"[57] might prevent straightforward Mendelian interpretations.

The next paper was delivered by Darbishire. This was essentially the same as his summary paper of March 15, 1905 (see Sect. 11.18). After having described how Mendelian proportions held from crosses between hybrids obtained from albinos and Japanese waltzing mice, he stuck to his newly adopted neutral position:

QUOT. 753 Some of the facts seem to confirm the Mendelian interpretation, while others may be described in terms of either Galton's or Pearson's formula of ancestral inheritance. (Anonymous, 1904b, p. 538)[58]

[56] See Sect. 3.3.

[57] Later to be called "linkage".

[58] All citations that follow in this section are the *Times* paraphrase of the speakers' true words.

The paper that followed was by the Mendelian C. C. Hurst, who described the results obtained when an inbred pair of albino Agora rabbits was crossed reciprocally with an inbred pair of Belgian hares. The hybrids from these crosses were then intercrossed for two generations. Hurst noted that the results of the crosses were in accordance with Mendelian expectations.

It was now Weldon's turn to intervene. Referring to the *Lychnis* experiments he had already commented on before (see Sect. 9.3.3), he again raised the issue of ambiguity of Mendelian categories:

QUOT. 754 ...*Lychnis dioico* had a hairy and a glabrous form. If these were crossed, the resulting plants were hairy; and the offspring of such hybrids were hairy or glabrous in Mendelian proportions. Now those hairy plants which were apparently "pure" bred, were said to be like the hairy plant used in making the original cross; but we were not told how hairy either plant was.[59]

With this example in mind, Weldon deplored the "looseness of Mendelian descriptions" (*ibidem*). According to him, the only phenomenon presented by the Mendelians and capable of definite interpretation was reversion. He argued that this phenomenon was in fact completely inconsistent with the Mendelians' claim that the characters of hybrids were combinations of pure determinants derived from parents. On the other hand, it was compatible with the ancestral law. Weldon concluded as follows:

QUOT. 755 What was wanted was more experiment, and, above all, a description of results so careful that the questions raised by the Mendelian hypothesis could be answered, instead of being masked and evaded by the adoption of loose and vague descriptive categories. Till such work, had been done it was, surely, better to use only the purely descriptive statements of Galton and Pearson, without invoking the cumbrous and demonstrable gametic mechanism on which Mendel's hypothesis rested. (*ibidem*)

In the afternoon, after R C Punnett (a Mendelian) and Professor Minot (from Harvard University) spoke, Bateson rose to rebut Weldon's arguments. His language was harsh.[60] He maintained that it was through the Mendelian hypothesis that a vast number of hitherto incoherent facts could be explained. Bateson scoffed at the ancestrians' claim that Mendelism was overrated; on the contrary, he maintained that:

QUOT. 756 [t]o the Mendelians, it appeared that the hypotheses of the Ancestrians were disposed of, and that the voluminous works based on these hypotheses had no scientific value. (*ibidem*)

Next, Bateson accused Weldon of passing over the fact that the Mendelian versus ancestrian issue had already been settled through increasing experimental evidence for gametic purity:

[59]"The British Association," *Times* (London), 20 August 1904, p. 8.

[60]In a September 6, 1904 letter to his niece (Pearson, 1930, p. 528, Vol. IIIB), Galton said that Bateson had used "some rather savage phrases," which were omitted in the *Times* article.

QUOT. 757 None of the various schemes of the schemes of the Ancestrians had ever contemplated such purity and all were totally unable to deal with it. (*ibidem*)

Bateson ended his reply mockingly with this comparison:

QUOT. 758 Without doubt in the old Regent House of the University disputants in the post had maintained the flatness of the earth before applauding crowds, much as Professor Weldon had to-day upheld the view of the Ancestrians. The paths of the heavenly bodies had been harmonized with the theory of a flat earth, as some of the facts of heredity had been with the law of Ancestry; but as the theory of gravitation had brought together great ranges of facts into one coherent whole, so had Mendelian theory begun to co-ordinate the facts of heredity, till then utterly incoherent and apparently contradictory. (*ibidem*)

Given the polemical tone of Bateson's reply, the intervention that followed by Pearson was relatively tame. He praised Galton for the introduction of mathematical methods into biology and for thus providing precision. He deplored the absence of such exactness in the work of the Mendelians thus far. Pearson then claimed that he had "elaborated the most complete Mendelian system ever worked out," which had produced "general principles which were singularly like those proposed by Galton from observation" (*ibidem*). Presumably, he was here referring to his 1904 paper (Pearson, 1904b). Pearson noted that, on the other hand, the Mendelians' "theories could not be grasped; they were always changing in form and definition." He concluded with these conciliatory words:

QUOT. 759 If both sides would work quietly without controversy for three or four years the world would know a great deal more at the end of that time. No hostile criticism of the Mendelians should appear in the journal of the Biometricians for the next four years if Mr. Bateson could see his way to going on quietly with his own work. Surely it was possible to discuss the question free from personal feeling. (*ibidem*)

11.21 Pearson Attacks Darbishire (1904)

Pearson's call for peace was short-lived for, soon after the British Association Meeting of August 1904, he launched a virulent attack on Darbishire. The biometricians were taken aback both by Darbishire's last paper (Darbishire, 1904a)[61] and his speech at the meeting. Darbishire had been unequivocally anti-Mendelian in all his four reports thus far, but now seemed to be admitting both ancestral and Mendelian theories. Pearson was especially irked by a particular contradiction in Darbishire's writings. In his fourth report, Darbishire had observed waltzing in 97 out of 555 mice (see QUOT. 733) and had statistically argued that this observed proportion (.175) could not be regarded as a Mendelian quarter. However, in the paper that soon followed (Darbishire, 1904a), he reversed his position, now claiming that he had found 25% waltzing mice (see QUOT. 738). Pearson scoffed at this change of position in a *Nature* article suggestively entitled "The Mendelian Quarter" (Pearson, 1904c):

[61] See Sect. 11.18.

QUOT. 760 Mr. Darbishire failed to see that 97 in 555 was a reasonable quarter, or 20 in 555 a reasonable sixteenth, but that he has learnt in Manchester, or perhaps in Cambridge from some dominant anaesthetist, that these things really are so? (*ibid.*, p. 530)

The "dominant anaesthetist" in the quotation above was an obvious reference to Bateson.

11.22 Castle in 1905: Darbishire's Fourth Report Contains a Key Error: His Extracted Hybrids Are Not All Heterozygotes and His Observed Proportions Are in Accordance with Mendelian Expectations

Darbishire was now in a singular position—the Mendelians were annoyed with him because they believed he had unduly undermined Mendelism and the biometricians were irked that he had unduly admitted that theory. How Darbishire could have done both things at the same time is a mystery but, as if he was not in enough trouble, Castle published a memoir in 1905 (Castle, 1905a),[62] which pointed to a fatal error in Darbishire's fourth report.

Castle's key observation concerned the F_2 pigmented mice that Darbishire had called "extracted hybrids" (Fig. 11.5). Although these pigmented mice were indeed hybrids (i.e., *DR* heterozygotes) for the crosses in (iii), this was not the case for the crosses in (i) and (ii). For the crosses in (i), it was possible for the pigmented mice to be either *DD* or *DR* obtained from the grandparental crosses *DR* × *DR*. For the crosses in (ii), the pigmented mice from the grandparental cross *DR* × *RR* were *DR*, but from the grandparental cross *DR* × *DR* they were either *DD* or *DR*. Darbishire's mistake was that he had assumed that *all* such pigmented mice were *DR*:

QUOT. 761 An examination of these records makes it evident that his so-called "hybrids" were not all hybrids. In mating (1) [(iii) in Fig. 11.5] the animals used undoubtedly were hybrids, since each had, as stated, one albino and one pigmented parent. It is noteworthy that in that case the Mendelian expectation of 25 per cent albinos is realized as perfectly as is possible in an odd number of young, there being 30 albinos in a total of 121 young. In cases (2) [(ii) in Fig. 11.5] and (3) [(i) in Fig. 11.5] Darbishire gives us no evidence that the animals which he mated were really hybrids in the sense that they contained recessive albinism. He calls them "extracted hybrids" because they had dark eyes, as did all his primary hybrids obtained by crossing pink-eyed with albino mice. The latter unquestionably were hybrids in the sense that they contained recessive albinism, for in every case one of the parents was an albino. The former may or may not have contained albinism recessive; Darbishire's experiments indicate that in some cases they did, and in other cases they did not. (*ibid.*, p. 19)

The particular experiments alluded to in the last sentence of the quotation above that provoked Castle's suspicion were the crossing of Darbishire's F_2 pigmented mice with albinos. If the pigmented mice were indeed *DR*, crossing them with albinos

[62]This paper was referred to in Sect. 10.6.

(*RR*) should have produced *both* pigmented mice and albinos. However, in Darbishire's Table F (Darbishire, 1904b, p. 26):

> **QUOT. 762** Five out of nineteen such matings failed to produce *any albinos whatever*, though they produced a total of 19 young, all pigmented. If the pigmented parents in these matings had really been hybrids, half their young should have been albinos. Such was precisely the proportion of albinos produced by the fourteen matings in which the pigmented parent, though of the same ancestry as in the foregoing cases, showed itself to be really a hybrid, for these fourteen matings produced 36 pigmented young and 36 albinos. (Castle, 1905a, p. 19)

However, the greatest confirmation that Castle was right in his criticism and that Darbishire had indeed erred is the following: when the fact that not all of Darbishire's "extracted hybrids" are heterozygotes is taken into account, then the percentages observed in Table 11.4 fit Mendelian expectations remarkably well. This will now be shown.

In Fig. 11.5(i), under Mendelian theory, each grandparental cross *DR* × *DR,* gives *DD*, *DR*, *RR* in the ratio 1:2:1. The pigmented types are *DD* and *DR*, and these occur in the ratio 1:2. Now, two *DR* mice (one from each grandparental cross) mate with probability $(2/3)^2 = 4/9$, and albinos *RR* are produced 1/4 of the time. Thus, the expected proportion of albino among the grandchildren, under a Mendelian model, is $4/9 \times 1/4 = 1/9 = .1111$. This is extremely close to the observed proportion of .1075 in Table 11.4.

Similarly, in Fig. 11.5 (ii), under Mendelian theory, a grandparental cross *DR* × *DR* gives the pigmented types *DD* and *DR* in the ratio 2:1 as before, while a grandparental cross *DR* × *RR* gives *DR* and *RR* in the ratio 1:1 and all pigmented types here are *DR*. Now, two *DR* mice (one from each grandparental cross) mate with probability $2/3 \times 1 = 2/3$ and albinos *RR* are produced 1/4 of the time. Thus, the expected proportion of albino among the grandchildren, under a Mendelian model, is $2/3 \times 1/4 = 1/6 = .1667$. This is reasonably close to the observed proportion of .1869 in Table 11.4.

Thus, as Castle said:

> **QUOT. 763** ...in fact Darbishire's results, when carefully analyzed, prove just the opposite of his deduction. (*ibid.*, p. 17)

Darbishire (1905) was now forced to make a major concession.

11.23 Darbishire's Key Concession to the Mendelians (1905): "The Table on p. 6 Therefore Is in No Wise a Refutation of Mendelian Theory"

Darbishire did not directly respond to the attacks from either the Mendelians or the biometricians, but wrote a paper (Darbishire, 1905) that laid bare his initial motivations for the mice experiments and his subsequent position on the matter.

Table 11.5 Table on p. 6 of Darbishire (1905)

Type of parentage	No. of young	Mendelian expectation (%)	Galtonian expectation (%)	Actual result (%)
HH × HH	93	25	9.37	10.75
HH × HA	107	25	17.96	18.69
HA × HA	121	25	26.56	24.79

Based on the key experiment in his fourth report. [HH here stands for DR × DR in Fig. 11.5, and HA for DR × RR]

At the start of his paper, Darbishire noted that there were two methods of scientific criticism. The first one was criticism arising simply out of the desire to crush a theory, and the second was criticism only when the theory had been thoroughly tested and found to be defective. He then observed that:

QUOT. 764 ...[t]he most flagrant example of the first kind of criticism, on which I can lay hands, flowed from the pen of a writer who, after having misrepresented the theory he was attacking by declaring that it was "an essential part of the Mendelian hypothesis that the (so-called 'extracted') recessive individual which is produced by pairing two first crosses, is in every respect similar to the original pure recessive,"* concludes with these words: "This mouse is clearly not a pure dominant, because it produces albinos; it is not a dominant hybrid because it has pink eyes; and it cannot be a recessive because when paired with an albino it produces some black-eyed forms." It is evident from this quotation that the stimulus which actuated the author was a desire to stultify and refute Mendelian theory at all costs, and that he did not make the smallest attempt to discover what Mendelian theory really was or to put himself in the position of those who held it to be true. (*ibid.*, p. 2)

Now, the asterisk in the quotation above points to Darbishire's own third report (Darbishire, 1903b, p. 282). What was just read is therefore Darbishire's confession that he had all along been biased against Mendelism, in spite of his January 3, 1903, reassurance to Bateson that this was not the case (see QUOT. 691). This was also a poor reflection on Weldon's motives since the mice experiments had been started under his direction.

Darbishire next considered the key experiment and result of his fourth report (Darbishire, 1904b) (see Fig. 11.5 and Table 11.4). Recall that this experiment consisted of three series of crossings, each involving the same parental types (extracted hybrids) but different grandparental types. Recall also that Castle (1905a) had quite convincingly questioned the authenticity of Darbishire's extracted hybrids. The latter reproduced the numbers he had obtained along with Galtonian expectations he had now calculated (see Table 11.5)

Fully aware of the point Castle had made, Darbishire gave three reasons why he had thought his extracted hybrids were indeed heterozygotes:

QUOT. 765 In the first place these putative hybrids bore the same features of coat and eye colour as the F_1 hybrids exhibited; secondly, they formed 50% of the F_2 generation, and lastly at least two upholders of Mendelian theory[63] had asserted that the heterozygote was represented in my experiment by the coloured mice with pigmented eyes. (Darbishire, 1905, p. 6)

[63] Castle and Allen, *Proc. Amer. Acad. Arts and Sci.*, Vol. 38; No. 21. (*Darbishire's footnote*).

Darbishire now addressed Castle's criticism. He reproduced a communication he previously had with the latter:

> **QUOT. 766** . . .these results were brought before the notice of a student of heredity [Castle] whose first question was, if there had ever resulted from the union of two hybrids a family in which there were no albinos; a query to which an affirmative answer was given. Then the following argument was used by my critic: 'The only proof that a given individual is a hybrid is one which is based on an examination of its gametic constitution; in the case of your mice you have no right to say that a grey mouse with black eyes is a hybrid until you have mated it with an albino and obtained albino young in the litter thus produced: until this has been done there is no evidence that it is not a dominant. In the case you have just shewn me you mated coloured mice with dark eyes without making this test, and by this neglect many dominants may have been included among them, and you see that this suggestion, against the truth of which you have no evidence, accounts just as well as Galtonian theory for the difference in the proportions of albinos in the three kinds of matings.' (*ibid.*, p. 7)

In the above, Castle had thus urged Darbishire to cross his extracted hybrids with albinos: the extracted hybrids would indeed be heterozygotes only if an albino was ever obtained from such a cross. Darbishire reported that he performed the requested test and found that no albino was produced in a few cases. He therefore conceded that, as Castle had predicted, his extracted hybrids consisted of both heterozygotes and pure dominants:

> **QUOT. 767** . . .it began to look very much as if these apparent hybrids really were dominants. This fact in itself pointed to the truth of the suggestion that mice with coloured coats and eyes were of two kinds—hybrids and dominants. (*ibid.*, p. 8)

Having conceded this point from Castle, Darbishire was left with no choice other than to concede the latter's other, more fundamental point, namely that:

> **QUOT. 768** The Table on p. 6 therefore is in no wise a refutation of Mendelian theory; at any rate the suggested Mendelian interpretation of the facts cannot be regarded as disproved. (*ibid.*, p. 9)

This was a major statement coming from a biometrician and marked a turning point in the biometry-Mendelism debate.

11.24 Was Darbishire Converted to Mendelism?

Several authors have used QUOT. 768 and similar statements made by Darbishire to argue that he "converted" to Mendelism. For example, Kim has spoken of the "case of Darbishire's conversion" (Kim, 1994, p. 63). Gillham has similarly stated that Dabishire's "conversion was engineered not only by Bateson, but independently by a brash young Mendelian named W. E. Castle" (Gillham, 2002, p. 319), and Provine has claimed that the 1905 paper (Darbishire, 1905) "represented a transition stage for Darbishire, who became an avowed Mendelian." (Provine, 2001, p. 80)

However, if Darbishire's alleged conversion is to be interpreted as an embrace of Mendelism together with the abandonment of his ancestrian beliefs, then no such conversion seems to have actually happened.[64] For, Darbishire's very next statement after QUOT. 768 was that:

QUOT. 769 Nevertheless the facts are no less in accord with Galtonian theory. (Darbishire, 1905, p. 9)

Thus, instead of being converted to Mendelism, Darbishire adopted a viewpoint that admitted both Mendelian and ancestrian theories. In fact, this new viewpoint was already evident in his summary paper delivered on March 15, 1904, and his speech at the British Association Meeting in August 1904 (see Sects. 11.18 and 11.20). Moreover, in the 1905 paper, Darbishire argued that the two theories were complementary: each provided an alternative way of describing the same phenomenon. He used the analogy of kinetic theory whereby the velocity of particles could be described either statistically by adopting an *en masse* approach, or mechanistically by focusing on the individual molecules:

QUOT. 770 The biometrician, with a new weapon of observation, is only concerned with mass phenomena; the individuals which go to swell his correlation tables are, like the atoms of the physicist, units of which no knowledge is required to attain the result at which he aims...

The Mendelian, on the other hand, with a new application of experiment, is a biological demon who, "perceiving" and "handling the separate" units themselves, tries to find out their properties by mating them with other units. (*ibid.*, p. 9)

Darbishire likened the biometrician's approach to heredity to the physicist's macroscopic attitude toward the second law of thermodynamics. This law, which states that heat cannot flow from a cold to a warmer body,[65] holds only statistically, i.e., it might not hold for some particles (although the probability of these events is quite low). To illustrate how this is possible, imagine an all-knowing entity (the so-called Maxwell's demon) that can follow the path of each particle. Suppose this demon opens and closes a tiny partition between a cold and a warmer body to allow the fastest molecules to move from the cold to the warmer body, and the slowest molecules to move in the opposite direction. It is thus possible for the second law not to hold microscopically (although it does hold macroscopically). Darbishire likened the Mendelian to a Maxwellian demon such that:

QUOT. 771 [t]he point of view of the demon is so different from that of the physicist that one of the truest generalizations of the latter would be declared absolutely false by the former: yet no one remains blind for a moment to the fact that the contradiction of their respective statements is only apparent, and is due to the radical difference in their points of view. (*ibid.*, p. 14)

Similar analogies were made in Darbishire's next paper (Darbishire, 1906). In that paper, he pointed out the well-known fact that the Mendelian theory was

[64] Ankeny (2000) is one of the few papers which has also recognized that Darbishire did not actually convert to Mendelism.

[65] The second law is more commonly stated as: "The entropy of a closed thermodynamic system can never decrease."

physiological whereas the ancestral theory was statistical. Toward the end, however, Darbishire made his position more precise:

> **QUOT. 772** If the true function of the biometrician is to give us statistics of average conduct where we cannot predict individual conduct it seems to me that to deal by the biometric method with cases where we can is not only un profitable, but likely to lead men to think that where there are two methods dealing with the same material of which the one can predict while the other cannot, the latter is fallacious... Whereas if the biometrician confined himself to the non-predicable and the Mendelian to the predicable, the general conclusion would be that each had his proper sphere—which indeed, in my belief, he has.

Darbishire's position was thus based on practicality: the biometric method was applicable in cases where individual predictions were not possible; otherwise, the Mendelian method was appropriate. That is, the biometrician and Mendelian each "had his proper sphere." It is hard to conclude from this that Darbishire had been converted to Mendelism.

Likewise, no conversion to Mendelism can be seen in Darbishire's review of the second edition of Punnett's "Mendelism" (Darbishire, 1907, p. 73). In fact, Darbishire here revealed that he had doubts about gametic purity, which was an essential component of Mendelism:

> **QUOT. 773** At the same time, no good can be done by refusing to face the fact that the truth of the Mendelian doctrine is not universally admitted. No one denies the extraordinary interest of these discoveries...But we hold that he must be a very rash man who accepts without further question the doctrine of gametic purity. (*ibidem*)

Finally, Darbishire's 1911 book "Breeding and the Mendelian Discovery" (Darbishire, 1911) was an exposition of the practical applications of Mendelism to animal breeding. Instead of a conversion to Mendelism, one finds that, in the words of a reviewer, "Dr. Darbishire gives an exposition of Mendelism and refrains alike from defence and criticism of Mendelian phenomena" (J.L.T., 1912, p. 173).

11.25 Adoption of Mendelism by Schuster (1905)

Since the biometricians could no longer rely on Darbishire to promote their agenda, Weldon turned to his student Edgar Shuster to repeat some of Darbishire's experiments.

E. H. J. Schuster (1897–1969) was born in a distinguished family[66] and he is described by Paton and Phillips as "one of that select group of men who devote themselves, not to personal research, but to helping the research of others" (Paton & Phillips, 1973, p. 111). Schuster obtained a B.A. with first class in Natural Science in 1901. At that point, he was awarded a Biological Scholarship by New College,

[66]His father, Ernest Schuster, was a banker and founder of the firm of Schuster Son & Co. and his uncle, Sir Arthur Schuster, was Secretary of the Royal Society and Copley Medalist.

Oxford, and worked under Weldon. In 1904, Schuster became the first holder of the Galton Research Fellowship in National Eugenics.

Kim (1994, p. 79) reported that in 1904, at around the time Darbishire started admitting both Mendelian and ancestrian theories,[67] Schuster invited the latter to lunch with Bateson. It is very probable that during this meeting, Bateson shared several Mendelian ideas with them and informed them of Cuénot's latest research in the field. Schuster's acquaintance with Cuénot's chromogene theory[68] was pivotal and he made full use of it in his first paper on albinos (Schuster, 1905).

Schuster started his paper with the following admission:

QUOT. 774 THE experiments here recorded were commenced in the department of comparative anatomy in Oxford in the autumn of 1902, but various personal circumstances have delayed the publication of any account of them for more a year after their completion.

They were made at the suggestion and with the advice and help of Professor Weldon, and I have therefore numerous reasons for thanking him. (*ibid.*, p. 1)

He then outlined Cuénot's theory according to which the fur on the gray mouse was determined by two enzymes (each an oxidizing diastase) that can react on a substance called the *chromogene*. These three substances were located on three "mnenons" (genes) in the germ cells. Regarding the black mouse, only one enzyme and one chromogene were located on three mnenons, while the albino mouse possessed two enzymes but no chromogene.

Using *C* for the chromogene, *A* for albino, and *G, B* and *Y* for the colors gray, black, and yellow, respectively, then:

CG represents an ordinary house mouse
CB represents a pure black mouse
CY represents a pure yellow mouse
AG represents an albino mouse with gray as latent color
AB represents an albino mouse with black as latent color
AY represents an albino mouse with yellow as latent color

Thus, since yellow was dominant over gray and black, and gray was dominant over black, a gray *CG* crossed with a white *AY* produced yellow *CG-AY* mice, a black *CB* crossed with a gray *CG* produced gray *CB-CG* mice, and a black *CB* crossed with an albino *AG* produced gray *CB-AG* mice.

Schuster next made a statement that seemed to be a direct jab at the ancestrians:

QUOT. 775 All this is not a modification nor an extension of Mendel's laws, but merely an application of them to a special case; Cuénot's theory as to the two factors which determine coat colour renders such an application possible. The further behaviour of the hybrids when crossed among themselves or with pure mice can therefore be predicted by any one conversant with these laws and possessed of a knowledge of the gametic formula of the hybrids used. (*ibid.*, p. 2)

[67] See Sect. 11.18.

[68] See Sect. 11.2.

This remarkable statement was aimed at the ancestrians who had often accused the Mendelians of arbitrarily modifying Mendel's theory to explain any kind of observation (e.g., see QUOT. 720). It also signified that Schuster's paper, contrary to what Weldon might have expected, was going to be anything but a defense of ancestrianism.

The following are the first three results from Schuster's experiments:

(a) *Crosses between albino and gray parents:* Schuster crossed 106 albinos with gray mice, obtaining 342 F_1 offspring, of which 329 were gray, 7 yellow, and 6 chinchilla. He explained the appearance of yellow due to parental crosses of the type $AY \times CG$. Similarly, gray mice could be explained by crosses of the type $AB \times CG$ or $AG \times CG$.
(b) *Intercrossing of* F_1 *yellow mice:* Schuster intercrossed the F_1 yellow mice AY-CG. From all 16 pairwise combinations, the Mendelian expectations of F_2 albinos, yellows, and grays are in the ratio 4:9:3. Since Schuster obtained only 13 F_2 mice he was unable to verify if his observations were close to Mendelian expectations.
(c) *Crossing of* F_1 *mice with albinos*: Crossing of the F_1 mice AY-CG with albinos produce albinos and colored mice in equal Mendelian proportions. Schuster obtained 261 albinos and 276 colored mice, in close agreement with Mendelian expectations.

Schuster performed several other crossings, all yielding results fairly close to Mendelian expectations. He thus concluded:

QUOT. 776 ...it is hardly necessary to point out that with regard to characters, colour-productiveness and albinism the mice under consideration here behave in complete accordance with Mendel's laws, both with regard to dominance and to segregation. (*ibid.*, p. 16)

Does this mean that Shuster was converted to Mendelism in his very first mouse paper? A possible inference may be made from an exchange that Schuster had in 1910 with Geo Mudge (the editor of the newly created *Mendel Journal*), who had written under the pen name "Ardent Mendelian."

QUOT. 777 "ARDENT MENDELIAN" is correct in supposing that I am a biometrician; but I am, at the same time, a believer in Mendelism, and I hold that the main aims of the *Mendel Journal* and *Biometrika* are not opposed. To me it appears that people who are studying the same problem by different methods should work in sympathy with one another... (Schuster, 1910)

Thus, as late as 1910, Schuster still regarded himself as a biometrician, although he did believe in the correctness of the Mendelian theory. His conciliatory language in the quotation above is reminiscent of Darbishire's similar equipoise (see Sect. 11.24).

Schuster's statements in his 1912 book *Eugenics* are even more enlightening:

QUOT. 778 The study of inheritance in recent years has been pursued vigorously along two different paths. First, by the statistical summarisation of the facts, a process introduced by Sir Francis Galton; and, secondly, by the experimental hybridisation of animals and plants. The latter method, when carried out on the lines laid down by Gregor Mendel, has alone led to some real understanding of the underlying physiological processes, and it appears capable of leading to more. (Schuster, 1912, p. 83)

However, despite Schuster's endorsement of the Mendelian theory, he still held there was a place for biometrical methods:

> **QUOT. 779** Though its [i.e. the Mendelian] advance during the last twelve years has been one of surprising rapidity, it has up to the present left many fields unworked, and even if it may be expected to increase a hundredfold the knowledge which we owe to it, there is as yet no indication that its methods of analysis will ever be able to touch some of the problems of inheritance. They may fail for two reasons. First, there may be characters which are not inherited according to Mendel's laws; and, secondly, the somatic character may be determined by so large a number of gametic factors that, although each factor is transmitted unchanged in all its purity, the somatic character might appear to be completely blended or modified in many other ways by inheritance. It is not in any way in disagreement with Mendel's theory to suppose that a character may depend for its full development on the presence of a hundred factors... (*ibid.*, pp. 107–108)

In the above, Schuster made two points that would soon be an integral part of the modern synthesis: the realization that continuous characters were the result of many Mendelian factors and the use of biometrical methods within Mendelian theory.

Chapter 12
Round 8: The Johannsen Breakthrough - Pure-Line Theory (1903), Gene and Genotype–Phenotype Distinction (1909)

Abstract In 1903, Johannsen performed a key experiment that had a significant influence in the biometry-saltationism debate. He planted seeds from the brown "Princess" beans and grouped the F_1 offspring according to the weight classes of the parent seeds. He found that, as the law of regression predicted, the deviation in the mean weight of a particular group of offspring from the mean F_1 weight was less extreme than the deviation in the weight class of the parents from the parental mean. However, when Johannsen divided the beans according to pure lines (i.e., different homogenous groups), the offspring in a given pure line showed complete regression to their own mean, irrespective of the deviations of their parents. This showed that natural selection was ineffective in pure lines, i.e., was unable to act on individual differences. In 1909, Johannsen introduced the terms phenotype, gene, and genotype, and made the genotype–phenotype distinction.

Keywords Genealogical selection · Law of regression · Pure lines · Complete regression · Displacement of center · Random fluctuations · Phenotype · Gene · Genotype

12.1 Introduction

Wilhelm Johannsen[1] (Fig. 12.1) was born in Denmark. In his younger years, he taught himself chemistry and worked as an apprentice pharmacist. He left his position at the Carlsberg laboratory in 1887 to do research in botany. He became a lecturer in 1892 and Professor of Botany and Plant Physiology in 1903 at the Copenhagen Agricultural College. Although he did not have any university education, Johannsen became Professor of Plant Physiology at the University of Copenhagen in 1905.

[1]Johannsen's life and work are also discussed in Roll-Hansen (1979, 2009, 2014), Churchill (1974), Kim (1994, Chap. 6), Provine (2001, pp. 92–100), Gayon (1998, pp. 260–271), and Tort (1996, pp. 2404–2406, Vol. 2: F-N).

P. Gorroochurn, *The Development of Evolutionary Genetics*,
https://doi.org/10.1007/978-3-031-69374-8_12

Fig. 12.1 Wilhelm Johannsen (1857–1927)

In the 1890s, Johannsen developed an interest in heredity, variation, and evolution. During the same period, the Frenchman Louis de Vilmorin[2] (Fig. 12.2) recognized the importance of selecting for exceptional characters in a single plant (as opposed to mass selection) and its lineage. After many generations, improved varieties could be obtained. The principles of this method, called genealogical selection (or pedigree selection), were described in an 1856 article entitled "Création d'une nouvelle race de betterave à sucre et considérations sur l'hérédité des végétaux" (Vilmorin, 1856). In another later article, Vilmorin summarized the method as follows:

> **QUOT. 780** Only when operating on unique individuals that one can obtain a nobility in plants, that a series of individuals in which individual characters are transmitted without alteration in successive generations. (Vilmorin, 1858, p. 77)

---

[2]From 1840 till 1920, the Vilmorin family ran the most important seed company in the world. During the 1830s, Vilmorin's father Phillipe-André de Vilmorin (1713–1780) grew wild carrots and then planted the best out of these. Within three years, he obtained extremely good carrots. For more details about the Vilmorin dynasty, see Chap. 5 of Kingsbury (2011).

Fig. 12.2 Louis de Vilmorin (1816–1860)

The method of genealogical selection was independently discovered at around 1890 by the Swedish Nils Hjalmar Nilsson (1856–1925), who was the director of the Agricultural Experimental Station at Svalöf. It was also referred to in Johannsen's 1896 small book entitled *Om Arvelighed og Variabilitet*[3] (Johannsen, 1903), which presaged the latter's interest in pure-line theory.

Om Arvelighed og Variabilitet represented Johannsen's views on heredity and variation at the time. He reviewed the pioneering work of Galton and stressed the fundamental importance of heredity in the understanding of evolution:

QUOT. 781 Evolution needs theory of heredity; but not vice versa. (*ibid.*, p. 12)

Johannsen was critical of Weismann's doctrine of continuity of the germ plasm[4] and traced its origin to Galton's stirp theory.[5] He had not endorsed the saltationist viewpoint yet because he admitted both de Vries and Bateson's mutations, and Weldon and Pearson's continuous variation as mechanisms of evolution.

[3] *On Heredity and Variation.*

[4] See Sect. 5.2.5.

[5] See Sect. 4.5.2.

The direction of Johannsen's future research was stated in the fourth edition of the book *Den almindelige Botanik*[6] (Warming & Johannsen, 1900), in which he was coauthor. In the last chapter entitled "Naturlig Slægtskab; Variabilitet, Arveligehd og Fylogenese",[7] Johannsen wrote that a method for distinguishing fluctuating variation from mutations was through regression to the mean. In the spring of 1901, Johannsen started his ground-breaking pure-line experiments, as will be explained in the next section.

Johannsen learned of Mendel's work in November 1903, but his 1902 paper "Organismernes formrigdom"[8] (Johannsen, 1902) shows that he had already endorsed the saltationist viewpoint. The following paragraph is so mutationist in spirit that it might also have been written by either de Vries or Bateson:

> **QUOT. 782** The theoretical importance of the mutation theory is exceedingly great. It frees us from the very difficult matter of having to explain the successive emergence of new types by a long-lasting selection of individual variants. All natural scientists are certainly convinced that natural selection takes place. Individuals that do not suit the conditions of the habitat or that have overwhelming enemies or competitors in a given location succumb in the struggle for existence. But this does not create new types! These are formed by mutation, and then natural selection can eradicate the types that cannot cope in the given place, under the given conditions. Naturally, weak individuals of any type would most easily perish. But for the development of species through the ages, for the composition of the entire flora and fauna of a land area—the epitome of all the existing systematic units, it is the relationships of new types arising from mutation in the struggle for existence that become decisive. (*ibid.*, p. 564)

Reading the paragraph above, it will be remembered[9] that one of the key tenets of the saltationists was that natural selection was incapable of *creating* new types—this was done by mutations.

Johannsen made his entry into the biometry-saltationism fray in 1903 as an ardent defender of the mutationist viewpoint. His research had two distinctive features, which made his results even more persuasive:

(a) Whereas previously Mendelians had focused their research on *qualitative* traits (such as seed color or form, mice color, etc.), Johannsen used a *quantitative* trait, namely, the seed weight of "Princess" beans.
(b) Whereas Mendelians before had refrained from using biometrical methods on the grounds that they were too complex and often obscured the heredity process, Johannsen used both experimental and biometrical methods.

Johannsen made another breakthrough in 1909, by introducing the term "gene" and making a key distinction between genotype and phenotype. The latter was a much-awaited conceptual advance and, together with the Hardy–Weinberg law of 1908,[10] marked the beginning of population genetics.

[6] *General Botany*.

[7] Natural Kinship; Variability, Heredity, and Phylogenesis.

[8] The plurality of forms of organism.

[9] See QUOT. 387 and QUOT. 516.

[10] See Chap. 10.

Although not all of Johannsen's conclusions in the 1903 and 1909 publications are now regarded as accurate, Johannsen remains a pioneering figure in the historical development of population genetics.

12.2 Pure-Line Theory in 1903

In the groundbreaking small book *Ueber Erblichkeit in Poulationen und in reinen Linien*[11] (Johannsen, 1903, 1955), Johannsen started by referring to the biometry-saltationism debate. He reviewed the biometricians' approach and drew attention to their reliance on Galton's regression law (later generalized to the ancestral law) as representing the process of heredity in Darwinian evolution. While Johannsen praised the mathematical contributions of Pearson, he noted that before a population could be analyzed as a mathematical unit, as the biometricians did, it first needed to be studied as a biological unit so that its elements could be understood. With this objective in mind, Johannsen referred to Louis Vilmorin's method of genealogical selection,[12] which he called the "isolation principle" and which he explained as follows:

> **QUOT. 783**... "isolation principle", a principle that can best be denoted as the principle of the individual estimation of progeny[13]. According to this principle, one must, by means of isolation—keeping separate the seeds of every single mother plant—determine whether the "force of heredity" (force héréditaire), as Vilmorin expresses it, is great or small. (*ibid.*, p. 175)

Johannsen now announced the primary aim of his work, namely, to use pure lines to verify the law of regression to the mean and thus establish whether there was a real difference between individual variations and mutations:

> **QUOT. 784** My task here then is this: by means of investigations after Galton's and Pearson's methods, but using Vilmorin's principle, to test the range of the law of regression–first of all only in the relations·between parents and offspring. By this it will be possible to show whether a real difference exists between mutation and fluctuating variability. (*ibid.*, p. 177)

Johannsen's reasoning was that a verification of regression to the mean would downplay the importance of mutations and instead favor individual variations.

As mentioned in Sect. 12.1, two characteristics of Johannsen's methodology that demarcated him from other saltationists and made his results even more credible were: (i) his use of quantitative (rather than qualitative) traits, and (ii) his adoption of biometrical methods.

---

[11] *Concerning Heredity in Populations and in Pure Lines.*

[12] See Sect. 12.1.

[13] All underlines in this and future quotations of this section are Johannsen's.

Table 12.1 Johannsen's experiment on "Princess" beans grouped in weight classes (Johannsen, 1903, 1955, p. 183)

	Weight classes of mother seeds (P) in mg					
	250–350	350–450	450–550	550–650	650–750	750–900
No of seeds sown (P)	25	40	161	24	12	25
Average weight of crop (F_1) in mg	374	388	401	434	446	457
No of seeds in the crop (F_1)	606	914	4476	604	370	598

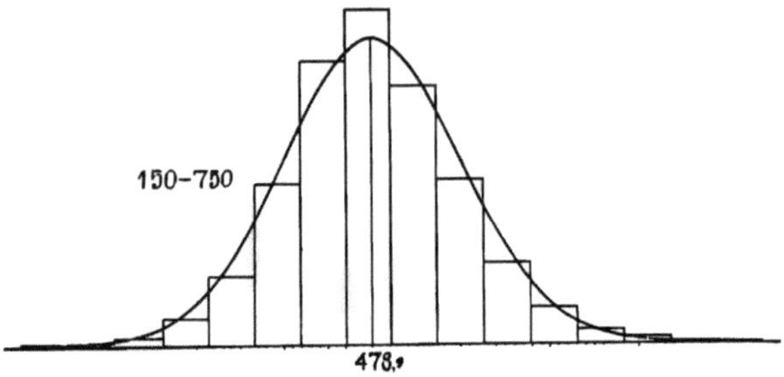

Fig. 12.3 Histogram of the 1902 daughter crops (F_1') (5494 beans) grouped together in Johannsen's pure line experiment

Johannsen's experiment started when, in the spring of 1901, he bought 8 kg (approximately 16,000 beans) of brown "Princess" beans that had been harvested in the previous year on the island of Fühnen. The average weight of the seeds, based on a sample of 5000, was 495 mg. He then sowed a sample of parent seeds that had been grouped into weight classes and examined the harvest in the fall of 1901 (see Table 12.1).

Table 12.1 shows that the law of regression holds fairly well when weights are analyzed *en masse*: parents with large deviations in a given direction from the parental mean (495 mg) result in offspring that deviated in the same direction to a lesser degree from their own mean (which can be calculated to be approximately 406 mg). For example, the 25 parent seeds that belonged to the 250–350 mg class deviated by approximately +200 mg from their mean; on the other hand, the 606 seeds in F_1 deviated by approximately only +30 mg from their mean.

Now Johannsen wanted to know if the law of regression also applied to pure lines:

> **QUOT. 785** And we must ask therefore whether or not, through selection of plus- and minus-variants within the pure lines, a type displacement, or rather a Galtonian regression, could be achieved. (*ibid.*, p. 187)

In the spring of 1901, Johannsen planted 19 "well-characterized" beans, each from one plant of the 1900 crop, such that each bean gave rise to a pure line (A, B, C, ..., S, T). In the summer of 1901, a first generation of crops was obtained. These parental crops (P′) were planted in the spring of 1902 and the daughter crops (F_1') were collected in the same year. When analyzed *en masse*, the F_1' crops yielded a Gaussian distribution as shown in Fig. 12.3.

Table 12.2 The tenth (*J*) of 19 pure lines in Johannsen's experiments. For example, 40 F_1' seeds were obtained in 1902, which belonged to parents (P′) in the 250–300 mg class. These F_1' seeds had an average weight of 445 mg and a standard deviation (SD) of 84.0 Mg. The P′ seeds for this pure line weighed 400 mg on average

Mean weight of parent (P′) seeds = 400 mg			
	F_1' offspring		
Weight classes of mother seeds (P′)	Average weight	No. of seeds	SD
250–300	445	40	84.0
300–350	498	53	64.8
350–400	453	164	69.9
400–450	447	155	65.5
450–500	434	103	74.9
500–550	468	102	79.1
550–600	458	95	79.9
Entire pure line *J*	454.4	712	74.0

Figure 12.3 appears to show that the underlying population is homogenous, subject to a similar regression to the mean as in Table 12.1. However, when the F_1' crops were kept separate according to their pure lines and categorized into weight classes with each class corresponding to a weight class of the parents (P′), a totally different picture emerged. As an example, the weights for pure line *J* are shown in Table 12.2.

Whereas Table 12.1 shows Galtonian regression to the mean, Table 12.2 shows that the offspring (F_1') weights fluctuate *randomly* around their mean of 454.4 mg, irrespective of the deviations of the parents from their own means. For example, the parents in the 300–350 mg weight class deviate by approximately −75 mg from their mean, but their offspring deviate by +43.6 mg from their mean! The regression to the mean in a population as whole and the random variation in pure lines mean that:

(a) *In the language of regression*: *En masse*, there is *partial regression to the mean*, but within pure lines, there is *complete regression to the mean of the pure line*.
(b) *In the language of natural selection: En masse*, there is a *displacement of the center and natural selection is effective*, but within pure lines there is *no displacement of center and natural selection is completely ineffective*.

Similar conclusions hold for each of the 19 pure lines used in the experiment. Johannsen used the conclusion in (a) to take a strong anti-ancestrian stand:

QUOT. 786 The personal nature of the parents, grandparents, or any ancestor whatsoever has, so far as my experience extends, no influence on the mean character of the offspring. But it is the type of the line which determines the mean character of the individuals. (*ibid.*, p. 210)

On the other hand, although (b) is valid for pure lines, his next argument that natural selection would also be ineffective in real populations is incorrect:

QUOT. 787 The usual well-known result of selection - successive progress in the direction of selection in the course of a few generations—depends accordingly on the progressive

> purification with each generation of the deviating line concerned. And it will now be easily understood that the action of selection cannot be carried out beyond fixed limits—it must indeed cease when the purification, the isolation of the particular most strongly deviating line, practically speaking, is carried to completion. (*ibid.*, p. 207)

Johannsen's erroneous inference is seen to rely on the unrealistic assumption that a real population would eventually be "purified," i.e., be similar to one of his pure lines.

Given the inefficacy of natural selection in the presence of chance fluctuations, Johannsen argued for the importance of mutations in evolution:

> **QUOT. 788** If my investigations are accurate, and if their significance spreads beyond the particular special cases, the general results of this work will form a not unimportant support for the doctrine at present especially represented by Bateson and de Vries, of the great significance of "discontinuous" variation or "mutation" for the theory of heredity. For a selection in populations acts in my cases only in so far as it chooses representatives of already existing types. These types are not successively formed, perhaps through the protection of those individuals which fluctuatingly vary in the specific direction; but they are found and isolated. (*ibid.*, p. 212)

Johannsen's endorsement of the mutationist viewpoint follows directly from the primary aim he had set himself in QUOT. 784. His differentiation between chance fluctuations and mutations was a direct challenge to the biometricians' position that such a distinction was futile (e.g., see QUOT. 394).

The fundamental importance of Johannsen's pure-line experiments lies in that they pointed to the importance of mutations in evolution: natural selection by itself does not produce any change; it needs variations produced by mutations. As such, the experiments do not in fact contradict or undermine the Darwinian theory. However, many biologists used them to attack Darwinian selection on two fronts:

1. They argued, as Johannsen did, that a cross-fertilization population would eventually be decomposed into pure lines whence natural selection would stop.
2. They took the apparent inability of natural selection to act on small variations to mean that the Darwinian theory was incorrect.

However, it was soon realized that:

(1′) Cross-fertilizing populations are exposed to mutations and recombination, and are *not* eventually decomposed into pure lines;

(2′) Random fluctuations arising through changes in somatic cells are not heritable and thus are *not* acted upon by natural selection, but there are small variations that can arise through mutations (i.e., changes in germ cells): these are heritable and *are* acted upon natural selection. Johannsen was correct that, *in his pure lines*, the individual differences were not heritable and thus natural selection was ineffective. However, this does not mean that natural selection was ineffective with respect to small variations *in general*, as the critics in (2) would argue, since some individual differences may be caused by mutations. Johannsen was again correct in resorting to mutations in QUOT. 788 but the large, species-forming and discontinuous mutations of de Vries would soon prove to be the exception than the rule.

12.3 The Biometricians' Reaction to Johannsen's 1903 Book

12.3.1 Weldon and Pearson (1903)

Naturally, the biometricians published a sharp critique of Johannsen's 1903 book. Weldon and Pearson published a paper in 1903 (Weldon and Pearson, 1903) in which they first claimed that Johannsen's observation of regression to the mean of pure lines had already been referred to by the biometricians. For example, referring to Pearson (1898), they reproduced the following statement from p. 402 of that paper:

QUOT. 789 We now see that with the law of ancestral heredity ...... a race with six generations of "selection will breed within 1–2 per cent. of truth ever afterwards. (Weldon and Pearson, 1903, p. 499)

However, the above is, in fact, a case of *partial regression* to the mean of the type, not a case of *complete regression*, as demonstrated by Johannsen.

Weldon and Pearson next put forward two arguments that were meant to undermine Johannsen's conclusions from his experiments:

(a) *Correlation between the weights of* F_1' *beans and the mean weight of parental* (P') *beans.* Weldon and Pearson argued that if the offspring within a pure line bred true to the type of the line, "then when the whole filial generation is compared with the whole parental generation, correlation between two must necessarily be perfect." They calculated the regression coefficient of filial (F_1') weight on mean parental (P') weights (obtained from Table 12.1 and pp. 188–192, respectively, of Johannsen (1955), obtaining 0.591 ± 0.125, which they took to "closely represent the true parental correlation, being a relation between mean filial character and mean parental character." Since this was far from the perfect correlation expected under Johannsen's pure line theory, Weldon and Pearson concluded that:

QUOT. 790...there is a regression from parent to offspring, leading to the inference that parental correlation has for *Phaseolus vulgar*is a value closely identical with that found other animals and plants, when we compare mean parental and mean filial characters. (Weldon and Pearson, 1903, p. 503)

However, there are two major problems with Weldon and Pearson's calculations: (i) a regression coefficient is equal to a correlation coefficient only when the standard deviations of the two concerned quantities are the same; here, the standard deviations of the *mean* parental (P') weights and of the offspring (F') weights must be assumed to be the same, which is highly questionable; (ii) even if the standard deviations are assumed to be the same, a regression of F' weights on mean P' weights yields a correlation between the weights of F' individual offspring and the mean P' weights, and this is *not* "a relation between mean filial character and mean parental character," as Weldon and Pearson had claimed.

(b) *Regression of* F_1' *bean weight on maternal bean weight and on grandmaternal bean weight.* By assigning the weight of a bean to the middle of the weight class to which it belonged, Weldon and Pearson calculated the following three regression equations:

$$\begin{aligned}\text{Estimated mean weight of offspring } (F_1') \text{ bean} &= 538.31 + 0.2691 \\ &\quad \times \text{weight of grandmaternal bean (1900 crop)}\end{aligned}$$

$$\begin{aligned}\text{Estimated mean weight of offspring } (F_1') \text{ bean} &= 417.41 + 0.1074 \\ &\quad \times \text{weight of maternal bean } (P').\end{aligned}$$

$$\begin{aligned}\text{Estimated mean weight of offspring } (F_1') \text{ bean} &= 330.71 + 0.2373 \\ &\quad \times \text{weight of maternal bean } (P') + 0.0731 \\ &\quad \times \text{weight of grandmaternal bean (1900 crop)}\end{aligned}$$

By referring to the first two regression coefficients in the above, Weldon and Pearson concluded:

QUOT. 791 This shows us at once that from Professor Johannsen's own data the maternal bean is more than twice as influential as the grandmaternal bean in settling the weight of the filial gene. (*ibidem*)

This was a direct challenge to Johannsen's statement that the mean weight of F_1' offspring depended on the type of the pure line only, irrespective of other ancestral influences.[14] Weldon and Pearson claimed that they had in effect proven the efficacy of natural selection even in pure lines:

QUOT. 792 ...when we compare the characters of individual seeds in successive generations the correlation between a seed and its parental seed is so much greater than that between a seed and its grandparental seed (both belonging to the same pure ancestral line) as to give strong evidence that characters arising in one generation within the line are inherited, and do therefore afford a basis on which selection may act. (*ibidem*)

However, Weldon and Pearson's regression equations do not necessarily support an ancestrian view. The first two equations are compatible with both a Mendelian and an ancestrian view. The third equation seems to show that, given the maternal contribution, the grandmaternal one is negligible—precisely the Mendelian view. Of course, a definite answer may be obtained only by testing if the observed regression coefficient (0.0731) is not significantly different from zero.

Overall, current opinion does not side with Weldon and Pearson's conclusions in QUOT. 790 and QUOT. 792.

A similar, but shorter, review to the one presented here was anonymously published by Pearson in the December 17, 1903, issue of *Nature* (Pearson, 1903a).

---

[14] See QUOT. 786.

12.3.2 Yule (1903)

In contrast to the sharp tone of Weldon and Pearson (1903), Yule's review was more measured (Yule, 1904). In fact, Yule's objective was not to review Johannsen's book but to comment on Weldon and Pearson's review, Pearson's anonymous review in *Nature* (Pearson, 1903a), and to make some remarks on the book.

Because of a lack of clarity on what specific correlation Weldon and Pearson wished to show was different from unity,[15] Yule pointed out that a zero correlation between parent and offspring in a pure line in no way implied a perfect correlation for parent and offspring across generations.

Yule took exception to the last paragraph of the *Nature* review, which had claimed that Johannsen had not stated his problem clearly. On the contrary, Yule thought that:

> **QUOT. 793** Prof. Johannsen undertook the definite task, clearly stated, of elucidating the nature of intra-racial heredity by the study of heredity within the pure line, i.e. the offspring of one self-fertilised individual. (Yule, 1904, p. 223)

However, Yule did not completely agree with Johannsen's claim of no natural selection within a pure line:

> **QUOT. 794** He [Johannsen] has shown that the intensity of heredity between the first two generations sprung from such a single individual may be vanishingly small, although it is quite sensible within the race at large. (*ibidem*)

Yule concluded his short paper by again disagreeing with Weldon and Pearson regarding the merits of Johannsen's work:

> **QUOT. 795** . . .the methods he [Johannsen] has used are legitimate and sufficient for his immediate purpose, and, in my opinion, the work as a whole is one of the most stimulating contributions to the study of intra-racial heredity published in recent years. (*ibid.*, p. 224)

Although Yule's comments sound sensible, they ought to be understood in the context of a disagreement he had been having since 1900 with Pearson regarding the 2×2 table. Yule's short note certainly did not help to appease a controversy that escalated in 1906 and raged until 1913. See pp. 244–252 of Gorroochurn (2016a).

12.4 Genes and Genotype–Phenotype Distinction in 1909

12.4.1 The Necessity of a Genotype–Phenotype Distinction

Genotype refers to the internal hereditary factors of an organism whereas *phenotype* refers to its external appearance. The first occurs as a result of germinal cell division whereas the second as a result of somatic cell division. The phenotype is usually a consequence of the genotype and environmental factors.

---

[15] See Sect. 12.3.1 for the different wordings Weldon and Pearson used for the desired correlation.

It is important to distinguish genotypes and phenotypes because there is a many-to-one relationship between the two. To the modern biologist, the distinction is essential since Mendel's two laws refer only to the internal constitution, and the external appearance still needs to be specified. After all, natural selection acts on the phenotype, *not* the genotype.

Mayr (1973, p. 130) has indicated that the failure to make a genotype–phenotype distinction was the most damaging confusion responsible for some to the greatest controversies in genetics.

One of the most famous theories that emerged and prevailed for a long time as a direct consequence of genotype–phenotype confusion was the blending theory of inheritance, whereby maternal and paternal hereditary materials were assumed to fuse and form an offspring that consequently had characters midway between those of the parents. Blending inheritance was a working assumption in Darwin's works (see QUOT. 109 and QUOT. 173) and was also a key tenet of the biometricians (see Sect. 7.1.2 (c)).

The saltationists were not spared by the genotype–phenotype confusion when they constantly emphasized the discontinuity of traits as proof of the discontinuity of hereditary factors. In fact, the biometry-saltationism war was largely rooted in this confusion. In their thinking and writing, the two parties regularly used the word "variation" to mean both the hereditary material and the trait.

Several writers tried to point out the distinction between genotype and phenotype, but their warnings were not sufficiently headed. For example, as early as 1866, Mendel had already warned that:

QUOT. 796 One sees how risky it can sometimes be to draw conclusions about the internal kinship of hybrids from their external similarity. (Mendel, 1866, 1966a, p. 40)

A similar distinction was made by Bateson and Saunders in 1901 (see QUOT. 606) and especially by Yule in 1902 (see Sect. 10.1.4.3). However, it was with Johannsen (1909) that this distinction was made explicit, and its importance realized.

12.4.2 Phenotype, Gene, and Genotype

12.4.2.1 Phenotype

Johannsen's definition of the phenotype was based on Quetelet's essentialist (or typological)[16] notion of average type.[17] In his book *Elemente der exakten*

---

[16] See Sect. 2.9 for more information on the typological tradition in scientific thinking.

[17] In the memoir "Recherches sur le penchant au crime aux différent âges," Quetelet said: "If the average man were determined for a nation he would present the type of that nation; if he could be determined from the ensemble of men, he would then present the type of type of the entire human species." (Quetelet, 1833, p. 1)

erblichkeitslehre (Johannsen, 1909), Johannsen arranged the lengths of beans in classes and showed that they nicely fit a binomial distribution. He then noted:

> **QUOT. 797** In reality, however, there were three biologically distinct types in relation to length, namely the short bean type, the long bean type and a type which can be called the hybrid type here. (*ibid*., p. 122)

Johannsen thus concluded that the statistical average (or type in Quetelet's sense) often gave the impression of homogeneity but in fact contained several biologically different types. He called this statistical average the "phenotype" (*Phaenotypus*):

> **QUOT. 798** Then we recognize that the "type" in QUETELET's sense is only an *appearance of a superficial nature*, which can deceive; only through further investigations will it be decided whether a single or several biologically different types are present. Therefore one could use the statistically derived type appropriately to designate as appearance type, or, briefly and clearly, as "*Phenotype*"). Such *phenotypes* are in themselves measurable realities: just what can be *observed* as typical; that is, in the case of variation series, the centers around which the variants are grouped. The word phenotype only takes the necessary reservation that no further-reaching conclusion may be drawn from the *appearance* itself. A given phenotype may express a biological entity; but he need not be. The phenotypes found in nature by statistical investigations are probably not in most cases! (*ibid*., p. 123)

The current understanding of the phenotype does not rely on Johannsen's essentialist concept of a statistically derived type—it only retains the "appearance" part of his definition.

12.4.2.2 Gene

Of the three key concepts (phenotype, gene, and genotype) introduced by Johannsen in his 1909 book, his description of the gene is furthest from today's beliefs. This is how he adduced the term gene (*gen*):

> **QUOT. 799** There has never been any doubt that the sex cells, the gametes, as is now called by a common name for egg and sperm cells, contain "something" which determines or very significantly influences the character of the organism formed by fertilization...This "something" in the gametes or in the zygote, which is of essential importance for the character of the organism, is usually called, with a very ambiguous expression, "dispositions". Many other expressions have been proposed, mostly, unfortunately, in precise connection with certain hypothetical conceptions. The word "pangene" introduced by DARWIN is probably used most frequently instead of "Anlage". However, the word "pangene" is not a happy choice, being a combination that contains the root Pan (all, everything) and Gen (to become). Only the meaning of the latter is considered here; only the simple idea is expressed that a property of the developing organism is or can be conditioned or co-determined by "something" in the gametes. No hypothesis about the nature of this "something" should be advanced or supported. Therefore, it seems easiest to extract from DARWIN'S well-known word, the root "gene" in which only we are interested, to replace the inappropriate, ambiguous word "plant." We will thus simply say "gene" and "genes" for "pangene" and "pangenes." (*ibid*., pp. 123–124)

Although it is widely believed that Johannsen derived the word "gene" from de Vries' 'pangen',[18] there is no mention of the latter in this connection in the whole book.[19] Instead, Johannsen had Darwin's gemmules (derived from the provisional hypothesis of pangenesis[20]) in mind, although he mistakenly ascribed to the latter the name "pangene." In the second edition of *Elemente*, this oversight was addressed as follows:

QUOT. 800 The word "pangene" introduced by DE VRIES following DARWIN has probably been used most frequently in place of "Anlage". (Johannsen, 1913, p. 143)

More importantly, Johannsen's description of the gene completely precluded any material nature to its existence. In this sense, it was "hypothesis-free" and completely different from Darwin's gemmules, Mendel's factors (Sect. 6.2.7), Galton's stirp (Sect. 4.5.2), Nägeli's micellae (Sect. 5.1.2), Weismann's biophors (Sect. 5.2.8), and de Vries pangens (Sect. 5.3.2). In fact, today's conception of the gene is closer to de Vries' pangens than to Johannsen's gene.

The reason for Johannsen's denial of a morphological property for the gene was his adherence to a physicalist doctrine, according to which all physical, chemical, and biological phenomena needed to be explained in terms of *forces*.[21]

In the 1911 English paper "The genotype conception of heredity," one can further read:

QUOT. 801 The "gene" is nothing but a very applicable little word, easily combined with others, and hence it may be useful as an expression for the "unit-factors," "elements" or 'allelomorphs' in the gametes, demonstrated by modern Mendelian researches. (Johannsen, 1911, p. 132)

QUOT. 802 As to the nature of the "genes" it is as yet of no value to propose any hypothesis; but that the notion "gene" covers a reality is evident from Mendelism. (*ibid*., p. 133)

Johannsen's introduction of the "gene" was thus so vague that the geneticist and historian E.A. Carlson has claimed that:

QUOT. 803 Johannsen did not define the gene; he couldn't. He merely said it was convenient to have a word for the "something" in reproductive cells that eventually led to traits that could be studied in breeding tests. (Carlson, 1967, p. 2)

---

[18] See Sect. 5.3.2.

[19] Van der Pas (1970, p. 205) is one of the few authors to have pointed this out.

[20] See Sect. 3.7.2.

[21] See Mayr (1982, p. 737). Johannsen's physicalism also lead him to deny, like Bateson (see Sect. 7.2.3.1), that chromosomes were the bearers of hereditary material: "These two ideas: that 'elements' in the zygote correspond to *special organs*, and that *discrete particles of the chromosomes* are 'bearers' of special parts of the whole inheritance in question are neither corollaries of, nor premises for, the stirp- or genotype-conception. Those special ideas may have some interest as expressions of the searching mind, but they have no support in experience; the first of them is evidently erroneous, the second a purely speculative morphological view of heredity without any suggestive value." (Johannsen, 1911, p. 132)

12.4.2.3 Genotype

In contrast to Johannsen's introduction of the term "gene," his definition of "genotype" (*genotypus*) was much closer to today's beliefs. In his definition, he also made a key distinction between genotype and phenotype:

> **QUOT. 804** The manner in which the phenotypes manifest themselves, whether they are evidenced by qualitative or quantitative properties, predicts absolutely nothing about the genes. There can be very obvious phenotypic differences where there is no genotypic difference; and there are also cases where, despite genotypic differences, the phenotypes are the same. That is why it is of the greatest importance to clearly separate the concept of *phenotype* (appearance type) from the concept of *genotype* (one could say predisposition type). However, we will not be able to operate with this last term—a genotype appears—not purely in appearance; however, the derived concept of genotypic difference will be of use to us in many ways. (Johannsen, 1909, p. 130)

In the 1911 paper, one can further read:

> **QUOT. 805** A "genotype" is the sum total of all the "genes" in a gamete or in a zygote. When a monohybrid is formed by cross fertilization, the "genotype" of this F_1-organism is heterozygotic in one single point and the "genotypes" of the two "genodifferent" gametes in question differ in one single point from other. (Johannsen, 1911, p. 133)

Johannsen's genotype is thus the genetic constitution of the zygote. In QUOT. 804, Johannsen explicitly pointed out that genotype and phenotype needed to be distinguished because they did not have a one-to-one relationship.

Although the genotype–phenotype distinction was first explicitly spelled out in the 1909 book, it was already implied in Johannsen's 1903 pure line paper[22]: the phenotype corresponded to the external appearance of the population of beans, and the genotypes corresponded to the separate pure lines.

---

[22] See Sect. 12.2.

Part IV
The Modern Synthesis

Chapter 13
Stages Leading to The Modern Synthesis

Abstract During the period 1915–1930, three key events occurred that enabled the reconciliation of Mendelism and Darwinism: (i) Castle and Norton proved the effectiveness of natural selection acting on small differences; (ii) Nilsson-Ehle and East showed that Mendelian heredity could explain quantitative traits, and (iii) Morgan and his co-workers demonstrated that most mutations were small rather than drastic changes. The reconciliation of Mendelism and Darwinism was done primarily by Fisher, Wright, and Haldane, and gave rise to the field of mathematical evolutionary (or population) genetics. The results obtained from this discipline constituted the modern synthesis, which was further strengthened by the investigation of field biologists working in related disciplines.

Keywords Modern synthesis · Mathematical evolutionary (population) genetics · Field biologists · Hardening of natural selection · Mosaic inheritance · Gametic contamination · Multifactor hypothesis · Sex-linked inheritance · Recombination · Linkage

13.1 What Is the Modern Synthesis?

During the period 1915–1930, three key events occurred that enabled the union of Mendelism with Darwinism:

I. Despite attacks by the saltationists, *the effectiveness of natural selection acting on small differences (as conceived by Darwin) was eventually demonstrated*[1].
II. The various rounds of conflict between the biometricians and saltationists (described in Part III) resulted in an increasing realization among the biological community of the validity of Mendelian heredity. Although the experiments thus far had involved qualitative traits, *a Mendelian explanation was soon put forward even for quantitative traits*[2].

[1] See Sect. 13.2.

[2] See Sect. 13.3.

P. Gorroochurn, *The Development of Evolutionary Genetics*,
https://doi.org/10.1007/978-3-031-69374-8_13

III. Finally, Johannsen's pure-line experiments[3] had highlighted the importance of mutations. *However, it was soon realized that the latter ought to be conceived as small genetic changes rather than the drastic morphological changes of de Vries*[4].

The scene was thus set for a reconciliation of Mendelian heredity with Darwinian selection: evolution was in a large part driven by natural selection according to Darwin's continuous model, the heredity component of this theory was supplied by Mendel's laws, and the raw materials on which natural selection acted were (usually small) mutations. The tools used for the reconciliation were borrowed from the biometricians' armory, and the new field that emerged as a result was called *mathematical evolutionary (or population) genetics*.

Thus, mathematical evolutionary genetics is the study of Darwinian selection within the framework of Mendelian heredity. The results obtained from this new discipline constitute the so-called *modern synthesis*[5] and its main architects were Fisher, Wright, and Haldane. The results of the modern synthesis were later strengthened by field biologists who united concepts and results from several different fields, namely:

- Genetics: Dobzhansky (1937)
- Animals and Systematics: Mayr (1942)
- Paleontology: Simpson (1944), and
- Plants: Stebbins (1950)

The approach adopted here mirrors that of Provine (1977), who has also highlighted the decisive influence of mathematical population genetics. Likewise, in the latest edition of his *Evolution*, Ridley has stated that:

> **QUOT. 806** The next step was to show that natural selection could operate with Mendelian genetics. The theoretical work was mainly done, independently, by R.A. Fisher, J.B.S. Haldane, and Sewall Wright.... Their synthesis of Darwin's theory of natural selection with the Mendelian theory of heredity established what is known as neo- Darwinism, or the synthetic theory of evolution, or the modern synthesis. (Ridley, 2003, p. 15)

However, it must be stressed that this interpretation of the modern synthesis is by no means universally accepted. Several biologists, most especially Mayr, have stressed the importance of field biology and downplayed the role of mathematical population genetics. This first happened during the "beanbag genetics" debate.[6] In his *The Growth of Biological Thought*, Mayr stated:

---

[3]See Sect. 12.2.

[4]See Sect. 13.4.

[5]The modern synthesis is sometimes also called the neo-Darwinian theory. However, in this book, neo-Darwinism shall be used only to denote Weismann's version of Darwin's theory (i.e., Darwin's theory without any Lamarckian component; see Sect. 5.2.1). The modern synthesis is also discussed in Dickins (2021), Mayr & Provine (1980), Provine (1977), Gould (1983; 2002, Chap. 7), Beatty (1986), Mayr (1982, pp. 556–570), Reid (1985), Smocovitis (1997), Ishida (2017), Huneman & Walsh (2017), Gayon (1990), Gayon & Huneman (2019), and Eldredge (1985).

[6]See Sect. 16.4.

QUOT. 807 The difficulties and misunderstandings were finally resolved during the period between 1936–1947, resulting in a unified evolutionary theory often referred to as the "evolutionary synthesis" (Mayr and Provine, 1980). Dobzhansky, Rensch, Mayr, Huxley, Simpson, and Stebbins, among others, showed that the major evolutionary phenomena such as speciation, evolutionary trends, the origin of evolutionary novelties, and the entire systematic hierarchy could be explained in terms of the genetic theory as matured during the 1920s and 30s. Except for shifts in emphasis and for a far more precise analysis of all the various mechanisms, the synthetic theory of evolution is the paradigm of today. (Mayr, 1982, pp. 119–120)

There was not much room, if any, for mathematical population genetics in Mayr's view of the modern synthesis.

On the other hand, Huxley, who coined the term "modern synthesis" (Huxley, 1942), regarded the work of the mathematical population geneticists as an integral component of the synthesis. In particular, he praised Fisher for having "radically transformed our outlook on the subject" and stated that:

QUOT. 808 Evolution may lay claim to be considered the most central and the most important of the problems of biology. For an attack upon it we need facts and methods from every branch of the science-ecology, genetics, paleontology, geographical distribution, embryology, systematics, comparative anatomy—not to mention reinforcements from other disciplines such as geology, geography, and mathematics.

Biology at the present time is embarking upon a phase of synthesis after a period in which new disciplines were taken up in turn and worked out in comparative isolation. (*ibid.*, p. 13)

Similarly, Dobzhansky fully acknowledged the pioneering works of Fisher, Wright, and Haldane,[7] and called them (together with Chetverikov) "the creators of the modern biological theory of evolution" (Dobzhansky, 1970, p. 25).

The writings of Huxley, Dobzhansky, and several others who followed them are indicative of a modern synthesis consisting of two phases:

- *Phase I,* consisting of the theoretical framework created by the mathematical population geneticists.
- *Phase II,* consisting of several biological disciplines united under the framework created by Phase I.

The paleontologist and evolutionary biologist Stephan Jay Gould (1941–2002) provided an interesting variation to this two-phase version (Gould, 2002, Chap. 7). His first phase of the modern synthesis was the same as Phase I above; however, his Phase II was a *hardening* of Phase I. By hardening, Gould meant that, whereas in Phase I evolution was understood as an interplay of natural selection, genetic drift, and other random phenomenon, in Phase II there was an increasing emphasis on the role of natural selection:

QUOT. 809 . . .the key methodological claim of the original synthesis: Small-scale genetic changes within natural populations are the stuff of all evolution. I have called this original version "pluralistic" because it admitted a range of theories about evolutionary change, Darwinian and otherwise, and insisted only that explanations at all levels be based upon known genetic causes operating within populations and laboratory stocks.

---

[7] See QUOT. 1011.

> This pluralistic version was slowly and subtly altered, primarily during the 1940s...as the intent of explanation by known genetics shifted to the content of one particular theory—Neo-Darwinism and its insistence that cumulative natural selection leading to adaptation be granted pride of place as the mechanism of evolutionary change. (Gould, 1983, p. 75)

Gould provided several examples of the hardening of the initial phase of the modern synthesis, including even that in the work of Sewall Wright (*ibid.*, pp. 83–86). Gould's interpretation of the modern synthesis has been adopted by one of the most recent works on the subject (Dickins, 2021, p. 64).

The three major stages, indicated at the start of this section, that led to the modern synthesis, as understood in this text, will now be described.

13.2 Stage One: The Rehabilitation of Natural Selection: The Hooded-Rat Experiments (Castle, 1906, 1911), Norton's 1915 Table

13.2.1 Introduction

Recall that Darwinian selection had been relentlessly attacked by the saltationists. In the early 1900s, de Vries' mutationism was still popular. Mendelism seemed to have gained the upper hand over Darwinism in the various rounds of fight between the saltationists and biometricians (see Part III), especially in the wake of the Darbishire controversy (see Chap. 11). Several biometricians had thus defected to the Mendelian camp. As explained before, the saltanionists thought that evolution was caused mainly by large variations and natural selection had only a minor role in the process. Furthermore, Johannsen (see Chap. 12) seemed to have delivered a death blow to the Darwinian paradigm. Castle's demonstration of the efficacy of natural selection was therefore paramount in its rehabilitation and eventual synthesis with Mendelism.

William Castle's life and early acceptance of Mendelism was described in Sect. 10.2.1. His 1903 derivation of the Hardy-Weinberg equilibrium law was explained in Sect. 10.2.3. In Sects. 11.8.1 and 11.22, it was shown that Castle took a pro-Mendelian stance in the Darbishire controversy. However, Sect. 11.8.2 also showed that, as early as 1903, Castle's version of Mendelism was peculiar: he approved of the particulate nature of that theory, but he also believed that Mendelian factors were altered (contaminated) during fertilization. His version of Mendelism was described as *mosaic inheritance* in Sect. 11.8.2, and one of its highlights was the violation of a key tenet of Mendelism, namely, *gametic purity*.

Just as Castle believed that Mendelian factors were contaminated during fertilization, so too he believed that these factors were altered under the action of natural selection. There is no doubt that Castle's work was fundamental in the rehabilitation of natural selection, but this rehabilitation was carried out within the framework (or under the burden?) of gametic contamination. Even after multiple objections, Castle vociferously stuck to his theory and relented only in 1919. These events will be described next.

Early in his career, Castle embraced both Mendelism (at least his own version of it) and de Vries' mutationism. In a 1905 paper, he referred to his own experiments on guinea pigs. He explained that he had found one male that had a supernumerary fourth digit in its hind foot. By breeding offspring with extra toes from the male, he was able to create a race with a well-developed forth toe on either hind foot within three generations. Castle then declared:

QUOT. 810 This race was not *created* by selection, though it was *improved* by that means. Like the poet, in the proverb, it was *born*, not made. Any amount of selection practised on other families of my guinea-pigs would probably never produce a four-toed race, for though carefully, watched through as many as seven generations no four-toed pig has appeared among them. (Castle, 1905b, p. 523)

Later, he added:

QUOT. 811 On the whole, it appears that the formation of new breeds begins with the discovery of an exceptional individual, or with the production of such an individual by means of cross-breeding. Such exceptional individuals are mutations. (*ibid.*, p. 525)

However, this viewpoint changed very quickly.

13.2.2 The Hooded-Rat Experiments (1906, 1911)

In 1906, Hansford MacCurdy finished some breeding experiments on "hooded" rats under Castle's supervision. The hooded rat is piebald with a dark dorsal stripe, and the piebald pattern is recessive to the gray color of wild rats. By selecting for increased or decreased size of the dorsal stripe, Castle and MacCurdy were able to permanently change the stripe size in either direction within five generations. They contrasted de Vries' and Darwin's positions on the efficacy of natural selection as follows:

QUOT. 812 De Vries maintains that all species-forming variations are of this sort [i.e. due to mutations]; that selection is unable to form new species, because it can neither call into existence mutations nor permanently modify a race by cumulation of abmodal fluctuations. Darwin, on the other hand, and the great majority of his followers, while admitting that races are occasionally produced by discontinuous or "sport" variation, ascribe evolutionary progress chiefly to the cumulation through long periods of time of slight individual differences, such as de Vries calls fluctuations. The issue between the two views is sharp and clear. According to de Vries, if we rightly understand him, selection is not a factor in the *production*[8] of new species, but only in their *perpetuation*, since it determines merely what species shall survive; according to the Darwinian view, new species arise through the direct agency of selection, which leads to the cumulation of fluctuating variations of a particular sort. (MacCurdy and Castle, 1907, p. 3)

Which of these two opposite views (de Vries' large mutations vs. Darwin's small differences) did Castle and MacCurdy endorse? Their answer could not be clearer:

[8] All italics in this quotation are MacCurdy and Castle's.

QUOT. 813 So far, then, as these experiments go, they support the Darwinian view rather than that of de Vries. (*ibid.*, p. 4)

Castle and MacCurdy's demonstration of the efficacy of Darwinian selection was an important achievement. However, when it came to the underlying mode of inheritance:

QUOT. 814 ...though the inheritance is clearly Mendelian, when hooded and Irish rats are crossed, the gametes formed by cross-breds are not pure, but modified, each extracted pattern being changed somewhat in the direction of that pattern with which it was associated in the cross-bred parent. This means simply that the inheritance, though in the main alternative, is to some extent blending. (*ibid.*, p. 34)

That is, the inheritance was Mendelian, but at the same time, the gametes formed were modified by natural selection!

What were the reasons for Castle's belief in gametic contamination? One reason seems to be a genotype–phenotype confusion:

QUOT. 815 By unit-character was understood (1) any visible character of an organism which behaves as an indivisible unit in Mendelian inheritance and (2) by implication, that thing in the germ-cell which produces the visible character. (Castle, 1919b, p. 127)

Castle could later have used a multifactor hypothesis to explain variation in the visible characters (phenotypes). However, he believed in natural selection and was opposed to Johannsen's 1903 pure line theory, which seemed to show that constant Mendelian factors made natural selection ineffective.

With the assistance of Dr John C Phillips, Castle started an extensive set of experiments with hooded rats in 1907. The rats were assigned different grades depending on the amount of pigmentation (Fig. 13.1). The grades of the initial rats ranged from −2 to +3. The extreme variants in either direction were then selected and bred separately. This process was repeated across eight generations, resulting in two categories of rats, the *minus* series (narrow striped rats) and the *plus* series (wide striped rats) (see Table 13.1). These data were provided in a 1910 paper (Castle, 1910, p. 100). Referring to the numbers in Table 13.1, Castle observed:

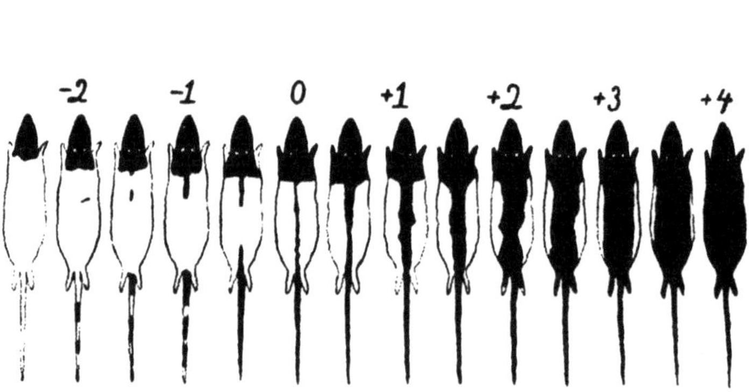

Fig. 13.1 Grades assigned by Castle to various degrees of pigmentation in the hooded rat (Fig. 18 on Plate 4 of Castle (1910))

Table 13.1 Average grades of hooded rats across eight generations of repeated selection

	Generation	Average grade of parents	Average grade of offspring	Number of offspring
Plus series	1	2.50	2.05	150
	2	2.51	1.92	471
	3	2.73	2.51	341
	4	3.09	2.72	444
	5	3.33	2.90	610
	6	3.51	3.09	834
	7	3.53	3.14	874
	8	3.65	3.30	91
				3815
Minus series	1	1.46	1.00	55
	2	1.41	1.07	132
	3	1.56	1.18	195
	4	1.69	1.28	329
	5	1.73	1.41	701
	6	1.86	1.56	1252
	7	2.00	1.70	1544
	8	2.03	1.78	713
				4921

QUOT. 816 The interesting feature of this experiment is the production, as a result of selection, of wholly new grades; in the narrow series, of animals having less pigment than any known type other than the albino; in the wide series, of animals so extensively pigmented that they would readily pass for the "Irish type"...By selection we have practically obliterated the gap which originally separated these types, though selected animals still give regression toward the respective types from which they came. But this regression grows less with each successive selection and ultimately should vanish, if the story told by these statistics is to be trusted. As yet there is no indication that a limit to the effects of selection has been reached. (*ibid.*, p. 101)

This statement was followed by an unequivocal endorsement of Darwinian selection, i.e., natural selection acting on small differences:

QUOT. 817 From the evidence in hand we conclude that Darwin was right in assigning great importance to selection in evolution; that progress results not merely from sorting out particular combinations of large and striking unit-characters, but also from the selection of slight differences in the potentiality of gametes representing the same unit-character combinations. (*ibidem*)

However, Castle still stuck to his previous position on the modifiability of Mendelian factors (*ibidem*). In a 1912 article, he also attacked Johannsen's genotype concept[9]:

[9] See Sect. 12.4.2.3.

QUOT. 818 I therefore regard as unsubstantiated the genotype conception of heredity in which unit-characters are regarded as indestructible and unmodifiable entities. Organisms are not devoid of variability; neither are the unit-characters which they manifest devoid of variability, nor yet is the germinal basis of such unit-characters devoid of variability. Unit-characters may arise gradually as the result of repeated selection in a particular direction. (Castle, 1912, p. 280)

13.2.3 Problems with Castle's Theory of Gametic Contamination

In 1914, Castle and Phillips published a paper (1914) where Castle first realized that there could be a serious problem with his theory of gametic contamination. In their paper, Castle and Phillips noted that when a rat in the minus series[10] was crossed with one in the plus series, the offspring was less negative (the authors gave as an example a -2-gamete uniting with a +6-gamete, giving a +4-gamete). On the other hand, they observed that a rat in the plus series that mated with another rat in the plus series gave an offspring that was *less positive,* "which is the opposite of what a contamination theory would demand" (*ibid*, p. 23). Castle and Phillips noted that this discrepancy could be explained by East's multifactor hypothesis,[11] which they referred to as the (gene) modifier hypothesis. However, since the latter was based on the genotype concept, they rejected it:

QUOT. 819 At this point the hypothesis of modifiers encounters serious difficulty, if one holds the prevalent or "genotype" conception as to the nature of Mendelian factors, viz, that they are fixed and unchangeable and not subject to quantitative variation, but only to combination in different ways with other factors…Some objections to this view had previously been stated by Castle (1912) and need not here be repeated. (*ibid.*, p. 24)

Castle and Phillips' paper elicited a sharp response from H. J. Muller (1914). The latter belonged to Morgan's school of genetics and took exception to Castle and Phillips' theory of gametic contamination:

QUOT. 820 A conclusion so radical and so opposed to previous work should not be accepted, however, as long as it remains at all reasonably possible to use instead an explanation in harmony with the results of Johannsen and other investigators. (*ibid.*, p. 568)

Thereupon, Muller explained how each of several results in Castle and Phillips' paper could be accounted for by a multifactor hypothesis for hoodedness, rather than by gametic contamination:

1. *The production of rats of any desired grade.* Muller explained the various grades of rats through a multifactor hypothesis as follows:

[10] See Sect. 13.2.2.

[11] See Sect. 13.3.2. Nilsson-Ehle was the first to experimentally demonstrate the multifactor hypothesis.

QUOT. 821 ...the production of animals of desired grade by selection may perhaps be explained as a mere sorting out, into different lines of descent, of different combinations of the various factors for hoodedness originally present in the heterozygous ancestors. (*ibid.*, p. 567)

2. *Return selection.* Castle and Phillips were able to show that an original race could be successively selected for the more pigmented members to give a highly pigmented race. These changes were permanent even when natural selection was relaxed but if the less pigmented members were now successively selected, the original race could be retrieved. They called this phenomenon *return selection.* Muller explained it through heterozygosity at various loci after the first series of selection so that recombination made return selection possible (*ibid.*, p. 571).
3. *Decrease in variability after continued selection.* Muller explained this through an increase in homozygosity by the direct action of natural selection (*ibid.*, p. 572)
4. *Variability in F_2 hooded rats when hooded and wild (or Irish) rats were crossed.* Castle and Phillips had observed that when hooded rats were crossed with a wild or Irish[12] rat, the F_1 hybrids were all pigmented ("selfs"), and the F_2 consisted of selfs and hooded in the ratio 3:1. However, the hooded rats in F_2 varied in intensity and Muller asked whether this was because of gametic contamination or modifying factors. He then stated (*ibid.*, p. 573) that there was strong evidence for the latter possibility since, when light-hooded rats were crossed with wild or Irish rats, the hooded rats in F_2 were darker and showed more variation. According to Muller, this could only be because the light-hooded rats differed from the wild or Irish at more modifiers than did the normal hooded rats, so in the former case, the hooded rats in F_2 had more modifiers that were heterozygous; hence the greater variation and darker intensity.
5. *Plus crossed with plus gives less positive.* Muller used the discrepancy mentioned by Castle and Phillips at the start of this section to deliver a major blow to the contamination theory. This theory predicted *darker* F_2 hooded-rats when wild or Irish rats were crossed with rats in the plus series. However, multifactor theory predicted *lighter* F_2 hooded rats (see (4) above), which agreed with actual observations. Muller concluded:

QUOT. 822 It was fatal to the idea that the difference between the P1 strain of hooded rats and the F_2 hoodeds was due to contamination of the allelomorph for hooded with that for self, since such contamination should have resulted in F_2 hooded rats darker than those of P1, not lighter. (*ibid.*, p. 574)

Muller ended his paper by making the following critical assessment of Castle's theory of genetic contamination:

[12]The wild rat is pigmented all over and the "Irish" rat is pigmented almost all over.

QUOT. 823 To thus suppose that independent genes *fuse*[13] or induce changes in one another, merely because they happen to produce similar *end effects* upon the organism, and in spite of the fact that they usually lie in different chromosomes and are apt to differ from each other as much as do other genes, is utterly teleological. (*ibid.*, p. 575)

Castle (1915a) shot back immediately with an article trying to defend against the various criticisms raised by Muller. However, in the meantime, Pearl published a paper detailing 17 years of selection for winter egg production on the Barred Plymouth Rock fowls (Pearl, 1915). Pearl was categorical: Castle and Phillips (1914) had only shown that natural selection was effective—there was no proof at all that natural selection had changed any hereditary factor:

QUOT. 824 A careful study of the very interesting and valuable work of Castle and Phillips with rats leaves the writer with the feeling that those experiments prove no more than do the experiments here reported: namely that the composition of a population may be altered by selection. It does not appear to be proven that selection has essentially altered the constitution of the germ-plasm of any particular individual as compared with the germ-plasm of that individual's ancestors, making due allowance of course for the phenomenon of segregation. (*ibid*, p. 608)

Castle (1915b) vehemently rebutted Pearl's comments, but it was inevitable that his position was becoming weaker. In the same paper, Castle also reported that his experiments on hooded rats had reached 17 generations with a total of 33,249 rats. This was followed by one further back-and-forth between Pearl (1916) and Castle (1916), as well as one between MacDowell (1916) and Castle (1917).

However, a resolution of the issue of modifying factors versus gametic contamination was soon in coming. Sewall Wright, who was Castle's student, suggested that the plus and minus series should each be repeatedly crossed with a pure third race (e.g., the wild rat). If multiple modifiers were responsible for hoodedness, then the repeated crossing in the plus series should remove the plus modifiers and the repeated crossing in the minus series should remove the minus modifiers. Eventually, the members in the two series should be similar. This strategy was first attempted in Castle and Wright (1916) with seemingly negative results:

QUOT. 825 Previous experiments had shown that when a race of hooded rats, whose character has been modified by selection (either plus or minus), is crossed with wild rats, the extracted hooded animals obtained in F_2 as recessives show regression toward the mean condition of the recessive race before selection began. This result suggested that the regression observed might be due to removal by the cross of modifying factors, which selection had accumulated in the hooded race. If this view was correct, it was thought that further crossing of the extracted hooded animals with the same wild race should result in further regression, and that if this further regression was not observed a different explanation must be sought for the regression already noted.

The entire experiment has accordingly been repeated from the beginning, with the same result as regards regression in the first F_2 generation, but with no regression of the same sort in a second F_2 containing twice-extracted hooded animals. So far from observing further regression as a result of the second cross with wild rats, we have unmistakable evidence that

[13] All italics in this quotation are Muller's.

Table 13.2 Castle's key experiment to test for multiple modifiers (Castle, 1919a, p. 372)

		Mean grade	Standard deviation	Number of hooded young
Plus series	Control, uncrossed plus race, generation 10	+3.73	0.36	776
	Once extracted hooded F_2 young	+3.17	0.73	73
	Twice extracted hooded F_2 young	+3.34	0.50	256
	Thrice extracted hooded F_2 young	+3.04	0.64	19
Minus series	Control, uncrossed plus race, generation 16	−2.63	0.27	1980
	Once extracted hooded F_2 young	−0.38	1.25	121
	Twice extracted hooded F_2 young	+1.01	0.92	49
	Thrice extracted hooded F_2 young	+2.55	0.66	104

> the movement of the mean, mode, and range of the hooded character has been in the reverse direction. So the hypothesis of modifying factors to account for the regression and for the progressive changes observed under selection becomes untenable. (*ibid.*, p. 163)

In the meantime, Sturtevant published a very comprehensive and balanced paper that concluded that:

> **QUOT. 826** Modification of factors by selection, crossing, fractionation, or similar means is undemonstrated in any given case, and has been shown not to occur in other cases that are typical of the results usually obtained. (Sturtevant, 1918, p. 51)

The turning point for Castle finally occurred in 1919. Wright had suggested that the extracted hooded rats in F_2 needed to be repeatedly crossed with the wild rats across several generations before meaningful results could be obtained. This new strategy was adopted by Castle (1919a) for both the plus and minus series giving the results shown in Table 13.2.

The table shows that, in the plus series, starting with rats in the 10th generation with an average grade of +3.73, thrice extracted hooded F_2 offspring yielded an average grade of +3.04. For the minus series, starting with rats in the 16th generation with an average grade of −2.63, thrice extracted hooded F_2 offspring yielded an average grade of +2.55. Additionally, one family in the minus series yielded 14 thrice-extracted offspring with a mean grade of +3.05.

Given the proximity of the mean final grades in the plus and minus series, Castle had no other option than to admit that:

> **QUOT. 827** ...the changes which had occurred in the hooded character as a result of selection were detachable changes and are probably in nature independently inherited modifying factors. (*ibid.*, p. 373)

It is true that Castle had carried the burden of gametic contamination for all these years but, as he made sure to mention in his paper, his experiments had clarified the fundamental importance of Darwinian selection in the process of evolution:

QUOT. 828 I should feel like apologizing for my own obtuseness in not reaching a similar conclusion sooner, did I not recall with satisfaction how much clearer the rôle of selection now stands revealed than it did when these experiments were begun, and to the clearing up of the situation I shall at least hope that this rat work has contributed something, if only by provoking investigation. (*ibid.*, p. 374)

13.2.4 Norton's 1915 Table

In 1915, Reginald Punnett,[14] a Mendelian, published a book (Punnett, 1915) that was meant to show that mimicry in butterflies had developed through mutation rather than natural selection:

QUOT. 829 If we assume that sudden and readily appreciable variations of the nature of "sports" turn up from time to time, and if these variations happen to resemble a form protected by distastefulness so closely that the two can be confused by an enemy which has learned to avoid the latter, then there would appear to be good grounds for the mimicking sport becoming established as the type form of the species.

...

Instead of building up a mimetic likeness bit by bit it [natural selection] will merely have conserved and rendered numerically preponderant a likeness which had turned up quite independently. The function of natural selection in respect of a mimetic likeness lies not in its formation but in its conservation. It does not bring about the likeness, neither does it accentuate it: it brings about the survival of those forms which happen to shew the likeness. (*ibid.*, pp. 143–144)

In Chap. VIII of his book, Punnett considered a population that was in Hardy-Weinberg equilibrium[15] with respect to a gene with two alleles, G (dominant) and g (recessive). He wanted to investigate what would happen if the dominant forms (GG and Gg) were given a selective advantage. Punnett enlisted the help of the mathematician H. T. J. Norton[16] (1886–1937) (Fellow of Trinity College, Cambridge), who showed, for example, that if $GG{:}Gg{:}gg = 1{:}4{:}4$ initially, then after 12 generations the new ratio would be 1:2:1 under a selection advantage of 10% for the dominant form. The calculations were not shown, but Norton also computed generation times for several selection advantages and initial genotype frequencies,

[14]See Sect. 10.9 for Punnett's role in Hardy's derivation of what later came to be known as the Hardy-Weinberg equilibrium.

[15]Weinberg's derivation of the equilibrium was unknown at the time, so Punnett referred only to Hardy.

[16]H. T. J. Norton was Hardy's student and, motivated by the latter's 1908 paper (Hardy, 1908), wrote the first thesis in population genetics in 1910. In that same year, Norton communicated his results to Punnett. In 1922, Norton also showed some of his work to Haldane, who had already independently derived them (Haldane, 1927a, p. 607). Due to ill health, Norton published his results only in 1928 (Norton, 1928).

and the results were tabulated in Appendix I of Punnett's book. In that table (Table 13.3), a selection advantage of 100/99 under Column A, for example, means that the chances of the dominants *GG* and *Gg* surviving to reproduction against those of *gg* are in the ratio 100:99. To obtain the 12 generations cited above, the procedure is as follows: since the new variety is dominant, Column A needs to be considered, and since the selection advantage is 10%, the column under 100/90 is relevant. Now the proportion of the new variety (i.e., dominant) moves from 5/9 = 55.6% to 3/4 = 75%. Since there are no such numbers in the column for the new variety, the column for the old variety (recessives) needs to be considered: the proportions then go from 44.4% to 25%. The 5th and 6th rows of numbers in the leftmost column correspond to these proportions and reading the 6th row in Column A for 100/90 gives the required 12 generations.

Norton's calculations were important for two reasons:

(a) They represented the first attempt to mathematically model natural selection within a Mendelian framework. Punnett remarked:

> **QUOT. 830** The development of Mendelian studies has given us a method, rough perhaps but the best yet found, of testing for the presence, and of measuring the intensity, of natural selection. (*ibid.*, p. 103)

(b) They showed the rapidity with which even a relatively weak selection advantage can lead to the substitution of one gene for another. As Punnett himself acknowledged:

> **QUOT. 831** . . .a selective advantage of 10% operating against the recessives will reduce their numbers in 70 generations from nearly one-half of the population to less than one-fortieth. . .it is remarkable in how brief a space of time a form which is discriminated against, even lightly, is bound to disappear. Evolution, in so far as it consists of the supplanting of one form by another, may be a very much more rapid process than has hitherto been suspected, for natural selection, if appreciable, must be held to operate with extraordinary swiftness where it is given established variations with which to work. (*ibid.*, p. 96)

Although, in his book, Punnett wished to highlight the importance of mutations over natural selection, Norton's calculations revealed the efficacy of the latter. This had a big influence on several key contributors to the modern synthesis, especially J. B. S. Haldane[17] and S. S. Chetverikov.[18]

[17] See Sect. 16.2.1.

[18] See Sect. 16.5.2.4.

Table 13.3 Table compiled by H.T.J. Norton showing the number of generations required to transition between different genotype frequencies under varying selection intensities (Punnett, 1915, p. 155)

Percentage of total population formed by the old variety	Percentage of total population formed by the hybrids	Percentage of total population formed by the new variety	Number of generations taken to pass from one position to another as indicated In the percentages of different individuals in left-hand column							
			A. Where the new variety is dominant				B. Where the new variety is recessive			
			$\frac{100}{50}$	$\frac{100}{75}$	$\frac{100}{90}$	$\frac{100}{99}$	$\frac{100}{50}$	$\frac{100}{75}$	$\frac{100}{90}$	$\frac{100}{99}$
99.9	.09	.000								
98.0	1.96	.008	4	10	28	300	1920	5740	17,200	189,092
90.7	9.0	.03	2	5	15	165	85	250	744	8,160
69.0	27.7	2.8	2	4	14	153	18	51	149	1,615
44.4	44.4	11.1	2	4	12	121	5	13	36	389
25.	50.	25.0	2	4	12	119	2	6	16	169
11.1	44.4	44.4	4	8	18	171	2	4	11	118
2.8	27.7	69.0	10	17	40	393	2	4	11	120
.03	9.0	90.7	36	68	166	1,632	2	6	14	152
.008	1.96	98.0	170	333	827	8,243	2	6	16	165
.000	.09	99.9	3840	7653	19,111	191,002	4	10	28	299

13.3 Stage Two: The Multifactor Hypothesis: Mendelian Inheritance Is Compatible with a Continuous Model of Evolution (Nilsson-Ehle, 1908; East, 1910)

13.3.1 Nilsson-Ehle (1908)

Both Mendel[19] and Yule[20] had previously argued that continuous traits could be explained through Mendelian heredity by a multifactor hypothesis, i.e., by allowing for the action of several factors (genes). However, this was first experimentally demonstrated by the Swedish biologist Nilsson-Ehle. Herman Nilsson-Ehle (Fig. 13.2) was born in Skurup, Sweden. He did research in plant taxonomy and physiology in 1894 and became an assistant at the plant breeding station in Svalöf in 1900. At that time, the station was led by Nils Hjalmar Nilsson, who had introduced the method of "genealogical selection," i.e., repeated selection of exceptional characters.[21]

A key event occurred in Nilsson-Ehle's career when Erich von Tschermak, one of the three co-rediscovers of Mendelism,[22] delivered a lecture at Svalöf in 1901 on Mendel's laws. This event started Nilsson-Ehle's interest in Mendelism and prompted his subsequent hybridization experiments, much to Nilsson's dissatisfaction.

Based on hybridization experiments he started in 1903, Nilsson-Ehle published a paper in 1907 (Nilsson-Ehle, 1907), where he first argued hybridization was a better method than genealogical selection for obtaining a combination of desirable qualities. In that paper, he also provided a first indication of the multifactor hypothesis:

> **QUOT. 832** It seems that it can happen that what appears to us as a characteristic in the work itself constitutes a composition of several characteristics, and that on the other hand several seemingly separate characteristics are followed together and thus in the work itself constitute an expression of a property. (*ibid.*, p. 214)

In Nilsson-Else's next paper (Nilsson-Ehle, 1908), the multifactor factor was much more explicit. In crosses that were started in 1900 between black and yellow-grained oat, he reported that white grains were produced in F_2. However, he remarked that the white was not latent or atavistic but rather the result of a two-factor Mendelian model:

> **QUOT. 833** It is easy to show that the white color is not latent or atavistic, but simply has to arise if one maintains the assumption of the feature pairs as the presence and absence of the units. The black and the yellow color do not form a pair of characteristics, but each designate

[19] See Sect. 6.2.8.

[20] See Sect. 10.1.4.3.

[21] See Sect. 12.1.

[22] See Sect. 6.6.4.

Fig. 13.2 Herman Nilsson-Ehle (1873–1949). (From: https://commons.wikimedia.org/wiki/File:Herman_Nilsson-Ehle.jpg)

a unit. Black (S)[23] forms a feature pair with its absence (s), and yellow (G)[24] with its absence (g). (*ibid.*, p. 262)

Using the notation in the quotation above, Nilsson-Ehle represented the gamete associated with a black parent grain by Sg and a yellow parent grain by sG. The white color then appears in the offspring if its gamete is sg. Nilsson-Ehle also reported some yellowish offspring resulting from crosses of the type sG × sg or sg × sG. It followed from his reasoning that a continuum of color could be obtained if more factors (genes at different loci) were allowed.

Having given a few similar examples, Nilsson-Ehlse made the following important statement:

QUOT. 834 This simple case gives a key to the interpretation of quantitative properties and allows bridging the certainly more imaginary than real gap between so-called discontinuous and continuous variability seem very possible. The true gulf is probably instead between hereditary variations (whether large or small) on the one hand, and non-hereditary variations caused by underlying circumstances on the other. However, the small hereditary variations

[23]The German for black is "schwarz."

[24]The German for yellow is "gelb."

> can be quite continuous and are therefore often thrown together with the non-hereditary variations caused by external factors, which are also usually continuous, as individual variability. Conversely, modifications brought about by external conditions can appear quite discontinuous. (*ibid.*, p. 267)

Nilsson-Ehle's key to the interpretation of quantitative traits was then through Mendelian heredity of several factors considered together. This would be one of the key tenets of the modern synthesis.

In the 1908 paper, Nilsson-Ehle gave an example that was meant to show that "such color gradations are the result of several independent units even clearer than the last-mentioned oat crossing" (*ibid.*, p. 268). This example, involving white and red wheat grain, has been referred to by many subsequent geneticists and is also discussed in Nilsson-Ehle's doctoral dissertation (Nilsson-Ehle, 1909).

In the crosses of red and white wheat grains, all F_1 hybrids were red (red was therefore dominant over white). When the F_1 hybrids were intercrossed, 75 F_2 offspring were obtained, none of which were white. When the F_2 offspring were intercrossed, some crosses yielded F_3 red:white offspring in the ratio 3:1, but ratios such as 15:1 and 63:1 were also obtained (Nilsson-Ehle, 1908, p. 269). Nilsson-Elhe explained these ratios through a three-factor Mendelian model.[25] For example, in cases where one of the three factors always has recessive gametes, the 15:1 ratio can be accounted for, as follows (in modern notation):

$$\underbrace{(3R_1+r_1)}_{\text{1st factor}}\underbrace{(3R_2+r_2)}_{\text{2nd factor}}\underbrace{r_3}_{\text{3rd factor}}=\underbrace{9R_1R_2r_3+3R_1r_2r_3+3r_1R_2r_3}_{\text{15 red}}+\underbrace{r_1r_2r_3}_{\text{1 white}} \qquad (13.1)$$

where the capital R stands for a dominant genotype and the small r for a recessive one. From (13.1), the 15:1 ratio is really a 9:3:3:1 ratio, with the 15 red being made of different shades of red depending on the number of dominant alleles.

Similarly, Nilsson-Ehle explained the 63:1 ratio when the three factors could be heterozygous, as follows (in modern notation):

$$\underbrace{(3R_1+r_1)}_{\text{1st factor}}\underbrace{(3R_2+r_2)}_{\text{2nd factor}}\underbrace{(3R_3+r_3)}_{\text{3rd factor}}$$

$$=\underbrace{27R_1R_2R_3+9R_1R_2r_3+9R_1r_2R_3+9r_1R_2R_3+3R_1r_2r_3+3r_1R_2r_3+3r_1r_2R_3}_{\text{63 red}}$$

$$+\underbrace{r_1r_2r_3}_{\text{1 white}}$$

As mentioned above, different shades of red could be obtained depending on the number of dominant gametes:

[25] Referring to similar numbers in Nilsson-Ehle's doctoral thesis (Nilsson-Ehle, 1909) where 78 F_2 plants had no white at all, Clausen and Hiesey (1958, p. 162) have argued that a Mendelian model with at least four factors might have been more appropriate.

QUOT. 835 It is also to be remarked that among the beds of the third generation with only red-grained plants, some showed on the average a deeper, others a lighter red color distinctly, and in still others a division into deeper and lighter red-grained individuals was quite conspicuous... (*ibidem*)

Thus, quantitative traits could arise by the action of several Mendelian factors (genes):

QUOT. 836 ...the crossings now mentioned have provided reliable experimental proof that quantitative hereditary characteristics can come about through the interaction of different independent, qualitatively inseparable units. The gradations designate different combinations and can therefore become much more numerous than the units themselves and therefore of course also bring with them a greater continuity in the variations than the units themselves can do. So there is no such thing as a unit for every gradation. The gradations are not independently arising variations, as pre-Mendelian research should think, but only different combinations of a relative, i.e. in relation to the innumerable gradations, small but absolutely possibly large number of existing units. (*ibid*., p. 271)

The above is one of the clearest articulations of the multifactor hypothesis (with "units" replaced by "genes at different loci" for the modern reader).

13.3.2 East (1910)

Shortly after Nilsson-Ehle's publication in German, East published his own English version of similar experiments he had independently conducted on corn. Edward Murray East (1879–1938) was born in DuQuoin, Illinois. He graduated in Chemistry from the University of Illinois in 1900 and soon after joined C. G. Hopkins' corn-breeding laboratory, where he developed an interest in corn-breeding. His first acquaintance with Mendel's laws seems to have occurred in 1905 during the second meeting of the American Breeders' Association. A key event occurred in East's life when, during that meeting, he was invited (upon Hopkins' recommendation) to join the Connecticut Experiment Station by its director, E. H. Jenkins. The latter was a chemist and had decided to expand research in plant breeding. At the Connecticut Station, East started experiments to improve corn varieties and learned of hybrid vigor (the selective superiority of heterozygotes), a practice that had been known well before him. East's attendance of a speech by G. H. Shull in 1908 and his subsequent correspondence with the latter marked the start of a close collaboration between the two scientists.

Of interest here is East's paper with the self-explanatory title "A Mendelian interpretation of variation that is apparently continuous" (East, 1910). In that paper, East referred to Nilsson-Ehle's doctoral thesis (Nilsson-Ehle, 1909) but claimed that he had reached similar conclusions prior to his colleague's publication.

He first reported crosses between 15 yellow varieties of maize and white varieties, which yielded the 3 yellow:1 white ratio in F_2. Next, he explained that when one variety of yellow endosperm was used to pollinate a white variety, the F_1 hybrids consisted of 159 medium yellow and 149 light yellow hybrid kernels. From this, East surmised that the yellow pollen parents were always homozygous for one yellow

(Y_1) and always heterozygous for another yellow (Y_2). This is a two-factor Mendelian model, with each factor associated with a different yellow:

$$\underbrace{Y_1}_{\text{1st factor}}\ \underbrace{(Y_2 + y_2)}_{\text{2nd factor}} = \underbrace{Y_1 Y_2}_{\text{medium yellow}} + \underbrace{Y_1 y_2}_{\text{light yellow}}$$

Now the medium yellow segregated in the ratio 15 yellow:1 white in F_2 (similar to Nilsson-Ehle's (13.1)), while the light yellow segregated in the ratio 3:1.

East presented many other similar examples in his papers (some also involving the number of rows per ear), leading him to conclude:

> **QUOT. 837** . . .it would appear that we may have several allelomorphic pairs each of which is inherited independently of the others, and each of which is separately capable of forming the same character. When present in different numbers, in different individuals, these units simply form quantitative differences. (East, 1910, p. 71)

In a follow-up paper, Emerson and East (1913) proved similar Mendelian interpretations for many other quantitative characters of maize, such as the length and diameter of ears, weight and breadth of seeds, and so on.

13.4 Stage Three: Small Variations Can Be Created by Mutations (Morgan, 1909)

13.4.1 Introduction to Thomas Hunt Morgan

Biographers of Morgan rightly hail him as "a pioneer of genetics" but his path to earning this title was all but linear. Thomas Hunt Morgan[26] (Fig. 13.3) was born in Lexington, Kentucky, and obtained his bachelor's degree in science in 1886 (the same year de Vries made his observations on the evening primroses—see footnote 41 of chap. 5). He then joined Johns Hopkins University to study zoology. He studied comparative anatomy under William Brooks, the renowned morphologist who had previously influenced Bateson's evolutionary views (see Sect. 7.2.3). Brooks was a Darwinian and no doubt also influenced Morgan. In 1890, Morgan obtained his doctorate with a thesis entitled "A contribution to the embryology and phylogeny of the Pycnogonids." He then left for the Marine Biology Station in Naples, where he developed an interest in the experimental study of growth. In 1900, Morgan visited de Vries' laboratory in Holland and then attended a meeting in Naples, where he first learned of the newly rediscovered Mendelian theory.

[26]Thomas Morgan's life and work are also discussed in Allen (1966; 1968; 1975, pp. 52–65; 1978; 1980), Moore (1983), Cuny (1969), Carlson (2004, Chap. 13), Kohler (1994), Sturtevant (1965, Chaps. 7 and 8), Schwarz (2008, pp. 165–208), Weinstein (1980), Shine & Wrobel (1976), and Tort (1996, pp. 3055–3059, Vol 2: F-N).

Fig. 13.3 Thomas Hunt Morgan (1866–1945). (From: https://commons.wikimedia.org/wiki/File:Thomas_Hunt_Morgan_sketch_1931.png)

13.4.2 Initial Disapproval of Natural Selection (1903–1909)

Morgan's 1903 book *Evolution and Adaptation* (Morgan, 1903) is an invaluable source of information on his early views on evolution (especially with respect to *Darwinian* evolution) and heredity (especially with respect to *Mendelian* heredity). The book as a whole is quite consistent with de Vries' views on evolution at the time.

The primary motivation for Morgan's 1903 work was the relevancy of Darwinian evolution to the problem of *regeneration* (the natural replacement of an individual's lost parts). There was no doubt in Morgan's mind that the Darwinian theory was inadequate for this problem and this situation further provoked his curiosity about the generality of the theory:

> **QUOT. 838** One of the general questions that I have always kept before me in my study of regenerative phenomena is how such a useful acquirement as the power to replace lost parts has arisen, and whether the Darwinian hypothesis is adequate to explain the result. The conclusion that I have reached is that the theory is entirely inadequate to account for the *origin* of the power to regenerate; and it seemed to me, therefore, desirable to reexamine the whole question of adaptation, for might it not prove true here, also, that the theory of natural selection was inapplicable? This was my starting-point. (*ibid.*, p. ix–x)

As suggested by the title of his book, Morgan's key consideration in any theory of evolution was *adaptation*. He thus spoke of the origin and evolution of *adaptations*:

QUOT. 839 Darwin's theory of natural selection is preeminently a theory of adaptation. It appears, in fact, better suited to explain this phenomenon than that of the "origin of species.". (*ibid*., p. 91)

The reason Morgan did not speak of the origin and evolution of *species* is because, following the nominalist[27] tradition, he denied that the species concept had any reality and maintained that it was invented solely for the purpose of classification:

QUOT. 840 We should always keep in mind the fact that the individual is the only reality with which we have to deal, and that the arrangement of these into species, genera, families, etc., is only a scheme invented by man for purposes of classification. Thus there is no such thing in nature as a species, except as a concept of a group of forms more or less alike. In nature there are no genera, families, orders, etc. These are inventions of man for purposes of classification. (*ibid*., p. 33)

Morgan believed that evolution was true and did not totally reject Darwin's theory. However, he had serious reservations about certain aspects of it:

QUOT. 841 . . .we can profitably reject, as I believe, much of the theory of natural selection, and more especially the idea that adaptations have arisen because of their usefulness. . . (*ibid*., p. ix)

Morgan's major objections to Darwin's theory of natural selection, in the order they appear in the book, are as follows:

(i) *Natural selection may have only a negative impact in evolution, with the primary driver of evolution being variation*. This was one of the major arguments of the mutationists against Darwin's theory: ". . .if definite variation can go on accumulating without selection, then possibly we could account for evolution without supposing any other process to intervene. Under these circumstances all that could be claimed for selection would be the destruction of those variations incapable of living, or of competing with other forms. Hence the process of selection would have an entirely negative value." (*ibid*., p. 97)
(ii) *Darwin's fluctuating variations may not lead to the formation of new species*. Here, Morgan argued that, given the scant evidence for the efficacy of fluctuating variation in artificial selection, there was no reason to suppose that natural selection could effectively act on such variations: "If it be admitted that the evidence from artificial selection is far from convincing, in showing that selection of fluctuating variations could have been the main source, even in the formation of new races, we need not be prejudiced in favor of such a

[27] Nominalism is the theory that only individuals exist, and classes and essences are artifacts. Its roots are theological: it can be traced to William of Ockham who was opposed to human categories taking precedence over Divine omnipotence.

process, when we come to examine the formation of species in nature." *(ibid.*, p. 101*)*

(iii) *Darwin's concept of struggle for existence may not be generally true.* Morgan accepted the struggle-for-existence concept for higher organisms but questioned its validity for lower ones. For example, "We cannot follow the individuals in...a culture of bacteria, but there is nothing to be seen that suggests a struggle for existence, if this idea conveys the impression of the destruction of certain individuals by competition with others. In fact, the results are in some respects exactly the reverse. Millions of individuals are present at the time when the food supply becomes exhausted, and they all pass into a protected resting stage." *(ibid.*, p. 112*)*

(iv) *The absence of transitional forms.* Although Darwin had explicitly addressed this issue in Chapter VI of his book (Darwin, 1859, p. 173),[28] Morgan reiterated a criticism that had most vehemently been expressed by Mivart previously[29]: "Why, if species have descended from other species by fine gradations, do we not everywhere see innumerable transitional forms? Why is not all nature in confusion, instead of the species being, as we see them, well defined?" (Morgan, 1903, p. 129)

(v) *Darwinian selection cannot account for organs of extreme perfection* (e.g., *the eye) or for organs of little apparent importance.* Morgan used Darwin's own halting explanation for the extreme perfection of the eye to undermine the theory of natural selection (*ibid.*, p. 131). Similarly, he dismissed Darwin's explanations for organs of little apparent importance (*ibid.*, p. 133).

(vi) *Doubtful evidence for the inheritance of acquired characters.* Morgan criticized Darwin's theory for its reliance on Lamarckian principles: "By falling back on the theory of inheritance of acquired characters Darwin tacitly admits the incompetence of natural selection to explain the evolution of the flatfish." (*ibid.*, p. 137). Regarding the validity of the Lamarckian theory, this is how Morgan expressed his skepticism: "Until it can be shown that more of the sense-organs develop in any special part, as the result of the increased use of the part, we have no real basis on which to establish, even as probable, the Lamarckian view." (*ibid.*, p. 245)

(vii) *The beginning stages of organs cannot be accounted for by natural selection.* This was a well-known criticism previously most vehemently expressed by Mivart.[30] Morgan echoed the same criticism and disapproved of Darwin's explanations (*ibid.*, p. 138).

In Morgan's criticism of Darwinism, Weismann's neo- (or ultra-) Darwinism was also not spared:

[28] See also paragraph before QUOT. 176.

[29] See Sect. 2.14.5 (e).

[30] See Sect. 2.14.5 (a).

QUOT. 842 Thus Weismann has piled up one hypothesis on another as though he could save the integrity of the theory of natural selection by adding new speculative matter to it. (*ibid*., p. 165)

What Morgan rejected was Darwin's natural selection and Weismann's neo-Darwinism, and what he endorsed was de Vries' mutation theory as a theory of evolution:

QUOT. 843 De Vries gives a critical analysis of a number of cases in which new races have been formed under domestication. He shows very convincingly that, whenever the result has been the outcome of the selection of fluctuating variations, the product that is formed can only be kept to its highest point of development by the most rigid and ever watchful care. If selection ceases for only a few generations, the new form sinks back at once to its original level. Many of our cultivated plants have really arisen, not by selection of this sort, but by mutations; and there are a number of recorded cases where the first and sudden appearance of a new form has been observed. In such cases as these there is no need for selection, for if left to themselves there is no return to the original form. If, however, after a new mutation has appeared in this way, we subject its fluctuating variations to selection, we can keep the new form up to its most extreme limit, but can do nothing more. (*ibid*., p. 291)

13.4.3 Initial Criticism of Mendelism (1903–1909)

Morgan's endorsement of mutationism in 1903 should not be confused with an endorsement of Mendelism. He correctly recognized that the de Vries' mutationism was a theory of evolution while:

QUOT. 844 [t]he chief value of Mendel's results lies in their relation to the theory of inheritance rather than to that of evolution. (*ibid*., p. 286)

Morgan believed that Mendelism could not be a viable theory of heredity because of swamping. Given that he had already discarded the importance of individual differences, what remained was discontinuous variation. However:

QUOT. 845 ...the evolution of animals and plants cannot rest primarily on the appearance of discontinuous variations, because they occur rarely and would be swamped by intercrossing. If Mendel's law applies to such cases, that is, if a cross were made between such a sport and the original form, the hybrid in this case, if self-fertilized, would begin to split up into the two original forms. But, on the other hand, it could very rarely happen that the hybrid did fertilize its own eggs, and, unless this occurred, the hybrid, by crossing with the parent forms in each generation, would soon lose all its characters inherited from its "sport" ancestor. (*ibid*., p. 286)

Morgan's swamping argument, which he attributed to Darwin, was thus different from Jenkin's.[31] For the latter, the argument was directed to undermine Darwinian selection and swamping occurred through the numerical inferiority of a sport. For the former, however, the argument was specifically directed to undermine Mendelism,

[31] See Sect. 2.14.4.3.

and swamping occurred when a hybrid with an inherited sport was intercrossed with a normal ancestor across generations.

Even as late as 1907, presumably influenced by Castle's views,[32] Morgan doubted gametic purity. In a paper published in *Science,* he discussed the possible Mendelian inheritance of sex and gave two versions of Mendelian heredity:

QUOT. 846 [Preformationist Interpretation of Mendelism] The currently accepted interpretation of Mendelian inheritance is strictly one of preformation. Alternative characters are treated as entities in the germ cells that may be shuffled, but seldom get mixed.

[Epigenetic Interpretation of Mendelism] If we take the opposite point of view, that of epigenetic development, the outcome, wherever alternative or contrasted characters are involved, is not due to separation, but to alternative dominance and recession, which need not give three types, but only two, if selective fertilization occurs, or, if only the egg or the sperm contains the internal factor that determines, that one or the other of the alternative sex characters shall predominate. (Morgan, 1907, p. 384)

The first interpretation is the one that accords with gametic purity. In the second interpretation, the germ cell is a mixture that contains the potential for both maleness and femaleness. Which interpretation did Morgan lean toward? In the concluding parts of the 1907 paper, one can read:

QUOT. 847 My own preference - or prejudice, perhaps - is for the epigenetic interpretation, but the whole truth may lie somewhere between these two forms of thought that are the Scylla and Charybdis of biological speculation. (*ibidem*)

Morgan's position on gametic purity would only change when he completely embraced Mendelian inheritance in 1910.

13.4.4 Breakthrough with Drosophila *(1910)*

13.4.4.1 Introduction

Morgan became Professor of Zoology at Columbia University in 1904, and started experiments on mice, poultry, pigeons, and so on, in an attempt to find mutations. However, these experiments were expensive to maintain and the progeny were few. Before that, in 1900, William Castle at Harvard University had started using the fruit fly (*Drosophila*) to study the effects of inbreeding. Castle suggested *Drosophila* to W.J. Moenkhaus who wanted to work on sex ratios. Soon after, following Moenkhaus, F.E. Lutz also performed experiments on the fruit fly in Cold Spring Harbor. Morgan was aware of his contemporaries' work, especially Castle's, and he obtained fruit flies from Lutz and elsewhere. In 1907, he started experimenting with *Drosophila*. He instructed a graduate student (Fernandus Payne) to perform a Lamarckian experiment: *Drosophila* were reared in the dark for several generations and then tested for atrophy due to disuse. The experiment was (not surprisingly)

[32] See Sect. 11.8.2.

unsuccessful, but Morgan realized that the fruit fly would be ideal for his experiments since it was inexpensive to maintain and proliferated rapidly (about 300 offspring for each pair of parents, with a reproduction time of approximately 10 days).

In 1908, Morgan and Payne started a second experiment in which they attempted to induce artificial mutations in *Drosophila*. Morgan used a method that had been recommended in 1904 by de Vries[33]:

QUOT. 848 . . .the rays of Roentgen and Curie, which are able to penetrate into the interior of living cells, [may] be used in an attempt to alter the hereditary particles in the germ cells. (Blakeslee, 1936, p. 36)

As his experiments progressed, Morgan started to recognize that certain heritable variations could be quite small and made Darwin's continuous evolution plausible:

QUOT. 849 It is not without interest to note in this connection that Darwin often assumed that *fluctuating variations*[34] are transmitted to the offspring . . . But we have discovered the additional fact that *some small* variations are inherited. Let us call these *definite variations*, and if these be the material with which evolution is concerned. Darwin's assumption in regard to the nature of variation will be, in part, justified. (Morgan, 1909, p. 375)

However, a breakthrough occurred in 1910 (probably in January[35]) when a male *Drosophila* offspring with *white* eyes[36] was discovered by Morgan.[37] This first report of this ground-breaking discovery was briefly stated in a somewhat perfunctory manner on May 18, 1910:

QUOT. 850 Several other mutants of these flies are under observation. A male in which the red pigment of the eye is totally absent has produced in the first generation, by his sisters, 162 red eyed to one white eyed fly. The results show that the character is germinal, but

[33] And that would later be used to full effect by H. J. Muller.

[34] All italics in this quotation are Morgan's.

[35] "Probably" in January because the month reported in Morgan and Bridges (1916, p. 25) and used by some historians, is May. However, Green (1996) has made a strong case that the month should be January. Green has used several pieces of evidence, the most compelling of which is based on Morgan's wife's, Lilian Morgan (herself an accomplished geneticist), recollection of the discovery. E.g., see Kohler (1994, p. 42). Carlson has also endorsed the January month (Carlson, 2011, p. 36).

[36] The common form of the fruit fly has *red* eyes, as were the parents Morgan used.

[37] In his 1941 book, Lutz claimed that he had discovered the white-eyed fruit fly first: "Professor T. H. Morgan visited the Station and I told him that a white-eyed Drosophila had appeared in one of the pedigreed strains but that I was too busy with abnormal veins to attend to it. He took live descendants of this white-eyed "sport"and bred from them. Eventually he got the white eye back. On working out the inheritance of this character he discovered an important addition to the Mendelian Laws. . . This little story is no particular credit to me. If I had realized how valuable that white-eyed mutant was destined to be, I would not have been happy to give it away. However, it fell into good hands and, really, *Drosophila melanogaster* should be called Morgan's Heredity Fly." (Lutz, 1941, p. 238). This claim was repeated by Ramsay Spillman (1942) in his review of Lutz's book. The review prompted Morgan to clarify that the flies he obtained from Lutz could not have been the descendants of the white-eyed fly since their matings did not yield any white-eyed offspring (Morgan, 1942). Another author, Hilaire Cuny, has stated that the white-eyed fly was actually discovered by Bridges, one of Morgan's students (Cuny, 1969, p. 39).

reappears at present in only a very small percentage of the first generation. (Morgan, 1910b, p. 161)

Morgan's classic paper on *Drosophila* appeared in *Science* on July 22, 1910 (Morgan, 1910c), shortly after the first report. The *Science* paper was the first where:

(i) Morgan adopted Mendelism.
(ii) The phenomenon of *sex-linkage* (in particular, X-linkage) was illustrated.
(iii) The term "mutation" was used by Morgan for a small change in genetic material within a Mendelian framework, thus distancing himself from de Vries' mutation as a non-Mendelian species-forming change in the nature of pangens (see QUOT. 382).

Each of the above will be explained next.

13.4.4.2 Adoption of Mendelism (1910)

The 1910 *Science* paper started in more or less the same way as the first report had ended:

QUOT. 851 In a pedigree culture of Drosophila which had been running for nearly a year through a considerable number of generations, a male appeared with white eyes. The normal flies have brilliant red eyes. (Morgan, 1910c, p. 120)

The following results were reported by Morgan:

(a) When the white-eyed male was crossed with red-eyed females, the offspring consisted of:

$$\left.\begin{array}{l}1{,}237 \text{ red-eyed offspring} \\ 3 \text{ white-eyed males}\end{array}\right\} (F_1)$$

(b) When the white-eyed male was crossed with some of the F_1 females, the following was obtained:

$$\begin{array}{c}129 \text{ red-eyed females,} \\ 132 \text{ red-eyed males,} \\ 88 \text{ white-eyed females,} \\ 86 \text{ white-eyed males.}\end{array}$$

(c) When the F_1 hybrids were intercrossed, the offspring consisted of:

$$\left.\begin{array}{l}\text{2,459 red-eyed females}\\ \text{1,011 red-eyed males}\\ \text{782 white-eyed males}\end{array}\right\}\ (\mathrm{F}_2)$$

However, there were no white-eyed females in F_2.

To explain these results, Morgan used a two-factor Mendelian model as follows: he assumed that the eye factor was either *R* (red) or *W* (white), and the sex factor was *X* such that *XX* (i.e., when both gametes carried the *X*) were females and *X* (i.e., when only one gamete carried the *X*) were males. A red-eyed female was then represented by *RRXX* (or *RWXX*, since *R* is dominant over *W*) and her gametes by *RX-RX* (or *RX-WX*), and a white-eyed male by *WWX* and his gametes by *WX-W*.

(a′) A Mendelian representation of the crosses in (a) is:

$$\underbrace{WX-W}_{\text{white male}}\times\underbrace{RX-RX}_{\text{red female}}=\underbrace{WX-RX}_{\text{red female}}+\underbrace{W-RX}_{\text{red male}}$$

The F_1 thus consists of an equal number of males and females, but all are red (Morgan attributed the three white males in F_1 to "further sporting").

(b′) For the crosses in (b),

$$\underbrace{WX-W}_{\text{white male}}\times\underbrace{WX-RX}_{\text{F1 red female}}=\underbrace{WX-WX}_{\text{white female}}+\underbrace{WX-RX}_{\text{red female}}+\underbrace{WX-W}_{\text{white male}}+\underbrace{RX-W}_{\text{red male}}$$

Morgan explained the cross above as a third verification of his two-factor Mendelian model.

Morgan also noted that wild-type red females were usually homozygous for red (i.e., *RRXX*) while wild-type red males were heterozygous for red (i.e., *RWX*).

(c′) The crosses in (c) will be explained in the next section because they are related to a new concept, namely, *sex-linkage.*

13.4.4.3 "Sex-Limited Inheritance" (1910), Recombination (1911), and Linkage (1912)

For the crosses in 13.4.4.2 (c), Morgan used his two-factor Mendelian model as follows:

$$\underbrace{WX-RX}_{\text{red female}}\times\underbrace{W-RX}_{\text{red male}}=\underbrace{WX-W}_{\text{white male}}+\underbrace{WX-RX}_{\text{red female}}+\underbrace{RX-W}_{\text{red male}}+\underbrace{RX-RX}_{\text{red female}}$$

From this and the numerical results he obtained, Morgan noted that:

QUOT. 852 . . .the outcome is Mendelian in the sense that there are three reds to one white. But it is also apparent that all of the whites are confined to the male sex. (*ibid.*, p. 121)

The fact that no white-eyed females appeared in F_1 prompted Morgan to hypothesize that the white-eyed character was sex-limited. In the next sentence, he rightly modified this hypothesis since it was also possible, in some rare cases, to obtain *white-eyed females*:

$$\underbrace{W - WX}_{\text{white male}} \times \underbrace{WX - RX}_{\text{F1 red female}} = \underbrace{WX - WX}_{\text{white female}} + \underbrace{WX - RX}_{\text{red female}} + \underbrace{W - WX}_{\text{white male}} + \underbrace{W - RX}_{\text{red male}}.$$

Therefore:

QUOT. 853 The results show that the new character, white eyes, can be carried over to the females by a suitable cross, and is in consequence in this sense not limited to one sex.

Thus, contrary to the title of his paper, Morgan had *not* discovered "sex-limited inheritance" (which refers to a trait expressed only in one sex, not necessarily because the gene for that trait resides on a sex chromosome). What he *had* discovered was "sex-linked inheritance" (which refers to a trait due to a gene residing on a sex chromosome). Morgan changed to the more appropriate "sex-linked inheritance" in 1912 (Morgan and Cattell, 1912, p. 80).

Morgan also made some comments on what would soon be called linkage.[38] He observed that:

QUOT. 854 In order to obtain these results it is necessary to assume, as in the last scheme, that, when the two classes of the spermatozoa are formed in the F_1 red male (RWX), R and X go together-otherwise the results will not follow (with the symbolism here used. (*ibid.*, p. 121)

Later in the paper, he called the phenomenon by its common name at the time, namely, "coupling' (see footnote 97 of chap. 6):

QUOT. 855 It now becomes evident why we found it necessary to assume a coupling of R and X in one of the spermatozoa of the red-eyed F_1 hybrid (RXO). The fact is that this R and X are combined, and have never existed apart. (*ibid.*, p. 122)

Readers will recognize that the association between *R* and *X* is not due to linkage: the latter term refers to genes lying on the same chromosome. As explained above, Morgan had discovered sex-linkage, a gene for color lying on the X chromosome. However, early in 1910, Morgan still doubted that chromosomes could contain Mendelian factors[39]:

[38]The phenomenon was first explicitly described by Correns (1900a) (see footnote 97 of chap. 6). It was also referred to in Bateson and Saunders (1901) (see Sect. 9.2.4 (e)), and in Bateson, Saunders & Punnett (1905). Its true explanation was first given in Morgan (1911) and "linkage" was first used in Morgan and Lynch (1912, p. 177).

[39]This hypothesis had been made as early as 1903 independently by Sutton (1903) and Boveri (1903). See QUOT. 508.

QUOT. 856 Since the number of chromosomes is relatively small and the characters of the individual are very numerous, it follows on the theory that many characters must be contained in the same chromosome. Consequently many characters must Mendelize together. Do the facts conform to this requisite of tie hypotheses? It seems to me that they do not. (Morgan, 1910a, p. 467)

Morgan was also not yet convinced that his *X* factor was, in fact, the *X* chromosome. This was despite his knowledge of the work of both Stevens (1905) and Wilson (1905), who had shown that female insects had two X chromosomes and males had only one.

However, later in 1910, Morgan discovered several other mutations, two of which (yellow body and miniature wings) were "sex-limited." He was now forced to acknowledge the presence of Mendelian factors on the *X* chromosome. However, his experiments revealed a recombination of factors when *Drosophila* with white eyes and miniature wings were crossed with *Drosophila* with white eyes and a yellow body color. Drawing on the work of the Belgian cytologist F.A. Janssens (1909), who had observed physical intertwining between homologous chromosomes during the early stages of meiosis, Morgan correctly hypothesized that the factors for body color and wings resided on the same *X* chromosome (i.e., were *linked*) such that the homologous pairs of the latter crossed over during meiosis to give recombinants:

QUOT. 857 In place of attractions, repulsions and orders of precedence, and the elaborate systems of coupling, I venture to suggest a comparatively simple explanation based on results of inheritance of eye color, body color, wing mutations and the sex factor for femaleness in *Drosophila*. If the materials that represent these factors are contained in the chromosomes, and if those factors that "couple" be near together in a linear series, then when the parental pairs (in the heterozygote) conjugate like regions will stand opposed. There is good evidence to support the view that during the strepsinema stage homologous chromosomes twist around each other, but when the chromosomes separate (split) the split is in a single plane, as maintained by Janssens. In consequence, the original materials will, for short distances, be more likely to fall on the same side of the split, while remoter regions will be as likely to fall on the same side as the last, as on the opposite side. In consequence, we find coupling in certain characters, and little or no evidence at all of coupling in other characters; the difference depending on the linear distance apart of the chromosomal materials that represent the factors...The results are a simple mechanical result of the location of the materials in the chromosomes, and of the method of union of homologous chromosomes, and the proportions that result are not so much the expression of a numerical system as of the relative location of the factors in the chromosomes. *Instead of random segregation in Mendel's sense we find "associations of factors" that are located near together in the chromosomes. Cytology furnishes the mechanism that the experimental evidence demands.*[40] (Morgan, 1911, p. 384)

The above describes linkage and recombination as it is understood today, although the term "linkage" is not used yet. This term first appeared in Morgan and Lynch (1912), where linkage was described for factors that were not sex-linked.

By using the idea that recombination fractions increased with distance along the chromosome (as explained in the quotation above), Morgan's student Sturtevant constructed the first genetic map in 1913 (Sturtevant, 1913). This work culminated in

[40] Italics are Morgan's.

the landmark book *The Mechanism of Mendelian Heredity* (Morgan et al., 1915), where the field of genetics was fully harmonized with chromosome theory.

Morgan's incorporation of chromosome theory into Mendelian heredity made him the founder of the *classical school of genetics*. Other prominent members of this school included his co-workers Alred Sturtevant, Calvin Bridges, and Herman Muller.

13.4.4.4 Mutations as *Small* Changes in Genetic Material (1910), Acceptance of Darwinian Selection (1916)

A third aspect of Morgan's 1910 *Science* paper (Morgan, 1910c) was his definition of a mutation as a small change in genetic material and subject to Mendel's laws:

> **QUOT. 858** Since the wild males (RWX) are heterozygous for red eyes, and the female (RXRX) homozygous, it seems probable that the sport arose from a change in a single egg of such a sort that instead of being RX (after reduction) the red factor dropped out, so that RX became WX or simply OX[41]. If this view is correct it follows that the mutation took place in the egg of a female from which a male was produced by combination with the sperm carrying no X, no R (or W in our formulae). (*ibid.*, p. 122)

This was a marked departure from de Vries' species-changing mutations. Following the 1910 *Science* paper, Morgan and his team discovered even smaller heritable variations (i.e., mutations) in *Drosophila*. By 1912, approximately forty *Drosophila* mutants had been identified. In that same year, he reviewed de Vries' *Die Mutationen in der Erblichkeitslehre* (de Vries, 1912) and even cast some doubt on whether the variations de Vries had observed on *Oenothera lamarckiana* were indeed mutations:

> **QUOT. 859** The only serious doubt that has arisen is whether the plant Oenothera Lamarckiana, on which the mutation theory so largely rested, is a hybrid or a "good" species. (Morgan, 1912, p. 543)

As Morgan discovered more small mutations, the plausibility of Darwinian selection increased. By 1916, he had accepted evolution by natural selection and shown that it was compatible with Mendelian inheritance:

> **QUOT. 860** The evidence shows clearly that the characters of wild animals and plants, as well as those of domesticated races, are inherited both in the wild and in the domesticated forms according to Mendel's Law. (Morgan, 1916, p. 193)
>
> The question still remains: Does selection play any role in evolution, and, if so, in what sense?...
>
> If through a mutation a character appears that is neither advantageous nor disadvantageous, but indifferent, the chance that it may become established in the race is extremely small, although by good luck such a thing may occur rarely. It makes no difference whether the character in question is a dominant or a recessive one, the chance of its becoming established is exactly the same. If through a mutation a character appears that has an

[41] The change of WX to OX is in line with Bateson's notion that red is due to a gene for color and white is due to its *absence our note.*

> injurious effect, however slight this may be, it has practically no chance of becoming established. If through a mutation a character appears that has a beneficial influence on the individual, the chance that the individual will survive is increased, not only for itself, but for all of its descendants that come to inherit this character. It is this increase in the number of individuals possessing a particular character, that might have an influence on the course of evolution. This gives a better chance for improvement by several successive steps; but not because the species is more likely to mutate again in the same direction. (*ibid.*, pp. 187–190)

The above explanation of natural selection soon became one of the most prevalent and influential interpretations of the process.

Chapter 14
Modern Synthesis (1): The Contributions of Fisher

Abstract Fisher was the first of the founding fathers of the modern synthesis. The two tenets of his evolutionary work were that natural selection acted slowly on populations that were large and panmictic, and that the effects of natural selection could be gauged by focusing on individual genes. In 1918, Fisher wrote a fundamental paper on biometrical (quantitative) genetics. His 1922 paper introduced the concepts of balancing selection and diffusion equations. It also investigated equilibria under different evolutionary scenarios. Fisher first clashed with Wright in 1928 on the issue of the evolution of dominance. One of Fisher's most controversial contributions to evolutionary genetics is the fundamental theorem of natural selection, which he first enunciated in 1930.

Keywords Biometrical (quantitative) genetics · Genotypic variance · Additive genetic variance · Dominance variance · Correlation between relatives · Infinitesimal model · Dominance ratio · Balancing selection · Diffusion equation · Wright-Fisher model · Hagedoorn effect · Evolution of dominance · Fundamental theorem of natural selection · Average effect · Average excess · Change in mean fitness due to natural selection · Partial change in mean fitness

14.1 Introduction

Fisher's contributions to the modern synthesis as a mathematical population geneticist were as fundamental as his work in statistics.[1] Sir Ronald Aylmer Fisher (Fig. 14.1) was born in London on February 17, 1890, and was a surviving twin. His father, George Fisher, was a fine-art auctioneer. Fisher entered Harrow School in 1904 and was good at mathematics right from the start. However, his eyesight was poor so he could not work in artificial light. This also meant that he often had to

[1]Ronald Fisher's life and work in evolutionary genetics are discussed in Box (1978), Edwards (1990, 1993, 2014, 2017), Provine (2001, pp. 140–154), Skipper (2007), Thompson (1990), Mather (1963, 1964), Karlin (1992), Owen (1962), Ford (2005), Piegorsch (1990), Pence (2021, Chap. 6), and Crow (1990). Fisher's work in statistics is discussed in Gorroochurn (2016a, Chap. 5), Fienberg & Hinkley (1980), Stigler (2006), Rao (1992), Street (1990), and Healy (2003).

P. Gorroochurn, *The Development of Evolutionary Genetics*,
https://doi.org/10.1007/978-3-031-69374-8_14

Fig. 14.1 Sir Ronald Fisher (1890–1962) in 1912. (From: https://commons.wikimedia.org/wiki/File:RonaldFisher1912.jpg)

resort to mental visualization to understand mathematics. In 1909, Fisher entered Gonville and Caius College, Cambridge, and read Karl Pearson's series of papers entitled "Mathematical contributions to the theory of evolution." It was undoubtedly through these papers that Fisher recognized the value of biometrical methods and adopted the Darwinian view of evolution. At the same time, Fisher was probably exposed to Mendelian ideas through Punnett, a member of Caius College. Finally, Edwards (2013) has produced evidence that Fisher was also greatly influenced by the second edition of Robert Heath *Lock's Recent Progress in the Study of Variation, Heredity, and Evolution* (Lock, 1909).

The purpose of this chapter is to show how, starting in 1918, Fisher became one of the founding fathers of modern evolutionary genetics. The entirety of Fisher's work in this field rested on two key assumptions:

(a) *Evolution occurs mainly through natural selection acting slowly on existing variation.* Natural populations are panmictic and large enough that the effects of random variation (genetic drift) are minimized. Fisher's evolutionary theory is thus known as *large population size theory* (LPST).

(b) *The effects of natural selection can be gauged by focusing on individual genes.*[2] Interaction effects between a given allele with alleles across multiple loci are pooled so that selection operates on the *average effect* of the given allele. In the words of Wright, Fisher's view of natural selection is *genic*[3] (Wright, 1980).

14.2 Foundational Biometrical Genetics Paper in 1918: "The Correlation Between Relatives on the Supposition of Mendelian Inheritance"

14.2.1 Background and Introduction to the 1918 Paper

As was common with evolutionists and geneticists of the time, Fisher developed an early interest in eugenics. In fact, during his final year as an undergraduate in 1911, he helped found the Cambridge University Eugenics Society and became the Chairman of its student committee. On November 10, 1911, during a meeting of the Eugenics Society, he read a paper on "Heredity" to a group of students and expressed his desire to reconcile Mendelism and biometry:

QUOT. 861 I have almost entirely devoted myself to the two lines of modern research which are of particular interest in Eugenics, that is to Biometrics and Mendelism: and perhaps experts and professionals will forgive the absence of more complicated details in both branches, if I explain that my object has been to give a fair view of the merits of the two methods, whose advocates have shown so little appreciation of the other school. (Bennett, 1983, p. 51)

Fisher's "Heredity" paper was a precursor to his later foundational paper on biometrical (or quantitative) genetics (Fisher, 1918). The latter was written at the suggestion of Leonard Darwin (1850–1943) (Charles Darwin's second youngest son and close friend of Fisher) and completed in 1916, at which time Fisher wrote a letter to Karl Pearson (who was editor of *Biometrika*) critiquing a recent paper by Kirstine Smith (who had used the minimization of χ^2 as a method of estimation) and informing him of his most recent work:

QUOT. 862 I have recently completed an article on Mendelism and Biometry which will probably be of interest to you. I find on analysis that the human data is as far as it goes, not inconsistent with Mendelism. But the argument is rather complex. (Pearson, 1968, p. 454)

---

[2]Fisher's model of natural selection acting on individual genes in a large population is also known as *mass selection*, a term borrowed from animal breeders, who often tried to select a single advantageous character in a large population.

[3]While Wright's own view is *organismic*.

Fisher had recently published a paper on the exact distribution of the sample correlation coefficient (Fisher, 1915)[4] in *Biometrika* and thought that his new work might also interest Pearson. However, in his June 26, 1916, reply letter, Pearson not only disagreed with Fisher's assessment of Smith's work but also made no mention of Fisher's biometrical genetics paper. He obviously was not interested.

Fisher then submitted his paper to the *Royal Society*, where it was reviewed by Pearson and Punnett. Although the two men came from diametrically opposite sides of the evolutionary debate, both found issues with Fisher's paper. Pearson's main criticism was that:

QUOT. 863 The author [Fisher] adopts a special hypothesis for determining the somatic character of an individual dropping the Mendelian phenomenon of dominance. It appears to me that his hypothesis is only one of a very large number that would lead to similar results, and it is not supported by any observational or experimental evidence that could differentiate it from them. (Norton and Pearson, 1976, p. 153)

Punnett was skeptical about Fisher's multifactor assumption:

QUOT. 864 And with regard to (3) [multifactor assumption] there are already cases well worked out in which the clearest interpretation is based on three independent factors acting in the same direction with regard to a given character. Such cases of course require great labour to work out & I should doubt therefore if assumption (3) could ever be tested by experiment. It may be true or it may not be and at present I don't see how one can say any more. (*ibid.*, p. 155)

Neither reviewer recommended outright rejection of Fisher's paper, but the reviews were negative, and the paper was rejected. At this point, Leonard Darwin intervened and financed its publication in the *Transactions of the Royal Society of Edinburgh* in 1918. When Fisher sent a copy to Pearson, the latter replied dismissively:

QUOT. 865 Many thanks for your memoir which I hope to find time for. I am afraid I am not a believer in cumulative Mendelian factors as being the solution of the heredity puzzle. (Pearson, 1968, p. 456)

It is no wonder that the relationship between Fisher and Pearson soured very quickly.

Fisher's paper (1918) is notoriously difficult because of its condensed style, but an attempt will be made here to delineate the main ideas it contains. The objective of the paper was clearly stated at the very start:

QUOT. 866 Several attempts have already been made to interpret the well-established results of biometry in accordance with the Mendelian scheme of inheritance. It is here attempted to ascertain the biometrical properties of a population of a more general type than has hitherto been examined, inheritance in which follows this scheme. (*ibid.*, p. 399)

[4]Pearson's handling of this paper was controversial. He accepted it and decided to write a "cooperative" paper (Soper et al., 1917) refuting the alleged use of inverse probability in Fisher's paper. This cooperative paper was supposedly an appendix to Fisher's paper, but Fisher was not consulted about it nor was his reply to the cooperative paper accepted by Pearson. See Gorroochurn (2016a, pp. 264–267) for more details.

14.2.2 Introduction to the Concept of Variance

At the start of his paper, Fisher introduced the variance as a more appropriate measure of variability than the standard deviation, which has been introduced previously by Karl Pearson (1894, p. 80). The variance was defined as the square of the standard deviation, and Fisher's preference was based on the additive property of the variance:

> **QUOT. 867** When there are two independent causes of variability capable of producing in an otherwise uniform population distributions with standard deviations σ_1 and σ_2, it is found that the distribution, when both causes act together, had a standard deviation $\sqrt{\sigma_1^2 + \sigma_2^2}$. It is therefore desirable in analyzing the causes of variability to deal with the square of the standard deviation as the measure of variability. We shall term this quantity the Variance of the normal population to which it refers, and we now ascribe to the constituent causes fractions and percentages of the total variance which they together produce. (Fisher, 1918, p. 399)

Apart from introducing the variance, the above also contains the germ of the idea of analysis of variance (ANOVA), which was later formally introduced by Fisher and Mackenzie (1923).

14.2.3 Genotypic Variance = Additive Genetic Variance + Dominance Variance

Fisher developed his methodology through a motivating example. The observed correlation coefficient between brothers was approximately 0.54. Using the formula $V/\sigma^2 = 1 - r$ (where r is the fraternal correlation, V is the variance of each fraternity, and σ^2 is the variance of the sons), it is seen that approximately 46% of the variance is attributable to other causes. Fisher argued that such causes cannot be due to the environment only, as had previously been pointed out by both Galton and Pearson. Furthermore, Galton had found that the unexplained variance was proportionately much less in identical twins. Therefore:

> **QUOT. 868** The simplest hypothesis, and the one which we shall examine, is that such features as stature are determined by a large number of Mendelian factors, and that the large variance among children of the same parents is due to the segregation of those factors in respect to which the parents are heterozygous. Upon this hypothesis we will attempt to determine how much more of the variance, in different measurable features, beyond that which is indicated by the fraternal correlation, is due to innate and heritable factors. (Fisher, 1918, p. 400)

Fisher's aim was therefore to find the "innate and heritable factors" that would account for some of the 46% unexplained variance.

Table 14.1 Fisher's basic quantitative model of 1918

Zygote	A_1A_1	A_1A_2	A_2A_2
Scaled phenotypic values	a	d	$-a$
Frequency	P	$2Q$	R

Fisher's model consisted of first considering a particular locus with two alleles A_1 and A_2 such that the three genotypes A_1A_1, A_1A_2, A_2A_2 had the scaled phenotypic values[5] and frequencies shown in Table 14.1.

His next approach was quite innovative: he proceeded to fit a weighted (by the genotype frequencies) least-squares line of the type

$$y = c + b(x - \overline{x}) \quad \text{for} \quad x = 0, 1, 2, \tag{14.1}$$

where y is the genotypic value, x is the number of A_1 alleles, and $\overline{x} = 1$. The task then was to minimize the sum of squares of deviations from linearity:

$$P(c + b - a)^2 + 2Q(c - d)^2 + R(c - b + a)^2 \tag{14.2}$$

(*ibid.*, p. 404) with respect to c and b. The least-squares solutions are[6]

$$c = \frac{Q(P + R)d}{PQ + 2PR + QR}, \qquad b = a - \frac{Q(P - R)d}{PQ + 2PR + QR}$$

When these values of c and b are substituted back in (14.2), the minimum sum of squares is obtained as

$$\delta^2 = \frac{4PQRd^2}{PQ + 2PR + QR} \tag{14.3}$$

(*ibidem*),[7] which Fisher called "the contribution to the variance of the irregular behaviour of the soma" and which is now called the *dominance variance*.

Next, when the phenotype effects in (14.2) are replaced by the *mean genotypic value*

$$m = Pa + 2Qd - Ra, \tag{14.4}$$

the resulting sum of squares becomes

[5]The scaled phenotypic value associated with a genotype A_iA_j is also called its *genotypic value*.

[6]Fisher's solution for b below (on p. 404 of his paper) mistakenly has a plus after the a (see Moran & Smith (1966, p. 11)).

[7]Under random mating, $P = p^2$, $Q = pq$ and $R = q^2$ so that $\delta^2 = 4p^2q^2d^2$.

$$P(c+b-m)^2 + 2Q(c-m)^2 + R(c-b-m)^2.$$

When the least squares solutions for c and b are substituted, the above becomes

$$\beta^2 = 2a^2(PQ + 2PR + QR) - 4Q(P-R)ad + \frac{2Q^2(P-R)^2 d^2}{PQ + 2PR + QR} \tag{14.5}$$

(*ibidem*),[8] which Fisher called "the contribution of the additive part" and is now known as the *additive genetic variance.*

Finally, the *genotypic variance* is

$$\alpha^2 = P(a-m)^2 + 2Q(d-m)^2 + R(-a-m)^2, \tag{14.6}$$

where m is given in Eq. (14.4).

Fisher used the quantities in Eqs. (14.3)–(14.6) to write

$$\alpha^2 = \beta^2 + \delta^2$$

(*ibid.*, p. 405). This corresponds to the decomposition of the total sum of squares (α^2) in Eq. (14.1) into the regression sum of squares (β^2) and the residual sum of squares (δ^2). By adding across several loci (which are assumed not to interact), Fisher wrote $\sigma^2 = \Sigma\alpha^2$, $\tau^2 = \Sigma\beta^2$ and $\varepsilon^2 = \Sigma\delta^2$, so that the above becomes the following fundamental equation of biometrical genetics:

$$\underbrace{\text{total genotypic variance}}_{\sigma^2} = \underbrace{\text{total additive genetic variance}}_{\tau^2} + \underbrace{\text{total dominance variance}}_{\varepsilon^2} \tag{14.7}$$

14.2.4 Correlation Between Relatives: Simple Cases

Fisher now proceeded to show how the theory he developed in the last section could be used to obtain the correlation between relatives. His technique was to construct 3 × 3 association tables, the entries of which denoted, under Hardy–Weinberg equilibrium (i.e., $P = p^2$, $Q = pq$, $R = q^2$), the probabilities of different relative–relative combinations. For example, the association table between a parent and an offspring is shown in Table 14.2.

[8]Under random mating, $P = p^2$, $Q = pq$ and $R = q^2$ so that $\beta^2 = 2pq\{a - (p-q)d\}^2$.

Table 14.2 Association table between parent and offspring (Fisher, 1918, p. 406)

		Offspring		
		A_1A_1	A_1A_2	A_2A_2
Parent	A_1A_1	p^3	p^2q	—
	A_1A_2	p^2q	$pq(p+q)$	pq^2
	A_2A_2	—	pq^2	q^3

The probability of a parent–offspring combination being, say, (A_1A_1, A_1A_2) is $p^2 \times q = p^2q$ (since, if the parent is A_1A_1 (with probability p^2), the offspring will have one A_1 allele, and the probability of its other allele being A_2 is q).

Next, Fisher wrote the genotypic effects i, j, and k as follows:

$$a-m=i, \quad d-m=j, \quad -a-m=k.$$

From the association table, Fisher calculated the covariance between the parental and offspring deviations. However, the term covariance had not yet been invented,[9] so he simply called it a "quadratic," written as follows:

$$p^3i^2 + 2p^2qij + pq(p+q)j^2 + 2pq^2jk + q^3k^2$$

(*ibid.*, p. 406). Since the genotypic mean is $m = Pa + 2Qd - Ra$, then $iP + 2jQ + kR = 0$. Substituting for j in Fisher's quadratic, the covariance becomes[10]

$$\frac{1}{4pq}\left(ip^2 - kq^2\right)^2 = \frac{1}{2}\beta^2.$$

Assuming the same genotypic variance α^2 for each relative (see Eq. (14.6)), the parent-offspring correlation for a given locus is thus $\beta^2/(2\alpha^2)$. Summing across loci (which are assumed not to interact) in both the numerator and denominator, the correlation becomes

$$\frac{1}{2}\left(\frac{\tau^2}{\sigma^2}\right). \tag{14.8}$$

Note that, if dominance is not assumed, then $\varepsilon^2 = 0$ in Eq. (14.7) and the correlation above becomes 1/2. Hence:

QUOT. 869. . .the effect of dominance is simply to reduce certain relationship correlations in the ratio $\frac{\tau^2}{\sigma^2}$. (*ibid.*, p. 405)

[9]The term "covariance" was first used on p. 195 of Fisher (1930c) and first defined on p. 198 of the same book.

[10]The bracketed expression on the left in the equation below is mistakenly cubed instead of squared on p. 406 of Fisher's paper (see Moran & Smith (1966, p. 15)).

By using a similar procedure, Fisher derived correlations between several other relatives. For example, he showed that the fraternal (brother–brother) correlation is

$$\frac{1}{2\sigma^2}\left(\tau^2+\frac{1}{2}\varepsilon^2\right) \tag{14.9}$$

(*ibid.*, p. 406). If dominance is not assumed, the correlation becomes 1/2, as before.

Furthermore, the reduction in correlation in parent–offspring due to dominance is

$$\frac{1}{2}-\frac{1}{2}\left(\frac{\tau^2}{\sigma^2}\right)=\frac{1}{2}\left(\frac{\varepsilon^2}{\sigma^2}\right).$$

The corresponding reduction between brothers is

$$\frac{1}{2}-\frac{\tau^2}{2\sigma^2}-\frac{\varepsilon^2}{4\sigma^2}=\frac{1}{4}\left(\frac{\varepsilon^2}{\sigma^2}\right).$$

Therefore:

QUOT. 870 The effect of dominance is to reduce the fraternal correlation to only half the extent to which the parental correlation is reduced. This allows us to distinguish, as far as the accuracy of the existing figures allows, between the random external effects of environment and those of dominance. (*ibidem*)

Other correlations obtained by Fisher were those between:

- uncle-nephew: $\tau^2/(4\sigma^2)$(*ibid.*, p. 407),
- double cousins: $\left(\tau^2+\frac{1}{4}\varepsilon^2\right)/(4\sigma^2)$(*ibidem*),
- first cousins: $\tau^2/(8\sigma^2)$ (*ibid.*, p. 408).

Now, consider the example given at the start of Sect. 14.2.3. Given a fraternal correlation of 0.54, Fisher had stated that 54% of the total variation was due to ancestry and that the remaining 46% was due to other causes. He had also claimed that the 46% could not be due to environmental effects alone. Based on the biometrical genetics model he had developed, Fisher justified this claim as follows.

Using the sib–sib correlation in Eq. (14.9), the variance within a sibship is

$$\begin{aligned}V_s&=\sigma^2\left\{1-\frac{1}{2\sigma^2}\left(\tau^2+\frac{1}{2}\varepsilon^2\right)\right\}\\&=(\tau^2+\varepsilon^2)-\frac{1}{2}\tau^2-\frac{1}{4}\varepsilon^2\\&=\frac{1}{2}\tau^2+\frac{3}{4}\varepsilon^2.\end{aligned}$$

Fisher's calculations (*ibid.*, p. 424) showed that $\tau^2/2 = 31\%$ and $3\varepsilon^2/4 = 15\%$. Since 46% of the total variation had previously been attributed to other causes, he therefore concluded:

Table 14.3 Fisher's treatment of two-locus interactions

	$B_1B_1(i')$	$B_1B_2(j')$	$B_2B_2(k')$
A_1A_1 (i)	a_{11}	a_{12}	a_{13}
A_1A_2 (j)	a_{21}	a_{22}	a_{23}
A_2A_2 (k)	a_{31}	a_{32}	a_{33}

QUOT. 871 These determinations are subject, as we have seen, to considerable errors of random sampling, but our figures are sufficient to show that, on this hypothesis, it is very unlikely that so much as 5 per cent. of the total variance is due to causes not heritable, especially as every irregularity of inheritance would, in the above analysis, appear as such a cause. (*ibidem*)

Furthermore, Fisher calculated the genotypic variance to be 83% of the total variation. Of this, 21% was due to dominance deviations; hence, the dominance ratio was $21/83 = 0.253$. Later in the paper (*ibid.*, p. 428), Fisher compared the dominance ratios (ranging from 0.25 to 0.38) he had found for various quantitative traits to Pearson's theoretical value of 1/3.

The issues of dominance and dominance ratios would be major themes in Fisher's later research.

14.2.5 Correlation Between Relatives Under Epistasis

One of the several innovations in Fisher's 1918 paper was his mathematical treatment of two-locus epistasis[11] ('dual epistacy'). He considered two unlinked loci, A and B, with alleles A_1, A_2 and B_1, B_2, and frequencies p, q and p', q', respectively (see Table 14.3).

In addition to i, j, and k, Fisher also introduced i', j', and k', all of which represented the actual contributions to the phenotypic effects of the individual genotypes at each locus. These values could be estimated if the weighted sum of squares of their deviations from the observed phenotypic effects (a_{IJ}) inside the table was minimized. That is, by writing the individual deviations as:

$$\begin{aligned}
e_{11} &= a_{11} - (i + i'), \quad e_{21} = a_{21} - (j + i'), \quad e_{31} = a_{31} - (k + i'),\\
e_{12} &= a_{12} - (i + j'), \quad e_{22} = a_{22} - (j + j'), \quad e_{32} = a_{32} - (k + j'),\\
e_{13} &= a_{13} - (i + k'), \quad e_{23} = a_{23} - (j + k'), \quad e_{33} = a_{33} - (k + k'),
\end{aligned}$$

the sum of squares to be minimized is

[11] Epistasis is an *inter-locus* interaction, as opposed to dominance, which is an *intra-locus* interaction.

$$S_1 = p^2q'^2e_{11}^2 + 2p^2p'q'e_{12}^2 + p^2q'e_{13}^2 + 2pqp'^2e_{21}^2 + 4pqp'q'e_{22}^2 + 2pqq'^2e_{23}^2 + q^2p'^2e_{31}^2 + 2q^2p'q'e_{32}^2 + q^2q'^2e_{33}^2.$$

Differentiating with respect to i, j, k, i', j', and k', the following least-square equations are obtained:

$$\begin{aligned} &p^2e_{11} + 2pqe_{21} + q^2e_{31} = 0, \quad && p'^2e_{11} + 2p'q'e_{21} + q'^2e_{31} = 0, \\ &p^2e_{12} + 2pqe_{22} + q^2e_{32} = 0, \quad && p'^2e_{12} + 2p'q'e_{22} + q'^2e_{32} = 0, \\ &p^2e_{13} + 2pqe_{23} + q^2e_{33} = 0, \quad && p'^2e_{13} + 2p'q'e_{23} + q'^2e_{33} = 0, \end{aligned} \tag{14.10}$$

(*ibid.*, p. 408). Of the six equations above, only five are independent and there are nine unknowns $e_{11}, e_{12}, \ldots, e_{33}$, so the system cannot be uniquely determined (it has $9 - 5 = 4$ *free* variables).

Now, the phenotypic effect x due to loci A and B in an offspring can be written as

$$x = x_1 + x_2 + x_3 + x'_1 + x'_2 + x'_3 + x''_4,$$

where the effects of alleles A_1, A_2 are x_1, x_2, the dominance effect at locus A is x_3, the effects of alleles B_1, B_2 are x'_1, x'_2, the dominance effect at locus B is x'_3, and the epistatic effect is x''_4.[12] These seven effects are uncorrelated, and

$$\text{var}x = \sum_{u=1}^{3} \text{var}x_u + \sum_{u=1}^{3} \text{var}x'_u + \text{var}x''_4. \tag{14.11}$$

Similarly, for the parent

$$X = X_1 + X_2 + X_3 + X'_1 + X'_2 + X'_3 + X''_4.$$

Then,

$$\text{var}X = \sum_{u=1}^{3} \text{var}X_u + \sum_{u=1}^{3} \text{var}X'_u + \text{var}X''_4 \tag{14.12}$$

and

[12] Although Fisher used a single term (x''_4) for the epistatic effect, the latter is in fact made up of four components: additive x additive, additive x dominance, dominance x additive, and dominance x dominance (e.g., see Cockerham (1954, p. 863).

$$\operatorname{cov}(x, X) = \sum_{u=1}^{3} \operatorname{cov}(x_u, X_u) + \sum_{u=1}^{3} \operatorname{cov}\left(x'_u, X'_u\right) + \operatorname{cov}\left(x''_4, X''_4\right), \tag{14.13}$$

since $\operatorname{cov}(x_I, X_J) = 0$ for $I \neq J$. In Eqs. (14.11) and (14.12), the eight variances with subscripts 1 and 2 are each half their respective additive genetic variances, and the two variances with subscript 3 are each their respective dominance variance. It remains to find $\operatorname{var}x''_4$, $\operatorname{var}X''_4$, and $\operatorname{cov}\left(x''_4, X''_4\right)$.

Fisher next returned to the association table of the type shown in Table 14.2, except that now the table had nine rows and nine columns (A_1B_1/A_1B_1, $A_1B_1/A_1B_2, \ldots, A_2B_2/A_2B_2$). He noted that, from each of the five independent equations in Eq. (14.10), one e_{IJ} could be eliminated (e.g., $e_{21} = -(p^2e_{11} + q^2e_{31})/(2pq)$ from the first equation), so that the initial 9 × 9 table can be reduced to a 4 × 4. The four deviations he retained were e_{11}, e_{13}, e_{31}, and e_{33}.

The variance $\operatorname{var}x''_4$, written in terms of the original nine e_{IJ}'s, is

$$\begin{aligned}\operatorname{var}x''_4 = {} & p'^2 \underbrace{\left(p^2e_{11}^2 + 2pqe_{21}^2 + q^2e_{31}^2\right)}_{Q_1} + 2p'q' \underbrace{\left(p^2e_{12}^2 + 2pqe_{22}^2 + q^2e_{32}^2\right)}_{Q_2} \\ & + q'^2 \underbrace{\left(p^2e_{13}^2 + 2pqe_{23}^2 + q^2e_{33}^2\right)}_{Q_3}\end{aligned} \tag{14.14}$$

and will now be reduced to an expression in terms of e_{11}, e_{13}, e_{31}, and e_{33}. Now, the first bracketed expression in Eq. (14.14) can be written as a quadratic form:

$$Q_1 = [e_{11}\ e_{21}\ e_{31}] \begin{bmatrix} p^2 & 0 & 0 \\ 0 & 2pq & 0 \\ 0 & 0 & q^2 \end{bmatrix} \begin{bmatrix} e_{11} \\ e_{21} \\ e_{31} \end{bmatrix}.$$

But since $e_{21} = -(p^2e_{11} + q^2e_{31})/(2pq)$,

$$\begin{aligned} Q_1 &= \left(p^2 + \frac{p^3}{2q}\right)e_{11}^2 + pqe_{11}e_{31} + \left(q^2 + \frac{p^3}{2q}\right)e_{31}^2 \\ &= [e_{11}\ e_{31}] \underbrace{\begin{bmatrix} \frac{p^2}{2q}(p + 2q) & \frac{1}{2}pq \\ \frac{1}{2}pq & \frac{q^2}{2p}(2p + q) \end{bmatrix}}_{\mathbf{T}(p,q)} \begin{bmatrix} e_{11} \\ e_{31} \end{bmatrix}. \end{aligned} \tag{14.15}$$

Similarly, the quadratic forms Q_2 and Q_3 in Eq. (14.14) can be simplified, resulting in

Table 14.4 Parent–offspring covariance under epistasis

Offspring	Parent e_{1n}	Parent e_{2n}	Parent e_{3n}
e_{1m}	p^3	p^2q	0
e_{2m}	p^2q	pq	pq^2
e_{3m}	0	pq^2	q^3

Table 14.5 Simplification of Table 14.4

	e_{1n}	e_{3n}
e_{1m}	$p^3/4q$	$-pq/4$
e_{3m}	$-pq/4$	$q^3/4p$

$$\text{var}x_4'' = \left(p^2 + \frac{p^3}{2q}\right)\left(p'^2e_{11}^2 + 2p'q'e_{12}^2 + q'^2e_{13}^2\right) + pq\left(p'^2e_{11}e_{31} + 2p'q'e_{12}e_{32} + q'^2e_{13}e_{33}\right) + \left(q^2 + \frac{q^3}{2p}\right)\left(p'^2e_{31}^2 + 2p'q'e_{32}^2 + q'^2e_{33}^2\right).$$

By simplifying the three quadratic forms in the above (as was done for Q_1), the variance was written in terms of four deviations (e_{11}, e_{13}, e_{31}, and e_{33}) as

$$\begin{aligned}\text{var}x_4'' = {} & \left(p^2 + \frac{p^3}{2q}\right)\left\{\left(p'^2 + \frac{p'^3}{2q'}\right)e_{11}^2 + p'q'e_{11}e_{13} + \left(q'^2 + \frac{q'^3}{2p'}\right)e_{13}^2\right\} + \\ & pq\left\{\left(p'^2 + \frac{p'^3}{2q'}\right)e_{11}e_{31} + \frac{1}{2}p'q'e_{11}e_{33} + \frac{1}{2}p'q'e_{13}e_{31} + \left(q'^2 + \frac{q'^3}{2p'}\right)e_{13}e_{33}\right\} + \\ & \left(q^2 + \frac{q^3}{2p}\right)\left\{\left(p'^2 + \frac{p'^3}{2q'}\right)e_{31}^2 + p'q'e_{31}e_{33} + \left(q'^2 + \frac{q'^3}{2p'}\right)e_{33}^2\right\}\end{aligned} \quad (14.16)$$

(*ibid.* p. 409).[13] As Moran and Smith (1966, p. 25) have indicated, going from Eqs. (14.14) to (14.16) involves taking the direct product of the matrix $\mathbf{T}(p, q)$ in Eq. (14.15) and the matrix $\mathbf{T}(p', q')$.

The variance $\text{var}X_4''$ can now be written down by interchanging p and p', and q and q' in Eq. (14.16). To find $\text{cov}\left(x_4'', X_4''\right)$, Fisher used an association table (Table 14.4) similar to Table 14.2:

Since $e_{2m} = -(p^2e_{1m} + q^2e_{3m})/(2pq)$ and $e_{2n} = -(p^2e_{1n} + q^2e_{3n})/(2pq)$, Table 14.4 simplifies to Table 14.5.

By using a similar procedure to the one used to go from (14.14) to (14.16), Fisher finally obtained

[13] This formula is given on p. 25 of Moran and Smith (1966) and corrects some of the errors on p. 409 of Fisher's paper.

$$\mathrm{cov}\left(x_4'', X_4''\right) = \frac{1}{16pqp'q'}\left(p^2p'e_{11} - p^2q'e_{13} - q^2p'e_{31} + q^2q'e_{33}\right)^2$$

(Fisher, 1918, p. 409).

Although Fisher's treatment of correlations under epistasis is nothing short of trailblazing, his method is complex and seldom used. Modern derivations can be found in Cockerham (1954) and Kempthorne (1954). For a recent lucid demonstration, see the book by Ewens (2004, pp. 75–77).

14.2.6 Infinitesimal Model of Assortative Mating

14.2.6.1 Frequency of Mating

Fisher's analysis of assortative mating[14] is based on the *infinitesimal model* assumption he had made earlier in his paper (*ibid.*, p. 402), namely, that a quantitative trait is the contribution of many Mendelian factors and is thus normally distributed. If such a trait has mean zero and variance V, then the probability that it lies in the interval $(x, x + dx)$ is

$$M = \frac{1}{\sqrt{2\pi V}}\exp\left(-\frac{x^2}{2V}\right)dx$$

and the probability that it lies in the interval $(y, y + dy)$ is

$$N = \frac{1}{\sqrt{2\pi V}}\exp\left(-\frac{y^2}{2V}\right)dy.$$

The probability of a mating between the two groups is then

$$\begin{aligned}&\frac{1}{2\pi V\sqrt{1-\mu^2}}\exp\left\{-\frac{x^2-2\mu xy+y^2}{2(1-\mu^2)V}\right\}dxdy\\ &= \frac{MN}{\sqrt{1-\mu^2}}\exp\left\{-\frac{\mu^2x^2-2\mu xy+\mu^2y^2}{2(1-\mu^2)V}\right\},\end{aligned}$$

where μ is the marital correlation. For the purpose of assortative mating, Fisher now considered two groups with variance V each, but with different means a and b in a stationary population. Their probability of mating is

[14] Assortative mating is nonrandom mating based on phenotype (as opposed to inbreeding, which is nonrandom mating between individuals who are genetically similar).

$$\iint\left[\frac{MN}{\sqrt{1-\mu^2}}\exp\left\{-\frac{\mu^2x^2-2\mu xy+\mu^2y^2}{2V(1-\mu)^2}\right\}\right.$$
$$\left.\times\frac{1}{2\pi V}\exp\left\{-\frac{(x-a)^2}{2V}-\frac{(y-b)^2}{2V}\right\}\right]dxdy \tag{14.17}$$
$$=MN\exp\left(\frac{\mu ab}{V}\right)$$

(*ibid.* p. 410).

14.2.6.2 Deviation from HWE

Fisher wrote *D*, *H*, and *R* for A_1A_1, A_1A_2, and A_2A_2, respectively, and noted that for the six possible matings ($D \times D$, $D \times H$, $D \times R$, $H \times H$, $H \times R$, $R \times R$), only $D \times R$ and $H \times H$ resulted in a change in genotypic proportions. For example, $D \times H$ resulted in *D/2* and *H/2* (same as in parents), but

$$D\times R=H,$$
$$H\times H=\frac{1}{4}D+\frac{1}{2}H+\frac{1}{4}R.$$

However, if the second mating is twice as frequent as the first one, then the genotypic frequencies will be the same in parents and offspring (i.e., equilibrium will be achieved). If *I, J, K* denote the mean trait values for the genotypes A_1A_1, A_1A_2, A_2A_2, respectively, then using Eq. (14.17), the equilibrium condition becomes

$$4Q^2e^{\mu J^2/V}=4PRe^{\mu IK/V}$$

Since the contributions of the individual loci are small, $I/\sqrt{V}, J/\sqrt{V}, K/\sqrt{V}$ are all small. Ignoring terms of third-degree smallness, the above becomes

$$PR-Q^2\approx\frac{\mu}{V}\left(Q^2J^2-PRIK\right)\approx\mu\left(\frac{Q^2}{V}\right)\left(J^2-IK\right), \tag{14.18}$$

where $PR - Q^2$ is assumed to be of second-order of smallness. Writing the derivations from HWE as δ_1, δ_2, δ_3, where $P = p^2 + \delta_1$, $Q = pq + \delta_2$ and $R = q^2 + \delta_3$, this implies $\delta_1 = \delta_3 = -\delta_2$. Substituting in Eq. (14.18),

$$\delta_1=\delta_3=-\delta_2=p^2q^2\mu\frac{J^2-IK}{V}$$

(*ibid.*, p. 411).

14.2.6.3 Increase in Additive Genetic Variance

Fisher now considered two Mendelian factors and wrote D, H, and R for A_1A_1, A_1A_2, and A_2A_2 at the first locus (as in (b) above). The corresponding quantities at the second locus were denoted by $D^{'}$, $H^{'}$, and$R^{'}$. The frequencies of the nine genotypic combinations are as follows:

$$\begin{aligned} \text{freq}(DD') &= PP'(1 + f_{11}), \\ \text{freq}(DH') &= 2PQ'(1 + f_{12}), \\ &\cdots \\ \text{freq}(RR') &= RR'(1 + f_{33}), \end{aligned} \qquad (14.19)$$

under the assumption of no epistasis. Now, the mean value I of the genotype D can be found by averaging those of $DD^{'}$, $DH^{'}$, and $DR^{'}$. The same rule applies for the values J and K of the genotype H and R, respectively, and Fisher obtained

$$\begin{aligned} I &= i + \sum (P'i'f_{11} + 2Q'j'f_{12} + R'k'f_{13}), \\ J &= j + \sum (P'i'f_{21} + 2Q'j'f_{22} + R'k'f_{23}), \\ K &= k + \sum (P'i'f_{31} + 2Q'j'f_{32} + R'k'f_{33}) \end{aligned} \qquad (14.20)$$

(*ibid*., p. 412), where i, j, k, i', j', k' are, respectively, the effects of the genotypes D, H, R, D', H', R' if the genes at other loci were fixed, and the summation extends over all other loci.

From (14.19),

$$\begin{aligned} P'f_{11} + 2Q'f_{12} + R'f_{13} = 0, &\qquad Pf_{11} + 2Qf_{21} + Rf_{31} = 0, \\ P'f_{21} + 2Q'f_{22} + R'f_{23} = 0, &\qquad Pf_{12} + 2Qf_{22} + Rf_{32} = 0, \\ P'f_{31} + 2Q'f_{32} + R'f_{33} = 0, &\qquad Pf_{13} + 2Qf_{23} + Rf_{33} = 0. \end{aligned}$$

By using the above equations in (14.20),

$$\begin{aligned} I &= i + \sum \{P'(i' - j')f_{11} - R'(j' - k'f_{13})\}, \\ K &= k + \sum \{P'(i' - j')f_{31} - R'(j' - k'f_{33})\}. \end{aligned}$$

Therefore,

$$\begin{aligned} IP - KR &= iP - kR + \sum \frac{\mu}{V}(IP - KR)(I'P' - K'R')\{p'(i' - j') + q'(j' - k')\} \\ &= iP - kR + A(IP - KR), \end{aligned}$$

where

$$A = \frac{\mu}{V} \sum (IP - KR)(I'P' - K'R')\{p'(i' - j') + q'(j' - k')\}$$

and

$$A(1-A) = \frac{\mu}{V} \sum \frac{(iP - kR)^2}{2Q} = \mu \frac{\tau^2}{V} \tag{14.21}$$

(*ibid.*, p. 413). By the infinitesimal model assumption, a quantitative trait X can be written as $\sum X_i$ where each X_i has mean zero. Furthermore, the genotypic variance can be written as

$$\begin{aligned} V &= \sum \mathrm{E}X_i^2 + \sum_{i \neq j} \mathrm{E}X_i X_j \\ &= \sum (Pi^2 + 2Qj^2 + Rk^2) + 2\sum_{i \neq j} \{PP'(1+f_{11})ii' + 2Q'P(1+f_{12})ij' + PR'(1+f_{13})ik' + \\ &\quad 2QP'(1+f_{21})ji' + 4QQ'(1+f_{22})jj' + 2QR'(1+f_{23})jk' + RP'(1+f_{31})ki' + \\ &\quad 2RQ'(1+f_{33})kj' + RR'(1+f_{33})kk'\}. \end{aligned}$$

By using Eq. (14.20), the above can be written as

$$V = \sum (PiI + 2QjK + RkK)$$

(*ibid.*, p. 412), where the summation is across all component factors. Fisher was able to simplify the above to

$$V = \sum \alpha^2 + \frac{A}{1-A} \sum (iP - kR)\{p(i-j) + q(j-k)\} = \sigma^2 + \frac{A}{1-A}\tau^2 \tag{14.22}$$

(*ibid.*, p. 414). The value of A can be obtained from the following quadratic equation:

$$VA(1-A) = \sigma^2 A(1-A) + A^2\tau^2 = \mu\tau^2 \quad \Rightarrow \quad A^2\varepsilon^2 - A\sigma^2 + \mu\tau^2 = 0,$$

where $\varepsilon^2 = \sigma^2 - \tau^2$. From Eq. (14.21), $A < 1$. Furthermore, by rewriting the above equation as $(\mu - A^2)\tau^2 = (A - A^2)\sigma^2$ gives $A < \mu$.

From Eq. (14.22), the genotypic variance under assortative variance is

$$\begin{aligned} V &= \sigma^2 + \frac{A}{1-A}\tau^2 = \sigma^2 + \frac{\tau^2}{1-A} - \tau^2. \\ \therefore\ V &= \frac{\tau^2}{1-A} + \varepsilon^2. \end{aligned} \tag{14.23}$$

Thus, the effect of assortative mating is to increase the genotypic variance through inflation of the additive component by a factor $1/(1 - A)$.

The quantity A may be interpreted as the correlation between the additive values of two mates and

$$A = \mu \frac{\tau^2}{V_P}, \tag{14.24}$$

where V_P is the phenotypic variance.

Crow and Kimura have disagreed with Fisher's conclusion that assortative mating affects only the additive variance. They have argued that the dominance variance also changes, albeit slightly:

> **QUOT. 872** According to Fisher (1918), assortative mating will increase V_g [additive variance], but not V_d [dominance variance] and V_e [environmental variance]. However, it is not strictly true, for V_d does change. But, as noted earlier, with a large number of genes there is very little change in heterozygosity under assortative mating, and therefore V_d is not expected to change very much. (Crow and Kimura, 1970, p. 156)

However, Vetta (1976b) has convincingly argued that, under the model used by Fisher, the dominance deviations of one factor are independent of both the dominance deviations and additive effects of another factor. Therefore:

> **QUOT. 873** Fisher's model of assortative mating and to the approximation used by him, the dominance variance does not change under assortative mating. (*ibid.*, p. 453)

14.2.6.4 Correlation Between Relatives Under Assortative Meeting

Fisher took the example of height as a typical quantitative trait and stated the three ways it could be measured:

> **QUOT. 874** if we call x the actual height of the individual, y what his height would have been under some standard environment, and z what his height would have been if in addition, without altering the extent to which different factors are associated, each phase is given its representative value. (*ibid.*, p. 419)

Thus, x, y, and z are the phenotypic, genotypic, and additive values, respectively. Then, since $x = y + e$, where e is the environmental effect, and y and e are independent,

- the regression of x on y has regression coefficient $\frac{\text{cov}(x,y)}{\text{var}\,y} = \frac{\text{var}\,y}{\text{var}\,y} = 1$;
- the regression of y on x has regression coefficient

$$c_1 = \frac{\text{cov}(x,y)}{\text{var}\,x} = \frac{\text{var}\,y}{\text{var}\,x} = \frac{V_G}{V_G + V_E} = \frac{V_G}{V_P},$$

where V_P, V_G, and V_E are the phenotypic, genotypic, and environmental variance, respectively.

Furthermore, writing the genotypic effect as the sum of the additive and dominance values, i.e., $y = z + d$ (where z and d are independent)

- the regression of y on z has regression coefficient $\frac{\text{cov}(y,z)}{\text{var}\,z} = \frac{\text{var}\,z}{\text{var}\,z} = 1$;
- the regression of z on y has regression coefficient (see Eq. (14.23))

$$c_2 = \frac{\text{cov}(z,y)}{\text{var}\,y} = \frac{\text{var}\,z}{\text{var}\,y} = \frac{\tau^2}{(1-A)V} = \frac{\tau^2}{\sigma^2 - A\varepsilon^2}$$

(*ibid.*, p. 420).

First, assuming that the association occurs through parental phenotypes, Eq. (14.24) implies that the correlation between the additive values of two parents is

$$A = \mu\frac{\tau^2}{V_P} = \left(\frac{\sigma^2 - A\varepsilon^2}{V_P}\right)\left(\frac{\tau^2}{\sigma^2 - A\varepsilon^2}\right)\mu = c_1c_2\mu.$$

If one parent deviates one unit from the population, the other parents will deviate by μ, and the mean deviation of the parents is $(1 + \mu)/2$. The correlation between the parent and offspring under assortative mating is then[15]

$$\frac{\tau^2}{V_P}\frac{(1+\mu)}{2} = c_1c_2\frac{(1+\mu)}{2}.$$

The grandparental–offspring correlation is the product of the above and $(1 + A)/2$, since A now replaces μ:

$$c_1c_2\frac{(1+\mu)}{2}\frac{(1+A)}{2}.$$

Similarly, the correlation between the offspring and the $(n + 1)^{\text{th}}$ ancestral parent is

$$c_1c_2\frac{(1+\mu)}{2}\frac{(1+A)^n}{2}$$

(*ibid.*, p. 421). From the above, Fisher claimed that he had shown:

> **QUOT. 875**...the Law of Ancestral Heredity as a necessary consequence of the factorial [i.e., Mendelian] mode of inheritance. (*ibidem*)

[15] Vetta (1976a, p. 317) has shown that the parent-offspring correlation below is incorrect and should be $\frac{1}{2}c_1c_2\{1 + A(1 - A)^2\} + \frac{1}{2}c_1(1 - c_2)A(1 - A)\}$.

However, Fisher had not shown that the Law of Ancestral Heredity was a consequence of Mendelian theory: what he had shown is that the geometrically decreasing correlations predicted by the ancestral law (see Sect. 4.6) could be derived from Mendelian theory. The ancestral law cannot be a consequence of Mendelian theory since the two are fundamentally incompatible: as indicated in Sect. 9.2.4(2), the ancestral law admits contributions from ancestors in addition to those of parents while Mendelian theory admits no contribution from ancestors once those of parents are known.

Next, Fisher assumed that the association was through parental genotypes. Since the correlation between parental phenotypes is μ and the regression of genotype on phenotype is c_1, the correlation between parental genotypic values must be μ/c_1. Furthermore, the correlation between the additive values of two parents is

$$A = \frac{\mu}{c_1}\frac{\tau^2}{V_G} = \frac{\mu}{c_1}c_2.$$

Reasoning as above, the parent–offspring correlation is

$$\frac{(1+\mu/c_1)}{2}\frac{\tau^2}{V_P} = \frac{(1+\mu/c_1)}{2}c_1c_2 = \frac{c_1c_2 + Ac_1}{2}$$

and the correlation between the offspring and the $(n + 1)^{\text{th}}$ancestral parent is

$$\frac{c_1c_2 + Ac_1}{2}\left(\frac{1+A}{2}\right)^n$$

(*ibidem*).

Finally, assuming that the association is though parental additive values, it can be similarly shown that $A = \mu/(c_1c_2)$ and the correlation between the offspring and the $(n + 1)^{\text{th}}$ ancestral parent is

$$c_1c_2\left(\frac{1+A}{2}\right)^{n+1}$$

(*ibidem*).

14.2.7 Weinberg's Prior Work in Quantitative Genetics (1910)

Although Weinberg's fame rests mainly on his co-discovery with Hardy of one of the most fundamental relationships in population genetics, namely, the Hardy–

Weinberg equilibrium,[16] he wrote a pioneering paper in 1910 (Weinberg, 1910, 1984) that contained some of the results in Fisher (1918). Weinberg's 1910 paper was an extension of two papers he had recently published (Weinberg, 1908b, 1909).

The aim of Weinberg's paper was to examine:

QUOT. 876....the characteristic effects certain modes of inheritance have on the composition of successive generations and on the relationships between composition, means, and correlations among various categories of relatives. (Weinberg, 1910, 1984, p. 42)

At the very start of his paper, Weinberg referred to his and Hardy's almost simultaneous discovery of the equilibrium law. He stated the law as follows. Consider a population such that a generation G_0 consists of the genotypes A_1A_1, A_1A_2, A_2A_2 with relative frequencies u, $2v$, w:

$$G_0 = uA_1A_1 + 2vA_1A_2 + wA_2A_2.$$

Then, under panmixia, "the composition of all following generations can be expressed as a second-degree binomial":

$$G_x = (u+v)^2A_1A_1 + 2(u+v)(v+w)A_1A_2 + (v+w)^2A_2A_2 = m^2A_1A_1 + 2mnA_1A_2 + n^2A_2A_2.$$

Weinberg called the mean genotypic value "the generation mean," defined as:

$$G = \sum_{x=1}^{s} n_x T_x$$

(*ibid.*, p. 44), where n_x is the relative frequency of the xth genotype and T_x is the corresponding value of the trait (i.e., the corresponding phenotypic value). He then defined the genotypic variance (see Eq. (14.6)), which he called the *endogenous variability*, as follows:

$$E^2 = \sum_{x=1}^{s} n_x (T_x - G)^2 = \sum_{x=1}^{s} n_x T_x^2 - G^2$$

(*ibid.*, p. 45). Using this definition, Weinberg obtained the parent–offspring correlation as

$$\alpha_p = \alpha_f = \frac{1}{2}\,\frac{\sum_{x=1}^{s} n_x T_x^2 - G^2 - z^2U}{E^2} = \frac{1}{2}\left(1 - \frac{z^2U}{E^2}\right) \tag{14.25}$$

[16] See Sect. 10.10.

(*ibidem*), where $U = m^2n^2(A_1A_1 - A_2A_2)^2$ is a positive constant ($A_1A_1 - A_2A_2$ being the difference in phenotypic values of the two homozygotes) and z is the degree of dominance (i.e., $z = \pm 1$ with complete dominance and $z = 0$ with no dominance). The quantity z^2U turns out to be Fisher's dominance variance δ^2 in Eq. (14.3), so that the above equation becomes

$$\alpha_p = \alpha_f = \frac{1}{2}\left(1 - \frac{\delta^2}{\alpha^2}\right) = \frac{1}{2}\left(\frac{\beta^2}{\alpha^2}\right).$$

Summing across loci, Fisher's Eq. (14.8) is obtained.

Weinberg also obtained the sib–sib correlation as

$$\alpha_c = \frac{\frac{1}{2}\left(\sum_{x=1}^{s} n_x T_x^2 + G^2\right) - \frac{z^2}{2}U - G^2}{E^2} = \frac{1}{2}\left(1 - \frac{z^2U}{2E^2}\right) \tag{14.26}$$

(*ibid.*, p. 46).[17] Setting $z^2U = \delta^2$ and $E^2 = \alpha^2$ as before, the above becomes

$$\alpha_c = \frac{1}{2}\left(1 - \frac{\delta^2}{2\alpha^2}\right) = \frac{1}{2\alpha^2}\left(\beta^2 + \frac{1}{2}\delta^2\right).$$

Summing across loci, Fisher's Eq. (14.9) is obtained.

Of final interest is Weinberg's decomposition of the phenotypic variance V^2 into a genotypic component E^2 ("endogenous component of variability") and an environmental component Z^2 ("exogenous component due to the effect of different environments"):

$$V^2 = E^2 + Z^2 \tag{14.27}$$

(*ibid.*, p. 49).

14.3 Foundational Population Genetics Paper in 1922: "On the Dominance Ratio"

14.3.1 Introduction

Fisher's foundational population genetics paper "On the Dominance Ratio" (Fisher, 1922)[18] contained many of the ideas that would constitute the backbone of his later

[17] The E^2 on the right side is missing in Weinberg's formula on p. 46.

[18] This paper has also been analyzed by Charlesworth (2022).

evolutionary views. The paper illustrates Fisher's gene-centered view of evolution and had a deep influence on several population geneticists, most notably Sewall Wright and J.B.S. Haldane, and set the agenda for future investigations.

Before the key topics in Fisher's paper are investigated, it is important to recall[19] that Fisher considered his populations to be very large so that:

> **QUOT. 877**. . .the whole investigation may be compared to the analytical treatment of the Theory of Gases, in which it is possible to make the most varied assumptions as to the accidental circumstances, and even the essential nature of the individual molecules, and yet to develop the general laws as to the behaviour of gases, leaving but a few fundamental constants to be determined by experiment. (*ibid.*, pp. 321–322)

Although this is the model Fisher usually had in mind in his treatment of evolutionary questions, the 1922 paper also contained an important section on finite populations, heralding what later came to be known as the Wright–Fisher model.

14.3.2 Balancing Selection

Although natural selection was usually thought of as acting to replace a less favorable variant by a more favorable one, Fisher demonstrated that this was not always the case and that natural selection could, in fact, lead to an equilibrium between two existing variants. H.J. Muller had previously shown the existence of balanced lethals (Muller, 1918) but Fisher should be credited as the first to explicitly derive the conditions for such an equilibrium.

Fisher considered a bi-allelic factor such that its "three phases" are in the ratio

$$P : 2Q : R$$

in any generation. Suppose the three phases are denoted by the genotypes A_1A_1, A_1A_2 and A_2A_2. Then,

$$\begin{aligned}\text{freq}(A_1) &= P + Q \equiv p,\\ \text{freq}(A_2) &= Q + R \equiv q,\end{aligned}$$

where $p + q = 1$. Now let natural selection act in such a way that the frequencies of A_1A_1, A_1A_2, A_2A_2 parents are in the ratio $aP : 2bQ : cR$.[20] Then, assuming random mating, the genotypic proportions in the next generation are

[19]See Sect. 14.1.

[20]*a, b,* and *c* represent the proportional contributions of A_1A_1, A_1A_2, A_2A_2 and are known as the (Darwinian) *fitnesses* of the corresponding genotypes. In general, if N_t is the number of individuals in generation *t*, then the Darwinian fitness *W* is implicitly defined by $N_t = WN_{t-1}$ or $N_t = W^tN_0$.

$$(aP + bQ)^2 : 2(aP + bQ)(bQ + cR) : (bQ + cR)^2.$$

Equilibrium holds when

$$\begin{aligned}(aP + bQ)^2 &= P,\\ (aP + bQ)(bQ + cR) &= Q,\\ (bQ + cR)^2 &= R\end{aligned}$$

(Fisher, 1922, p. 324). The above three equations imply $PR = Q^2$, or $P = p^2$, $Q = pq$, $R = q^2$, so that

$$\begin{aligned}p &= aP + bQ,\\ q &= bQ + cR.\end{aligned}$$

Fisher finally re-parametrized the fitnesses as follows: $a \equiv 1 + \alpha$, $b \equiv 1 + \beta$, $c \equiv 1 + \gamma$.[21] The equation $p = aP + bQ$ then becomes

$$p = (1 + \alpha)p^2 + (1 + \beta)pq \Rightarrow p\alpha + q\beta = 0 \tag{14.28}$$

and $q = bQ + cR$ becomes

$$q = (1 + \beta)pq + (1 + \gamma)q^2 \Rightarrow p\beta + q\gamma = 0. \tag{14.29}$$

By combining Eqs. (14.28) and (14.29), the conditions for equilibrium under selection become

$$\frac{\beta}{pq} = -\frac{\alpha}{q^2} = -\frac{\gamma}{p^2} \tag{14.30}$$

(*ibidem*).[22] From Eq. (14.30), $\beta > 0 \Rightarrow \alpha < 0, \gamma < 0$, i.e., $b > 1$ and $a < 1, c < 1$. Thus, one condition for equilibrium is $b > a, c$, i.e., the fitness of the heterozygote is higher than that of either homozygote. This phenomenon is now called *overdominance*.[23] Similarly, $\beta < 0 \Rightarrow \alpha > 0, \gamma > 0$, i.e., $b < 1$ and $a > 1, c > 1$. Thus, an alternative condition for equilibrium is $b < a, c$, i.e., the fitness of the heterozygote is lower than that of either homozygote (*underdominance*).

---

[21] $-\alpha$, $-\beta$, and $-\gamma$ are now known as *selection coefficients.*

[22] Fisher's equilibrium equations on p. 324 wrongly have the p^2 and q^2 in the denominators interchanged.

[23] The concept of *overdominance* goes back to the independent works of Shull (1908) and East (1908). The term overdominance was first used by Hull (1946).

Fisher then declared that overdominance corresponds to a stable equilibrium and underdominance to an unstable equilibrium:

> **QUOT. 878** If selection favours the homozygotes, no stable equilibrium will be possible, and selection will then tend to eliminate whichever gene is below its equilibrium proportion; such factors will therefore not commonly be found in nature: if, on the other hand, the selection favours the heterozygote, there is a condition of stable equilibrium, and the factor will continue in the stock. (*ibidem*)

The above statement was made without any proof but was in accordance with Fisher's observations (*ibid.*, p. 323) from Muller's 1918 paper.

14.3.3 The Survival of an Initial Mutation

Here, Fisher used the powerful method of probability generating functions (PGF) to investigate the probability that a particular number of copies of a parental mutant gene are transmitted to future generations. For example, if Y denotes the number of copies of an initial mutation in the next generation, then the PGF of Y is

$$f(t) \equiv \mathrm{E}t^Y \equiv p_0 + p_1 t + p_2 t^2 + \ldots,$$

where $p_i \equiv \Pr\{Y = i\}$ $(i = 0, 1, 2, \ldots)$ is the coefficient of t^i in $f(t)$.

If Y_1 and Y_2 denote the number of copies in the next generation of two initial mutations, then the PGF of Y_1+Y_2 is

$$\mathrm{E}t^{Y_1+Y_2} = \mathrm{E}t^{Y_1}t^{Y_2} = \mathrm{E}t^{Y_1}\mathrm{E}t^{Y_2} = \{f(t)\}^2,$$

where Y_1 and Y_2 are assumed to be independent and identically distributed, and the probability of i total copies in the next generation of the two initial mutations is the coefficient of t^iin $\{f(t)\}^2$.

Finally, suppose the number of copies of an initial mutation in the next generation is n. Then, by the above, the PGF of the number of copies in the third generation of the initial mutation is $\{f(t)\}^n$. The unconditional PGF is, by the definition of $f(t)$,

$$\sum_{n=0}^{\infty} \{f(t)\}^n \Pr\{Y = n\} = f(f(t)) \tag{14.31}$$

and the probability of i total copies in the third generation of the initial gene is the coefficient of t^i in$f(f(t))$.[24]

[24] Fisher was, in effect, using the Galton—Watson process (Watson and Galton, 1875), a key result of which had been announced before by Bienaymé (1845). See Heyde and Seneta (1972).

Fisher next took the specific example of a species of constant size, i.e., each parent pair produces on average two offspring that can procreate. In terms of an initial gene, the mean number of copies (m) in the next generation is one. This corresponds to a neutral mutant (no natural selection). Fisher assumed that the actual number of copies (Y) has a Poisson distribution, i.e., $Y \sim \text{Po}(1)$ and

$$f(t) = \sum_{y=0}^{\infty} \frac{t^y e^{-1}}{y!} = e^{t-1}.$$

Using the formalism thus developed, Fisher stated one of the most important results of his paper:

> **QUOT. 879** While it is rare, its [a mutation's] survival will be at the mercy of chance, even if it is well fitted to survive...
>
> ...it may be seen that only about 2 per cent. will survive 100 generations...
> (*ibid.*, p. 326)

Fisher's survival probability of 0.02 may be obtained as follows. In analogy with Eq. (14.31), the PGF of the number of copies of an initial mutation in the 100th generation is the 98th iteration of $f(t)$:

$$f_{100}(t) = \underbrace{f(f.\ldots(f(t)))}_{99\,f's}$$

Using the fact that: (i) $f_i(0)$ is the probability of zero copies of an initial mutation in generation i, (ii) $f_i(t) = f(f_{i-1}(t))$, and (iii) $f(t) = e^{t-1}$, the extinction probabilities can be recursively calculated as follows:

$$\begin{aligned} f_2(0) &= f(0) = e^{-1} = .3679, \\ f_3(0) &= f(f_2(0)) = f(.3679) = e^{.3679-1} = .5315, \\ &\ldots \\ f_{100}(0) &= f(f_{99}(0)) = f(.9803) = e^{.9803-1} = .9805. \end{aligned}$$

The survival probability in the 100th generation of an initial mutation is thus $1 - .9805 \approx .0195$. Similarly, if $m = 1.1$, the survival probability in the 100th generation is 0.1761.

It is remarkable that Fisher's results are independent of population size. They show that, for rare neutral or nearly neutral mutations, random fluctuations in gene frequencies[25] may be a powerful evolutionary force even in large populations.

Lest it be thought from QUOT. 879 that Fisher was emphasizing the superior evolutionary power of random fluctuations over natural selection even in large populations, he was quick to add:

[25] Later called (genetic) drift by Wright (1929c, p. 559).

Table 14.6 Survival probability in the 100th generation of an initial mutation with selection advantage $m - 1$

Selective advantage $(m - 1)$	0.0	0.1	0.2	0.5	0.75	1.0
Survival probability	0.0195	0.1762	0.3137	0.5828	0.7127	0.7968

QUOT. 880 Only when the number of individuals affected becomes large will the effect of selection predominate over that of random survival, though even then only a very small minority of the population may be affected. (*ibidem*)

For example, if $m = 1.5$, the PGF is $f(t) = e^{1.5(x - 1)}$ and $f_{100}(0) = .5828$, i.e., the survival probability has jumped from 0.0195 to 0.5828. If $m = 2.0$, the survival probability is 0.7968 (see Table 14.6).

14.3.4 Diffusion Equation for Neutral Alleles in a Finite Population

14.3.4.1 The Wright–Fisher Model *Avant La Lettre*

Fisher used what later came to be known as the Wright–Fisher model to obtain a diffusion equation for the distribution of neutral alleles under genetic drift. In the simplest version of the Wright–Fisher model, a diploid population of constant size is made up of alleles of two types, A_1 and A_2, such that the number of alleles of interest, say A_1, in the next generation is obtained by binomial sampling with replacement from the previous one, i.e., if $X_T$is the number of A_1 alleles in generation T (=1, 2, 3,...) and n is the population size, then

$$\Pr\{X_{T+1} = j | X_T = i\} = \binom{2n}{j}\left(\frac{i}{2n}\right)^j\left(1 - \frac{i}{2n}\right)^{2n-j} \quad i, j = 0, 1, 2, \ldots, 2N. \quad (14.32)$$

Although this formula was not given in Fisher's paper, a binomial distribution was implied, since Fisher wrote:

QUOT. 881...if a second generation of n individuals be now formed at random, the standard departure of p from its previous value will be

$$\sigma_p = \sqrt{\frac{pq}{2n}}$$

(*ibid*., p. 327)

To remove the dependence of the standard deviation on p, Fisher used the following transformation:

$$\cos\theta = 1 - 2p, \qquad 0 \leq \theta \leq \pi, \tag{14.33}$$

so that

$$\sigma_\theta = \sqrt{\frac{pq}{2n}}\frac{d\theta}{dp} = \sqrt{\frac{pq}{2n}}\cdot\frac{2}{\sin\theta} = \frac{1}{\sqrt{2n}}.$$

14.3.4.2 Fisher's Diffusion Equation and The Error in It

Fisher next obtained a diffusion equation for the density of θ. As his "derivation" was done in two lines, an attempt will be made here to fill in the missing steps.[26] If the density of θ in any given generation is $y(\theta)$, then the density in the next generation is

$$y + \delta y = \int_0^\pi y(\theta - \delta\theta) f(\theta - \delta\theta, \theta) d(\delta\theta), \tag{14.34}$$

where $f(\theta - \delta\theta, \theta)$ is the density of a change from $\theta - \delta\theta$ to θ in one generation. Fisher assumed

$$\delta\theta \sim N(M_{\delta\theta}, V_{\delta\theta}) \quad \text{where} \quad M_{\delta\theta} = 0, V_{\delta\theta} = \sigma_\theta^2 = \frac{1}{2n}. \tag{14.35}$$

Equation (14.34) becomes

$$\begin{aligned} y + \delta y &= \int_0^\pi \left\{ yf - \delta\theta \frac{\partial}{\partial\theta}(yf) + \frac{(\delta\theta)^2}{2}\frac{\partial^2}{\partial\theta^2}(yf) + \ldots \right\} d(\delta\theta) \\ &= y - \frac{\partial}{\partial\theta}\left\{ y \int_0^\pi (\delta\theta) f d(\delta\theta) \right\} + \frac{1}{2}\frac{\partial^2}{\partial\theta^2}\left\{ y \int_0^\pi (\delta\theta)^2 f d(\delta\theta) \right\} + \ldots \end{aligned}$$

By ignoring terms containing $(\delta\theta)^3$ and higher powers,

[26] For an excellent introduction to diffusion theory in population, see Kimura (1964) and Gale (1990, Chaps. 8, 9, and 10). The standard modern treatment can be found in Ewens (2004, Chaps. 4 and 5).

$$\frac{\partial y}{\partial T} = -\frac{\partial (yM_{\delta\theta})}{\partial \theta} + \frac{1}{2}\frac{\partial^2 (yV_{\delta\theta})}{\partial \theta^2}. \tag{14.36}$$

Using the means and variances in Eq. (14.35), Fisher thus obtained the following diffusion equation for $y(\theta, T)$, i.e., for the density of θ in generation T:

$$\frac{\partial y}{\partial T} = \frac{1}{4n}\frac{\partial^2 y}{\partial \theta^2} \tag{14.37}$$

(*ibid.*, p. 327). The solution is

$$y = A_0 e^{-T/(4n)} \sin\theta \tag{14.38}$$

(*ibidem*). The diffusion approach represents a trailblazing attempt by Fisher to model the process of evolution under genetic drift. The only problem is that Fisher's Eq. (14.37), and hence his Eq. (14.38), suffer from a key error related to the zero mean assumption ($M_{\delta\theta} = 0$) in Eq. (14.35).

The whole problem came to light after Fisher first met Sewall Wright in the USA in 1924 and expressed his interest in Wright's 1921 papers on path coefficients (Wright, 1921b, c). Later, Fisher sent his paper on the dominance ratio to Wright. Now, as will be explained in the next section, Fisher had used his Eq. (14.38) to argue that the rate of decay of mutations was proportional to $1/(4n)$. However, using the method of path coefficients, Wright independently obtained a rate of decay proportional to $1/(2n)$. This was developed in his big paper on population genetics (which was published in 1931[27]). The discrepancy was communicated to Fisher in a July 24, 1929, letter:

> **QUOT. 882** I have a manuscript on the subject which was stimulated by your extremely interesting 1922 paper which I studied carefully at the time when you sent me a reprint shortly after your return from this country [summer 1924]. A discrepancy between your result for the rate of decay ($1/4N$) of factor frequency (no mutation) and the figure for the decrease of heterozygosis by my path coefficient method ($1/2N$) led me to look for the cause. I reached the somewhat different formula for factor frequency which I mentioned. It is of course a matter of two approximate formulae. I am not wholly clear yet as to the cause of the discrepancy. I would like very much to get your comment on it. I will try to send a copy in a few days[28]. (Provine, 1989, p. 253)

Fisher acknowledged his error in an October 15, 1929, letter to Wright (Bennett, 1983, p. 274) and gave the correct version of (14.37), as next explained.

[27] "Evolution in Mendelian Populations" (Wright, 1931, pp. 108–109). See Sect. 15.5.2.

[28] A copy was sent to Fisher on August 13, 1929.

14.3.4.3 Correction of the Error in Fisher's Diffusion Equation (Fisher, 1930)

The correction of Eq. (14.37) was first published in a 1930 paper (Fisher, 1930a).[29] The essential observation was that since $1 - 2p = \cos\theta$,

$$-2\delta p = -\delta\theta\sin\theta - \frac{(\delta\theta)^2}{2}\cos\theta + \ldots$$

Therefore, the term involving $(\delta\theta)^2$ *cannot* be ignored in the calculation of $M_{\delta\theta}$ because, like the variance $V_{\delta\theta}$, it is of the order $1/n$. Therefore, the above becomes

$$\begin{aligned} \delta\theta\sin\theta &\approx 2\delta p - \frac{(\delta\theta)^2}{2}\cos\theta \\ \delta\theta &\approx \frac{2\delta p}{\sin\theta} - \frac{(\delta\theta)^2}{2}\cot\theta \\ &= \frac{\delta p}{\sqrt{pq}} - \frac{(\delta p)^2}{2pq}\cot\theta \\ \therefore\quad M_{\delta\theta} &= 0 - \frac{pq/(2n)}{2pq}\cot\theta = -\frac{\cot\theta}{4n} \end{aligned} \tag{14.39}$$

(*ibid*., p. 205). Eq. (14.36) then becomes

$$\frac{\partial y}{\partial T} = \frac{1}{4n}\left\{\frac{\partial(y\cot\theta)}{\partial\theta} + \frac{\partial^2 y}{\partial\theta^2}\right\} \tag{14.40}$$

(*ibidem*). The above corrects the erroneous Eq. (14.37). It is known as the Fokker–Planck equation in physics (Fokker, 1914; Planck, 1917). In mathematics, it is called a Kolmogorov forward equation (Kolmogorov, 1931). It is also worth noting that modern writers have not found it necessary to perform the angular transformation $\cos\theta = 1 - 2p$[30] and have written the diffusion equation directly as

$$\frac{\partial y}{\partial T} = -\frac{\partial(yM_{\delta p})}{\partial p} + \frac{1}{2}\frac{\partial^2(yV_{\delta p})}{\partial p^2},$$

where

---

[29] It was also included in Fisher's fundamental book *The General Theory of Natural Selection* (Fisher, 1930c, pp. 87–88).

[30] The angular transformation is especially useful for data on drift over a relatively short time or when $p \approx 0.5$. See Kimura (1964, pp. 200–202) for more details.

$$M_{\delta p} = 0 \quad \text{and} \quad V_{\delta p} = \frac{p(1-p)}{2n}$$

so that

$$\frac{\partial y}{\partial T} = \frac{1}{4n} \frac{\partial^2}{\partial p^2} \{p(1-p)y\}. \tag{14.41}$$

The derivation shows that the normality assumption in Eq. (14.35) is unnecessary.

Having obtained the correct diffusion equation in Eq. (14.40), Fisher gave its solution as follows:

$$y(\theta, T) = A \sin\theta = A_0 e^{-T/(2n)} \sin\theta \tag{14.42}$$

(Fisher, 1930a, p. 206). This gives the correct rate of decay of mutations as $1/(2n)$, as opposed to Eq. (14.38), which gives an incorrect rate of $1/(4n)$.[31]

14.3.4.4 The Hagedoorn Effect

An important component of Fisher's 1922 paper (Fisher, 1922) was that the rate of decay of mutations due to genetic drift was proportional to $1/n$. This is because the Dutch geneticists Arend Lourens Hagedoorn (1880–1953) and his wife Anna Cornelia Hagedoorn-Vorstheuvel La Brand had published a book in 1921 where they underplayed the role of natural selection compared to chance effects in determining the level of genetic variation in a population:

> **QUOT. 883** For Darwin, natural selection was the only cause for specific stability. If a species, by continual selection, were wrought up to the point where it [w]as perfectly adapted to its environment, natural selection would keep it there. Darwin did not think that ever a group of organisms could be really pure, really stable, in such a way, that in new surroundings it would not be able to change and adapt itself to them. It appeared to him, that only continual selection could make a species pure, and only continuous selection would keep it pure. At present we know, that all closed groups of organisms, groups which are in some way protected from admixture, speedily become stable automatically. The total geno-variability of every group tends to diminish even in the absence of every selection. This makes it unnecessary to assume that every quality for which we find a group stable, must be necessarily useful, and have selection-value. (*ibid.* 129–130)

Fisher was, of course, a staunch Darwinist, and he responded vigorously to the Hagedoorns' book. In a 1921 review, he wrote:

> **QUOT. 884** A second point, which is much emphasised, is that the authors believe (ii) that the random selection of individuals of one generation to become the parents of the next generation is a more important factor in reducing the variability of a species than is the

[31] However, Fisher's solution in Eq. (14.42) is not the most general one. The latter was first obtained for the transformed diffusion equation in Eq. (14.41) by Kimura (1964, p. 194, Eq. (4.9)).

natural selection of advantageous characters, while they also believe that crossing is a far more important factor in increasing variability is mutation. The whole question is worthy of a thorough discussion, but the authors evidently lack the statistical knowledge necessary for its adequate treatment. By taking simple examples of species of *only two or three individuals* it is easy to show that by random selection the variability will be rapidly reduced. It by no means follows that the reduction will be equally rapid in an interbreeding group of 2 or 3 million individuals. The present reviewer has examined this particular point, and finds that in the absence of mutation or crossing, a perceptible reduction in variability will take place in a number of generations equal to four times the number of the species breeding...

The number of robins in a county even in winter must be many thousands, and in a thousand generations interbreeding will have taken place to an extent which will have spread the blood of any local group over the whole country. In cases of this kind, therefore, the Hagedoorn effect (the reduction of variability by random selection) will be very slow in its action; while the effect of selection will be as rapid for a large population as for a small one. In a population of a million, for example, it will require four million generations for random selection to reduce the variability of a group by a certain proportion, but even so minute an advantage as 1 per cent. in the rate of increase of a variety will in a thousand generations change it from a condition of great rarity to one of complete predominance. (Fisher, 1921, pp. 467–468)

The factor of $4n$ referred to at the end of the first paragraph in the above quotation is a consequence of Fisher's erroneous Eq. (14.37), as explained before, and should have been $2n$. Even then, it shows the weak power of the "Hagedoorn effect"[32] compared to natural selection in a large population.

14.3.5 Equilibrium Under Rare Mutations

In the 1922 paper (Fisher, 1922), Fisher also investigated the stationary distribution under the opposing effects of genetic drift and new mutations. Fisher did not give the details, but this involves the use of the probability flux[33] $\Psi(\theta, T)$, which is the rate of flow of probability (from left to right) across the point θ at time T, such that

$$\frac{\partial}{\partial\theta}\Psi(\theta,T) = -\frac{\partial}{\partial T}y(\theta,T) \tag{14.43}$$

and Eq. (14.36) gives the probability flux as

$$\Psi(\theta,T) = yM_{\delta\theta} - \frac{1}{2}\frac{\partial}{\partial\theta}(yV_{\delta\theta}). \tag{14.44}$$

Using the incorrect mean in Eq. (14.35), the above becomes

[32] Later called (genetic) drift by Wright (1931, p. 249).

[33] See Watterson (1962) for a formal introduction to the concept of probability flux, but the idea was already known to Fisher, e.g. (Fisher, 1930a, p. 205). An accessible introduction can be found in Gale (1990, Chap. 8).

$$\Psi(\theta, T) = -\frac{1}{4n}\frac{\partial y}{\partial \theta}. \tag{14.45}$$

When the distribution is stable, $\Psi(\theta, T) = 0$ so that $y(\theta, T) = C_1$, a constant. Since $\int_0^\pi y dy = 1$, we have $C_1 = 1/\pi$. Fisher wrote the symmetric stationary solution as

$$y(\theta, T) = \frac{A}{\pi}, \tag{14.46}$$

where A is the number of "factors" (heterozygous loci). The uniform distribution above is wrong since it is based on the incorrect (14.37). In the 1930 paper (Fisher, 1930a), to obtain the *symmetric* stationary solution, Eq. (14.44) was corrected to (using the correct mean in Eq. (14.39)):

$$0 = \frac{y \cot \theta}{4n} + \frac{1}{2}\frac{\partial}{\partial \theta}\left(y \frac{1}{2n}\right)$$

giving $dy/d\theta + y \cot \theta = 0$ with solution $y = A \operatorname{cosec} \theta$ (*ibid.*, p. 206). In terms of p, the stationary solution is

$$\begin{aligned} y(p, T) &= y(\theta, T)\left|\frac{d\theta}{dp}\right| \\ &= A \cos \mathrm{ec}\theta . \frac{2}{\sin \theta} \\ &= \frac{A}{2} p^{-1}(1-p)^{-1}, \qquad 0 < p < 1 \end{aligned} \tag{14.47}$$

14.3.6 Equilibrium Under Uniform Genetic Selection and Uniform Genotypic Selection

Using the fitnesses a, b, and c for the genotypes A_1A_1, A_1A_2, and A_2A_2 as in Sect. 14.3.2, Fisher defined uniform genetic selection to be selection such that a, b, and c are in geometric progression. The common ratio is $r = b/a = c/b$ and b is the geometric mean of a and c (i.e., by $b = \sqrt{ac}$). Note that this corresponds neither to overdominance or underdominance (see Sect. 14.3.2), in which cases equilibrium could be found. Under random mating, the ratio of the frequencies of A_1 and A_2 in generation n as a function of the ratio initially is

$$\frac{p}{q} = r^n \frac{p_0}{q_0}$$

(Fisher, 1922, p. 331). Fisher considered the case when $r \approx 1$ and defined $\alpha = \log_e r$ as a measure of the strength of selection. To obtain the equilibrium distribution from

Eq. (14.44), he assumed $V_{\delta\theta} = 1/(2n) + (\sigma_\alpha^2 \sin^2\theta)/4$ (the first term being due to drift, the second due to selection, σ_α^2 is the mean square value of α for different factors) and $M_{\delta\theta} = 0$. However, his $M_{\delta\theta} = 0$ assumption is again wrong since, in fact, $M_{\delta\theta} = \frac{\alpha}{4}\sin^2\theta$ (or $M_{\delta p} = \alpha p(1-p)$) and Fisher's equilibrium solution on p. 332 of his book is erroneous.

In spite of errors in his diffusion equation, Fisher's overall conclusion was correct:

> **QUOT. 885** The existence of even the slightest selection is in large populations of more influence in keeping variability in check than random survival. (*ibid.*, p. 334)

Fisher also considered the case when the fitnesses a, b, and c for the genotypes A_1A_1, A_1A_2 and A_2A_2 are such that $a = b \neq c$. After one generation of natural selection, the initial freq(A_1) : freq(A_2) ratio of p: q becomes

$$\frac{p}{q}\frac{a}{ap + cq} = \frac{p}{q}\frac{1+\beta}{1+p\beta},$$

where $\beta = a/c - 1$ (*ibid.*, p. 427). Fisher considered the case $\beta \approx 0$, so that

$$\frac{p}{q}\frac{1+\beta}{1+p\beta} = \frac{p}{q}\frac{1+\beta}{1+\beta-q\beta} = \frac{p}{q}\left(1 - \frac{q\beta}{1+\beta}\right)^{-1} \approx \frac{p}{q}(1+q\beta),$$

so that the strength of selection here is $\alpha = q\beta$. The variance is

$$V_{\delta\theta} = 1/(2n) + \sin^2(\theta/2)\cos^6(\theta/w)\sigma_\beta^2,$$

where σ_β^2 is the mean square value of β for different factors. Again, Fisher's assumption of $M_{\delta\theta} = 0$, and consequently his equilibrium distribution on p. 335, are both incorrect (in fact, $M_{\delta\theta} = \beta\sin^2\theta(1 + \cos\theta)/8$).

14.3.7 The Dominance Ratio

Fisher attached considerable importance to the dominance ratio.[34] To him, its relatively high value was evidence that natural selection was more important than chance effects in limiting the variability of a species:

> **QUOT. 886** The decay in the variance of a species breeding at random without selection, and without mutation, is almost inconceivably slow: a moderate supply of fresh mutations will be sufficient to maintain the variability. When selection is at work even to the most trifling extent, the new mutations must be much more numerous in order to maintain equilibrium. That such is the actual state of the case in mankind may be inferred from the fact that the frequency distribution of the numerical proportion of the allelomorphs,

[34] See end of Sect. 14.2.4.

calculated on the assumption of selection maintained in equilibrium by occasional mutation, leads to the value of the Dominance Ratio which is actually observed. (Fisher, 1922, pp. 323–324)

Thereupon, Fisher calculated the dominance ratio for each of the evolutionary scenarios he had considered in his paper.

For example, for the Hagedoorn model (i.e., no selection or mutation, only genetic drift, see Sect. 14.3.4), Fisher wrote the density of θ in generation T (see Eq. 14.38) as

$$y(\theta, T) = \frac{1}{2} A \sin \theta, \quad 0 \leq \theta \leq \pi \tag{14.48}$$

(*ibid.*, p. 336), where $A = 2A_0 e^{-T/(4n)}$. As noted before, the expression for A is incorrect (the correct one is given in Eq. (14.42)), but the calculations below show that the value of A cancels out so that Fisher was still able to obtain the correct dominance ratio.

Fisher set $\phi \equiv \theta/2$ so that, from Eq. (14.33), $p = \sin^2\phi$. Using the density in Eq. (14.48) and the formulas for δ^2 and β^2 under random mating (see footnotes 7 and 8),

$$\begin{aligned}
\varepsilon^2 &= \sum \delta^2 = 8A\overline{d^2} \int_0^{\pi/2} \sin^5\phi \cos^5\phi d\phi, \\
\sigma^2 &= \sum \alpha^2 = \sum (\delta^2 + \beta^2) = 8A\overline{d^2} \int_0^{\pi/2} (\sin^5\phi \cos^5\phi + 2\sin^3\phi \cos^7\phi) d\phi.
\end{aligned}$$

Therefore, the dominance ratio is

$$\frac{\varepsilon^2}{\sigma^2} = \frac{8A\overline{d^2} \int_0^{\pi/2} \sin^5\phi \cos^5\phi d\phi}{8A\overline{d^2} \int_0^{\pi/2} (\sin^5\phi \cos^5\phi + 2\sin^3\phi \cos^7\phi) d\phi} = \frac{1/60}{1/60 + 2(1/40)} = .25$$

(*ibidem*). This value was later confirmed by Wright (1931, p. 18) by using a different approach.

For the case of equilibrium between genetic drift and rare mutations (see Sect. 14.3.5), Fisher used the incorrect stationary distribution in Eq. (14.46), namely

$$y(\theta, T) = \frac{A}{\pi}.$$

Proceeding as above, he calculated the dominance ratio to be

$$\frac{\varepsilon^2}{\sigma^2} = \frac{\int_0^{\pi/2} \sin^4\phi \cos^4\phi d\phi}{\int_0^{\pi/2} \left(\sin^4\phi \cos^4\phi + 2\sin^2\phi \cos^6\phi\right) d\phi} = \frac{\frac{3}{256}}{\frac{3}{256} + 2\left(\frac{5}{256}\right)} = 0.2308$$

(*ibid*., p. 337). The correct value was later obtained by Wright (1931, p. 36) as 1/5.

For the case of equilibrium between genetic drift and unform genetic selection (see Sect.14.3.6), Fisher obtained $\varepsilon^2/\sigma^2 = 0.2$. For uniform genotypic selection, $\varepsilon^2/\sigma^2 = 1/3$.

Based on his calculations, Fisher concluded:

QUOT. 887 In the light of the above discussion in which we have deduced the distribution of allelomorphic ratios from the conditions of equilibrium with selective influences, from which condition it is probable that natural species do not widely depart, we find that the value $^1/_3$ for the dominance ratio is produced by the asymmetry of the distribution, and in such a manner as to be independent of the activity of the selective agencies, provided that this exceeds a certain very low level. When differential survival to the extent of only about 1 per cent. in a generation affects the different Mendelian factors, in a population of only a million, and far more for more powerful selection, or a larger population, the dominance ratio will be very close to its characteristic value of $^1/_3$. (*ibid*., p. 338)

As mentioned before, Fisher's "On the Dominance Ratio" was a groundbreaking paper. However, in retrospect, its fame does not lie as much in Fisher's preoccupation with the dominance ratio than with the innovations he made with respect to balancing selection, genetic drift, and diffusion theory.

14.4 Fisher's Theory of the Evolution of Dominance (1928): First Controversy with Wright[35]

14.4.1 Fisher's Theory of Dominance

Fisher first clashed with Wright on the issue of dominance.[36] The phenomenon had been known well before Mendel's time, but it was the latter who first delineated it most clearly.[37] Mendel's definition correctly characterized dominance as a property of the trait (phenotype) rather than that of the factor (allele).[38] A first theory on the

[35]The controversy around the evolution of dominance is also discussed in Sheppard (1975, Chap. 9), Sved & Mayo (1970, pp. 289–316), Ewens (1965a, 1965b, 1967), Gale (1980, Chap. 9), Cornish-Bowden & Nanjundiah (2006), Charlesworth (1979), Billiard & Castric (2011), and Bourguet (1999).

[36]See footnote 9 of chap. 3 on the first usage of 'dominance'.

[37]See QUOT. 408.

[38]Although, for a given trait in a given environment, the alleles themselves are often called dominant and recessive.

nature of dominance, the so-called "presence-absence" theory, was put forward by Bateson in 1909:

QUOT. 888 All observations point to a conclusion of great importance, namely that a dominant character is the condition due to the *presence*[39] of a definite factor, while the corresponding recessive owes its condition to the *absence* of the same factor. (Bateson, 1909, pp. 53–54)

The "presence-absence" theory was soon found to be deficient, the main reasons behind which have been nicely summarized by Merrell as follows:

QUOT. 889 In a system of multiple alleles, the intermediate alleles posed a problem; also, known deficiencies like Haplo-IV and Notch, which should have been recessive under this theory, were dominant. The discovery of reverse mutations from recessive to dominant meant that nothing was capable of mutating to something. Finally, if evolution occurs by the gradual substitution of mutants for existing wild-type genes, and the mutants represent losses, then evolution must occur as a series of losses from the ancestral genotypes, which seems highly improbable. (Merrell, 1981, pp. 92–93)

In 1928, drawing on Castle's work on modifying factors[40] and his own recent work on mimicry (Fisher, 1927a), Fisher proposed an evolutionary theory for the origin of dominance (Fisher, 1928a).[41] Fisher first noted that, since new mutations were often deleterious, this posed a problem for the theory of evolution, which works through the gradual replacement of wild-types by mutants. He argued that this problem loses its importance if, as a response to a new mutant, the wild-type *evolved to become dominant* through the action of modifiers on the heterozygotes. Thus, according to Fisher's theory, a new mutation is not completely recessive and the fitness of the heterozygote *Aa* (*A*: wild-type, *a*: mutant) is intermediate between that of the two homozygotes *AA* and *aa*; the tendency for natural selection to eliminate *Aa* is countered by the action of those modifying genes that gradually increase the fitness of *Aa* to that of *AA*; and if the mutation *a* is recurrent, in the course of time natural selection favors a constellation of modifiers that make *A* dominant to *a*.

A key assumption in Fisher's theory is that modifiers act almost exclusively on the heterozygotes *Aa* rather than the homozygotes *aa* since the latter are much less frequent than the former:

QUOT. 890...if the heterozygote is at any appreciable disadvantage compared to the wild type, it will be so enormously more frequent than the homozygote that any selection of modifiers which is in progress will be determined by the reaction of the heterozygote[42]. (*ibid.*, p. 120)

Fisher backed this claim by giving the equilibrium frequency (under the opposite forces of mutation and natural selection) of the mutant *a* as

---

[39] All italics are Bateson's.

[40] See Sect. 13.2.3, especially QUOT. 827.

[41] Interestingly, 1 year earlier, Wright also gave examples in the guinea pig where dominance at one locus was affected by other loci (Wright, 1927).

[42] Italics are Fisher's.

$$p = \frac{k}{1 - v}, \tag{14.49}$$

where k is the mutation rate (typically ~10^{-6}) and v is the viability of Aa compared to aa. Then[43]

$$\frac{\text{freq}(Aa)}{\text{freq}(aa)} = \frac{2(1-p)p}{p^2} \approx \frac{2}{k}, \tag{14.50}$$

a very large number. Fisher concluded his paper as follows:

> **QUOT. 891** It may be calculated that with mutation rates of the order of one in a million the corresponding selection in the state of nature, though extremely slow, can not safely be neglected in the case of the heterozygotes. (*ibid.*, p. 128)

Fisher's paper was soon followed by a second one (Fisher, 1928b) that cited the *Crinkled Dwarf* mutant, found in the Sea Island cottons, and poultry as examples that supported his theory.

14.4.2 Wright's Reaction (1929) to Fisher's Theory of Dominance: Selective Advantage of Modifier Gene Is Too Small, Wright's Physiological Theory

Wright (1929b) promptly reacted to Fisher's theory of dominance and took exception to the latter's main conclusion, namely, that selection was effective in promoting modifiers that make the wild-type dominant. Wright expressed his disagreement with Fisher as follows:

> **QUOT. 892** Fisher does not present his calculation of the selection rate due to this cause. The present writer reaches a figure so "extremely slow" as to make its efficacy seem highly questionable. (*ibid.*, p. 275)

Wright supported his conclusion through the bi-allelic model shown in Table 14.7, which assumes a random mating population. The quantities q, s, and h represent the frequency of the allele A, the selection coefficient against aa, and the degree of dominance, respectively.

The frequencies of the three genotypes after selection are given by the array

$$\frac{q^2 AA + 2(1-hs)q(1-q)Aa + (1-s)(1-q)^2 aa}{1 - 2hsq(1-q) - s(1-q)^2}.$$

The frequency of A after selection is

[43] For complete dominance of the A allele, Fisher also gave $p = \sqrt{k/(1-w)}$ (p. 123). Both formulas were later proved by Wright, see Sect. 14.4.2.

Table 14.7 Wright's model for the evolution of dominance

	AA	*Aa*	*aa*
Frequency	q^2	$2q(1-q)$	$(1-q)^2$
Fitness	1	$1-hs$	$1-s$

$$q_1 = \frac{q^2 + (1-hs)q(1-q)}{1-2hsq(1-q)-s(1-q)^2} = q\left\{\frac{1-hs(1-q)}{1-2hsq(1-q)-s(1-q)^2}\right\} \quad (14.51)$$

(*ibidem*). Because of the higher fitnesses of *AA* and *Aa*, $q_1 > q$ but this increase is counterbalanced by the recurrent mutation of *A* to *a* (at a rate u), until an equilibrium is reached. Using $q_1 - q = uq_1$ and Eq. (14.51), the equilibrium frequency of *A* is

$$q \approx 1 - \frac{u}{hs} \quad \text{for} \quad h > \sqrt{\frac{u}{s}}, \quad (14.52)$$

which is the same as Fisher's Eq. (14.49). In the case of complete dominance ($h = 0$), the equilibrium frequency becomes

$$q \approx 1 - \sqrt{\frac{u}{s}} \quad \text{for} \quad h = 0. \quad (14.53)$$

Wright next introduced a modifier allele *M* that increases the fitness of *Aa* from $1 - hs$ to $1 - h's$ ($h' < h$). Then, the mean fitness of the *M*- genotype (i.e., one that contains at least one *M* allele) is

$$q^2 + 2q(1-q)(1-h's) + (1-q)^2(1-s) \approx 1 - 2\left(\frac{h'}{h}\right)u,$$

since $q \approx 1 - u/(hs)$. Similarly, if *m* is another allele at the modifier locus, then the mean fitness of the *mm* genotype is

$$q^2 + 2q(1-q)(1-hs) + (1-q)^2(1-s) \approx 1 - 2u.$$

Therefore,

$$\frac{\text{mean fitness of } M-}{\text{mean fitness of } mm} \approx \frac{1-2\left(\frac{h'}{h}u\right)}{1-2u} \approx 1 + 2u\left(1 - \frac{h'}{h}\right)$$

(*ibid.*, p. 276). This ratio is hardly larger than unity, i.e., the selective advantage of *M*- is so small that it makes the evolution of dominance insignificant. Moreover:

QUOT. 893 Even if it [the modifier gene] is able at once to bring the heterozygote *Aa* to equality with the wild type in reproductive rate ($h' = 0$) the pressure per generation toward its fixation (which is $2up^2(1-p)$, where p is the proportion of m's) is never more than half that which would result from recurrent mutation from *m* to *M* at the same rate as that from *A* to *a* (i.e., *up*). (*ibidem*)

The above criticism is especially relevant since a modifier would likely be subject to its own selection forces and thus it would be unlikely that a selection force of the order of the mutation rate would have much of an effect:

QUOT. 894. . . the hypothesis that a selection pressure, of the order calculated here, can be the general factor making for dominance of wild type, depends on the assumption that modifiers of dominance (assumed to be sufficiently abundant) are in general so nearly indifferent to selection on their own account that a force of the order of mutation pressure is the major factor controlling their fate. With the prevalence of multiple effects in mind it seems doubtful to the present writer whether there are many such genes. (*ibid*., p. 277)

Wright stressed that his overall conclusion on Fisher's theory was based on the low frequency of heterozygous mutants: if the latter frequency approaches that of the wild-type, then Fisher's evolutionary dominance would become more plausible.

Having rejected Fisher's evolutionary theory and consistent with his early training in physiological genetics,[44] Wright offered a physiological explanation for dominance:

QUOT. 895 On the view that genes act as catalysts and largely through bringing about the production of catalysts of a second order, it is easy to show that increase in the activity of a gene should soon lead to a condition in which even doubling of its immediate effect brings about little or no increase in the ultimate effects. Under such conditions, there would be dominance of the active phase. (*ibid*., p. 278)

Wright's physiological model thus posited that genes function as catalysts and if a trait is fully expressed by the activity of only one allele, then the allele is dominant; on the other hand, a recessive allele is one that is inactive.

14.4.3 Back-and-Forth Between Fisher and Wright on the Evolution of Dominance

Fisher did not deny Wright's point that the selection pressure on the modifier genes was small, but he was convinced that even if selection was weak, given enough time it would have the same effect as a large selection force acting for a proportionally shorter time. In a June 6, 1929, letter to Wright, he wrote:

QUOT. 896 What I mainly want to know, however, is whether you agree with me that a very slight selective effect acting for a correspondingly long time will be equivalent to a much greater effect acting for a proportionately shorter time. Or, whether, on the other hand, you think I have underestimated the ratio of the selective intensities, or overestimated the ratio of the times. I cannot see how a conclusion can be reached without considering the latter. (Provine, 1989, p. 247)

[44] See Sect. 15.1.

Wright made a key counterargument against Fisher's argument above, namely that the modifier gene would be in a state of equilibrium under the action of several evolutionary forces, thus nullifying the time factor. In a June 28, 1929, letter, he wrote:

QUOT. 897 The main question at issue seems to be clearly whether a very slight selective effect acting for a correspondingly long time will be equivalent to a much greater effect acting for proportionately shorter time. I assumed that the relative frequencies of any factor pair would be affected by many evolutionary pressures of varying orders of magnitude. As the resultant of all such pressures there would be a certain equilibrium point but this point would practically be dependent (in this sense of approximate fixation of *MM* or of *mm*) on only the one or two most important pressures. Thus it seemed to me that selection and mutation pressures of a lower order would have virtually no effect, however long the period of time. (*ibid.*, p. 248)

At this point, Fisher decided to formally respond to Wright's (1929b) comments on his dominance paper. He sent a copy of his response to Wright, informing the latter that he was submitting it to the *American Naturalist.* In his response, Fisher (1929) admitted that he found Wright's comments to be "thoughtful" and that their disagreement was not on the numerical values of their main formulas but rather on their respective interpretations of these values:

QUOT. 898...it is perhaps worthwhile to point out that the numerical values of his [Wright's] primary formulas differ in no essential respects from my own, and that the selective intensity which inclines Professor Wright to reject the theory is in fact the same that originally led me to adopt it. (*ibid.*, p. 553)

Having said that much, Fisher went on to criticize Wright's 1929 comments on two grounds:

(a) Wright's claim in QUOT. 893 that the rate of change per generation of the frequency of the modifier gene was always less than or equal to half the mutation rate was incorrect because it did not account for the effect of the frequency of the modifier on the frequency of the modified heterozygote.
(b) Even when the effect of the modifier gene is taken into account, the proper quantity to be used should not be, as Wright had assumed, the rate of change of modifier frequency per generation ($\delta p'$), but should instead be the selective intensity (i). Using Wright's formulation, $i = 2u(1 - p')$, but when the frequency of the modifier gene (M) is taken into account, the correct formula becomes $i = 2u/(1 - p')$, which shows that the selection intensity "increases without limit as p' increases towards unity"[45] (*ibid.*, p. 554).

---

[45]This result may be obtained as follows: first, $\delta p' = 2up'$ when the modifier gene M is dominant over m (the case considered by Fisher); then, the selection intensity i satisfies $\delta p' = ip'(1 - p')$ so that $i = 2u/(1 - p') \to \infty$ as $p' \to 1$. However, Ewens (1965b) has shown that the whole reasoning is flawed since both $\delta p'$ and i are, in fact, bounded above by very small numbers, so that $i \not\to \infty$ as $p' \to 1$. More specifically, when M is dominant, $\delta p' \leq p'(1 - p')\left(1.24h^{1/2}u^{3/4}s^{1/4}\right)$ and $i \leq 1.24h^{1/2}u^{3/4}s^{1/4}$(where h is the degree of dominance of A over a) (Ewens, 2004, p. 223). A similar conclusion holds for other dominance values of M.

Fisher next took on Wright's argument in QUOT. 897 that, since the modifier gene would probably be in an equilibrium state, hardly any change in its frequency could occur even if a large amount of time was allowed. Fisher disagreed:

QUOT. 899 I have myself attached a good deal of importance, in other connections, to gene ratios held in stable equilibrium, though it never occurred to me that such a state was universal. Even if it were nearly so, the conditions on which the stability rests must be liable to change; stability may turn to instability, a transient state, admittedly, but one in which a minute, but steadily increasing, selective intensity would be well fitted to tip the balance. (*ibid.*, p. 556)

14.4.4 Wright's Introduction of Two Key Evolutionary Concepts in the Dominance Debate: Genetic Drift and Effective Population Size

At this stage of the debate, Wright had already written (but not yet published) his major population genetics paper (Wright, 1931). In his last latter, Fisher had urged him to have "another go" in the *American Naturalist* on the issue of dominance. Wright now decided to make a major counterattack on Fisher's argument for the primacy of natural selection on the evolution of dominance. He sent his comment (Wright, 1929c) on Fisher's reply on July 23, 1929. In his comment, Wright first conceded Fisher's criticism that he had not taken the frequency of the modified heterozygote into account when calculating the rate of change of the frequency of the modifier gene.[46] However, even when this was conceded, Wright noted that the speed of selection was still low.

Next, for the first time in his debate with Fisher, Wright pointed to the importance of genetic drift and the effective population size (n):

QUOT. 900 In a freely interbreeding population of limited size (n) gene frequency shows random variation with a standard deviation $\sqrt{\frac{q(1-q)}{2n}}$. Being random, such variations largely neutralize each other, but there is a second order drift which can not be ignored. Dr. Fisher himself has calculated the probability distribution resulting from this cause under various conditions (*ibid.*, p. 559).

. . .

Selection controls the situation if s is larger than $1/(2n)$ but is of little importance below this figure. In small inbred populations ($1/(2n)$ large) even vigorous selection is ineffective in keeping injurious factors from drifting into fixation, and in the histories of a number of such lines one can follow in the laboratory the course of evolution under extreme isolation. In the case of Fisher's modifiers of dominance with selection coefficients at best of the order of mutation rate, the latter must be greater than $1/(2n)$ if the gene is not to drift back and forth in the course of logic time from one state of approximate fixation to the other and practically as freely in the face of the selection pressure as with it. (*ibid.*, p. 560)

[46]See Sect. 14.4.3 (a).

Wright's argument was that for small enough effective population sizes, genetic drift would overpower the effect of natural selection. Regarding the effective population size below which genetic drift predominates:

QUOT. 901 Unfortunately it is difficult to estimate *n* in animal and plant populations. In the calculations, it refers to a population breeding at random, a condition not realized in natural species as wholes. In most cases random interbreeding is more or less restricted to small localities. These and other conditions such as violent seasonal oscillation in numbers may well reduce *n* to moderate size, which for the present purpose may be taken as anything less than a million. If mutation rate is of the order of one in a million per locus, an interbreeding group of less than a million can show little effect of selection of the type which Dr. Fisher postulates even though there be no more important selection process and time be unlimited. (*ibidem*)

Having submitted his manuscript on July 23, the next day Wright sent a letter to Fisher informing the latter of the submission. The letter also contained a handwritten note[47] that mentioned Wright's yet unpublished big population genetics paper (Wright, 1931) and highlighted the importance he attached to genetic drift and the effective population size:

QUOT. 902 The question of the effect of random variation in factor frequency is one of the reasons that has long made me doubtful of the effect of feeble selection. I did not go into the matter in my first discussion of your hypothesis as being rather complicated, involving an unpublished formula and especially because of the uncertainty in the application of the *N* in my formula. I am sorry that I forgot this point at the moment in answering your previous letter as you may wish to add to your note on this subject now that I have discussed it. (Provine, 1989, p. 253)

Fisher's response on August 13, 1929 showed that he had not yet grasped the full meaning and implications of the concept of the effective population size *N*:

QUOT. 903 I am not sure that I agree with you as to the magnitude of the population number *N*. To reduce it to the number in a district requires that there shall be *no* diffusions even over the number of generations considered. For the relevant purpose I believe *N* must usually be the total population on the planet, enumerated at sexual maturity, and at the minimum of the annual or other periodic fluctuation. For birds twice the number of nests would be good. I am glad, however, that you stress the importance of this number. (*ibid.*, p. 255)

The concept of effective population size was formally introduced in Wright's big paper (Wright, 1931) in the form of "the population number":

QUOT. 904 It will be well to discuss more fully, before going on, what is to be understood by the symbol N used here for population number. The conception is that of two random samples of gametes, N sperms and N eggs, drawn from the total gametes produced by the generation in question (N/2 males and N/2 females each with a double representation from each series of allelomorphs). Obviously N applies only to the breeding population and not to the total number of individuals of all ages. If the population fluctuates greatly, the effective N is much closer to the minimum number than to the maximum number. If there is a great difference between the number of mature males and females, it is closer to the smaller number than to the larger. In fact, as just shown, a population of N, males and an indefinitely

[47] See also QUOT. 882.

large number of females is approximately equivalent to a population of 4N, individuals equally divided between males and females. (*ibid.*, pp. 110–111)

The effective population size is thus the breeding size of the population, i.e., the number of individuals who contribute to the gene pool. In other words, the effective size of a real population is the size of a breeding population, which results in the same amount of genetic drift.

14.4.5 Haldane (1930) Joins the Dominance Debate: Fisher's Modifier Theory Works Only in Rare Cases

Together with Fisher and Wright, John Burdon Sanderson Haldane (1892–1964) was one of the founders of evolutionary genetics.[48] In 1930, he subjected Fisher's modifier theory to his own mathematical analysis (Haldane, 1930). He considered a random mating population that consisted mostly of *AAmm*. The genotypic combinations *Aamm* and *AAMm* are assumed to have fitnesses $1 - k$ and $1 - k'$, where $k' < k$. The proportion of *Aa* in the population is approximately $2u/k$ (see Eq. 14.52), where u is the mutation rate from A to a.

Now Haldane argued that the genotype *AAMm* had a disadvantage of k' compared to *AAmm* (i.e., in most cases) and an advantage of k when combined with *Aa* (i.e., in a proportion $2u/k$ of cases). Therefore, the genotype *AAMm* spreads if and only if

$$k\left(\frac{2u}{k}\right) > k' \quad \Rightarrow \quad k' < 2u$$

(*ibid.*, p. 87). Next, Haldane assumed $u = 10^{-6}$ and the modifier M increases the fitness of *Aa* to that of *AA*, estimated to be approximately 0.99. Then, the proportion of *Aa* that are selected is approximately

$$\frac{2(10^{-6})}{1-0.99} = \frac{1}{5000}.$$

Hence, Haldane concluded that:

QUOT. 905...it is only in rare cases that suitable modifiers will be found. (*ibid.*, p. 87–88)

If the action of modifiers on heterozygotes was not the main mechanism for the origin of dominance, how then could the latter be explained? Haldane favored Wright's physiological theory, which he attributed to Goldschmidt (1927):

QUOT. 906 I suspect that the Fisher effect has operated in a different manner. Adopting Goldschmidt's (27) view that genes are catalysts acting at a definite rate, there is, as pointed out elsewhere (Haldane, 27b) no obvious way of distinguishing those which act at more than

[48] See Chap. 16.

a certain rate. E.g., if an enzyme can oxidize a certain substance as quickly as it is formed, no visible result arises from doubling the amount of that enzyme. Hence, while a minus mutation (diminution of activity) of a normal gene may yield a recessive type, a plus mutation is often unobservable. (Haldane, 1930, p. 88)

Haldane also explained the occurrence of dominance in several linked genes (e.g., in studies by Nabours (1917, 1925)) through gene duplication. Fisher (1930b) replied that gene duplication could, in some cases, result in dominance but this did not explain the *origin* of dominance.

14.4.6 Breakdown of the Fisher-Wright Relationship (1933)[49]

Fisher and Wright had hitherto remained cordial to each other and exchanged several fruitful letters. However, an incident in 1933 caused the relationship to sour suddenly. Thereupon, their differences became irreconcilable, and all communication ceased.

On June 21, 1933, Wright presented a paper entitled "Physiological and evolutionary theories of dominance" at the Symposium of the American Society of Naturalists at the Century of Progress meeting of the American Association for the Advancement of Science in Chicago. This paper was later published, with the mathematical parts extended, in the *American Naturalist* (Wright, 1934a).

In his paper, Wright recapitulated his physiological theory for the origin of dominance. He also repeated his main criticism of Fisher's evolutionary theory of dominance, namely that the force of selection was too weak to enable the fixation of the modifier gene. He disagreed with Fisher that, given enough time, a weak selection force would have the same effect as a large selection force acting for a proportionately smaller amount of time.

None of what Wright had said would have irked Fisher if Wright next had not criticized Fisher's very conception of evolution:

QUOT. 907 Fisher used the observed frequency of dominance as evidence for his conception of evolution as a process under complete control of selection pressure, however small the magnitude of the latter.

My interest in his theory of dominance was based in part on the fact that I had reached a very different conception of evolution (1931) and one to which his theory of dominance seemed fatal if correct. As I saw it, selection could exercise only a loose control over the momentary evolutionary trend of populations. A large part of the differentiation of local races and even of species was held to be due to the cumulative effects of accidents of sampling in populations of limited size. Adaptive advance was attributed more to intergroup than intragroup selection. (*ibid.*, p. 51)

---

[49]The Fisher-Wright dispute on dominance and other topics to be considered later in the text are described in Provine (1985; 1989, Chap. 8), Skipper (2000, 2009), and Hodge (1992).

In the above, Wright demoted Fisher's vision of the evolutionary process as being driven mainly by natural selection and promoted his own conception in which the major role belonged to the cumulative effects of genetic drift within groups, followed by some role for selection between groups.

Fisher (1934) replied immediately, and his irritation could hardly be concealed. He accused Wright of giving a selective history of the dominance problem in his 1934 paper. Fisher pointed to the error in Wright's initial model, which implied a selection intensity of the same order as the mutation rate[50]; when Fisher corrected that error, he showed the selection intensity "increasing without limit as the dominant modifier becomes more and more numerous" (*ibid.*, p. 371).[51] Fisher maintained his previous position that the frequency of mutant heterozygotes, maintained by the balancing action of mutation and natural selection, would be relatively high and that selection would be "proportionately more rapid" (*ibidem*). He concluded his retort as follows:

> **QUOT. 908** If...Professor Wright's [physiological] views can only be made plausible, by the exclusion of all alternatives, he must find other objections to the selection theory more weighty than those he has revived. (*ibid.*¸ p. 374)

Wright was no doubt taken aback by the tone of Fisher's paper and decided to respond tit-for-tat (Wright, 1934b). Concerning Fisher's charge of selective history, Wright explained that he had acknowledged Fisher's correction of his error (Wright, 1929c). He then complained that it was Fisher who had not "always been careful in historical matters" (Wright, 1934b, p. 563). He recalled how Fisher had made an error in the calculation of the rate of loss of heterozygosity in 1922[52] (Fisher, 1922). Wright had communicated this error to Fisher in a July 24, 1929, letter[53] and the latter had used this correction to derive the distribution of gene frequencies in three cases (Fisher, 1930a, c). Wright charged that although Fisher had acknowledged his error in print, he had not admitted that two of his distributions of gene frequencies had already been derived by Wright.

Wright also claimed that he had not overlooked the importance of natural selection on the issue of dominance. He concluded his paper with the following conciliatory words:

> **QUOT. 909** Finally in view of the tone of Dr. Fisher's note let me say here that I have the greatest admiration for his contributions to statistical genetics and biometry. If I have devoted considerable space to their criticism it is only because his views have seemed to me to be worthy of exhaustive study. (Wright, 1934b, p. 564)

---

[50] See Sect. 14.4.3 (b).

[51] But see footnote 45.

[52] See Sects. 14.3.4.2 and 14.3.4.3.

[53] See QUOT. 882.

14.4.7 Concluding Remarks

Fisher's theory of the evolution of dominance by gene modification has been thoroughly criticized since its inception, although it did garner occasional support (e.g., Ford (1930)). Kimura (1983, p. 9)[54] has more recently made a strong argument against Fisher's theory. On several occasions, Fisher stressed that the selection intensity (modification of dominance) on the mutant heterozygote (*Aa*) increased as its fitness became closer to that of the normal homozygote (*AA*) (e.g., Fisher (1928a, p. 122)). This means that if the fitnesses of *AA, Aa*, and *aa* are denoted by $1, 1 - hs, 1 - s$, then Fisher's theory predicts that a small h should correspond to a small s. However, most studies have shown that there exists an inverse relationship between h and s. But see also Billiard and Castric (2011) for evidence in favor of Fisher's theory.

14.5 The "Fundamental Theorem of Natural Selection" (1930, 1958)[55]: Concepts and Controversies

14.5.1 Introduction

In 1930, Fisher released a landmark book entitled *The Genetical Theory of Natural Selection* (Fisher, 1930c), which represented the culmination of his research in evolutionary genetics thus far. The book is now regarded as the first population genetics book but did not include any material from the groundbreaking 1918 paper (Fisher, 1918). This is understandable since the latter paper was not concerned with natural selection, which was the current topic of Fisher's interest.

Fisher's motivation for writing the 1930 book was that, since the times of Darwin and Wallace, there had been much interest in both the theories of natural selection and of evolution by natural selection. An unfortunate consequence of this was that the two theories were often confused and utilized interchangeably. In the Preface, Fisher thus started with the following clarification:

> **QUOT. 910** NATURAL Selection is not Evolution. Yet, ever since the two words have been in common use, the theory of Natural Selection has been employed as a convenient abbreviation for the theory of Evolution by means of Natural Selection, put forward by Darwin and Wallace. This has had the unfortunate consequence that the theory of Natural Selection itself has scarcely ever. if ever, received separate consideration. (Fisher, 1930c)

[54] See also Charlesworth (1979).

[55] 'The Fundamental Theorem of Natural Selection' is also discussed in Ewens (1989, 1995, 2004, pp. 64–67, 259–261; 2014; 2024), Castilloux and Lessard (1995), Lessard (1997), Turner (1970), Frank & Slatkin (1992), Frank (1997), Okasha (2018, pp. 84–96), Walsh & Lynch (2018, Chap. 6), Ågren (2021, pp. 31–35), Rice (2004, pp. 33–37), Bürger (2000, pp. 33–35, 41–44, 89–92), Gayon (1998, pp. 331–340), Li (1967b), and Reiss (2009, pp. 177–182).

Of all the topics discussed in Fisher's book, none has generated as much debate and research than his main conclusion of Chap. 2, which Fisher called "The Fundamental Theorem of Natural Selection" (FTNS). Briefly, this states that the rate of increase in mean fitness is equal to the additive genetic variance in fitness. The title of Fisher's chapter is proof of the importance he attached to this statement. Indeed, he compared it to the second law of thermodynamics, claiming that:

QUOT. 911. . .so similar a law should hold the supreme position among the biological sciences. (*ibid.*, p. 37)

The task of Sect. 14.5 will be to describe Fisher's development of this work and to trace the historical sequence of events that lead to a better understanding of his 'FTNS', with the ultimate realization that his claim that the theorem was exact was inaccurate.[56] It will also be explained how researchers were able to "extract" a result from Fisher's 'FTNS' that is indeed correct, although this adaptation of the 'FTNS' showed that there was nothing truly fundamental involved.

14.5.2 *Malthusian Parameter of Fitness, Average Excess, and Average Effect (1938, 1940, 1958)*

14.5.2.1 Malthusian Parameter of Fitness

To state his 'FTNS', Fisher started with some definitions. Since his model was *continuous in time with overlapping generations*, Fisher needed an alternative to the usual Darwinian fitness, which is the proportional contribution of a given genotype from one generation to the other. He chose the Malthusian parameter m as a measure of fitness. This parameter is "implicit in any given system of rates of death and reproduction, and measures the relative rate of increase or decrease of a population when in the steady state appropriate to any such system" (Fisher, 1930c, p. 26). Thus, if $N(t)$ is the number of individuals in a population at time t, then

$$m = \frac{1}{N(t)}\frac{dN(t)}{dt} = \frac{d}{dt}\log N(t). \tag{14.54}$$

Furthermore, m will be different for different genotypes.

The equation above also implies $N(t) = N(0)e^{mt}$, which, when compared to the one for the Darwinian fitness W, namely $N(t) = N(0)W^{t}$,[57] gives

[56]This is why, in this book, Fisher's claim will be enclosed by single inverted commas, i.e. his FTNS will be written as 'FTNS'.

[57]See footnote 20.

$$m = \log W$$

Alternatively, the Malthusian parameter can also be viewed as the root of the equation

$$\int_0^{\infty} e^{-mx} l_x b_x dx = 1$$

(*ibidem*), where l_x is the probability of survival from birth to age x, and b_x is the rate of reproduction at age x. The above equation is similar to the one used previously by Lotka (1913, 1918; Dublin and Lotka, 1925) and was first used by Fisher in 1927 (Fisher, 1927b).

14.5.2.2 Average Excess Associated with a Gene Substitution

The next concept is that of the *average excess* associated with a gene substitution:

QUOT. 912 Let us now consider the manner in which any quantitative individual measurement, such as human stature, may depend upon the individual genetic constitution. We may imagine, in respect of any pair of alternative genes, the population divided into two portions, each comprising one homozygous type together with half of the heterozygotes, which must be divided equally between the two portions. The difference in average stature between these two groups may then be termed the average excess (in stature) associated with the gene substitution in question. (Fisher, 1930c, p. 30)

Thus, given that the genotypes at a locus are *GG*, *Gg*, and *gg*, with frequencies P, $2Q$, and R, respectively, and fitnesses w_{11}, w_{12}, and w_{22}, the average excess associated with the substitution of the allele g by G is defined by

$$a = \frac{Pw_{11} + Qw_{12}}{p} - \frac{Qw_{12} + Rw_{22}}{q}, \tag{14.55}$$

where p and q are the frequencies of the alleles G and g, respectively. This formula was given in a later paper (Fisher, 1941, p. 53).

The expression for the average excess changes when fitnesses are expressed in terms of the Malthusian parameter. Let n_{ij} be the number of individuals with the ordered genotype A_iA_j and let N_i be the number of alleles A_i. Then using the definition of the Malthusian fitness in Eq. (14.54),

$$\frac{dN_i}{dt} = Nm_i = \sum_j n_{ij} m_{ij},$$

where m_{ij} is the Malthusian fitness of A_iA_j. Therefore,

$$\frac{dN}{dt} = \sum_{i,j} n_{ij} m_{ij}$$

and

$$\begin{aligned}\frac{dp_i}{dt} &= \frac{d}{dt}\left(\frac{N_i}{N}\right)\\ &= \frac{1}{N}\frac{dN_i}{dt} - \frac{N_i}{N^2}\frac{dN}{dt}\\ &= \frac{1}{N}\sum_j n_{ij} m_{ij} - p_i\left(\frac{1}{N}\frac{dN}{dt}\right)\\ &= \sum_j P_{ij} m_{ij} - p_i \sum_j P_{ij} m_{ij}\\ &= p_i(m_i - \overline{m}),\end{aligned}$$

where $\overline{m}$ is the mean Malthusian fitness and $m_i - \overline{m} = a_{mi}$ is the average excess of Malthusian fitness of the allele A_i in a multi-locus system. Therefore, the average excess (of Malthusian fitness) associated with a substitution of A_2 by A_1 is

$$a_m = a_{m1} - a_{m2} = \frac{1}{p_1}\frac{dp_1}{dt} - \frac{1}{p_2}\frac{dp_2}{dt} = \frac{1}{p_1 p_2}\frac{dp_1}{dt}, \tag{14.56}$$

which is the same as Fisher's $a = d\{\log(p/q)\}/dt$ for a bi-allelic system (*ibid.*, p. 34).

In the second edition of *The Genetical Theory of Natural Selection*, Fisher ((1958)) extended his treatment of Darwinian fitnesses to include the average excess a_i of fitness associated with a given allele A_i ($i = 1, 2, \ldots, s$) at a multi-allelic locus:

$$a_i = \frac{P_{ii} w_{ii} + \sum_{j \neq i} P_{ij} w_{ij}}{p_i} \tag{14.57}$$

(*ibid.*, p. 34), where P_{ii} is the frequency of the homozygotes A_iA_i, $2P_{ij}$ is the frequency of the heterozygotes A_iA_j, and w_{ij} is the centered (on the mean) Darwinian fitness of the genotype A_iA_j. For uncentered (i.e., raw) fitnesses, the formula is

$$a_i = \frac{P_{ii} w_{ii} + \sum_{j \neq i} P_{ij} w_{ij}}{p_i} - \overline{w} = \frac{P_{ii} w_{ii} + \sum_{j \neq i} P_{ij} w_{ij}}{p_i} - \sum_{i,j} P_{ij} w_{ij}. \tag{14.58}$$

In the 1941 paper, Fisher explained that the concept of average excess was not enough since the difference in Eq. (14.55) could *not* be attributed to the substitution of one allele by another, ceteris paribus.

QUOT. 913 This difference [the average excess] is not, however, to be ascribed to the gene substitution, as though it were necessarily an effect of it, and of it only: for by reason of such common phenomena as homogamy, or the mating of like with like, or a variety of similar causes, it may well be, and is probably the case in all real populations, that the two moities [*sic*], of which the averages are compared, differ not only in the one gene substitution selected, but in the frequencies of a number of other genes affecting the measurement in question. (Fisher, 1941, p. 53)

14.5.2.3 Average Effect of a Gene Substitution

Fisher was thus led to the concept of the *average effect* of a gene substitution. Although a preliminary attempt was made to explain this in the 1930 book, it is best to refer to the 1941 paper as this is where the concept is more clearly delineated:

QUOT. 914 We require, therefore, to form an equally definite conception, explicitly defined, of what is meant by the *average effect* of a given gene substitution. It is natural to conceive this as the actual increase in the total of the measurements of a population, when without change in the environment, or in the mating system, the gene substitution is experimentally brought about, as it might be by mutation. If, however, we think thus in experimental terms it is necessary, for the idea, to be applicable to all cases, to draw a further distinction. A change in the proportion of any pair of genes itself constitutes a change in the environment in which individuals of the species find themselves. In mammals, for example, a gene substitution may affect milk yield, and thus directly influence infantile nutrition. To find the effect of the gene substitution in a constant environment, any effect of the gene substitution itself on the environment must be discounted. In other words, the effect that is wanted is only that due to the change in the frequencies of the different possible genotypes, not including any change in the average measurement of a given genotype, which the change in gene frequency may in fact bring about. (*ibid.*, pp. 53–54)

The average effect of a gene on fitness is thus the mean change in fitness experienced by the addition of an extra copy of the gene.

To determine the average effect, α, of substituting the allele g by G, the genotypic fitness (w) of GG, Gg, and gg is regressed on the number of G alleles[58]:

$$w = \overline{w} + \alpha(x - \overline{x}) \quad \text{for} \quad x = 0, 1, 2, \tag{14.59}$$

where $\overline{w}$ is the mean fitness, x is the number of G alleles, and $\overline{x} = 1$. The parameter α is obtained by least-squares, i.e., by minimizing

$$P(w_{11} - \overline{w} - \alpha)^2 + 2Q(w_{12} - \overline{w})^2 + R(w_{22} - \overline{w} + \alpha)^2, \tag{14.60}$$

where, as before, the genotypes GG, Gg, and gg, have fitnesses w_{11}, w_{12}, and w_{12}, and frequencies P, $2Q$, and R, respectively. The least-squares solutions for α can be written alongside the expression for a in Eq. (14.55) as:

[58] This technique was already employed in Fisher's 1918 paper (Fisher, 1918). See Eq. (14.1).

$$a = \frac{P(Q+R)(w_{11} - w_{12}) + R(P+Q)(w_{12} - w_{22})}{pq} \quad \text{(average excess)},$$
$$\alpha = \frac{P(Q+R)(w_{11} - w_{12}) + R(P+Q)(w_{12} - w_{22})}{PQ + QR + 2PR} \quad \text{(average effect)} \tag{14.61}$$

(*ibid*., p. 55). Fisher noted that the average excess and average effect were different quantities unless $Q^2 = PR$(random mating or HWE). This is indeed the case in one-locus models.

Fisher's treatment of the average effect in a multi-locus model occurred in the second edition of *The Genetical Theory of Natural Selection* (Fisher, 1958, p. 35). He used the method of Lagrange multipliers to minimize the sum of squares of dominance deviations subject to a constraint on the additive effects. More specifically, following the more accurate treatment of Castilloux & Lessard (1995), consider an n-locus model such that the kth locus ($k = 1, 2, \ldots, n$) has s_k different alleles and the fitness of the genotype $A_{i_k}A_{j_k}$ ($i_k, j_k = 1, 2, \ldots, s_k$) is given by

$$w_{ij} = \overline{w} + \sum_k \left(\alpha_{ki_k} + \alpha_{kj_k}\right) + \gamma_{ij},$$

where α_{ki_k} is the average effect of allele A_{i_k} at locus k and γ_{ij} is the dominance deviation of the genotype A_iA_j. The additive effects are estimated by minimizing the dominance deviations subject to the constraint $\sum_{i_k} \alpha_{ki_k} p_{ki_k} = 0$. Let

$$L = \sum_{i,j} g_{ij} \left\{ w_{ij} - \overline{w} - \sum_k \left(\alpha_{ki_k} + \alpha_{kj_k}\right) \right\}^2 + \sum_k \lambda_k \sum_{i_k} \left(\alpha_{ki_k} p_{ki_k}\right).$$

By differentiating with respect to α_{ki_k} and setting to zero[59]

$$-4 \sum_j \sum_{i \ni i_k} g_{ij} \left\{ w_{ij} - \overline{w} - \sum_k \left(\alpha_{ki_k} + \alpha_{kj_k}\right) \right\} + \lambda_k p_{ki_k} = 0.$$

Summing on i_k for each k,

$$\lambda_k = 4 \sum_{i,j} g_{ij} \left\{ w_{ij} - \overline{w} - \sum_h \left(\alpha_{hi_h} + \alpha_{hj_h}\right) \right\} = -8 \sum_h \sum_{i_h} \alpha_{hi_h} p_{hi_h} = 0.$$

Using the last two equations, the average effects are implicitly defined through

[59] Fisher's equation (1958, p. 35) is erroneous as it omits the genotypes g_{ij} in front of the sum of squares. It also uses a single Lagrange multiplier λ, instead of one for each locus.

$$\sum_j \sum_{i \ni i_k} g_{ij} \left\{ w_{ij} - \overline{w} - \sum_k \left(\alpha_{ki_k} + \alpha_{kj_k} \right) \right\} = 0.$$

14.5.3 Fisher's "Fundamental Theorem of Natural Selection" (1930, 1958)

Having given the necessary background on the Malthusian fitness, average excess and average effect, Fisher proceeded to the stated aim in Chap. 2 of his book:

QUOT. 915 [t]o state the principle of Natural Selection in the form of a rigorous mathematical theorem, by which the rate of improvement of any species of organisms in relation to its environment is determined by its present condition. (Fisher, 1930c, p. 22)

Fisher's demonstration and statement of the 'FTNS' were as follows:

QUOT. 916...any increase *dp* in the proportion of one type of gene at the expense of the other will be accompanied by an increase αdp in the average fitness of the species, where α may be negative; but the definition of *a* requires that the ratio *p*:*q* must be increasing in geometrical progression at a rate measured by *a*, or in mathematical notation that

$$\frac{d}{dt}\log\left(\frac{p}{q}\right) = a$$

which may be written

$$\left(\frac{1}{p} + \frac{1}{q}\right) dp = a dt,$$

or $dp = pqadt$
whence it follows that[60],

$$\alpha dp = pqa\alpha dt$$

and, taking all factors into consideration, the total increase in fitness[61],

$$\sum \alpha dp = \sum (pqa\alpha) dt = W dt. \tag{14.62}$$

If therefore the time element *dt* is positive, the total change of fitness *Wdt* is also positive, and indeed the rate of increase in fitness due to all changes in gene ratio is exactly equal to the genetic variance of fitness *W* which the population exhibits. We may consequently state the fundamental theorem of Natural Selection in the form:

---

[60]The equation below is more properly written as $2\alpha dp = 2pqa\alpha dt$ since the additive genetic variance in fitness is $2pqa\alpha$ (see Price (1972, p. 135)).

[61]The equation below is incorrect: it should have been $\Sigma 2\alpha dp = \Sigma 2pqa\alpha dt = Wdt$ (see Price (1972, p. 135)).

> *The rate of increase in fitness of any organism at any time is equal to its genetic variance in fitness at that time*[62].
> (*ibid.*, pp. 34–35)

Fisher's 'FTNS' assumed a *continuous-time* model for a *given environment.* The *genetic variance* should not be confused with the total genetic (or genotypic) variance: Fisher's genetic variance is, in fact, the *additive genetic variance*[63] (or *genic* variance). Similarly, the *fitness* in the statement should be interpreted (for now) as the *mean fitness.*

Despite Fisher's claim that he had actually "proved" the 'FTNS' (*ibid.*, p. 36), the content of QUOT. 916 is no proof. At best, it is an *informal argument* leading to a deficient theorem, and it is tersely written. The following is a reconstruction of his argument.

At a given locus, let the genotypes GG, Gg, and gg have frequencies P, $2Q$, and R, and Malthusian fitnesses m_{11}, m_{12}, and m_{22}, respectively. From (14.60), the additive genetic variance is

$$V_A = R(-\alpha)^2 + P(\alpha)^2 - \alpha^2(P-R)^2 = \alpha^2\left\{(P+R) - (P-R)^2\right\},$$

where α is the average effect (on fitness) of substituting the allele g by B, obtained by regressing the genotypic fitness on the number of G alleles. Substituting for the average excess (a) and average effect (α) in Eqs. (14.61),

$$V_A = 2pqa\alpha,$$

where p and q are the frequencies of the alleles G and g, respectively. Next the change in the allele frequency of p in $(t, t + dt)$ is given from Eq. (14.56) as:

$$dp = apqdt.$$

Therefore, $d\overline{m}/dt = (d\overline{m}/dp)(dp/dt) = (d\overline{m}/dp)(pqa)$, where $\overline{m}$ is the mean Malthusian fitness. Fisher wrote $d\overline{m}/dt = 2pqa\alpha$, thus obtaining

$$d\overline{m}/dt = V_A.$$

To obtain the above result, Fisher had thus used $d\overline{m}/dp = 2\alpha$. However, this holds either if there is random mating or no dominance (Turner, 1970, p.41, Eq. (14.33); p. 45, Eq. (14.57)).

The reconstruction above shows that, even though Fisher wanted his 'FTNS' to be a very general result and did not assume *random mating*, his result for a single locus is valid only under the latter condition (or under non-dominance).

---

[62] Italics are Fisher's.

[63] See Sect. 14.2.3.

Furthermore, even though Fisher claimed that his result was valid across all loci in the genome, the multi-locus result in Eq. (14.62) is incorrect because it assumes average effects are additive across loci. However, this is true if and only if there is no linkage disequilibrium between their corresponding alleles (see Ewens (2004, pp. 248, 260)).

The last shortcoming was addressed in the second edition of *The Genetical Theory of Natural Selection* (Fisher, 1958). Fisher here also extended Eq. (14.62) to multiple alleles. He used a procedure essentially equivalent to that shown at the start of Sect. 14.5.5.4 to obtain the multi-locus version of Eq. (14.81):

$$V_a = \sum \sum{}' (2pa\alpha)$$

(*ibid.*, p. 37), where V_a is the additive genetic variance, the summation $\sum'$ is across alleles at a given locus, and the summation $\sum$ is across loci in the genome. Fisher's 'FTNS' then becomes

$$\sum \sum{}' 2\alpha dp = \sum \sum{}' (2pa\alpha)dt = Wdt \tag{14.63}$$

(*ibidem*). However, despite properly addressing the issue of average effects across loci in 1958, Fisher's 'FTNS' is still incorrect[64]: in most cases it is only approximate, and in some cases, the mean fitness can be shown to decrease across generations.[65]

The 'FTNS' was meant to hold when the "environment" is constant. In such a case, the rate of increase of m is positive and the value of m could greatly exceed zero, resulting in population growth without bounds. However, environments are usually not constant (e.g., because of increased competition) and the increase in m is countered by a deteriorating environment:

> **QUOT. 917** The balance left over when from the rate of increase in the mean value of m produced by Natural Selection, is deducted the rate of decrease due to deterioration in environment, results not in an increase in the average value of m, for this average value cannot greatly exceed zero, but principally in a steady increase in population. (*ibid.*, p. 42)

This is expressed as:

$$\frac{dM}{dt} = W - \left(D + \frac{M}{C}\right) \tag{14.64}$$

(*ibidem*), where M is the mean value of the Malthusian parameter, C is a constant expressing the relationship between fitness and population increase, and "defined as the increase in the natural logarithm of the population, supposed stationary at each

[64] This is despite Fisher's insistence that, if correctly interpreted, his 'FTNS' is an exact result: "The rigor of the demonstration requires that the terms employed should be used strictly as defined" (1930c, p. 35). On the incorrectness of Fisher's 'FTNS', Gillespie has said: "Thus, it [Fisher's 'FTNS'] is neither fundamental nor a theorem" (Gillespie, 2004, p. 69).

[65] See Sect. 14.5.4.

stage, produced by unit increase in the value of *M*'', *W* is the (additive) genetic variance in fitness (defined here as the rate of actual increase in fitness determined by natural selection, because of the 'FTNS'), and *D* is the rate of loss due to the deterioration of the environment. See Sect. 14.5.4.3 for Kimura's treatment of Fisher's 'environment'.

Scholars spent several decades trying to understand what the 'FTNS' meant. It was only in 1972 that a breakthrough was made by Price toward a proper understanding of the fitness component of the 'FTNS' (Price, 1972). With this new understanding, Ewens (1989) showed that the 'FTNS' held exactly in all its generality, but its significance remained elusive. The Price-Ewens contributions will be fully discussed in Sect 14.5.6.

14.5.4 The Behavior of the Mean Fitness

14.5.4.1 Wright (1948): The Mean Fitness of a Single-Locus System May Decrease Under Frequency-Dependent Selection

Owing to the terseness of Fisher's demonstration of the 'FTNS' in QUOT. 916, it was widely misunderstood, even after Fisher revisited it on two additional occasions (Fisher, 1941, 1958). Since variances are nonnegative, many researchers naturally interpreted the 'FTNS' to imply that the mean fitness of a population could never decrease in a constant environment.[66]

In 1948, Wright[67] provided a simple example of frequency-dependent selection at one locus and demonstrated a decrease in mean fitness (Wright, 1948, 1986).

In his paper, Wright considered a bi-allelic system such that each allele (A_1 and A_2) was favored when rare but selected against when abundant. Let the fitnesses of A_1A_1, A_1A_2, and A_2A_2 be $1 - s + t(1 - p)$, 1, and $1 + s - t(1 - p)$, respectively, where $p = \text{freq}(A_1)$ and $s, t > 0$. Assuming random mating, the change in frequency from one generation to the next is

$$\Delta p = -p(1-p)\frac{(s-t+tp)}{\overline{w}} \tag{14.65}$$

(*ibid*., p. 541), where

[66] Although, in 1941, Fisher gave an example of "a model involving powerful selection, in which the fitness of the species as a whole, judged by external criteria is entirely inoperative" (Fisher, 1941, p. 58). Fisher's argument was that selection acted on individuals and not for the "general good."

[67] Sewall Wright (1889–1988), together with R. A. Fisher and J. B. S. Haldane, was one of the key founders of the modern synthesis. More will be said about him in Chap. 15.

Fig. 14.2 Motoo Kimura (1924–1994). (Photo obtained with permission from Joseph Felsenstein)

$$\overline{w} = 1 - (s - t + pt)(2p - 1). \tag{14.66}$$

From Eq. (14.65), it is seen that for $s < t$, there is a point of stable equilibrium at $\widetilde{p} = 1 - s/t$ and from Eq. (14.66), $\overline{w}$ is maximized at $p^* = \frac{1}{2}\left(\frac{1}{2} + \widetilde{p}\right)$. This implies a steady decrease in mean fitness from p^* to $\widetilde{p}$.

14.5.4.2 Kimura (1958): The Mean Fitness in a Multi-Locus System May Decrease

Motoo Kimura (Fig. 14.2) was one of the most influential population geneticists of his time. He was born in Okazaki, Japan. In 1944, he entered Kyoto Imperial University and was advised by the geneticist Hitoshi Kihara to study botany. Kimura learned population genetics after joining Kihara's laboratory after the war. In 1949, he joined the National Institute of Genetics in Mishima, Shizuoka. His first population genetics paper, published in 1953, was very influential and described the stepping-stone model (Kimura, 1953). Kimura entered graduate school at Iowa State College in 1953 and soon moved to the University of Wisconsin, where he became associated with James Crow (1916–2012). Kimura's most influential contribution to population genetics was his proposal of the neutral theory of molecular evolution (Kimura, 1968a).

In 1958, Kimura indirectly showed that the mean fitness could also decrease in a multi-locus system even when selection is not frequency-dependent. For a two-locus system, consider two loci A and B such that when the locus A has genotype A_iA_j and the locus B has genotype B_kB_l, the fitness of the combined genotype $A_iA_jB_kB_l$ can be written as

$$a_{ijkl} = \overline{a} + A_{ij} + B_{kl} + \varepsilon_{ijkl}$$

(Kimura, 1958, p. 151), where $\overline{a}$ is the mean fitness, A_{ij} is the additive effect of the first locus, B_{kl} is the additive effect of the second locus, and ε_{ijkl} is the epistatic deviation. Next, Kimura wrote

$$\begin{aligned} A_{ij} &= \alpha_i + \alpha_j + d_{ij}^{(a)}, \\ B_{kl} &= \beta_k + \beta_l + d_{kl}^{(b)}, \end{aligned}$$

where α_i, α_j are the additive effects at locus A and $d_{ij}^{(a)}$ is the dominance deviation (similar definitions apply at locus B).

Furthermore, if the frequencies of alleles are A_i and B_k, then the deviations from HWE at each of loci A and B can be quantified through

$$\theta_{ij}^{(a)} = \frac{P_{ij}^{(a)}}{x_i x_j}, \qquad \theta_{kl}^{(b)} = \frac{P_{kl}^{(b)}}{y_k y_l},$$

where $P_{ij}^{(a)}, P_{kl}^{(b)}$ are the genotype frequencies of A_iA_j and B_kB_l, respectively. The departure from random association across loci can similarly be defined by

$$\theta_{ijkl} = \frac{P_{ijkl}}{P_{ij}^{(a)} P_{kl}^{(b)}}.$$

In analogy with his previous one-locus derivation of the 'FTNS' in the same paper,[68] Kimura was able to write

$$\frac{d\overline{a}}{dt} = V_g + \overline{\frac{da}{dt}} + \sum_{ij} P_{ij}^{(a)} d_{ij}^{(a)} \frac{d}{dt} \log \theta_{ij}^{(a)} + \sum_{kl} P_{kl}^{(b)} d_{kl}^{(b)} \frac{d}{dt} \log \theta_{kl}^{(b)}$$
$$+ \sum_{kl} P_{ijkl} \varepsilon_{ijkl} \frac{d}{dt} \log \theta_{ijkl}$$

(*ibid*., p. 152), where V_g is the additive genetic (or genic) variance, $d\overline{a}/dt$ is the rate of change of mean fitness, and $\overline{da/dt}$ is the mean rate of change of individual fitness.

[68] See Sect. 14.5.5.4 and, in particular, Eq. (14.85).

Using an analogous notation, the two-locus result above can be extended to a multi-locus system as follows, which represents a more general version of the 'FTNS':

$$\frac{d\bar{a}}{dt} = V_g + \overline{\frac{da}{dt}} + \overline{\sum \varepsilon \frac{d}{dt} \log \theta} \tag{14.67}$$

(*ibidem*), where "ε stands for an interaction effect (dominance or epistatic) and θ is the corresponding coefficient of departure from random combination." Now since the last two terms of Eq. (14.67) can be negative, the implication is that the mean fitness in a multi-locus system may decrease with time.

14.5.4.3 Kimura (1958): The Meaning of Fisher's 'Environment'

In his 1958 paper (see the last section), Kimura defined the Malthusian parameter M by

$$M = \bar{a} - f(N)$$

(Kimura, 1958, p. 164), where $\bar{a}$ is the intrinsic rate of population increase, N is the population number, and f is a function that represents the "resistance against population increase due to overcrowding." Therefore,

$$\begin{aligned}\frac{dM}{dt} &= \frac{d\bar{a}}{dt} - \frac{df}{dt} \\ &= \frac{d\bar{a}}{dt} - \frac{df}{dN} \cdot \frac{dN}{dt} \\ &= \frac{d\bar{a}}{dt} - MN\frac{df}{dN}\end{aligned}$$

since $dN/dt = NM$ from Eq. (14.54). Kimura's (14.67) becomes

$$\frac{dM}{dt} + MN\frac{df}{dN} = V_g + \overline{\frac{da}{dt}} + \overline{\sum \varepsilon \frac{d}{dt} \log \theta}$$

In the case of a relative weak resistance $f(N) = K \log N$, where K is a constant, the above becomes

$$\frac{dM}{dt} = V_g + \overline{\frac{da}{dt}} + \overline{\sum \varepsilon \frac{d}{dt} \log \theta} - \frac{M}{C}$$

(*ibid.*, p. 165), where $C = 1/K$. Comparing the above with Fisher's Eq. (14.64), $W = V_g$ and

$$-D = \overline{\frac{da}{dt}} + \overline{\sum \varepsilon \frac{d}{dt} \log \theta}$$

Thus, Fisher's environmental factor D depends not only on changes in individual fitnesses ($\overline{da}/dt$) as a result of external conditions, but also on the system of mating (θ), and on dominance and epistatic effects (ε).

14.5.4.4 Scheuer and Mandel (1959), Kingman (1961): The Mean Fitness Is Non-Decreasing Under Random Mating at a Single-Locus

A direct proof that the mean fitness increases at a single locus under random mating was given in 1959 by Scheuer and Mandel (1959). The authors assumed discrete time and considered the $K = 2$ allele case first. Let p_i ($i = 1, 2$) be the frequency of allele A_i and let a_{ij} ($i, j = 1, 2$) be the relative viability of the genotype A_iA_j. Under random mating, the mean viability of the population then is

$$\overline{W} = a_{11}p_1^2 + 2a_{12}p_1p_2 + a_{22}p_2^2$$

(*ibid.*, p. 519). The frequencies of the alleles A_i and A_i in the next generation are, respectively,

$$\begin{aligned} p_1' &= \frac{a_{11}p_1^2 + a_{12}p_1p_2}{\overline{W}}, \\ p_2' &= \frac{a_{22}p_2^2 + a_{12}p_1p_2}{\overline{W}}. \end{aligned} \tag{14.68}$$

The mean viability now is

$$\overline{W}' = a_{11}p_1'^2 + 2a_{12}p_1'p_2' + a_{22}p_2'^2.$$

After some algebra, Scheuer and Mendel were able to show that

$$\begin{aligned} \overline{W}' - \overline{W} = \frac{1}{\overline{W}^2} &\left\{p_1\left(a_{11}p_1 + a_{12}p_2 - \overline{W}\right)^2 + p_2\left(a_{12}p_1 + a_{22}p_2 - \overline{W}\right)^2\right\} \\ &\times \left(\overline{W} + a_{11}p_1 + a_{22}p_2\right) \end{aligned}$$

(*ibid.*, p. 520). Therefore,

$$\overline{W}' - \overline{W} \geq 0.$$

The authors noted that no such simplification occurred for $K \geq 3$ alleles and resorted to matrix algebra for a proof in that case. See Edwards (2000, pp. 17–18, 40–41).

The most elegant proof for the general K-allele case under random mating was given by Kingman (1961). Sir John Kingman was born in 1939 in Beckenham, Kent. He graduated in 1960 and did his PhD under Peter Whittle. Kingman studied queuing theory and stochastic processes, but never completed his PhD. His greatest contribution to population genetics was the development of coalescent theory (1982a, b).

In his proof, Kingman extended the notation above so that the frequency of allele A_i ($i = 1, 2, \ldots, K$) in the next generation is, in analogy with Eq. (14.68), given by

$$p_i' = \frac{a_{i1}p_ip_1 + a_{i2}p_ip_2 + \ldots + a_{ii}p_i^2 + \ldots . a_{iK}p_ip_K}{\overline{W}} = \frac{p_i}{\overline{W}}\sum_{j=1}^{K} a_{ij}p_j = \frac{p_ia_i}{\overline{W}},$$

where $a_i = \sum_j a_{ij}p_j$ is the marginal fitness of allele A_i. Then, the mean fitness in the next generation is

$$\overline{W}' = \sum_{i,j} a_{ij}p_i'p_j' = \frac{1}{\overline{W}^2}\sum_{i,j} a_{ij}(a_ip_i)(a_jp_j) = \frac{1}{\overline{W}^2}\sum_{i,j,k} a_{ij}a_{ik}a_jp_ip_jp_k$$

By interchanging j and k, the above becomes

$$\overline{W}' = \sum_{i,j} a_{ij}p_i'p_j' = \frac{1}{\overline{W}^2}\sum_{i,j,k} a_{ij}a_{ik}a_kp_ip_jp_k$$

Averaging the last two expressions,

$$\overline{W}' = \frac{1}{2}\frac{1}{\overline{W}^2}\sum_{i,j,k} a_{ij}a_{ik}(a_j + a_k)p_ip_jp_k$$

Using the fact that the arithmetic mean is greater than or equal to the geometric mean, the above becomes

$$\begin{aligned}\overline{W}' &\geq \frac{1}{\overline{W}^2}\sum_{i,j,k} a_{ij}a_{ik}a_j^{1/2}a_k^{1/2}p_ip_jp_k \\ &= \frac{1}{\overline{W}^2}\sum_i p_i\left(\sum_j p_ja_{ij}a_j^{1/2}\right)^2 \\ &\geq \frac{1}{\overline{W}^2}\left(\sum_i p_i\sum_j p_ja_{ij}a_j^{1/2}\right)^2,\end{aligned}$$

the last step being due to Jensen's inequality. Therefore,

Table 14.8 Fitness values for various genotypic combinations are two loci A and B

		Locus A		
		AA	*Aa*	*aa*
Locus B	*BB*	W_{22}	W_{12}	W_{02}
	Bb	W_{21}	W_{11}	W_{01}
	bb	W_{20}	W_{10}	W_{00}

$$\begin{aligned}\overline{W}' &\geq \frac{1}{\overline{W}^2}\left(\sum_j p_j a_j^{1/2} \sum_i p_i a_{ji}\right)^2\\ &= \frac{1}{\overline{W}^2}\left(\sum_j p_j a_j^{3/2}\right)^2\\ &\geq \frac{1}{\overline{W}^2}\left(\sum_j p_j a_j\right)^3\\ &= \overline{W}\end{aligned}$$

(*ibid.*, p. 78), the last but one step being due to Jensen's inequality again.

14.5.4.5 Kojima and Kelleher (1961): Analysis of Mean Fitness in Bi-Allelic Two-Locus Models Under Random Mating

Kojima and Kelleher provided a thorough analysis of a bi-allelic two-locus model under random mating in 1961 (Kojima and Kelleher, 1961). The fitness values of the various genotypic combinations are shown in Table 14.8.

The haplotypes *AB, Ab, aB*, and *ab* are assumed to have frequencies x, y, z, and u, and marginal mean fitnesses W_x, W_y, W_z, and W_u, respectively, such that:

$$\begin{aligned}W_x &= xW_{22} + yW_{21} + zW_{12} + uW_{11},\\ W_y &= xW_{21} + yW_{20} + zW_{11} + uW_{10},\\ W_z &= xW_{12} + yW_{11} + zW_{02} + uW_{01},\\ W_u &= xW_{11} + yW_{10} + zW_{01} + uW_{00}.\end{aligned} \tag{14.69}$$

By referring to the Kimura (1956, Eq. (1) on p. 279), Kojima and Kelleher wrote the following differential equations:

$$\begin{aligned}\frac{dx}{dt} &= x\left(W_x - \overline{W}\right) - rW_{11}D,\\ \frac{dy}{dt} &= y\left(W_y - \overline{W}\right) + rW_{11}D,\\ \frac{dz}{dt} &= z\left(W_z - \overline{W}\right) + rW_{11}D,\\ \frac{du}{dt} &= u\left(W_u - \overline{W}\right) - rW_{11}D,\end{aligned} \tag{14.70}$$

where r is the recombination fraction between the two loci, $D = xu - yz$ is the linkage disequilibrium parameter, and $\overline{W} = xW_x + yW_y + zW_z + uW_u$ is the mean fitness.[69] The rate of change in mean fitness is then

$$\frac{d\overline{W}}{dt} = \frac{dx}{dt}W_x + \frac{dy}{dt}W_y + \frac{dz}{dt}W_z + \frac{du}{dt}W_u + x\frac{dW_x}{dt} + y\frac{dW_y}{dt} + z\frac{dW_z}{dt} + u\frac{dW_u}{dt}.$$

By differentiating the mean marginal fitnesses in Eqs. (14.69), the above simplifies to

$$\frac{d\overline{W}}{dt} = 2\left(\frac{dx}{dt}W_x + \frac{dy}{dt}W_y + \frac{dz}{dt}W_z + \frac{du}{dt}W_u\right)$$

(*ibid.*, p. 528). Furthermore, $\frac{dx}{dt} + \frac{dy}{dt} + \frac{dz}{dt} + \frac{du}{dt} = 0$ from Eqs. (14.70), so the above becomes

$$\begin{aligned}\frac{d\overline{W}}{dt} &= 2\left(\frac{dx}{dt}W_x + \frac{dy}{dt}W_y + \frac{dz}{dt}W_z + \frac{du}{dt}W_u\right) - 2\overline{W}\left(\frac{dx}{dt} + \frac{dy}{dt} + \frac{dz}{dt} + \frac{du}{dt}\right)\\ &= 2\left\{\left(W_x - \overline{W}\right)\frac{dx}{dt} + \left(W_y - \overline{W}\right)\frac{dy}{dt} + \left(W_z - \overline{W}\right)\frac{dz}{dt} + \left(W_u - \overline{W}\right)\frac{du}{dt}\right\}.\end{aligned}$$

Substituting the derivatives from Eqs. (14.70), the authors finally obtained

$$\begin{aligned}\frac{d\overline{W}}{dt} &= 2\left\{x\left(W_x - \overline{W}\right)^2 + y\left(W_y - \overline{W}\right)^2 + z\left(W_z - \overline{W}\right)^2 + u\left(W_u - \overline{W}\right)^2\right\}\\ &\quad - 2\left(W_x - W_y - W_z + W_u\right)W_{11}rD\end{aligned}$$

(*ibid.*, p. 529). Now, the term in the curly brackets {} is the variance, denoted by V_m, among the gametic marginal means and $(W_x - W_y - W_z + W_u)$ is the additive x additive epistatic deviation, denoted by α_{AB}. Hence the last equation can be written as

$$\frac{d\overline{W}}{dt} = 2(V_m - \alpha_{AB}W_{11}rD), \tag{14.71}$$

which shows that $\overline{W}$ is an increasing function of t unless at least one of α_{AB}, W_{11}, r, or D is zero. If none of the latter quantities are zero, $\overline{W}$ may decrease. In particular, even if the loci are unlinked ($r = 1/2$), $\overline{W}$ may still decrease, which is the opposite of what occurs in a one-locus model under random mating (see Sect. 14.5.4.4).

[69] Kojima and Kelleher's version of (14.70) mistakenly has $-rW_{11}D$ instead $+rW_{11}D$ in the 2nd and 3rd equations.

14.5.4.6 Ewens (1969): The Mean Fitness in Non-Decreasing When Fitnesses Are Additive Across Multiple Loci Under Random Mating

Ewens is an influential population geneticist. More will be said on him in Sect. 14.5.6.2. In 1969, he provided a remarkable sufficient condition for the mean fitness to increase across multiple loci, namely, that the fitnesses are additive across loci, i.e., the fitness of the combined genotype at several loci is the sum of the fitnesses of the corresponding genotype at each locus (Ewens, 1969b). He first considered the bi-allelic two-locus case. Here, the fitness of a combined genotype at the two loci A and B is the sum of the fitness v of the genotype at locus A and the fitness u at locus B. This is depicted in Table 14.9.

Some background information will now be given before Ewens' proof is described. In the absence of natural selection, the frequency of the gamete, say A_1B_1, in generation $t + 1$ is

$$g'_{A_1B_1} = (1 - R)g_{A_1B_1} + Rp_{A_1}p_{B_1} = g_{A_1B_1} - RD,$$

where $D = g_{A_1B_1} - p_{A_1}p_{B_1} = g_{A_1B_1}g_{A_2B_2} - g_{A_1B_2}g_{A_2B_1}$ is the linkage disequilibrium parameter, R is the recombination fraction between the two loci, and the g without a prime refers to a gamete frequency in generation t. Denoting the frequencies of the gametes $A_1B_1, A_2B_1, A_1B_2, A_2B_2$ in generation t by c_1, c_2, c_3, c_4, respectively, in analogy with the equation above, the gamete frequencies in generation $t + 1$ are

$$c'_i = c_i + \delta_i RD \quad \text{for } i = 1, 2, 3, 4,$$

where $\delta_1 = \delta_4 = 1$ and $\delta_2 = \delta_3 = -1$.

Coming back to Ewens' proof, natural selection is assumed to act according to the fitness scheme in Table 14.9. The above equation becomes

$$c'_i = \frac{c_i w_i + \delta_i(u_2 + v_2)RD}{\overline{w}} \quad \text{for } i = 1, 2, 3, 4 \tag{14.72}$$

where $w_i = \sum_j c_j w_{ij}$ (the w_{ij}'s are given in Table 14.9) is the mean fitness of gamete i and $\overline{w}$ is the mean fitness of the population, i.e., under random mating

Table 14.9 Fitness scheme for Ewens' 1969 model

		Locus A		
		A_1A_1	A_1A_2	A_2A_2
Locus B	B_1B_1	$w_{11} = u_1 + v_1$	$w_{12} = u_1 + v_2$	$w_{22} = u_1 + v_3$
	B_1B_2	$w_{13} = u_2 + v_1$	$w_{14} = u_2 + v_2$	$w_{24} = u_2 + v_3$
	B_2B_2	$w_{33} = u_3 + v_1$	$w_{34} = u_3 + v_2$	$w_{44} = u_3 + v_3$

$$\overline{w} = u_1(c_1 + c_2)^2 + 2u_2(c_1 + c_2)(c_3 + c_4) + u_2(c_3 + c_4)^2 + v_1(c_1 + c_3) + 2v_2(c_1 + c_3)(c_2 + c_4) + v_3(c_2 + c_4)^2$$

(*ibid.*, p. 1076). The mean fitness in generation $t + 1$ is then

$$\overline{w}' = u_1(c_1' + c_2')^2 + 2u_2(c_1' + c_2')(c_3' + c_4') + u_2(c_3' + c_4')^2 + v_1(c_1' + c_3') + 2v_2(c_1' + c_3')(c_2' + c_4') + v_3(c_2' + c_4')^2.$$

Thus, for given fitness values, $\overline{w}'$ depends solely on the frequencies of allele A_1 $(=c_1' + c_3')$, allele B_1 $(=c_1' + c_2')$, and the alternative of each of these alleles (whose frequencies are also functions of $c_1' + c_3'$ and $c_1' + c_2'$). However, Eqs. (14.72) show that $c_1' + c_3'$ and $c_1' + c_2'$ do not depend on R, and hence $\overline{w}'$ also does not depend on R. Therefore, $\overline{w}'$ has the same value for any R, in particular when $R = 0$. In the latter case, the bi-allelic two-locus model is equivalent to a single-locus model with four alleles. Hence, using Sheuer and Mendel's result in Sect. 14.5.4.4, $\overline{w}' \geq \overline{w}$, i.e., the mean fitness is non-decreasing.

By using a similar argument, Ewens extended the result above to a multi-locus model (with additive fitnesses) in a follow-up paper (Ewens, 1969a).

14.5.5 *Interpretations and Proofs of Fisher's Fundamental Theorem of Natural Selection*

14.5.5.1 Wright's (1930) Influence on Later Interpretations and Proofs of the Fundamental Theorem

Although Fisher and Wright already clashed on the issue of the evolution of dominance, they were still in communication at the time Fisher released his book (Fisher, 1930c). In a March 19, 1930 letter, Fisher informed Wright that he was sending him a complimentary copy of his book.[70] In his reply, dated June 10, 1930, Wright admitted that the book was "extremely interesting and stimulating" and informed Fisher that he had been asked to review the book for the *Journal of Heredity* (Bennett, 1983, pp. 275–276).

In his review, Wright described Fisher's book as follows:

> **QUOT. 918** It is a book which is certain to take rank as one of the major contributions to the theory of evolution. (Wright, 1930, p. 349)

After commenting on several points where he and Fisher agreed, Wright brought about the issue of the evolution of dominance. The two men had disagreed on the efficacy of natural selection on this topic. To further clarify the power of natural

[70] However, the book was forwarded to Wright in May 26, 1930. The latter received a second copy from *Journal of Heredity* with a request to review it.

selection, Wright turned his attention to the 'FTNS'. Wright stated the theorem verbatim and showed that he understood that Fisher had used "genetic variance" in the special sense of "genic variance" (or additive genetic variance). However, he next stated that the 'FTNS' was not exact since it ignored dominance and epistasis:

QUOT. 919 Obviously there could be exact agreement in all cases only if dominance and epistatic relationships were completely lacking. Actually, dominance is very common and with respect to such a character as fitness, it may safely be assumed that there are always important epistatic effects. (*ibid.*, p. 353)

Wright also believed that the 'FTNS' needed revision since it also ignored mutations (and migrations):

QUOT. 920 The formula itself seems to need revision in the case of another important class of genes, ones slightly deleterious in effect but maintained at a certain equilibrium in frequency by recurrent mutation (or migration). These contribute to the genetic variance of the species, but not to the increase in fitness. Terms involving mutation (and migration) rates seem to be omitted in the formula as given. (*ibidem*)

In his January 19, 1931 reply, Fisher tried to explain that he had, in fact, taken mutation into account in his more general formula (14.64): mutation was incorporated into the factor D, the rate of loss due to the deterioration of the environment:

QUOT. 921 I had not intended to take up any special point in this letter, but I am tempted to mention this one, (p. 353) 'The formula itself seems to need revision in the case of another important class of genes, ones slightly deleterious in effect but maintained at a certain equilibrium in frequency by recurrent mutation' (I can leave migration aside here). The point here is that the average fitness *is* continually being increased by selection, at exactly the same rate as it is being decreased by mutation. This cause of deterioration of adaptedness, due to mutations of the organism, is, in my treatment, classed with the parallel deterioration due to changes in the environments. This supplies an amendment to the corresponding statement on p. 352, 'The only effective offset to undeviating increase in fitness, which he recognizes, is change of environment'. I think, if you happen to re-read p. 41 [GTNS, p. 44), you will see that I class deleterious mutations equally as an offset. (Bennett, 1983, p. 279)

Wright replied on February 3, 1931, but his response showed that he remained unconvinced that the 'FNTS' was accurate. He proposed his own version of the 'FTNS':

QUOT. 922 I think that if the theorem on page 35 had been stated with its qualifications: "The rate of increase in fitness of any population at any time is equal to its genetic variance in fitness at that time, except as affected by mutation, migration, change of environment and the effects of random sampling," I would not have been confused by the discussion in the following pages. (Provine, 1989, p. 272)

There are two important aspects to Wright's interpretation of the 'FTNS' above, and these would permeate future interpretations and proofs of the theorem:

1. To Wright, Fisher's 'FTNS' was only an approximation since it ignored the effects of mutation, migration, change in environmental factors, and drift.
2. To Wright, Fisher's rate of increase in fitness referred to the increase in *total fitness*.

Elsewhere, Wright also claimed that the 'FTNS' held only under the assumption of random mating. For example:

QUOT. 923 As noted, Fisher's theorem holds strictly only under the assumption of random combination of loci. (Wright, 1977, p. 425)

Although no such assumption was made by Fisher, Wright was incorrect on two fronts: (i) in general, the 'FTNS' holds only approximately, and (ii) for a one-locus model, the 'FTNS' holds exactly under HWE (or in the absence of dominance). See the discussion in Sect. 14.5.3.

14.5.5.2 Li (1955): First Attempted Proof of Fisher's 'FTNS'

It was in C.C. Li's famous population genetics book (Li, 1955, pp. 288–289) that most probably the first attempt was made to rigorously prove the 'FTNS'. However, Li's proof used a discrete model, unlike the continuous model used by Fisher.

Chin Chun Li (1912–2003) was born in Tianjin, China. He received his B.S. degree in agronomy in 1936 from the University of Nanking, an American missionary school. He obtained his PhD in Plant Breeding and Genetics from Cornell University, where he had studied from 1937 to 1940. In 1946, he became professor and chair of the Agronomy Department at National Peking University. There, he published his famous textbook in English, entitled *Introduction to Population Genetics* (Li, 1948). When the communist government took over mainland China in 1949, they forced Li to resign and asked him to denounce Mendelism in favor of Lysenkoism (which was promoted by Stalin).[71] Li and his family fled to Hong-Kong. After H.J. Muller wrote a strong letter of recommendation for him, Li joined the University of Pittsburgh in 1951. He served as the president of the American Society of Human Genetics in 1960 and was widely regarded as a pioneer in the field until his death in 2003.

In his proof, Li did not assume random mating and considered the fitness scheme depicted in Table 14.10. Here, f refers to a genotypic frequency *before* natural selection occurs.

The mean and variance of the fitness are, respectively,

$$\overline{W} = \sum f_i W_i, \quad \sigma_W^2 = \sum f_i (W_i - \overline{W})^2 = \sum f_i W_i^2 - \overline{W}.$$

Note that Li's σ_W^2 is the genotypic variance in fitness, not the *additive* genetic variance in fitness. Once natural selection has taken place, the frequency of genotype G_i is $f_i W_i / \sum f_i W_i = f_i W_i / \overline{W}$ and the new mean fitness is

[71] See Sect. 16.5.1.

Table 14.10 Li's fitness scheme for his "proof" of the 'FTNS' (Li, 1955, p. 272)

Genotype G	G_1	G_2	...	G_i	...
Proportions f	f_1	f_2	...	f_i	...
Fitness W	W_1	W_2	...	W_i	...

$$\overline{W}' = \frac{\sum f_i W_i^2}{\overline{W}}.$$

The change in mean fitness is therefore

$$\Delta \overline{W} = \overline{W}' - \overline{W} = \frac{\sum f_i W_i^2}{\overline{W}} - \overline{W} = \frac{\sigma_W^2}{\overline{W}}$$

(*ibid.*, p. 289).

Li's proof of the 'FTNS' suffers from two serious problems:

(a) It confuses the genotypic variance with the additive genetic variance and formulates the theorem in terms of the former rather than the latter.
(b) It ignores the round of mating so that it is a result *for a given generation and not across generations*, as specified by Fisher. This fact makes Li's result lack the evolutionary importance of Fisher's.[72]

14.5.5.3 Bennett (1956): First Proof in the Spirit of Fisher's 'FTNS'

Bennett (1956) was the first to attempt a proof of the 'FTNS' in the same spirit that Fisher had enunciated. Like Fisher, Bennett considered a continuous-time model but, unlike Fisher's, his result pertained to a *single locus*. Suppose the latter has two alleles A_1 and A_2, with respective frequencies p_1 and p_2, and genotypes A_1A_1, A_1A_2 and A_2A_2 with respective frequencies P_{11}, $2P_{12}$ and P_{22}. The average excesses[73] of alleles A_1 and A_2 are

$$a_1 = m_1 - \overline{m} = \frac{1}{p_1}\frac{dp_1}{dt}, \qquad a_2 = m_2 - \overline{m} = \frac{1}{p_2}\frac{dp_2}{dt}, \tag{14.73}$$

where m_1, m_2 are the marginal (Malthusian) fitnesses of the A_1, A_2 alleles, respectively, and $\overline{m}$ is the mean genotypic fitness. The average excess associated with a substitution of A_2 by A_1 is

[72] On this point, Turner has correctly observed: "Fisher has proposed a theorem, not of natural selection, but of evolution" (Turner, 1970, p. 34).

[73] See Sect. 14.5.2.2.

$$a = a_1 - a_2 = \frac{1}{p_1}\frac{dp_1}{dt} - \frac{1}{p_2}\frac{dp_2}{dt} = \frac{1}{p_1 p_2}\frac{dp_1}{dt} \tag{14.74}$$

(*ibid.*, p. 303). Bennett then used the following argument in Fisher (1941, p. 55). Suppose, in an infinitesimal time δt, the frequencies of genotypes A_1A_1, A_1A_2 and A_2A_2 change by δP_{11}, $2\delta P_{12}$ and δP_{22}, respectively. Since $P_{11} + 2P_{12} + P_{22} = 1$ and $p_1 = P_{11} + P_{12}$, then $\delta P_{11} + 2\delta P_{12} + \delta P_{22} = 0$ and $\delta p_1 = \delta P_{11} + \delta P_{12} = (\delta P_{11} - \delta P_{22})/2$. Thus, for a given δp_1, if one of δP_{11}, δP_{12}, δP_{22} is fixed, then the other two can be completely determined. Therefore, Fisher set

$$\delta P_{11} = \frac{2p_2P_{11}}{p_2P_{11} + p_1P_{22}}\delta p_1 \quad \Rightarrow \quad \begin{aligned} \delta P_{22} &= -\frac{2p_1P_{22}}{p_2P_{11} + p_1P_{22}}\delta p_1, 2\delta P_{12} \\ &= \frac{2P_{12}(P_{22} - P_{11})}{p_2P_{11} + p_1P_{22}}\delta p_1. \end{aligned} \tag{14.75}$$

Writing the mean genotypic fitness as $\overline{m} = m_{11}P_{11} + 2m_{12}P_{12} + m_{22}P_{22}$, Bennett used the equations above to obtain

$$\begin{aligned} \frac{d\overline{m}}{dp_1} &= \frac{\partial \overline{m}}{\partial P_{11}}\frac{dP_{11}}{dp_1} + \frac{\partial \overline{m}}{\partial P_{12}}\frac{dP_{12}}{dp_1} + \frac{\partial \overline{m}}{\partial P_{22}}\frac{dP_{22}}{dp_1} \\ &= 2\frac{p_2P_{11}(w_{11} - w_{12}) + p_1P_{22}(w_{12} - w_{22})}{p_2P_{11} + p_1P_{22}} \\ &= 2\alpha \end{aligned} \tag{14.76}$$

(*ibid.*, p. 301), where α is the average effect of substituting A_2 by A_1 (see Eq. (14.61)). From the above and Eq. (14.74),

$$\frac{d\overline{m}}{dt} = \frac{d\overline{m}}{dp_1}\frac{dp_1}{dt} = 2p_1p_2\alpha a. \tag{14.77}$$

Finally, using the values of the average excess in (14.61), namely

$$a = \frac{P_{11}p_2(m_{11} - m_{12}) + P_{22}p_1(m_{12} - m_{22})}{p_1p_2},$$

Bennett wrote the additive genetic variance (i.e., the variance of the genic values $\mu + \alpha$, μ, $\mu - \alpha$ with probabilities P_{11}, $2P_{12}$, P_{22}) as

$$\sigma_A^2 = 2p_1p_2a\alpha. \tag{14.78}$$

From Eqs. (14.77) and (14.78), Fisher's ‘FTNS’ follows: $d\overline{m}/dt = \sigma_A^2$. Moreover, Bennett noted that the equations in Eq. (14.75) imply that

$$\frac{\delta P_{11}}{P_{11}} + \frac{\delta P_{22}}{P_{22}} = \frac{2\delta P_{12}}{P_{12}}.$$

Integrating on both sides, $\log P_{11}$ + $\log P_{22} = 2\log_{12}(kP_{12})$, where k is a constant of integration, so that the 'FTNS' is valid as long as $P_{12}^2/(P_{11}P_{22})$ is constant (in particular, it holds under random mating). As was indicated in Sect. 14.5.3, the relationship in Eq. (14.76), and hence Bennett's proof, is also valid if there is no dominance (see also Turner (1970), p. 45, Eq. (57)).

14.5.5.4 Kimura (1958): First Generalization of Fisher's 'FTNS'

Kimura (1958) was the first to successfully obtain a correct version of Fisher's 'FTNS'. His model was continuous in time, multi-locus, did not assume random mating, and was multi-allelic. In the latter respect, it therefore extended Fisher's 1930 formulation (Fisher, 1930c, pp. 34–35) and anticipated Fisher's final enunciation of the 'FTNS' (Fisher, 1958, p. 37). Kimura first considered a single locus A with n alleles. The fitness of the genotype A_iA_j (whose frequency is $2P_{ij}$) is given by

$$a_{ij} = \overline{a} + \alpha_i + \alpha_j + d_{ij},$$

where α_i, α_j are, respectively, the average effects of the alleles A_i, A_j, d_{ij} is the dominance deviation, and $\overline{a}$ is the mean genotypic fitness. The average effects are obtained by minimizing the weighted sum of squares

$$Q = \sum_{i,j} P_{ij}\left(a_{ij} - \overline{a} - \alpha_i - \alpha_j\right)^2.$$

Differentiating with respect to α_k and setting to zero,

$$\sum_j P_{kj}\left(a_{kj} - \overline{a} - \alpha_k - \alpha_j\right) + \sum_i P_{ik}\left(a_{ik} - \overline{a} - \alpha_i - \alpha_k\right) = 0$$

The normal equations then are

$$\alpha_k \sum_i P_{ki} + \sum_i P_{ki}\alpha_i = \sum_i P_{ki}\left(a_{ki} - \overline{a}\right) \tag{14.79}$$

(Kimura, 1958, p. 149). By defining the average excess of allele A_k as follows:

$$\widetilde{a}_{k.} = a_k - \overline{a} = \frac{\sum_i P_{ki}a_{ki}}{x_k} - \overline{a},$$

(a_k is the marginal fitness of allele A_k), then multiplying by α_k, and finally summing over k, Eq. (14.79) becomes

$$\sum_{k,i} \alpha_k^2 P_{ki} + \sum_{k,i} P_{ki} \alpha_i \alpha_k = \sum_k x_k \tilde{a}_{k.} \alpha_k. \tag{14.80}$$

Now the genic (or additive genetic) variance is the sum of squares of average effects:

$$V_g = \sum_i P_{ii} (2\alpha_i)^2 + \sum_{k<i} 2P_{ki} \alpha_i \alpha_k,$$

where P_{ii} is the frequency of the homozygote A_iA_i. The right side of the above is twice the left side of Eq. (14.80), so that

$$V_g = 2\sum_k x_k \tilde{a}_{k.} \alpha_k \tag{14.81}$$

(*ibidem*). Using the formula for the average excess in (14.73),[74] the above can also be written as

$$V_g = 2\sum_i \alpha_i \frac{dx_i}{dt}. \tag{14.82}$$

Kimura next introduced the quantity θ_{ij}, which “may be called the coefficient of departure from random combination of alleles and is unity if mating is random”:

$$\theta_{ij} = \frac{P_{ij}}{x_i x_j}.$$

Substituting the above into Eq. (14.79) and dividing by x_k,

$$\sum_j x_j \theta_{ij} (a_{ki} - \bar{a}) = \alpha_i + \sum_j \alpha_j \theta_{ij} x_j. \tag{14.83}$$

Regarding the rate of change of population fitness under natural selection, the mean genotypic fitness is

[74]Note that the formula is based on the *Malthusian* fitness.

$$\begin{aligned}\bar{a} &= \sum_{i,j} a_{ij} P_{ij} \\ \therefore \quad \frac{d\bar{a}}{dt} &= \sum_{i,j} \frac{da_{ij}}{dt} P_{ij} + \sum_{i,j} a_{ij} \frac{dP_{ij}}{dt}.\end{aligned} \tag{14.84}$$

The first term on the right side is the average rate of change of individual fitness and may be written as $\overline{da/dt}$ and the second term on the right side is shown to be

$$\sum_{i,j} a_{ij} \frac{dP_{ij}}{dt} = V_g + \sum_{i,j} d_{ij} P_{ij} \frac{d}{dt} \log \theta_{ij}.$$

Therefore, Eq. (14.84) becomes

$$\frac{d\bar{a}}{dt} = V_g + \overline{\frac{da}{dt}} + \sum_{i,j} d_{ij} P_{ij} \frac{d}{dt} \log \theta_{ij} \tag{14.85}$$

(*ibid.*, p. 150), which is a generalized version of Fisher's 'FTNS'. Thus, for a single locus, Fisher's statement (i.e., $d\bar{a}/dt = V_g$) holds exactly if and only if (i) there is no change in individual fitness, and (ii) either there is no dominance or the deviation from HWE as measured by θ_{ij} is constant (HWE is therefore sufficient but not necessary for Fisher's 'FTNS' to hold exactly).

The multi-locus version of the above was covered in Sect. 14.5.4.2,[75] where it was argued that the presence of the terms after V_g implied that the mean fitness may decrease with time. The general equation is

$$\frac{d\bar{a}}{dt} = V_g + \overline{\frac{da}{dt}} + \overline{\sum \varepsilon \frac{d}{dt} \log \theta}\ , \tag{14.86}$$

where "ε stands for an interaction effect (dominance or epistatic) and θ is the corresponding coefficient of departure from random combination."[76]

Upon receiving a draft of Kimura's paper, Fisher wrote back on May 3, 1956:

> **QUOT. 924** In answer to my question about in what respect you thought the fundamental theorem needed extension, you say that your original purpose was 'to obtain the general expression for the rate of change of population fitness'. Now, of course I purported to give such a general expression, and I should like to know whether your expression differs from mine in substance rather than only in form, and in what respect you think that my expression is erroneous. Of course I had developed the multiple allele case actually before the book was published, and have put it into the Dover Publications edition, which I hope will soon appear. (Bennett, 1983, pp. 229–230)

[75] See Eq. (14.67).

[76] See Eq. (14.67).

Kimura made sure that his answer to Fisher's question was included in his paper:

QUOT. 925 The above formulation establishes the roles of additive, dominance and epistatic components on the rate of increase of population fitness. It is the reformulation of the results published by Fisher in 1930 under the title *The Fundamental Theorem of Natural Selection*, and in no way differs from it in substance. (Kimura, 1958, p. 166)

It is also noteworthy that Kimura's 1958 proof was included for the two-allele case in Crow and Kimura (1970). In their book, the authors made a most remarkable observation, the importance of which would be realized just 2 years later (Price (1972)):

QUOT. 926 One interpretation of the theorem is to say that it measures the rate of increase of fitness that would occur if the gene frequency changes took place, but nothing else changed. The theorem thus gives the effect of gene frequency changes alone, isolated from the other things that are happening. Fitness becomes a rather abstract quantity that continually increases while the population size is roughly stabilized by the various factors that cause the environment to change. (Crow and Kimura, 1970, p. 209)

14.5.6 The Price-Ewens Fundamental Theorem of Natural Selection

14.5.6.1 Price (1972): Change in Mean Fitness Due to Natural Selection

George Robert Price[77] (Fig. 14.3) was born in the state of New York. He graduated in chemistry in 1943 from the University of Chicago and obtained his doctorate from the same university in 1946. He worked various jobs, including consultant to Argonne National Laboratory and chemist at Bell Laboratories, and was treated for thyroid cancer in 1966. He left for the UK in 1967, and it was there that he made fundamental contributions to evolutionary genetics during the period 1967–1974. These included: (i) the formulation of the Price equation, which describes how a trait or allele changes with time; (ii) working in conjunction with John Maynard Smith to introduce the concept of evolutionary stable strategy, and (iii) a new formulation of the 'FTNS'.

Despite his pioneering work, Price spent his last months homeless and committed suicide in 1975.

Price had already derived the famous Price equation (Price, 1970) when he wrote a paper in 1972 that gave a novel interpretation to the 'FTNS' (Price, 1972). Using this interpretation, he was able to show that Fisher's theorem held exactly and with the full generality that Fisher had claimed:

[77]George Price's life and work are described in Harman (2010), Frank (1995), and Walsh and Lynch (2018, Chap. 6).

Fig. 14.3 George Robert Price (1922–1975). (From: Harman (2010, p. 282), photo obtained with permission from the author)

George Price in his University College, London, office, 1973

QUOT. 927 ...many authors (not reviewed here) have maintained that the theorem holds only under very special conditions, while only a few (e.g. Edwards, 1967) have thought that Fisher may have been correct—if only we could understand what he meant!

It will be shown here that this latter view is correct. Fisher's theorem does indeed hold with the generality that he claimed for it. The mystery and the controversy result from incomprehensibility rather than error. (*ibid.*, p. 129)

Price's key insight was that, in Fisher's statement about the rate of change in (mean) fitness, the total change in mean fitness (dM) was actually made up of two components, one due to natural selection ($\partial_{NS}M$), the other due to the 'environment'[78] ($\partial_{EC}M$):

$$dM = \partial_{NS}M + \partial_{EC}M \tag{14.87}$$

(*ibidem*). Price argued that previous authors had not been able to prove the 'FTNS' in its full generality *because they had focused on the total change* (dM) *instead of the*

[78] Fisher's "environment" included both dominance and epistatic effects.

change due to natural selection ($\partial_{NS}M$). He contended that Fisher had $\partial_{NS}M$ in mind when he stated his fundamental theorem, which should therefore be interpreted as follows:

$$\frac{\partial_{NS}M}{\partial t} = W,$$

where W is the additive genetic variance in fitness, or more generally (to emphasize the dependence of the various quantities on time):

$$\frac{\partial_{NS}M_t}{\partial t} = W_t \tag{14.88}$$

(*ibidem*).

Price's interpretation of the first term on the right of (14.87) as the change due to natural selection is problematic. In his paper, Price himself later acknowledged that:

> **QUOT. 928** One defect is the device of treating nonadditive gene effects as 'environment'." (*ibid.*, p. 139)

Since nonadditive effects are included in the second term on the right side of Eq. (14.87) and potentially involve natural selection, calling the first term "the change due to natural selection" is thus inaccurate.[79]

To prove Eq. (14.88) Price first obtained an expression for $\partial_{NS}M_t$. He considered the regression:

$$M = c + \sum_{l,k} \beta_{m,lk} Q_{lk},$$

where c is a constant, Q_{lk} is the frequency of allele k at locus l, and $\beta_{m,\ lk}$ is a regression coefficient. The above can be rewritten as $M = \sum_{l,\ k} b_{m,\ lk} Q_{lk}$, or more simply as

$$M = \sum bQ.$$

Now let $T = t + dt$ so that

$$M_t = \sum b_t Q_t, \quad M_T = \sum b_T Q_T \tag{14.89}$$

and

[79] Some authors have also interpreted the first term as the change in mean fitness "due to gene frequency changes only." But this interpretation is also problematic. See the discussion following Eq. (14.90).

$$dM_t = M_T - M_t = \sum b_T Q_T - \sum b_t Q_t.$$

Using the equation above, Price wrote the decomposition in Eq. (14.87) as follows:

$$dM_t = \sum b_T Q_T - \sum b_t Q_t = \left(\sum b_t Q_T - \sum b_t Q_t\right) + \left(\sum b_T Q_T - \sum b_t Q_T\right).$$

Price next equated the first and second terms on the right side of the equation above to $\partial_{NS} M_t$ and $\partial_{EC} M_t$, respectively:

$$\sum b_t Q_T - \sum b_t Q_t = \partial_{NS} M_t \tag{14.90}$$

and $\sum b_T Q_T - \sum b_t Q_T = \partial_{EC} M_t$ (*ibid.*, p. 130). Eq. (14.90) explains why some authors have also described Price's $\partial_{NS} M$ as the change in mean fitness due to changes in gene frequency only: the first term on the left *seems* to represent the effect of only gene frequency change (from Q_t to Q_T) on mean fitness and keeps all else (through b_t) fixed. However, as Eq. (14.89) shows, a major problem with the "change in gene frequencies only" interpretation is that changes in gene frequencies also change b_t to b_T. A more appropriate description of Eq. (14.90) will be discussed in Sect. 14.5.6.2.

It still remained for Price to show that: (i) W_t was the additive genetic variance, and (ii) $\partial_{NS} M_t = W_t dt$.

He defined the *average excess* $a_{\phi,\,lk}$ of allele lk for the trait ϕ as

$$a_{\phi,lk} = \frac{\sum_{i=1}^{N} \left\{ \left(\phi_i - \overline{\phi}\right)\left(q_{lki} - Q_{lk}\right) \right\} / N}{Q_{lk}} = \frac{\text{cov}(\phi, q_{lk})}{Q_{lk}} \tag{14.91}$$

(*ibid.*, p. 137), where q_{lki} is the frequency of allele lk in individual I_i, Q_{lk} is the population frequency of allele lk and ϕ_i is the trait value in individual I_i. He next wrote the *average effect* $\alpha_{\phi,\,lk}$ of allele lk for the trait ϕ as

$$\alpha_{\phi,lk} = \frac{1}{2}\left\{ \beta_{\phi,lk} - \sum_{k=1}^{s(l)} \beta_{\phi,lk} Q_{lk} \right\} \tag{14.92}$$

(*ibidem*). Now, using (14.91) and (14.92), he was able to prove that

$$\sum_{l,k} 2 Q_{lk} a_{\phi,lk} \alpha_{\phi,lk} = W_\phi \tag{14.93}$$

(*ibidem*). Referring to the second edition of *The Genetical Theory of Natural Selection* (Fisher, 1958, p. 37), Price noted that the above was simply Fisher's $\sum'(2pa\alpha)$, which "represent[s] the contribution of each factor to the genetic variance in fitness." Thus, (i) above is proved.

Next, Price interpreted Fisher's next statement that "any increase dp in the frequency of the chosen gene will be accompanied by an increase $2\alpha dp$ in the average fitness of the species" as

$$\frac{\partial_{NS} M_t}{\partial Q_{lk}} = 2\alpha_{m,lk} \tag{14.94}$$

(Price, 1972, p. 138), and showed that

$$\partial_{NS} M_t = \sum_{k,l} 2\alpha_{m,lk} a_{m,lk} Q_{lk} dt.$$

Using (14.93), this means that

$$\partial_{NS} M_t = W_t dt$$

(*ibidem*) and (ii) is proved. This concludes Price's proof of Fisher's 'FNTS.'

Having proven the 'FTNS' in its full generality, Price thought there were two aspects of the theorem that undermined its utility:

(a) Fisher's treatment of nonadditive effects as environmental: "Still one feels disappointed that it does not say more. One defect is the device of treating nonadditive gene effects as 'environment'." (*ibid.*, p. 139).[80]
(b) Fisher's theorem was about the rate of change in the mean fitness (due to natural selection) while, at the same time, the mean fitness always stayed close to zero: "A much graver defect is the matter of the shifting standard of 'fitness' that gives the paradox of *M* tending always to increase and yet staying generally close to zero. Much more interesting would be a theorem telling of increase in 'fitness' defined in terms of some fixed standard" (*ibid.*, pp. 139–140).

Price's final words were thus of disappointment:

QUOT. 929 Thus there is a challenge here to find a deeper definition of this elusive concept 'fitness' and to give a deeper and sharper explanation of why it increases and under what conditions. (*ibid*, p. 140)

However, A. W. F. Edwards (Fig. 14.4), who has spent countless hours in the exegesis of Fisher's work (both in population genetics and statistics) and who was one of the first to realize the importance of Price's paper (Edwards, 1990a, p. 281), has acknowledged that:

QUOT. 930 Price has got closer than anyone in divining Fisher's thoughts. (*ibidem*)

[80]This limitation was already mentioned in QUOT. 928.

Fig. 14.4 Anthony William Fairbank Edwards (1935). (Photo taken in October 2023 and obtained with permission from Anthony Edwards)

14.5.6.2 Ewens (1989): Partial Change in Mean Fitness

Warren Ewens (1937) (Fig. 14.5) was born in Australia and majored in mathematical statistics at the University of Melbourne in 1960. The next year, he moved to the Australia National University and decided to study genetics. He did his PhD under P. A.P. Moran (1917–1988) in stochastic processes in population genetics. In 1964, having completed his PhD, Ewens did a 6-month postdoc at Stanford and joined the Department of Biology at the University of Pennsylvania in 1972. His contribution to population genetics has been enormous, including the well-known Ewens' Sampling formula (Ewens, 1972).

Using Price's insight (namely, that the total change in mean fitness ought to be thought of as the sum of two components, only the first of which needed to be considered in the 'FTNS'), in 1989 Ewens provided what may be considered to be the definite proof of the 'FTNS' (Ewens, 1989) under different evolutionary scenarios. Ewens' proof was adopted by later researchers (e.g., see Edwards (1994, 2002)) since it eschewed some of the abstractness in Price's proof, provided a more appropriate interpretation of the first term on the right side of Eq. (14.90), and was written in standard population genetics language.

Before giving his treatment of the 'FTNS', Ewens enumerated five important clarifications regarding the theorem (*ibid.*, p. 168):

(i) *Nonnecessity of random mating:* "It is usually claimed that the Fundamental Theorem requires the assumption of random mating for its validity. Fisher

Fig. 14.5 Warren Ewens (1937). (Photo obtained with permission from Warren Ewens)

nowhere makes this assumption, and is quite explicit in stating that, for the genes defining important characters, random mating would not hold ... As we note later, the Fundamental Theorem is true for any mating scheme, a fact also stated explicitly by Fisher."

(ii) *Nonnecessity of an explicit evolutionary model or mating scheme*: "...it is often claimed that an explicit evolutionary model is necessary for the development of theorem [*sic*]. No evolutionary model or mating scheme is necessary, provided only that one quite simple and reasonable evolutionary assumption (described later) is made."

(iii) *Validity in both discrete and continuous time*: "...it is often debated whether the theorem assumes discrete or continuous time. It will be argued here that this debate is vacuous and that the theorem is true in both cases."

(iv) *Validity for multiple alleles and multiple loci:* "...a formal extension of Fisher's analysis will be given to show that the theorem is true no matter how many loci fitness depends on and no matter how many alleles arise at these loci. This is despite the fact that even under random mating, mean fitness can decrease in the multi-locus case."

(v) *Exactness of the 'FTNS'*: "...it is often stated that Fisher intended the theorem merely as a close approximation. It will be argued here that there is substantial evidence that he intended it as a precise and exact result."

Thereupon, Ewens systematically addressed the following scenarios: (a) one-locus in discrete time under general mating; (b) one-locus in discrete time under random mating; (c) evolution in continuous time; and (e) evolution of multiple loci. He showed that, under Price's decomposition in Eq. (14.87), the 'FTNS' held in each case. Each of these proofs will now be described.

(a) *Single locus in discrete time under general mating*. Ewens considered alleles A_1, A_2, ..., A_m at a locus such that the genotype A_iA_j has (viability) fitness w_{ij} and frequency $2P_{ij}$ or P_{ii} depending on whether $i \neq j$ or $i = j$, respectively. Before natural selection has acted, the frequency of allele A_i is

$$p_i = \sum_j P_{ij}. \tag{14.95}$$

Once natural selection has occurred, the frequency of allele A_i in the adults is

$$p_i' = \frac{\sum_j P_{ij} w_{ij}}{\overline{w}} \tag{14.96}$$

(*ibid*., p. 169), where $\overline{w}$ is the mean fitness

$$\overline{w} = \sum_{i,j} P_{ij} w_{ij}.$$

Ewens then noted that, although a complete dynamical model is not necessary for the 'FTNS', it will be sufficient to assume that p_i' is the frequency of A_i in the next generation.

The next important definition is of the *average excess* a_i of allele A_i:

$$a_i = \frac{1}{p_i} \sum_j P_{ij} \left(w_{ij} - \overline{w} \right) \tag{14.97}$$

(*ibidem*). Subtracting (14.95) from (14.96) and substituting the above, the change in the frequency of A_i from one generation to the next is

$$\begin{aligned} \Delta p_i = p_i' - p_i = \frac{\sum_j P_{ij} w_{ij}}{\overline{w}} - \sum_j P_{ij} = \frac{1}{\overline{w}} \sum_j P_{ij} \left(w_{ij} - \overline{w} \right) \\ \therefore \quad \Delta p_i = \frac{a_i p_i}{\overline{w}} \end{aligned} \tag{14.98}$$

(*ibidem*). Having defined the average excess, the next concept is that of the *average effect*. Suppose

$$w_{ij} = \overline{w} + \alpha_i + \alpha_j,$$

where $\alpha_1, \ldots, \alpha_m$ are constants. Multiplying the above by P_{ij} and summing across all i and j shows that $\sum_i p_i \alpha_i = 0$. Then the average effect on fitness when an allele A_j is replaced by an allele A_i in a genotype is defined by

$$\alpha = \alpha_i - \alpha_j.$$

On the other hand, when a randomly chosen allele is replaced by an allele A_i, then the average effect of the allele A_i on fitness is defined by

$$\alpha_i - \sum_j \alpha_j p_j \equiv \alpha_i.$$

The average effects are estimated through least-squares, i.e., by minimizing $\sum_{i,j} P_{ij} \left(w_{ij} - \overline{w} - \alpha_i - \alpha_j\right)^2$. Differentiating with respect to α_k and setting to zero, the normal equations are

$$\begin{aligned} \alpha_k \sum_i P_{ki} + \sum_i P_{ki}\alpha_i &= \sum_i P_{ki}(w_{ki} - \overline{w}) \\ \therefore \quad \alpha_k p_k + \sum_i P_{ki}\alpha_i &= p_k a_k \end{aligned} \tag{14.99}$$

The additive genetic variance is then

$$\begin{aligned} \sigma_A^2 &= \sum_{i,j} P_{ij}\left(\alpha_i + \alpha_j\right)^2 \\ &= 2\sum_i p_i \alpha_i^2 + 2\sum_{i,j} P_{ij}\alpha_i \alpha_j. \end{aligned}$$

Using (14.99), the above becomes

$$\begin{aligned} \sigma_A^2 &= 2\sum_i p_i \alpha_i^2 + 2\sum_{i,j} P_{ij}\alpha_i \alpha_j \\ &= 2\sum_i p_i \alpha_i^2 + 2\sum_i p_i (a_i - \alpha_i)\alpha_i \\ &= 2\sum_i p_i a_i \alpha_i \end{aligned} \tag{14.100}$$

(*ibid.*, p. 170). Substituting from (14.98), the additive genetic variance may be written as

$$\sigma_A^2 = 2\overline{w}\sum_i (\Delta p_i)\alpha_i. \tag{14.101}$$

Ewens next considered changes in the mean fitness. He argued that, in view of the *transmission value*[81] of a genotype, Fisher's mean fitness is more appropriately written as

$$\sum_{i,j} P_{ij}\left(\overline{w} + \alpha_i + \alpha_j\right) \tag{14.102}$$

instead of $\overline{w}$. It is in the computation of the change in mean fitness next that Price's insight comes into play, and Ewens introduced the concept of "partial change in mean fitness":

> **QUOT. 931** . . .Fisher considers the change in mean fitness only through changes of the P_{ij} factors . . .and not through changes in $\overline{w}$, α_i, and α_j, and the Fundamental Theorem relates only to this change. It is central to the understanding of the statement of the Fundamental Theorem to note this, as Price (1972) (and perhaps no one else) observes. Fisher presumably saw a significance in this change, which I shall call the partial change in mean fitness, and Price, in describing and in part supporting Fisher's viewpoint, describes this partial change as the change in mean fitness due to natural selection, as opposed to the change caused by the environment. (*ibid.*, p. 171)

The essence of the "partial change in fitness" argument, therefore, is to consider only changes in P_{ij} in (14.102), leaving $\overline{w}$, α_i, and α_j unchanged. This term provides a more appropriate description of Eq. (14.90), instead of "change in mean fitness due to natural selection" or "change in mean fitness due to gene frequency changes only."

Denoting the frequency of the ordered genotype A_iA_j in the next generation by P_{ij}, the partial change in mean fitness is

$$\begin{aligned}\sum_{i,j}\left(P'_{ij} - P_{ij}\right)\left(\overline{w} + \alpha_i + \alpha_j\right) &= \sum_{i,j}\left(P'_{ij} - P_{ij}\right)\left(\alpha_i + \alpha_j\right)\\ &= 2\sum_i \alpha_i \sum_j \left(P'_{ij} - P_{ij}\right)\\ &= 2\sum_i \alpha_i \Delta p_i\end{aligned} \tag{14.103}$$

Using Eq. (14.101), the above becomes

[81]The transmission value from parent to offspring is in terms of *average effects of alleles* rather than *fitnesses of genotypes*, since parents transmit alleles, not genotypes.

$$\sum_{i,j}\left(P'_{ij}-P_{ij}\right)\left(\overline{w}+\alpha_i+\alpha_j\right)=\frac{\sigma_A^2}{\overline{w}} \tag{14.104}$$

(*ibidem*), which is the 'FTNS' for one locus in discrete time (the mean fitness $\overline{w}$ may be assumed to be unity without any loss of generality). The *partial* change in mean fitness $\sum_{i,j}\left(P'_{ij}-P_{ij}\right)\left(\overline{w}+\alpha_i+\alpha_j\right)$ should be contrasted with the *total* change in mean fitness $\sum_{i,j}P'_{ij}w_{ij}-\overline{w}$, which is how the 'FTNS' had been interpreted until Price's paper (Price, 1972).

(b) *Single locus in discrete time under random mating*. The reader may have noted that no assumption of a mating scheme was made in the derivation in Eq. (14.104). Therefore, that equation should continue to hold under random mating, i.e., when $P_{ij}=p_ip_j$ for all i, j.
(c) *Evolution in continuous time.* Ewens considered a proportion P_{ij} of the ordered genotype A_iA_j at time t. Since w_{ij} is the viability fitness, a proportion $P_{ij}(1-w_{ij})\delta t$ of ordered genotypes A_iA_j will not survive to time $t+\delta t$. During the same time interval $(t, t+\delta t)$, the total proportion of non-survivors will be replaced by newborns, a proportion P^*_{ij} of whom will be of ordered genotype A_iA_j. Therefore,

$$\frac{d}{dt}P_{ij}=-P_{ij}\left(1-w_{ij}\right)+\left(1-\overline{w}\right)P^*_{ij}$$

(*ibid*., p. 174), and the partial rate of change in mean fitness becomes

$$\begin{aligned}\sum_{i,j}\frac{dP_{ij}}{dt}\left(\overline{w}+\alpha_i+\alpha_j\right) &=\sum_{i,j}\left\{-P_{ij}\left(1-w_{ij}\right)+\left(1-\overline{w}\right)P^*_{ij}\right\}\left(\overline{w}+\alpha_i+\alpha_j\right)\\ &=2\sum_i\alpha_i\sum_j\left(P_{ij}w_{ij}-P_{ij}+P^*_{ij}-\overline{w}P^*_{ij}\right)\\ &=2\sum_i\alpha_i\left(\sum_j P_{ij}w_{ij}-p_i+p^*_i-\overline{w}p^*_i\right)\end{aligned}$$

Assuming that "that the frequency of A_i among newborns at time t is equal to that among individuals alive at that time", i.e., $p^*_i=p_i$, the above becomes

$$\begin{aligned}\sum_{i,j}\frac{dP_{ij}}{dt}\left(\overline{w}+\alpha_i+\alpha_j\right) &=2\sum_i\alpha_i\left(\sum_j P_{ij}w_{ij}-p_i\overline{w}\right)\\ &=2\sum_i p_i\alpha_i\left(\frac{\sum_j P_{ij}w_{ij}}{p_i}-\overline{w}\right)\\ &=2\sum_i p_i\alpha_i a_i,\end{aligned}$$

where a_i is the average excess in Eq. (14.97). Using Eq. (14.100), the 'FTNS' in continuous time is obtained:

$$\sum_{i,j} \frac{dP_{ij}}{dt}\left(\overline{w} + \alpha_i + \alpha_j\right) = \sigma_A^2.$$

(d) *Multiple loci*. For the many-loci case, Ewens assumed that each locus was made up of two gametes *i* and *j*, which made up an (*i*, *j*) genotype. Suppose that, before natural selection acts, the genotype (*i*, *j*) has frequency g_{ii} if $i = j$ and frequency g_{ij} if $i \neq j$. Furthermore, let the allele *u* at locus *k* be denoted A_{ku} and have frequency p_{ku}. Then, the mean fitness is

$$\overline{w} = \sum_{i,j} g_{ij} w_{ij},$$

where w_{ij} is the fitness of the genotype (*i*, *j*). The change in genotype frequency in the current generation once natural selection has acted is

$$\Delta g_{ij} = \frac{g_{ij} w_{ij}}{\overline{w}} - g_{ij} = \frac{g_{ij}\left(w_{ij} - \overline{w}\right)}{\overline{w}}.$$

Therefore, the corresponding change in the frequency of allele A_{ku} is

$$\Delta p_{ku} = \sum \sum \frac{g_{ij}\left(w_{ij} - \overline{w}\right)}{\overline{w}}$$

(*ibid*., p. 176), where the summation is over all ordered genotypes (i, j) such that the gamete *i* contains A_{ku}. In analogy with the single-locus case, the average effect a_{ku} of allele A_{ku} is found by minimizing

$$\sum_{i,j} g_{ij} \left\{ w_{ij} - \overline{w} - \sum_{(i,j)} \alpha_{ku} \right\}^2$$

subject to $\sum_k \alpha_{ku} p_{ku} = 0$ and such that α_{ku} is counted twice if A_{ku} appears twice in the genotype. The normal equations are

$$p_{ku}\alpha_{ku} + \sum_{\substack{l,v \\ l \neq k}} P_{ku,lv}\alpha_{lv} = (\Delta p_{ku})\overline{w}$$

(Ewens, 2004, p. 247). Using the values of α_{ku} above, the additive genetic variance is (in analogy with (14.101))

$$\sigma_A^2 = 2\overline{w}\sum_{k,u} \alpha_{ku}(\Delta p_{ku}). \tag{14.105}$$

In analogy with (14.102), the mean fitness to be considered is written as

$$\sum_{i,j} g_{ij}\left\{\overline{w} + \sum_{(i,j)} \alpha_{ku}\right\}$$

and the partial change in mean fitness due to changes in g_{ij} values only is

$$\sum_{i,j} \Delta^* g_{ij}\left\{\overline{w} + \sum_{(i,j)} \alpha_{ku}\right\} = \sum_{i,j} \Delta^* g_{ij}\left\{\sum_{(i,j)} \alpha_{ku}\right\} = 2\sum_{k,u} \alpha_{ku}(\Delta^* p_{ku})$$

(Ewens, 1989, p. 176), where $\Delta^* p_{ku}$ is the change in the frequency of allele A_{ku}. Using Eq. (14.105), the above may be written as

$$\sum_{i,j} \Delta^* g_{ij}\left\{\overline{w} + \sum_{(i,j)} \alpha_{ku}\right\} = 2\sum_{k,u} \alpha_{ku}(\Delta p_{ku}) = \frac{\sigma_A^2}{\overline{w}},$$

which is the 'FTNS' for the multi-locus case.

Ewens has expressed the same disappointment as Price regarding the purview of the 'FTNS':

> **QUOT. 932** It is therefore important to assess whether the partial change in mean fitness is that component due to gene frequency changes. Price's concluding view is in the negative, and he is "disappointed that [the theorem] does not say more", regarding "the device of treating non-additive gene effects as 'environment' as a defect". These views are in line with my own negative assessment of the theorem as a biological statement. In the expression (12) [Eq. 14.102] for mean fitness, $\overline{w}$, α_i, and α_j, all depend on gene frequency, and thus all change with changes in gene frequency. Thus there appears to be no justification for singling out the partial change (13) [Eq. 14.103] as isolating the "natural selection" or "change in gene frequency" component of the total change in mean fitness, since the first component in each term in (12) [Eq. 14.102] is no more involved with gene frequency than the second component. A calculus analogy is the following. The rate of change of the function $f(x) = x - x^2$ with respect to changes in x cannot be found by writing $f(x) = x(1 - x)$, considering the rate of change of the first term (x) only, and arriving at a rate of change $1 - x$. The expression $(1 - x)$ is a function of x just as x is itself, and changes in $1 - x$ must equally be taken account of in calculating the change in $f(x)$. (*ibid.*,p. 179)

Ewens' point is that a partial change in mean fitness requires a change in P_{ij} while keeping $\overline{w}, \alpha_i$ and α_j fixed; however, it is not at all obvious how such a change can actually occur, since $\overline{w}, \alpha_i$ and α_j are all functions of P_{ij}.

A proper resolution of the 'FTNS', in the full generality that Fisher intended, has still not been achieved.

Chapter 15
Modern Synthesis (2): The Contributions of Wright

Abstract In contrast to Fisher's genic view of natural selection, Wright's approach focused on gene combinations. One of Wright's most useful inventions was the method of path coefficients. Using the latter, he defined the inbreeding coefficient in 1922. Wright's biggest achievement was his shifting balance theory (SBT) of evolution (1930, 1931), which was most efficient when a large population was divided into isolated (or almost isolated) groups. The SBT is a three-phase theory: (i) genetic drift creates various gene combinations within groups, (ii) intragroup selection creates local groups with high fitness, (iii) intergroup selection results in the local groups with highest fitness making the most offspring contribution, followed by gene flow from the fittest to the less fit groups, thus increasing the overall fitness of the population. In a subsequent paper, Wright used a visual metaphor called the adaptive landscape to illustrate his SBT.

Keywords Path coefficients · Tracing rule · "system of mating" papers · Correlation between uniting gametes · Inbreeding coefficient · Shifting balance theory · Adaptive landscape · The Sewall Wright equation" · Stochastic theory · Cotte · man's derivatives · Malécot's probability of identity.

15.1 Introduction

Together with Fisher and Haldane, Wright shares the distinction of being one of the founding fathers of the synthetic theory of evolution. Sewall Wright[1] (Fig. 15.1) was born in Massachusetts on December 21, 1889. In high school, he was exposed to evolutionary theory through Darwin's *Origin of Species*. While doing his BSc at Lombard College, he learned mathematics but soon became interested in biology under the influence of the geneticist Wilhelmine "Minnie" Marie Enteman Key (1872–1955). Wright completed his master's thesis on the anatomy of *Microphallus*

[1] Sewall Wright's life and work are also discussed in Provine (1989, 2001, pp. 154–167), Crow (1988, 1990b), Hill (1990), Tort (1996, pp. 4686–4698, Vol. 3: O-Z), Dronamraju (1990), and Russell (1989).

P. Gorroochurn, *The Development of Evolutionary Genetics*,
https://doi.org/10.1007/978-3-031-69374-8_15

Fig. 15.1 Sewall Wright (1889–1988) at the age of 96. (Photo obtained with permission from Joseph Felsenstein)

opacus in 1912, at which time he enrolled at Harvard University to work under William Castle. Under the latter's supervision, Wright completed his PhD thesis in 1915. In his dissertation, Wright demonstrated the existence of several loci and alleles that explained coat color in guinea pigs, and he provided physiological explanations for the observed patterns of coat color. Working with Castle, Wright also made key contributions to the latter's hooded-rat experiments, resulting in Castle's acceptance of modifiers and the abandonment of his controversial contamination theory.[2]

[2]See Sect. 13.2.3.

Wright's early experience in the areas of physiological genetics, multiple loci, and their interactions would later permeate his entire approach to evolutionary theory.

In 1915, Wright became a senior animal husbandman at the U.S. Department of Agriculture (USDA). He maintained that position until 1925 and joined the University of Chicago the next year. However, even before joining Chicago, Wright had already made three major contributions: (1) the invention of the method of path coefficients in 1918; (2) the development of the inbreeding coefficient in a series of papers in 1921; and (3) the completion of the first draft of his most important paper, "Evolution in Mendelian Populations",[3] in 1925.

Wright retired in 1955 to become the Leon J. Cole Professor of Genetics at the University of Wisconsin. He wrote his last paper in 1988, the same year he passed away (at the age of 98).

15.2 Wright's Invention of the Method of Path Coefficients (1918, 1920, 1921)

15.2.1 Introduction

Wright's method of path coefficients[4] was one of his greatest inventions. It turned out to be a powerful quantitative tool that could compete with some of Fisher's advanced mathematics. He was able to use his method to decompose phenotypic variances into genotypic and environmental components, similar to what Fisher had done. Wright went even further by defining a key concept in population genetics, namely, the inbreeding coefficient, through the use of path coefficients. In fact, much of Wright's evolutionary theory was built on path analysis.

According to Provine (1989, p. 126), Wright invented the method of path coefficients while he was a graduate. He first put it to use, without naming the method, in the 1918 paper "On the nature of size factors" (Wright, 1918). The term "path coefficient" was first used in Wright (1920), and the method was fully developed in Wright (1921a).

A good description of the method of path coefficients was provided by Wright in a March 15, 1924, letter to Henry A. Wallace:

QUOT. 933...Assume that we have a complex system of variables in which each one is represented as a linear function of others, as indicated by arrows in such a diagram as below:

[3]Wright continually revised this paper until its publication in 1931 (Wright, 1931).

[4]Wright's method of path coefficients is also described in Provine (1989, Chap. 5), Li (1955, Chaps. 12–13, 1975), and Wright (1960b). See also Wolfle (1999) for an annotated bibliography of Wright's work on path coefficients.

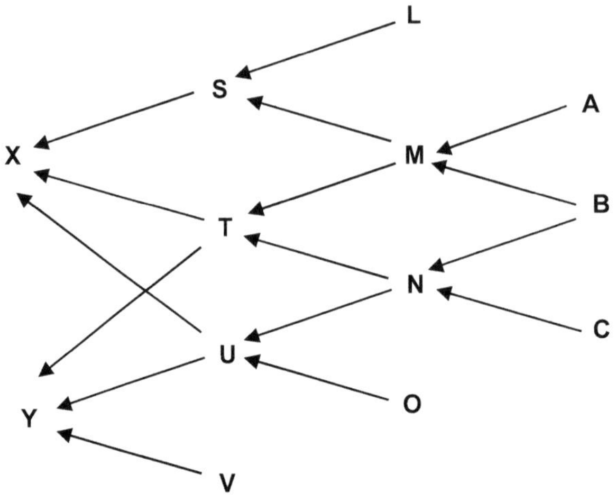

> The problem in which I was interested was to find a coefficient which would measure the influence along each path and would be related in a simple way to the correlations…
>
> …The result is a diagrammatic representation of the course of events, in which each line of influence is measured by a coefficient which gives the portion of the variation of the dependent variable for which the causal factor is directly responsible. This coefficient is identical with the partial regression coefficient if the standard deviations are the units of measurement.
>
> … The apparent simplicity of the idea of partial correlation, however, may be dangerous I think as likely to lead to an interpretation in terms of causal relation which is not justified. It goes without saying, however, that there are hundreds of cases in which multiple correlation and regression, and partial correlation (carefully interpreted) can be used for one in which there is sufficient knowledge of causal relations to make this use of path coefficients profitable.
>
> (Provine, 1989, pp. 148–149)

In the above, Wright explained that his method of path coefficients was an extension of multiple regression and represented the relationships within a network of variables, such that the arrow linking two variables in the network was assigned a path coefficient equal to the proportion of variation (measured in standard deviations) in the dependent variable due to variation in the causal variable.

To see why Wright's method is a set of regressions, consider the figure in QUOT. 933 with path coefficients *a, b, c,…* assigned to some of the paths (Fig. 15.2)

Then, the path model in Fig. 15.2 is equivalent to the following equations:

$$\begin{aligned} X &= S + T + U \\ Y &= T + U + V \\ S &= L + M \\ &\cdots \end{aligned} \tag{15.1}$$

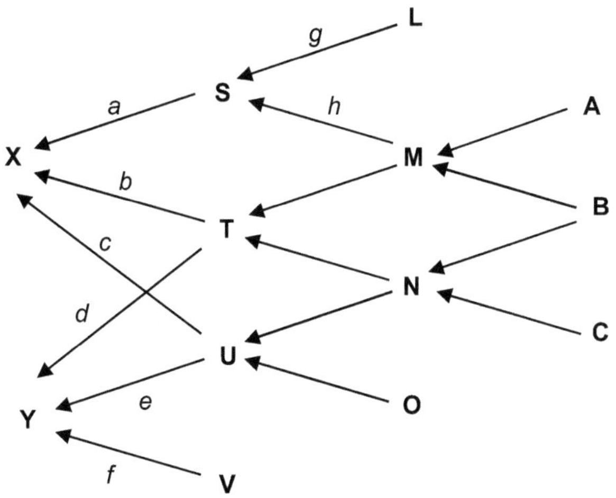

Fig. 15.2 Wright's path model in QUOT. 933 with path coefficients assigned to some of the paths

When the variables are standardized (by subtracting the sample mean of each variable and dividing by its standard deviation) to X^*, Y^*, S^*, T^*, ... the equations above become

$$\begin{aligned} X^* &= aS^* + bT^* + cU^* \\ Y^* &= dT^* + eU^* + fV^* \\ S^* &= gL^* + hM^* \\ &\ldots \end{aligned}$$

The system of equations above shows why path coefficients are equivalent to partial regression coefficients (measured in standard deviations).

In modern parlance, any variable that has at least one arrow pointing into it is said to be *endogenous* (dependent) and any variable that has no such arrow is said to be *exogenous* (independent). Thus, in Fig. 15.2, the variables *L, A, B, C, O,* and *V* are exogenous, and all other variables are endogenous.

In his letter, Wright rightly warned that path analysis could not *prove* causation but could efficiently be used to strengthen assumed causal relationships or theories. Thus, the method was not meant to discover *new* causal factors; rather, it operated on *assumed* ones.

Since Wright's invention of path coefficients, this method has blossomed, especially in the social sciences, into the powerful technique of *structural equation modeling* (SEM). The strength of SEM derives from its ability to accommodate both measured (as in Wright's case) and unmeasured (or latent) variables.

15.2.2 First Application of the Method of Path Coefficients Avant La Lettre *(1918)*

In 1914, Castle wrote an appendix entitled "The Nature of Size Factors as Indicated by a Study of Correlation" to a paper by his student MacDowell on the inheritance of size in rabbits (MacDowell, 1914). In his appendix, Castle considered the relative importance of factors that affected the growth of the body as a whole (i.e., *general factors*) versus factors that affected only specific parts of the body (i.e., *specific factors*). Based on the correlations (on rabbits) that Wright and Fisher had calculated for him, Castle declared:

> **QUOT. 934** In view of the high correlations obtaining between one skeletal dimension and another (and these agree closely with those observed in the case of man by Pearson and others), it follows that to a large extent the factors which determine size are general factors affecting all parts of the skeleton simultaneously. When the skull is long, the legs are long and the skull is wide, and every other part varies in proportion (or within 65 to 85 per cent of the same proportion). Whatever special factors (if any) there are, which are concerned in limiting the size of particular bones, these can play only a subordinate part in determining size. (*ibid.* p. 51)

The correlations Castle had presented (e.g., occipital to maxilla and zygoma posterior, length of zygoma posterior and humerus) were all high, ranging between .7 and .9.

Three years later, Davenport (1917) considered human parts and argued that the measurement pairs for rabbits considered by Castle were not relevant for humans. He reported several low correlations; for example, the correlation between stature and the femur was .37, and that between the clavicle and scapula was .12–.16. Based on the totality of the data he had considered, Davenport concluded:

> **QUOT. 935** . . .general factors control growth only to a degree that may be estimated as less than half. On the other hand, special factors are present that control, independently, the growth of the various elements that go to make up stature. And the graduated nature of the variations of stature must be largely due to the number of these independently varying units. (*ibid.*, p. 349)

It was in this context, namely, the discrepant conclusions reached by Castle and Davenport, that Wright decided to publish his trailblazing paper, "On the nature of size factors" (Wright, 1918). This paper was meant to be

> **QUOT. 936** . . .a further study of Prof. CASTLE'S results, designed to illustrate a method of analysis as well as to bring out certain conclusions. (*ibid.*, p. 367)

In the above, Wright wished "to illustrate a method of analysis." This was his method of path coefficients, although neither the name was used nor any path coefficient was explicitly calculated in 1918.

Before introducing his new method, Wright displayed several partial correlations that he had already calculated in 1914. For example, Castle's initial correlations and a new set of correlations between two factors while holding three other factors constant are shown in Table 15.1.

Table 15.1 Bivariate and partial correlations (holding three factors constant) on rabbits (Wright, 1918, p. 368)

Pairs of factors	Bivariate correlation $R \pm$ SE	Partial Correlation (holding three factors constant)
OM – *Zp*	.750 ± .015	.448
H	.743 ± .016	.172
F	.760 ± .015	.224
T	.701 ± .018	.022
Zp – *H*	.675 ± .019	.119
F	.674 ± .019	.004
T	.658 ± .020	.136
H – *F*	.857 ± .009	.463
T	.791 ± .013	.161
F – *T*	.858 ± .009	.517

OM = length of skull (occipital to maxilla), *Zp* = width of skull (posteriorly across the zygomatic arch), *H* = length of humerus, *F* = length of femur, *T* = length of tibia

Now, if the size of the body parts was determined by general factors, the partial correlations would be expected to decrease considerably. Table 15.1 shows that this is precisely what occurs, except for the OM-*Zp*, *H*-*F*, and *F*-*T* partial correlations. Therefore,

> **QUOT. 937** These three correlations suggest the existence of growth factors which affect the size of the skull independently of the body, others which affect similarly the length of homologous long bones apart from all else, and others which affect similarly bones of the same limb. (*ibid.*, p. 369)

Wright thus inferred separate specific growth factors for skull, homologous long bones, and bones of the same limb. Other variations in bone size were determined by general growth factors.

Similar conclusions held when partial correlations based on one and then two constant factors were calculated.

Having done that much, Wright next wished to "assign definite values to the different classes of growth factors which are indicated." It was for the accomplishment of this task, namely that of gauging the relative importance of different growth factors (as measured by the relative proportion of the variation in the observed values each factor explained) that Wright deployed his method of path coefficients. Accordingly, he posited separate factors for each of the following: (a) general size (all parts of the skeleton alike), (b) size of the skull only, but all skull bones alike, (c) length of leg bones only, but these alike, (d) length of bones of hind limbs only, but these alike, (e) length of homologous leg bones only, but these alike, (f) each part independently.

To illustrate the method of path coefficients, Wright considered the hypothetical example of two measurements, *X* and *Y*, which were determined by six independent causes as shown in Table 15.2.

Wright explained that the statistical basis of his method is the decomposition of variation:

Table 15.2 Wright's hypothetical example of two measurements X and Y determined by six independent causal factors [The variables a, b, c,... stand for the proportion of the variance in X explained by A, B, C,..., respectively; the variables a^1, b^1, c^1,... stand for the proportion of the variance in Y explained by A, B, C,..., respectively]

Effects	Causes					
	A	B	C	D	M	N
X	a	b	c	d	m	
Y	a^1	b^1	c^1	d^1		n^1

QUOT. 938 The extent to which a cause determines the variation in an effect is measured by the proportion of the squared standard deviation of the latter for which it is responsible. This follows from the proposition that the squared standard deviations due to single causes acting alone may be combined by simple addition to find the squared standard deviation of an array in which all causes are acting, provided the causes are independent of each other, i.e., $\sigma^2_{A+B+C} = \sigma^2_A + \sigma^2_B + \sigma^2_C$.

Like Fisher (see Sect. 14.2.2), Wright had thus recognized that the advantage of using the square of the standard deviation (which Fisher had called the "variance") as a measure of variation was the additive property it enjoyed. Both men were unknown to each other, but both published their papers in 1918 and both were concerned with the decomposition of the variance!

Furthermore, since a, b, c,... stand for the proportion of the variance in X explained by A, B, C,..., respectively, and a^1, b^1, c^1,... stand for the proportion of the variance in Y explained by A, B, C,..., respectively,

$$\begin{aligned} a + b + c + d + \ldots + m &= 1, \\ a^1 + b^1 + c^1 + d^1 + \ldots n^1 &= 1. \end{aligned} \tag{15.2}$$

Wright ended his hypothetical example by giving the correlation between X and Y as

$$r_{XY} = \pm\sqrt{aa^1} \pm \sqrt{bb^1} \pm \sqrt{cc^1} + \ldots, \tag{15.3}$$

where the coefficient of $\sqrt{aa^1}$ is positive if A produces effects in the same direction in X and Y, and negative otherwise, and similarly for the other terms.

The correlation above can be obtained by first representing the relationships in Table 15.2 in terms of standardized variables[5]

$$\begin{aligned} X^* &= \pm\sqrt{a}A^* \pm \sqrt{b}B^* \pm \ldots \pm \sqrt{m}M^*, \\ Y^* &= \pm\sqrt{a^1}A^* \pm \sqrt{b^1}B^* \pm \ldots \pm \sqrt{n^1}N^*. \end{aligned}$$

Then, since A^*, B^*, C^*, ... are independent of each other and are standardized,

[5]The equations below are in terms of the *path coefficients* which are $\pm\sqrt{\text{variance proportion}}$ (when the causes are independent).

Table 15.3 Wright's assumed model for the factors responsible for the five measured effects in rabbits

	Growth factors					
Effects	General	Skull	Legs	Hind Legs	Proximal leg bones	Special
OM	a	b	...	...	...	m
Zp	a^{I}	b^{I}	...	...	...	n^{I}
H	a^{II}	...	c^{II}	...	e^{II}	o^{II}
F	a^{III}	...	c^{III}	d^{III}	e^{III}	p^{III}
T	a^{IV}	...	c^{IV}	d^{IV}	...	q^{IV}

$$\mathrm{cov}(X^*, Y^*) = \pm\sqrt{aa^1}\,\mathrm{var}A^* \pm \sqrt{bb^1}\,\mathrm{var}B^* \pm \ldots$$

$$\frac{\mathrm{cov}(X^*, Y^*)}{\sqrt{\mathrm{var}X^*}\sqrt{\mathrm{var}Y^*}} = \pm\sqrt{aa^1}\frac{\mathrm{var}A^*}{\sqrt{a\mathrm{var}A^* + b\mathrm{var}B^* + \ldots}\sqrt{a^1\mathrm{var}A^* + b^1\mathrm{var}B^* + \ldots}} \pm$$

$$\sqrt{bb^1}\frac{\mathrm{var}B^*}{\sqrt{a\mathrm{var}A^* + b\mathrm{var}B^* + \ldots}\sqrt{a^1\mathrm{var}A^* + b^1\mathrm{var}B^* + \ldots}} \pm \ldots$$

$$\therefore\ r_{XY} = r_{X^*Y^*} = \pm\sqrt{aa^1}\frac{1}{\sqrt{a+b+\ldots}\sqrt{a^1+b^1+\ldots}}$$

$$\pm\sqrt{bb^1}\frac{1}{\sqrt{a+b+\ldots}\sqrt{a^1+b^1+\ldots}} \pm \ldots$$

By using Eq. (15.2), the correlation in Eq. (15.3) is obtained.

A simpler derivation can be achieved by using either the tracing rule or the first law of path analysis.[6]

Returning to his study of stature, Wright's Table 15.2 took the form shown in Table 15.3. The latter was based on his knowledge of human anatomy and the correlations he had calculated.

Wright's aim now was to estimate variance proportions represented by a, a^{I}, a^{II}, ..., b, b^{I}, ..., q^{IV}. The usual practice is to write equations of correlations in terms of path coefficients and solve for the latter, but in his initial paper Wright resorted to an approximate solution for the variance proportions. This approximation method may be summarized as follows:

(I) Using Eq. (15.3) and the bivariate correlations in Table 15.1, the following squared correlations were obtained:

(i) $r^2_{OM-H} = aa^{\mathrm{II}} = .5520$
(ii) $r^2_{Zp-H} = a^{\mathrm{I}}a^{\mathrm{II}} = .4556$
(iii) $r^2_{OM-F} = aa^{\mathrm{III}} = .5776$
(iv) $r^2_{Zp-F} = a^{\mathrm{I}}a^{\mathrm{III}} = .4543$
(v) $r^2_{OM-T} = aa^{\mathrm{IV}} = .4914$
(vi) $r^2_{Zp-T} = a^{\mathrm{I}}a^{\mathrm{IV}} = .4330$

[6] See QUOT. 944 and footnote 8.

(II) By dividing ii. by i., iv. by iii., and vi. by v., three estimates of a^{I}/a were obtained and averaged to .8310. Similarly, the average value for $a^{\mathrm{III}}/a^{\mathrm{II}}$ is 1.0218 and for $a^{\mathrm{IV}}/a^{\mathrm{II}}$ is .9203.

(III) Wright had calculated the partial correlation between OM and *Zp*, holding *H* constant, to be .502. Therefore,

$$ {}_{H}r_{OM-Zp} = \frac{r_{OM-Zp} - \sqrt{aa^{\mathrm{II}}}\sqrt{a^{\mathrm{I}}a^{\mathrm{II}}}}{\sqrt{(1-r^2_{OM-H})(1-r^2_{Zp-H})}} = .502 \quad \Rightarrow \quad \sqrt{aa^{\mathrm{I}}} = \frac{.5021}{a^{\mathrm{II}}} $$

Similarly, ${}_{F}r_{OM-Zp}$ was calculated to be .496. Using the above definition of the partial correlation and $a^{\mathrm{III}}/a^{\mathrm{II}} = 1.0218$ in (b), Wright obtained $\sqrt{aa^{\mathrm{I}}} = .5010/a^{\mathrm{II}}$. Finally, using ${}_{T}r_{OM-Zp} = .538$ and $a^{\mathrm{IV}}/a^{\mathrm{II}} = .9203$ in (b), $\sqrt{aa^{\mathrm{I}}} = .5010/a^{\mathrm{II}}$. Hence the average value of $\sqrt{aa^{\mathrm{I}}}$ is $.5014/a^{\mathrm{II}}$. Similarly, the average (based on two estimates) value of $\sqrt{a^{\mathrm{II}}a^{\mathrm{IV}}}$ was calculated to be $.5264/a$.

(IV) From Table 15.3, Eq. (15.3) yields $r_{OM-Zp} = \sqrt{aa^{\mathrm{I}}} + \sqrt{bb^{\mathrm{I}}} = .750$ (Table 15.1). Wright reasoned that the maximum value of $\sqrt{aa^{\mathrm{I}}}$ is .750 (when $\sqrt{bb^{\mathrm{I}}} = 0$). From (b), $a^{\mathrm{I}}/a = .831$, hence the maximum value of a is $.750/\sqrt{.831} = .823$. Similarly, the maximum value of a^{II} is $.791/\sqrt{.920} = .825$.

(V) From (c), $\sqrt{aa^{\mathrm{I}}} = .5014/a^{\mathrm{II}}$, and the minimum value of $\sqrt{aa^{\mathrm{I}}}$ is $.5014/.825 = .609$

(VI) From (d) and (e), the maximum and minimum values $\sqrt{aa^{\mathrm{I}}}$ are .750 and .638, respectively, so Wright took $\sqrt{aa^{\mathrm{I}}} = (.750 + .638)/2 = .679$ as a good approximation.

(VII) Once the value $\sqrt{aa^{\mathrm{I}}} = .679$ is adopted, all variance proportions can be estimated. For example, from (b) $a^{\mathrm{I}}/a = .8310$ so that $a = .679/\sqrt{.831} = .745$ and $a^{\mathrm{I}} = (.8310)(.745) = .619$, and so on

Wright was now able to fill in Table 15.3 to give Table 15.4. The latter showed a predominance of general growth factors and he concluded:

QUOT. 939 Analysis of the relations between five bone lengths in a population of rabbits shows that most differences between individuals are those which involve the size of the body as a whole. (*ibid.*, pp. 373–374)

Table 15.4 Wright's table of variance proportions for his assumed model on rabbits

	Growth factors					
Effects	General	Skull	Legs	Hind Legs	Proximal leg bones	Special
OM	.746	.077	. . .	. . .	. . .	.177
Zp	.620	.065	. . .	. . .	. . .	.315
H	.739	. . .	.083	. . .	.030	.148
F	.755	. . .	.085	.064	.030	.066
T	.680	. . .	.076	.058	. . .	.186

15.2.3 First Path Diagram: Path Coefficients First Explicitly Defined and Used (1920)

The reader may have noticed that Wright's 1918 paper (Wright, 1918) not only lacked any path diagram but also did not explicitly use path coefficients, preferring instead variance proportions. Both shortcomings were addressed in Wright's 1920 paper (Wright, 1920), which also contained the first explicit definition of the path coefficient.

In his paper, Wright reported that brother–sister matings had been performed at the USDA in 23 families of guinea pigs since 1906. A second stock where even second cousins were prevented from mating was also maintained. Wright remarked that, surprisingly, substantial variation existed in the coat colors within families of the inbred stock. His aim was therefore as follows:

> **QUOT. 940** As a supplement to more direct attempts to determine the cause of these variations in the pattern, it seemed desirable to find the relative importance of heredity and environmental factors. (*ibid.*, p. 321)

Wright first produced a table of correlations between members of a given family, as shown in Table 15.5.

He then noted that:

> **QUOT. 941** The negligible correlations between the parents (+0.019 in the control stock and +0.029 in the inbred family) show that matings actually were made at random as far as color is concerned. In the control stock, both the average correlation between parent and offspring

Table 15.5 Correlations between parents, between parent and offspring and between litter mates in random-bred stock and inbred family no. 35, during 1916, 1917, and 1918 (Wright, 1920, p. 327)

	Experiment B (random-bred)		Family (inbred)	
	No.	Correlation	No.	Correlation
Sire-Dam	105	+0.019 ± 0.066	73	+0.029 ± 0.079
Sire-Son	492	+0.231 ± 0.029	235	+0.013 ± 0.044
Sire-Daughter	484	+0.194 ± 0.030	236	+0.082 ± 0.044
Dam-Son	498	+0.251 ± 0.028	235	+0.042 ± 0.044
Dam-Daughter	488	+0.165 ± 0.030	236	−0.080 ± 0.044
Average, Parent-Offspring	**1962**	**+0.211 ± 0.015**	**942**	**+0.014 ± 0.022**
Brother-Brother	390	+0.219 ± 0.033	182	+0.090 ± 0.050
Brother-Sister	437	+0.228 ± 0.031	203	+0.062 ± 0.047
Sister-Sister	406	+0.180 ± 0.032	194	+0.064 ± 0.048
Average, Litter Mates	**1233**	**+0.214 ± 0.018**	**579**	**+0.069 ± 0.028**

The matings in the correlation between sire and dam are each weighted by the number of offspring. The probable errors in this case are based merely on the number of matings. The probable errors in the other cases are based on the number of entries in the tables. Owing to repetition of individuals, they are probably somewhat too small. The correlation between litter mates in experiment B is based on 894 individuals in litters in which two or more were graded. The probable error based on this number is ±0.022 instead of ±0.018 as given in the table. Similarly there were 426 individuals in such litters in family 35, giving a probable error of ±0.031 instead of ±0.028

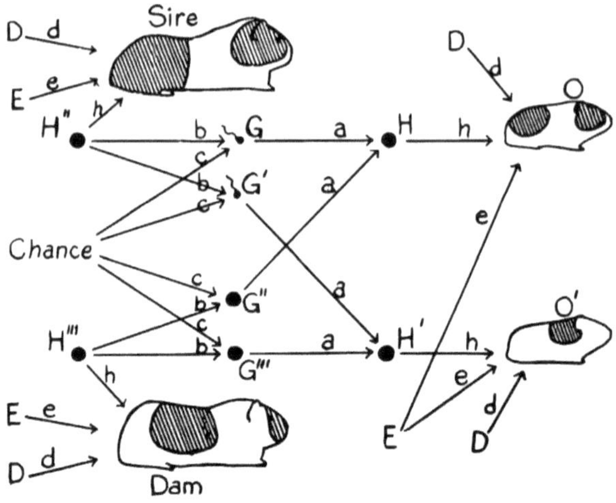

Fig. 15.3 Diagram illustrating the casual [sic] relations between litter mates (O, O′) and between each of them and their parents. H, H′, H″, H‴ represent the genetic constitutions of the four individuals, G, G′, G″, and G‴ that of four germ cells. E represents such environmental factors as are common to litter mates. D represents other factors, largely ontogenetic irregularity. The small letters stand for the various path coefficients (Wright, 1920, p. 328)

(+0.211) and that between litter mates (+0.214) are unquestionably significant. In the inbred family, there is virtually no correlation between parent and offspring (0.014) while that between litter mates is of doubtful significance (+0.069). (*ibid.*, p. 327)

Having given the necessary motivation, Wright was now ready to introduce his path diagram:

QUOT. 942 There is evidently very little genetic variability left in family 35, but is a good deal in the control stock. A closer analysis can be made by considering the relations between parent and offspring as shown diagrammatically in figure 5 (Fig. 15.3). (*ibid.*, pp. 327–328)

Wright's first path diagram is shown in Fig. 15.3. The small letters *a, b, c,. . .*, *h* represent the path coefficients. There are three causal factors, namely, *H* (hereditary), *E* (environmental conditions), and *D* (other factors such as developmental irregularities). Wright's aim was "to determine the degree of determination by each of these factors."

Before returning to this problem, Wright provided some details about the mechanics of path analysis through the simple example shown in Fig. 15.4. Here, two effects (*X* and *Y*) were determined in part by two common, correlated causes (*X* and *Y*). The path coefficients are represented by the small letters *a, b, c,. . ., d′*.

Assuming that all relations are approximately linear and that the various causes are additive, Wright provided an explicit definition of the path coefficient for the first time:

QUOT. 943 The path coefficient, measuring the importance of a given path of influence from cause to effect, is defined as the ratio of the variability of the effect to be found when all causes are constant except the one in question, the variability of which is kept unchanged, to the total variability. Variability is measured by the standard deviation. (*ibid.*, p. 329)

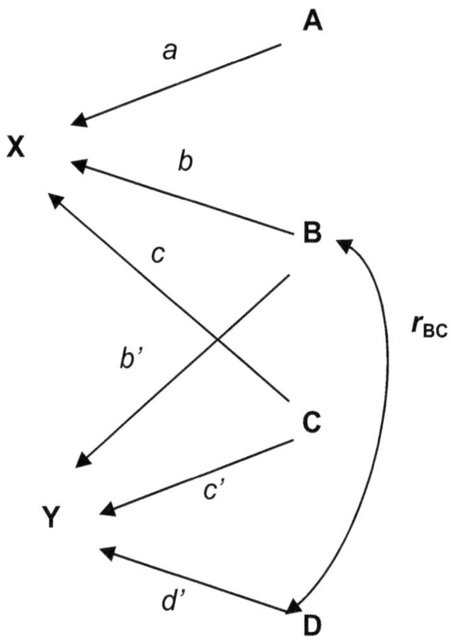

Fig. 15.4 Wright's example of two effects (*X*, *Y*) determined in part by two correlated causes (*B*, *C*)

For example, in Fig. 15.4, the path coefficients from *A* to *X* is

$$p_{X.A} = a = \frac{\sigma_{X.A}}{\sigma_X},$$

where σ_X is the standard deviation of *X* and $\sigma_{X.A}$ is the standard deviation of *X* due to variation in *A* only (i.e., holding *B* and *C* constant).[7] Moreover, the squares of the path coefficients are equal to the proportion of the variance in an effect measured by a cause, but "if the causes are correlated, terms representing joint determination must be recognized." Therefore, counting all the variance contributions for *X*,

$$a^2 + b^2 + c^2 + 2bcr_{BC} = 1. \tag{15.4}$$

Similarly, for *Y*,

$$b'^2 + c'^2 + d'^2 + 2b'c'r_{BC} = 1. \tag{15.5}$$

[7]The path coefficient is also the partial regression coefficient when variables are measured in standard deviations. See Fig. 15.2 and the discussion below it.

Next, Wright enunciated an important rule (the so-called *tracing rule*)[8] for determining the correlation between any two variables:

QUOT. 944 The correlation between two variables can be shown to equal the sum of the products of the chains of path coefficients along all of the paths by which they are connected. (*ibid.*, p. 330)

Applying this rule to Fig. 15.4,

$$r_{XY} = bb' + cc' + bc' r_{BC} + b' c r_{BC}$$

(*ibidem*). Similar equations can be written for the known correlations in terms of the unknown path coefficients. These equations, combined with Eqs. (15.4) and (15.5), can be used to uniquely solve for the unknowns if there are as many equations as there are unknowns.

Returning to Wright's original aim, namely the determination of the relative contributions of heredity and environment to the variation in coat color of guinea-pigs, in analogy with Eqs. (15.4) or (15.5), Fig. 15.3 yields[9]

$$h^2 + e^2 + d^2 = 1. \tag{15.6}$$

Using the tracing rule, the correlation between an offspring and its parent is

$$r_{O,\text{Sire}} = hbah = abh^2, \tag{15.7}$$

and the correlation between the two offspring O and O' is

$$r_{OO'} = habbah + habbah + ee = 2a^2b^2h^2 + e^2. \tag{15.8}$$

Since half of a parent's genetic constitution is passed on to any one germ cell, which in turn contributes half of an offspring's genetic constitution,

$$r^2_{GH''} = r^2_{GH} = a^2 = b^2 = \frac{1}{2} \quad \Rightarrow \quad a = b = \frac{1}{\sqrt{2}}$$

Using the value above and the correlations in Table 15.5, Eqs. (15.6)–(15.8) reduce to, for the randomly-bred family,

[8]The tracing rule holds if: (i) the same variable is not entered twice, and (ii) a variable is not left through an arrowhead and entered through an arrowhead at the same time (Kenny, 1979, p. 30). The tracing rule needs to be modified in *nonhierarchical models*, which are characterized by *feedback loops* (e.g., *X* causes *Y* and *Y* causes *X*, or *X* causes *Y*, which causes *Z*, which causes *X*). A more general method is the *first law* of path coefficients: $r_{YZ} = \Sigma_i p_{YX_i} r_{X_iZ}$, where *Y* is an endogenous variable, X_i is a causal variable for *Y*, p_{YX_i} is the path coefficient from X_i to *Y*, and r_{X_iZ} is the correlation between X_i and *Z*. The first law was enunciated in Wright (1934c, p. 163).

[9]The quantity h^2 would later be written as H^2 and called the "broad-sense heritability."

$$\left.\begin{aligned} h^2 + e^2 + d^2 &= 1 \\ \frac{1}{2}h^2 &= .211 \\ \frac{1}{2}h^2 + e^2 &= .214 \end{aligned}\right\} \Rightarrow h^2 = .422, e^2 = .003, d^2 = .575$$

and for the inbred family,

$$\left.\begin{aligned} h^2 + e^2 + d^2 &= 1 \\ \frac{1}{2}h^2 &= .014 \\ \frac{1}{2}h^2 + e^2 &= .069 \end{aligned}\right\} \Rightarrow h^2 = .028, e^2 = .055, d^2 = .917$$

Hence:

> **QUOT.** 945 Thus in the control stock, variations in pattern are determined about 42% by heredity, and 58% by irregularity in development, leaving nothing for tangible environmental factors. In the inbred family the corresponding figures are 3% for heredity, 5% for tangible environment and 92% for irregularity in development. (*ibid.*, p. 332)

Since inbreeding increases homozygosity, the contribution of heredity to pattern variability is, as expected, much less in the inbred family than in the control stock.

15.2.4 Synthesis of Results on Path Coefficients (1921)

In 1921, Wright wrote his famous paper entitled "Correlation and causation" (Wright, 1921a). The first part was a synthesis of the results of his 1918 and 1920 papers (especially the latter), and the second part contained two applications. The main results were summarized as follows:

(i) *Definitions of path coefficient and coefficient of determination.* In a path model where the variation of *X* is completely determined by, say, *A*, *B*, and *C*, the path coefficient from *A* to *X* is given by

$$p_{X.A} = \frac{\sigma_{X.A}}{\sigma_X}$$

(*ibid.*, p. 562), where σ_X is the standard deviation of *X* and $\sigma_{X.A}$ is the standard deviation of *X* due to variation in *A* only (i.e., holding *B* and *C* constant). Correspondingly, the proportion of variation in *X* explained by *A* only is $d_{X.A}$.

(ii) *Systems of independent causes*. If X is the sum of two independent causes A and B, then $\sigma_{X.A} = \sigma_A$, the standard deviation of A. Therefore,

$$p_{X.A} = \frac{\sigma_A}{\sigma_X}, \quad d_{X.A} = \frac{\sigma_A^2}{\sigma_X^2}.$$

Furthermore, the correlation between X and A is

$$\begin{aligned} r_{XA} &= \frac{\text{cov}(X, A)}{\sigma_X \sigma_A} \\ &= \frac{\text{cov}(A + B, A)}{\sigma_X \sigma_A} \\ &= \frac{\text{cov}(A, A) + \text{cov}(B, A)}{\sigma_X \sigma_A} \\ &= \frac{\text{var}A + 0}{\sigma_X \sigma_A} \\ &= \frac{\sigma_A}{\sigma_X} \\ &= p_{X.A} \end{aligned}$$

(*ibidem*).

(iii) *Chains of causes*. If $X = M + N$, where M and N are independent, and $M = A + B$, where A and B are independent, then

$$d_{X.M} = \frac{\sigma_M^2}{\sigma_X^2}, d_{M.A} = \frac{\sigma_A^2}{\sigma_M^2}, d_{X.A} = \frac{\sigma_A^2}{\sigma_X^2} \quad \Rightarrow \quad d_{X.A} = d_{X.M} d_{M.A}.$$

Similarly,

$$p_{X.A} = p_{X.M} p_{M.A}$$

(*ibid*., p. 563).

(iv) *Nonadditivity and nonlinearity*. Wright considered the example of $X = AB$, where A and B are independent, nonadditive factors. He showed that $d_{X.A} = r_{XA}^2$ still held, with $r_{XA}^2 = M_B^2 \sigma_A^2 / \sigma_X^2$, where M_B is the mean of B. Wright's conclusion was that, unless the variation of A and B was large in comparison with their mean values,

QUOT. 946 …[i]t is safe to deal with path coefficients and degrees of determination in the case of multiplying factors just as in the case of additive factors. (p. 564)

On the other hand, if X is a nonlinear function of A, then although $d_{X.A} = \eta_{X.A}^2 = p_{X.A}^2$, where $\eta_{X.A}^2$ is the correlation ratio between A and X, Wright warned that

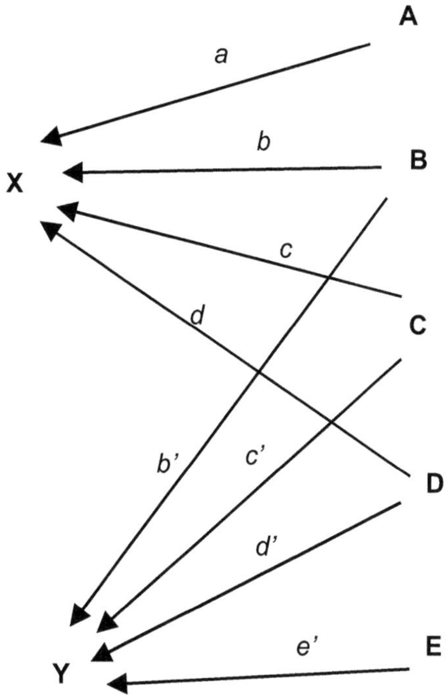

Fig. 15.5 Effects of common causes

QUOT. 947 [u]nfortunately we can not deal with chains of factors which involve nonlinear relations by mere multiplication of the path coefficients of the component links. (p. 565)

(v) *Effects of common causes*. Wright considered two effects *X* and *Y* which are due to common causes *B, C,* and *D*, as shown in Fig. 15.5. Assuming the common causes to be independent, the successive partial correlations are:

$$_B r_{XY} = \frac{r_{XY} - bb'}{\sqrt{(1-b^2)(1-b'^2)}},$$

$$_{CB} r_{XY} = \frac{_B r_{XY} - {_B r_{XC}} \cdot {_B r_{YC}}}{\sqrt{(1 - {_B r_{XC}^2})(1 - {_B r_{YC}^2})}} = \frac{r_{XY} - bb' - cc'}{\sqrt{(1-b^2-c^2)(1-b'^2-c'^2)}},$$

$$_{DCB} r_{XY} = \frac{r_{XY} - bb' - cc' - dd'}{\sqrt{(1-b^2-c^2-d^2)(1-b'^2-c'^2-d'^2)}}.$$

Since *B, C,* and *D* explain all the variation in *X* and *Y*, $_{DCB} r_{XY} = 0$ so that

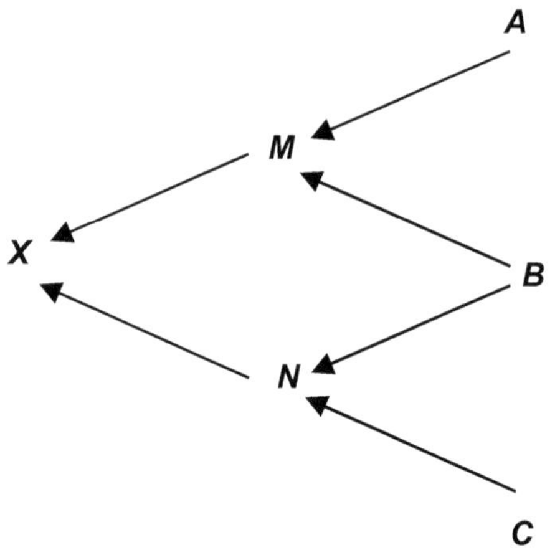

Fig. 15.6 Systems of correlated cases

$$r_{XY} = bb' + cc' + dd' = \sum_{I=B,C,D} p_{X.I}p_{Y.I}$$

(*ibidem*).

(vi) *Systems of correlated causes*. This is an extension of (c) for the case where B and C are correlated. Wright set $X = M + N$, where $M = A + B$ and $N = B + C$ as in

(vii) Figure 15.6.

Then $\sigma_{X.\ M} = \sigma_M$ and $\sigma_{X.\ N} = \sigma_N$ so that

$$d_{X.M} = \frac{\sigma_M^2}{\sigma_X^2}, \quad d_{X.N} = \frac{\sigma_N^2}{\sigma_X^2}.$$

However, $d_{X.\ M} + d_{Y.\ M} < 1$ since M and N are correlated. The remaining variation is explained by M and N jointly as

$$d_{X.\overline{MN}} = 2p_{X.M}p_{X.N}r_{MN}$$

(*ibid*., p. 566).

(viii) *General formula*. Here, Wright restated the tracing rule for the path coefficient method (see QUOT. 944). Applying this rule to Fig. 15.7 yields

$$r_{XY} = mm' + nn' + mbb'n' + nb'bm'$$

(*ibid*., p. 567).

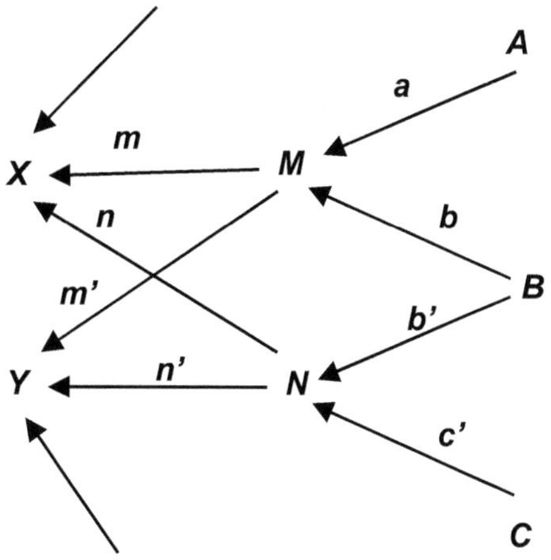

Fig. 15.7 General formula for the path coefficient method

15.3 Wright's Invention of the Inbreeding Coefficient (1921, 1922): An Application of Path Coefficients

15.3.1 Introduction

Wright's use of the method of path coefficients to define the inbreeding coefficient was another of his great achievements in evolutionary genetics. Among many other applications, he was able, with this newly defined quantity, to bypass much of Fisher's complicated mathematics by providing a simpler derivation of the correlation between relatives.[10]

Inbreeding or consanguinity (i.e., mating between close relatives) has been practiced since very ancient times. For example, the Pharaohs often married their own sisters, believing that this would prevent their divine blood from being diluted. Similar practices existed in many royal families, although various religious organizations (such as the Roman Catholic Church) placed restrictions on marriages between relatives who were too close. However, beyond religious and moral reasons, consanguineous marriages soon tended to be viewed negatively because of the observed deleterious effects on offspring born from such marriages.

The first systematic studies of the effects of inbreeding started with Jennings, Pearl, and others (Jennings, 1912, 1914, 1916, 1917; Pearl, 1913, 1914a, b, 1917a; Fish, 1914; Robbins, 1917, 1918b). Jennings (1912) studied continuous self-fertilization and showed that, starting with the genotype *Aa*, the proportion of heterozygotes y after n fertilizations is given by

[10] See Sect. 14.2.

$$y = \left(\frac{1}{2}\right)^n \quad n = 0, 1, 2, \ldots$$

(*ibid.*, p. 488), i.e., y decreases by a factor of (1/2) after each fertilization. Pearl (1913, 1914a, b), Fish (1914), Jennings (1914), and Robbins (1917) studied continued brother–sister mating.

Pearl (1917a) was probably the first to define an inbreeding coefficient. He first equated inbreeding with "the condition or state in which an organism has in fact fewer different ancestors than the maximum number possible" (*ibid.*, p. 545) and then defined the total amount of inbreeding by $Z_1 + Z_2 + \ldots + Z_n$, where

$$Z_i = \frac{100(p_{i+1} - q_{i+1})}{p_{i+1}}$$

(*ibidem*) is the inbreeding coefficient for the $(i + 1)^{\text{th}}$ ancestral generation, p_{i+1} is the maximum number of different individuals involved in the matings of the $(i + 1)^{\text{th}}$ generation, and q_{i+1} is the actual number of different individuals involved in the matings. However, a major problem with Pearl's definition is that the same values of $Z_1, Z_2, \ldots, Z_n$ may be obtained for different degrees of inbreeding of individuals in various generations.[11]

Most researchers modeled mating across successive generations by calculating the probabilities of the actual genotypic crosses. Robbins (1917, 1918a, b) made some advances by using recurrence relations and induction but, like others before him, he could not proceed beyond brother–sister matings.

15.3.2 *The "Systems of Mating" Papers (1921)*

15.3.2.1 Introduction

Against this background, Wright saw an opportunity to apply his method of path coefficients to investigate the effects of inbreeding on heterozygosity. He quickly wrote a series of five papers under the general title "Systems of Matings." They were all submitted at the same time to *Genetics* and published as one continuous series on March 1921 (Wright, 1921b, c, d, e, f). The purpose of the series was as follows:

> **QUOT. 948** In the problems to be dealt with in the present series of papers, we will assume that the original population is already in equilibrium under random mating. The question to be investigated in each case is the change in composition of the population brought about by a given system of mating. In some cases a new equilibrium point is approached, in others there is no equilibrium until complete homozygosis is reached. (Wright, 1921b, p. 113)

[11] Pearl attempted to address this issue in a subsequent paper (Pearl, 1917b) but, as Wright later explained, "[t]here was no assurance, however, that it did this in a way that was quantitatively related to experimental results." (Wright, 1969, p. 169)

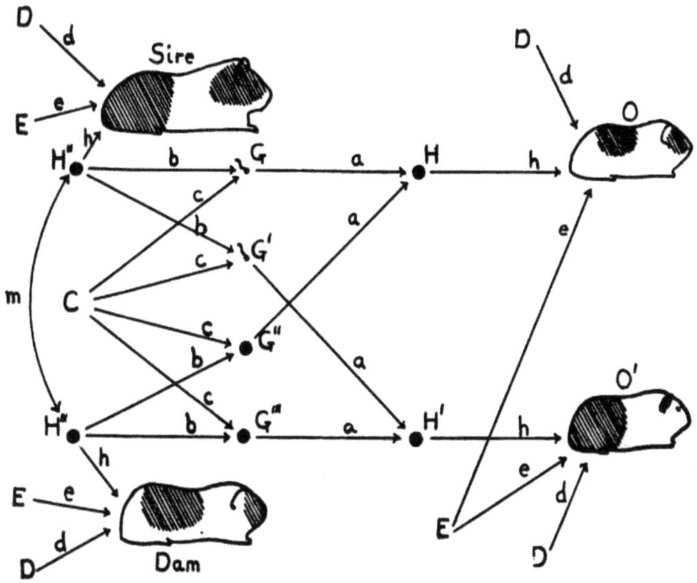

Fig. 15.8 Wright's reproduction of the path diagram in Fig. 15.3 with a correlation m now introduced between the two parents

15.3.2.2 First Paper: The Correlation f Between Uniting Gametes

In his first paper of the series, Wright reproduced the path diagram he had used in 1920 (Fig. 15.3), but with one major addition: a correlation m was now introduced between the two parents of an offspring, as shown in Fig. 15.8.

Furthermore, Wright introduced the fundamental quantity f,[12] defined as follows:

QUOT. 949 If there is assortative mating from any cause, there will be some correlation between the gametes which unite. Represent this correlation by f. (*ibid.*, p. 118)

Wright used the above definition to address one of the major aims of his current work, namely, the investigation of the heterozygosity under inbreeding. He noted that the percentage of heterozygotes was 50% under random mating in a bi-allelic system (A, a) with equal allele frequencies. However, under inbreeding:

QUOT. 950 Assume that there are x cases in which A unites with A, an equal number in which a unites with a, and $1 - 2x$ cases in which A unites with a. The product-moment correlation between the gametes, which we have called f, comes out $4x - 1$. Thus for the percentage of heterozygosis we have

$$p = 1 - 2x = ½\,(1 - f)$$

(*ibid.*, p. 120)

[12] Soon to be called the inbreeding coefficient.

Since the frequencies of the two alleles are each .5 in the above, it is instructive to see how the definition of f as the correlation between uniting gametes leads to a heterozygosity of $p = (1 - f)/2$. Let

$$X_A = \begin{cases} 1 & \text{if male gamete is } A \\ 0 & \text{if male gamete is } a, \end{cases}$$
$$Y_A = \begin{cases} 1 & \text{if female gamete is } A \\ 0 & \text{if female gamete is } a. \end{cases}$$

Then,

$$\begin{aligned} f &= \text{corr}(X_A, Y_A) \\ &= \frac{\text{E}X_A Y_A - \text{E}X_A \text{E}Y_A}{\sqrt{\text{E}X_A^2 - \text{E}^2 X_A}\sqrt{\text{E}Y_A^2 - \text{E}^2 Y_A}} \\ &= \frac{x - (1/2)(1/2)}{\sqrt{1/2 - 1/4}\sqrt{1/2 - 1/4}} \\ &= \frac{x - 1/4}{1/4}. \end{aligned}$$

Hence, the heterozygosity is[13]

$$p = 1 - 2x = \frac{1}{2}(1 - f). \tag{15.9}$$

Thus, as expected, under increased inbreeding ($f \to 1$), $p \to 0$ and under random mating ($f = 0$), $p = 1/2$.

Wright next showed how the path coefficients in Fig. 15.8 could be obtained. Since the sum of the relative contributions to the total variation in H (or H') is unity and the correlation between G and G'' (or G' and G'') is f, he obtained

$$2a^2 + 2a^2 f = 1 \quad \Rightarrow \quad a = \frac{1}{\sqrt{2(1 + f)}} \tag{15.10}$$

(*ibidem*), where a is the path coefficient from gamete to zygote.

To determine the path coefficient b from zygote to gamete, Wright considered hypothetical path coefficients into H''. Using the same principle as above,

[13]This formula was first generalized for unequal allele frequencies in Wright (1927, p. 40) to give $p = 2xy\,(1 - f)$, where x is the frequency of A and $y = 1 - x$ is the frequency of a.

$$2p^2_{H''.G} + 2p^2_{H''.G}g = 1 \quad \Rightarrow \quad 2p^2_{H''.G}(1+g) = 1,$$

where g is the correlation between two gametes (G and G') from a given individual (H''). Moreover, applying the tracing rule,[14]

$$r_{H''G} = p_{H''.G} + p_{H''.G}g = p_{H''.G}(1+g) = b.$$

Eliminating $p_{H''.G}$ from the last two equations,

$$b^2 = \frac{1}{2}(1+g) = \frac{1}{2}\left(1+f'\right)$$

since g is also the correlation between uniting gametes from the previous generation (i.e., $g = f'$). Hence,

$$b = \sqrt{\frac{1+f'}{2}} \tag{15.11}$$

(*ibid.*, p. 119). Finally, the relationship between f and m can be obtained from

$$r_{G'G'''} = f = bmb \quad \Rightarrow \quad f = mb^2 \tag{15.12}$$

Wright now proceeded to the calculation of correlations between relatives. Denoting the male parent (Sire) in Fig. 15.8 by P and writing the path coefficient from H'' to P as h' (to distinguish it from the h from H to O in the next generation), the parent-offspring correlation is obtained from the tracing rule as

$$r_{PO} = habh' + habmh' = abhh'(1+m) \tag{15.13}$$

and the offspring–offspring correlation is

$$r_{OO'} = 2habbah + 2habmbah + e^2 = 2a^2b^2h^2(1+m) + e^2 \tag{15.14}$$

(*ibid.*, p. 120).[15]

Wright's formulas were summarized (see Table 15.6) for consanguine mating ($f > 0$), population in equilibrium (m, g, f, a^2, b^2 constant) and random mating ($f = 0$).

Having done that much, Wright set the agenda for the second paper in the series:

---

[14] See QUOT. 944.

[15] Wright also gave correlations for the case of dominance but these were erroneous. He acknowledged and addressed this issue in Wright (1931, pp. 115–116).

Table 15.6 Summary of formulas used by Wright for consanguine mating ($f > 0$), population in equilibrium (m, g, f, a^2, b^2 constant), and random mating ($f = 0$)

Consanguine mating	Equilibrium	Random mating
$h^2 + d^2 + e^2 = 1$	$h^2 + d^2 + e^2 = 1$	$h^2 + d^2 + e^2 = 1$
$m = \phi(a'b'm')$	$m = \text{constant}$	$m = 0$
$g = f'$	$g = f' = f = \dfrac{m}{2-m}$	$g = f' = f = 0$
$b^2 = \dfrac{1}{2}(1+f')$	$b^2 = \dfrac{1}{2-m}$	$b^2 = \dfrac{1}{2}$
$f = b^2 m$	$f = \dfrac{m}{2-m}$	$f = 0$
$a^2 = \dfrac{1}{2(1+f)}$	$a^2 = \dfrac{1}{4}(2-m)$	$a^2 = \dfrac{1}{2}$
$ab = \dfrac{1}{2}\sqrt{\dfrac{1+f'}{1+f}}$	$ab = \dfrac{1}{2}$	$ab = \dfrac{1}{2}$
$p = \dfrac{1}{2}(1-f)$	$p = \dfrac{1-m}{2-m}$	$p = \dfrac{1}{2}$
$h^2 = \dfrac{2h_0^2(1-p)}{h_0^2(1-2p)+1}$	$h^2 = \dfrac{2h_0^2}{2-m\left(1-h_0^2\right)}$	$h^2 = h_0^2$
$r_{pp} = mh'^2$	$r_{pp} = mh^2$	$r_{pp} = 0$
$r_{po} = abhh'(1+m)$	$r_{po} = \dfrac{1}{2}h^2(1+m)$	$r_{po} = \dfrac{1}{2}h^2$
$r_{oo} = 2a^2b^2h^2(1+m) + e^2$	$r_{oo} = \dfrac{1}{2}h^2(1+m) + e^2$	$r_{oo} = \dfrac{1}{2}h^2 + e^2$

> **QUOT. 951**...it is possible to express the correlations and path coefficients in each generation in terms of those in the preceding generation, provided that an expression can be found for *m*, the correlation between the genetic constitutions of the parents. This can easily be done for many systems of consanguine mating as will be brought out in a subsequent paper. (*ibid.*, p. 121)

15.3.2.3 Second Paper: Inbreeding

As explained in QUOT. 951, Wright's major aim in the second paper (Wright, 1921c) was to obtain an appropriate value of *m* for a given system of mating. This value then enabled him to calculate *f* and various path coefficients. Wright did this first for the then well-known example of continued brother–sister mating and was thus able to verify his method. Using ', ", ''',... to denote one, two, three, ... generations prior to the present, he gave the following recurrence relation for *m* (correlation between two parents):

$$m = 2a'^2 b'^2 \left(1 + m'\right) \tag{15.15}$$

(*ibid.*, p. 126). This formula can be obtained by applying the tracing rule to Fig. 15.9, which is a simplification of Fig. 15.8.

Using Eqs. (15.10)–(15.15) and assuming that initially $m = f = 0$, Table 15.7 can be generated for successive values of *m*, *f*, and *p*.

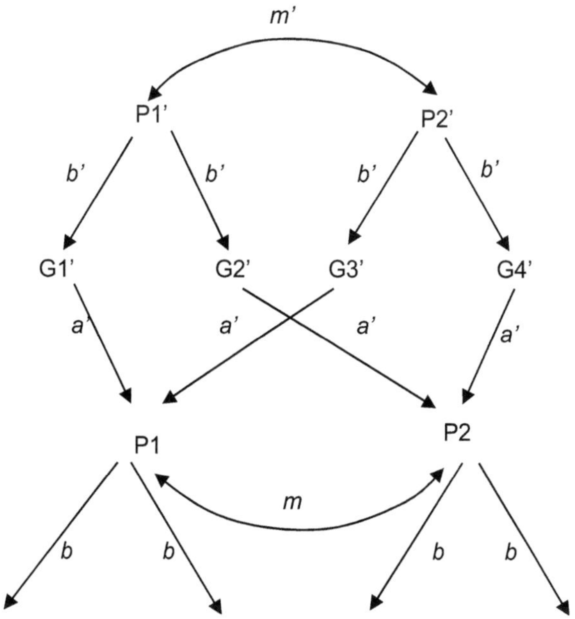

Fig. 15.9 Recursive relationship between *m* and *m*′ for continued brother–sister mating. *P*1, *P*2,... are the parents, and *G*1,...,*G*4 are the gametes, spring. The dash (′) refers to one previous generation

Table 15.7 Wright's calculation of successive values of *m, f,* and *p* for repeated brother–sister matings

Gen	m	f'	$(1+f')/2$ (b^2)	mb^2 (f)	$\{2(1+f)\}^{-1}$ (a^2)	$2a^2b^2(1+m)$ (r_{OO})	$ab(1+m)$ (r_{OP})	$p=(1-f)/2$
0	0	0	1/2	0	1/2	1/2	1/2	2/4
1	1/2	0	2/4	1/4	2/5	3/5	$\frac{3}{2}\sqrt{\frac{1}{5}}$	3/8
2	3/5	1/4	5/8	3/8	4/11	8/11	$\frac{8}{5}\sqrt{\frac{5}{22}}$	5/16
3	8/11	3/8	11/16	8/16	8/24	19/24	$\frac{19}{11}\sqrt{\frac{11}{48}}$	8/32
4	19/24	8/16	24/32	19/32	16/51	43/51	$\frac{43}{24}\sqrt{\frac{4}{17}}$	13/64
5	43/51	19/32	51/64	43/64	32/107	94/107	$\frac{94}{51}\sqrt{\frac{51}{214}}$	21/128

Wright next showed how a recursion for *f* could be obtained. Using Eqs. (15.10)–(15.12), Eq. (15.15) becomes

$$m=\frac{1+2f^{/}+f^{//}}{2\left(1+f^{/}\right)}.$$

Using Eqs. (15.11) and (15.12), the above becomes

$$\frac{f}{\left(1+f^{/}\right)/2}=\frac{1+2f^{/}+f^{//}}{2\left(1+f^{/}\right)}\quad\Rightarrow\quad f=\frac{1}{4}\left(1+2f^{/}+f^{//}\right)$$

(*ibid.*¸p. 127). Using $p=(1-f)/2$ (Eq. 15.9), the corresponding recursive equation for the heterozygosity p is

$$p=\frac{p^{/}}{2}+\frac{p^{//}}{4}$$

(*ibid.*, p. 128). Wright noted that his values of p were compatible with Jennings' calculations of the homozygosity (Jennings, 1914, p. 694).

Wright provided another verification of his method by considering self-fertilization. He put $m=1$ in Eq. (15.12) and used Eq. (15.11), so that

$$f=\frac{1}{2}\left(1+f^{/}\right).$$

Using Eq. (15.9), he also obtained

$$p=\frac{1}{2}p^{/}$$

(*ibid*., p. 130), in agreement with Jennings' formula $(1/2)^n$, $n = 1, 2, \ldots$ (Jennings, 1912, p. 488).

Several other mating systems were considered by Wright. For example:

(a) *Parent–Offspring Mating* [$m = \sqrt{\frac{1}{2}a' + a'b'm'}$]:

$$f = \frac{1}{4}\left(1 + 2f'\right), \qquad p = \frac{1}{8}\left(1 + 4p'\right)$$

(*ibid*., p. 131)

(b) *Double first Cousin Mating* [$m = a'^2b'^2(2m' + 4a''^2b''^2(m'' + 1))$]:

$$f = \frac{1}{8}\left(4f' + 2f'' + f''' + 1\right), \qquad p = \frac{1}{8}\left(4p' + 2p'' + p'''\right)$$

(*ibid*., p. 134)

15.3.2.4 Third Paper: Assortative Mating

In the third paper of the "Systems of Mating" series, Wright (1921d) investigated the effect of assortative mating on heterozygosity and the correlation f between uniting gametes. Since the case of assortative mating (where the phenotype was determined by alleles at one locus) had already been investigated (e.g., Jennings (1916, 1917); Wentworth and Remick (1916)), Wright provided an example of a trait determined by genes at two loci. He numerically showed how assortative mating led to a decrease in heterozygosity and an increase in the correlation between uniting gametes (f).

He then considered the general case of n loci A, B, C,... and used the path diagram in Fig. 15.10 to show the assortative mating between two individuals.

Using a dash (') to denote quantities in the previous generation, let:

r_{pp} = correlation between mated individuals based on their somatic resemblance,

h^2 = relative contribution of heredity to total phenotypic variation,

m = correlation between zygotic constitutions of two mated individuals,

f_u = correlation between uniting alleles (e.g. A and A') at the same locus,

k_u = correlation between alleles (e.g. A and B) at different loci on the same uniting gamete,

j_u = correlation between alleles (e.g. A and B'') at different loci on different uniting gametes.

Now,

> ...with assortative mating based wholly on external characters, there is no reason why an individual of formula *AAbb* should have a greater tendency to mate with his like than with *aaBB*.

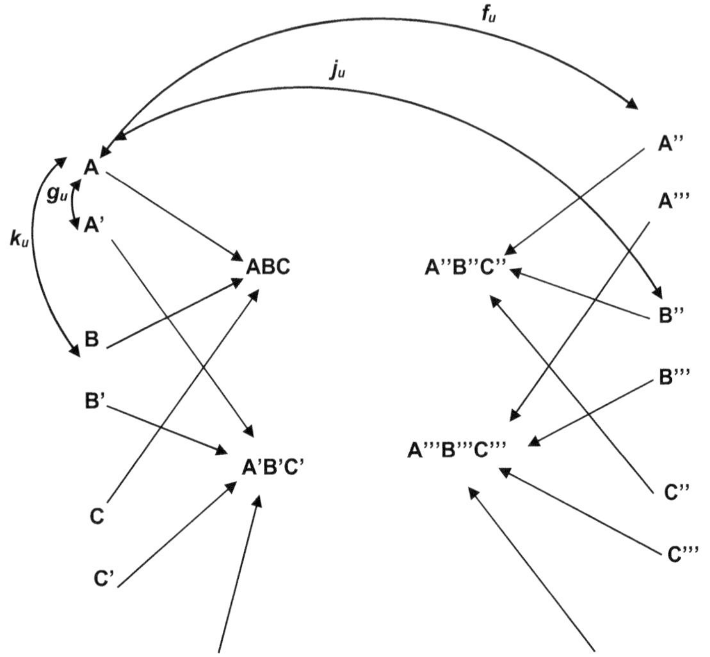

Fig. 15.10 Wright's path diagram for assortative mating when the phenotype is determined by genes at multiple loci A, B, C,.... ABC, A′B′C′,... represent uniting gametes

> Thus, $j_u = f_u$
> (*ibid.*, p. 147)

Wright's key formulas for assortative mating were as follows:

$$
\begin{aligned}
m &= h'^2 r_{pp}, \\
g_u &= f'_u, \\
k_u &= \frac{1}{2}\left(f'_u + k'_u\right), \\
f_u &= \frac{m}{2n}\left\{1 + f'_u + 2(n-1)k_u\right\}, \\
g &= \frac{f'_u + (n-1)k_u}{1 + (n-1)k_u}, \\
f &= \frac{m}{2}(1 + g), \\
p &= \frac{1}{2}(1 - f_u)
\end{aligned}
$$

(*ibid.*, p. 151). For one locus ($n = 1$), these formulas give

$$m = 1, \quad f_u = \frac{1}{2}\left(1 + f'_u\right), \quad p = \frac{1}{2}\left(1 - f'_u\right) = \frac{1}{2}p',$$

so that, as expected, the heterozygosity is halved after each generation of assortative mating. For two loci ($n = 2$),

$$m = 1, \quad k_u = \frac{1}{2}\left(f_u' + k_u'\right), \quad f_u = \frac{1}{4}\left(1 + f_u' + 2k_u\right).$$

Starting with $f_u = k_u = 0$, Wright obtained

$$\begin{aligned} k_u &= 0, 0, \frac{1}{8}, \frac{1}{4}, \frac{23}{64}, \frac{29}{64}, \ldots, 1, \\ f_u &= 0, \frac{1}{4}, \frac{3}{8}, \frac{30}{64}, \frac{35}{64}, \frac{157}{256}, \ldots, 1. \end{aligned}$$

Therefore,

$$\begin{aligned} g &= 0, 0, \frac{1}{3}, \frac{1}{2}, \frac{53}{87}, \frac{64}{93}, \ldots, 1, \\ f &= \frac{1}{2}, \frac{1}{2}, \frac{2}{3}, \frac{3}{4}, \frac{79}{87}, \frac{157}{186}, \ldots, 1, \\ p &= \frac{1}{2}, \frac{3}{8}, \frac{5}{16}, \frac{17}{64}, \frac{29}{128}, \frac{99}{512}, \ldots, 0. \end{aligned}$$

Wright noted several differences between assortative mating and inbreeding:

(a) In the case of assortative mating, the decrease in heterozygosity depends on the number of loci involved. On the other hand, in the case of inbreeding, the decrease in heterozygosity is independent of the number of loci involved.
(b) The relative amount of variation (measured in standard deviation) due to heredity in the case of assortative mating, compared to random mating, increases with the number of loci. On the other hand, this variation is constant for the case of inbreeding.
(c) With perfect assortative mating at, say, two bi-allelic loci, A and B, the population tends to split into only two extreme types *AABB* and *aabb*, whereas with inbreeding all homozygous types *AABB, AAbb, aaBB,* and *aabb* tend equally to fixation.

15.3.2.5 Fourth Paper: Natural Selection

Remarkably, one year before Fisher wrote his "On the Dominance Ratio" paper (Fisher, 1922), where he established that natural selection does not necessarily lead to the fixation of a favorable allele,[16] Wright stated a similar result in the fourth of the "Systems of Mating" series (Wright, 1921e).

---

[16] See Sect. 14.3.2.

Wright started his paper by referring to the works of Jennings (1916) and Wentworth & Remick (1916). These authors had established that, in an original population consisting of ¼ *AA*, ½ *Aa*, and ¼ *aa* (where *A* is dominant over *a*), selection for dominant genotypes led to the following series for heterozygotes:

$$\left(\frac{1}{2}\right), \frac{4}{9}, \frac{6}{16}, \frac{8}{25}, \ldots, \frac{2(n+1)}{(n+2)^2}, \ldots 0$$

If selection was on recessives, they were immediately fixed. Now, if there was no dominance:

QUOT. 952. . .selection of either homozygous type means immediate fixation. Selection of heterozygotes leads to no fixation whatever, however long it may be continued. The color of the Blue Andalusian fowl is the classical example of an unfixable characteristic of this kind. The roan color of Shorthorn cattle, the cream color of guinea-pigs and the yellow color of mice are other cases. (*ibid.*, p. 163)

Wright also provided an example of a two-locus model with alleles (*A*, *a*) at one locus and alleles (*B*, *b*) at the other locus (the capital denotes the wild-type and the lowercase denotes the mutant type). The number of possible wild-type alleles ("positive factors") in an individual is then 0, 1, 2, 3, or 4. He argued that, if natural selection acted on intermediate individuals (i.e., those with only two positive factors—AAbb, AaBb, or aaBB), the proportion of positive factors would remain constant at ½, no matter how long natural selection acted. Furthermore, the standard deviation decreases to only 72.8% of its original value after continued selection.

It is noteworthy that, as early as 1921, Wright had doubts about the power of natural selection in small populations:

QUOT. 953 In a population of limited size, such selection would, of course, tend to fix some intermediate type, but the inbreeding consequent on small numbers is a necessary condition for this fixation. (*ibid.*, pp. 164–165)

15.3.2.6 Fifth Paper: Summary

Wright's last paper (Wright, 1921f) in the series was a summary of the results of his four previous papers:

QUOT. 954 It will be seen that all of the systems of mating have their advantages and disadvantages. Close inbreeding automatically brings about fixation of type and prepotency. Intermediate types are fixed as readily as extremes. It is the only method of bringing to light hereditary differences in characters which are determined largely by factors other than heredity. On the other hand, close inbreeding is likely to lead toward reduced fertility, size and vigor.

Matings between relatives more remote than first cousins have little significance as inbreeding, except in so far as there is continued breeding within a population of small size.

Assortative mating and selection can lead to fixation of extreme types only and are not very efficient in this respect. Selection, however, is the only means of permanently changing the relative proportions of the various genes present in the original stock. It is an essential adjunct of the other systems as means of improvement.

Assortative mating leads to the greatest diversification of the population as a whole, and thus is practically always accompanied by selection either in nature or in live-stock breeding. Under conditions such that all progeny are to be saved for breeding, this diversification of the population is a disadvantage. Disassortative mating is the method which best holds the whole population together, pending the fixation of the average type by close inbreeding. (*ibid.*, pp. 177–178)

In the above, Wright again downplayed the power of natural selection: although it was the only force of permanent change, it was "not very efficient." He reemphasized the importance of population structure, especially close breeding:

QUOT. 955 Straight selection unaccompanied by close breeding is thus not an effective method for permanently fixing characteristics. (*ibid.*, p. 176)

15.3.3 The Correlation f Named "Inbreeding Coefficient," The Pedigree Formula (1922)

15.3.3.1 Development of the Inbreeding Coefficient f

Building on his 1921 series, Wright (1922a) was now ready to define a quantity that could measure the extent of inbreeding—he called such a quantity an *inbreeding coefficient*. He first explained:

An inbreeding coefficient to be of most value should measure as directly as possible the effects to be expected on the average from the system of mating in the given pedigree.

There are two classes of effects which are ascribed to inbreeding: First, a decline in all elements of vigor, as weight, fertility, vitality, etc., and second, an increase in uniformity within the inbred stock, correlated with which is an increase in prepotency in outside crosses.... it may readily be shown that the decrease in vigor on starting inbreeding in a previously random-bred stock should be directly proportional to the increase in the percentage of homozygosis... As for the other effects of inbreeding, fixation of characters and increased prepotency, these are of course in direct proportion to the percentage of homozygosis. Thus, if we can calculate the percentage of homozygosis which would follow on the average from a given system of mating, we can at once form the most natural coefficient of inbreeding. (*ibid.*, pp. 331–332)

In the above, Wright argued that a first requirement of the "most natural coefficient of inbreeding" was that it should be an increasing function of homozygosity since the latter was an inevitable consequence of inbreeding. Next:

QUOT. 956 Taking the typical case in which there are an equal number of dominant and recessive genes (A and a) in the population, the random-bred stock will be composed of 25 per cent. *AA*, 50 per cent. *Aa* and 25 per cent. *aa*. Close inbreeding will tend to convert the proportions to 50 per cent. *AA*, 50 per cent. *aa*, a change from 50 per cent. homozygosis to 100 per cent. homozygosis. For a natural coefficient of inbreeding, we want a scale which runs from 0 to 1, while the percentage of homozygosis is running from, 50 per cent. to 100 per cent. (*ibid.*, pp. 332–333)

Wright's second requirement was that, for equal frequencies of two alleles at a locus, the measure of inbreeding should vary from 0 to 1 as the homozygosity varied from .5 (random mating) to 1 (complete inbreeding).

He next claimed that the quantity f, which he had defined[17] as the correlation between uniting gametes in the first of the 1921 series (Wright, 1921b), satisfied these two requirements. From Eq. (15.9), for equal allele frequencies, the formula for f is

$$f = 1 - 2p = 2h - 1 \tag{15.16}$$

(*ibid.*, p. 333), where p is the heterozygosity and $h = 1 - p$ is the homozygosity.

Wright's correlation f was thus an inbreeding coefficient "of general application."

In a paper (Wright, 1922b) published the same year, Wright used the correlation definition of f to generalize (15.9) for unequal allele frequencies to

$$f = 1 - \frac{p}{2xy} \tag{15.17}$$

(*ibid.*, p. 40), where x and y are the frequencies of alleles A and a, respectively. The formula above gives the following frequencies for the genotypes Aa, AA, and aa:

$$\begin{aligned} P_{Aa} &= p = 2xy(1 - f), \\ P_{AA} &= x^2 + xyf, \\ P_{aa} &= y^2 + xyf. \end{aligned}$$

Thus, although these formulas are usually attributed to Bernstein (1930) (e.g., see Cavalli-Sforza and Bodmer (1971, p. 349)), Wright had essentially derived them in 1922.

15.3.3.2 The Pedigree Formula

Having developed the inbreeding coefficient f, Wright used it to determine the path coefficient from a male parent (S) to an offspring (O) as

$$p_{o.s} = \frac{1}{2}\sqrt{\frac{1 + f_s}{1 + f_o}}$$

(*ibid.*, p. 333), where f_s, f_o are the inbreeding coefficients of S and O, respectively. This formula may be obtained by modifying Fig. 15.8 as shown in Fig. 15.11.

From (15.11) and (15.10), the path coefficient from S to O is

[17] See QUOT. 949.

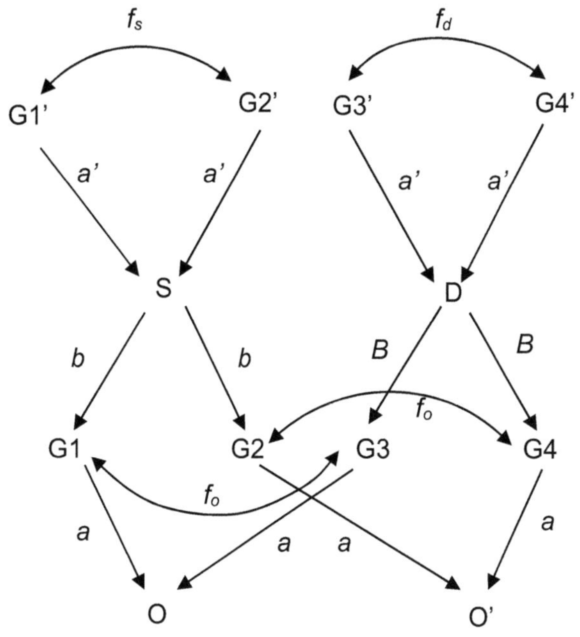

Fig. 15.11 Wright's development of the pedigree formula

$$p_{o.s} = ba = \sqrt{\frac{1}{2}(1+f_s)}\sqrt{\frac{1}{2(1+f_o)}} = \frac{1}{2}\sqrt{\frac{1+f_s}{1+f_o}}.$$

By using a similar reasoning, the path coefficient from S's father (G) to O is

$$\begin{aligned} p_{o.g} &= p_{o.s}p_{s.g} \\ &= \frac{1}{2}\sqrt{\frac{1+f_s}{1+f_o}}\cdot\frac{1}{2}\sqrt{\frac{1+f_g}{1+f_s}} \\ &= \frac{1}{4}\sqrt{\frac{1+f_g}{1+f_o}} \end{aligned}$$

and the path coefficient from the Nth great-grandfather (A) to O is

$$p_{o.g} = \left(\frac{1}{2}\right)^N \sqrt{\frac{1+f_a}{1+f_o}},$$

where f_g and f_a are the inbreeding coefficients of the grandfather and Nth great-grandfather, respectively. By applying the tracing rule,[18] the correlation between S and D is the sum of the products of all path coefficients connecting s and t:

[18] See QUOT. 944.

$$
\begin{aligned}
r_{sd} &= \sum p_{s.a} p_{d.a} \\
&= \sum \left(\frac{1}{2}\right)^{n+n'} \sqrt{\frac{1+f_a}{1+f_s}} \sqrt{\frac{1+f_a}{1+f_d}} \\
&= \sum \left(\frac{1}{2}\right)^{n+n'} \frac{1+f_a}{\sqrt{1+f_s}\sqrt{1+f_d}},
\end{aligned}
\tag{15.18}
$$

where n and n' are the number of generations from S and D, respectively, to the ancestor A. Now, from Fig. 15.11,

$$
\begin{aligned}
f_o &= b r_{sd} B \\
&= r_{sd} \sqrt{\frac{1}{2}(1+f_s)} \sqrt{\frac{1}{2}(1+f_d)} \\
&= \frac{1}{2} r_{sd} \sqrt{(1+f_s)(1+f_d)}.
\end{aligned}
$$

Substituting for r_{sd} in Eq. (15.18), the above becomes

$$
f_o = \sum \left(\frac{1}{2}\right)^{n+n'+1} (1+f_a) \tag{15.19}
$$

(*ibid.*, p. 334), which is the famous *pedigree formula*. If it can be assumed that the ancestor is not inbred, the formula above simplifies to

$$
f_o = \sum \left(\frac{1}{2}\right)^{n+n'+1}. \tag{15.20}
$$

As an example, Wright considered the pedigree of the famous Shorthorn sire, Roan Gauntlet (RG) (Fig. 15.12). This yields Table 15.8 for the calculation of f_o for RG.

15.4 Wright's Theory of Evolution: The Shifting Balance Theory (1930, 1931), The Adaptive Landscape (1931, 1932)[19]

15.4.1 Introduction

Wright's exposition of the Shifting Balance Theory (SBT) was a landmark in evolutionary genetics. In essence, this theory emphasizes the importance of

[19]Wright's Shifting Balance Theory and adaptive landscape are also discussed in Gayon (1998, pp. 340–351), Ruse (1996), Svensson and Calsbeek (2012), Coyne et al. (1997, 2000), Wade and Goodnight (1998), Goodnight and Wade (2000), Okasha (2018, pp. 74–84), Provine (1989, Chap. 9), Ridley (2003, pp. 216–220), Futuyma (1998, pp. 408–409), Reiss (2009, pp. 166–177), and Wright (1970).

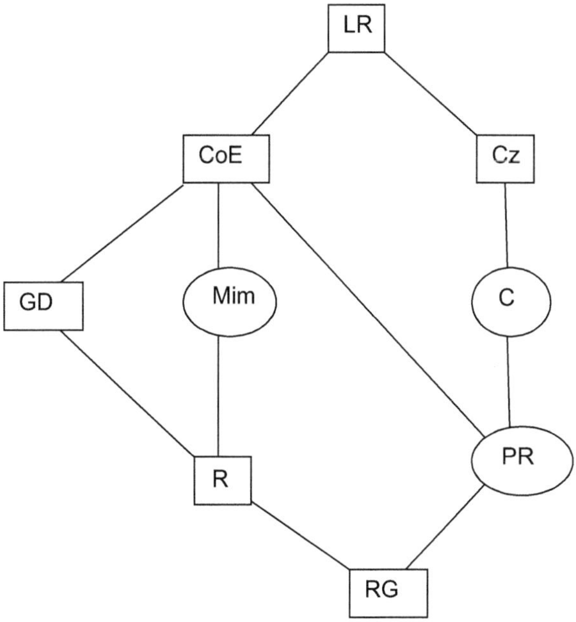

Fig. 15.12 Wright's illustration of the pedigree formula (LR Lord Raglan, CoE Champion of England, Cz The Czar, GD Grand Duke of Gloster, Mim Mimulus, Car Carmine, RD Royal Duke of Gloster, PR Princess Royal, RG Roan Gauntlet)

Table 15.8 Calculation of f_o for Roan Gauntlet (RG) in Fig. 15.12

Common ancestor	Path	f_a	n	n'	$(½)^{n+n'+1} \times (1+f_a)$
CoE	RD→GD→CoE→PR	0	2	1	$(½)^4 = .0625$
CoE	RD→Mim→CoE→PR	0	2	1	$(½)^4 = .0625$
LR	RD→GD→CoE→LR→Cz→Car→Pr	0	3	3	$(½)^7 = .007812$
LR	RD→Mim→CoE→LR→Cz→Car→Pr	0	3	3	$(½)^7 = .007812$
					$f_o = .140625$

population structure, gene combinations, and gene flow in the evolution of populations. It argues that genetic drift and gene flow can produce changes that cannot be accounted for by natural selection alone. As such, Wright's SBT is at odds with Fisher's (more Darwinian) theory of evolution by mass selection, although, to be fair, the two theories have many more similarities than differences. The latter fact was recognized by both men on several occasions.

15.4.2 Influences on Wright's Views on Evolution and Development of the SBT

Next to Darwin's *Origin,* which Wright read while in high school, the book that had a strong early influence on him was *Darwinism Today* (Kellogg, 1907), written by the American entomologist Vermon Lyman Kellogg (1867–1937). In this book,

Kellogg discussed the universality of natural selection. In one revealing section, he compared the amount of variation in two groups of insects, one collected "after having lived for more or less a free life exposed, that is, to the rigor of the individual struggle for existence," the other collected "just at the time of reaching maturity but before enjoying any opportunity to be weeded out (on a basis of disadvantageous variation) by the rigour of the life-struggle." Kellogg reported his results as follows:

QUOT. 957 Just as many varying individuals, with variations of just as much extent and variety, were found in series exposed to the struggle...as among series not yet exposed. (*ibid.*, p. 83)

To Kellogg, these observations implied that natural selection was not always effective.

Kellogg quoted Conn (1900) to emphasize the role of *chance* in evolution:

QUOT. 958 Indiscriminate destruction occurs constantly, and certainly influences the problem of survival. Of the hundreds of individuals that are produced where few can live, many are destroyed indiscriminately, independent of the principle of survival of the fittest, and of these that are thus killed doubtless some are superior to those that survive. This principle of indiscriminate elimination does not in the slightest deny the force of the principle of survival of the fittest, but only indicates that its action is not absolutely rigid. (*ibidem*)

The importance of chance was that it often led to *nonadaptive* features:

QUOT. 959 Whether an individual shall live or die in the struggle is largely a matter of accident. Many a well-equipped individual will die, while many another, even though handicapped by decidedly unfavourable characters, will continue to live and produce offspring because of some specially favourable conditions. (*ibid.*, pp. 83–84)

In addition to chance, Kellogg also emphasized the role of *isolation* in species formation:

QUOT. 960 Selection needs help from isolation. To my mind, also, these means of isolation actually exist, and the resulting isolation is actually a very potent factor in species-forming. The proofs seem to me obvious. (*ibid.*, p. 233)

Chance, isolation, and occasionally ineffective natural selection would soon become trademarks of Wright's evolutionary views. Kellogg and similar field naturalists such as John Thomas Gulick (1832–1923), David Starr Jordan (1867–1937), Henry Fairfield Osborn (1857–1935), and Henry Edward Crampton (1875–1956) were later often cited by him.

In a 1978 retrospective article, Wright (1978) referred to four further influences that led him to the development of the SBT while he was working at the USDA during the period 1915–1925:

(a) *Castle's hooded-rat experiments (1915)* From Castle's hooded-rat experiments,[20] Wright learned about the ability of mass selection to produce permanent change in organisms. However, Wright also threw the following qualification:

QUOT. 961 My recognition of the efficacy of mass selection was, however, qualified by recognition that this process is likely to lead to deleterious side effects because of the

[20] See Sect. 13.2.

interference of selection in any one respect with selection of the others." (*ibid.*, pp. 1197–1198)

(b) *Wright's PhD thesis (1916)* Wright's doctoral dissertation was entitled "An intensive study of the inheritance of color and of other coat characters in guinea pigs with especial reference to graded variation" (Wright, 1916). In it, Wright showed that alleles at several loci were responsible for the coat color of guinea pigs. Therefore:

QUOT. 962 From my studies of gene combinations, . . ., I recognized that an organism must never be looked upon as a mere mosaic of "unit characters," each determined by a single gene, but rather as a vast network of interaction systems. (Wright, 1978, p. 1198)

(c) *Study of inbred guinea pigs (1915)*. Wright started this work when he joined the USDA in 1915. From his experiments on closely inbred lines of guinea pigs, Wright found that genetic drift caused great differentiation between the lines and could even overcome the effects of natural selection to decrease the mean fitness of the population. It was also possible, by mere chance, for genetic drift to create favorable gene combinations whose frequency would then be increased by natural selection. These observations provided Wright with the elements of a theory (namely the SBT) that explained how evolution could proceed at a higher rate than by mass selection alone.
(d) *Clarification of the roles of inbreeding and selection in breeding of livestock (1921)* Wright's observations on the breeding history of Shorthorn cattle added support to the SBT:

QUOT. 963 There were always many herds at any given time, but only a few were generally perceived as distinctly superior; . . . These herds successively made over the whole breed by being principal sources of sires. (*ibid.*, p. 1199)

Wright's early viewpoint on evolution may therefore be summarized as a qualified acceptance of the power of mass selection together with an explicit recognition of the importance of population structure and gene interactions across loci.

15.4.3 The SBT (1929, 1930, 1931): A Theory in Three Phases

All the elements of Wright's SBT were put together in a trailblazing paper whose first version (no longer extant) was completed in 1925. Wright continually revised the paper until its publication in 1931 under the title "Evolution in Mendelian populations" (Wright, 1931). Starting in 1924, he benefited greatly from communications with Fisher. This enabled Wright to iron out a few discrepancies between his and Fisher's results.[21] On July 24, 1929, during the dominance debate between the two, he informed Fisher of his big paper, which was sent to the latter on August

---

[21] For example, see end of Sect. 14.3.4.2.

13, 1929. At around the same time, Wright started to express his ideas on the importance of population structure and genetic drift both in print[22] and in his communication with Fisher.[23]

Wright submitted his big paper for publication in *Genetics* on January 20, 1930. A few weeks before, in December 1929, he delivered a short summary (Wright, 1929a) of his paper at the annual meeting of the *American Association for the Advancement of Science* (AAAS). The summary argued that natural selection was ineffective in very small, very large, and intermediate-sized populations:

> **QUOT. 964** In too small a population, there is nearly complete random fixation, little variation, little effect of selection and thus a static condition, modified occasionally by chance fixation of a new mutation, leading to degeneration and extinction. In too large a freely interbreeding population, there is great variability, but such a close approach of all gene frequencies to equilibrium that there is no evolution under static conditions. Changed conditions cause a usually slight and reversible shift of the gene frequencies to new equilibrium points. With intermediate size of population, there is continual random shifting of gene frequencies and consequent alteration of all selection coefficients, leading to relatively rapid, indefinitely continuing, irreversible and large fortuitous but not degenerative changes even under static conditions. The absolute rate, however, is slow, being limited by mutation pressure. (*ibid.*, p. 287)

Wright's argument in the above was that: (i) for a very small population, random effects are very strong but the force of natural selection is weak since there is little variation; (ii) for a very large population, there is much variation for natural selection to act, but a quick approach to equilibrium limits its efficacy; and (iii) for a population of intermediate size, random variation and selection may both be relatively strong, but will be limited by mutation pressure.

According to Wright, the ideal condition under which natural selection was effective was that of a large population subdivided into smaller groups. This is because:

> **QUOT. 965** ...in a large but subdivided population, there is continually shifting differentiation among the local races, even under uniform static conditions, which through intergroup selection, brings about indefinitely continuing, irreversible, adaptive and much more rapid evolution of the species as a whole. (*ibid.*, p. 287)

Thus, genetic drift *within* groups created continual genetic differences *between* groups, followed by intergroup selection on the most favorable gene combinations. This was Wright's preliminary description of the SBT.

However, the description was incomplete (there is no mention of gene flow between groups) and improved upon in Wright's 1930 review of Fisher's *Genetical Theory of Natural Selection*. Commenting on the slowness of Fisher's model of mass selection acting on mutations, Wright stated:

> **QUOT. 966** A much more favorable condition would be that of a large population, broken up into imperfectly isolated local strains. The probability array for genes within such a local strain has been given on a previous page. The rate of evolutionary change depends primarily on the balance between the effective size of population in the local strain (n) and the amount

[22] See QUOT. 901.

[23] See QUOT. 902.

of interchange of individuals with the species as a whole (*m*) and is therefore not limited by mutation rates. The consequence would seem to be a rapid differentiation of local strains, in itself non-adaptive, but permitting selective increase or decrease of the numbers in different strains and thus leading to relatively rapid adaptive advance of the species as a whole. Thus I would hold that a condition of subdivision of the species is important in evolution not merely as an occasional precursor of fission, but also as an essential factor in its evolution as a single group. (Wright, 1930, p. 355)

The three phases of the SBT were given even more completely in Wright's big paper of 1931:

QUOT. 967. . .in a large population, divided and subdivided into partially isolated local races of small size, there is a continually shifting differentiation among the latter (intensified by local differences in selection but occurring under uniform and static conditions) which inevitably brings about an indefinitely continuing, irreversible, adaptive, and much more rapid evolution of the species. Complete isolation in this case, and more slowly in the preceding, originates new species differing for the most part in nonadaptive respects but is capable of initiating an adaptive radiation as well as of parallel orthogenetic lines, in accordance with the conditions. (Wright, 1931, p. 158)

The description above has all three phases of Wright's SBT for a large population subdivided into isolated (or almost isolated) groups:

Phase I: *Genetic drift creates various gene combinations by chance in the groups and enables each group to discover new gene combinations.*
Phase II: *Intragroup selection occurs for the most favorable local combinations, resulting in the establishment of various gene combinations with high fitness within each group.*
Phase III: *Intergroup selection occurs so that the groups with the highest mean fitness contribute more offspring to the total population; this is followed by gene flow from the fittest to the less fit groups, thus rapidly increasing the mean fitness of the total population by further local selection.*

Thus, Wright's SBT is not just about the sole power of genetic drift: all of genetic drift, natural selection and gene flow play vital roles. The process, however, *is* initiated by genetic drift.

Although the SBT was initially well received, subsequent scrutiny raised various issues with this theory. These will be examined in Sect. 15.4.6.

15.4.4 Wright's Adaptive Landscape (1931, 1932): A Visual Metaphor of the SBT

15.4.4.1 Introduction

Although Wright's 1931 paper established him as a first-class population geneticist, it was quite technical. The SBT and other novel ideas it contained were undoubtedly not accessible to all geneticists. However, the situation changed when Wright presented a nontechnical follow-up paper (Wright, 1932) at the 1932 International Congress of Genetics in Ithica, New York. The highlight of Wright's presentation

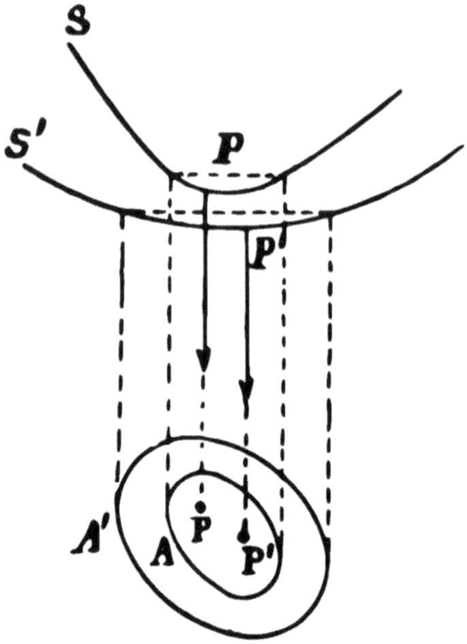

Fig. 15.13 Janet's (1895, p. 144) depiction of two selection curves (*S* and *S*′). *S* represents stronger selection than *S*′, and the corresponding area *A* represents the variability in the population and is smaller than *A*′

was what later came to be known as the *adaptive landscape*—a pictorial representation of the SBT consisting of peaks and valleys to denote gene combinations of high and low adaptive values, respectively.

15.4.4.2 Janet's Adaptive Landscape (1895)

In a relatively recent article, McCoy (1979) indicated that the first adaptive landscape had been constructed by a French engineer (with an interest in entomology) named Armand Janet (1895). Janet's construct was explicitly based on the notion of Newtonian forces in mechanics. In its simplest version, a two-dimensional curve (*S* or *S*′ in Fig. 15.13) is used to represent the mean phenotype at various points. The latter are held in equilibrium by the action of several forces, represented by vectors, such as natural selection and environmental factors. Natural selection acts like gravity and pulls a population at a local maximum or peak (low fitness) toward a local minimum or valley (high fitness). Underneath the curve, an area is constructed to represent the variability in the population—the stronger the force of selection (as indicated by the concavity of the curve), the lower the variability in the population, and the smaller its area.

However, as Dietrich and Skipper (2012) have argued, Wright's metaphor was independent and fundamentally different from Janet's. In the former case, gene combinations rather than phenotypes were represented, the peaks and valleys were the opposite of those of Janet's, and Wright's landscape had a topographical interpretation that Janet's did not have.

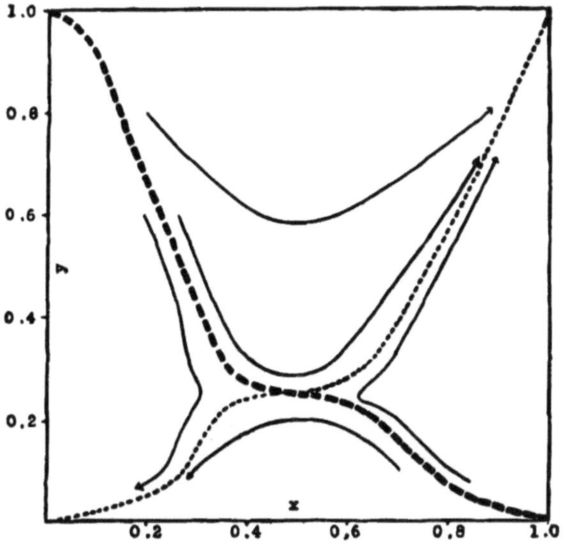

Fig. 15.14 Adaptive landscape drawn by C. H. Waddington for Haldane (1931b, p. 140)

15.4.4.3 Haldane's Adaptive Landscape (1931)

Much closer to Wright's times was J.B.S. Haldane's (1931b) eighth installment of the "Mathematical theory of natural and artificial selection" series of ten papers. With the help of C.H. Waddington, Haldane produced an adaptive landscape similar to Wright's. He considered several genes that were disadvantageous individually but advantageous in combination. For example, suppose that *A*, *B* are dominant alleles at two loci and that *a, b* are their recessive counterparts. Assume that the fitnesses of the four phenotypes are [*AB*]: 1, [*aB*]: $1 - k_1$, [*Ab*]: $1 - k_2$, and [*ab*]: $1 + K$, where k_1, $k_2 > 0$ and $|K| < k_1, k_2$. Let $x = \text{freq}(a)$ and $y = \text{freq}(b)$. Then, Haldane argued that such a system had two points of stable equilibrium, namely, at (0, 0) and (1, 1), as shown in Fig. 15.14.

The broken curve joining (0, 0) and (1, 1) and that joining (0, 1) and (1, 0) split the (*x*, *y*)-space into four partitions, such that starting from any point (*x*, *y*), the continuous arrows show the trajectories from this point to either of the stable points (0, 0) or (1, 1).

Next, Haldane considered a population initially consisting of *AABB* only. Allowing for rare mutations of *A* to *a* and of *B* to *b*, this population soon reaches an equilibrium point (x_0, y_0) close to (0, 0). However, chance fluctuations often cause *AABB* genotypes to move from (x_0, y_0) to either the region on the right of the intersection of the two broken curves or the region above that intersection, whence these *AABB* genotypes become fixed to *aabb* through natural selection. Haldane called (x_0, y_0) a "metastable" equilibrium point, which, when ruptured, resulted in the formation of two species *AABB* and *aabb*. He further noted that:

QUOT. 968. . .such a rupture will be specially likely where small communities are isolated. (*ibid.*, p. 142)

Haldane's adaptive landscape is similar to Wright's but does not have the latter's sophistication and comprehensiveness, as will be explained in the next section.

15.4.4.4 Wright's Adaptive Landscape (1931, 1932)

Haldane's metastable paper was published in January 1931. One month later, on February 3, Wright wrote a letter to Fisher and explained the elements of his SBT graphically[24]:

QUOT. 969 Think of the field of visible joint frequencies of all genes as spread out in a multidimensional space. Add another dimension measuring degree of fitness. The field would be very humpy in relation to the latter because of epistatic relations, groups of mutations which were deleterious individually producing a harmonious result in combination. In the figure below this field (very imperfectly represented by a single line) is plotted against fitness.

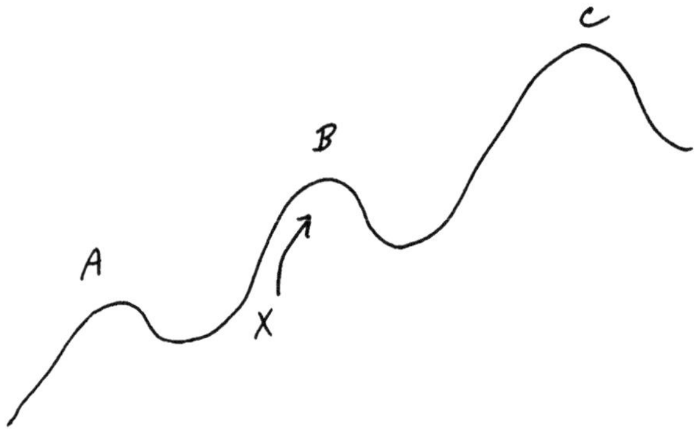

A species, at point X with respect to its system of gene frequencies, will tend to move steadily toward some point near B (not B exactly if there is recurrent mutation) under the influence of selection, but will tend to stay there even though C would be a position of greater fitness if it could be reached. Something other than the steady pressure of selection is needed. I recognized four factors in my review. First, an irregularly changing environment. This by continually changing the system of humps themselves would keep the species moving, the immediate effect being downward as a rule but in the long run permitting some species to reach positions of adaptation which could never have been reached by a direct process. Life as a whole would advance at least in complexity of adaptation. Second, novel mutations in indefinitely extended multiple series, creating new dimensions and occasionally new paths of advance upwards. I interpreted your idea of the mechanism for

[24] In a letter dated June 5, 1931, to Fisher, Wright indicated that he had recently received Haldane's metastable paper (Provine, 1989, p. 275), so it is unlikely that Wright got his adaptive landscape ideas from Haldane.

> indefinite continuance of the evolutionary process as being of this nature. Third, limitation in the size of the population to such a figure that random variation in gene frequency becomes important (but not overwhelmingly important as under close inbreeding). The position of the species would move at random about B and occasionally might go so far that the upward pressure of selection would tend to carry it toward A or C instead of to B. Since it could more easily escape from A than C, the general tendency would be toward increasing adaptation even though the process would be very slow. Fourth, subdivision of an indefinitely large species into many small, not quite completely isolated groups. The positions of these groups would move at random about B. Some would reach such positions as A, others C. The latter being better adapted would tend to increase in population and become the major source of migrants to other groups. The position of the mean of the species should thus approach C.
>
> I certainly do not profess to have any complete mathematical analysis of this situation, but as far as my figures go they have seemed to indicate that the fourth scheme is the most effective. In a sense selection plays as great a role as in your theory, the selection of individuals to which I have attributed less complete control being supplemented by selection between local races. (Provine, 1989, pp. 272–273)

Wright's adaptive curve in the quotation above has a gene frequency on the *x*-axis and the corresponding fitness on the *y*-axis.[25] The curve clearly illustrates the key issue: how can the species move from *X* to the highest peak *C*? Natural selection is not sufficient because, although it will allow the species to go from *X* to *B*, it will not permit movement from *B* to the valley on the right as this entails a decrease in fitness.

In Haldane's scheme, the key insight was that genetic drift had the ability to temporarily decrease the mean fitness of a population to enable selection to reach a higher mean fitness. This corresponds to Wright's third scheme in QUOT. 969 (i.e., limitation of population size), but Wright argued that natural selection would be too slow in this case. Wright's best scenario for natural selection was the fourth scheme (i.e., the subdivision of a large population into isolated or partially isolated groups)—genetic drift enabled a population to go down a valley, natural selection then established several local peaks, and gene flow finally took place from the highest peaks.

Wright's adaptive landscape was formally introduced in a paper (Wright, 1932) presented at the 1932 International Congress of Genetics in Ithaca, New York. The meeting was organized by Edward East who invited Wright, Fisher, and Haldane to share a session and present a simplified version of their latest work to the audience.

Wright started his paper by pointing out the exponential increase in the number of possible genetic combinations with the number of alleles at a given locus and the number of loci. For example, with five bi-allelic loci, there are $2^5 = 32$ homozygous combinations and these can be represented in five dimensions with an extra dimension for fitness. With 1000 loci each with ten possible alleles, the number of homozygous combinations is 10^{1000} and the number of dimensions required to represent them is 9000. Wright then argued that the two-dimensional diagram in Fig. 15.15, although inadequate, could be used to represent such a field, with the contour lines representing gene combinations of constant adaptive value.

[25] On the other hand, in Haldane's Fig. 15.14, the *x*- and *y*-axes are the gene frequencies at two loci, and a third axis of fitness (projecting from the *xy*-plane) is also implied.

Fig. 15.15 Wright's adaptive landscape (Wright, 1932, p. 358). The dotted lines represent gene combinations of constant adaptive value. A + denotes a local maximum (peak) and a—denotes a local minimum (valley)

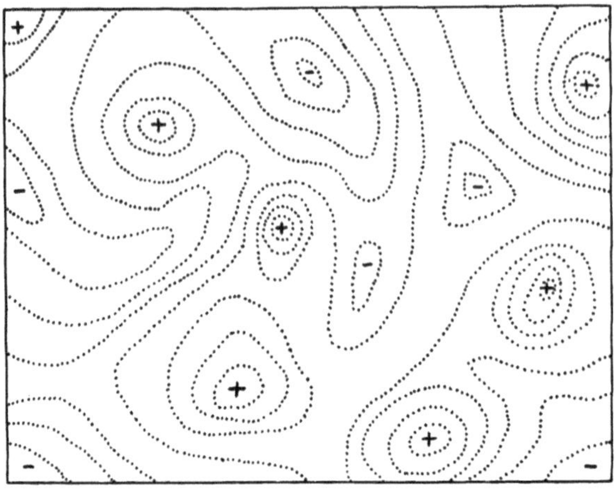

The reader should at once note that this second representation of the adaptive landscape in terms of gene combinations is fundamentally different from the first representation to be found in Wright's February 3, 1931, letter, which was in terms of gene frequencies.[26] But more on this later.[27]

In any case, Wright continued his presentation and indicated that the number of local gene combinations with maximum fitness (peaks) was enormous, making the adaptive landscape very rugged:

> **QUOT. 970** With something like 10^{1000} possibilities [of homozygous gene combinations]... it may be taken as certain that there will be an enormous number of widely separated harmonious combinations. The chance that a random combination is as adaptive as those characteristic of the species may be as low as 10^{-100} and still leave room for 10^{800} separate peaks, each surrounded by 10^{100} more or less similar combinations. In a rugged field of this character, selection will easily carry the species to the nearest peak, but there may be innumerable other peaks which are higher but which are separated by "valleys." (*ibid.*, p. 358)

According to Wright, the main problem of evolution was that

> **QUOT. 971** ...of a mechanism by which the species may continually find its way from lower to higher peaks in such a field. In order that this may occur, there must be some trial and error mechanism on a grand scale by which the species may explore the region surrounding the small portion of the field which it occupies. To evolve, the species must not be under strict control of natural selection. Is there such a trial and error mechanism? (*ibid.*, p. 359)

Thereupon, Wright presented adaptive landscapes (Fig. 15.16) for the various population scenarios he had considered in his 1929 summary (Wright, 1929a)[28] and 1931 big paper (Wright, 1931). Which one of these would be more likely to result in movement from a lower to higher peak?

---

[26] See first sentence of QUOT. 969.

[27] See Sect. 15.4.6.

[28] See QUOT. 964.

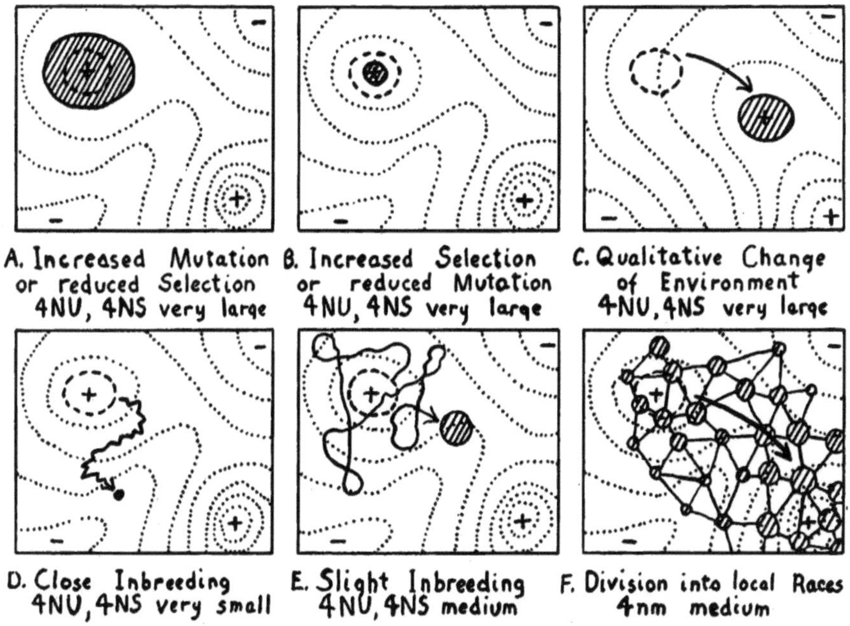

Fig. 15.16 Wright's (1932, p. 361) depiction of the adaptive landscapes for various population scenarios (see text). *N:* population size, *n:* subpopulation size, *U:* mutation rate, *S:* selection coefficient, *m:* migration rate. Wright argued that only the scenario in F is conducive to any significant evolution

Wright considered each scenario in turn and, in accordance with his earlier view in QUOT. 966 (or QUOT. 967) concluded that F was the only scenario that would be conducive to any significant evolution. Wright's translation of this view in the language of adaptive landscapes is as follows:

> **QUOT. 972** Finally [scenario F in Fig. 15.16], let us consider the case of a large species which is subdivided into many small local races, each breeding largely within itself but occasionally crossbreeding. The field of gene combinations occupied by each of these local races shifts continually in a nonadaptive fashion (except in so far as there are local differences in the conditions of selection). The rate of movement may be enormously greater than in the preceding case since the condition for such movement is that the reciprocal of the population number be of the order of the proportion of crossbreeding instead of the mutation rate. With many local races, each spreading over a considerable field and moving relatively rapidly in the more general field about the controlling peak, the chances are good that one at least will come under the influence of another peak. If a higher peak, this race will expand in numbers and by crossbreeding with the others will pull the whole species toward the new position. The average adaptiveness of the species thus advances under intergroup selection, an enormously more effective process than intragroup selection. The conclusion is that subdivision of a species into local races provides the most effective mechanism for trial and error in the field of gene combinations. (*ibid.*, p. 363)

The following are the three phases of the SBT given after QUOT. 967, now translated into the language of adaptive landscapes. In a large population subdivided into isolated (or almost isolated) groups:

Phase I: *Genetic drift creates novel local peaks and valleys, and enables a group at an existing local peak to explore valleys around it.*

Phase II: *If a group is in the valley between two local peaks, intragroup selection will take the group to the higher of the two local peaks. Repetition of phases I and II will eventually move a group to its highest local peaks.*

Phase III: *Intergroup selection occurs so that the groups with the highest local peaks contribute more offspring to the total population; this is followed by gene flow from the highest peaks to the lower peaks, thus rapidly increasing the mean fitness of the total population by further local selection.*

Wright's adaptive landscape proved to be an immediate success. His Fig. 15.15 is one of the most well-known metaphors ever used in biology. However, despite the popularity of Wright's SBT, it also has had its fair share of ambiguities and controversies, as will be explained in Sect. 15.4.6.

15.4.5 Mathematizing the Adaptive Landscape: The Sewall Wright Equation (1935, 1937)

In 1935, Wright published two papers (which were originally written as one) where he took the extra step of mathematizing his adaptive landscape by assigning an adaptive value (H) to each point on it (Wright, 1935a, b). Let $\overline{H}$ be the "mean adaptive value," (a_i, A_i) be the possible alleles at locus i ($i = 1, 2, \ldots, n$), q_i be the frequency of allele a_i, and γ_i be the average effect[29] of substituting a_i by A_i. Then,

$$\frac{\partial \overline{H}}{\partial q_i} = 2\gamma_i \tag{15.21}$$

(*ibid.*, p. 258).[30] Now, at a given locus, for two alleles a and A with (marginal) adaptive values $1 + s$ and 1, respectively, the change in frequency of a after one round of selection is

$$\Delta q = \frac{1+s}{\overline{H}} q - q,$$

where $\overline{H} = q(1+s) + 1 - q = qs + 1$ is the mean adaptive value. Therefore,

[29]This is denoted by α in Eq. (14.61).

[30]See Eq. (14.76).

$$\Delta q = \frac{qs(1-q)}{1+qs} \approx sq(1-q).$$

Since, in general, "the selective disadvantage must be proportional to the momentary net effect of gene replacement on adaptive value," Wright replaced s in the above by $\gamma_i/\overline{H}$ so that

$$\Delta q_i = \frac{\gamma_i q_i(1-q_i)}{\overline{H}}. \tag{15.22}$$

From Eqs. (15.21) and (15.22), it follows that

$$\Delta q_i = \frac{q_i(1-q_i)}{2\overline{H}} \frac{\partial \overline{H}}{\partial q_i}.$$

The equation above does not appear in Wright (1935a), but is implied by Eqs. (15.21) and (15.22). Explicit formulas first appeared in Wright (1937), both for the bi-allelic case at a locus and for the multi-allelic case. The latter case is the most general and was written as

$$\Delta q_i = \frac{q_i(1-q_i)}{2\overline{W}} \frac{\partial \overline{W}}{\partial q_i} \tag{15.23}$$

(*ibid.*, p. 310), where $\overline{W}$ was now called the mean selective value (later the mean fitness).

Equation (15.23) is known as the *Sewall Wright equation* and holds for random mating in the bi-allelic case. In the latter case, the equation implies that a strong force of selection (as measured by the slope $\partial\overline{W}/\partial q_i$ of the adaptive curve) leads to a greater change in allele frequency, i.e., the population climbs the curve until it reaches the top. In the case of random mating and multiple alleles at a locus, the "hill-climbing" will occur but not necessarily along the steepest path (Edwards, 1971, p. 333). In the case of multiple loci, the mean fitness may decrease and hill-climbing may not occur (see Sect. 14.5.4).

The above are but a few of the difficulties associated with Wright's adaptive landscape. More controversies will be examined in the next section.

15.4.6 *The SBT and Adaptive Landscape: Difficulties and Controversies*

15.4.6.1 Introduction

Wright's SBT, the adaptive landscape, and the associated equation in Eq. (15.23) have all been subjected to scrutiny. Not surprisingly, some of the strongest criticisms were first made by Fisher, although Coyne et al. (1997) later also raised some particularly salient points.

15.4.6.2 Gene Frequencies or Gene Combinations?

As pointed out in Sect. 15.4.4.4, Wright made use of two diametrically different adaptive landscapes to illustrate his theory—in the February 3, 1931, letter to Fisher,[31] he spoke of "the field of visible joint frequencies of all genes as spread out in a multidimensional space," with an extra dimension for the mean fitness. However, in the 1932 paper (Wright, 1932), his landscape[32] was a two-dimensional plane consisting of "an enormous number of widely separated harmonious combinations" and the meaning of the axes was unclear. Fisher was the first to point out this inconsistency:

> **QUOT. 973** Prof. Wright here confuses the number of genotypes, e.g. 3^{1000}, which may be distinguished among individuals, with the continuous field of variation of gene frequencies. (Fisher, 1941, p. 58

Provine has been much harsher and claimed that Wright's Figs. 15.15 and 15.16 were meaningless (!):

> **QUOT. 974** The first and most important thing to notice about Wright's first published version of his fitness "surface" is that his construction does not in fact produce a continuous surface at all. Each axis is simply a gene combination; there are no gradations along the axis. There is no indication of what the units along the axis might be or where along the axis the gene combination should be placed. No intelligible surface can be generated by this procedure. By no stretch of the imagination can Wright's famous diagrams of the 1932 paper be constructed by his method of utilizing gene combinations. The diagrams represent a nicely continuous surface of selective value of individual genotypic combinations; the method Wright used to generate this surface actually yields an unintelligible result. Thus the famous diagrams of Wright's 1932 paper, certainly the most popular of all graphic representations of evolutionary biology in the twentieth century, are meaningless in any precise sense. (Provine, 1989, p. 310)

However, Provine's criticism is somewhat unfair, as it would have been impossible for Wright to represent a fitness surface with gene frequencies in thousands of dimensions, which is why he resorted to a heuristic device to illustrate the fitness surface. His graph in QUOT. 969 is a mathematical depiction with well-defined axes. On the other hand, his Figs. 15.15 and 15.16 are topographical depictions that should be interpreted non-mathematically.[33] Ruse (1996, p. 336) is of a similar opinion and has concluded that "Wright's adaptive landscapes have played a crucial role in evolutionary thought in this century."

[31] See QUOT. 969.

[32] See Fig. 15.15.

[33] A good analogy to Wright's topological depiction is a surface (weather) map with isobars.

15.4.6.3 Is $\overline{W}$ a Potential Function?

Wright's use of an adaptive surface that was climbed by a deme to achieve a state of stable equilibrium implied that his $\overline{W}$ in Eq. (15.23) behaved like a potential function. In physics, such potential functions are common; for example, in a (conservative) gravitational field, the potential energy function of a body tends toward a minimum and the derivative of such a potential function gives the (negative of the) gravitational force. However, Fisher had a good understanding of mechanics and was categorically against such an interpretation and disapproved of Wright's Eq. (15.23):

> **QUOT. 975** Wright's conception embodied in equation (6) [Eq. 15.23] that selective intensities are derivable, like forces in a conservative system, from a simple potential function dependent on the gene ratios of the species as a whole, has led him into extensive but untenable speculations. (Fisher, 1941, p. 58)

Fisher's disapproval was based on the fact that the mean fitness, being a function of fitnesses of discrete genotypes, was a discontinuous function of the continuous allele frequencies and therefore the interpretation of the derivative $\partial W/\partial q_i$ was problematic. Despite Fisher's very pertinent protest, the potential function view of $\overline{W}$ has been embraced by several authors (e.g., Crow and Kimura (1970, p. 181), Barrows (2011, p. 6)).

15.4.6.4 Multi-Peaked Landscape?

In a May 31, 1931, letter to Wright, Fisher voiced one of the earliest difficulties with the adaptive landscape. With the high-dimensional nature of Wright's model, Fisher doubted the existence of numerous peaks and valleys in the adaptive landscape:

> **QUOT. 976** In one dimension, a curve gives a series of alternate maxima and minima, but in two dimensions two inequalities must be satisfied for a true maximum, and I suppose that only about one fourth of the stationary points will satisfy both. Roughly I would guess that with n factors only 2^{-n} of the stationary points would be stable for all types of displacement, and any new mutation will have a half chance of destroying the stability. This suggests that true stability in the case of many interacting genes may be of rare occurrence, though its consequence when it does occur is especially interesting and important. (Provine, 1989, p. 274)

According to Fisher, there should exist a single adaptive peak with ridges along it, so that selection alone can take the population to the top.

In his reply on June 5, Wright stated that Haldane had also devised a similar landscape as him and reached similar conclusions:

> **QUOT. 977** He [Haldane] points out that with m genes it would be possible for alternate apices in his hypercube to be stable. (*ibid.*, p. 275)

Moreover, Wright noted that, with as many as 2^{1000} gene combinations, even if the maxima were sparse, there would practically be an infinity of them.

As explained below, Coyne et al.'s (1997) extensive study on the SBT concluded that several peaks and valleys may indeed exist but that genetic drift was not necessary for peak shifts.

15.4.6.5 What Does the Evidence Say Regarding the Three Phases of the SBT?

In an important article, Coyne et al. (*ibidem*) performed a comprehensive review of the evidence regarding the three phases of Wright's SBT and reported that:

(i) Phases 1 and 2 are possible, but genetic drift is often unnecessary for movement between peaks within a group.
(ii) Phase 3 is unlikely to spread higher peaks (from the previous two phases) across demes.
(iii) Although there is evidence for each of the three phases of the SBT, there is no known example of the SBT going through all three phases.

Coyne et al.'s paper was countered by a defense of the SBT by Wade and Goodnight (1998), but Wright's three-phase theory is still not generally agreed upon.

15.4.7 Influence of Wright's SBT on Dobzhansky

One of the most ardent initial promoters of Wright's SBT was Theodosius Dobzhansky. Dobzhansky (Fig. 15.17) was born in 1900 Nemirov in Ukraine (then part of the Russian Empire). He went to New York in 1927 to join the *Drosophila* Group of T H Morgan and soon became one of the foremost experimental evolutionary geneticists of his time. In 1937, he wrote the first post-synthetic book, entitled *Genetics and the Origin of Species* (Dobzhansky, 1937). In that same year, he became a naturalized citizen of the United States. He passed away in California in 1975.

In *Genetics and the Origin of Species*, Dobzhansky (1937) introduced Wright's adaptive landscape as follows:

> **QUOT. 978**...one can hardly eschew trying to sketch some sort of a general picture of evolution. A very interesting attempt in this direction has been made by Wright (1931, 1932), whose lead we may partly follow. (*ibid.*, p. 186)

The next paragraph is so close to Wright's beliefs that it might also have been written by the latter:

> **QUOT. 979** In an organism possessing only 1,000 genes each capable of producing ten allelomorphs, the number of the possible gene combinations that may be formed is 10^{1000}. Some, probably a great majority, of these combinations are discordant and have no survival value, but still very numerous ones may be supposed to be harmonious in the different ecological niches of the same environment, as well as in different environments. If the entire ideal field of possible gene combinations is graded with respect to the adaptive value, we may find numerous "adaptive peaks" separated by "valleys." The "peaks" are the groups of related gene combinations that make their carriers fit for survival in a given environment; the "valleys" are the more or less unfavorable combinations. Each living species or race may be thought of as occupying one of the available peaks in the field of gene combinations...a shift from one adaptive peak to another, which may be even higher than the first but separated

Fig. 15.17 Theodosius Dobzhansky (1900–1975). (From: https://commons.wikimedia.org/wiki/File:Dobzhansky_no_Brasil_em_1943.jpg)

from it by an adaptive "valley," must involve a trial and error mechanism on a grand scale which will enable the species to "explore" the region around its own peak in order to finally encounter a gradient leading toward the other peak or peaks. (*ibid.*, p. 187)

Dobzhansky next reproduced Wright's Fig. 15.16 and the latter's argument for why the scenario in F is the optimal one for progressive evolution:

QUOT. 980 Since in a semi-isolated colony of a species the fixation or loss of genes is to a certain degree independent of their adaptive values (owing to the restriction of the population size), a colony may become different from others simultaneously in several characters. One or a few of the latter may be adaptive, and may enable the population to conquer new territories or ecological stations. (*ibid.*, p. 191)

It is also in Dobzhansky's *Genetics and the Origin of Species* that the modern definition of evolution was first explicitly stated and the agenda of population genetics was set:

QUOT. 981 Since evolution is a change in the genetic composition of populations, the mechanisms of evolution constitute problems of population genetics (*ibid.*, pp. 11–12)

This reductionist definition was almost universally adopted by population geneticists: the investigation of evolution was performed by studying *changes in allele frequencies*. However, Mayr voiced his opposition to such an approach, calling it "beanbag genetics." See Sect. 16.4.

15.5 The Wright–Fisher Model: Wright's Stochastic Theory

15.5.1 Introduction

Starting with his big paper (Wright, 1931), Wright produced a considerable body of stochastic work in population genetics. Although the pioneer in the field was Fisher, Wright's work was more comprehensive since it also incorporated the effects of population structure.

15.5.2 Rate of Decrease of Heterozygosity (1931)

In the big paper, Wright (1931) investigated the rate of loss of heterozygosity in a *finite* population of breeding size N (consisting of N_m males and N_f females). Using the path diagram in Fig. 15.9, he obtained the correlation M between two parents by considering matings between full brothers and full sisters (with probability $1/(N_m N_f)$), half brothers and sisters (with probability $(N_m + N_f - 2)/(N_m N_f)$), and less related individuals (with probability $(N_m - 1)(N_f - 1)/(N_m N_f)$). Using a similar reasoning to the one used to obtain Eq. (15.15), Wright wrote

$$M = a'^2 b'^2 \left\{ \frac{1}{N_m N_f}\left(2 + 2M'\right) + \frac{N_m + N_f - 2}{N_m N_f}\left(1 + 3M'\right) + \frac{(N_m - 1)(N_f - 1)}{N_m N_f} 4M' \right\}$$

(*ibid.*, p. 108), where a is the path coefficient from gamete to zygote,[34] b is the path coefficient from zygote to gamete,[35] and $'$ denotes one generation before the present. In terms of the inbreeding coefficient F (or correlation between uniting gametes, see Eq. (15.12)), the equation above becomes

$$F = F' + \left(\frac{N_m + N_f}{8N_m N_f}\right)\left(1 - 2F' + F''\right),$$

where F'' is the value of F two generations before, or

[34] See Eq. (15.10).

[35] See Eq. (15.11).

$$P = P^{/} - \left(\frac{N_m + N_f}{8N_m N_f}\right)\left(2P^{/} - P^{//}\right),$$

where P is the heterozygosity. Since "[i]t is to be expected that the proportional change per generation will reach approximate constancy," Wright set $P/P^{/} = P^{/}/P^{//}$ so that the above becomes

$$\frac{P}{P^{/}} = \frac{N_e - 1 + \sqrt{N_e^2 + 1}}{2N_e},$$

where $N_e = 4N_m N_f/(N_m + N_f)$. Hence,

$$-\frac{\Delta P}{P^{/}} = \frac{P - P^{/}}{P^{/}} = \frac{1}{2}\left(1 + \frac{1}{N_e}\right) - \frac{1}{2}\sqrt{1 + \frac{1}{N_e^2}}$$

(*ibidem*). If the numbers of males and females are the same, i.e., $N_m = N_f = N/2$, and large, then the rate of loss of heterozygosity is given by

$$-\frac{\Delta P}{P^{/}} = \frac{1}{2}\left(1 + \frac{1}{N_e}\right) - \frac{1}{2} \approx \frac{1}{2N_e}.$$

Wright referred to Fisher's dominance ratio paper (Fisher, 1922), where a wrong rate of $1/(4N)$ was obtained.[36]

15.5.3 Introducing the Wright–Fisher Model: Distribution of Gene Frequencies After One Generation of Random Mating Under Genetic Drift (1931)

In his very first foray into the distribution of gene frequencies, Wright devised an ingenious method that bypassed some of the intricacies of Fisher's diffusion equation approach. Instead, he used what later came to be known as the Wright–Fisher model.[37]

Wright's first investigation was the distribution of gene frequencies after one generation of random mating in a finite population. No mutation, migration, or selection was assumed. Consider an initial population of breeding size N with a frequency q of alleles A. If the initial absolute (i.e., non-normalized) probability mass

[36] See Eq. (14.38) and the discussion after.

[37] Fisher's first implicit use of that model was explained in Sect. 14.3.4.1.

function of the number of A alleles is f, then the mass function of the number of A alleles after one generation of random mating, evaluated at $2Nq_1$, is given by using the classic equation of the Wright–Fisher model[38] as

$$\binom{2N}{2Nq_1} q^{2Nq_1}(1-q)^{2N(1-q_1)} f$$

Summing across q gives the total absolute mass function:

$$\binom{2N}{2Nq_1} \sum_q q^{2Nq_1}(1-q)^{2N(1-q_1)} f$$

Now, Wright had shown that the absolute distribution decreased by an amount $1/(2N + 1)$ from one generation to the next, so that

$$\frac{2N}{2N+1} f = \binom{2N}{2Nq_1} \sum_q q^{2Nq_1}(1-q)^{2N(1-q_1)} f.$$

In the continuous limit, using $\phi(q)$ to denote the density function of q, Wright wrote $f = \phi(q)/(2N) = \phi(q)dq$ so that the above equation becomes

$$\frac{\phi(q_1)}{2N+1} = \binom{2N}{2Nq_1} \int_0^1 q^{2Nq_1}(1-q)^{2N(1-q_1)} \phi(q)dq \qquad (15.24)$$

(Wright, 1931, p. 113). Next, by studying the cases of small N, Wright found that the uniform distribution $\phi(q) = C$, where C is a constant, satisfied the equation above. Since $\int_0^1 \phi(q)dq = 1$, the final solution is

$$\phi(q) = 1 \text{ for } 0 < q < 1. \qquad (15.25)$$

Thus, after repeated random mating in a finite population, a "steady" state is eventually reached whereby, although the heterozygosity decreases by $1/(2N)$ from one generation to the other, the frequency of A is uniform in (0, 1) *within* a generation.

Kimura (1955, p. 148) has shown that the stationary distribution in Eq. (15.25) may not be relevant since, by the time this distribution is reached, the allele A may already be fixed or lost. See also the discussion in Ewens (2004, pp. 23, 161).

[38] See Eq. (14.32).

15.5.4 Stationary Distribution of Gene Frequencies Under Genetic Drift and Nonrecurrent Mutations (1931)

It is possible for rare, nonrecurrent mutations to occur such that an equilibrium state is eventually reached whereby the rate of increase in heterozygosity due to mutations is exactly counterbalanced by the rate of decrease in heterozygosity due to genetic drift. Wright modified Eq. (15.24) as follows:

$$\frac{\phi(q_1)}{2N}=\binom{2N}{2Nq_1}\int_0^1 q^{2Nq_1}(1-q)^{2N(1-q_1)}\phi(q)dq$$

(Wright, 1931, p. 116). By trial and error, he obtained the general solution $\phi(q)=C_1q^{-1}+C_2(1-q)^{-1}$, where C_1, C_2 are arbitrary constants. Since the solution must be symmetric in q, $C_1=C_2=C$ and

$$\phi(q)=Cq^{-1}(1-q)^{-1}$$

is the general solution. This formula matches Fisher's Eq. (14.47) but Wright's method is less mathematically demanding.

To determine the constant C, Wright went back to the mass function f, where

$$f(q)=\frac{1}{2N}\phi(q)=\frac{C}{2N}q^{-1}(1-q)^{-1} \tag{15.26}$$

such that $\sum f=1$. This gives

$$C=\frac{1}{2\{.577+\log(2N-1)\}}\approx\frac{1}{2\log(3.6N)}.$$

Furthermore, Wright let $K/2$ denote the rate of change of allele frequencies (i.e., flux) at a boundary ($q=0$ or 1) and showed that $K=C$. Now, the frequency of a subterminal class (i.e., at $q=1/(2N)$ or $q=1-1/(2N)$) is

$$f_1=f_{2N-1}=2NC\left(\frac{1}{2N}\right)\left(1-\frac{1}{2N}\right)=C\frac{2N-1}{2N}=K$$

for large N. This gives the important result:

QUOT. 982 The proportion exchanged at each extreme is thus about half the corresponding subterminal class where N is large. (*ibid.*, p. 118)

15.5.5 Stationary Distribution of Gene Frequencies Under Reversible Recurrent Mutations (1931)

Wright next considered the case when A mutates to a at a rate u and the reverse rate is v. Then Eq. (15.24) needs to be modified to

$$\begin{aligned}\phi(q_1)\left(\frac{1-K}{2N}\right) &= \binom{2N}{2Nq_1}\int_0^1 \{q+v(1-q)-uq\}^{2Nq_1}\{(1-q)+uq-v(1-q)\}^{2N(1-q_1)}\phi(q)dq \\ &= \binom{2N}{2Nq_1}\int_0^1 \{q(1-u-v)+v\}^{2Nq_1}\{1-q(1-u-v)-v\}^{2N(1-q_1)}\phi(q)dq\end{aligned} \tag{15.27}$$

(Wright, 1931, p. 121), where $K/(2N)$ is the decrease in the absolute distribution from one generation to the next. By trial and error, Wright obtained a general solution of the form $Cq^s(1 - q)^t$. By substituting the latter in the equation above and making use of the identity

$$\int_0^1 x^{\alpha-1}(1-x)^{\beta-1}dx = \frac{\Gamma(\alpha)\Gamma(\beta)}{\Gamma(\alpha+\beta)}, \tag{15.28}$$

where $\Gamma()$ is the gamma function and $x = q(1 - u - v) + v$, Eq. (15.27) becomes

$$\begin{aligned}(1-K)q_1^s(1-q_1)^t = \frac{1}{1-(u+v)(s+t+1)}\Bigg[&\frac{4N}{4N+(s+t+1)(s+t+2)}q_1^s(1-q_1)^t \\ &+\left\{\frac{s(s+1)}{4N+(s+t+1)(s+t+2)} - \frac{4Nsv}{4N+(s+t)(s+t+1)}\right\}q_1^{s-1}(1-q_1)^t \\ &+\left\{\frac{t(t+1)}{4N+(s+t+1)(s+t+2)} - \frac{4Ntu}{4N+(s+t)(s+t+1)}\right\}q_1^s(1-q_1)^{t-1}\Bigg].\end{aligned}$$

For the equality above to hold, the coefficients of $q_1^{s-1}(1-q_1)^t$ and $q_1^s(1-q_1)^{t-1}$ must be zero, so that, assuming $s, t \ll N$,

$$\begin{aligned} s &= 0 \quad \text{or} \quad s = 4Nv - 1, \\ t &= 0 \quad \text{or} \quad t = 4Nu - 1. \end{aligned}$$

Wright thus obtained the most general solution as

$$\phi(q) = \frac{\Gamma(4Nu+4Nv)}{\Gamma(4Nu)\Gamma(4Nv)} q^{4Nv-1}(1-q)^{4Nu-1} \tag{15.29}$$

(*ibid.*, p. 123).

15.5.6 Stationary Distribution of Gene Frequencies Under Two-Way Symmetric Migration (1931)

Suppose a subgroup is partially isolated from a main population such that the frequency of allele A is q_m in each population. If each exchanges alleles A at a rate m per generation, then the equilibrium distribution of alleles A is obtained from (15.29) by replacing $u \to mq_m$ and $v \to m(1 - q_m)$:

$$\phi(q) = \frac{\Gamma(4Nm)}{\Gamma\{4Nmq_m\}\Gamma\{4Nm(1-q_m)\}} q^{4Nmq_m - 1}(1-q)^{4Nm(1-q_m)-1} \tag{15.30}$$

(Wright, 1931, p. 126).

15.5.7 Stationary Distribution of Gene Frequencies Under Selection and Rare Mutations: Wright's Error and Correction (1931)

Wright considered a model of rare mutations and genic selection such that the alleles a and A after one generating of mating are in the ratio $1 - s : 1$. Then, starting with frequencies of $1 - q$ and q, respectively, the frequencies of allele A after selection is

$$\frac{q}{q + (1-s)(1-q)} = \frac{q}{1 - s(1-q)} \approx q\{1 + s(1-q)\}$$

for small s. Then, Eq. (15.24) becomes

$$\frac{\phi(q_1)}{2N} = \binom{2N}{2Nq_1} \int_0^1 [q\{1+s(1-q)\}]^{2Nq_1}[(1-q)(1-sq)]^{2N(1-q_1)}\phi(q)dq$$

$$= \binom{2N}{2Nq_1} \int_0^1 q^{2Nq_1}(1-q)^{2N(1-q_1)} \underbrace{\{1+s(1-q)\}^{2Nq_1}\{1-sq\}^{2N(1-q_1)}}_{\text{selection terms}} \phi(q)dq$$

(*ibid.*, p. 129). On approximating the selection terms above, Wright obtained

$$\frac{\phi(q_1)}{2N+1} = \binom{2N}{2Nq_1} \int_0^1 q^{2Nq_1}(1-q)^{2N(1-q_1)} e^{2Ns(q_1 - q)}\{1 - Ns^2 q_1(1-q_1)\}\phi(q)dq$$

$$= \binom{2N}{2Nq_1}\{1 - Ns^2 q_1(1-q_1)\} \int_0^1 q^{2Nq_1}(1-q)^{2N(1-q_1)} e^{2Ns(q_1 - q)}\phi(q)dq.$$

On trying a solution of the form $\phi(q) = e^{2Nsq}q^{-1}(1 - q)^{-1}(a + bq + cq^2 + dq^3 + \ldots)$, where $a, b, c, d, \ldots$ are constants, and using the identity in Eq. (15.28), he obtained

$$\phi(q) = 2e^{2Nsq}q^{-1}(1 - q^{-1})(C_1\cosh 2Nsq + C_2 \sinh 2Nsq)$$

for constants C_1 and C_2. From symmetry arguments, another solution can be obtained by substituting q by $(1 - q)$ and s by $-s$, so that a more general solution is

$$\phi(q) = q^{-1}(1 - q^{-1})\left[C_1(e^{4Nsq} + 1) + C_2(e^{4Nsq} - 1) + C_3\left\{1 + e^{-4Ns(1-q)}\right\} + C_4\left\{1 - e^{-4Ns(1-q)}\right\}\right]$$

By using the result that the mutation rate at each boundary is half the subterminal frequencies,[39]

$$\begin{aligned} f_1 &= 2C_1 + 2sC_2 + \left(1 + e^{-4Ns}\right)C_3 + \left(1 - e^{-4Ns}\right)C_4 \\ f_{2N-1} &= \left(e^{4Ns} + 1\right)C_1 + \left(e^{4Ns} - 1\right)C_2 + 2C_3 + 2sC_4. \end{aligned}$$

At equilibrium, the flux due to selection is balanced by that due to mutation, so that

$$\int_0^1 \phi(q)\Delta q dq - \frac{f_{2N-1}}{4N} + \frac{f_1}{4N} = 0,$$

where $\Delta q = sq(1 - q)$, resulting in $C_1 - C_2 + C_3 + C_4 = 0$. Using this relationship, the general solution simplifies to

$$\phi(q) = Ce^{4Nsq}q^{-1}(1 - q)^{-1} \tag{15.31}$$

(*ibid.*, p. 131).

It is noteworthy that Wright had previously given an incorrect selection factor of $2Nsq$ (instead of the correct $4Nsq$ in Eq. (15.31)) in his exchange on the evolution of dominance[40] with Fisher (Wright, 1929c, p. 559). In an October 15, 1929 letter (where Fisher had acknowledged his own error (of a different kind)[41]), the latter had insinuated that there was an error in Wright's own formula:

> **QUOT. 983** With this correction I find myself in entire agreement with your value 2N for the time of relaxation and *with your corrected distribution for factors in the absence of selection.*[42] (Provine, 1989, p. 259)

---

[39] See QUOT. 982.

[40] See Sect. 14.4 for a description of this debate.

[41] See Sect. 14.3.4.2.

[42] Italics are ours.

Provine (*ibidem*) reported that Wright might have become increasingly distrustful of Fisher, given that the latter did not explicitly point out the error to Wright, but had instead gone ahead and published a paper with the correct selection factor (Fisher, 1930a, p. 216) as well as in Fisher's book (Fisher, 1930c, p. 92). Fortunately, Wright realized and corrected his error while his big paper was in proof.

15.5.8 The General Case: Stationary Distribution of Gene Frequencies Under Selection, Mutation, and Migration (1931)

By combining the equations in (15.29), (15.30), and (15.31), Wright wrote the stationary distribution under the balancing effects of selection, mutation, and migration in a finite population as

$$\phi(q) = Ce^{4Nsq} q^{4N(mq_m+v)-1}(1-q)^{4N\{m(1-q_m)+u\}-1}$$

(Wright, 1931, p. 134).

15.5.9 Wright's General Formula for the Stationary Distribution of Gene Frequencies (1937, 1938)

In 1937, Wright gave an elegant demonstration of a formula for the calculation of stationary gene frequencies (Wright, 1937). The advantage of this formula was that it avoided the need to guess the form of $\phi(q)$ in Eq. (15.24) and all similar equations in Sects. 15.5.4, 15.5.5, 15.5.6, 15.5.7, and 15.5.8.

Wright's starting point was Eq. (15.23), which, when modified to take into account mutations from A to A' at a rate u and from A' to A at a rate v, becomes

$$\Delta q = Zq(1-q) - uq + v(1-q) \tag{15.32}$$

(*ibid*., p. 315), where $Z = \frac{1}{2}d(\log \overline{w})/dq$, q is now the frequency of A and $\overline{w}$ is the mean fitness. The stationary solution is assumed to be of the form

$$\phi(q) = \psi q^{4Nv-1}(1-q)^{4Nu-1},$$

where ψ is a function of the selection coefficient. At a mode of the distribution ϕ, $d\{\log\phi(q)\}/dq = 0$ so that

$$\begin{aligned} &\frac{d}{dq}\log\psi+\frac{4Nv-1}{q}-\frac{4Nu-1}{1-q}=0.\\ \therefore\quad &\frac{q(1-q)}{4N}\frac{d}{dq}\log\psi+v(1-q)-uq+\frac{2q-1}{4N}=0. \end{aligned} \tag{15.33}$$

From Eq. (15.31), equilibrium implies

$$Zq(1-q)+v(1-q)-uq=0. \tag{15.34}$$

By ignoring the term $(2q-1)/(4N)$ and comparing Eqs. (15.33) and (15.34),

$$Z=\frac{1}{4N}\frac{d}{dq}\log\psi \;\Rightarrow\; \psi=Ce^{4N\int Zdq}$$

$$\Rightarrow\; \phi=Ce^{4N\int Zdq}q^{4Nv-1}(1-q)^{4Nu-1}.$$

Since $Z=\Delta q/\{q(1-q)\}+u/(1-q)-v/q$ from Eq. (15.34), the equation above becomes

$$\phi(q)=\frac{C}{q(1-q)}e^{4N\int\frac{\Delta q}{q(1-q)}dq} \tag{15.35}$$

(*ibid*., p. 316). Wright verified that the solution above implied constancy of both the mean and variance of the distribution.

As examples of the application of (15.35), the stationary solution in Eq. (15.26) can be obtained by setting $\Delta q=0$, that in Eq. (15.29) can be obtained by setting $\Delta q=v(1-q)-uq$, and that in Eq. (15.31) can be obtained by setting $\Delta q=sq(1-q)$.

By imposing the conditions of the constancy of the mean and variance of ϕ, Wright re-derived Eq. (15.30) in 1938 in the form

$$\phi(q)=\frac{C}{\sigma_{\Delta q}^2}e^{2\int\frac{\Delta q}{\sigma_{\Delta q}^2}dq}$$

(Wright, 1938, p. 254), where $\sigma_{\Delta q}^2=q(1-q)/(2N)$.

Wright (1952, pp. 320–321) later derived the stationary solution in Eq. (15.35) under the condition of constancy of all moments.

15.6 Cotterman's "Derivatives" (1940) and Malécot's "Probability of Identity" (1941): Probabilistic Definitions of Wright's Inbreeding Coefficient (f)

15.6.1 Introduction

The reader will recall that Wright's definition of the inbreeding coefficient (see QUOT. 949) was in terms of *correlation*, which is a *statistical* measure, i.e., can be evaluated once data are obtained. Moreover, his use of path coefficients to derive the inbreeding coefficient assumed linearity. In the early 1940s, Cotterman and Malécot independently developed a *probabilistic* definition of the inbreeding coefficient (by using the concept of identity by descent of two alleles) so that the inbreeding coefficient could now be calculated a priori.[43] Moreover, their approach avoided Wright's assumption of linearity. It is thus the probabilistic approach that is now used to define the inbreeding coefficient.

15.6.2 Cotterman's "Derivatives" (1940)

Charles Cotterman (1914–1989) stands as a singular figure in the history of evolutionary genetics. He was born in Dayton, Ohio and showed early interest in science. While still an undergraduate at Ohio State University, he published several papers in human genetics. Working under Laurence Snyder, he submitted his PhD thesis entitled *A Calculus for Statistico-Genetics* in 1940 (Cotterman, 1940). When it was eventually published in 1974 in *Benchmark Papers in Genetics* (Cotterman, 1974), the editor Paul Ballonoff had this to say on Cotterman's thesis:

> **QUOT. 984** Had it been properly published at that time [in 1940], it would have advanced the progress of genetics by a decade or more on several fronts. (Ballonoff, 1974, p. 156)

Despite the brilliance and abundance of his work, Cotterman was not interested in publishing and only a couple of dozen papers have appeared under his name. He passed away in 1989, at the age of 74.

In his 1940 thesis, Cotterman sought a probabilistic interpretation of Wright's inbreeding coefficient. The aim of the thesis was as follows:

> **QUOT. 985** One of the objects of the present paper is to show that the solution of many statistico-genetic problems is greatly simplified when the analysis proceeds from a consideration of the probabilities of the various genic resemblances limited to the types of relatives

[43] In an earlier paper on inbreeding in Mendelian populations, Haldane and Moshinsky (1939) used probability to derive Wright's correlation-based formulas for the inbreeding coefficient, but their approach lacked a fundamental concept that is present in Cotterman and Malécot's approach, namely that of probability of identity by descent of two alleles.

> involved. To the knowledge of the author, no systematic use of such procedure has heretofore been undertaken, with the result that earlier derivations of many genetic principles have often been made unnecessarily complex. (Cotterman, 1940, 1974, p. 267)

Because the concept of gene identity was important for his probabilistic interpretation, Cotterman distinguished two types of identity:

(a) Two genes could be identical "in the sense that, if they could be exchanged, there is reason to believe that no change would be wrought in the phenotypes or breeding behaviors of their possessors." (*ibid.*, p. 167) These genes (e.g., both *A* or both *a*) were called simply "identical" by Cotterman. Crow (1954, p. 544) later used the term "alike in state"—the current term is *identical by state (IBS)*
(b) Two genes could be identical "if they result one from the other or both from a common gene through the process of gene reproduction." These genes were said to be "derivative" (Cotterman, 1940, 1974, p. 168). Crow (1954, p. 544) later used the term "*identical by descent*" *(IBD)* and Cotterman (1983, p. 397) used "*autozygote*" (as opposed to "allozygote"). Both IBD and autozygote are currently used.

Since all genes were ultimately derived from a single gene, Cotterman recommended that the term derivative (i.e., IBD) be applied when the common ancestral gene was within five to six generations ago. Moreover, since mutations were rare, it was *usually* the case that genes that were IBD were also IBS.

Now, consider two individuals *X* and *Y* at a locus with genotypes $[\alpha_1\alpha_2]$ and $[\alpha_3\alpha_4]$, respectively. Cotterman wrote the genotypes in terms of their IBD relationships. For example, if α_1 and α_2 are IBD (i.e., are copies of the same ancestral allele ξ_1), he represented the genotype pair by $[\xi_1\xi_1][\xi_2\xi_3]$, as shown in Fig. 15.18a. On the other hand, if α_1 and α_3 are IBD, the genotypic pair is $[\xi_1\xi_2][\xi_1\xi_3]$, as shown in Fig. 15.18b.

It can be shown that there is a total of 15 such configurations, as first shown by Gillois (1965, p. 29), although Cotterman thought there were only 12.

Cotterman noted that Wright's inbreeding coefficient f (see Eqs. 15.16 and 15.17), being the correlation between two uniting gametes, was also a "coefficient of relationship" between the two contributing individuals. He then defined this coefficient of relationship as the probability that two individuals "will share derivatives of a single gene, irrespective of whether the second gene of the pair is shared" (Cotterman, 1940, 1974, p. 189). *That is, Cotterman defined f as the probability that two individuals will share at least one pair of IBD alleles.* In terms of genotypic pairs,

$$f = \underbrace{\Pr\{[\xi_1\xi_2][\xi_1\xi_3]\}}_{\text{one allele IBD}} + \underbrace{\Pr\{[\xi_1\xi_2][\xi_1\xi_2]\}}_{\text{two alleles IBD}}$$

(*ibid.*, p. 188). In his thesis, Cotterman also defined the probabilities of two individuals sharing 0, 1, and 2 IBD alleles, respectively, as

Fig. 15.18 The four alleles (*) of individuals X (top) and Y (bottom). A line joining any two or more alleles means the latter are IBD. In (**a**), alleles α_1 and α_2 of X are IBD and the genotypic pair is represented by $[\xi_1\xi_1][\xi_2\xi_3]$. In (**b**), allele α_1 of X and allele α_3 of Y are IBD and the genotypic pair is represented by $[\xi_1\xi_2][\xi_1\xi_3]$

$$
\begin{aligned}
c_0 &= \Pr\{[\xi_1\xi_2][\xi_3\xi_4]\},\\
c_1 &= \Pr\{[\xi_1\xi_2][\xi_1\xi_3]\} = \Pr\{[\xi_1\xi_2][\xi_2\xi_3]\},\\
c_2 &= \Pr\{[\xi_1\xi_2][\xi_1\xi_2]\}
\end{aligned}
$$

(*ibid.*, p. 195). These later came to be known as the *k*-coefficients.

15.6.3 Malécot's "Probability of Identity" (1941)

Gustave Malécot (1911–1998) was a revered figure in population genetics, mainly because of the originality, depth, and clarity of his research in this field.[44] Malécot was born in a small village near St. Étienne. In 1832, he attended the École Normale Supérieure to study mathematics and was taught by Émile Borel, Maurice Fréchet, and Georges Darmois. He graduated in 1935 and joined the Institut Henri Poincaré at the University of Paris. Under the supervision of Darmois, Malécot completed his PhD thesis entitled *Théorie Mathématique de l'Hérédite Mendélienne Genéralisée* (Malécot, 1939), which generalized Fisher's 1918 treatment of correlation between relatives (Fisher, 1918). Subsequently, Malécot published many important papers and the results were summarized in his book *Les Mathématiques de l'Hérédité* (Malécot, 1948).[45] Malécot remained professor of applied mathematics at the University of Lyon from 1945 until 1981, when he retired. He passed away in 1988.

Unaware of Cotterman's work one year ago, in 1941 Malécot published a breakthrough paper where he provided a probabilistic definition of Wright's inbreeding coefficient (Malécot, 1941). Just three years earlier, Malécot had used the *statistical* definition of Wright's inbreeding coefficient as a correlation between uniting gametes to show that the correlation between two relatives X and Y is given by

[44] Malecot's life and work are also discussed in Nagylaki (1989), Slatkin and Veuille (2002), Tort (1996, pp. 2769–2785, Vol. 2: F:N), and Epperson (1999).

[45] Published in English under the title *The Mathematics of Heredity* (Malécot, 1969).

$$r_{XY} = \frac{r+\rho}{2}\frac{\tau^2}{\tau^2+\varepsilon^2} + r\rho\frac{\varepsilon^2}{\tau^2+\varepsilon^2}$$

(Malécot, 1938, p. 404), where r is the correlation between the alleles of X and Y on one chromosome, ρ is the correlation on the other chromosome, τ^2 is the additive genetic variance, and ε^2 is the dominance variance. The above formula generalizes Fisher's formulas in Sect. 14.2.4. For example, for two brothers, $r = \rho = 1/2$ and the above formula reduces to Fisher's Eq. (14.9).

In the 1941 paper, Malécot considered the union of two gametes H and $\overline{H}$ such that the correlation between them is f. If the gametes H and $\overline{H}$ are contributed by the parents H and $\overline{\text{H}}$, respectively, then Malécot noted that f was also:

(i) The *coefficient of parentage* ($\Phi_{\text{H}\overline{\text{H}}}$) between H and $\overline{\text{H}}$.
(ii) The *coefficient of consanguinity* ($f_{\text{H}\times\overline{\text{H}}}$) of a child of H and $\overline{\text{H}}$.

That is, $\Phi_{\text{H}\overline{\text{H}}} \equiv f_{\text{H}\times\overline{\text{H}}}$.

Now, suppose that each gamete can contain an allele A with probability p or an allele a with probability q ($= 1 - p$). Malécot then rewrote the three equations implied by (14.17) as follows:

$$\begin{aligned} P &= p^2 + pqf, \\ 2Q &= 2pq(1-f), \\ R &= q^2 + pqf \end{aligned} \tag{15.36}$$

(Malécot, 1941, p. 48), where P, $2Q$, and R are the frequencies of AA, Aa, and aa, respectively. Malécot now observed:

> **QUOT. 986** A remarkable interpretation of f is as follows: from the point of view of the stochastic relationship between H and $\overline{H}$, everything takes place as if these random quantities had the probability f of being identical and the probability $1 - f$ of being independent. (*ibid.*, p. 49)

Using Malécot's interpretation, if the uniting gametes are identical (with probability f), then the frequency of AA is p; on the other hand, if the uniting gametes are independent (with probability $1 - f$), then the frequency of AA is p^2. Thus,

$$P = f \times p + (1-f) \times p^2 = p^2 + pqf.$$

Similarly, for the two other genotypes,

$$\begin{aligned} 2Q &= f \times 0 + (1-f) \times 2pq = 2pq(1-f), \\ R &= f \times q + (1-f) \times q^2 = q^2 + pqf. \end{aligned}$$

Malécot thus defined Wright's f *as the probability that two uniting gametes are identical* and showed that, with this definition, the equations in (15.36) are obtained.

Note that, in the 1941 paper, Malécot simply used identity, without any further explanation as to the nature of the identity (i.e., *IBD* vs. *IBS*).[46]

Malécot also showed how the coefficient of parentage of an individual A with itself could be calculated. This is the probability that two gametes from the individual are identical. This event can happen either if the two gametes are from the same parent (with probability 1/2) or if the two gametes are from two different parents (with probability 1/2), which themselves have a probability f_A of bearing identical alleles. Hence, the coefficient of parentage of an individual with itself is

$$\Phi_{\mathrm{AA}} = \frac{1 + f_A}{2}.$$

This formula was fundamental for Malécot's proof of Wright's pedigree formula in Eq. (15.19). Suppose H and $\overline{\mathrm{H}}$ have a common ancestor A such that there are $n - 1$ intermediate parents between H and A, and $p - 1$ intermediate parents between $\overline{\mathrm{H}}$ and A (e.g., see Fig. 15.12). This implies a total n line segments or segregating events from H through A so that the probability that a particular allele is transmitted from A to H is $(1/2)^n$. Similarly, the probability that a particular allele is transmitted from A to $\overline{\mathrm{H}}$ is $(1/2)^p$. Now, Malécot's coefficient of parentage between H and $\overline{\mathrm{H}}$ is $(1/2)^{n+p}$ multiplied by the probability that the two particular alleles in A are identical, which is $(1 + f_A)/2$, where f_A is the inbreeding coefficient of the common ancestor *A*. Hence the coefficient of parentage between H and $\overline{\mathrm{H}}$ is $(1/2)^{n+p}(1 + f_A)/2$. In the case of several common ancestors, Malécot finally obtained the following expression for the coefficient of parentage between H and $\overline{\mathrm{H}}$:

$$\Phi_{\mathrm{H}\overline{\mathrm{H}}} = \sum \left(\frac{1}{2}\right)^{n+p} \left(\frac{1 + f_A}{2}\right)$$

(*ibid*., p. 52), which is Wright's (15.19). Here, $n + p$ is the total number of line segments from H to $\overline{\mathrm{H}}$ via A.

As mentioned above, in the 1941 paper, Malécot did not offer any further information on the nature of the *identical* gametes he used in his definition of *f*. This changed a few years later when he made use of a common ancestral gene in his definition:

> **QUOT. 987** The *mean consanguinity coefficient*[47] f_n of an individual A_n chosen at random in F_n [the nth generation]; it is also the *coefficient of parentage* of its two parents A_{n-1} and A'_{n-1} chosen at random in F_{n-1}; it is the probability that two homologous genes H_{n-1} and H'_{n-1}, taken at random from A_{n-1} and A'_{n-1} and uniting in A_n, descend from a single gene in common ancestor. (Malécot, 1946, p. 841)

The above fills in the IBD component that was previously missing in Malécot's definition of the inbreeding coefficient.

[46] But more on this at the end of the section.

[47] All italics in this citation are Malécot's.

Chapter 16
Modern Synthesis (3): The Contributions of Haldane

Abstract Haldane was the third founding father of the modern synthesis. In one of his early papers, he derived a fundamental equation for the selection intensity, which he applied to industrial melanism. His estimate of a large value for the selection intensity was later confirmed by Kettlewell. In another paper, he introduced mutation in his selection model and showed that the probability of survival of an initial mutation is approximately twice the selection intensity when the latter was small. In 1957, Haldane formally introduced the concept of the cost of natural selection. This was one of the influences in Kimura's later development of the neutral model, although both Haldane's and Kimura's calculations of the cost have been criticized. In 1964, Haldane defended the works of the population geneticists against Mayr's beanbag genetics criticism.

Keywords Selection intensity · Industrial melanism · Cost of natural selection · Genetic load · Classical vs. balance hypothesis · Segregational load · Neutral theory · Beanbag genetics.

16.1 Introduction

Together with Fisher and Wright, Haldane is recognized as the third person to have made the most significant contributions to the modern synthesis. John Burdon Sanderson Haldane[1] (Fig. 16.1) was born in Oxford. In 1911, he won a scholarship to study mathematics at New College, Oxford, and graduated with first-class honors. In his short autobiography, he later recalled:

> **QUOT. 988** But as nobody can study mathematics intensively for more than about 5 hours daily and retain sanity, I also attended the final honors course in zoology in my first year. (Haldane, 1966, p. 477)

[1] John Haldane's life and work are also discussed in Dronamraju (1968, 1985, 1986, 2016), Clark (1968), Majumder (1993), Provine (2001, pp. 168–177), Edwards (2017), Zimmermann (1975), Tort (1996, pp. 2124–2127, Vol. 2: F-N), and Indian Statistical Institute (1992).

P. Gorroochurn, *The Development of Evolutionary Genetics*,
https://doi.org/10.1007/978-3-031-69374-8_16

Fig. 16.1 John Burdon Sanderson Haldane (1892–1964). (From: https://commons.wikimedia.org/wiki/File:Haldane.jpg)

Haldane also won first-class honors in classics and philosophy in 1914. He had no formal training in science but was exposed to it from an early age through his father, the physiologist John Scott Haldane. Thus, at the age of 8, Haldane reportedly attended with his father a talk on Mendelism by Darbishire. In 1919, Haldane became a fellow in New College, and in 1922, he moved to Cambridge University until 1932. He then joined University College, London, where he became Chair of Biometry in 1937. In 1957, Haldane joined the Indian Statistical Institute, and in 1961, he became a naturalized Indian citizen. He passed away in 1964 in Bhubaneseswar, India.

Fisher, Wright, and Haldane all worked independently of each other, although they did exchange letters on their work. While the relationship between Wright and Fisher eventually broke down, Haldane remained on speaking terms with his two colleagues. His scientific interests were also broader. Haldane first met Wright in 1932, during the International Congress of Genetics in Ithaca, New York. He may have met Fisher at around 1923 in Cambridge but definitely met him in April 1930 when they each gave a talk (Edwards, 2017, p. 748). Haldane's approach to population genetics was closer to Fisher's than to Wright's, in the sense that he shared the former's gene-centered approach.

Haldane's first paper, co-authored with C. G. Douglas and his father, was on physiology (Douglas et al., 1912) and his first genetics paper, co-authored with A. D. Sprunt and N. M. Haldane (his sister), was on linkage (Haldane et al., 1915). In 1919, Haldane published two further papers on linkage (Haldane, 1919a, b), the second of which gave a formula for the map distance as a function of the recombination fraction.

16.2 The "Mathematical Theory of Natural and Artificial Selection" Papers (1924–1934)

16.2.1 Introduction

Haldane's entry into the population genetics scene occurred through a succession of ten groundbreaking papers written between 1924 and 1934 under the generic title "A mathematical theory of natural selection." These represent his most important contributions to population genetics. It is very likely that Haldane was influenced by H. T. J. Norton, whom he met in Cambridge and who showed him results that Norton had derived as far back as 1910. Recall that[2] Punnett had solicited Norton's help in constructing a key table in Appendix I of his *Mimicry in Butterflies* (Punnett, 1915). This table provided generation times for specific changes in genotype frequencies under varying selection intensities and different modes of inheritance.

16.2.2 Part I: Fundamental Equation for Selection Intensity, Application to Industrial Melanism

In his first paper, Haldane (1924b) devised explicit mathematical models to explain the results in Punnett's table as well as for a host of other models of inheritance. Before these are explained, however, it is instructive to quote in full the opening paragraph of the first paper:

> **QUOT. 989** A satisfactory theory of natural selection must be quantitative. In order to establish the view that natural selection is capable of accounting for the known facts of evolution we must show not only that it can cause a species to change, but that it can cause it to change at a rate which will account for present and past transmutations. In any given case we must specify the following.

[2]See Sect. 13.2.4.

(1) The mode of inheritance of the character considered.
(2) The system of breeding in the group of organisms studied.
(3) The intensity of selection.
(4) Its incidence (e.g. on both sexes or only one).
(5) The rate at which the proportion of organisms showing the character increases or diminishes.

It should then be possible to obtain an equation connecting (3) and (5).
(*ibid.*, p. 19)

The above is an announcement of a *quantitative* agenda to address questions of evolutionary interest, with an emphasis on the *rate* of evolution due to natural selection.

In one of the classic inheritance models considered by Haldane, a Mendelian character is assumed to be due to a single locus consisting of two alleles A (dominant) and a (recessive). Let u_n be the ratio of the number of A to the number of a alleles in the nth generation so that, under the assumption of Hardy–Weinberg equilibrium, the initial proportions of genotypes in the nth generation are given by

$$u_n^2 AA : 2u_n Aa : 1aa.$$

Now, assuming that a proportion $(1 - k)$ of recessives survive (i.e., the selection intensity against aa is k), the proportions of survivors in the nth generation are given by

$$u_n^2 AA : 2u_n Aa : (1-k)aa$$

so that

$$u_{n+1} = \frac{u_n(u_n+1)}{u_n+1-k} \tag{16.1}$$

(*ibid.*, p. 23). Haldane next devised an ingenious method to solve the difference equation above. Let

$$\Delta u_n \equiv u_{n+1} - u_n = \frac{ku_n}{u_n+1-k}.$$

Haldane assumed that k is small, so u_n increases continuously and $\Delta u_n \approx du_n/dn$. Then, approximately,

$$\frac{du_n}{dn} = \frac{ku_n}{u_n+1} \quad \Rightarrow \quad kn = \int_{u_0}^{u_n} \frac{1+u}{u}\,du,$$

where u_0 is the initial A: a allele ratio. Hence,

$$kn = u_n - u_0 + \log\left(\frac{u_n}{u_0}\right) \tag{16.2}$$

(*ibid.*, p. 24) is the fundamental equation relating the number of generations n required for the $A{:}a$ allele ratio to change from u_0 to u_n for small selection intensities k against the recessive. In terms of recessive proportions y_n [$= (1 + u_n)^{-2}$], the equation above can be written as[3]

$$kn = y_n^{-1/2} - y_0^{-1/2} + \log\left(\frac{y_n^{-1/2} - 1}{y_0^{-1/2} - 1}\right). \tag{16.3}$$

As an example, Haldane considered the following well-known case of industrial melanism[4]:

> **QUOT. 990** The dominant melanic form *doubledayaria* of the peppered moth *Amphidasys betularia* first appeared at Manchester in 1848. Some time before 1901 when Barrett described the case, it had completely ousted the recessive variety in Manchester. It is required to find the least intensity of natural selection[5] which will account for this fact. (*ibid.*, p. 26)

Haldane assumed that "there were not more than 1% of dominants in Manchester in 1848, nor less than 99% in 1898." This gives $y_0 = .99$, $y_n = .01$ and $n = 50$. Haldane devised two sets of tables, one to read off y_n from kn in Eq. (16.3), and the other to read off kn from y_n. However, in the current case, k can be calculated directly to give $k = .330$ (compared to Haldane's table value of $k = .332$).

Despite Haldane's assurance that the calculated k was *not* a "very intense degree of natural selection," its value is indeed surprisingly large since Darwinian selection is usually thought of as a gradual process with only small (~.01) fitness differences between different phenotypes. Although Haldane's results were initially viewed with some suspicion, they were later vindicated by Kettlewell (1956) who performed systematic experiments on the peppered moth and found that the selection coefficient could, in fact, be even higher.

Haldane's first paper also contained several extensions of his basic model, and these are summarized below:

---

[3]Using this formula, the numbers calculated by H. T. J. Norton and cited by Punnett in QUOT. 831 may be obtained. Setting $y_0 = .5$, $y_n = .025$ and $n = 70$ in Eq. (16.3) gives $k = .1066$, which is close to the cited 10%.

[4]This well-known phenomenon refers to the appearance of melanic (or dark) variants of animals living in industrial areas. The most famous example is that of the peppered moth. Industrial melanism and the peppered-moth story are also discussed in Majerus (1998), Kettlewell (1961, 1973), Ford (1975, Chap. 14), Hooper (2002), Cook and Saccheri (2013), Neal (2004, pp. 171–179), and Numbers & Kampourakis (2015, pp. 171–177).

[5]The subsequent preponderance of the melanic (dark) form was ascribed to natural selection first by Tutt (1896) who argued that the dark coloration in a polluted environment offered camouflage against predators (*our note*).

(i) *Familial selection* (p. 27): Haldane introduced this term to denote selection due to competition between family members, some of whom may possess an advantageous gene. Haldane reported that, for slow selection against the recessives, the selection intensity was 1/2 the value in Eq. (16.2) if each family had two parents in common (as in mammals), and 3/4 the value in Eq. (16.2) if each family had one parent in common (as in cross-pollinated seed plants).

(ii) *Sex-limited selection* (p. 29): Haldane considered the case when a deleterious allele affects only one sex. He showed that for slow selection, the selection intensity was half the value in Eq. (16.2).

(iii) *Selection of an alternatively dominant character* (p. 30). It was possible for an allele to be dominant for a character in one sex and recessive for that character in another sex. Haldane cited Wood (1905), who argued that horns in Dorset and Suffolk sheep were dominant in males and recessive in females. Haldane found that the selection intensity in this case was higher than that in Eq. (16.2).

(iv) *Bisexual selection of a sex-linked character* (p. 31). The next case considered was that of a sex-linked character such that, in each sex, the recessives were selected against. Haldane found that, for slow selection in the homozygous sex, selection acted more rapidly on a sex-linked character than on an autosomal character, whereas in the heterozygous sex, selection acted at approximately the same rate in the two cases.

(v) *Bisexual familial selection of a sex-linked character* (p. 33). This was the same situation as above, except that selection occurred due to competition between family members (see (i) above). Haldane considered the case of two parents in common within mixed families and found that, for slow selection, selection acted at half to a third of the rate in Eq. (16.2).

(vi) *Selection of a sex-linked character in the homozygous sex only* (p. 34). Haldane considered the case where sex-linked factors had an effect on the homozygous rather than heterozygous sex. For slow selection, he found that, in the former case, the selection intensity was 1.5 times smaller than that in Eq. (16.2), while in the latter case, it proceeded more slowly.

(vii) *Selection of a sex-linked character in the heterozygous sex only* (p. 34). This is the same case as above except that the effect was on the heterozygous sex only. Haldane showed that, for slow selection, the selection intensity in the heterozygous sex was 1/3 times that in an asexual model, while in the homozygous sex it proceeded slightly faster.

(viii) *Gametic selection of an autosomal character* (p. 35). Haldane now considered the case when selection acted on gametes. He found that for slow selection in heterozygotes, the selection intensity was half that when selection was slow and on recessives.

(ix) *Gametic selection on a sex-linked character* (p. 36). As an example of this type of selection, Haldane considered selection among the gametes of the homozygous sex, with no selection in the heterozygous sex. He found that the selection intensity was twice that in (vii).

16.2.3 Parts II–IV: Further Extensions of the Basic Selection Model

In the second paper of the series, Haldane (1924a) continued extending the mode of inheritance and calculated the selection intensity for the following cases: (a) partial self-fertilization, (b) partial inbreeding, (c) partial assortative mating, and (d) selective fertilization. His conclusion was that inbreeding and self-fertilization greatly increased the selection intensity.

In Part III of the series, Haldane (1926) further considered: (a) selection of an incompletely dominant autosomal character, (b) selection of an incompletely dominant sex-linked character, (c) selection for a multiple recessive character when the loci are unlinked, (d) selection of a multiple sex-linked recessive character, (e) selection for a multiple autosomal dominant character when the loci are unlinked, (f) two linked loci in the absence of selection (here the linkage disequilibrium results of Robbins (1918b, p. 376) are independently rederived), and (g) selection in a tetraploid organism.

In the fourth installment of the series, Haldane (1927a) considered overlapping generations. He cited H. T. J. Norton, who had realized, as early as 1910,[6] that this case needed to be approached through integral rather than difference equations. Norton showed his results to Haldane in 1922, although the latter had independently derived them before.

16.2.4 Part V: Mutations Introduced in Selection Models

In Part V, Haldane (1927b) introduced mutations in his selection models and followed the stochastic approach that had been previously used by Fisher (1922) (see Sect. 14.3.3). As in the latter paper, let $f(x)$ be the probability generating function (PGF) of the number of copies of an initial gene in the next generation. Then, the PGF of the number of copies of an initial gene after n generations is $f(f(f..f(x)...))$, the operation being repeated n times.

In the case of a dominant gene, Haldane assumed a selection advantage of k. The mean number of copies of an initial gene in the next generation is $f'(1) = 1 + k$. If the distribution of the number of copies is assumed to be Poisson, the PGF is $f(x) = e^{(1+k)(x-1)}$ and, for $k > 0$, the probability of ultimate extinction is the smallest nonnegative root of

$$x = f(x) = e^{(1+k)(x-1)}.$$

Writing $x = 1 - y$, where y is the probability of ultimate survival, the above becomes

[6] See footnote 16 of chap. 13.

$$1 - y = e^{-(1+k)y}.$$

Therefore,

$$(1 + k)y = -\log(1 - y) \Rightarrow k = \frac{y}{2} + \frac{y^2}{3} + \frac{y^3}{4} + \ldots$$

For small k, Haldane obtained

$$y \approx 2k$$

(Haldane, 1927b, p. 839), i.e., for a small selective advantage of k, the probability of survival of a dominant gene is approximately $2k$, irrespective of the population size. This agrees reasonably well with Fisher's second and third values in Table 14.6.

Haldane next considered the case of a recessive gene with selective advantage k. The survival probability was much lower than that in the previous case, being now given by

$$y \approx \sqrt{\frac{k}{N}},$$

where N is the population size.

Haldane also considered a two-way mutation model in a random mating population. Suppose that, in each generation, a proportion p of alleles A mutates to a, a proportion q of alleles a mutates to A, and the selection intensity against recessives is k. Then, Eq. (16.1) becomes

$$u_{n+1} = \frac{(1 - p)\left(u_n^2 + u_n\right) + q(u_n + 1 - k)}{(1 - q)(u_n + 1 - k) + p\left(u_n^2 + u_n\right)},$$

where u_n is the ratio of the number of A to a alleles in the nth generation, so that

$$\Delta u_n = u_{n+1} - u_n \approx \frac{ku_n}{u_n + 1} - pu_n(u_n + 1) + q(u_n + 1)$$

for small p, q, and k. At equilibrium, $\Delta u_n = 0$ and $u_{n+1} = u_n = u$, so that

$$pu^3 + (2p - q)u^2 + (p - 2q - k + kq)u - q + kq = 0.$$

If $k, q >> p$,

$$u = \frac{\text{freq}(A)}{\text{freq}(a)} \approx \sqrt{\frac{k}{p}}$$

(*ibid.*, p. 841). This formula was subsequently also derived by both Fisher (1928a)[7] and Wright (1929b).[8]

Haldane summarized his views about the relative importance of natural selection and mutation as follows:

> **QUOT. 991** To sum up, if selection acts against mutation, it is ineffective provided that the rate of mutation is greater than the coefficient of selection. Moreover, mutation is quite effective where selection is not, namely in causing an increase of recessives where these are rare. It is also more effective than selection in weeding out rare recessives provided that it is not balanced by back mutation of dominants. Mutation therefore determines the course of evolution as regards factors of negligible advantage or disadvantage to the species. It can only lead to results of importance when its frequency becomes large. (Haldane, 1927b, p. 842)

Haldane thus believed that mutation was an important factor in evolution. Its effects surpassed those of natural selection when the mutation rate was greater than the coefficient of selection and when the frequency of mutants became large enough.

16.2.5 Part VI–X: Gene Flow, Varying Selection Intensity, Metastable Equilibrium, Large Selection Coefficients, and Artificial Selection

In Part VI, Haldane (1930a) introduced gene flow in his selection models. His basic model was one of an isolated population of size N, in which a new type with selection coefficient k is favored. In each generation, a number $l \times N$(where l is a constant) of immigrants of the old type were admitted to the population.

Under the above assumptions, Haldane considered ten different models. For example, when the dominants are favored and the recessives immigrate, Eq. (16.1) becomes

$$u_{n+1} = \frac{u_n(u_n + 1)}{u_n + 1 - k + l(u_n + 1)^2},$$

so that

$$\Delta u_n = \frac{u_n}{u_n + 1}\left\{k - l(u_n + 1)^2\right\}.$$

Equilibrium occurs when

---

[7] See footnote 43 of chap. 14.

[8] See Eq. (14.53).

$$u = \sqrt{\frac{k}{l}} - 1 \quad \text{for} \quad k > l$$

(*ibid.*, p. 221). On the other hand, when $k < l$, $\Delta u_n < 0$ so that the dominants disappear, and selection is ineffective.

In Part VII, Haldane (1931a) investigated selection models for which the selection intensity varied with the mortality rate. He showed that the selection intensity may diminish and become negative at high elimination rates, and under the best conditions the intensity may increase at an extremely small rate.

In Part VIII, Haldane (1931b) considered a multi-locus system in which a special type of equilibrium, which he called a *metastable* equilibrium, existed. He showed that the action of genetic drift could rupture this equilibrium and lead to different species. Haldane illustrated the two-locus case with a diagram that almost mirrored Wright's adaptive landscape. His "metastable" paper has been described in more detail in Sect. 15.4.4.3.

In Part IX, Haldane (1932a) extended his basic selection model in Part I to accommodate large selection intensities. The difference equation in (16.1) could no longer be solved by using $\Delta u_n \approx du_n/dn$. Instead, Haldane proceeded by first writing $x_n = 1 + 1/u_n$, so that Eq. (16.1) becomes

$$x_{n+1} = x_n - \frac{k(x_n - 1)^2}{x_n}.$$

By appealing to a result he had previously proven in Haldane (1932b), he obtained the following for the difference equation above:

$$\begin{aligned} n = \int_{x_0}^{x_n} \Bigg\{ &-\frac{x}{k(x-1)^2} + \frac{x+1}{2x(x-1)} + \frac{k(x^2+2x+3)}{12x^3} + \frac{k^2(x^2-1)(x^2+2x+5)}{24x^5} + \\ &\frac{k^3(x-1)^2(19x^4+76x^3+220x^2+360x+105)}{720x^7} + \ldots \Bigg\} dx \\ \approx \frac{u_n - u_0}{k} &+ \frac{\log\left(\dfrac{1+u_n^{-1}}{1+u_0^{-1}}\right)}{\log(1-k)} + \frac{1-k}{k}\log\left(\frac{1+u_n}{1+u_0}\right) \quad \text{for} \quad k < 1 \end{aligned}$$

(*ibid.*, p. 245). With this formula, Haldane was able to verify Elton's (1927) suggestion that occasional intense selection (such as during periodic famines and plagues) was more consequential than less intense selection acting in every generation. Haldane applied his formula to the following two situations: (i) selection of intensity k acting in every generation, and (ii) selection occurring with intensity mk (where $|mk| < 1$) in every m^{th} generation. He showed that Elton's suggestion was true as long as $u_n > u_0$, i.e., when dominants were favored.

Table 16.1 Genotype frequencies and fitnesses in Haldane's dominant model (allele *a* is a dominant mutant)

Genotype	*AA*	*Aa*	*aa*
Freq.	$1 - x$	x	0
Fitness	1	f	f

In the tenth and final installment of the series, Haldane considered various theorems in artificial selection.

Several of the results in the ten-paper series were included in Chap. V and the Appendix of Haldane's landmark book *The Causes of Evolution* (Haldane, 1932c).

16.3 Haldane's Cost of Natural Selection (1937, 1957): Genetic Loads and Associated Controversies[9]

16.3.1 Haldane's Principle and the "Price" of Evolution (1937)

In 1937, Haldane published an important paper (Haldane, 1937) containing the germ of an idea (the genetic load) that later became one of the major influences for a new theory of evolution (the neutral or non-Darwinian theory), which directly challenged Darwinian evolution.

Haldane's aim in 1937 was simple enough, namely, to quantify the effect of deleterious genes on the fitness of a species. He had already noted earlier in the paper that the initial mean fitness must be close to unity:

> **QUOT. 992** It is clear that the mean fitness of all members of a species must always be very close to unity, if we average over any length of time. If the fitness were 1.01 the population would increase 20,959 times in 1,000 generations. In almost all species the mean fitness over 1,000 generations must vary from unity by far less than one per cent. (*ibid.*, pp. 339–340)

He now first considered the case of a normal allele, *A*, which mutates to a dominant allele, *a*, at a rate μ per locus per generation. He also assumed that the homozygote *aa* is so rare that it may be ignored. Let x be the frequency of the mutant *Aa* and let $f(<1)$ be its fitness (see Table 16.1). Then, the increase in frequency of the allele *a* in the next generation due to mutation only is $2\mu(1 - x) + \mu x = (2 - x)\mu$ and the decrease in frequency due to natural selection only is $x - xf = x(1 - f)$. At equilibrium the two changes are the same in magnitude, so that

[9]The concepts of cost of selection and genetic load are also discussed in Wills (1980, Chap. 3), Crow (1970, pp. 128–177), Ewens (1993), Wallace (1970, 1991), Williams (1992, pp. 143–148), Nei (2013, pp. 35–37, 60–65), and Tort (1996, pp. 1622–1628, Vol. 1: A-E).

Table 16.2 Genotype frequencies and fitnesses in Haldane's recessive model (allele *a* is a recessive mutant)

Genotype	*AA*	*Aa*	*aa*
Freq.	$1-2y-x$	$2y$	x
Fitness	1	1	f

$$x = \frac{2\mu}{1-f+\mu} \approx \frac{2\mu}{1-f} \quad \text{since } \mu \ll f.$$

Since the initial mean fitness is assumed to be unity, the decrease in the mean fitness of the population is

$$\begin{aligned} 1-\{(1-x)+xf\} &= x(1-f) \\ &\approx \frac{2\mu}{1-f}(1-f) \\ &= 2\mu \end{aligned} \tag{16.4}$$

(*ibid*., p. 342).

Haldane next considered the case of a normal allele *A* that mutates to a recessive allele *a* at a rate μ per locus per generation. Let the frequencies of *aa* and *Aa* be x and $2y$, respectively, and the fitnesses be as shown in Table 16.2. Then, the increase in the frequency of alleles *a* due to mutation is $2\mu(1-2y-x) + 2\mu y = 2\mu(1-x-y)$ and the decrease in frequency due to natural selection is $2x(1-f)$.[10]

At equilibrium,

$$x = \frac{\mu - \mu y}{1-f+\mu} \approx \frac{\mu}{1-f} \quad \text{since } \mu \ll f \text{ and } y \text{ is small}$$

The decrease in mean fitness is now

$$\begin{aligned} 1-\{(1-2y-x)+2y+xf\} &\approx x(1-f) \\ &\approx \frac{\mu}{1-f}(1-f) \\ &= \mu \end{aligned} \tag{16.5}$$

As Haldane noted, it is remarkable that the decrease in mean fitness in both (16.4) and (16.5) is independent of the selection intensity (f): instead, it depends only on the mutation rate (μ). This result is known as *Haldane's principle*. Moreover, the decrease in mean fitness in the first case is twice that in the second because, in the first case (dominant mutant), a single mutant is sufficient to cause detrimental effects, while in the second case (recessive mutant), two mutants are necessary.

[10] The factor 2 before $x(1-f)$ occurs because *two* alleles *a* are lost for each loss of *one aa* genotype.

Not only were Haldane's results remarkable but also the interpretation he gave to the loss of fitness was noteworthy:

QUOT. 993 It is suggested that this loss of fitness is the price paid by a species for its capacity for further evolution. (*ibid.*, p. 349)

Haldane's reasoning was that the effectiveness of natural selection depended on the presence of genetic variation. However, genetic variation implied a fitness difference in genotypes, i.e., some genotypes would have lower fitnesses than others. This decrease in fitness and consequent loss of the associated genotypes was the cost incurred by the species so that evolution by natural selection could proceed.

The novelty here was the idea of a *price* or *cost* associated with natural selection.

16.3.2 Muller (1950): "Genetic Load" Introduced

Haldane's 1937 paper was not particularly influential. However, in 1949, it played a key role in H. J. Muller's presidential address to the American Society of Human Genetics.

Hermann Joseph Muller (Fig. 16.2) was born in New York in 1890. He showed interest in genetics at the early age of 16. His initial experiments on spontaneous gene mutations were performed under the guidance of T. H. Morgan. He became one of the foremost geneticists of his time, winning the Nobel Prize in 1946 for his much earlier discovery of the use of X-rays to induce genetic mutations.

In his presidential address, Muller read a paper entitled "Our load of mutations," (Muller, 1950) which was meant to highlight, contrary to prevailing opinion, the significant effect of mutations on human health. The paper introduced the term "genetic load,[11]" which was a rephrasing of Haldane's price of selection in terms of the load or burden that deleterious mutations placed on the species. Muller referred to Haldane's calculations of 1937 as pioneering and admitted that he had independently reached a similar conclusion. Using a mutation rate per gene per generation of 1 in 50,000 and assuming that the total number of loci was 5000, he concluded that at least 1 in 10 gametes in humans carried a mutation (*ibid.*, p. 128). Moreover, using an average of eight detrimental genes per person, with each gene having a selective disadvantage of 2.5%, he concluded that there was:

QUOT. 994 ...nearly 20% chance of death or of reproductive inefficacy, from genetic causes. (*ibid.*, p. 142)[12]

---

[11]More specifically, Muller's genetic load of 1949 was, like Haldane's, a *mutational* load.

[12]Muller's calculation was later generalized by Morton et al. (1956, p. 857) to give the concept of a *lethal equivalent:* "A lethal equivalent is a group of mutant genes of such number that, if dispersed in different individuals, they would cause on the average one death, e.g., one lethal mutant, or two mutants each with 50 per cent probability of causing death, etc." Thus, the number of lethal equivalents is the number of deleterious recessive alleles in heterozygous condition multiplied by the probability the recessive allele would cause death when homozygous. For example, if there are 10 recessive alleles in heterozygous condition and each recessive homozygote has a probability of .8 of causing death, the number of lethal equivalents is 8.

Fig. 16.2 Hermann Joseph Muller (1890–1967). (From: https://commons.wikimedia.org/wiki/File:HJ_Muller_1952.jpg)

These are relatively large numbers and implied that mutations produced a significant load on human beings.

16.3.3 Dobzhansky (1955): The Classical versus Balance Hypothesis, Two Types of Genetic Load

Following Muller's presidential address, Dobzhansky[13] (1955) wrote a paper that made a fundamental contribution to the understanding of how genetic variation was maintained in natural populations.

Dobzhansky's 1955 paper started by reviewing the *classical hypothesis (or model)* of genetic variation, according to which alleles at a given locus were either wild-type or mutant. The wild-type was the normal form of the gene; it was

[13] See Sect. 15.4.7 for more on Dobzhansky.

favored by natural selection and therefore abundant. On the other hand, the mutant was the abnormal form of the gene; it was deleterious and rare. Moreover:

QUOT. 995. . .evolutionary changes consist in the main in gradual substitution and eventual fixation of the more favorable, in place of the less favorable, gene alleles and chromosomal structures. Superior alleles are established by natural selection, and supplant inferior ones.

Most individuals in a Mendelian population should, then, be homozygous for most genes. Heterozygous loci will be a minority. (*ibid.*, p. 3)

As indicated by Dobzhansky, the classical hypothesis goes back to Plato's Theory of Forms (Sect. 1.1.2) and the associated typological (or essentialist) thinking (Sect. 2.9). However, the model is best exemplified by the work of Morgan and his colleagues.[14] Fisher, Wright and Haldane also used the classical model, but Dobzhansky had Muller's paper (Muller, 1950) in mind and referred to the latter on several occasions in his paper.

Dobzhansky next expressed doubts about the ubiquity of the classical hypothesis as follows:

QUOT. 996 An array of related genotypes, consonant with the demands of the environment is the adaptive norm of a population. The "norm" is, thus, neither a single genotype nor a single phenotype. It is not a transcendental constant standing above or beyond the multiform reality. The "Norm" of *Drosophila melanogaster* has as little reality as the "Type" of *Homo sapiens*. (Dobzhansky, 1955, p. 3)

Basing on the selection models of Mather (1943) and Lerner (1954), Dobzhansky pointed to another hypothesis worth considering, namely, the *balance hypothesis*. Under this hypothesis, there was no wild-type or mutant. Instead, a locus consisted of (usually) two alleles, such that the heterozygote had a higher fitness than the two homozygotes (overdominance), and thus both alleles were maintained at appreciable frequencies:

QUOT. 997 . . .the adaptive norm is an array of genotypes heterozygous for more or less numerous gene alleles, gene complexes, and chromosomal structures. Homozygotes for these genes and gene complexes occur in normal outbred populations only in a minority of individuals, and make these individuals more or less inferior to the norm in fitness. (*ibidem*)

Under the classical hypothesis, selection was purifying, working to reduce variation (average heterozygosity ~5–10%), while under the balance hypothesis, selection was stabilizing, working to maintain variation (average heterozygosity ~50%).

Dobzhansky acknowledged Chetverikov's influence on his development of the balance hypothesis. In the late 1920s and early 1930s, the Russian geneticist and his students (Timofeeff-Ressovsky and Dubinin) had experimented with *Drosophila* and found relatively high proportions of heterozygotes containing recessive mutants.[15] Dobzhansky thus noted:

QUOT. 998 Many of the classical mutants which grace the pages of genetics manuals were thus shown to exist concealed in natural populations. (*ibid.*, p. 4)

---

[14] See Sect. 13.4.4.

[15] See Sect. 16.5.2.3.

Dobzhansky believed that the classical and balance hypotheses were *not* mutually exclusive, since some loci could behave according to the first, while others according to the second. However, his 1955 paper had already drawn the battlelines of a contentious debate between the two schools of variation, with mainly Dobzhansky and his student Bruce Wallace on the balance side, and mainly Muller on the classical side.[16] The subsequent discovery by Lewontin and Hubby (Hubby & Lewontin, 1966; Lewontin & Hubby, 1966) of extensive electrophoretic variation should have settled the issue in favor of the balance hypothesis, but it instead gave rise to a new debate between the balance school (now called the selectionists) and the neutralists.[17]

In any case, having defined two broad hypotheses (classical versus balance) that could explain genetic variation, Dobzhansky next made a case for two types of mutational load, the first of which was compatible with the classical hypothesis and the second of which was compatible with the balance hypothesis:

> **QUOT. 999** One load is what the classical hypothesis of population structure leads us to expect; it consists of dominant, semidominant, and incompletely recessive genetic variants which are deleterious to their carriers in most environments in which the population lives. The other load accords better with the balance hypothesis. The components of this load are deleterious when in double dose, but they give rise to heterosis in heterozygotes. (Dobzhansky, 1955, p. 5)

16.3.4 Crow (1958): The Genetic Load Mathematically Defined, The Segregational Load

A quantitative definition of the genetic load was first given by Crow (1958). James F. Crow (1916–2012) was born in Pennsylvania and was one of the foremost population geneticists during the period from the modern synthesis until the genomic era. In his paper, Crow wrote:

> **QUOT. 1000** The *genetic load* of a population is defined as the proportion by which the population fitness (or whatever other trait is being considered) is decreased in comparison with an optimum genotype. Thus, if the character being measured is fitness, the genetic load is a measure of the total genotypic selection intensity. (*ibid.*, p. 768)

If $W_{\max}$ is the fitness of the optimal genotype and $\overline{W}$ is the mean fitness of the population, then Crow's definition of the genetic load L amounts to:

$$L = \frac{W_{\max} - \overline{W}}{W_{\max}}.$$

The usual practice is to set $W_{\max} = 1$ so that $L = 1 - \overline{W}$.

---

[16]See Beatty (1987) for a description of the Dobzhansky-Muller debate.

[17]See Sect. 16.3.6.

Table 16.3 Crow's model for the calculation of the mutational load

Genotype	*AA*	*AA'*	*A'A'*
Frequency	p^2	$2pq$	q^2
Relative fitness	1	$1 - hs$	$1 - s$

Table 16.4 Crow's model for the calculation of the segregational load

Genotype	*AA*	*AA'*	*A'A'*
Frequency	p^2	$2pq$	q^2
Relative fitness	$1 - s$	1	$1 - t$

The first type of genetic load considered by Crow was the mutational load. Consider a locus with two alleles, *A* and *A'*, and genotype frequencies and fitnesses as in Table 16.3.

The genotypes *A'A'* and *AA'* are selected against, resulting in a decrease in the frequency of the allele *A'*. This decrease is counter-balanced by the mutation of allele *A* to *A'* at a rate u.[18] In every generation, the decrease in frequency of genotype *AA'* due to selection is $2pq(hs)$ and the increase in its frequency due to mutation of *AA* is $2u$. At equilibrium,

$$u = pqhs$$

To calculate the mutational load L_m, Crow noted that the fitness of *AA'* is lowered by hs. Since the fraction of *AA'* is $2pq$,

$$L_m = (hs)(2pq) = 2u$$

(*ibid.*, p. 770), in accordance with Haldane's (16.4).

Next, Crow introduced the *segregational load*, which was the same as the second type of mutational load proposed by Dobzhansky (see QUOT. 999), except that it precluded balanced polymorphisms "due to factors that are favored in different niches, or are frequency-dependent." The segregational load thus denoted the load due to segregation only, such that the heterozygote was favored over both homozygotes in a two-allele model (i.e., when there was overdominance).

Using the model in Table 16.4, Crow reasoned that, in each generation, a number proportional to p^2s of *A* alleles was eliminated because of natural selection and, for the same reason, a number proportional to q^2t of *A'* alleles was eliminated. An equilibrium is reached when the *proportions* of the two types of alleles eliminated are the same, i.e.,

$$\frac{p^2 s}{p} = \frac{q^2 t}{q} \quad \Rightarrow \quad p = \frac{t}{s+t}, q = \frac{s}{s+t}.$$

The segregational load is then

[18]In his paper, Crow used a mutation rate m for both *A* alleles in *AA*, i.e., $m = 2u$.

Table 16.5 Haldane's (1957b) diploid model for the calculation of the cost of selection

Genotype	AA	Aa	aa
Frequency in nth generation	p_n^2	$2p_nq_n$	q_n^2
Relative fitness	1	$1-\lambda K$	$1-K$

$$\begin{aligned} L_s &= sp^2 + tq^2 \\ &= s\left(\frac{t}{s+t}\right)^2 + t\left(\frac{s}{s+t}\right)^2 \\ &= \frac{st}{s+t} \end{aligned}$$

(*ibid*., p. 771).

16.3.5 Haldane (1957): The Cost of Natural Selection (Substitutional Load)

Haldane's "Cost of Natural Selection" (Haldane, 1957b) was a follow-up to his paper published two decades ago (Haldane (1937)) and had a major influence on the way evolutionary genetics subsequently unfolded.

Whereas Haldane had investigated the cost associated with mutation in 1937, the current paper examined the cost associated with natural selection due to "the substitution of one gene for another at the same locus" (Haldane, 1957b, p. 517).[19] He first investigated a haploid model but here, his second model of a diploid autosomal locus, as shown in Table 16.5, will be considered (the conclusions in the two cases are similar).

The fraction of deaths due to natural selection in the nth generation is

$$d_n = 2\lambda K p_n q_n + K q_n^2 = K q_n\{2\lambda + (1-2\lambda)q_n\}$$

(*ibidem*). The frequency of a in the $(n+1)$th generation is

$$q_{n+1} = \frac{p_nq_n(1-\lambda K) + q_n^2(1-K)}{p_n^2 + 2p_nq_n(1-\lambda K) + q_n^2(1-K)} = \frac{p_nq_n(1-\lambda K) + q_n^2(1-K)}{1 - 2\lambda K p_n q_n - K q_n^2}$$

so that the change in the frequency of a from generation n to $n+1$ is

[19]Haldane's cost of natural selection is thus usually known as the *substitutional load*, a name first given by Kimura (1960, p. 23).

$$\begin{aligned}\Delta q &= q_{n+1} - q_n \\ &= \frac{-p_n q_n \{\lambda K(p_n - q_n) + K q_n\}}{1 - 2\lambda K p_n q_n - K q_n^2} \\ &= -K p_n q_n \{\lambda + (1 - 2\lambda) q_n\},\end{aligned}$$

by assuming the denominator is close to 1. Taking one generation as the unit of time, the above difference equation becomes

$$\frac{dq}{dt} = -Kpq\{\lambda + (1 - 2\lambda)q\}$$

approximately (Haldane, 1957b, p. 515). Now the total fraction of deaths due to natural selection, i.e., *the cost of natural selection*, is

$$\begin{aligned}D &= \sum_{n=0}^{\infty} d_n \\ &= K \sum_{n=0}^{\infty} q_n \{2\lambda + (1 - 2\lambda) q_n\} \\ &\approx K \int_0^{\infty} q\{2\lambda + (1 - 2\lambda)q\} dt \\ &= K \int_0^{q_0} q\{2\lambda + (1 - 2\lambda)q\} \frac{dt}{dq} dq \\ &= \int_0^{q_0} \frac{\{2\lambda + (1 - 2\lambda)q\}}{(1 - q)\{\lambda + (1 - 2\lambda)q\}} dq.\end{aligned}$$

For $\lambda < 1$ (A is partially dominant), the above simplifies to

$$D = \frac{1}{1 - \lambda}\left\{ - \log p_0 + \lambda \log\left(\frac{1 - \lambda}{\lambda}\right)\right\}, \tag{16.6}$$

while, for $\lambda = 1$ (A is recessive advantageous),

$$D = \frac{1}{p_0} - \log p_0. \tag{16.7}$$

Thus, the cost of natural selection (D) is *independent of the selection intensity K*.

Using the expressions in (16.6) and (16.7), Haldane calculated values of D for various degrees of dominance λ and initial frequencies p_0. From his calculations, he concluded:

QUOT. 1001. . .D usually lies between 10 and 100, with 30 as a representative value. (*ibid.*, p. 517)

Just so that the reader realizes it, $D = 30$ represents an enormous cost. It implies that the price the population must pay for selection to substitute just one allele with another is equal to a number of deaths that is 30 times the size of the population in a generation!

Naturally, the cost cannot be paid within a single generation. If a population can afford to lose 10% of its size by natural selection in a generation (i.e., the selection intensity is .1), then it will take 300 generations for a single gene substitution.

Moreover, substitutions may occur across several loci. Suppose that the genetic deaths are spread across n generations and that the decrease in fitness at locus i is δ_i per generation. Then, the mean number of loci transformed per generation is $\sum_i \delta_i / D$ and the number of generations required is $n = D / \sum_i \delta_i$. By assuming the decrease in fitness at these loci occurs *independently*, the fitness due to substitutions at these loci decreases to

$$\prod_i (1 - \delta_i) \approx 1 - \sum_i \delta_i \approx e^{-\sum_i \delta_i} \tag{16.8}$$

From the above, the selection intensity is

$$I = \sum_i \delta_i = \frac{D}{n} \tag{16.9}$$

(*ibid.*, p. 520). Using $D = 30$ and an intensity of $I = .1$, Haldane obtained $n = 300$ generations. Both the single-locus and multiple-loci calculations indicate a very slow rate of evolution by natural selection.

Haldane placed much importance on his concept of the cost of selection:

QUOT. 1002 . . . I am quite aware that my conclusions will probably need drastic revision. But I am convinced that quantitative arguments of the kind here put forward should play a part in all future discussions of evolution. (*ibid.*, p. 523)

However, his calculations show a serious conundrum associated with natural selection. This has been called *Haldane's dilemma*, a term first used by van Valen (1963). The dilemma is that natural selection is usually associated with increasing fitness in a population,[20] yet the formula in Eq. (16.8) indicates an extremely low fitness, given that the number of loci is usually very large. For example, even if the decrease in average fitness at a locus is as small as $\delta_i = .01$, then with 1000 loci, the fitness of the population decreases to $e^{-10} \approx 4.5 \times 10^{-5}$.

[20] Haldane had himself previously acknowledged that the mean fitness should be close to unity. See QUOT. 992.

16.3.6 *Influence of Haldane on Kimura's Neutral Theory of Evolution (1968)*

Haldane's paper had a large impact on several population geneticists, the foremost of whom was probably Motoo Kimura.[21] Kimura (1968b, p. 137) declared that he shared with Haldane the belief expressed in the second sentence of QUOT. 1002. In that same year, Kimura wrote his landmark paper on neutral evolution (Kimura, 1968a), according to which most genetic variation between individuals (or species) was due to DNA differences that did not affect or hardly affected the reproductive success of the individuals.[22] The neutral theory is, of course, contrary to classical Darwinian selection.

Shortly afterward, in a paper (Kimura & Ohta, 1972, p. 51) co-written with Tomoko Ohta, Kimura explained how Haldane's paper motivated his neutral theory.

Kimura had taken the average substitution rate per acid site (per year) in cistrons to be 1.6×10^{-9}, which implies a substitution rate per nucleotide of approximately 6.3×10^{-10}. Assuming the total number of nucleotides to be 3×10^9, this gives two substitutions per haploid DNA per year. Taking one generation to be three years on average in mammals, the rate of substitution is approximately 6 per generation. Compared to Haldane's estimate of one substitution for every 300 generations, Kimura's number was "unbelievably high." More specifically, using Eq. (16.9) (which assumes independence across loci) and taking a cost (substitution load) for one substitution of $D = 30$, Kimura obtained a selection intensity (or substitution load *per generation*) of

$$I = \sum_i \delta_i = \frac{D}{n} = \frac{30}{1/6} = 180. \tag{16.10}$$

According Eq. (16.8), the fitness of the population decreases to $e^{-\sum_i \delta_i} = e^{-180}$, so that to maintain a constant population size under such a substitution load, each parent would need to produce $e^{180} \approx 10^{78}$ offspring for one to survive. In the authors' own words:

> **QUOT. 1003** This means that in order to carry out independent gene substitutions at this rate and still maintain the same population number, each parent must leave e^{180} or about 10^{78} offspring for only one of them to survive. It is obvious that no mammalian species can stand such a heavy load of substitution. This reasoning led one of us [i.e. Kimura] to put forward

[21] See Sect. 14.5.4.2 for more on Motoo Kimura.

[22] In a way, the neutral theory is a return to the classical hypothesis (as opposed to the balance hypothesis, see Sect. 16.3.3) because it postulates that the genetic variations detected by electrophoretic methods are due to DNA changes that maintain all or almost all of the functionality of the wildtype. A similar theory was proposed by King and Jukes (1969) under the name "non-Darwinian evolution" since, under the neutral theory, evolution at the molecular level occurs mainly through genetic drift.

for the first time the neutral mutation-random drift theory as the main cause of molecular evolution. For neutral mutants, there is no selection, and hence, no genetic load. (*ibid.*, p. 52)

The reasoning behind the numbers in the quotation above has been criticized by several scholars, especially Ewens (see Sect. 16.3.7).

* * *

A conclusion similar to Haldane's and Kimura's regarding the segregational load was reached by Lewontin and Hubby (1966), who found that a typical population was polymorphic for approximately 33% of all loci in the genome, giving about 2000 polymorphic loci. Assuming that there is a 5% loss in reproductive potential at a locus because of overdominance and assuming independence across loci, the fitness of the population is reduced to

$$(.95)^{2000} \approx 10^{-46} \tag{16.11}$$

of its maximum. This is an enormous segregational load:

QUOT. 1004 We then have a dilemma. If we postulate weak selective forces, we cannot explain the observed variation in natural populations unless we invoke much larger mutation and migration rates than are now considered reasonable. If we postulate strong selection, we must assume an intolerable load of differential selection in the population. (*ibid.*, p. 607)

16.3.7 Critique of Haldane's Cost of Natural Selection

Before answering the question as to whether the neutral theory does indeed solve Haldane's dilemma (as claimed by Kimura in the last sentence of QUOT. 1003), it may perhaps be more appropriate to ask, *was there truly a dilemma to start with?* In other words, are Haldane's (and Kimura's) calculations regarding the cost of selection correct?

Indeed, soon after Haldane published his 1957 paper, biologists scrutinized his analysis and found much to disagree with. One of the earliest critiques came from the renowned evolutionary biologist Ernst Mayr (1959b), who questioned the appropriateness of Haldane's assumption that the fitnesses of the different genotypes remained constant:

QUOT. 1005 Haldane (1957a, b) has recently pointed out that selection places a considerable strain upon populations. Too rapid a rate of simultaneous selection against too many genes might eliminate the entire population. By these considerations, Haldane has called attention to a neglected aspect of selection. But is this really a serious problem? Is this not a situation where the new genetic theory of relativity should be applied? Haldane himself emphasizes correctly that we are dealing with relative values . . .The 'goodness' or 'badness' of a gene will be relative depending on the fitness of the averages of all the individuals in the population. And as the population density drops, even rather poor phenotypes may survive by not being exposed to competition from better ones. (*ibid.*, pp. 25–26)

An equally powerful criticism concerning the appropriateness of Haldane's independence assumption for fitnesses across loci in Eq. (16.8) was made by Mayr a few years later:

QUOT. 1006 The contributions to genetic death made by each gene substitution are treated as if they were independent of each other. It is quite likely, however, that they are partially synergistic, when coming together in a genotype. (Mayr, 1963, p. 260)

Similarly, the independence assumption made by Kimura in Eq. (16.10) and by Lewontin and Hubby in (16.11) is questionable.

Ewens (1979, p. 254, 1993, p. 10, 2004, p. 83) has also criticized the independence assumption, especially the one behind the reasoning in QUOT. 1003, which Kimura used to justify the neutral theory:

QUOT. 1007 Our calculations show that the fitness e^{180} is arrived at by assuming that fitnesses are multiplicative over loci. This is a quite unreasonable assumption, and the large offspring requirement of the most fit individual is a direct consequence of it. It is certainly true that in nature substantial epistasis occurs, and if this is so there will be a considerable reduction to the load from that calculated formally by using marginal fitnesses and multiplicativity. . .(*ibidem*)

Mention should also be made of the population geneticist Bruce Wallace (1920–2015), one of the fiercest critics of the genetic load concept. In the late 1960s, he (Wallace, 1968, pp. 427–435) proposed the concepts of *hard* and *soft* selection to counter the genetic load calculations that had been made by Haldane, Kimura, Crow, and others. Wallace's hard selection occurs when the fitness of an individual is an absolute property with respect to its environment and in no way depends on the competing individuals. For example, hard selection occurs when an individual has a lethal dominant allele in a given environment: the individual will die irrespective of other individuals who may or may not possess the lethal allele. Hard selection thus constitutes a genuine genetic load. On the other hand, soft selection occurs when the fitness of an individual may depend on the type, frequency, or distribution of other phenotypes in the population. For example, soft selection occurs when competition occurs between various genotypes for a limited resource in an environment. Soft selection does not constitute a genetic load.[23] Wallace countered the load argument by reasoning that Haldane and Kimura had considered only hard selection, while selection is usually of the soft type and does not entail any genetic load.

Taking into account these criticisms, it may be concluded that genetic loads *do exist*, but it is unlikely that their values are as high as the models of Haldane, Kimura, and others would suggest. Likewise, there is a place for the neutral theory but the arguments now used to defend it are not based on genetic load.

[23] Mayr's argument in QUOT. 1005, in fact, uses the concept of soft selection.

Fig. 16.3 Ernst Mayr (1904–2005). (From: https://commons.wikimedia.org/wiki/File:Ernst_Mayr_PLoS.jpg)

16.4 Haldane versus Mayr: Is Mathematical Population Genetics "Beanbag Genetics"[24]?

16.4.1 Introduction

In 1959 and 1963, Mayr raised important questions about the validity of the mathematical approach adopted by the three founders of the evolutionary synthesis (Fisher, Wright, and Haldane) and, on both occasions, compared their approach to "beanbag genetics." That is, genes were treated as if they were independent units, like colored beans placed inside a bag.

Ernst Mayr (Fig. 16.3) was born in Kempten, Germany. He completed his doctorate in ornithology at the University of Berlin in 1926, and in the same year became assistant at the Berlin Museum. In 1927, at the International Zoological Congress at Budapest, he was introduced to the naturalist Walter Rothschild who asked him to collect in New Guinea for the Rothschild Museum at Tring (UK) and the American Museum of Natural History (AMNH). In 1931, Mayr joined the AMNH where he became curator in 1953. He subsequently joined Harvard University as Alexander Agassiz Professor of Zoology (1957–1975) and as director of the Museum of Comparative Zoology (1961–1970). Mayr died in 2005 in Bedford, Massachusetts.

In 1942, Mayr wrote one of the first post-synthetic books entitled *Systematics and the Origin of Species* (Mayr, 1942), where he presented evidence for evolution from the point of view of modern systematics. To explain the process of speciation, Mayr gave a new definition of a species:

[24]"Beanbag genetics" and the Haldane-Mayr controversy are also discussed in Dronamraju (2011), Rao and Nanjundiah (2011, 2016), de Winter (1997), Provine (1977), Ewens (1993, 2008), Sarkar (2007, 2016), Crow (2008), Morton (2008), Borges (2008), and Nei (2013, pp. 56–58).

QUOT. 1008 Species are groups of actually or potentially interbreeding natural populations, which are reproductively isolated from other such groups. (*ibid.*, p. 120)[25]

This definition came to be known as the *biological species concept* and marked a departure from the Linnaean-Darwinian *morphological species concept.*

Among many other accomplishments, the 1942 book also introduced the *founder principle* (p. 237), whereby low genetic variation could be observed in a population not only by random gene loss, but also when the population was formed by a few founding members originating from a parental population.

16.4.2 Waddington's Initial Attack on the Mathematical Approach of the Three Founders (1952)

Before examining Mayr's beanbag genetics comments, it is worth noting that the British geneticist Conrad Hal Waddington (1905–1975) had, many years before, already voiced discontent with the highly mathematical approach of the three founders. Thus, in 1952, at the seventh meeting of the Symposia of the Society for Experimental Biology, Waddington alleged that the mathematical approach had:

QUOT. 1009. . .the peculiar character of not having achieved either of the two results which one normally expects from a mathematical theory. It has not, in the first place, led to any noteworthy quantitative statements about evolution. The formulae involve parameters of selective advantage, effective population size, migration and mutation rates, etc., most of which are still too inaccurately known to enable quantitative predictions to be made or verified. But even when this is not possible, a mathematical treatment may reveal new types of relation and of process, and thus provide a more flexible theory, capable of explaining phenomena which were previously obscure. It is doubtful how far the mathematical theory of evolution can be said to have done this. Very few qualitatively new ideas have emerged from it. (Waddington, 1953, p. 186)

The criticism above was repeated verbatim in Waddington's book *The Strategy of the Genes* (Waddington, 1957, pp. 59–60).

Waddington's attacks were refuted by Haldane, who had written the foreword to the 1952 symposium volume. Referring to the last sentence in QUOT. 1009, Haldane retorted that:

---

[25] Mayr indicated that this definition was based on one he had given in a previous paper: "A species consists of a group of populations which replace each other geographically or ecologically and of which the neighboring ones intergrade or hybridize wherever they are in contact or which are potentially capable of doing so (with one or more of the populations) in those cases where contact is prevented by geographical or ecological barrier" (Mayr, 1940, p. 256). Mayr's definition had, in fact, been anticipated by Dobzhansky (1935, p. 354): "the species represents that stage of evolutionary divergence, at which the once actually or potentially interbreeding array of forms becomes segregated into two or more separate arrays which are physiologically incapable of interbreeding." This was repeated in Dobzhansky (1937, p. 312). However, the definition seems to have originated with Chetverikov in 1926 (see Sect. 16.5.2.2).

QUOT. 1010. . .some of them have emerged so completely that their origin is forgotten. (Haldane, 1953, p. ix)

He then enumerated some of the qualitative ideas that had emerged from the mathematical approach:

(i) Darwin had resorted to environmental (i.e., Lamarckian) effects for the maintenance of variation, presumably because he had thought that such variation would be lost relatively quickly. However, Fisher had shown "clearly that in large populations it diminished with extreme slowness apart from the effects of natural selection."
(ii) According to common opinion, human races could be purified from lethal genes by negative eugenics. However, Penrose and Haldane had shown that lethal genes may be maintained by recurrent mutation.
(iii) Mathematical analysis had shown that, for a locus with two alleles, if a heterozygote had higher fitness than the two homozygotes, then both alleles would be maintained in the population by balancing selection.
(iv) A mathematical approach was required to dispel "ideas with a superficial appeal to common sense, for example, that dominants must oust recessives."

16.4.3 Mayr's Renews Waddington's Challenge (1955): The "Beanbag Genetics" Accusation (1959, 1963)

Mayr first reiterated Waddington's attacks during the 1955 Cold Spring Harbor Symposia on Quantitative Biology. The first paper was by Dobzhansky who gave an unequivocal endorsement of the mathematical approach of the three founders:

QUOT. 1011 The foundations of population genetics were laid chiefly by mathematical deduction from basic premises contained in the works of Mendel and Morgan and their followers. Haldane, Wright, and Fisher are the pioneers of population genetics whose only research equipment was paper and ink rather than microscopes, experimental fields, *Drosophila* bottles, or mouse cages. Theirs is theoretical biology at its best, and it has provided a guiding light for rigorous quantitative experimentation and observation. (Dobzhansky, 1955, pp. 13–14)

On the other hand, Mayr's paper showed that he had a diametrically opposite viewpoint:

QUOT. 1012 Although the models and the formulae of the mathematicians may contain symbols and expressions for all evolutionary factors, they are nevertheless of sharply limited usefulness since the factual equivalents of these mathematical expressions are largely unknown. (Mayr, 1955, p. 327)

Mayr's point in the above was that mathematical expressions often could not be explained in practical terms. Continuing his argument, Mayr stated that there was nevertheless promise with the new empirical approaches of field naturalists, since the latter were "not dealing merely with the genetic analysis of simple Mendelian characters or simple frequencies of genes in natural populations but rather with entire genetic complexes, locked up in chromosome sections, whole chromosomes or genotypes."

At this stage, there was neither any mention of the derogative term "beanbag genetics" nor any finger-pointing at Fisher, Wright and Haldane. However, the situation changed in 1959. Sarkar (2007, p. 74) has reported that Mayr had read Waddington's book *The Strategy of the Genes* (Waddington, 1957) and agreed with the latter's argument. Mayr may have also been emboldened by Waddington's remarks and thus sharpened his attacks during the 1959 Cold Spring Symposia on Quantitative Biology. In the paper that he read at the meeting ("Where are we?"), Mayr (1959b, p. 2) argued that there were two approaches to population genetics: the first one was mathematical and exemplified by the works of Fisher, Wright, and Haldane; the second one was based on systematics and exemplified by the works of Sumner, Timofeeff-Ressovsky, and Dobzhansky. Mayr complained that the first approach had thus far overshadowed the second. Citing Waddington (1953), who had questioned why the mathematical approach enjoyed such "immense importance and prestige," Mayr explained that the mathematical approach had "restored the prestige of natural selection... and prepared the ground for a treatment of quantitative characters." However, he argued that this approach was inherently deficient:

QUOT. 1013 Yet, this period [of mathematical modeling] was one of gross oversimplification. Evolutionary change was essentially presented as an input or output of genes, as the adding of certain beans to a beanbag and the withdrawing of others. This period of "beanbag genetics" was a necessary step in the development of our thinking, yet its shortcomings became obvious as a result of the work of the experimental population geneticists, the animal and plant breeders, and the population systematists, which ushered in a third era of evolutionary genetics. (Mayr, 1959b, p. 2)

Thus, Mayr's accusation of "beanbag genetics" was based on his perception that, in their mathematical works, Fisher, Wright, and Haldane had treated genes as independent units without considering their interactions. Next, Mayr asked:

QUOT. 1014 But what, precisely, has been the contribution of this mathematical school to the evolutionary theory, if I may be permitted to ask such a provocative question? (*ibidem*)

This was not simply a rhetorical question, for Mayr left it to "Fisher, Wright, and Haldane to point out for themselves what they consider their major contributions."

As an example of the "beanbag genetics" approach, Mayr (p. 5) cited Haldane's recent work on the cost of natural selection (Haldane, 1957b).[26] By calculating the substitutional load at one locus, and then treating different loci to be independent of each other, Haldane had concluded that there was "a considerable strain on populations" Mayr questioned the validity of Haldane's approach (see QUOT. 1005) and asked if the substitutional load was truly a problem.

Neither Fisher nor Haldane seemed to have taken notice of Mayr's challenge. Fisher had already migrated to Australia in 1957, and Haldane was living in India since 1956. Only Wright (1960a) responded to Mayr.[27] However, Wright's defense went unheeded, and Mayr renewed the "beanbag genetics" attacks in his book

[26] See Sect. 16.3.5.

[27] See Sect. 16.4.4.

Animal Species and Evolution. After having criticized the genetic load concept in Chap. 9 of the book, Mayr opened the next chapter as follows:

> **QUOT. 1015** The procedure of the classical Mendelian genetics, of studying each gene locus separately and independently, was a simplification necessary to permit the determination of the laws of inheritance and to obtain basic information on the physiology of the gene. When dealing with several genes, the geneticist was inclined to think in terms of their relative frequencies in the population. The Mendelian was apt to compare the genetic contents of a population to a bag full of colored beans. Mutation was the exchange of one kind of bean for another. This conceptualization has been referred to as "beanbag genetics."
>
> Work in population and developmental genetics has shown, however, that the thinking of beanbag genetics is in many ways quite misleading. To consider genes as independent units is meaningless from the physiological as well as the evolutionary viewpoint. Genes not only *act*[28] (with respect to certain aspects of the phenotype) but also *interact*. (Mayr, 1963, p. 263).

16.4.4 Wright's Response (1960) to Mayr's 1959 Criticism

Mayr's accusation of "beanbag genetics" did have some merits in so far as Fisher and Haldane were concerned, since these two population geneticists had explicitly adopted a gene-centered approach.[29] However, if there was one person against whom the charge of "beanbag genetics" was meritless, that would have to be Wright who, as early as 1916, had "recognized that an organism must never be looked upon as a mere mosaic of 'unit characters,' each determined by a single gene, but rather as a vast network of interaction systems".[30] Indeed, Wright's most important theory in population genetics, namely, his Shifting Balance Theory,[31] and the associated adaptive landscape[32] were built around the notion of gene interactions.

This was precisely the main point in Wright's rebuttal paper (Wright, 1960a) when he was asked to review the proceedings of the 1959 symposium. In an eight-page review, Wright devoted one page to listing the various speakers, while the remaining seven pages were dedicated to countering Mayr's attacks. Wright took issue with Mayr's division of the history of evolutionary genetics into three periods—the Mendelian period (~1900 to ~1920) marked by disagreement between the mutationists and the biometricians, the period of classical population genetics (~1930s) marked by the "beanbag genetics" of the three founders, and the period of the "newer population genetics" (~1940s) marked by an increased attention of the

---

[28] All italics are Mayr's.

[29] This is despite the fact that both had, at one time or another, considered gene interactions: Haldane had explicitly considered two-locus interactions in 1931 (see Sect. 15.4.4.3) and Fisher had done so as early as 1918 (see Sect. 14.2.5).

[30] See QUOT. 962.

[31] See Sect. 15.4.3.

[32] See Sect. 15.4.4.

field naturalists to gene interactions. Wright argued that, in fact, the classical population geneticists had all considered gene interactions in their work. It is noteworthy here that Wright defended not only his own work, but also those of his two co-workers, Fisher and Haldane. Concerning the latter, Wright explained:

> **QUOT. 1016** Haldane (1924a, b) made a more fruitful attack on the mathematical theory of selection by considering the case of a single gene subject to selection and mutation. This was a very important contribution to the dynamics of populations. If, however, nothing further had been developed in mathematical population genetics, Mayr's criticism would be sound. As it was, Haldane himself discussed selection under many complicating conditions including factor interaction (summarized in his 1932 book). (*ibid.*, p. 369)

Wright next referred to Fisher's "Fundamental Theorem of Natural Selection",[33] which states that *the rate of increase in mean fitness of an organism is equal to its additive genetic variance in fitness*. Wright explained that, in his theorem, Fisher had considered average effects of genes, but:

> **QUOT. 1017**...[t]o demonstrate that progress by selection is restricted to the net effects of the genes in the combinations in which they enter is not to ignore interaction as Mayr seems to suppose. (*ibidem*)

Although he did not make the next point, Wright could have indicated that the 1958 version of Fisher's 'FTNS' (see Eq. 14.63) explicitly took into account all loci and their interactions across the genome.[34]

Finally, Wright considered his own work. Recall that, in the first phase of his Shifting Balance Theory,[35] genetic drift creates various gene combinations in each of several (almost isolated) demes, so that, in the second phase, natural selection selects for the most favorable combinations in each deme. The third stage involves interdemic selection for the best gene combination followed by gene flow from the various demes to the optimal combination. Gene combinations thus formed an integral part of Wright's theory, which he explained in his review. However, in his 1959 article, Mayr had understood the fitness W to apply to a single gene.[36] Wright concluded that Mayr had misunderstood his work:

> **QUOT. 1018** It may be noted parenthetically that Mayr states flatly that the symbol W for selective value in this theory represented an absolute value assigned to a given gene. Actually it has always been defined as applying to a total genotype in the system under consideration, thus involving whatever interaction effects there may be among the component genes... (*ibid.*, p. 369)

Wright then took Mayr's query, regarding the contribution of the mathematical school to evolutionary theory (see QUOT. 1014), and simply turned the question around and asked:

---

[33] See Sect. 14.5.3.

[34] As argued in Sect. 14.5.3, both the 1930 and 1958 versions of Fisher's 'FTNS' turn out to be inexact.

[35] See Sect. 15.4.3.

[36] See Mayr (1959b, p. 6).

QUOT. 1019 What then has been the contribution of "the newer population genetics"? (*ibid.*, p. 371)

Wright argued that Mayr had obscured the answer to this question by treating mathematical population genetics and population genetics of field naturalists as alternative methods. Instead, they complemented each other: "each plays a role in the synthetic theory that the other cannot possibly play" (*ibidem*). Having been one of the major contributors to the field, Wright had a precise understanding of the role of mathematical theory, namely:

QUOT. 1020 . . .that of an intermediary between the bodies of factual knowledge discovered at two levels, that of the individual and that of the population. It must deduce from the postulates at the level of the individual and from models of population structure what is to be expected in populations, and then modify its postulates and models on the basis of any discrepancies with observation and so on. (*ibidem*)

In the quotation above, the knowledge at the "individual level" is of the Mendelian type and that at the "population level" is of the biometric type. *Mathematical evolutionary genetics thus reconciles Mendelism with biometry.*

It is remarkable that Wright's measured, pointed rebuttal has virtually been ignored by scholars, who have instead focused on Haldane's later refutation. This will be described next.

16.4.5 Haldane's Response (1964) to Mayr's 1959 and 1963 Criticisms

Haldane does not seem to have been aware of Mayr's 1959 article when the latter visited him in Calcutta in 1959, a few months after the Cold Spring Symposia on Quantitative Biology. The two men exchanged friendly letters thereafter, and on February 25, 1963, Mayr informed Haldane that he had arranged for a set of page proofs of his forthcoming book *Animal Species and Evolution* to be sent to the latter (Dronamraju, 2011, p. 252). It was while going through the proofs that Haldane most likely first became aware of Mayr's criticism, at which point he must have also availed himself of Mayr's 1959 paper. Haldane's April 6, 1963 letter to Mayr was meant to inform the latter that he was *now* aware of the beanbag criticism:

QUOT. 1021 Thank you for the proofs, which I am reading. Whether people agree with your conclusions or not, your book will be an invaluable guide to the literature. As a "bean-bag geneticist," I think your view of a species may be a little too physiological and not historical enough. (*ibid.*, p. 253)

Haldane's May 8, 1963 letter, apart from being more formal than his previous friendlier letters to Mayr, gave a first indication that his take on the beanbag genetics affair was going to be distinctly different from Wright's:

QUOT. 1022 I am an unrepentant "beanbag geneticist". Non-mathematicians often fail to realise the complexity of behavior, and sometimes the self-regulatory capacity, of material systems composed of simple components. (*ibid.*, p. 254)

Thus, while Wright had tried to argue that the founders had *not* practiced beanbag genetics and were all along concerned with genetic interactions, Haldane embraced the beanbag genetics tag and lauded the "complexity of behavior" that can be exhibited by systems consisting of "simple components."

This would be the main theme of his formal response (Haldane, 1964) to Mayr.

On June 3, 1963, Haldane informed Mayr of his forthcoming paper entitled "A defense of beanbag genetics." The paper was published the following year. At the start of the paper, Haldane noted that he was left to "hold the fort" since Fisher had passed (in 1962) and Wright was not used to counterattacking. Thus, Haldane was unaware of Wright's response[37] to Mayr. Coming to the substance of his rebuttal, he responded to Mayr's description of the mathematical theory as being "impressive" by denying that the mathematics used had been overwhelming. In fact, he argued that many simplifying assumptions had been made to keep the mathematics elementary. Next, Haldane went on to highlight some of the accomplishments of "beanbag genetics":

1. Around 1930, he and Wright had shown that even in a population of several hundred thousand members, mutation could not be a driving force of evolution. Haldane explained that common sense would imply that, since a mutation rate was usually much lower than a typical selection intensity, mutation would be a "rate-determining process." However, his and Wright's mathematical analysis indicated the contrary. Hence, "only algebraical argument can be decisive in such a case." (*ibid.*, p. 346)
2. Although a verbal argument could lead to an estimation of mutation rates by assuming equilibrium between mutation and natural selection, it was another matter when it came to estimate the time for such an equilibrium to be reached. In two papers, Haldane had shown that:

   > **QUOT. 1023.**...while harmful dominants and sex-linked recessives reach equilibrium fairly quickly, the time needed for the frequency of an autosomal recessive to get halfway to equilibrium after a change in the mutation rate, the selective disadvantage, or the mating system, may be several thousand generations. (*ibid.*, p. 347)

   Haldane added that "the verbal argument is liable to be fallacious."
3. Haldane considered one of his most well-known contributions to population genetics, namely the investigation of the selection intensity in the moth *Biston betularia* in Manchester (Haldane, 1924b).[38] Haldane explained that he had calculated the selection intensity as $k = .5$,[39] although "few or no biologists accepted this conclusion" because k seemed to be too high. However, the experiments by Kettlewell (1956) vindicated his calculations. Haldane next took a passing shot at Mayr for inaccurately representing his work:

---

[37] See Sect. 16.4.4.

[38] See Sect. 16.2.2 and in particular QUOT. 990.

[39] The value reported on p. 26 of Haldane (1924b) is $k = .332$.

QUOT. 1024 I think that in this particular instance Professor Mayr may have unwittingly been a little less than fair to us beanbaggers. On his page 191 [1][40] he says that my "classical" calculations in a book published in 1932 were deliberately based on very small selective intensities and implies that I only reached the same conclusion for industrial melanism in 1957. (Haldane, 1964, p. 358)

Haldane also counterattacked by alleging that Mayr's tendency to use verbal, rather than algebraic, reasoning made his "verbal arguments hard to follow." He also blamed Mayr for showing "considerable ignorance of the earlier literature of beanbag genetics".

There is no doubt that Haldane provided some interesting examples of "beanbag genetics" in his rebuttal. His explanation of the role of mathematical theory in providing a support structure on which a theory can be built is especially illuminating:

QUOT. 1025 In the consideration of evolution, a mathematical theory may be regarded as a kind of scaffolding within which a reasonably secure theory expressible in words may be built up. (*ibid.*, p. 350)

However, Haldane's rebuttal appears weak when it passingly mentions work on genetic interactions by Fisher and Wright but fails to emphasize the importance of genetic interactions, which is what Mayr had argued. In his rebuttal, Haldane gave the impression that a gene-centered approach to population genetics was sufficient to understand evolution. This view of the self-sufficiency of beanbag genetics is inaccurate, as evidenced by Haldane's work on the cost of natural selection.[41] At one point, he even argued that the investigation of interactions was antithetical to mathematical population genetics:

QUOT. 1026 Of course, Mayr is correct in stating that beanbag genetics do not explain the physiological interaction of genes and the interaction of genotype and environment. If they did so they would not be a branch of biology. They would be biology. (*ibid.*, p. 353)

Granted population genetics cannot explain the *physiological* interaction of genes, it is still possible to uncover *mathematical (or statistical) interactions* that may, in some cases, be explained physiologically with the help of biology. Haldane himself had been a pioneer in investigating mathematical interactions.[42]

Haldane's apparent underplaying of the importance of interactions should be contrasted with Wright's (1960a) more reasonable stance that population genetics was *not*, in fact, beanbag genetics and was very much concerned with gene–gene interactions. However, compared to Wright's rebuttal, Haldane's was more spirited and perhaps more interesting to read, which could explain why it is the one usually remembered.

---

[40] Mayr (1963).

[41] See Sect. 16.3.7. Haldane's conclusion regarding the cost of natural selection was one of the reasons for Mayr's criticism of the beanbag genetics approach.

[42] See Sect. 15.4.4.3.

16.5 S. S. Chetverikov (1880–1959): A Forgotten Founder of the Modern Synthesis?

16.5.1 Introduction

Haldane was well-known to give due credit to his predecessors in his writings. One such predecessor was Chetverikov, whose name has often been associated with those of Fisher, Wright, and Haldane as the fourth founder of the modern synthesis (e.g., Dobzhansky (1970, p. 25), Weber (2013, p. 68), Ninio (2014, p. 48), Langridge (1987, p. 251)).

Sergei Chetverikov[43] (Fig. 16.4) was born in Moscow in 1880. In 1892, he studied in the excellent private school Voskresenskij in Moscow and came under the influence of the zoologist Vladimir Zykov (1855–1913). Chetverikov was introduced by the latter to Darwin's theory of evolution and decided to become "professor of Zoology" after leaving school. However, in 1898, Chetverikov was urged by his father, a liberal-minded industrialist, to move to Germany to study engineering. After two semesters, Chetverikov realized that his proper vocation was zoology and he returned to Russia. He studied at the Kiev-Pechersk Gymnasium in Kiev (now the capital of Ukraine) to prepare for his entry at university. In 1900, he enrolled in the Department of Natural Sciences at Moscow University. Chetverikov specialized in lepidopterology (the study of moths and the two superfamilies of butterflies) and published *A Manual for Collecting Lepidoptera* (Chetverikov, 1902).

In 1905, Chetverikov published an important article entitled "Waves of life" (Chetverikov, 1905), where he described chance fluctuations in the population size of *Lepidoptera*. The paper was a precursor of the later ideas of genetic drift, population bottlenecks, and founder effects. The following year, Chetverikov graduated in natural sciences and in 1911 he defended his thesis, entitled "Beiträge zur Anatomie der Wasserassel (Asellus aquaticus L.)," which was published in German in the *Bulletin of the Moscow Society of Naturalists*. In 1914, Chetverikov co-founded the Moscow Entomological Society.

The following years were marked by tumultuous events in Russia, and these events would have a significant impact on Chetverikov's life. The First World War broke out in 1914. On August 1, Germany declared war on Russia. However, despite the support of the opposition, the Russian government failed to react, leading to increasing political unrest. The first Russian revolution of 1917 occurred on March 8 of that year when strikes and riots broke out in Petrograd (now St. Petersburg), followed by a general mutiny of the troops 2 days later. A provisional government was established on March 12, and the Russian Emperor Nicholas II abdicated on March 15. Meanwhile, Lenin and other Bolshevik leaders arrived in Petrograd from Switzerland on April 16, 1917. Lenin's aim was to transfer power from the

[43] Chetverikov's life and work are also discussed in Adams (1968, 1980), Graham (1993, pp. 240–242), Ruse and Travis (2009, pp. 481–483), and Gall and Konashev (2008, pp. 504–510).

Fig. 16.4 Sergei Chetverikov (1880–1959). (From: https://en.wikipedia.org/wiki/Sergei_Chetverikov)

provisional government (the "bourgeois") to the working class. After a failed coup on July 16–18, the Bolsheviks seized power on November 6 (old style calendar: October 24). This was the second Russian revolution of 1917. Following Lenin's death in 1924, Stalin assumed total control of Russia by 1927 and continued his predecessor's Marxist-Leninist policies. In 1928, a first 5-year plan was set up by the government, with special emphasis on industry and agriculture. At the same time, Stalin took control of science, resulting in the purge of the intelligentsia. In the mid-1930s, Lysenko, a pseudoscientist, Lamarckian, and protégé of Stalin, took a strong anti-Mendelian stance and purged many geneticists, including Nikolai Vavilov (1887–1943), one of Russia's greatest geneticists.

Trofim Lysenko[44] (Fig. 16.5) was a disciple of Ivan Michurin (1855–1935), an influential (in Russia) plant-breeder with Lamarckian beliefs. Lysenko advocated the use of vernalization (cooling of seeds during germination to accelerate flowering when they are planted). Lysenko claimed that: (i) this acceleration would increase

[44]Trofim Lysenko's crusade against genetics is also described in Medvedev (1969), Soyfer (1994), Joravsky (1970), DeJong-Lambert (2012), Pollock (2006), Birstein (2001), and Graham (1993).

Fig. 16.5 Trofim Lysenko (1898–1976). (From: https://commons.wikimedia.org/wiki/File:Trofim_Lysenko_portrait.jpg)

the production of grain, and (ii) vernalization would not need to be repeated since it would become an acquired characteristic. He was wrong on both counts. Nonetheless, he took control of the VASKhNL[45] in 1940 and advocated a total rejection of gene theory. In the same year, Vavilov, who had been president of the VASKhNL until 1935 and who forcefully opposed Lysenko's pseudo-scientific ideas, was arrested. In 1941, Vavilov was sentenced to death, but the sentence was commuted to 20 years imprisonment in 1942. Vavilov died in prison in 1943 due to harsh conditions.

When Chetverikov joined the Kol'tsov Institute of Experimental Biology in 1921 and became the head of its genetics section, the political climate was favorable to science, particularly to genetics. The genetics station consisted of several members including Nikolai Timoféeff-Ressovsky (1900–1981), who then moved in 1925 to Germany and achieved fame through pioneering work in molecular biology. Nikolai Dubinin (1906–1908) also joined Chetverikov's group later. A key event occurred in

[45] VASKhNIL is the acronym for the Lenin All-Union Academy of Agricultural Sciences or the V.I. Lenin Academy of Agricultural Sciences.

August 1922 when H. J. Muller, one of Morgan's prominent students, brought *Drosophila melagonaster* strains to the genetics station.[46] Although Chetverikov was a trained naturalist rather than a geneticist, he soon started attacking genetic problems. This resulted in the 1926 publication of a foundational paper (Chetverikov, 1926), published in the "Zhurnal Eksperimental'noi Biologii," which has enshrined Chetverikov's name in the annals of population genetics.

Chetverikov and his team published several more experimental papers in the following years. In 1928, Chetverikov started to translate his 1926 paper in English for the journal *Genetics* and added a fifth chapter to the original paper. However, the paper was never published due to Chetverikov's arrest at the start of 1929, probably as a result of specious political denunciations by one of his co-workers. Chetverikov spent 2 months in a Moscow prison, and was then exiled for three years in Sverdlovsk. On April 13, 1932, he was further exiled for three years in Vladimir. These events meant that several articles in press when Chetverikov was arrested never saw the light of day. Nor was he able to ever resume his research on *Drosophila.*

After his exile in Vladimir, during the period 1935–1948, Chetverikov became head of the Department of Genetics and Selection at the University of Gorky. In 1948, a death blow was administered to Russian geneticists at the hands of Lysenko. During the infamous VASKhNL session held in August of that year, with the support of Stalin, Lysenko launched virulent attacks against "Mendelist-Morganist genetics" and Darwinian biology. These attacks were widely publicized, and geneticists were officially removed from their positions. Chetverikov was not spared and was forced to resign. He died on June 2, 1959, due to a brain hemorrhage.

Stalin died in 1953. Almost a decade later, in 1964, genetics was rehabilitated in Russia, under Leonid Brezhnev. The following year, Lysenko was formally removed as director of the VASKhNL.

But the harm inflicted on Soviet genetics at the hands of Lysenko and Stalin could hardly be undone.

16.5.2 Chetverikov's 1926 Foundational Paper

16.5.2.1 Introduction

As mentioned in Sect. 16.5.1, Chetverikov's English translation of the 1926 paper was never published owing to his arrest. However, he seems to have met Haldane at the Fifth International Congress of Genetics in Berlin in 1927 and given the latter a copy of his paper (Adams, 1980, p. 268). Haldane then apparently arranged for the paper to be translated and made available in his laboratory. The paper was referenced

[46]Chetverikov later gave some of the *Drosophila* strains to Dobzhansky, who was then working at the Kiev Agricultural Institute. Dobzhansky left for the United States in 1927 but always acknowledged the influence of Chetverikov's ideas.

by Haldane's student Cecil Gordon in 1936 (Gordon, 1936, p. 25). Furthermore, Dobzhansky included an English translation of parts of Chetverikov's paper as an appendix to his 1959 address at the Cold Spring Harbor Symposium on Quantitative Biology (Dobzhansky, 1959, pp. 27–30). A full English translation was done by Malina Barker and edited by I. Michael Lerner, and published by the *Proceedings of the American Philosophical Society* in 1961 (Chetverikov, 1961).

Chetverikov started his paper by acknowledging the validity of the Mendelian laws, which therefore formed the basis of modern genetics. He explained the hostility to Mendelism by "many outstanding evolutionists, both here and abroad" because genetics was often in "contradiction with conventional views of general evolutionary concepts." His paper therefore had the goal of

> **QUOT. 1027**...clarifying certain questions on evolution in connection with our current genetic concepts. (*ibid.*, p. 169)

Like Fisher, Wright, and Haldane, Chetverikov therefore wished to reconcile Mendelian heredity with Darwinian evolution.

Chapter I of Chetverikov's paper was devoted to the origin of mutations in nature. Chetverikov noted that the term "mutation" had previously been used to denote sports, so he preferred the term "genovariation" to denote a Morganian-type variation in the genotype.[47] Chetverikov stated that, while mutations were usually harmful and rarely beneficial, there was also "a series of 'neutral ones', not having any biological significances and therefore not subjected to selection." His general conclusion regarding the origin of mutations was that there was no method to produce mutations artificially. He was thus unaware of Muller's work on ionizing radiation, which was being done at around the same time Chetverikov's paper was published. Chetverikov was more accurate in his belief that mutations in the wild occurred naturally and at random.

16.5.2.2 Biological Definition of Species

In Chapter II of his paper, Chetverikov (1926, 1962) turned his attention to random mating. Before this could be done, he provided a definition of a species that broke from the Linnaean-Darwinian tradition (which was based on morphological similarity) and that anticipated Mayr's biological definition (see QUOT. 1008) by approximately 15 years:

> **QUOT. 1028** The definition of species, as an aggregate of individuals constituting a single freely intercrossing complex, corresponds most closely to our genetic and systematic ideas... potentially, all the individuals of a single species are able to cross freely...(*ibid.*, p. 174)

Interestingly, Chetverikov's "potentially" in the quotation above was also used by Mayr in his 1940[48] and 1942[49] definitions.

[47] Chetverikov's "genovariation" is now an obsolete term and will be referred to simply as a mutation in the text.

[48] See footnote 18 of chap. 16.

[49] See QUOT. 1008.

16.5.2.3 Solving Jenkin's Dilemma: "The Species, Like a Sponge, Soaks up Heterogeneous Mutations"

Having reviewed the Hardy–Weinberg law (which he ascribed to Hardy only, since Weinberg's 1908 derivation was unknown to most in 1926—see Sect. 10.10), Chetverikov next took on one of the most incisive criticisms that Darwin had faced regarding his theory of natural selection, namely, the swamping argument put forward by the engineer Fleeming Jenkin (see Sect. 2.14.4.3). As Chetverikov recalled, the latter had "demonstrated by a simple arithmetical calculation that as a result of free crossing every accidentally originating deviation, be it even beneficial, must be very quickly dissolved among the normal individuals" (*ibid.*, p. 175). Darwin was not able to appropriately counter Jenkin's criticism, with the result that he put more emphasis on individual differences, rather than sports, as the variation on which natural selection acted.

In Chapter II of his paper, Chetverikov proposed a simple argument that disproved the essence of Jenkin's argument. The word "essence" is used here since Jenkin had framed his argument using blending inheritance (which is now known to be incorrect), but Chetverikov pointed the error in Jenkin's argument by proceeding under Mendelian heredity; however, Chetverikov's analysis is still correct since Jenkin's argument is essentially the same under both modes of inheritance.

Chetverikov considered a large random mating population of (constant) size N consisting entirely of dominant wild-type homozygotes *AA*. Suppose that a recessive mutant homozygote *aa* is added to the population. This will temporarily disturb the equilibrium but when the *aa* homozygote crosses with an *AA* homozygote, the equilibrium is restored in the next generation (F_1), now with $(N - 1)$ *AA* homozygotes and two *Aa* heterozygotes. Chetverikov then asked, how likely was it for an *aa* mutant to reappear in F_2 as a result of an *Aa* x *Aa* cross in F_1? From the point of view of a particular *Aa* genotype, it may mate with any one of the 1 *Aa* or $(N - 1)$ *AA* genotypes. The probability that the *Aa* genotype will cross with a similar one in F_1 (so that an *aa* reappears in F_2) is therefore[50]

$$p = \frac{1}{1 + (N - 1)} = \frac{1}{N}$$

(*ibid.*, p. 177). Taking the example of $N = 1000{,}000$ gray (*AA*) cows and one albino (*aa*), Chetverikov obtained $p = 10^{-6}$. Thus, the albino would be "swallowed up" by free crossing, as argued by Jenkin. However, regarding the mutant *allele a*, Chetverikov warned that:

> **QUOT. 1029**... its fate will be completely different from what former evolutionists imagined. The mutation will not be destroyed, will not be dissolved in the mass of normal

[50] Note that this is the probability that a *particular Aa* will cross with the other *Aa*. The probability of an *Aa* x *Aa* cross, out of all possible crosses, is $1/\{N(N + 1)\}$.

individuals. It will exist in the heterozygous state remaining hidden from the eye, but in the form of a specific inheritable genotype generation after generation. (*ibidem*)

Thus, in a large random mating population, a recessive mutant allele may not be expressed phenotypically but may still persist in heterozygotic combination with a normal allele. Now, assuming there were initial mutations *bb*, *cc*, ... for other independent characters, Chetverikov concluded that:

QUOT. 1030. . .a *species, like a sponge, soaks up heterozygous mutations, while remaining from first to last externally (phenotypically) homogeneous.*[51] (*ibid.*, p. 178)

Chetverikov noted that his conclusion above was supported by data collected in *Drospohila* since 1925.[52]

It is for the conclusion in QUOT. 1030, more than anything else, that Chetverikov's paper is remembered. Moreover, there is no doubt that this conclusion had considerable influence in Dobzhansky's later development of the balance hypothesis.[53]

16.5.2.4 The Efficacity of Natural Selection

Chapter III of Chetverikov's paper addressed natural selection. Chetverikov reproduced the table (see Table 13.3) that H.T.J. Norton had prepared for Punnett (1915), which showed the evolution of a random mating population under different selection intensities. Chetverikov drew attention to the *completeness* of the process of natural selection, in the sense that

QUOT. 1031 . . .in both cases—when the selected trait is dominant as well as when it is recessive —the process of the *transformation of the species,*[54] that is, of the *complete replacement* of a former unadapted form by the more adapted one, proceeds, practically speaking, to an end. (Chetverikov, 1926, 1961, p. 182)

Chetverikov noted that the above was true even when selection was extremely slow. That is, "no matter how small the newly arisen improvement be, hundreds of thousands of generations, perhaps, will pass, but finally, sooner or later, it will emerge and gradually *become a property of all the individuals of the species*[55]" (*ibid.*, p. 184). This meant that, under Mendelian heredity, natural selection was able to proceed completely even with the slightest advantage of a genotype. Chetverikov argued that this fact overcame "one of the most substantial difficulties of Darwinism" and also meant that:

[51] Italics are Chetverikov's.

[52] This was later published in Chetverikov (1927).

[53] See Sect. 16.3.3.

[54] All italics in this quotation are Chetverikov's.

[55] Italics are Chetverikov's.

QUOT. 1032 Darwinism, in so far as natural selection and the struggle for existence are its characteristic features, received a completely unexpected and powerful ally in Mendelism. (*ibid.*, p. 183)

Comparing the selection rates for the cases of dominance and recessiveness, Chetverikov noted that the rates were approximately the same for small selection intensities. For higher selection intensities, a difference between the two cases became apparent—for a selection intensity of 50%, the rate of selection for the dominant case was twice that for the recessive case.

16.5.2.5 Importance of Isolating Mechanisms in Speciation

In Chapter II (Random Mating) of his paper, Chetverikov noted the importance of isolation in the creation of genetic differentiation both within and between a species:

QUOT. 1033. . .*isolation, under the conditions of a process of continuous accumulation of mutations becomes, by itself, a cause of intraspecific (and consequently, eventually, also of interspecific) differentiation.*[56] (*ibid.*,p. 179)

He then reviewed various forms of isolation, including geographic isolation (i.e., the subdivision of a species into colonies living in different regions), isolation in time (i.e., the subdivision of a species into colonies living in the same region but isolated because of nonoccurrence of reproductive periods in each colony), and ecological isolation (i.e., the subdivision of a species into colonies living in the same geographic region and reproducing at the same time but living in different conditions of habitat).

In Chapter III (Natural Selection), Chetverikov compared the relative roles of natural selection and isolation in the process of species formation or *speciation*. He maintained that the initiation of speciation was the result of isolation only:

QUOT. 1034 But it is absolutely necessary to appreciate clearly the fact that never can an adaptive change, related to the action of selection in the struggle for existence but not related to isolation, initiate speciation. *Not selection, but isolation is the actual source, the real cause of the origin of species.*[57]

Note that Chetverikov did *not* deny that natural selection had *any* role in species formation; he only denied that it could *initiate* speciation, leaving the possibility for natural selection to later act on the different species.

Current evolutionists do not deny that speciation can be initiated by isolation, but also argue that isolation itself is usually the result of adaptive divergence between colonies; thus, natural selection retains a role in the causal chain of speciation initiation. Such a model (involving two interactive loci) was proposed by Dobzhansky (1936) and Muller (1939).

[56] Italics are Chetverikov's.

[57] Italics are Chetverikov's.

16.5.2.6 Importance of the "Genotypic Milieu": Gene Interactions and Pleiotropy

In the last chapter (Chap. IV) of his paper, Chetverikov first denied Castle's theory of gametic contamination[58] and accepted the purity and constancy of genes. He then stressed the importance of the *genetic milieu* as follows:

QUOT. 1035 Each gene does not act isolatedly from the whole genotype, is not independent of it, but acts, manifests itself, within it, in relation to it. *The very same gene will manifest itself differently, depending on the complex of the other genes in which it finds itself.*[59] For it, this complex, this genotype, will be the *genotypic milieu*, within the surroundings of which it will be externally manifested. (*ibid.*, p. 190)

In the above, Chetverikov warned that the gene could *not* be studied in isolation but had to be looked through the lens of gene–gene interactions. This would be precisely Mayr's "beanbag genetics" argument, expressed in 1959,[60] although to be fair, Wright had also held views similar to both Chetverikov and Mayr as early 1916 (see QUOT. 962). By paying attention not only to the gene, but also its milieu ("the complex of the other genes it which it finds itself"), one is led to the notion of *pleiotropic effects*:

QUOT. 1036 The concept of pleiotropic action of genes consists of the idea that every gene may influence not only the specific character corresponding to it, but a whole series of others; generally speaking, the entire soma. (*ibid.*, p. 189)

---

[58] See Sect. 11.8.2.

[59] All italics in this quotation are Chetverikov's.

[60] See Sect. 16.4.3.

Bibliography

A Fellow of the International Education Board. (1926). William Bateson, 1861–1926: His life and work. *Journal of Heredity, 17*(12), 433–449.
Adams, A. B. (1969). *Eternal quest: The story of the great naturalists*. Putnam.
Adams, M. B. (1968). The founding of population genetics: Contributions of the Chetverikov school 1924–1934. *Journal of the History of Biology*, 23–39.
Adams, M. B. (1980). Sergei Chetverikov, the Kol'tsov Institute, and the evolutionary synthesis. In *The evolutionary synthesis: Perspectives on the unification of biology* (pp. 242–278). Harvard University Press.
Agassiz, L. J. R. (1860). Prof. Agassiz on the origin of species. *American Journal of Science and Arts, 30*, 142–154.
Ågren, J. A. (2021). *The gene's-eye view of evolution*. Oxford University Press.
Allen, G. E. (1968). Thomas Hunt Morgan and the problem of natural selection. *Journal of the History of Biology, 1*(1).
Allen, G. E. (1969). Hugo de Vries and the reception of the "mutation theory". *Journal of the History of Biology, 2*(1), 55–87.
Allen, G. E. (1975). *Life science in the twentieth century*. Wiley.
Allen, G. E. (1978). *Thomas Hunt Morgan: The man and his science*. Princeton University Press.
Allen, G. E. (1980). The evolutionary synthesis: Morgan and natural selection revisited. In E. Mayr & W. B. Provine (Eds.), *The evolutionary synthesis: Perspectives on the unification of biology* (pp. 356–382). Harvard University Press.
American Academy. (1860). *Proceedings of the American Academy of Arts Vol. IV (May 1857 to May 1860)*.
Ankeny, R. A. (2000). Marvelling at the marvel: The supposed conversion of AD Darbishire to Mendelism. *Journal of the History of Biology, 33*, 315–347.
Anonymous. (1670). Review of historia generalis insectorum. *Philosophical Transactions of the Royal Society, 5*, 2078–2080.
Anonymous. (1859a). Darwin's origin of species. *Saturday Review*, 775–776.
Anonymous. (1859b). (Leifchild, J.R.) On the origin of species, by means of natural selection. *Athenaeum*, 659–660.
Anonymous. (1860a). (Asa Gray) Review of Darwin's theory on the origin of species by means of natural selection. *American Journal of Science, 29*(86), 153–184.
Anonymous. (1860b). (Owen, R.) Review of origin and other works. *Edinburgh Review, 111*, 487–532.

P. Gorroochurn, *The Development of Evolutionary Genetics*,
https://doi.org/10.1007/978-3-031-69374-8

Anonymous. (1860c). (Sedgwick, A.) Objections to Mr. Darwin's theory of the origin of species. *The Spectator, 33*, 285–286, 334–335.

Anonymous. (1895a). (Editor) The garden Cineraria. *The Gardener's Chronicle, 17*, 655.

Anonymous. (1895b). (Editor) The garden Cineraria. *The Gardener's Chronicle, 18*, 89–90.

Anonymous. (1895c). (Editor) The origin of the Cineraria. *The Gardener's Chronicle, 17*, 742.

Anonymous. (1895d). (Editor) The origin of the Cineraria. *The Gardener's Chronicle, 18*, 17.

Anonymous. (1895e). Origin of the Cineraria. *The Gardener's Chronicle, 18*, 186.

Anonymous. (1895f). Sports and seedlings. *The Gardener's Chronicle, 17*, 588.

Anonymous. (1896). Kew note: The origin of the garden Cineraria. *The Gardener's Chronicle, 20*, 239–240.

Anonymous. (1900). Societies: Royal horticultural lecture. *Gardeners Chronicle, 27*(III), 303.

Anonymous. (1903). Scientific serial: Biometrika. *Nature, 67*(1745), 550.

Anonymous. (1904a, August 20). The British Association. *Times*, p. 8.

Anonymous. (1904b). Zoology at the British Association. *Nature, 70*, 538–541.

Appel, T. A. (1987). *The Cuvier-Geoffroy debate: French biology in the decades before Darwin*. Oxford University Press.

Appleman, P. (Ed.). (1976). *An essay on the principle of population. Thomas Robert Malthus. A Norton critical edition*. Norton.

Armstrong, P. (2007). *All things Darwin: An encyclopedia of Darwin's world*. Greenwood Publishing Group.

Ayala, F. J. (2004). In William Paley's shadow: Darwin's explanation of design. *Ludus Vitalis, 12*(21), 53–66.

Ayala, F. J. (2012). The design of organisms: From Paley to Darwin. In K. R. Monroe (Ed.), *Science, ethics, and politics* (pp. 14–36). Routledge.

Bailey, E. B. (1962). *Charles Lyell*. Thomas Nelson.

Bailey, L. H. (1892). *Cross-breeding and hybridizing, the philosophy of the crossing of plants, considered with reference to their improvement under cultivation, with a brief bibliography of the subject*. Rural Publishing Company.

Bailey, L. H. (1895). *Plant breeding*. Macmillan.

Bailey, L. H. (1902). *Plant breeding* (2nd ed.). Macmillan.

Bailey, L. H. (1903). Some recent ideas on the evolution of plants. *Science (New York, N.Y.), 17*(429), 441–454.

Bakewell, R. (1815). *An introduction to geology* (2nd ed.). J. Harding.

Ballonoff, P. (1974). Editor's comments on paper 11. In P. Ballonoff (Ed.), *Genetics and social structure: Mathematical structuralism in population genetics and social theory* (pp. 156–156). Dowden, Hutchinson & Ross, Inc.

Barber, L. (1980). *The heyday of natural history, 1820–1870*. Doubleday.

Bardell, D. (2001). Erasmus Darwin expressed his concept of evolution in poetry. *Bios, 72*(2), 63–66.

Barkley, T. M. (1966). A review of the origin and development of the florists' Cineraria, Senecio cruentus. *Economic Botany, 20*(4), 386–395.

Barlow, N. (1958). *The autobiography of Charles Darwin, 1809–1882: With original omissions restored*. Norton.

Barlow, N. (1963). Darwin's ornithological notes. *Bulletin of the British Museum (Natural History). Historical Series, 2*(7), 201–278.

Barrett, P. H., et al. (1987). *Charles Darwin's notebooks, 1836–1844: Geology, transmutation of species, metaphysical enquiries*. Cornell University Press.

Barrows, E. M. (2011). *Animal behavior desk reference: A dictionary of animal behavior, ecology, and evolution* (3rd ed.). CRC Press.

Barthélemy-Madaule, M. (1982). *Lamarck the mythical precursor: A study of the relations between science and ideology*. MIT Press.

Bartholomew, M. (1973). Lyell and evolution: An account of Lyell's response to the prospect of an evolutionary ancestry for man. *The British Journal for the History of Science, 6*(3), 261–303.

Bateson, B. (1928). *William Bateson, FRS, naturalist: His essays & addresses, together with a short account of his life*. Cambridge University Press.

Bateson, W. (1884a). Memoirs: The early stages in the development of Balanoglossus (sp. incert.). *Journal of Cell Science, 2*(94), 208–236.

Bateson, W. (1884b). On the development of Balanoglossus. *Journal of Natural History, 13*(73), 65–67.

Bateson, W. (1885). Note on the later stages in the development of Balanoglossus Kowalevskii (Agassiz), and on the affinities of the enteropneusta. *Proceedings of the Royal Society of London, 38*(235–238), 23–30.

Bateson, W. (1886). Memoirs: Continued account of the later stages in the development of Balanoglossus Kowalevskii, and of the morphology of the Enteropneusta. *Journal of Cell Science, 2*(104), 511–534.

Bateson, W. (1889). On some variations of cardium edule apparently correlated to the conditions of life. *Philosophical Transactions of the Royal Society of London (Ser. B), 180*, 297–330.

Bateson, W. (1891). On variations in the floral symmetry of certain plants having irregular corollas. *Botanical Journal of the Linnean Society, 28*(196), 386–424.

Bateson, W. (1894). *Materials for the study of variation: Treated with especial regard to discontinuity in the origin of species*. Macmillan.

Bateson, W. (1895a). Home correspondence: The origin of the cultivated Cineraria. *The Gardener's Chronicle, 18*, 774.

Bateson, W. (1895b). Letter to the Editor: The origin of the cultivated Cineraria. *Nature, 51*(1330), 605–607.

Bateson, W. (1895c). Letter to the Editor: The origin of the cultivated Cineraria. *Nature, 52*(1332), 29.

Bateson, W. (1895d). The origin of the cultivated Cineraria. *Nature, 52*(1335), 103–104.

Bateson, W. (1897). Notes on hybrid Cinerarias produced by Mr Lynch and Miss Pertz. *Proceedings of the Cambridge Philosophical Society, IX*, 308–309.

Bateson, W. (1899). Hybridisation and cross-breeding as a method of scientific investigation. *Journal of the Royal Horticultural Society of London, 24*, 59–66.

Bateson, W. (1901). Problems of heredity as a subject for horticultural investigation. *Journal of the Royal Horticultural Society, 25*, 54–61.

Bateson, W. (1902a). Heredity, differentiation, and other conceptions of biology: a consideration of Professor Karl Pearson's paper "On the Principle of homotyposis". *Proceedings of the Royal Society of London, 69*(451–458), 193–205.

Bateson, W. (1902b). *Mendel's principles of heredity: A defence*. Cambridge University Press.

Bateson, W. (1903a). Letter to Editor: Mendel's principles of heredity in mice. *Nature, 67*(1742), 462–463.

Bateson, W. (1903b). Letter to Editor: Mendel's principles of heredity in mice. *Nature, 67*(1747), 585–586.

Bateson, W. (1903c). Letter to Editor: Mendel's principles of heredity in mice. *Nature, 68*(1750), 33–34.

Bateson, W. (1903d). The present state of knowledge of colour-heredity in mice and rats. *Proceedings of the Zoological Society of London, 2*, 71–99.

Bateson, W. (1904). Section D. Zoology. Opening address by William Bateson, M.A., F.R.S., President of the Section. *Nature, 70*(1817), 406–413.

Bateson, W. (1906). A text-book of genetics (Review of Lotsy's Vorlesungen uber Deszendenztheorien, I. Theil). *Nature, 74*(1911), 146–147.

Bateson, W. (1907). Facts limiting the theory of heredity. *Science, 26*(672), 649–660.

Bateson, W. (1909). *Mendel's principles of heredity* (2nd ed.). Cambridge University Press.

Bateson, W. (1922). Evolutionary faith and modern doubts. *Science, 55*(1412), 55–61.

Bateson, W., & Saunders, E. R. (1901). *Reports to the Evolution Committee of the Royal Society: Report I*. Harrison & Sons.

Bateson, W., Saunders, E. R., & Punnett, R. C. (1905). Experimental studies in the physiology of heredity. *Reports to the Evolution Committee of the Royal Society, II*, 1–131.

Beatty, J. (1986). The synthesis and the synthetic theory. In W. Bechtel (Ed.), *Integrating scientific disciplines: Case studies from life sciences* (pp. 125–135). Martinus Nijhoff.

Beatty, J. (1987). Weighing the risks: Stalemate in the classical/balance controversy. *Journal of the History of Biology, 20*(3), 289–319.

Beddall, B. G. (1968). Wallace, Darwin, and the theory of natural selection: A study in the development of ideas and attitudes. *Journal of the History of Biology*, 261–323.

Beddall, B. G. (1988a). Darwin and divergence: The Wallace connection. *Journal of the History of Biology, 21*(1), 1–68.

Beddall, B. G. (1988b). Wallace's annotated copy of Darwin's "Origin of Species". *Journal of the History of Biology*, 265–289.

de Beer, G. R. (1964). Mendel, Darwin, and Fisher (1865–1965). *Notes and Records of the Royal Society of London, 19*(2), 192–226.

de Beer, G. R. (1966). Mendel, Darwin, and Fisher: Addendum. *Notes and Records of the Royal Society of London, 21*(1), 64–71.

de Beer, G. R. (1967). *Charles Darwin: Evolution by natural selection*. Doubleday.

Bell, T. (1859). Presidential address to the members of the Linnean Society. *Journal of the Proceedings of the Linnean Society of London (Zoology), 4*, viii–ix.

van Beneden, E. (1875). La maturation de l'oeuf, la fécondation et les premières phases du développement embryonnaire des mammifères d'après des recherches faites chez le Lapin. *Bulletins de l'Académie Royale des Sciences, des Lettres et des Beaux-Arts de Belgique, 41*, 686–736.

van Beneden, E. (1883). Recherches sur la maturation de l'oeuf, la fécondation, et la division cellulaire. *Archives de Biologie, 4*, 265–640.

Bennett, A. W. (1870). The theory of natural selection from a mathematical point of view. *Nature, 3*, 30–33.

Bennett, J. H. (1956). Population genetics and natural selection. *Genetica, 28*, 297–307.

Bennett, J. H. (1983). *Natural selection, heredity, and eugenics: Including selected correspondence of R A Fisher with Leonard Darwin and others*. Clarendon Press.

Bernstein, F. (1930). Fortgesetzte untersuchungen aus der theorie der blutgruppen. *Zeitschrift für induktive Abstammungs-und Vererbungslehre, 56*(1), 233–273.

Berrill, N. J., & Liu, C. K. (1948). Germplasm, Weismann, and hydrozoa. *The Quarterly Review of Biology, 23*(2), 124–132.

Berry, A. (Ed.). (2003). *Infinite tropics: An Alfred Russel Wallace anthology*. Verso.

Besnard, A. (1872). Alphabetische Uebersicht der speciellen Literatur des Genus Hieracium. *Flora, 55*(25), 390–394.

Bibby, C. (1972). *Scientist extraordinary: The life and scientific work of Thomas Henry Huxley, 1825–1895*. Pergamon Press.

Bienaymé, I.-J. (1845). De la loi de multiplication et de la durée des familles. *Société philomathique de Paris Extraits*, Sefie 5, 37–39.

Billiard, S., & Castric, V. (2011). Evidence for Fisher's dominance theory: How many "special cases"? *Trends in Genetics, 27*(11), 441–445.

Biological Sciences Curriculum Study. (1963). *Biological science: An inquiry into life*. Harcourt, Brace & World.

Birkhead, T. (2014). *The red canary: The story of the first genetically engineered animal*. Bloomsbury.

Birstein, V. J. (2001). *The perversion of knowledge: The true story of soviet science*. Westview Press.

Blacher, L. I. (1982). *The problem of the inheritance of acquired characters: A history of a priori and empirical methods used to find a solution* (Eng. transl. ed. F. B. Churchill. Russian Original, Moscow: Nauka, 1971). Amerind Publishing Company.

de Blainville, H. (1825). Article: Nereide. *Dictionnaire des Sciences Naturelles, 34*, 408–455.

de Blainville, H.-M. D. (1890). *Cuvier et Geoffroy Saint-Hilaire: Biographies scientifiques*. Baillière.
Blakeslee, A. F. (1936). Twenty-five years of genetics, 1910–1935. *Brooklyn Botanic Garden Memoirs, 4*, 29–40.
Blanc, M. (1984). Gregor Mendel: la légende du génie méconnu. *La Recherche, 151*, 46–59.
Blanc, M. (1985). Darwin, Mendel, Morgan: le lancement de la génétique. *Diogène, 131*, 108–120.
Blaringhem, M. L. (1913). A propos de l'hérédité en mosaïque. *Bulletin de la Société Botanique de France, 60*(3), 282–283.
Blomberg, A. (1872). *Om hybridbildning hos de fanerogama växterna* (Dissertation). Uppsala University, Stockholm.
Blundell, D. J., & Scott, A. C. (1998). *Lyell: The past is the key to the present, Nature London*. Geological Society of London Special Publication 143.
Blunt, W. (1971). *The compleat naturalist: A life of Linnaeus, London: (First published 1971)*. The Viking Press.
Blyth, E. (1835). An attempt to classify the "varieties" of animals, with observations on the marked seasonal and other changes which naturally take place in various British species, and which do not constitute varieties. *Magazine of Natural History, 8*(1), 40–53.
Blyth, E. (1836). Observations on the various seasonal and other external changes which regularly take place in birds, more particularly in those which occur in Britain; with remarks on their great importance in indicating the true affinities of species; and upon the natural system of arrangement. *Magazine of Natural History, 9*(64), 393–409.
Blyth, E. (1837). On the psychological distinctions between man and all other animals; and the consequent diversity of human influence over the inferior ranks of creation, from any mutual or reciprocal influence exercised among the latter. *Magazine of Natural History, 1*(1), 977–985.
Boerhaave, H. (1744). *Praelectiones academicae in proprias institvtiones rei medicae (edited with notes by Albertus Haller)*. Vandenhoeck.
Bonnet, C. (1762). *Considerations sur les corps organisés: où l'on traite de leur origine, de leur développement, de leur réproduction, &c., & òu l'on a rassemblé en abrégé tout ce que l'histoire naturelle offre de plus certain & de plus intéressant sur ce sujet*. Chez Marc-Michel Rey.
Borges, R. M. (2008). Commentary: The objection is sustained: A defence of the defense of beanbag genetics. *International Journal of Epidemiology, 37*(3), 451–454.
Boston Society. (1861). *Proceedings of the Boston Society of Natural History, V.7 (1859–61)*.
Bourguet, D. (1999). The evolution of dominance. *Heredity, 83*(1), 1–4.
Boveri, T. (1887). Zellen-Studien. I. Die Bildung der Richtungskörper bei Ascaris megalocephala und Ascaris lümbricoides. *Jenaische Zeitschrift für Naturwissenschaft, 21*, 423–515.
Boveri, T. (1903). Über die Konstitution der chromatischen Kernsubstanz. *Verhandlungen der Deutschen Zoologischen Gesellschaft, 13*, 10–33.
Bower, F. O. (1919). *Joseph Dalton Hooker*. The Macmillan Company (Society for Promoting Christian Knowledge).
Bowlby, J. (1992). *Charles Darwin: A new life*. WW Norton & Co.
Bowler, P. J. (1976). Malthus, Darwin, and the concept of struggle. *Journal of the History of Ideas, 37*(4), 631–650.
Bowler, P. J. (1989). *The Mendelian revolution: The emergence of hereditarian concepts in modern science and society*. Bloomsbury publishing.
Bowler, P. J. (1996). *Charles Darwin: The man and his influence*. Cambridge University Press.
Bowler, P. J. (2003). *Evolution: The history of an idea* (3rd ed.). University of California Press.
Box, J. F. (1978). *R.A. Fisher: The life of a scientist*. Wiley.
Brackman, A. C. (1980). *A delicate arrangement: The strange case of Charles Darwin and Alfred Russel Wallace*. Times Books.
Brent, P. (1981). *Charles Darwin: A man of enlarged curiosity*. Hienemann.
Brooklyn Botanical Garden. (1950). The birth of genetics: Mendel, de Vries, Correns, Tschermak. *Supplement to Genetics, 36*(5 (part 2)), 33–47.

Brooks, J. L. (1984). *Just before the origin: Alfred Russel Wallace's theory of evolution*. Columbia U. Press.
Brooks, W. K. (1883). *The law of heredity: A study of the cause of variation, and the origin of living organisms*. J. Murphy.
Browne, J. E. (1995). *Charles Darwin: Voyaging*. Princeton University Press.
Browne, J. E. (1996). *Charles Darwin: The power of place*. Princeton University Press.
Browne, J. E. (2010). Asa Gray and Charles Darwin: Corresponding naturalists. *Harvard Papers in Botany, 15*(2), 209–220.
Brunnander, B. (2013). Did Darwin really answer Paley's question? *Studies in History and Philosophy of Biological and Biomedical Sciences, 44*(3), 309–311.
Buchenau, A., & Cassirer, E. (1924). Hauptschriften zur Grundlegung der Philosophie. In *Philosophische Bibliothek*. F. Meiner.
Buckland, W. (1836). *Geology and mineralogy considered with reference to natural theology*. William Pickering.
Buffon, G.-L. L. (1749a). *Histoire naturelle, générale et particulière avec la description du Cabinet du Roi (Tome 1)*. De l'imprimerie royale.
Buffon, G.-L. L. (1749b). *Histoire naturelle, générale et particulière avec la description du Cabinet du Roi (Tome 2)*. De l'imprimerie royale.
Buffon, G.-L. L. (1749c). *Histoire naturelle, générale et particulière avec la description du Cabinet du Roi (Tome 3)*. De l'imprimerie royale.
Buffon, G.-L. L. (1753). *Histoire naturelle, générale et particulière avec la description du Cabinet du Roi (Tome 4)*. De l'imprimerie royale.
Buffon, G.-L. L. (1761). *Histoire naturelle, générale et particulière avec la description du Cabinet du Roi (Tome 9)*. De l'imprimerie royale.
Buffon, G.-L. L. (1766). *Histoire naturelle, générale et particulière avec la description du Cabinet du Roi (Tome 14), VI (Paris: Imprimerie Royale, 1779)*. De l'imprimerie royale.
Buffon, G.-L. L. (1775). *Histoire Naturelle, Générale et Particulière, Servant de Suite à la Théorie de la Terre, & de préliminaire à l'histoire des Végétaux. Parties Expérimentale & Hypothétique. (Supplément, Tome Second)*. Imprimerie Royale.
Buican, D. (1984). *Histoire de la Génétique et de l'Évolutionnisme en France*. Presses Universitaires de France.
Bulmer, M. (1998). Galton's law of ancestral heredity. *Heredity, 81*(5), 579–585.
Bulmer, M. (2003). *Francis Galton: Pioneer of heredity and biometry*. Johns Hopkins Press.
Bulmer, M. (2004). Did Jenkin's swamping argument invalidate Darwin's theory of natural selection? *The British Journal for the History of Science, 37*(3), 281–297.
Burchfield, J. D. (2009). *Lord Kelvin and the age of the earth*. University of Chicago Press.
Bürger, R. (2000). *The mathematical theory of selection, recombination, and mutation*. Wiley.
Burian, R. M., Gayon, J., & Zallen, D. (1988). The singular fate of genetics in the history of French biology, 1900–1940. *Journal of the History of Biology, 21*, 357–402.
Burkhardt, F. (1974). England and Scotland: The learned societies. In T. F. Glick (Ed.), *The comparative reception of Darwinism*. University of Texas Press.
Burkhardt, F., et al. (Eds.). (1993). *The correspondence of Charles Darwin (Vol. 8:1860)*. Cambridge University Press.
Burkhardt, F., et al. (Eds.). (2004). *The correspondence of Charles Darwin (Vol. 14: 1866)*. Cambridge University Press.
Burkhardt, F., et al. (Eds.). (2015). *The correspondence of Charles Darwin (Vol. 23: 1875)*. Cambridge University Press.
Burkhardt, F., & Smith, S. (Eds.). (1987). *The correspondence of Charles Darwin (Vol. 3: 1844–1846)*. Cambridge University Press.
Burkhardt, F., & Smith, S. (Eds.). (1989). *The correspondence of Charles Darwin (Vol. 5: 1851–1855)*. Cambridge University Press.
Burkhardt, F., & Smith, S. (Eds.). (1991). *The correspondence of Charles Darwin (Vol. 7: 1858–1859)*. Cambridge University Press.

Burkhardt, R. W. (1977). *The spirit of system: Lamarck and evolutionary biology*. Harvard University Press.
Butler, S. (1879). *Evolution, old and new: Or, the theories of Buffon*. Hardwicke & Bogue.
Butler, S. (1880). *Unconscious memory*. David Bogue.
Butterfield, H. (1931). *The Whig interpretation of history*. G. Bells & Sons.
Çabej, N. (2012). *Epigenetic principles of evolution*. Elsevier.
Cahn, T. (1962). *La vie et l'oeuvre d'Etienne Geoffroy Saint-Hilaire*. Presses Universitaires de France.
Callender, L. A. (1988). Gregor Mendel: An opponent of descent with modification. *History of Science, 26*(1), 41–75.
Camerarius, R. J. (1694). *De sexu plantarum*. Typis Viduae Rommeii.
Campbell, M. (1976). Explanations of Mendel's results. *Centaurus, 20*(2), 159–174.
Canning, G., & Frere, J. H. (1800). *Poetry of the anti-Jacobin*. J. Wright.
Cannon, H. G. (1958). *The evolution of living things*. Manchester University Press.
Cannon, H. G. (1959). *Lamarck and modern genetics*. Manchester University Press.
Cannon, W. F. (1961). The impact of uniformitarianism: Two letters from John Herschel to Charles Lyell, 1836–1837. *Proceedings of the American Philosophical Society, 105*(3), 301–314.
Carey, N. (2012). *The epigenetics revolution: How modern biology is rewriting our understanding of genetics, disease, and inheritance*. Columbia University Press.
Carlson, E. A. (1967). *Gene theory*. Dickenson.
Carlson, E. A. (2004). *Mendel's legacy: The origin of classical genetics*. Cold Spring Harbor Laboratory Press.
Carlson, E. A. (2011). *Mutation: The history of an idea from Darwin to genomics*. Cold Spring Harbor Laboratory Press.
Carozzi, A. V. (1964). *Hydrogeology by J.B. Lamarck*. University of Illinois Press.
Castilloux, A.-M., & Lessard, S. (1995). The fundamental theorem of natural selection in Ewens' sense (case of many loci). *Theoretical Population Biology, 48*(3), 306–315.
Castle, W. E. (1903). The laws of heredity of Galton and Mendel, and some laws governing race improvement by selection. *Proceedings of the American Academy of Arts and Sciences, 39*(8), 223–242.
Castle, W. E. (1905a). Heredity of coat characters in guinea-pigs and rabbits. *Carnegie Institution of Washington Publication, 23*, 1–78.
Castle, W. E. (1905b). The mutation theory of organic evolution, from the standpoint of animal breeding. *Science, 21*(536), 521–525.
Castle, W. E. (1906). Yellow mice and gametic purity. *Science, 24*(609), 275–281.
Castle, W. E. (1910). The nature of unit characters. *The Harvey Lectures, 6*, 90–101.
Castle, W. E. (1912). Some biological principles of animal breeding. *Journal of Heredity, 3*(4), 270–282.
Castle, W. E. (1915a). Mr. Muller on the constancy of Mendelian factors. *The American Naturalist, 49*(577), 37–42.
Castle, W. E. (1915b). Some experiments in mass selection. *The American Naturalist, 49*(588), 713–726.
Castle, W. E. (1916). Can selection cause genetic change? *The American Naturalist, 50*(592), 248–256.
Castle, W. E. (1917). Piebald rats and multiple factors. *The American Naturalist, 51*(602), 102–114.
Castle, W. E. (1919a). Piebald rats and selection, a correction. *The American Naturalist, 53*(627), 370–376.
Castle, W. E. (1919b). Piebald rats and the theory of genes. *Proceedings of the National Academy of Sciences, 5*(4), 126–130.
Castle, W. E. (1951). The beginnings of Mendelism in America. In L. C. Dunn (Ed.), *Genetics in the twentieth century* (pp. 59–76). Macmillan.
Castle, W. E., & Allen, G. M. (1903). The heredity of albinism. *Proceedings of the American Academy of Arts and Sciences, 38*(21), 603–622.

Castle, W. E., & Little, C. C. (1909). The peculiar inheritance of pink eyes among colored mice. *Science, 30*(766), 313–314.
Castle, W. E., & Phillips, J. C. (1914). Piebald rats and selection: An experimental test of the effectiveness of selection and of the theory of gametic purity in Mendelian crosses. *Carnegie Institution of Washington, 195*, 1–56.
Castle, W. E., & Wright, S. (1916). Studies of inheritance in guinea-pigs and rats. *Carnegie Institution of Washington, 241*, 163–192.
Caudill, E. (1989). *Darwinism in the press: The evolution of an idea*. Erlbaum.
Cavalli-Sforza, L. L., & Bodmer, W. F. (1971). *The genetics of human populations*. W H Freeman.
Chambers, R. (1844). *Vestiges of the natural history of creation*. J. Churchill.
Charlesworth, B. (1979). Evidence against Fisher's theory of dominance. *Nature, 278*(5707), 848–849.
Charlesworth, B. (2022). Fisher's historic 1922 paper "On the dominance ratio". *Genetics, 220*(3), iyac006.
Chetverikov, S. S. (1902). A manual for collecting Lepidoptera. In G. A. Kozhevnikov (Ed.), *A manual for organizing zoological expeditions and collecting zoological material*. K.I. Tikhomirov.
Chetverikov, S. S. (1905). The waves of life (from lepidopterological observations in the summer of 1903). *Proceedings of the Zoological Section of the Imperial Society of Friends of Natural Sciences, Anthropology, and Ethnography, 3*(6), 106–111.
Chetverikov, S. S. (1926). O nekotorykh momentakh evolutsionnogo protsessa s tochki zreniya sovremennoi genetiki. *Zhurnal Eksperimental'noi Biologii. Seriya A, 2*(1), 1–54.
Chetverikov, S. S. (1927) Über die genetische Beschaffenheit wilder Populationen, *Zeitschrift fUr induktive Abstammungs- und Vererbungslehre, 46*(2), 1499–1500.
Chetverikov, S. S. (1961). On certain aspects of the evolutionary process from the standpoint of modern genetics (Eng. transl. by M. Baker and I. M. Lerner of Chetverikov (1926)). *Proceedings of the American Philosophical Society, 105*(2), 167–195.
Chomard-Lexa, A. (2004). *Lucien Cuénot: L'intuition naturaliste*. Editions L'Harmattan.
Churchill, F. B. (1968). August Weismann and a break from tradition. *Journal of the History of Biology*, 91–112.
Churchill, F. B. (1974). William Johannsen and the genotype concept. *Journal of the History of Biology*, 5–30.
Churchill, F. B. (2015). *August Weismann: Development, heredity, and evolution*. Harvard University Press.
Clark, R. W. (1968). *JBS: The life and work of JBS Haldane*. Hodder and Stoughton.
Clark, R. W. (1984). *The survival of Charles Darwin: A biography of a man and an idea*. Random House.
Clausen, J., & Hiesey, W. M. (1958). Experimental studies on the nature of species. IV. *Genetic Structure of Ecological Races, 615*, 1–312.
Cobb, M. (2006). *Generation: The seventeenth-century scientists who unraveled the secrets of sex, life, and growth*. Bloomsbury.
Cock, A. G., & Forsdyke, D. R. (2008). *Treasure your exceptions: The science and life of William Bateson*. Springer.
Cockerham, C. C. (1954). An extension of the concept of partitioning hereditary variance for analysis of covariances among relatives when epistasis is present. *Genetics, 39*(6), 859–882.
Cole, F. J. (1930). *Early theories of sexual generation*. Clarendon Press.
Coleman, W. (1964). *Georges Cuvier, zoologist: A study in the history of evolution theory*. Harvard University Press.
Coleridge, S. T. (1875). Notes on Stillingfleet. *The Athenaeum, 1*, 422–423.
Conn, H. W. (1900). *The method of evolution: A review of the present attitude of science toward the question of the laws and forces which have brought about the origin of species*. Putnam.
Cook, L. M., & Saccheri, I. J. (2013). The peppered moth and industrial melanism: Evolution of a natural selection case study. *Heredity, 110*(3), 207–212.

Corcos, A. F., & Monaghan, F. V. (1993). *Gregor Mendel's experiments on plant hybrids: A guided study*. Rutgers University Press.

Cornish-Bowden, A., & Nanjundiah, V. (2006). The basis of dominance. In R. A. Veitia (Ed.), *The biology of genetic dominance* (pp. 1–16). Landes Bioscience.

Correns, C. (1899). Untersuchungen über Xenien bei Zea mays. *Berichte der Deutschen Botanischen Gesellschaft, XVII*, 410–417.

Correns, C. (1900a). G. Mendel's Regel über das Verhalten der Nachkommenschaft der Rassenbastarde. *Berichte der Deutschen Botanischen Gesellschaft, 18*(4), 158–168.

Correns, C. (1900b). Gregor Mendel's "Versuche über Pflanzen-Hybriden" und die Bestätigung ihrer Ergebnisse durch die neuesten Untersuchungen. *Botanische Zeitung, 58*, 229–235.

Correns, C. (1900c). Über Levkojenbastarde: Zur Kenntnis der Grenzen der Mendelschen Regeln. *Botanisches Centralblatt, 84*, 97–113.

Correns, C. (1905). Gregor Mendels Briefe an Carl Nägeli 1866–1873. *Abhandlungen der Mathematisch-Physikalischen Klasse der Königlich Bayerischen Akademie der Wissenschaften, 29*, 189–265.

Correns, C. (1966a). G. Mendel's law concerning the behavior of progeny of varietal hybrids (Eng. transl. by Leonie K Piternick of Correns (1900)). In C. Stern & E. R. Sherwood (Eds.), *The origin of genetics: A Mendel source book* (pp. 119–132). W H Freeman & Co.

Correns, C. (1966b). Gregor Mendel's letters to Carl Nägeli: 1866-1873 (Eng. transl. by Leonie K Piternick & George Piternick of Correns (1905)). In C. Stern & E. R. Sherwood (Eds.), *The origin of genetics: A Mendel source book* (pp. 56–102). W H Freeman & Co.

Corsi, P. (1988). *The age of Lamarck: Evolutionary theories in France, 1790–1830*. University of California Press.

Costa, J. T. (2014). *Wallace, Darwin, and the origin of species*. Harvard University Press.

Cotterman, C. W. (1940). *A calculus for statistico-genetics* (Ph.D. thesis). Ohio State University, *Genetics and social structure*. Dowden, Hutchinson & Ross, Stroudsburg, PA, USA.

Cotterman, C. W. (1974). Calculus for statistico-genetics. In P. A. Ballonoff (Ed.), *Genetics and social structure: Mathematical structuralism in population genetics and social theory. Pennsylvania (First publication of Cotterman (1940))* (pp. 157–272). Dowden, Hutchinson & Ross, Inc.

Cotterman, C. W. (1983). Relationship and probability in Mendelian populations. *American Journal of Medical Genetics, 16*(3), 393–440 (Originally written in 1960).

Courrier, R. (1952). Notice sur la vie et les travaux de Lucien Cuénot. *Institut de France-Académie des Sciences*, 332–495.

Cowan, R. L. S. (1969). *Sir Francis Galton and the study of heredity in the nineteenth century* (Dissertation). Johns Hopkins University.

Cowan, R. S. (1972a). Francis Galton's contribution to genetics. *Journal of the History of Biology, 5*, 389–412.

Cowan, R. S. (1972b). Francis Galton's statistical ideas: The influence of eugenics. *Isis, 63*, 509–528.

Cowan, R. S. (1977). Nature and nurture: The interplay of biology and politics in the work of Francis Galton. *Studies in History of Biology, 1*, 132–208.

Cowan, R. S. (1981). Walter Frank Raphael Weldon. In C. C. Gillispie (Ed.), *Dictionary of scientific biography* (Vol. 14, pp. 251–252). Charles Scribner's Sons.

Coyne, J. A., Barton, N. H., & Turelli, M. (1997). Perspective: A critique of Sewall Wright's shifting balance theory of evolution. *Evolution, 51*(3), 643–671.

Coyne, J. A., Barton, N. H., & Turelli, M. (2000). Is Wright's shifting balance process important in evolution? *Evolution*, 306–317.

Crew, F. A. E. (1966). *The foundations of genetics*. Pergamon Press.

Crow, J. F. (1954). Breeding structure of populations. II. Effective population number. In O. Kempthorne et al. (Eds.), *Statistics and mathematics in biology* (pp. 543–556). Hafner Publishing Co.

Crow, J. F. (1958). Some possibilities for measuring selection intensities in man. *Human Biology, 30*(1), 1–13.

Crow, J. F. (1970). Genetic loads and the cost of natural selection. In K. Kojima (Ed.), *Mathematical topics in population genetics* (pp. 128–177). Springer.

Crow, J. F. (1988). Sewall Wright (1889–1988). *Genetics, 119*(1), 1–4.

Crow, J. F. (1990a). R. A. Fisher, a centennial view. *Genetics, 124*, 207–211.

Crow, J. F. (1990b). Sewall Wright's place in twentieth-century biology. *Journal of the History of Biology, 23*(1), 57–89.

Crow, J. F. (1993). Francis Galton: Count and measure, measure and count. *Genetics, 135*, 1–4.

Crow, J. F. (2008). Commentary: Haldane and beanbag genetics. *International Journal of Epidemiology, 37*(3), 442–445.

Crow, J. F., & Dove, W. F. (1987). Sewall Wright and physiological genetics. *Genetics, 115*(1), 1–2.

Crow, J. F., & Kimura, M. (1970). *An introduction to population genetics theory*. Harper & Row.

Crowther, J. G. (1935). *British scientists of the nineteenth century*. Kegan Paul, Trench, Trubner.

Cuénot, L. C. M. J. (1902a). La loi de Mendel et l'hérédité de la pigmentation chez les souris. *Archives de Zoologie Expérimentale et Générale 3eme Serie, 10*, XXVII–XXX.

Cuénot, L. C. M. J. (1902b). Sur quelques applications de la loi de Mendel. *Comptes rendus hebdomadaires des séances et mémoires de la Société de Biologie, 54*, 397–398.

Cuénot, L. C. M. J. (1903a). Hypothèse sur l'hérédité des couleurs dans les croisements des souris noires, grises et blanches. *Comptes rendus hebdomadaires des séances et mémoires de la Société de Biologie, 55*, 301–302.

Cuénot, L. C. M. J. (1903b). Transmission héréditaire de pigmentation par les souris albinos. *Comptes rendus hebdomadaires des séances et mémoires de la Société de Biologie, 55*, 299–301.

Cuénot, L. C. M. J. (1905). Les races pures et leurs combinaisons chez les souris. *Archives de Zoologie Experimentale, 3*(7), 123–132.

Cunningham, J. T. (1895). The statistical investigation of evolution. *Nature, 51*(1326), 510.

Cuny, H. (1969). *Thomas Hunt Morgan et la génétique*. Seghers.

Cuvier, G. (1799). Mémoire sur les espèces d'éléphans vivantes et fossiles. *Mémoire de l'Academie des Sciences Paris, 2*, 1–32.

Cuvier, G. (1812). *Recherches sur les ossemens fossiles de quadrupèdes, ou l'on rétablit les caractères de plusieurs espèces d'animaux que les révolutions du globe paroissent avoir détruites*. Deterville.

Cuvier, G. (1821). *Discours sur les revolutions de la surface du globe: recherches sur les ossements fossiles*. G. Dufour and E. d'Ocagne.

Cuvier, G. (1830). Considérations sur les mollusques, et en particulier sur les Céphalopodes. *Annales des Sciences Naturelles, 19*, 241–259.

Cuvier, G. (1835). Eloge de M. de Lamarck, lu à l'Académie des sciences, le 26 novembre 1832. *Mémoires de l'Académie des Sciences, 13*(2), i–xxx.

Dalrymple, G. B. (1994). *The age of the earth*. Stanford University Press.

Dana, J. D. (1853). *On the classification and geographical distribution of Crustacea: From the report on Crustacea of the United States exploring expedition under Captain Charles Wilkes, U.S.N., during the years 1838–1842*. C. Sherman.

Darbishire, A. D. (1902). Note on the results of crossing Japanese waltizing mice with European albino races. *Biometrika, 2*(1), 101–104.

Darbishire, A. D. (1903a). Second report on the result of crossing Japanese waltzing mice with European albino races. *Biometrika, 2*(2), 165–173.

Darbishire, A. D. (1903b). Third report on hybrids between waltzing mice and albino races: On the result of crossing Japanese waltzing mice with "extracted" recessive albinos. *Biometrika, 2*(3), 282–285.

Darbishire, A. D. (1904a). On the bearing of Mendelian principles of heredity on current theories of the origin of species. *Manchester Memoirs, 48*(24), 1–19.

Darbishire, A. D. (1904b). On the result of crossing Japanese waltzing with albino mice. *Biometrika, 3*(1), 1–51.
Darbishire, A. D. (1905). On the supposed antagonism of Mendelian to biometric theories of heredity. *Memoirs and Proceedings of the Manchester Literary and Philosophical Society, 49*(6), 1–19.
Darbishire, A. D. (1906). On the difference between physiological and statistical laws of heredity. *Memoirs and Proceedings of the Manchester Literary and Philosophical Society, 50*(11), 1–44.
Darbishire, A. D. (1907). Mendelism by RC Punnett. (Anonymous review). *Nature, 76*(7), 73–74.
Darbishire, A. D. (1911). *Breeding and the Mendelian discovery*. Cassell, Limited.
Darbishire, A. D. (1917). *An introduction to a biology and other papers*. Cassell & Company.
Darden, L. (1991). *Theory change in science: Strategies from Mendelian genetics*. Oxford University Press.
Darlington, C. D. (1959). *Darwin's place in history*. Blackwell.
Darlington, C. D. (1964). *Genetics and man*. Macmillan.
Darwin, C. (1840). *Journal of researches into the geology and natural history of the various countries visited by HMS beagle, under the command of Captain Fitzroy from 1832 to 1836*. H Colburn.
Darwin, C. (1846). *Journal of researches into the natural history and geology of the countries visited during the voyage of HMS beagle round the world, under the command of Capt. Fitz Roy, R.N.* Harper & brothers.
Darwin, C. (1859). *On the origin of species by means of natural selection*. John Murray.
Darwin, C. (1860a). Natural selection. *Gardeners' Chronicle and Agricultural Gazette*, 362–363.
Darwin, C. (1860b). *On the origin of species by means of natural selection* (2nd ed.). Murray.
Darwin, C. (1861). *On the origin of species by means of natural selection* (3rd ed.). Murray.
Darwin, C. (1866). *On the origin of species by means of natural selection* (4th ed.). Murray.
Darwin, C. (1868a). *The variation of animals and plants under domestication* (Vol. I). John Murray.
Darwin, C. (1868b). *The variation of animals and plants under domestication* (Vol. II). John Murray.
Darwin, C. (1869). *On the origin of species by means of natural selection* (5th ed.). Murray.
Darwin, C. (1871a). Pangenesis. *Nature, 3*(78), 502–503.
Darwin, C. (1871b). *The descent of man and selection in relation to sex* (Vol. I). D. Appleton and Co.
Darwin, C. (1872). *The origin of species by means of natural selection* (6th ed.). Murray.
Darwin, C. (1876). *The effects of cross and self fertilisation in the vegetable kingdom*. John Murray.
Darwin, C. (1877). *The different forms of flowers on plants of the same species*. John Murray.
Darwin, C., & Wallace, A. R. (1858). On the tendency of species to form varieties; and on the perpetuation of varieties and species by natural means of selection. *Journal of the Proceedings of the Linnean Society of London. Zoology, 3*(9), 45–62.
Darwin, E. (1791). The botanic garden: A poem. In *Two parts. Part I. Containing the economy of vegetation. Part II. The loves of the plants. With philosophical notes*. Printed for J. Johnson.
Darwin, E. (1794). *Zoonomia, or, the laws of organic life*. Printed for P. Byrne, and W. Jones.
Darwin, E. (1803). *The temple of nature; or, the origin of society: A poem, with philosophical notes*. J. Johnson.
Darwin, F. (1887). *The life and letters of Charles Darwin, including an autobiographical chapter* (Vol. I). John Murray.
Darwin, F. (1888a). *The life and letters of Charles Darwin, including an autobiographical chapter* (Vol. II). John Murray.
Darwin, F. (1888b). *The life and letters of Charles Darwin, including an autobiographical chapter* (Vol. III). John Murray.
Darwin, F. (1898). The Knight-Darwin law. *Nature, 58*, 630–632.
Darwin, F. (1909). *The foundations of the origin of species. Two essays written in 1842 and 1844 by Charles Darwin*. Cambridge University Press.

Darwin, F., & Seward, A. C. (1903a). *More letters of Charles Darwin: A record of his work in a series of hitherto unpublished letters* (Vol. I). D. Appleton and Company.
Darwin, F., & Seward, A. C. (1903b). *More letters of Charles Darwin: A record of his work in a series of hitherto unpublished letters* (Vol. II). D. Appleton and company.
Darwin, G. (1873). On beneficial restrictions to liberty of marriage. *The Contemporary Review, 22*, 412–426.
Daudin, H. (1926). *Cuvier et Lamarck: les classes zoologiques et l'idée de série animale (1790–1830)* (Vol. 2). Alcan.
Davenport, C. B. (1900). Review of von Guaita's experiments in breeding mice. *The Biological Bulletin, 2*(3), 121–128.
Davenport, C. B. (1901). Mendel's law of dichotomy in hybrids. *The Biological Bulletin, 2*(6), 307–310.
Davenport, C. B. (1917). Inheritance of stature. *Genetics, 2*, 313–389.
Davies, R. (2008). *The Darwin conspiracy*. Golden Square.
Davis, A. S. (1871). The "North British Review" and the origin of species. *Nature, 5*(113), 161.
Dean, D. R. (1992). *James Hutton and the history of geology*. Cornell University Press.
DeJong-Lambert, W. (2012). *The cold war politics of genetic research: An introduction to the Lysenko affair*. Springer.
Delange, Y. (1984). *Lamarck: Sa vie, son oeuvre*. Actes Sud.
Dennett, D. C. (1996). *Darwin's dangerous idea: Evolution and the meanings of life*. Simon & Schuster.
Depew, D. J., & Weber, B. H. (1994). *Darwinism evolving: Systems dynamics and the genealogy of natural selection*. The MIT Press.
Desmond, A. (1997). *Huxley: From Devil's disciple to evolution's high priest*. Addison-Wesley.
Desmond, A., & Moore, J. (1991). *Darwin*. Michael Joseph.
Dickins, T. E. (2021). *The modern synthesis: Evolution and the organization of information*. Springer Nature.
Dietrich, M. R., & Skipper Jr, R. A. (2012). A shifting terrain: A brief history of the adaptive landscape. In E. I. Svensson & R. Calsbeek (Eds.), *The adaptive landscape in evolutionary biology* (pp. 3–15). Oxford University Press.
van Dijk, P. J., & Ellis, T. H. N. (2016). The full breadth of Mendel's genetics. *Genetics, 204*(4), 1327–1336.
Disraeli, B. (1847). *Tancred, or, the new crusade*. Carey & Hart.
Dobzhansky, T. (1935). A critique of the species concept in biology. *Philosophy of Science, 2*(3), 344–355.
Dobzhansky, T. (1936). Studies on hybrid sterility. II. Localization of sterility factors in Drosophila pseudoobscura hybrids. *Genetics, 21*(2), 113–135.
Dobzhansky, T. (1937). *Genetics and the origin of species*. Columbia University Press.
Dobzhansky, T. (1955). A review of some fundamental concepts and problems of population genetics. *Cold Spring Harbor Symposia on Quantitative Biology, 20*, 1–15.
Dobzhansky, T. (1959). Evolution of genes and genes in evolution. In *Cold Spring Harbor symposia on quantitative biology* (pp. 15–30). Cold Spring Harbor Laboratory Press.
Dobzhansky, T. (1965). Mendelism, Darwinism, and evolutionism. *Proceedings of the American Philosophical Society, 109*(4), 205–215.
Dobzhansky, T. (1970). *Genetics of the evolutionary process*. Columbia University Press.
Douglas, C. G., Haldane, J., & Haldane, J. B. S. (1912). The laws of combination of haemoglobin with carbon monoxide and oxygen. *The Journal of Physiology, 44*(4), 275–304.
Douglas, J. (1895). Home correspondence: The origin of the cultivated Cineraria. *The Gardener's Chronicle, 18*, 774–776.
Dronamraju, K. (2011). *Haldane, Mayr, and beanbag genetics*. Oxford University Press.
Dronamraju, K. (2016). *Popularizing science: The life and work of JBS Haldane*. Oxford University Press.
Dronamraju, K. (2018). *A century of geneticists: Mutation to medicine*. CRC Press.

Dronamraju, K. R. (1985). *The life and work of JBS Haldane with special reference to India*. Aberdeen University Press.
Dronamraju, K. R. (1986). Possible worlds: Contributions of JBS Haldane to genetics. *Trends in Genetics, 2*, 322–324.
Dronamraju, K. R. (1990). Review: Sewall Wright (1889–1988). *The Japanese Journal of Genetics, 65*, 25–31.
Dronamraju, K. R. (Ed.). (1968). *Haldane and modern biology*. Johns Hopkins University Press.
Dröscher, A. (2015). Gregor Mendel, Franz Unger, Carl Nägeli and the magic of numbers. *History of Science, 53*(4), 492–508.
Drummond, J. (1827). On the cultivation and improvement of Cineraria cruenta. *The Gardener's Magazine, II*, 153–154.
Dublin, L. I., & Lotka, A. J. (1925). On the true rate of natural increase: As exemplified by the population of the United States, 1920. *Journal of the American Statistical Association, 20*(151), 305–339.
Dung, S. K., et al. (2019). Illuminating women's hidden contribution to historical theoretical population genetics. *Genetics, 211*(2), 363–366.
Dunn, L. C. (1965a). *A short history of genetics*. McGraw-Hill.
Dunn, L. C. (1965b). William Ernest Castle, October 25, 1867–June 3, 1962. In *Biographical memoirs* (Vol. 38, pp. 33–80). National Academy of Sciences (US).
Dupree, H. (1959). *Asa Gray, 1810–1888*. Harvard University Press.
Duzdevich, D. (2014). *Darwin's on the origin of species: A modern rendition*. Indiana University Press.
East, E. M. (1908). Inbreeding in corn. *Reports of the Connecticut Agricultural Experiment Station for Years 1907–1908*, 419–428.
East, E. M. (1910). A Mendelian interpretation of variation that is apparently continuous. *The American Naturalist, 44*(518), 65–82.
Edgeworth, F. Y. (1892). Correlated averages. *Philosophical Magazine, 5th Series, 34*, 190–204.
Edwards, A. W. F. (1971). Reviews – Evolution and the genetics of populations. Volume 2: The theory of gene frequencies. *Heredity, 26*(2), 332–338.
Edwards, A. W. F. (1986a). Are Mendel's results really too close? *Biological Reviews, 61*(4), 295–312.
Edwards, A. W. F. (1986b). More on the too-good-to-be-true paradox and Gregor Mendel. *Journal of Heredity, 77*(2), 138.
Edwards, A. W. F. (1990a). Fisher, W^-, and the fundamental theorem. *Theoretical Population Biology, 38*(3), 276–284.
Edwards, A. W. F. (1990b). R. A. Fisher twice professor of genetics: London and Cambridge or "a fairly well-known geneticist". *Biometrics, 46*(4), 897–904.
Edwards, A. W. F. (1993). Mendel, Galton, Fisher. *Australian Journal of Statistics, 35*(2), 129–140.
Edwards, A. W. F. (1994). The fundamental theorem of natural selection. *Biological Reviews, 69*(4), 443–474.
Edwards, A. W. F. (2000). *Foundations of mathematical genetics*. Cambridge University Press.
Edwards, A. W. F. (2002). The fundamental theorem of natural selection. *Theoretical Population Biology, 61*(3), 335–337.
Edwards, A. W. F. (2008). Anecdotal, historical and critical commentaries on genetics: GH Hardy (1908) and Hardy–Weinberg equilibrium. *Genetics, 179*(3), 1143–1150.
Edwards, A. W. F. (2013). Robert Heath Lock and his textbook of genetics, 1906. *Genetics, 194*(3), 529–537.
Edwards, A. W. F. (2014). R. A. Fisher's gene-centred view of evolution and the fundamental theorem of natural selection. *Biological Reviews, 89*(1), 135–147.
Edwards, A. W. F. (2017). Haldane and Fisher–scientific interactions. *Journal of Genetics, 96*(5), 747–752.
Eiseley, L. C. (1958). *Darwin's century: Evolution and the men who discovered it.* Doubleday.

Eiseley, L. C. (1959). Charles Darwin, Edward Blyth, and the theory of natural selection. *Proceedings of the American Philosophical Society, 103*(1), 94–158.

Eiseley, L. C. (1979). *Darwin and the mysterious Mr. X: New light on the evolutionists*. Dutton.

Eisenhart, C. (1974). Karl Pearson. In C. C. Gillispie (Ed.), *Dictionary of scientific biography* (pp. 447–473). Charles Scribner's Sons.

Eldredge, N. (1985). *Unfinished synthesis: Biological hierarchies and modern evolutionary thought*. Oxford University Press.

Ellegard, A. (1958). *Darwin and the general reader: The reception of Darwin's theory of evolution in the British Periodical Press, 1859–1872*. Goteborgs Universitets Arsskrift.

Elton, C. S. (1927). *Animal ecology*. Sidgwick & Jackson.

Emerson, R. A., & East, E. M. (1913). The inheritance of quantitative characters in maize. *Nebraska Agricultural Experiment Station Research Bulletin, 2*, 1–120.

Endersby, J. (2007). *A guinea pig's history of biology*. Random House.

Endersby, J. (2020). *Imperial nature: Joseph Hooker and the practices of Victorian science*. University of Chicago Press.

Epperson, B. K. (1999). Gustave Malécot, 1911–1998: Population genetics founding father. *Genetics, 152*(2), 477–484.

Everdell, W. R. (1997). *The first moderns: Profiles in the origins of 20th century thought*. University of Chicago Press.

Ewens, W. J. (1965a). A note on Fisher's theory of the evolution of dominance. *Annals of Human Genetics, 29*, 85–88.

Ewens, W. J. (1965b). Further notes on the evolution of dominance. *Heredity, 20*(3), 443–450.

Ewens, W. J. (1967). A note on the mathematical theory of the evolution of dominance. *The American Naturalist, 101*(917), 35–40.

Ewens, W. J. (1969a). A generalized fundamental theorem of natural selection. *Genetics, 63*(2), 531–537.

Ewens, W. J. (1969b). Mean fitness increases when fitnesses are additive. *Nature, 221*(5185), 1076.

Ewens, W. J. (1972). The sampling theory of selectively neutral alleles. *Theoretical Population Biology, 3*(1), 87–112.

Ewens, W. J. (1979). *Mathematical population genetics*. Springer.

Ewens, W. J. (1989). An interpretation and proof of the fundamental theorem of natural selection. *Theoretical Population Biology, 36*(2), 167–180.

Ewens, W. J. (1993). Beanbag genetics and after. In P. P. Majumder (Ed.), *Human population genetics: A centennial tribute to JBS Haldane* (pp. 7–29). Plenum Press.

Ewens, W. J. (1995). Addendum to "The fundamental theorem of natural selection in Ewens′ sense (case of many loci)" by Castilloux and Lessard. *Theoretical Population Biology, 48*(3), 316–317.

Ewens, W. J. (2004). *Mathematical population genetics. I Theoretical introduction* (2nd ed.). Springer.

Ewens, W. J. (2008). Commentary: On Haldane's "defense of beanbag genetics". *International Journal of Epidemiology, 37*(3), 447–451.

Ewens, W. J. (2014). Grafen, the Price equations, fitness maximization, optimisation and the fundamental theorem of natural selection. *Biology & Philosophy, 29*, 197–205.

Ewens, W. J. (2024). The fundamental theorem of natural selection: The end of a story. *Evolution, 78(5)*, 803–808.

Fairbanks, D. J., & Rytting, B. (2001). Mendelian controversies: A botanical and historical review. *American Journal of Botany, 88*(5), 737–752.

Falk, R., & Sarkar, S. (1991). The real objective of Mendel's paper: A response to Monaghan and Corcos. *Biology and Philosophy, 6*, 447–451.

Farmer, J. B., & Moore, J. E. S. (1905). On the maiotic phase (reduction divisions) in animals and plants. *Journal of Cell Science, 2*(192), 489–558.

Fellows, O. E., & Milliken, S. F. (1972). *Buffon*. Twayne Publishers.

Fichman, M. (2010). *An elusive Victorian: The evolution of Alfred Russel Wallace*. University of Chicago Press.

Fienberg, S. E., & Hinkley, D. V. (Eds.). (1980). *R. A. Fisher: An appreciation*. Springer.

Fish, H. D. (1914). On the progressive increase of homozygosis brother-sister matings. *The American Naturalist, 48*(576), 759–761.

Fisher, R. A. (1915). Frequency distribution of the values of the correlation coefficient in samples from an indefinitely large population. *Biometrika, 10*(4), 507–521.

Fisher, R. A. (1918). The correlation between relatives on the supposition of Mendelian inheritance. *Earth and Environmental Science Transactions of the Royal Society of Edinburgh, 52*(2), 399–433.

Fisher, R. A. (1921). Review of A.L. Hagedoorn and A.C. Hagedoorn "The relative value of the processes causing evolution". *The Eugenics Review, 13*, 467–470.

Fisher, R. A. (1922). On the dominance ratio. *Proceedings of the Royal Society of Edinburgh, 42*, 321–341.

Fisher, R. A. (1927a). On some objections to mimicry theory; statistical and genetic. *Transactions of the Entomological Society of London, 75*, 269–278.

Fisher, R. A. (1927b). The actuarial treatment of official birth records. *The Eugenics Review, 19*(2), 103–108.

Fisher, R. A. (1928a). The possible modification of the response of the wild type to recurrent mutations. *The American Naturalist, 62*(679), 115–126.

Fisher, R. A. (1928b). Two further notes on the origin of dominance. *The American Naturalist, 62*(683), 571–574.

Fisher, R. A. (1929). The evolution of dominance; reply to Professor Sewall Wright. *The American Naturalist, 63*(689), 553–556.

Fisher, R. A. (1930a). The distribution of gene ratios for rare mutations. *Proceedings of the Royal Society of Edinburgh, 50*, 204–219.

Fisher, R. A. (1930b). The evolution of dominance in certain polymorphic species. *The American Naturalist, 64*(694), 385–406.

Fisher, R. A. (1930c). *The genetical theory of natural selection*. Clarendon Press.

Fisher, R. A. (1934). Professor Wright on the theory of dominance. *The American Naturalist, 68*(717), 370–374.

Fisher, R. A. (1936). Has Mendel's work been rediscovered? *Annals of Science, 1*(2), 115–137.

Fisher, R. A. (1941). Average excess and average effect of a gene substitution. *Annals of Eugenics, 11*(1), 53–63.

Fisher, R. A. (1958). *The genetical theory of natural selection* (2nd ed.). Dover.

Fisher, R. A., & Mackenzie, W. A. (1923). Studies in crop variation. II. The manurial response of different potato varieties. *The Journal of Agricultural Science, 13*(3), 311–320.

Flannery, M. A. (2018). *Nature's prophet: Alfred Russel Wallace and his evolution from natural selection to natural theology*. University of Alabama Press.

Flemming, W. (1882). *Zellsubstanz, kern und zelltheilung*. FCW Vogel.

Flew, A. (1997). *Darwinian evolution* (2nd ed.). Transaction Publishers.

Focke, W. O. (1881). *Die Pflanzen-Mischlinge: Ein Beitrag zur Biologie der Gewächse*. Gebrüder Borntraeger.

Fokker, A. D. (1914). Die mittlere Energie rotierender elektrischer Dipole im Strahlungsfeld. *Annalen der Physik, 348*(5), 810–820.

Forbes, E. (1854). On the manifestation of polarity in the distribution of organized beings in time. *Notices of the Proceedings at the Meetings of the Members of the Royal Institution, 1*, 428–433.

Ford, E. B. (1930). The theory of dominance. *The American Naturalist, 64*(695), 560–566.

Ford, E. B. (1975). *Ecological genetics* (4th ed.). Chapman and Hall.

Ford, E. B., & Crow, J. F. (2005). R. A. Fisher: An appreciation. *Genetics, 171*(2), 415–417.

Fordyce, J. (1883). *Aspects of scepticism: With special reference to the present time*. Elliot Stock.

Forrest, D. W. (1974). *Francis Galton: The life and work of a Victorian genius*. Taplinger.

Fothergill, P. G. (1952). *Historical aspects of organic evolution*. Hollis & Carter.

Fancier, O. (1893). *Fancy mice, their varieties, management, and breeding* (3rd ed.). L. Upcott Gill.
Frängsmyr, T. (Ed.). (1983). *Linnaeus: The man and his work*. University of California Press.
Frank, S. A. (1995). George Price's contributions to evolutionary genetics. *Journal of Theoretical Biology, 175*(3), 373–388.
Frank, S. A. (1997). The Price equation, Fisher's fundamental theorem, kin selection, and causal analysis. *Evolution, 51*(6), 1712–1729.
Frank, S. A., & Slatkin, M. (1992). Fisher's fundamental theorem of natural selection. *Trends in Ecology & Evolution, 7*(3), 92–95.
Franklin, A., et al. (2008). *Ending the Mendel-Fisher controversy*. University of Pittsburgh Press.
Freeman, R. B. (1977). *The works of Charles Darwin: An annotated bibliographical handlist* (2nd ed.). Dawson.
Freeman, R. B. (1978). *Charles Darwin: A companion*. Dawson.
Froggatt, P., & Nevin, N. C. (1971a). Galton's "law of ancestral heredity": Its influence on the early development of human genetics. *History of Science, 10*(1), 1–27.
Froggatt, P., & Nevin, N. C. (1971b). The "law of ancestral heredity" and the Mendelian-ancestrian controversy in England, 1889–1906. *Journal of Medical Genetics, 8*(1), 1–36.
Futuyma, D. J. (1998). *Evolutionary biology* (3rd ed.). Sinauer.
Galbraith, G. (Ed.). (1965). *The Journal of the Rev. William Bagshaw Stevens*. Clarendon Press.
Gale, B. G. (1982). *Evolution without evidence: Charles Darwin and the origin of species*. University of New Mexico Press Albuquerque.
Gale, J. S. (1980). *Population genetics*. Wiley.
Gale, J. S. (1990). *Theoretical population genetics*. Unwin Hyman.
Gall, Y., & Konashev, M. B. (2008). The reception of Darwin's theory of evolution in Russia. In E.-V. Engels & T. F. Glick (Eds.), *The reception of Charles Darwin in Europe* (Vol. 2, pp. 502–521). Continuum.
Gallais, A. (2018). *Histoire de la Génétique et de l'Amélioration des Plantes*. Éditions Quae.
Gallesio, G. (1816). *Teoria della riproduzione vegetale*. Niccolò Capurro.
Galton, D. J. (2017). *Standing on the shoulders of Darwin and Mendel: Early views of inheritance*. CRC Press.
Galton, F. (1852). Recent expedition into the interior of South-Western Africa. *The Journal of the Royal Geographical Society of London, 22*, 140–163.
Galton, F. (1853). *Explorer of tropical South Africa*. John Murray.
Galton, F. (1865). Hereditary talent and character. *Macmillan's Magazine, 12*, 318–327.
Galton, F. (1869a). *Hereditary genius: An inquiry into its laws and consequences*. Macmillan.
Galton, F. (1869b). Hereditary genius: The judges of England between 1660 and 1865. *Macmillan's Magazine, 16*, 424–431.
Galton, F. (1871a). Experiments in Pangenesis, by breeding from rabbits of a pure variety, into whose circulation blood taken from other varieties had previously been largely transfused. *Proceedings of the Royal Society of London, 19*(123–129), 393–410.
Galton, F. (1871b). Pangenesis. *Nature, 4*(79), 5–6.
Galton, F. (1872). On blood-relationship. *Proceedings of the Royal Society of London, 20*, 394–402.
Galton, F. (1875). A theory of heredity. *Contemporary Review, 27*, 80–95.
Galton, F. (1876). A theory of heredity. *The Journal of the Anthropological Institute of Great Britain and Ireland, 5*, 329–348.
Galton, F. (1877). Typical laws of heredity. *Nature, 15*, 492–495, 512–514, 532–533.
Galton, F. (1883). *Inquiries into human faculty and its development*. Macmillan.
Galton, F. (1885a). Presidential address, section H, anthropology. *British Association Report, 55*, 1206–1214.
Galton, F. (1885b). Regression towards mediocrity in hereditary stature. *The Journal of the Anthropological Institute of Great Britain and Ireland, 15*, 246–263.
Galton, F. (1886). Family likeness in stature. *Proceedings of the Royal Society of London, 40*(242–245), 42–73.

Galton, F. (1888). Co-relations and their measurement, chiefly from anthropometric data. *Proceedings of the Royal Society of London, 45*, 135–145.

Galton, F. (1889). *Natural inheritance*. Macmillan.

Galton, F. (1891). *Hereditary genius: An inquiry into its Laws and Consequences* (2nd ed.). Macmillan.

Galton, F. (1892). *Finger prints*. Macmillan.

Galton, F. (1894). Discontinuity in evolution. *Mind, 3*(11), 362–372.

Galton, F. (1897a). Rate of racial change that accompanies different degrees of severity in selection. *Nature, 55*(1435), 605–606.

Galton, F. (1897b). The average contribution of each several ancestor to the total heritage of the offspring. *Proceedings of the Royal Society of London, 61*(369–377), 401–413.

Galton, F. (1901). Biometry. *Biometrika, 1*, 7–10.

Galton, F. (1908). *Memories of my life*. Methuen and Co.

Gärtner, C. F. (1849). *Versuche und Beobachtungen über die Bastarderzeugung im Pflanzenreich: mit Hinweisung auf die ähnlichen Erscheinungen im Thierreich*. Hering und Co.

Gärtner, K. F. (1826). Nachricht über Versuche, die Befruchtung einiger Gewächse betreffend. *Naturwissenschaftliche Abhandlungen. Herausgegeben von eine gesellschaft in Würtemberg, 1*, 35–66.

Gaudant, M., & Gaudant, J. (1971). *Les théories classiques de l'évolution*. Dunod.

Gayon, J. (1990). Critics and criticisms of the modern synthesis: The viewpoint of a philosopher. *Evolutionary Biology, 24*, 1–49.

Gayon, J. (1998). *Darwinism's struggle for survival: Heredity and the hypothesis of natural selection*. Cambridge University Press.

Gayon, J. (Ed.). (1992). *Buffon 88*. Librairie Philosophique VRIN.

Gayon, J., & Huneman, P. (2019). The modern synthesis: Theoretical or institutional event? *Journal of the History of Biology, 52*(4), 519–535.

Gee, H. (2004). *Jacob's ladder: The history of the human genome*. WW Norton & Company.

Geiringer, H. (1944). On the probability theory of linkage in Mendelian heredity. *The Annals of Mathematical Statistics, 15*(1), 25–57.

Geison, G. L. (1969). Darwin and heredity: The evolution of his hypothesis of pangenesis. *Journal of the History of Medicine and Allied Sciences, 24*(4), 375–411.

Geldart, H. D. (1879). Notes on the life and writings of Edward Blyth. *Transactions of the Norfolk and Norwich Naturalists Society, 3*, 38–46.

Geoffroy Saint-Hilaire, E. (1796). Mémoire sur les rapports naturels des makis Lemur L. et description d'une espèce nouvelle de Mammifere. *Magasin Encyclopedique, 2eme année, 1*, 20–50.

Geoffroy Saint-Hilaire, E. (1820a). Mémoire sur l'organisation des insectes: Second memoire, sur quelques regles fondamentales en philosophic naturelle (lu a l'Academie des Sciences le 17 janvier 1820). *Journal Complémentaire des Sciences Médicales, 6*, 31–36.

Geoffroy Saint-Hilaire, E. (1820b). Mémoires sur l'organisation des insectes: Premier mémoire sur un squelette chez les insectes, dont toutes les pièces, identiques entre elles dans les divers ordres du systeme entomologique, correspondent a chacun des os du squelette dans les classes super. *Journal Complementaire des Sciences Medicales, 5*, 340–351.

Geoffroy Saint-Hilaire, E. (1820c). Mémoires sur l'organisation des Insectes: Troisième mémoire, sur une colonne vertébrale et ses côtes dans les insectes apiropodes (lu à l'Académie des sciences, le 12 février 1820). *Journal Complémentaire des Sciences Médicales, 7*, 271–303.

Geoffroy Saint-Hilaire, E. (1825). Recherches sur l'organisation des Gavials, sur leurs affinités naturelles desquelles résulte la necessité d'une autre distribution générique: Gavialis, Teleosaurus, Steneosaurus. *Mémoires du Muséum National d'Histoire Naturelle, 12*, 97–155.

Geoffroy Saint-Hilaire, E. (1831). *Recherches sur de Grands Sauriens: Trouvés à l'État Fossile vers les Confins Maritimes de la Basse Normandie, Attribués d'Abord au Crocodile, Puis Déterminés Sous les Noms de Téléosaurus et Sténéosaurus*. Firmin Didot.

Geoffroy Saint-Hilaire, E. (1833). Le degré d'influence du monde ambiant pour modifier les formes animales. *Mémoires de l'Académie Royale des Sciences de l'Institut de France, 12*, 63–92.

Geoffroy Saint-Hilaire, E., & Cuvier, G. (1795a). Histoire naturelle des orangs-outangs. *Magasin Encyclopédique, ou Journal des Sciences, des Lettres et des Arts, 3*, 451–463.

Geoffroy Saint-Hilaire, E., & Cuvier, G. (1795b). Mémoire sur les rapports naturels du tarsier, (Didelphis macrotarsus Gm.), lu à la Société d'Histoire Naturelle, le 21 messidor, an troisièm. *Magasin Encyclopédique, ou Journal des Sciences, des Lettres et des Arts, 3*, 147–154.

George, W. (1964). *Biologist philosopher: A study of the life and writings of Alfred Russel Wallace, science and society*. Abelard-Schuman.

George, W. (1982). *Darwin*. Fontana Paperbacks.

Ghiselin, M. T. (1969). *The triumph of the Darwinian method*. University of California Press.

Ghiselin, M. T. (1975). The rationale of pangenesis. *Genetics (supp.), 79*, 47–57.

Gillespie, J. H. (2004). *Population genetics: A concise guide* (2nd ed.). Johns Hopkins University Press.

Gillham, N. W. (2002). *A life of Sir Francis Galton: From African exploration to the birth of eugenics*. Oxford University Press.

Gillispie, C. C. (1959). Lamarck and Darwin in the history of science. In B. Glass, O. Temkin, & W. L. Strauss Jr (Eds.), *Forerunners of Darwin, 1745–1859* (pp. 266–291). Johns Hopkins Press.

Gillois, M. (1965). Relation d'identité en génétique. *Annales de l'institut Henri Poincaré. Section B. Calcul des probabilités et statistiques, 2*(1), 1–94.

Gilman, D. (1899). *The life of James Dwight Dana: Scientific explorer, mineralogist, geologist, zoologist, professor in Yale University*. Harper & Brothers.

Glick, T. F. (Ed.). (1974). *The comparative reception of Darwinism, the American historical review*. University of Texas Press.

Goldschmidt, R. (1927). *Physiologische Theorie der Vererbung*. Springer.

Goldschmidt, R. (1951). L. Cuénot: 1866–1951. *Science, 113*(2934), 309–310.

Goldstein, H. (2012). Francis Galton, measurement, psychometrics and social progress. *Assessment in Education: Principles, Policy & Practice, 19*(2), 147–158.

Goodnight, C. J., & Wade, M. J. (2000). The ongoing synthesis: A reply to Coyne, Barton, and Turelli. *Evolution*, 317–324.

Gordon, C. (1936). The frequency of heterozygosis in free-living populations of Drosophila melanogaster and Drosophila subobsgura. *Journal of Genetics, 33*(1), 25–60.

Gorroochurn, P. (2016a). *Classic topics on the history of modern mathematical statistics: From Laplace to modern recent times*. Wiley.

Gorroochurn, P. (2016b). On Galton's change from "reversion" to "regression". *The American Statistician, 70*(3), 227–231.

Goss, J. (1824). On the variation in the colour of peas occasioned by cross impregnation. *Transactions of the Horticultural Society of London, 5*, 234–237.

Gould, S. J. (1983). The hardening of the modern synthesis. In M. Grene (Ed.), *Dimensions of Darwinism* (pp. 71–93). Cambridge University Press.

Gould, S. J. (1985). Fleeming Jenkin revisited. *Natural History (New York N.Y.), 94*(6), 14–20.

Gould, S. J. (1990). Darwin and Paley meet the invisible hand. *Natural History, 99*(11), 8–12.

Gould, S. J. (2002). *The structure of evolutionary theory*. Harvard University Press.

Graham, L. R. (1993). *Science in Russia and the Soviet Union: A short history*. Cambridge University Press.

Gray, A. (1859). Diagnostic characters of new species of phaenogamous plants, collected in Japan by Charles Wright, botanist of the U.S. North Pacific exploring expedition. *American Academy of Arts and Sciences, Memoirs, n.s., 6*, 377–452.

Gray, A. (1861). *Natural selection not inconsistent with natural theology*. Trubner.

Green, M. M. (1996). The "genesis of the white-eyed mutant"' in Drosophila melanogaster: A reappraisal. *Genetics, 142*, 329–331.

Greene, J. C. (1959). *The death of Adam: Evolution and its impact on Western thought*. Iowa State University Press.
di Gregorio, M. A. (1984). *T.H. Huxley's place in natural science*. Yale University Press.
Griffiths, A. J. F., et al. (2015). *An introduction to genetic analysis* (11th ed.). W.H. Freeman & Company.
Griggs, E. L. (1956). *Collected letters of Samuel Taylor Coleridge*. Clarendon Press.
Gruber, H. E. (1974). *Darwin on man: A psychological study of scientific creativity*. EP Dutton.
von Guaita, G. (1898). Versuche mit Kreuzungen von verschiedenen Rassen der Hausmaus. *Berichte der Naturforschenden Gesellschaft zu Freiburg, 10*, 317–332.
Le Guyader, H. (2004). *Geoffroy Saint-Hilaire: A visionary naturalist*. University of Chicago Press.
Guyénot, E. (1941). *Les Sciences de la vie aux XVIIè et XVIIIè siècles*. Albin Michel.
Haacke, W. (1895). Ueber Wesen, Ursachen und Verebung von Albinismus und Scheckung und ueber deren Bedeutung fuer vererbungstheoretische und entwicklungsmechanische Fragen. *Biologisches Centralblatt, 15*, 44–78.
Haeckel, E. (1866). *Generelle Morphologie der Organismen*. G. Reimer.
Hakewill, G. (1635). *An apologie or declaration of the power and providence of god in the government of the world* (3rd ed.). W. Turner.
Haldane, J. B. S. (1919a). The combination of linkage values and the calculation of distances between the loci of linked factors. *Journal of Genetics, 8*(29), 299–309.
Haldane, J. B. S. (1919b). The probable errors of calculated linkage values, and the most accurate method of determining gametic from certain zygotic series. *Journal of Genetics, 8*, 291–297.
Haldane, J. B. S. (1924a). A mathematical theory of natural and artificial selection. Part II. The influence of partial self-fertilisation, inbreeding, assortative mating, and selective fertilisation on the composition of Mendelian populations, and on natural selection. *Proceedings of the Cambridge Philosophical Society: Biological Sciences, 1*, 158–163.
Haldane, J. B. S. (1924b). A mathematical theory of natural and artificial selection—I. *Transactions of the Cambridge Philosophical Society, 23*, 19–41.
Haldane, J. B. S. (1926). A mathematical theory of natural and artificial selection. Part III. *Mathematical Proceedings of the Cambridge Philosophical Society, 23*(4), 363–372.
Haldane, J. B. S. (1927a). A mathematical theory of natural and artificial selection. Part IV. *Mathematical Proceedings of the Cambridge Philosophical Society, 23*(5), 607–615.
Haldane, J. B. S. (1927b). A mathematical theory of natural and artificial selection, Part V. Selection and mutation. *Proceedings of the Cambridge Philosophical Society, 23*(7), 838–844.
Haldane, J. B. S. (1930a). A mathematical theory of natural and artificial selection. Part VI. Isolation. *Proceedings of the Cambridge Philosophical Society, 26*(2), 220–230.
Haldane, J. B. S. (1930b). A note on Fisher's theory of the origin of dominance, and on a correlation between dominance and linkage. *The American Naturalist, 64*(690), 87–90.
Haldane, J. B. S. (1931a). A mathematical theory of natural and artificial selection. Part VII. Selection intensity as a function of mortality rate. In *Mathematical proceedings of the Cambridge Philosophical Society* (pp. 131–136). Cambridge University Press.
Haldane, J. B. S. (1931b). A mathematical theory of natural selection. Part VIII. Metastable populations. *Proceedings of the Cambridge Philosophical Society, 27*(1), 137–142.
Haldane, J. B. S. (1932a). A mathematical theory of natural and artificial selection. Part IX. Rapid selection. *Mathematical Proceedings of the Cambridge Philosophical Society, 28*(2), 244–248.
Haldane, J. B. S. (1932b). On the non-linear difference equation $\Delta xn = k\Phi$ (xn). *Mathematical Proceedings of the Cambridge Philosophical Society, 28*(2), 234–243.
Haldane, J. B. S. (1932c). *The causes of evolution*. Longmans Green and Co.
Haldane, J. B. S. (1937). The effect of variation of fitness. *The American Naturalist, 71*(735), 337–349.
Haldane, J. B. S. (1953). Foreword. "Evolution". *Symposia of the Society for Experimental Biology*, ix–xix.
Haldane, J. B. S. (1957a). Karl Pearson, 1857–1957. *Biometrika, 44*(3/4), 303–313.
Haldane, J. B. S. (1957b). The cost of natural selection. *Journal of Genetics, 55*, 511–524.

Haldane, J. B. S. (1964). A defense of beanbag genetics. *Perspectives in Biology and Medicine, 7*(3), 343–360.
Haldane, J. B. S. (1966). An autobiography in brief. *Perspectives in Biology and Medicine, 9*(4), 476–481.
Haldane, J. B. S., & Moshinsky, P. (1939). Inbreeding in Mendelian populations with special reference to human cousin marriage. *Annals of Eugenics, 9*(4), 321–340.
Haldane, J. B. S., Sprunt, A. D., & Haldane, N. M. (1915). Reduplication in mice (preliminary communication). *Journal of Genetics, 5*, 133–135.
Hall, T. S. (1969). Ideas of life & matter. In *From the enlightenment to the end of the nineteenth century* (Vol. 2). University of Chicago Press.
Hamblin, J. D. (2005). *Science in the early twentieth century: An encyclopedia*. ABC-CLIO.
Hamilton, E., & Cairns, H. (1961). *The collected dialogues of Plato*. Princeton University Press.
Hanks, L. (1966). *Buffon Avant l'Histoire Naturelle*. Presses Univ. France.
Hardin, G. J. (1959). *Nature and man's fate*. Rinehart.
Hardy, G. H. (1908). Mendelian proportions in a mixed population. *Science, 28*(706), 49–50.
Hardy, G. H. (1941). *A mathematician's apology*. Cambridge University Press.
Harman, O. (2010). *The price of altruism: George Price and the search for the origins of kindness*. W. W. Norton.
Harrison, J. (1971). Erasmus Darwin's view of evolution. *Journal of the History of Ideas, 32*(2), 247–264.
Harrow, W. (1895). The Cineraria. *The Gardener's Chronicle, 18*, 17.
Harrow, W. (1896). Home correspondence: The origin of the garden Cineraria. *The Gardener's Chronicle, 20*, 560.
Hartl, D. L., & Orel, V. (1992). What did Gregor Mendel think he discovered? *Genetics, 131*(2), 245–253.
Harvey, W. (1651). *Exercitationes de generatione animalium*. Jansson.
Harvey, W. (1653). *Anatomical exercitations, concerning the generation of living creatures*. Printed by James Young, for Octavian Pulleyn.
Haugen, P. (2014). *Biology: Decade by decade*. Infobase Publishing.
Healy, M. J. R. (2003). R. A. Fisher the statistician. *Journal of the Royal Statistical Society: Series D (The Statistician), 52*(3).
van Helvert, P., & van Wyhe, J. (2021). *Darwin: A companion*. World Scientific.
Hemsley, W. B. (1895). Letter to the Editor: The origin of the cultivated Cineraria. *Nature, 52*(1333), 54–55.
Henig, M. R. (2000). *The monk in the garden: The lost and found genius of Gregor Mendel, the father of genetics*. Houghton Mifflin.
Herbert, S. (1971). Darwin, Malthus, and selection. *Journal of the History of Biology*, 209–217.
Herbert, S. (2005). *Charles Darwin, geologist*. Cornell University Press.
Hercik, F., & Novak, L. (1952). Chybne zaklady mendelismu. *Cs. Biologie, 1*, 254–262.
Herrnstein, R. J., & Murray, C. A. (1994). *The Bell curve: Intelligence and class structure in American life*. Free Press.
Herschel, J. F. W. (1830). *Preliminary discourse on the study of natural philosophy*. Longman.
Hertwig, O. (1875). Beiträge zur Kenntnis der Bildung, Befruchtung und Teilung des tierishchen Eis. I. *Morphologisches Jahrbuch, 1*, 347–437.
Hertwig, O. (1890). Vergleich der Ei-und Samenbildung bei Nematoden. *Archiv für Mikroskopische Anatomie, 36*(1), 1–138.
Heyde, C. C., & Seneta, E. (1972). The simple branching process, a turning point test and a fundamental inequality: A historical note on IJ Bienaymé. *Biometrika, 59*, 680–683.
Hill, W. G. (1990). Sewall Wright: 21 December 1889-3 March 1988. *Biographical Memoirs of Fellows of the Royal Society, 36*, 569–579.
Himmelfarb, G. (1959). *Darwin and the Darwinian revolution*. Doubleday.
Hodge, J. (2010). The Darwin of pangenesis. *Comptes Rendus Biologies, 333*(2), 129–133.

Hodge, M. J. S. (1992). Biology and philosophy (including ideology): A study of Fisher and Wright. In S. Sarkar (Ed.), *The founders of evolutionary genetics: A century reappraisal* (pp. 231–293). Kluwer.
Hodge, M. S. J. (1994). Natural selection: Historical perspectives. In E. F. Keller & E. A. Lloyd (Eds.), *Keywords in evolutionary biology* (pp. 212–219). Harvard University Press.
Hoeveler, J. D. (2007). *The evolutionists: American thinkers confront Charles Darwin, 1860–1920*. Rowman & Littlefield.
Hoffmann, H. (1869). *Untersuchungen zur Bestimmung des Werthes von Species und Varietät: ein Beitrag zur Kritik der Darwin'schen Hypothese*. De Gruyter, Incorporated.
Holmes, S. J. (1916). *Studies in animal behavior*. Badger.
Holterhoff, K. (2014). The history and reception of Charles Darwin's hypothesis of pangenesis. *Journal of the History of Biology, 47*, 661–695.
Honenberger, P. (2018). Darwin among the philosophers: Hull and Ruse on Darwin, Herschel, and Whewell. *The Journal of the International Society for the History of Philosophy of Science, 8*(2), 278–309.
Hooker, J. D. (1853). *The botany of the Antarctic voyage. II. Flora Novae-Zealandiae*. Lovell Reeve.
Hooker, J. D. (1860). On the origination and distribution of species: Introductory essay to the flora of Tasmania. *American Journal of Science and Arts (1820–1879), 29*(85), 1–25, 305–326.
Hooker, J. D. (1868). Presidential address. *British Association Report (Norwich)*, lviii–lxxv.
Hooper, J. (2002). *Of moths and men: An evolutionary tale: The untold story of science and the peppered moth*. WW Norton & Company.
Hooykaas, R. (1963). *Natural law and divine miracle: The principle of uniformity in geology, biology and theology*. Brill.
Hoppe, B. (1971). Die Beziehung zwischen J. G. Mendel und C. W. Nägeli auf Grund neuer Dokumente. *Folia Mendeliana, 6*, 123–138.
Hubby, J. L., & Lewontin, R. C. (1966). A molecular approach to the study of genic heterozygosity in natural populations. I. The number of alleles at different loci in Drosophila pseudoobscura. *Genetics, 54*, 577–594.
Hull, D. L. (1965a). The effect of essentialism on taxonomy – two thousand years of stasis (I). *The British Journal for the Philosophy of Science, 15*, 314–326.
Hull, D. L. (1965b). The effect of essentialism on taxonomy – two thousand years of stasis (II). *The British Journal for the Philosophy of Science, 16*, 1–18.
Hull, D. L. (1973). *Darwin and his critics*. Harvard University Press.
Hull, D. L. (1974). *Philosophy of biological science*. Prentice-Hall.
Hull, D. L. (2010). *Science as a process: An evolutionary account of the social and conceptual development of science*. University of Chicago Press.
Hull, F. H. (1946). Regression analyses of corn yield data. *Genetics, 31*, 219–219.
Huneman, P., & Walsh, D. M. (Eds.). (2017). *Challenging the modern synthesis: Adaptation, development, and inheritance*. Oxford University Press.
Hutton, J. (1785). *Abstract of a dissertation read in the Royal Society of Edinburgh upon the seventh of March and fourth of April MDCCLXXXY, concerning the system of the earth, its duration and stability*. Edinburgh.
Hutton, J. (1788). Theory of the Earth; or an investigation of the laws observable in the composition, dissolution, and restoration of land upon the Globe. *Transactions of the Royal Society of Edinburgh, 1*(Part 2), 209–304.
Hutton, J. (1795). *Theory of the Earth: With proofs and illustrations*. Printed for Messrs. Cadell, Junior, and Davies, London, and William Creech, Edinburgh.
Huxley, J. (1942). *Evolution: The modern synthesis*. George Allen & Unwin, Ltd.
Huxley, L. (1900). *Life and letters of Thomas Henry Huxley* (Vol. I). Macmillan.
Huxley, L. (1918). *Life and letters of Sir Joseph Dalton Hooker* (Vol. II). D. Appleton & Co.
Huxley, T. H. (1854). Vestiges of the natural history of creation. *British and Foreign Medico-Chirurgical Review, 26*, 425–439.

Huxley, T. H. (1859, December 26). Darwin on the origin of species. *The Times*, p. 9.
Huxley, T. H. (1860). Darwin on the origin of species. *Westminster Review, 17*(2), 541–570.
Huxley, T. H. (1863). *Evidence as to man's place in nature*. Appleton & Co.
Huxley, T. H. (1864). Criticisms on "the origin of species". *Natural History Review, 4*, 566–580.
Huxley, T. H. (1869). President's annual address. *Quarterly Journal of the Geological Society of London, 25*, 305–339.
Huxley, T. H. (1871). Mr. Darwin's critics. *Contemporary Review, 18*, 443–476.
Iltis, H. (1932). *Life of Mendel*. George Allen and Unwin.
Indian Statistical Institute. (1992). *J. B. S. Haldane: A tribute*. Indian Statistical Institute.
Irvine, W. (1955). *Apes, angels, and Victorians*. Weidenfeld and Nicolson.
Ishida, Y. (2017). Sewall Wright, shifting balance theory, and the hardening of the modern synthesis. *Studies in History and Philosophy of Biological and Biomedical Sciences, 61*, 1–10.
Jablonka, E., & Lamb, M. J. (1995). *Epigenetic inheritance and evolution: The Lamarckian dimension*. Oxford University Press.
Jablonka, E., & Lamb, M. J. (2020). *Inheritance systems and the extended evolutionary synthesis*. Cambridge University Press.
Jackson, B. D. (1881). *Guide to the literature of botany: Being a classified selection of botanical works, including nearly 6000 titles not given in Pritzel's thesaurus*. Longmans, Green & Company.
Jackson, P. W. (2006). *The chronologers' quest: The search for the age of the earth*. Cambridge University Press.
Jäger, H. (1877). *Lehrbuch der gartenkunst: oder Lehr von der Anlage Ausschmückung und Künstlerischen unterhalung der Gärten und freien Anlagen*. Voigt.
Janet, A. (1895). Considérations méchaniques sur l'évolution et le problème des espèces. In *Comptes Rendue des 3me Congrès International de Zoologie. 3me Congres International de Zoologie, Leyden* (pp. 136–145).
Janssens, F. A. (1909). La Théorie de la Chiasmatypie. Nouvelle interprétation des cinèses de maturation. *La Cellule, 25*, 389–411.
Jenkin, F. (1867). The origin of species. *North British Review, 46*(92), 277–318.
Jennings, H. S. (1912). Production of pure homozygotic organisms from heterozygotes by self-fertilization. *The American Naturalist, 46*(548), 487–491.
Jennings, H. S. (1914). Formulae for the results of inbreeding. *The American Naturalist, 48*(575), 693–696.
Jennings, H. S. (1916). The numerical results of diverse systems of breeding. *Proceedings of the National Academy of Sciences, 2*(1), 45–50.
Jennings, H. S. (1917). The numerical results of diverse systems of breeding, with respect to two pairs of characters, linked or independent, with special relation to the effects of linkage. *Genetics, 2*, 97–154.
Johannsen, W. (1896). *Om Arvelighed og Variabilitet*. Copenhagen.
Johannsen, W. (1902). Om Organismernes Formrigdom. *Nordisk Tidskrift för Vetenskap (Letterstedske Tidskrift)*, 545–565.
Johannsen, W. (1903). *Ueber Erblichkeit in Populationen und in reinen Linien: ein Beitrag zur Beleuchtung schwebender* Selektionsfragen (Selections trans. in Johannsen (1955)). Johann. G. Fischer.
Johannsen, W. (1909). *Elemente der exakten Erblichkeithere mit Grundsatzen der biologische Variationsstatitik*. Gustav Fischer.
Johannsen, W. (1911). The genotype conception of heredity. *The American Naturalist, 45*(531), 129–159.
Johannsen, W. (1913). *Elemente der exakten Erblichkeitslehre* (Vol. XII, 2nd ed.). G. Fischer.
Johannsen, W. (1955). Concerning heredity in populations and in pure lines (Eng. transl. by H. Gall & E. Putschar of selected portions of Johannsen (1903)). In *Selected readings in biology for natural sciences* (Vol. 3, pp. 172–215). University of Chicago Press.

Johnson, C. N. (2019). *Darwin's historical sketch: An examination of the 'preface' to the origin of species*. Oxford University Press.
Jones, S. (2000). *Darwin's ghost: The origin of species updated*. Ballantine Books.
Jones, W. H. S. (1923). *Hippocrates' air, waters and places*. Heineman.
Joravsky, D. (1970). *The Lysenko affair*. Harvard University Press.
Jordanova, L. J. (1984). *Lamarck*. Oxford University Press.
Kant, I. (1781). *Kritik der reinen Vernunft*. Johann Friedrich Hartknoch.
Karlin, S. (1992). R. A. Fisher and evolutionary theory. *Statistical Science, 7*(1), 13–33.
Kastner, J. (1977). *A species of eternity*. Alfred A. Knopf.
Keeler, C. (1968). Some oddities in the delayed appreciation of "Castle's law". *Journal of Heredity, 59*(2), 110–112.
Keith, A. (1955). *Darwin revalued*. Watts.
Kellogg, V. L. (1907). *Darwinism to-day: A discussion of present-day scientific criticism of the Darwinian selection theories, together with a brief account of the principal other proposed auxiliary and alternative theories of species-forming*. H. Holt and Cpmpany.
Kempthorne, O. (1954). The correlation between relatives in a random mating population. *Proceedings of the Royal Society of London. Series B-Biological Sciences, 143*(910), 103–113.
Kenny, D. A. (1979). *Correlation and causality*. Wiley.
Kettlewell, B. (1973). *Evolution of Melanism: The study of a recurring necessity*. Clarendon Press.
Kettlewell, H. B. D. (1956). A resume of investigations on the evolution of melanism in the Lepidoptera. *Proceedings of the Royal Society of London. Series B-Biological Sciences, 145*(920), 297–303.
Kettlewell, H. B. D. (1961). The phenomenon of industrial melanism in Lepidoptera. *Annual Review of Entomology, 6*(1), 245–262.
Keynes, M. (Ed.). (1993). *Sir Francis Galton, FRS: The legacy of his ideas*. Macmillan.
Keynes, M., Edwards, A. W. F., & Peel, R. (Eds.). (2004). *A century of Mendelism in human genetics*. CRC Press.
Kim, K. M. (1994). *Explaining scientific consensus: The case of Mendelian genetics*. Guilford Press.
Kimura, M. (1953). "Stepping stone" model of population. *Annual Report of the National Institute of Genetics (Japan), 3*, 62–63.
Kimura, M. (1955). Solution of a process of random genetic drift with a continuous model. *Proceedings of the National Academy of Sciences, 41*(3), 144–150.
Kimura, M. (1956). A model of a genetic system which leads to closer linkage by natural selection. *Evolution*, 278–287.
Kimura, M. (1958). On the change of population fitness by natural selection. *Heredity, 12*(2), 145–167.
Kimura, M. (1960). Optimum mutation rate and degree of dominance as determined by the principle of minimum genetic load. *Journal of Genetics, 57*(1), 21–34.
Kimura, M. (1964). Diffusion models in population genetics. *Journal of Applied Probability, 1*(2), 177–232.
Kimura, M. (1968a). Evolutionary rate at the molecular level. *Nature, 217*(5129), 624–626.
Kimura, M. (1968b). Haldane's contributions to the mathematical theories of evolution and population genetics. In K. R. Dromanraju (Ed.), *Haldane and modern biology* (pp. 133–140). John Hopkins Press.
Kimura, M. (1983). *The neutral theory of molecular evolution*. Cambridge University Press.
Kimura, M., & Ohta, T. (1972). Population genetics, molecular biometry, and evolution. *Proceedings of the Sixth Berkeley Symposium on Mathematical Statistics and Probability, 5*, 43–68.
King, J. L., & Jukes, T. H. (1969). Non-Darwinian evolution: Most evolutionary change in proteins may be due to neutral mutations and genetic drift. *Science, 164*(3881), 788–798.
King, R. C., Stansfield, W. D., & Mulligan, P. K. (2006). *A dictionary of genetics* (7th ed.). Oxford University Press.
King-Hele, D. G. (1963a). Dr. Erasmus Darwin and the theory of evolution. *Nature, 200*, 304–306.

King-Hele, D. G. (1963b). *Erasmus Darwin*. Charles Scribner's.
Kingman, J. F. C. (1961). On an inequality in partial averages. *The Quarterly Journal of Mathematics, 12*(1), 78–80.
Kingman, J. F. C. (1982a). On the genealogy of large populations. *Journal of Applied Probability, 19*(A), 27–43.
Kingman, J. F. C. (1982b). The coalescent. *Stochastic Processes and their Applications, 13*(3), 235–248.
Kingsbury, N. (2011). *Hybrid: The history and science of plant breeding*. University of Chicago Press.
Kirwan, R. (1794). *Elements of mineralogy, with considerable improvements and additions* (Vol. 1, 2nd ed.). J. Nichols for P. Elmsly.
Klein, J. (2000). Johann Mendel's field of dreams. *Genetics, 156*(1), 1–6.
Klein, J., & Klein, N. (2013). *Solitude of a humble genius—Gregor Johann Mendel: Volume 1*. Springer.
Knight, T. A. (1799). An account of some experiments on the fecundation of vegetables. In a letter from Thomas Andrew Knight, Esq. to the right Hon. Sir Joseph Banks, KBPRS. *Philosophical Transactions of the Royal Society of London, 89*, 195–204.
Koerner, L. (1999). *Linnaeus: Nature and nation*. Harvard University Press.
Kohler, R. E. (1994). *Lords of the Fly: Drosophila genetics and the experimental life*. University of Chicago Press.
Kohn, D. (1981). On the origin of the principle of diversity. *Science, 213*(4512), 1105–1108.
Kohn, D. (Ed.). (1985). *The Darwinian heritage*. Princeton University Press.
Kojima, K.-I., & Kelleher, T. M. (1961). Changes of mean fitness in random mating populations when epistasis and linkage are present. *Genetics, 46*(5), 527.
Kolmogorov, A. N. (1931). Über die analytischen Methoden in der Wahrscheinlichkeitstheorie. *Mathematische Annalen, 104*, 415–458.
Kölreuter, J. G. (1761). *Vorläufige Nachricht von einigen das Geschlecht der Pflanzen betreffenden Versuchen und Beobachtungen*. Gleditsch.
Kölreuter, J. G. (1763). *Fortsetzung Der vorläufigen Nachricht von einigen das Geschlecht der Pflanzen betreffenden Versuchen und Beobachtungen*. Gleditsch.
Kölreuter, J. G. (1764). *Zweyte Fortsetzung der vorläufigen Nachricht von einigen das Geschlecht der Pflanzen betreffenden Versuchen und Beobachtungen, von Joseph Gottlieb Kölreuter*. Gleditsch.
Kölreuter, J. G. (1766). *Dritte Fortsetzung der vorläufigen Nachricht von einigen das Geschlecht der Pflanzen betreffenden Versuchen*. Gleditsch.
Kovalchuk, I., & Kovalchuk, O. (2012). *Epigenetics in health and disease*. FT Press.
Krumbhaar, E. B. (1931). Erasmus Darwin and his relation to the doctrine of evolution. *Annals of Medical History, 3*(5), 487–500.
Kruta, V., & Orel, V. (1974). Johann Gregor Mendel. In C. C. Gillispie (Ed.), *Dictionary of scientific biography* (pp. 277–283). Charles Scribner's.
Lamarck, J.-B.-P.-A. (1801). *Système des animaux sans vertèbres ou tableau général des classes, des ordres et des genres de ces animaux*. Author and Deterville.
Lamarck, J.-B.-P.-A. (1802). *Hydrogeologie, ou Recherches sur l'Influence qu'ont les Eaux sur la Surface du Globe Terrestre*. L'Auteur (Eng. transl. in Carozzi (1964)).
Lamarck, J.-B.-P.-A. (1809). *Philosophie Zoologique* (Eng. transl. by Hugh Elliot in Lamarck (1984)). G. Baillière.
Lamarck, J.-B.-P.-A. (1984). *Zoological philosophy: An exposition with regard to the natural history of animals (Eng. transl. by Hugh Elliot of Lamarck (1809))*. The University of Chicago Press.
Lamarck, J.-B.-P.-A., & Drouin, J. M. (1755). *Recherches sur l'organisation des corps vivants: précédé du Discours d'ouverture du cours de zoologie donné dans le Museum d'histoire naturelle*. Fayard.

Langridge, J. (1987). Old and new theories of evolution. In K. S. Campbell & M. F. Day (Eds.), *Rates of evolution* (pp. 248–262). Allen & Urwin.
Lankester, E. R. (1884). Mental evolution in animals (Reply to Romanes). *The Athenaeum, 1*(2944), 412.
Leibniz, G. W. (1765). Nouveaux essais sur l'entendement humain. In *Oeuvres philosophiques latines & françoises de feu Mr. de Leibnitz, tirées de ses manuscrits qui se conservent dans la Bibliothèque Royale à Hanovre, publiées par Rud. Eric Raspe, avec une préface de M. Kaestner* (pp. 1–496). Jean Schreuder.
Lenay, C. (2000). Hugo De Vries: From the theory of intracellular pangenesis to the rediscovery of Mendel. *Comptes Rendus de l'Académie des Sciences-Series III-Sciences de la Vie, 323*(12), 1053–1060.
Lerner, I. M. (1954). *Genetic homeostasis*. Oliver & Boyd.
Lessard, S. (1997). Fisher's fundamental theorem of natural selection revisited. *Theoretical Population Biology, 52*(2), 119–136.
Levin, S. M. (1966). Malthus and the idea of progress. *Journal of the History of Ideas, 27*(1), 92–108.
Lewontin, R. C., & Hubby, J. L. (1966). A molecular approach to the study of genic heterozygosity in natural populations. II. Amount of variation and degree of heterozygosity in natural populations of Drosophila pseudoobscura. *Genetics, 54*(2), 595–609.
Lewontin, R. C., & Kojima, K.-I. (1960). The evolutionary dynamics of complex polymorphisms. *Evolution*, 458–472.
Ley, W. (1968). *Dawn of zoology*. Prentice-Hall.
Li, C. C. (1948). *An introduction to population genetics*. National Peking University Press.
Li, C. C. (1955). *Population genetics*. University of Chicago Press.
Li, C. C. (1967a). Castle's early work on selection and equilibrium. *American Journal of Human Genetics, 19*(1), 70–74.
Li, C. C. (1967b). Fundamental theorem of natural selection. *Nature, 214*(5087), 505–506.
Li, C. C. (1975). *Path analysis: A primer*. The Boxwood Press.
Liang, G. H., et al. (1998). The discovery of Mendel's Laws. In *Discoveries in plant biology* (Vol. I, pp. 145–162). World Scientific.
Limoges, C. (1970). *La sélection naturelle. Étude sur la première constitution d'un concept (1837–1859)*. Presses Universitaires de France.
Lincoln, R., Boxshall, G., & Clark, P. (1982). *A dictionary of ecology, evolution and systematics*. Cambridge University Press.
Lindley, J. (1841). *Elements of botany, structural, physiological, systematical, and medical. Being a fourth edition of the outline of the first principles of botany*. Taylor & Watson.
Linnaeus, C. (1762). *Fundamentum fructificationis (JM Grâberg.)*. University of Upasala.
Linnaeus, C. V. (1735). *Systema naturæ regna tria, secundum classes, ordines, genera, species, cum characteribus, differentiis, synonymis, locis* (2nd ed.). Kiesewetter.
Linnaeus, C. V. (1736). *Fundamenta botanica quæ majorum operum prodromi instar theoriam scientiæ botanices per breves aphorismos tradunt*. apud S. Schouten.
Linnaeus, C. V. (1751). *Plantae hybridae*. J. J. Haartman.
Lints, F. A., & Delcour, J. (1968). Galton and the Mendelian ratios. *Heredity, 23*(1), 153–155.
Liu, Y. (2018a). *Darwin's pangenesis and its rediscovery part A, advances in genetics*. Academic Press.
Liu, Y. (2018b). *Darwin's pangenesis and its rediscovery part B, advances in genetics*. Academic Press.
Lock, R. H. (1909). *Recent progress in the study of variation, heredity, and evolution* (2nd ed.). J. Murray.
Locke, J. (1690). *An essay concerning human understanding*. William Coke, 1765.
Lonie, I. M. (1981). *The hippocratic treatises 'on generation', 'on the nature of the child', 'diseases IV': A commentary*. De Gruyter.

Lotka, A. J. (1913). A natural population norm. I. *Journal of the Washington Academy of Sciences, 3*(9), 241–248.
Lotka, A. J. (1918). The relation between birth rate and death rate in a normal population and the rational basis of an empirical formula for the mean length of life given by William Farr. *Quarterly Publications of the American Statistical Association, 16*(123), 121–130.
Loudon, J. (1842). On the history and culture of the greenhouse Cineraria. *The Ladies Magazine of Gardening, 1*, 111–113.
Loudon, J. C. (1829). *An encyclopaedia of plants*. Longman.
Lovejoy, A. O. (1936). *The great chain of being; a study of the history of an idea*. Harvard University Press.
Loveland, J. (2001). *Rhetoric and natural history: Buffon in polemical and literary context, studies on Voltaire and the eighteenth century*. Voltaire Foundation.
Lutz, F. E. (1941). *A lot of insects. Entomology in a suburban garden*. Putman.
Lyell, C. (1830). *Principles of geology: Being an attempt to explain the former changes of the earth's surface, by reference to causes now in operation* (Vol. I). J. Murray.
Lyell, C. (1832). *Principles of geology: Being an attempt to explain the former changes of the earth's surface, by reference to causes now in operation* (Vol. II). J. Murray.
Lyell, C. (1863). *The geological evidences of the antiquity of man*. J. Murray.
Lyell, C. (1881). *Life, letters and journals of Sir Charles Lyell, bart*. J. Murray.
Lynch, R. I. (1897). Origin of the cultivated Cineraria. *Nature, 55*(1424), 341–342.
Lynch, R. I. (1899). Hybrid Cinerarias. *Journal of the Royal Horticultural Society, 24*, 269–272.
MacCurdy, H., & Castle, W. E. (1907). Selection and cross-breeding in relation to the inheritance of coat-pigments and coat-patterns in rats and guinea-pigs. *Science, 70*, 1–50.
MacDowell, E. C. (1914). Size inheritance in rabbits. *Carnegie Institution of Washington, 196*, 1–49.
MacDowell, E. C. (1916). Piebald rats and multiple factors. *The American Naturalist, 50*(600), 719–742.
Magnello, E. (2001). Karl Pearson. In C. C. Heyde & E. Seneta (Eds.), *Statisticians of the centuries* (pp. 248–256). Springer.
Magnello, E. (2004a). The reception of Mendelism by the biometricians and the early Mendelians (1899–1909). In M. Keynes, A. W. F. Edwards, & R. Peel (Eds.), *A century of Mendelism in human genetics* (pp. 19–32). CRC Press.
Magnello, E. (2004b). Weldon, Walter Frank Raphael (1860–1906). In H. C. G. Matthew & B. Harrison (Eds.), *Oxford dictionary of national biography* (pp. 989–992). Oxford University Press.
de Maillet, B. (1748). *Telliamed: ou Entretiens d'un philosophe indien avec un missionnaire françois sur la diminution de la mer, etc* (Vol. 2). L'Honoré et Fils.
de Maistre, J. (1821) Les soirées de Saint-Pétersbourg ou entretiens sur le gouvernement temporel de la Providence: suivis d'un traité sur les sacrifices. Tome premier. : Librairie grecque, latine et française.
Majerus, M. E. N. (1998). *Melanism: Evolution in action*. Oxford University Press.
Majumder, P. P. (Ed.). (1993). *Human population genetics: A centennial tribute to JBS Haldane*. Plenum Press.
Malebranche, N. (1674). *De la recherche de la vérité* (Vol. I). Chez André Pralard.
Malécot, G. (1938). Sur les aléatoires mendéliennes et les corrélations de l'hérédité. *Comptes Rendus Académie des Sciences, 206*, 404–406.
Malécot, G. (1939). *Théorie mathématique de l'hérédité mendélienne généralisée*.
Malécot, G. (1941). Etude mathématique des populations "mendeliennes". *Annales de l'Universite de Lyon, Sciences, Section, 4*, 45–60.
Malécot, G. (1946). La consanguinité dans une population limitée. *Comptes Rendus Académie des Sciences, 222*(81), 1–81.
Malécot, G. (1948). *Les mathématiques de l'hérédité*. Maison et Cie.
Malécot, G. (1969). *The mathematics of heredity*. W H Freeman.

Malthus, T. R. (1798). *An essay on the principle of population, as it affects the future improvement of society*. J. Johnson.
Manier, E. (1978). *The young Darwin and his cultural circle*. Springer.
Marchant, J. (1916). *Alfred Russel Wallace: Letters and reminiscences* (Vol. 2). Harper.
Marchant, L., Cuvier, G., & Pfaff, C. M. (1858). *Lettres de Georges Cuvier a C.M. [sic] Pfaff sur l'histoire naturelle, la politique et la Littérature 1788–1792*. Masson.
Marsden, T. (1865). The sacred steps of creation; or, the revealed genetic theology illustrated by geology and astronomy. In *Twelve lectures*. Longman, Green, Longman, Roberts, and Green.
Martini, S. (1961). Giorgio Gallesio, pomologist and precursor of Gregor Mendel. *Fruit Varieties and Horticultural Digest, 15*(3), 53–54.
Mather, K. (1943). Polygenic inheritance and natural selection. *Biological Reviews, 18*(1), 32–64.
Mather, K. (1963). Fisher and genetics. *Journal of the Royal Statistical Society, Series A, 216*, 166–168.
Mather, K. (1964). R. A. Fisher's work in genetics. *Biometrics, 20*, 330–342.
Matthew, P. (1831). *On naval timber and arboriculture: With critical notes on authors who have recently treated the subject of planting*. Longman.
Matthew, P. (1860). Nature's law of selection. *Gardeners' Chronicle and Agricultural Gazette*, 312–313.
de Maupertuis, P.-L. M. (1751). *Systme de la nature. (In Oeuvres (1788), 2: 137-168. 4 vols.)*. Lyon: Jean-Marie Bruyset, 1756 (Originally published under the pseudonym Dr. Baumann of Erlangen and titled Dissertatio inauguralis metaphysica de universali naturae sytemate).
Mawer, S. (2006). *Gregor Mendel: Planting the seeds of genetics*. Abrams.
Mayer, W. V. (1987). Wallace and Darwin. *The American Biology Teacher, 49*(8), 406–410.
Mayr, E. (1940). Speciation phenomena in birds. *The American Naturalist, 74*(752), 249–278.
Mayr, E. (1942). *Systematics and the origin of species, from the viewpoint of a zoologist*. Harvard University Press.
Mayr, E. (1955). Integration of genotypes: Synthesis. *Cold Spring Harbor Symposia on Quantitative Biology, 20*, 327–333.
Mayr, E. (1959a). Typological versus population thinking. In B. J. Meggers (Ed.), *Evolution and anthropology: A centennial appraisal* (pp. 409–412). The Anthropological Society of Washington.
Mayr, E. (1959b). Where are we? *Cold Spring Harbor Symposia on Quantitative Biology*, 1–14.
Mayr, E. (1963). *Animal species and evolution*. Harvard University Press.
Mayr, E. (1973). The recent historiography of genetics. *Journal of the History of Biology, 6*(1), 125–154.
Mayr, E. (1982). *The growth of biological thought: Diversity, evolution, and inheritance*. Harvard University Press.
Mayr, E. (1985). Weismann and evolution. *Journal of the History of Biology, 18*, 295–329.
Mayr, E. (1990). When is historiography Whiggish? *Journal of the History of Ideas, 51*(2), 301–309.
Mayr, E., & Provine, W. B. (Eds.). (1980). *The evolutionary synthesis: Perspectives on the unification of biology*. Harvard University Press.
McCalman, I. (2010). *Darwin's Armada: Four voyages and the battle for the theory of evolution*. Pocket Books.
McCoy, J. W. (1979). The origin of the "adaptive landscape" concept. *The American Naturalist, 113*(4), 610–613.
McKinney, H. L. (1972). *Wallace and natural selection*. Yale University Press.
McNeil, M. (1987). *Under the banner of science: Erasmus Darwin and his age*. Manchester University Press.
Medvedev, Z. A. (1969). *The rise and fall of T D Lysenko*. Columbia University Press.
Mendel, G. (1866). Versuch Uber Pflanzen-Hybriden (Eng. transl. in Mendel (1966)). *Verhandlungen des naturforschenden Vereines in Brunn, 4*, 3–47.

Mendel, G. (1869). Uber einige aus künstlichen Befruchtung gewonnenen Hieracium-Bastarde. *Verhandlungen des naturforschenden Vereines in Brunn, 8*, 26–31.

Mendel, G. (1901). Experiments in plant hybridisation (Eng. transl. by the Journal of the Horticultural Society of Mendel (1866)). *Journal of the Horticultural Society, XXVI*(1), 1–32.

Mendel, G. (1966a). Experiments on plant hybrids (Eng. transl. by Eva R Sherwood of Mendel (1866)). In C. Stern & E. R. Sherwood (Eds.), *The origin of genetics: A Mendel source book* (pp. 1–48). W H Freeman & Co.

Mendel, G. (1966b). On Hieracium-Hybrids obtained by artificial fertilization (Eng. transl. from Bateson (1902) of Mendel (1869)). In C. Stern & E. R. Sherwood (Eds.), *The origin of genetics: A Mendel source book* (pp. 49–55). W H Freeman & Co.

Merrell, D. J. (1981). *Ecological genetics*. University of Minnesota Press.

Meyer, A. W. (1939). *The rise of embryology*. Stanford University Press.

Millhauser, M. (1959). *Just before Darwin: Robert Chambers and Vestiges*. Wesleyan University Press.

Milne-Edwards, H. (1851). *Introduction à la Zoologie Générale, ou, Considérations sur les Tendances de la Nature Dans la Constitution du Règne Animal*. Masson.

Milner, R. (2009). *Darwin's universe: Evolution from A to Z*. University of California Press.

Mivart, S. G. J. (1869). Difficulties of the theory of natural selection. *The Month, 11*, 35–53.

Mivart, S. G. J. (1871a). Darwin's descent of man. *Quarterly Review, 131*, 47–90.

Mivart, S. G. J. (1871b). Evolution and its consequences: A reply to Professor Huxley. *Contemporary Review, 19*, 168–197.

Mivart, S. G. J. (1871c). *On the genesis of species*. Macmillan.

Mivart, S. G. J. (1874). Primitive man: Tylor and Lubbock. *Quarterly Review, 137*, 40–77.

Mivart, S. G. J. (1876). *Lessons from nature as manifested in mind and matter*. Murray.

Mivart, S. G. J. (1897). Some reminiscences of Thomas Henry Huxley. *Nineteenth Century, 42*, 985–998.

Monaghan, F., & Corcos, A. (1985). Chi-square and Mendel's experiments: where's the bias? *Journal of Heredity, 76*(4), 307–309.

Monaghan, F. V., & Corcos, A. F. (1990). The real objective of Mendel's paper. *Biology and Philosophy, 5*, 267–292.

Monaghan, F. V., & Corcos, A. F. (1993). The real objective of Mendel's paper: A response to Falk and Sarkar's criticism. *Biology and Philosophy, 8*, 95–98.

Moore, J. A. (1983). Thomas Hunt Morgan – The geneticist. *Integrative and Comparative Biology, 23*, 862–863.

Moore, J. A. (1999). *Science as a way of knowing*. Harvard University Press.

Moran, P. A. P., & Smith, C. A. B. (1966). Commentary on RA Fisher's paper on the correlation between relatives on the supposition of Mendelian inheritance. *Eugenics Laboratory Memoirs, 41*, 1–62.

Morgan, T. H. (1903). *Evolution and adaptation*. Macmillan.

Morgan, T. H. (1907). Sex-determining factors in animals. *Science, 25*(636), 382–384.

Morgan, T. H. (1909). For Darwin. *Popular Science Monthly, 74*, 367–380.

Morgan, T. H. (1910a). Chromosomes and heredity. *The American Naturalist, 44*(524), 449–496.

Morgan, T. H. (1910b). Hybridization in a mutating period in Drosophila. *Proceedings of the Society for Experimental Biology and Medicine, 7*(5), 160–161.

Morgan, T. H. (1910c). Sex limited inheritance in Drosophila. *Science, 32*(812), 120–122.

Morgan, T. H. (1911). Random segregation versus coupling in Mendelian inheritance. *Science, 34*(873), 384.

Morgan, T. H. (1912). Some books on evolution (anonymous review). *Nation*, 542–544.

Morgan, T. H. (1916). *A critique of the theory of evolution*. Princeton University Press.

Morgan, T. H. (1942). Genesis of the white-eyed mutant. *Journal of Heredity, 33*(3), 91–92.

Morgan, T. H., & Bridges, C. B. (1916). Sex-linked inheritance in Drosophila. *Carnegie Institution of Washington Publication, 237*, 1–87.

Morgan, T. H., & Cattell, E. (1912). Data for the study of sex-linked inheritance in Drosophila. *Journal of Experimental Zoology, 13*(1), 79–101.
Morgan, T. H., & Lynch, C. J. (1912). The linkage of two factors in Drosophila that are not sex-linked. *The Biological Bulletin, 23*(3), 174–182.
Morgan, T. H., Sturtevant, A. H., & Bridges, C. B. (1915). *The mechanism of Mendelian heredity*. Holt.
Morris, S. W. (1994). Fleeming Jenkin and the origin of species: A reassessment. *The British Journal for the History of Science, 27*(3), 313–343.
Morton, N. E. (2008). Commentary: Growth of beanbag genetics. *International Journal of Epidemiology, 37*(3), 445–446.
Morton, N. E., Crow, J. F., & Muller, H. J. (1956). An estimate of the mutational damage in man from data on consanguineous marriages. *Proceedings of the National Academy of Sciences, 42*(11), 855–863.
Mukherjee, S. (2016). *The gene: An intimate history*. Scribner.
Müller, G. H. (1983). First use of biologie. *Nature, 302*(5911), 744.
Muller, H. J. (1914). The bearing of the selection experiments of Castle and Phillips on the variability of genes. *The American Naturalist, 48*(573), 567–576.
Muller, H. J. (1918). Genetic variability, twin hybrids and constant hybrids, in a case of balanced lethal factors. *Genetics, 3*(5), 422–499.
Muller, H. J. (1939). Reversibility in evolution considered from the standpoint of genetics. *Biological Reviews, 14*(3), 261–280.
Muller, H. J. (1950). Our load of mutations. *American Journal of Human Genetics, 2*, 111–176.
Müller-Wille, S., & Rheinberger, H.-J. (2012). *A cultural history of heredity*. University of Chicago Press.
Nabours, R. K. (1917). Studies of inheritance and evolution in orthoptera. II. *Journal of Genetics, 7*(1), 1–46.
Nabours, R. K. (1925). *Studies of inheritance and evolution in orthoptera. V. The grouse locust, Apotettix eurycephalus Hancock*. B.P. Walker.
Nägeli, C. (1842). Botanische Beitrage. *Linnaea, 16*, 237–285.
Nägeli, C. (1844). Zellenkerne, Zellenbildung und Zellenwachsthum bei den Pflanzen. *Zeitschrift für Wissenschaftliche Botanik, 1*, 34–134.
Nägeli, C. (1856). *Die Individualität in der Natur: mit vorzüglicher Berücksichtigung des Pflanzenreiches*. Meyer & Zeller.
Nägeli, C. (1858). *Die Stärkekörner: morphologische, physiologische, chemisch-physicalische und systematisch-botanische Monographie*. Friedrich Schulthess.
Nägeli, C. (1863). *Die Anwendung des Polarisationsapparates auf die Untersuchung vegetabilischer Elementartheile*. Leipzig.
Nägeli, C. (1865). *Entstehung und Begriff der naturhistorischen Art*. Königliche Akademie.
Nägeli, C. (1879). *Theorie der Gärung: ein Beitrag zur Molekularphysiologie*. De Gruyter Oldenbourg.
Nägeli, C. (1884). *Mechanisch-physiologische Theorie der Abstammungslehre*. Oldenburg.
Nägeli, C. (1898). *A mechanico-physiological theory of organic evolution* (English summary by V.A. Clark). Open Court Publishing Company.
Nägeli, C., & Peter, A. (1885). *Die Hieracien Mittel-Europas: Monographische Bearbeitung der Piloselloiden mit Besonderer Berücksichtigung der Mitteleuropäischen Sippen*. Oldenbourg.
Nägeli, C., & Schwendener, S. (1877). *Das Mikroskop, Theorie und Anwendung desselben*. Engelman.
Nagylaki, T. (1989). Gustave Malécot and the transition from classical to modern population genetics. *Genetics, 122*(2), 253–268.
Naudin, C. (1852). Considérations philosophiques sur l'espèce et la variété. *Revue Horticole, 1*(4), 102–109.
Naudin, C. (1863). Nouvelles recherches sur l'hybridité dans les végétaux. *Annales des Sciences Naturelles (Botanique)*, 180–203.

Naudin, C. (1865). Petunia violaceo-nyctaginiflora. Nouvelles Recherches sur l'Hybridité dans les Végétaux. *Nouvelles archives du Muséum d'histoire naturelle, 1*, 25–175.
Neal, D. (2004). *Introduction to population biology*. Cambridge University Press.
Needham, J. (1959). *A history of embryology* (2nd ed.). Abelard-Schuman.
Nei, M. (2013). *Mutation-driven evolution*. Oxford University Press.
Nilsson-Ehle, H. (1907). Om nordskandinaviska och andra tidiga hafresorter och försök till deras bättrande genom individualförädling och korsning. *Sveriges Utsädesförenings Tidskrift, 17*, 209–218.
Nilsson-Ehle, H. (1908). *Einige Ergebnisse von Kreuzungen bei Hafer und Weizen*. Berlingska Boktryckeriet.
Nilsson-Ehle, H. (1909). *Kreuzungsuntersuchungen an hafer und weizen*. H. Ohlssons buchdruckerei.
Ninio, J. (2014). *Molecular approaches to evolution*. Princeton University Press.
Noble, G. K. (1926). Kammerer's alytes. *Nature, 118*(2971), 518.
Nordenskiöld, E. (1928). *The history of biology: A survey*. A.A. Knopf.
Norman, A. (2013). *Charles Darwin: Destroyer of myths*. Pen and Sword.
Norton, B., & Pearson, E. S. (1976). A note on the background to, and refereeing of, RA Fisher's 1918 paper "On the correlation between relatives on the supposition of Mendelian inheritance". *Notes and Records of the Royal Society of London, 31*(1), 151–162.
Norton, H. T. J. (1928). Natural selection and Mendelian variation. *Proceedings of the London Mathematical Society, 2*(1), 1–45.
Novitski, C. E. (1995). Another look at some of Mendel's results. *The Journal of Heredity, 86*, 62–65.
Novitski, C. E. (2004a). Revision of Fisher's analysis of Mendel's garden pea experiments. *Genetics, 166*(3), 1139–1140.
Novitski, C. E. (2004b). On Fisher's criticism of Mendel's results with the garden pea. *Genetics, 166*(3), 1133–1136.
Numbers, R. L., & Kampourakis, K. (Eds.). (2015). *Newton's apple and other myths about science*. Harvard University Press.
Nuovo, V. (1992). Rethinking Paley. *Synthese, 91*, 29–51.
Nussbaum, M. (1880). *Zur differenzirung des geschlects im thierreich*. M. Cohen & Sohn.
Okasha, S. (2018). *Agents and goals in evolution*. Oxford University Press.
Olby, R. C. (1963). Charles Darwin's manuscript of pangenesis. *The British Journal for the History of Science, 1*(3), 251–263.
Olby, R. C. (1965). Francis Galton's derivation of Mendelian ratios in 1875. *Heredity, 20*(4), 636–638.
Olby, R. C. (1966). *Origins of Mendelism*. Constable and Company Ltd.
Olby, R. C. (1979). Mendel no mendelian? *History of Science, 17*(1), 53–72.
Olby, R. C. (1985). *Origins of Mendelism* (2nd ed.). University of Chicago Press.
Olby, R. C., & Gautrey, P. (1968). Eleven references to Mendel before 1900. *Annals of Science, 24*(1), 7–20.
Oldroyd, D. R. (1980). *Darwinian impacts*. The Open Univ. Press.
Orel, A. (1933). *Das Weltantlitz: eine gemeinverständliche Natur-, Kultur-, Religions-und Geschichtsphilosophie*. Matthias-Grünewald-Verlag.
Orel, V. (1996). *Gregor Mendel: The first geneticist*. Oxford University Press.
Osborn, H. F. (1896) A student's reminiscences of Huxley: Biological Lectures, Marine Biological Laboratory of Woods Hole, Ginn & Co. (pp. 29–42).
Ospovat, D. (1979). Darwin after Malthus. *Journal of the History of Biology, 12*, 211–230.
Ospovat, D. (1981). *The development of Darwin's theory: Natural history, natural theology, and natural selection, 1838–1859*. Cambridge University Press.
Ostoya, P. (1951). *Les théories de l'évolution*. Payot.
Ostwald, W. (Ed.). (1893). *Ostwald's Klassiker No. 41*. W. Engelmann.
Ostwald, W. (Ed.). (1899). *Ostwald's Klassiker No. 105*. W. Engelmann.

Outram, D. (1984). *Georges Cuvier: Vocation, science and authority in post-revolutionary France.* Manchester University Press.

Owen, A. R. G. (1962). An appreciation of the life and work of Sir Ronald Aylmer Fisher, F.R.S., F. S.S., Sc.D. *The Statistician, 12*(4), 313–319.

Owen, R. (1846). *Lectures on the comparative anatomy and physiology of the vertebrate animals: Delivered at the Royal College of Surgeons of England, in 1844 and 1846: Part 1 Fishes.* Longman.

Owen, R. (1848). *On the archetype and homologies of the vertebrate skeleton.* van Voorst.

Owen, R. (1849). *On parthenogenesis; or, the successive production of procreating individuals from a single ovum: A discourse introductory to the Hunterian lectures on generation and development, for the year 1849, delivered at the Royal College of Surgeons of England.* van Voorst.

Owen, R. (1851). On Dinornis (part IV.): Containing the restoration of the feet of that genus and of Palapteryx, with a description of the sternum in Palapteryx and Aptornis. *The Transactions of the Zoological Society of London, 4*(1), 1–20.

Owen, R. (1866). Professor Owen and the Darwinian theory. *London Review*, 516.

Packard, A. S. (1901). *Lamarck, the founder of evolution: His life and work.* Longmans, Green, and Company.

Paley, W. (1802). *Natural theology or evidences of the existence and attributes of the deity.* Printed for R. Faulder by Wilks and Taylor.

Paley, W. (1835). *The principles of moral and political philosophy.* R. Faulder.

van der Pas, P. W. (1970). The correspondence of Hugo de Vries and Charles Darwin. *Janus, 57*, 173–213.

van der Pas, P. W. (1976). Hugo de Vries and Gregor Mendel. *Folia Mendeliana, 11*, 3–16.

Paton, W. D. M., & Phillips, C. G. (1973). EHJ Schuster (1897–1969). *Notes and Records of the Royal Society of London, 28*(1), 111–117.

Pearl, R. (1913). A contribution towards an analysis of the problem of inbreeding. *The American Naturalist, 47*(562), 577–614.

Pearl, R. (1914a). On the results of inbreeding a Mendelian population: A correction and extension of previous conclusions. *The American Naturalist, 48*(565), 57–62.

Pearl, R. (1914b). Studies on inbreeding. IV. On a general formula for the constitution of the nth generation of a Mendelian population in which all matings are of brother X sister. *The American Naturalist, 48*(572), 491–494.

Pearl, R. (1915). Seventeen years selection of a character showing sex-linked Mendelian inheritance. *The American Naturalist, 49*(586), 595–608.

Pearl, R. (1916). Fecundity in the domestic fowl and the selection problem. *The American Naturalist, 50*(590), 89–105.

Pearl, R. (1917a). Studies on inbreeding. VII. Some further considerations regarding the measurement and numerical expression of degrees of kinship. *The American Naturalist, 51*(609), 545–559.

Pearl, R. (1917b). Studies on inbreeding. VIII. A single numerical measure of the total amount of inbreeding. *The American Naturalist, 51*(610), 636–639.

Pearson, E. S. (1938). *Karl Pearson: An appreciation of some aspects of his life and work.* Cambridge University Press.

Pearson, E. S. (1968). Some early correspondence between W S Gosset, R A Fisher and Karl Pearson, with notes and comments. *Biometrika, 55*(3), 445–457.

Pearson, K. (1892). *The grammar of science.* A. & C. Black.

Pearson, K. (1893a). Abstract of "Contributions to the mathematical theory of evolution". *Proceedings of the Royal Society, 54*, 329–333.

Pearson, K. (1893b). Contributions to the mathematical theory of evolution (Abstract of a paper read before the Royal Society, 1893). *Journal of the Royal Statistical Society, 56*(4), 675–679.

Pearson, K. (1894). Contributions to the mathematical theory of evolution. *Philosophical Transactions of the Royal Society of London, 185*(Ser. A), 71–110.

Pearson, K. (1898). Mathematical contributions to the theory of evolution. On the law of ancestral heredity. *Proceedings of the Royal Society of London, 62*, 386–412.
Pearson, K. (1900a). Mathematical contributions to the theory of evolution. VII. On the correlation of characters not quantitatively measurable. *Philosophical Transactions of the Royal Society of London (A), 195*, 1–47.
Pearson, K. (1900b). On the criterion that a given system of deviations from the probable in the case of a correlated system of variables is such that it can be reasonably supposed to have arisen from random sampling. *Philosophical Magazine, Series, 5*(50), 157–175.
Pearson, K., et al. (1901a). Mathematical contributions to the theory of evolution. IX. On the principle of homotyposis and its relation to heredity, to the variability of the individual, and to that of the race. Part I. Homotyposis in the vegetable kingdom (abstract). *Proceedings of the Royal Society of London, 68*, 1–5.
Pearson, K., et al. (1901b). Mathematical contributions to the theory of evolution: IX. On the principle of homotyposis and its relation to heredity, to variability of the individual, and to that of race. Part I: Homotyposis in the vegetable kingdom. *Philosophical Transactions of the Royal Society of London. Series A, 197*, 285–379.
Pearson, K. (1901). *The ethic of freethought: And other addresses and essays* (2nd ed.). A. and C. Black.
Pearson, K. (1902). On the fundamental conceptions of biology. *Biometrika, 1*(3), 320–344.
Pearson, K. (1903a). Prof. Johannsen on heredity (Anonymous). *Nature, 69*(17), 149–150.
Pearson, K. (1903b). The law of ancestral heredity. *Biometrika, 2*(2), 211–228.
Pearson, K. (1904a). A Mendelian's view of the law of ancestral inheritance. *Biometrika*, 109–112.
Pearson, K. (1904b). Mathematical contributions to the theory of evolution. XII.—On a generalised theory of alternative inheritance, with special reference to Mendel's laws. *Philosophical Transactions of the Royal Society of London. Series A, 203*, 53–86.
Pearson, K. (1904c). The Mendelian quarter. *Nature, 70*(1822), 529–530.
Pearson, K. (1906). Walter Frank Raphael Weldon, 1860–1906. *Biometrika, V*(1), 1–52.
Pearson, K. (1907a). Reply to certain criticisms of Mr G.U. Yule. *Biometrika, 5*, 470–476.
Pearson, K. (1907b). [Review of] On the theory of inheritance of quantitatively compound characters on the basis of Mendel's laws. *Biometrika, 5*, 481–482.
Pearson, K. (1911). *The grammar of science* (3rd ed.). A. & C. Black.
Pearson, K. (1914). *The life, letters and labours of Francis Galton* (Vol. I). Cambridge University Press.
Pearson, K. (1924). *The life, letters and labours of Francis Galton* (Vol. II). Cambridge University Press.
Pearson, K. (1930). The life, letters and labours of Francis Galton. In *IIIA & IIIB*. Cambridge University Press.
Pearson, K. (1934). *Speeches delivered at a dinner held in University College, London in honour of Professor Karl Pearson, 23 April, 1934*. Cambridge University Press.
Pearson, K., Lee, A., & Bramley-Moore, L. (1895). Mathematical contributions to the theory of evolution. III. Regression, heredity and panmixia. *Philosophical Transactions of the Royal Society of London. Series A, 187*, 253–318.
Peckham, M. (1959). *The origin of species: A variorum text*. University of Pennsylvania.
Pence, C. H. (2021). *The rise of chance in evolutionary theory: A pompous parade of arithmetic*. Academic Press.
Phipps, W. E. (2002). *Darwin's religious odyssey*. A&C Black.
Pichot, A. (2011). *Expliquer la vie: De l'âme à la molécule*. Quae.
Piegorsch, W. W. (1986). The Gregor Mendel controversy: Early issues of goodness-of-fit and recent issues of genetic linkage. *History of Science, 24*(2), 173–182.
Piegorsch, W. W. (1990). Fisher's contributions to genetics and heredity, with special emphasis on the Gregor Mendel Controversy. *Biometrics, 46*(4), 915–924.
Pilgrim, I. (1984). The too-good-to-be-true paradox and Gregor Mendel. *Journal of Heredity, 75*(6), 501–502.

Pilgrim, I. (1986). A solution to the too-good-to-be-true paradox and Gregor Mendel. *Journal of Heredity, 77*, 218–220.

Pilpel, A. (2007). Statistics is not enough: Revisiting Ronald A. Fisher's critique (1936) of Mendel's experimental results (1866). *Studies in History and Philosophy of Science Part C: Studies in History and Philosophy of Biological and Biomedical Sciences, 38*(3), 618–626.

Pinto-Correia, C. (1997). *The ovary of eve: Egg and sperm and preformation*. University of Chicago Press.

Pittendrigh, C. S. (1958). Adaptation, natural selection, and behavior. In A. Roe & G. G. Simpson (Eds.), *Behavior and evolution* (pp. 390–416). Yale University Press.

Plan, D. (1909). *Henri-Albert Gosse (1753–1816): Un Genevois d'Autrefois*. Fischbacher; Kundig.

Planck, M. (1917). Über einen Satz der statistischen Dynamik und seine Erweiterung in der Quantentheorie, Sitz. *Preuschen AkaD. Wissen, 24*, 324–341.

Platt, R. (1959). Mendel, Darwin and Galton. *Medical History, 87*, 87–99.

Pollock, E. (2006). *Stalin and the Soviet science wars*. Princeton University Press.

Porter, D. M. (1993). On the road to the origin with Darwin, Hooker, and Gray. *Journal of the History of Biology*, 1–38.

Porter, D. M., & Graham, P. W. (2015). *Darwin's sciences*. Wiley.

Porter, T. M. (2004). *Karl Pearson: The scientific life in a statistical age*. Princeton University Press.

Poulton, E. B. (1896). *Charles Darwin and the theory of natural selection*. Cassell.

Price, G. R. (1970). Selection and covariance. *Nature, 227*, 520–521.

Price, G. R. (1972). Fisher's "fundamental theorem" made clear. *Annals of Human Genetics, 36*(2), 129–140.

Pringsheim, N. (1856). *Zur Kritik und Geschichte der Untersuchungen über das Algengeschlecht*. A. Hirschwald.

Provine, W. B. (1977). Role of mathematical population geneticists in the evolutionary synthesis of the 1930's and 40's. In *Mathematical models in biological discovery* (pp. 2–31). Springer.

Provine, W. B. (1985). The R. A. Fisher-Sewall Wright controversy and its influence upon modern evolutionary biology. In R. Dawkins & M. Ridley (Eds.), *Oxford surveys in evolutionary biology* (Vol. 2, pp. 197–219). Oxford University Press.

Provine, W. B. (1989). *Sewall Wright and evolutionary biology*. University of Chicago press.

Provine, W. B. (2001). *The origins of theoretical population genetics: With a new afterword*. University of Chicago Press.

Prynne, W. (1630). *Anti-Arminianism, or the Church of Englands old antithesis to new Arminianisme* (2nd ed.). Imprinted, by Elizabeth Allde for Michael Sparke.

Punnett, R. C. (1908). Mendelism in relation to disease. *Proceedings of the Royal Society of Medicine, 1*, 135–168.

Punnett, R. C. (1911). *Mendelism* (3rd ed.). Macmillan.

Punnett, R. C. (1915). *Mimicry in butterflies*. Cambridge University Press.

Punnett, R. C. (1928). *Scientific papers of William Bateson*. Cambridge University Press.

Punnett, R. C. (1950). Early days of genetics. *Heredity, 4*, 1–10.

Quetelet, A. (1831). *Recherches sur le penchant au crime aux différens âges*. M. Hayez.

Raby, P. (2001). *Alfred Russel Wallace: A life*. Princeton University Press.

Radick, G. (2022). Mendel the fraud? A social history of truth in genetics. *Studies in History and Philosophy of Science, 93*, 39–46.

Radick, G. (2023). *Disputed inheritance: The battle over Mendel and the future of biology*. University of Chicago Press.

Rao, C. R. (1992). R A Fisher: The founder of modern statistics. *Statistical Science*, 34–48.

Rao, V., & Nanjundiah, V. (2011). JBS Haldane, Ernst Mayr and the beanbag genetics dispute. *Journal of the History of Biology, 44*, 233–281.

Rao, V., & Nanjundiah, V. (2016). Ernst Mayr's interactions with JBS Haldane. *History and Philosophy of the Life Sciences, 38*, 138–150.

Rauber, F. (1880). Formstörung in der Entwickelung von Wirbeltieren. *Morphologisches Jahrbuch, 5*, 661–705.

Reid, R. G. B. (1985). *Evolutionary theory: The unfinished synthesis*. Cornell University Press.

Reiss, J. (2009). *Not by design: Retiring Darwin's watchmaker*. University of California Press.

Remoff, H. (2016). Malthus, Darwin, and the descent of economics. *American Journal of Economics and Sociology, 75*(4), 862–903.

Repcheck, J. (2003). *The man who found time: James Hutton and the discovery of the earth's antiquity. Spectator-London-Weekly*. Perseus Publishing.

Rheinberger, H. J. (1995). When did Carl Correns read Gregor Mendel's paper? A research note. *Isis, 86*(4), 612–616.

Rheinberger, H.-J. (2003). Carl Correns' experiments with pisum, 1896–1899. In F. L. Holmes, J. Renn, & H.-J. Rheinberger (Eds.), *Reworking the bench: Research notebooks in the history of science (Archimedes, 7)* (pp. 221–252). Springer.

Rheinberger, H.-J. (2004). Carl Correns and the early history of genetic linkage. In H.-J. Rheinberger & J.-P. Gaudillière (Eds.), *Classical genetic research and its legacy: The mapping cultures of twentieth-century genetics* (pp. 21–33). Routledge.

Rice, S. (2004). *Evolutionary theory: Mathematical and conceptual foundations*. Sinauer Associates.

Richard, J. B. (1845). *Enrichissement de la langue Française: Dictionnaire de Mots Nouveaux* (2nd ed.). Léautey.

Richards, R. J. (1992). *The meaning of evolution: The morphological construction and ideological reconstruction of Darwin's theory*. University of Chicago Press.

Richardson, B. W. (1891). *Thomas Sopwith: With excerpts from his diary of fifty-seven years*. Longmans, Green.

Ridley, M. (2003). *Evolution. Genetics Research* (3rd ed.). Blackwell.

Robbins, R. B. (1917). Some applications of mathematics to breeding problems. *Genetics, 2*(5), 489–504.

Robbins, R. B. (1918a). Applications of mathematics to breeding problems II. *Genetics, 3*, 73–92.

Robbins, R. B. (1918b). Some applications of mathematics to breeding problems III. *Genetics, 3*, 375–389.

Roberts, H. F. (1929). *Plant hybridization before Mendel*. Princeton.

Robinson, D. C. (2019). Gravitation and general relativity at King's College London. *The European Physical Journal H, 44*, 181–270.

Robinson, G. (1979). *A prelude to genetics. Theories of a material substance of heredity: Darwin to Weismann*. Lawrence.

Roe, S. A. (1981). *Matter, life, and generation: Eighteenth-century embryology and the Haller-Wolff debate*. Cambridge University Press.

Roger, J. (1971). *Les sciences de la vie dans la pensée française au XVIIIe siècle* (2eme ed.). Armand Colin.

Roger, J. (1997). *Buffon: A life in natural history* (Trans. S. L. Bonnefoi). Cornell University Press.

Rolfe, R. A. (1888). The Cineraria. *The Gardener's Chronicle, 3*, 654.

Rolfe, R. A. (1898). Hybrid Cineraria. *The Gardener's Chronicle, 24*, 101–102.

Roll-Hansen, N. (1979). The genotype theory of Wilhelm Johannsen and its relation to plant breeding and the study of evolution. *Centaurus, 22*(3), 201–235.

Roll-Hansen, N. (2009). Sources of Wilhelm Johannsen's genotype theory. *Journal of the History of Biology, 42*, 457–493.

Roll-Hansen, N. (2014). Commentary: Wilhelm Johannsen and the problem of heredity at the turn of the 19th century. *International Journal of Epidemiology, 43*(4), 1007–1013.

Romanes, G. J. (1881). Hybridism. In *Encyclopedia Britannica* (9th ed., pp. 422–426). Encyclopaedia Britannica.

Romanes, G. J. (1888). Lamarckism versus Darwinism. *Nature, 38*(983), 413.

Romanes, G. J. (1895). *Darwin and after Darwin: Post-Darwinian questions: Heredity and utility* (Vol. II). The Open Court Publishing Company.

Root-Bernstein, R. S. (1983). Mendel and methodology. *History of Science, 21*(3), 275–295.

Rose, M. R. (1998). *Darwin's spectre: Evolutionary biology in the modern world*. Princeton University Press.

Ross, W. D. (Ed.). (1952a). *The works of Aristotle* (Vol. I). Encyclopedia Britannica.

Ross, W. D. (Ed.). (1952b). *The works of Aristotle* (Vol. II). Encyclopedia Britannica.

Rostand, J. (1945). *Esquisse d'une histoire de la biologie* (13th ed.). Gallimard.

Roux, W. (1883). *Über die Bedeutung der Zellteilungsfiguren*. Leipzig.

Ruse, M. (1996). Are pictures really necessary? The case of Sewall Wright's "adaptive landscapes". In *Picturing knowledge: Historical and philosophical problems concerning the use of art in science* (pp. 303–337). University of Toronto Press.

Ruse, M. (1999). *The Darwin revolution: Science red in tooth and claw* (2nd ed.). University of Chicago Press.

Ruse, M., & Travis, J. (2009). *Evolution: The first four billion years*. Harvard University Press.

Rushton, A. R. (2014). William Bateson and the chromosome theory of heredity: A reappraisal. *The British Journal for the History of Science, 47*(1), 147–171.

Russell, E. S. (1989). Sewall Wright's contributions to physiological genetics and to inbreeding theory and practice. *Annual Review of Genetics, 23*, 1–18.

Sageret, A. (1826). Considérations sur la production des hybrides, des variantes et des variétés en général, et sur celles de la famille des Cucurbitacées en particulier. *Annales des Sciences Naturelles, 8*, 294–314.

Sandler, L., & Novitski, E. (1957). Meiotic drive as an evolutionary force. *The American Naturalist, 91*(857), 105–110.

Sapp, J. (1991). The nine lives of Mendel. In H. E. Le Grand (Ed.), *Experimental Inquiries* (pp. 137–166). Kluwer.

Sargent, C. P. (1889). *Scientific papers of Asa Gray* (Vol. II). Houghton Mifflin.

Sarkar, S. (2007). Haldane and the emergence of modern evolutionary theory. In M. Matthen & C. Stephens (Eds.), *Philosophy of biology* (pp. 49–86). Elsevier.

Sarkar, S. (2016). Haldane and Mayr: A response to Rao and Nanjundiah. *History and Philosophy of the Life Sciences, 38*, 151–154.

Scheuer, P. A. G., & Mandel, S. P. H. (1959). An inequality in population genetics. *Heredity, 13*(4), 519–524.

Schleiden, M. J. (1838). Beitrage zür Phytogenesis. *Mullers Archiv fur Anatomie, Physiologie und wissen schaftliche Madizin, 5*, 137–176.

Schleiden, M. J. (1842). *Grundzüge der wissenschaftlichen Botanik: nebst einer Methodologischen Einleitung als Anleitung zum Studium der Pflanze* (Eng. transl. by E Lankester in Schleiden (1849)). Engelmann.

Schleiden, M. J. (1849). *Principles of scientific botany: Or, botany as an inductive science (Eng. transl. by E Lankester of Scheiden (1842))*. Longman.

Schmalhausen, I. F. (1874). *O rastitelnych pomesjach-nabljudenija iz petersburskoy flory* (Ph.D. dissertation). St. Petersburg University.

Schmitt, S. (2002). Lucien Cuénot et la théorie de l'évolution: un itinéraire hors norme. *La Revue Pour l'Histoire du CNRS, 7*, 10–16.

Schneer, C. J. (Ed.). (1969). *Toward a history of geology, New Hampshire inter-disciplinary conference on the history of geology (1967: Rye Beach)*. MIT Press.

Schuster, E. H. J. (1910). Reply to "The Mendel journal". *Nature, 82*(2102), 430.

Schuster, E. H. J. (1912). *Eugenics*. Collins.

Schuster, E. H. J. (1905). Results of crossing grey (house) mice with albinos. *Biometrika, 4*(1–2), 1–12.

Schwann, T. A. H. (1839). *Mikroskopische Untersuchungen über die Uebereinstimmung in der Struktur und dem Wachstum der Thiere und Pflanzen*. Reimer.

Schwartz, J. (2008). *In pursuit of the gene: From Darwin to DNA*. Harvard University Press.

Schwartz, J. S. (1974). Charles Darwin's debt to Malthus and Edward Blyth. *Journal of the History of Biology, 7*(2), 301–318.

Schwartz, J. S. (1984). Darwin, Wallace, and the descent of man. *Journal of the History of Biology, 17*(2), 271–289.

Sebright, J. (1809). *The art of improving the breeds of domestic animals, in a letter addressed to the right hon. Sir Joseph Banks, KB*. John Harding.

Sedgwick, A. (1845). Review of Vestiges (3rd ed.). *The Edinburgh Review, 82*, 1–85.

Sedgwick, A. (1850). *A discourse on the studies of the University of Cambridge* (5th ed.). JW Parker.

Senex (Robert Fitzroy). (1859, December 1). Review of origin of species. *The Times*, p. 8.

Serre, J.-L. (1984). La genèse de l'œuvre de Mendel. *La Recherche, 15*, 1072–1080.

Seton, A. (1823). On the variation in the colour of peas from cross-impregnation. *Transactions of the Horticultural Society of London, 5*, 236–237.

Seybold, K. S. (2016). *Explorations in neuroscience, psychology and religion*. Ashgate.

Shan, Y. (2020). *Doing integrated history and philosophy of science: A case study of the origin of genetics*. Springer.

Sheppard, P. M. (1975). *Natural selection and heredity* (4th ed.). Hutchinson.

Shermer, M. (2002). *In Darwin's shadow: The life and science of Alfred Russel Wallace: A biographical study on the psychology of history*. Oxford University Press.

Shine, I., & Wrobel, S. (1976). *Thomas Hunt Morgan: Pioneer of genetics*. University of Kentucky.

Shull, G. H. (1908). The composition of a field of maize. *Journal of Heredity, 1*, 296–301.

Silvers, W. K. (1979). *The coat colors of mice: A model for mammalian gene action and interaction*. Springer-Verlag.

Simpson, G. G. (1944). *Tempo and mode in evolution*. Columbia University Press.

Simunek, M., Hossfeld, U., & Breidbach, O. (2012). "Further development" of Mendel's legacy? Erich von Tschermak-Seysenegg in the context of Mendelian-biometry controversy, 1901–1906. *Theory in Biosciences, 131*(4), 243–252.

Simunek, M., Hossfeld, U., & Wissemann, V. (2011). "Rediscovery" revised – The cooperation of Erich and Armin von Tschermak-Seysenegg in the context of the "rediscovery" of Mendel's laws in 1899–1901. *Plant Biology (Stuttgart, Germany), 13*(6), 835–841.

Simunek, M. V., et al. (2017). Armin von Tschermak-Seysenegg (1870–1952): Physiologist and co-"rediscoverer" of Mendel's laws. *Theory in Biosciences, 136*(1–2), 59–67.

Skipper Jr, R. A. (2000). *The R. A. Fisher-Sewall Wright controversy in philosophical focus: Theory evaluation in population genetics* (Ph.D. dissertation). Department of Philosophy, University of Maryland, College Park, MD.

Skipper, R. A. (2007). Sir Ronald Aylmer Fisher. In M. Matthen & C. Stephens (Eds.), *Philosophy of biology* (pp. 37–48). Elsevier.

Skipper, R. A. (2009). Revisiting the Fisher-Wright controversy. *Transactions of the American Philosophical Society, 99*(1), 299–322.

Slatkin, M., & Veuille, M. (2002). *Modern developments in theoretical population genetics: The legacy of Gustave Malécot*. Oxford University Press.

Slotten, R. A. (2004). *The heretic in Darwin's court: The life of Alfred Russel Wallace*. Columbia University Press.

Smith, C. H., & Beccaloni, G. (Eds.). (2010). *Natural selection and beyond: The intellectual legacy of Alfred Russel Wallace*. Oxford University Press.

Smocovitis, V. B. (1997). *Unifying biology: The evolutionary synthesis and evolutionary biology*. Princeton University Press.

Soper, H. E., et al. (1917). On the distribution of the correlation coefficient in small samples. Appendix II to the papers of "Student" and R.A. Fisher. A cooperative study. *Biometrika, 11*, 328–413.

Sosna, M. (Ed.). (1966). *G. Mendel memorial symposium 1865–1965. Proceedings of a symposium held in Brno in August 4–7, 1965*. Academia, Prague.

Soyfer, V. N. (1994). *Lysenko and the tragedy of Soviet science*. Rutgers University Press.

Spencer, H. (1859). Illogical geology. *The Universal Review, 2*, 57–89.

Spencer, H. (1864). *The principles of biology* (Vol. I). Williams and Norgate.

Spencer, H. (1893). The inadequacy of "natural selection". *The Contemporary Review, 1866-1900*(63), 153–166.

van Speybroeck, L., van de Vijver, G., & de Waele, D. (Eds.). (2002). *From epigenesis to epigenetics: The genome in context*. The New York Academy of Sciences.

Spillman, R. (1942). Are insects people? *Journal of Heredity, 33*(1), 23–24.

Stafleu, F. A. (1971). *Linnaeus and the Linnaeans: The spreading of their ideas in systematic botany, 1735–1789*. Oosthoek.

Stamhuis, I. H. (2003). The reactions on Hugo de Vries's intracellular pangenesis; the discussion with August Weismann. *Journal of the History of Biology, 36*, 119–152.

Stamhuis, I. H., Meijer, O. G., & Zevenhuizen, E. J. (1999). Hugo de Vries on heredity, 1889–1903. Statistics, Mendelian laws, pangenes, mutations. *Isis, 90*(2), 238–267.

Stanfield, J. F. (1813). *An essay on the study and composition of biography*. G. Garbutt.

Stauffer, R. C. (1975). *Charles Darwin's natural selection: Being the second part of his big species book written from 1856 to 1858*. Cambridge University Press.

Stebbins, G. L. (1950). *Variation and evolution in plants*. Columbia University Press.

Stern, C. (1943). The Hardy-Weinberg law. *Science, 97*(2510), 137–138.

Stern, C. (1962). Wilhelm Weinberg (1862–1937). *Genetics, 47*, 1–5.

Stern, C. (1965). Mendel and human genetics. *Proceedings of the American Philosophical Society, 109*(4), 216–226.

Stern, C., & Sherwood, E. R. (Eds.). (1966). *The origin of genetics: A Mendel source book*. W H Freeman.

Stevens, N. M. (1905). Studies in spermatogenesis, with especial reference to the "accessory chromosome". *Carnegie Institution of Washington, 36*, 3–32.

Stigler, S. M. (2006). How Ronald Fisher became a mathematical statistician. *Mathematical Social Sciences, 176*, 23–30.

Stigler, S. M. (2010). Darwin, Galton, and the statistical enlightenment. *Journal of the Royal Statistical Society: Series A, 173*(3), 469–482.

Stomps, T. J. (1954). On the rediscovery of Mendel's work by Hugo de Vries. *Journal of Heredity, 45*(6), 293–294.

Strasburger, E. (1875). *Ueber Zellbildung und Zelltheilung*. G. Fischer.

Strasburger, E. (1880). *Zellbildung und zelltheilung*. G. Fischer.

Strasburger, E. (1884). *Neue Untersuchungen über den Befruchtungsvorgang bei den Phanerogamen als Grundlage für eine Theorie der Zeugung*. Fischer.

Street, D. J. (1990). Fisher's contributions to agricultural statistics. *Biometrics, 46*(4), 937–945.

Stubbe, H. (1972). *History of genetics: From prehistoric times to the rediscovery of Mendel's laws* (2nd ed.). MIT Press.

Sturtevant, A. H. (1913). The linear arrangement of six sex-linked factors in Drosophila, as shown by their mode of association. *Journal of Experimental Zoology, 14*(1), 43–59.

Sturtevant, A. H. (1918). An analysis of the effects of selection. *Carnegie Institution of Washington Publication, 264*, 1–68.

Sturtevant, A. H. (1965). *A history of genetics*. Harper & Row.

Sulloway, F. J. (1982a). Darwin and his finches: The evolution of a legend. *Journal of the History of Biology, 15*(1), 1–53.

Sulloway, F. J. (1982b). Darwin's conversion: The Beagle voyage and its aftermath. *Journal of the History of Biology, 15*(3), 325–396.

Sulloway, F. J. (1985). Darwin's early intellectual development: An overview of the Beagle voyage (1831–1836). In D. Kohn (Ed.), *The Darwinian heritage* (pp. 121–154). Princeton.

Sutton, W. S. (1903). The chromosomes in heredity. *Biological Bulletin, 4*(5), 231–251.

Sved, J. A., & Mayo, O. (1970). The evolution of dominance. In K. Kojima (Ed.), *Mathematical topics in population genetics* (pp. 289–316). Springer.

Svensson, E. I., & Calsbeek, R. (2012). *The adaptive landscape in evolutionary biology*. Oxford University Press.

Swammerdam, J. (1669). *Historia insectorum generalis*. M. Van Dreunen.

Swammerdam, J. (1682). *Histoire générale des insectes*. Chez Guillaume de Walcheren.
Sweeney, G. (2001). *Fighting for the good cause: Reflections on Francis Galton's Legacy to American Hereditarian Psychology*. American Philosophical Society.
Swinburne, R. G. (1965). Galton's law—Formulation and development. *Annals of Science, 21*(1), 15–31.
Szyfman, L. (1982). *Jean-Baptiste Lamarck et Son Époque*. Masson.
Tabery, J. G. (2004). The "evolutionary synthesis"' of George Udny Yule. *Journal of the History of Biology, 37*(1), 73–101.
Tayler, J. L. (1912). [Review of] "Breeding and the Mendelian discovery" by A. D. Darbishire. *The Sociological Review, 5*, 173–174.
Tétry, A. (1981). Lucien Cuénot. In C. C. Gillipsie (Ed.), *Dictionary of scientific biography* (pp. 492–494). Charles Scribner's Sons.
Thiselton-Dyer, W. T. (1895a). Letter to the Editor: The origin of the cultivated Cineraria. *Nature, 52*(1331), 3–4.
Thiselton-Dyer, W. T. (1895b). Letter to the Editor: The origin of the cultivated Cineraria. *Nature, 52*(1334), 78–79.
Thiselton-Dyer, W. T. (1895c). Letter to the Editor: The origin of the cultivated Cineraria. *Nature, 52*(1336), 128–129.
Thiselton-Dyer, W. T. (1895d). Variation and specific stability. *Nature, 51*(1324), 459–461.
Thiselton-Dyer, W. T. (1896). Home correspondence: Origin of garden Cineraria. *The Gardener's Chronicle, 20*.
Thompson, E. A. (1990). R. A. Fisher's contributions to genetical statistics. *Biometrics, 46*(4).
Thomson, W. (1862). On the age of the sun's heat. *Macmillan's Magazine, 5*, 288–375.
Thomson, W. (1863). On the secular cooling of the Earth. *Philosophical Magazine, 4*(25), 1–14.
Todhunter, I. (1876). *William Whewell, DD, Master of Trinity College, Cambridge: An account of his writings; with selections from his literary and scientific correspondence* (Vol. II). Macmillan.
Topham, J. R. (2010). Biology in the service of natural theology: Paley, Darwin, and the Bridgewater Treatises. In D. R. Alexander & R. L. Numbers (Eds.), *Biology and ideology: From Descartes to Dawkins* (pp. 88–113). University of Chicago Press.
Tort, P. (Ed.). (1996). *Dictionnaire du Darwinisme et de l'Évolution* (Vol. 3). Presses Universitaires de France.
Treviranus, G. R. (1802). *Biologie oder Philosophie der lebenden Natur für Naturforscher*. Röwer.
di Trocchio, F. (1991). Mendel's experiments: A reinterpretation. *Journal of the History of Biology, 24*(3), 485–519.
Tschermak, E. (1900a). Ueber künstliche Kreuzung bei Pisum sativum. *Berichte der Deutschen Botanischen Gesellschaft, 18*, 232–239.
Tschermak, E. (1900b). Ueber künstliche Kreuzung bei Pisum sativum. *Zeitschrift für das landwirtschaftliche Versuchswesen in Österreich, 3*, 465–555.
Tudge, C. (2002). *In Mendel's footnotes*. Vintage.
Tunlid, A., Kristoffersson, U., & Åström, F. (2020). A century of Hereditas: From local publication to international journal. *Hereditas, 157*, 1–19.
Turner, B. M. (2001). *Chromatin and gene regulation: Molecular mechanisms in epigenetics*. Blackwell Science.
Turner, J. R. G. (1970). Changes in mean fitness under natural selection. In K. Kojima (Ed.), *Mathematical topics in population genetics* (pp. 32–78). Springer.
Turrill, W. B. (1963). *Joseph Dalton Hooker: Botanist, explorer, and administrator*. Nelson.
Tutt, J. W. (1896). *British moths*. G. Routledge.
Unger, F. (1841). Genesis der Spiralgefässe. *Linnaea, 15*, 385–407.
Unger, F. (1854). *Botanical letters to a friend*. Lindsay and Blakiston (Translated by D. B. Paul).
Valadés, D. (1579). *Rhetorica christiana ad concionandi et orandi vsvm accommodata, vtrivsque facvltatis exemplis svo loco insertis; qvae qvidem ex Indorvm maximè deprompta svnt historiis*.

Vnde praeter doctrinam, svm˜a qvoqve delectatio comparabitvr. Pervsiae, apud Petrumiacobum Petrutium.
van Valen, L. (1963). Haldane's dilemma, evolutionary rates, and heterosis. *The American Naturalist, 97*(894), 185–190.
Vetta, A. (1976a). Correction to Fisher's correlations between relatives and environmental effects. *Nature, 263*(5575), 316–317.
Vetta, A. (1976b). Dominance variance in Fisher's model of assortative mating. *Annals of Human Genetics, 39*(4), 447–453.
Vickers, H. M. (1911). An apparently hitherto unnoticed "anticipation" of the theory of natural selection. *Nature, 85*(2155), 510–511.
de Vilmorin, L. (1856). Note sur la création d'une nouvelle race de betterave-considération sur l'hérédité des végétaux. *Comptes Rendus des séances hebdomadaires de l'Académie des Sciences, 43*(1856), 871–874.
de Vilmorin, L. (1858). Expériences sur la culture du colza. *Bulletin de la Société Impériale et Centrale d'Agriculture (2e série), 13*, 77–78.
Vorzimmer, P. (1969). Darwin, Malthus, and the theory of natural selection. *Journal of the History of Ideas, 30*(4), 527–542.
Vorzimmer, P. J. (1963). Charles Darwin and blending inheritance. *Isis, 54*(3), 371–390.
Vorzimmer, P. J. (1970). *Charles Darwin: The years of controversy*. Temple University Press.
de Vries, H. (1889). *Intracellulare pangenesis* (Eng. transl. by E.S. Granger (1910) as 'Intracellular pangenesis'). Gustav Fischer.
de Vries, H. (1894). Ueber halbe Galton – Kurven als Zeichnen diskontinuirlichen Variation. *Berichte der Deutschen Botanischen Gesellschaft, XII*, 197–207.
de Vries, H. (1895). Les demi-courbes Galtoniennes comme indice de variation discontinue. *Archives Neerlandaises, 28*, 442–457.
de Vries, H. (1897). Erfelijke monstrositeiten in den ruilhandel der botanische Tuinen. *Botanisch Jaarboek Dodonaea*, 62–79.
de Vries, H. (1899a). Hybridising of monstrosities. *Journal of the Royal Horticultural Society*, 69–75.
de Vries, H. (1899b). Sur le fécondation hybride de l'albumen. *Comptes Rendus Hebdomadaires des Seances de l'Académie des Sciences de Paris, 129*, 973–975.
de Vries, H. (1900a). Das Spaltungsgesetz der bastarde (Eng. transl. in de Vries (1966)). *Berichte der Deutschen Botanischen Gesellschaft, 18*, 83–90.
de Vries, H. (1900b). Sur la fécondation hybride de l'endosperme chez le Mais. *Revue générale de botanique, 12*, 129–136.
de Vries, H. (1900c). Sur la loi de disjonction des hybrides. *Comptes Rendus de l'Academie des Sciences, 130*, 845–847.
de Vries, H. (1900d). Sur les unités des caractères spécifiques et leur application à l'étude des hybrides. *Revue Générate de Botanique, 12*, 257–271.
de Vries, H. (1901). *Die Mutationstheorie. I. Die Entstehung der Arten durch Mutation* (Eng. transl. in de Vries (1909)). Von Veit & Co.
de Vries, H. (1903). *Die Mutationstheorie II. Elementare Bastardlehre* (Eng. transl. in de Vries (1910)). Von Veit & Co.
de Vries, H. (1909). *The mutation theory. I. The origin of species by mutations* (Eng. transl. by J B Farmer and A D Darbishire of de Vries (1901)). Open Court.
de Vries, H. (1910a). *Intracellular pangenesis* (Eng. transl. by C. Stuart Gager of de Vries (1889)). Open Court Publishing Company.
de Vries, H. (1910b). *The mutation theory. II. The origin of varieties by mutation* (Eng. transl. by J B Farmer and A D Darbishire of de Vries (1903)). Open Court.
de Vries, H. (1912). *Die Mutationen in der Erblichkeitslehre* (3rd ed.). Gebrüder Borntraeger.
de Vries, H. (1917). The origin of the mutation theory. *The Monist, 27*(3), 403–410.
de Vries, H. (1918). *Opera e periodicis collata* (Vol. V). Oosthoek.
de Vries, H. (1920). *Opera e periodicis collata* (Vol. VI). Oosthoek.

de Vries, H. (1950). Concerning the law of segregation of hybrids. *Genetics, 35*(5(Part 2)), 30–32.

de Vries, H. (1966). The law of segregation of hybrids (Eng. transl. by Evelyn Stern of de Vries (1900a)). In C. Stern & E. R. Sherwood (Eds.), *The origin of genetics: A Mendel source book* (pp. 107–117). W H Freeman & Co.

Waagen, W. (1869). Die Formenreihe des Ammonites subradiatus. *Benecke's Geognostiche-Paläontologhische Beitrage, 2*, 179–256.

Waddington, C. H. (1942). The epigenotype. *Endeavour, 1*, 18–20.

Waddington, C. H. (1953). Epigenetics and evolution. *Symposia of the Society for Experimental Biology, 7*, 186–199.

Waddington, C. H. (1957). *The strategy of the genes: A discussion of some aspects of theoretical biology*. Allen and Unwin.

Wade, N. J. (2003). *Destined for distinguished oblivion: The scientific vision of William Charles Wells* (1757–1817). Kluwer/Plenum.

Wade, M. J., & Goodnight, C. J. (1998). Perspective: The theories of Fisher and Wright in the context of metapopulations: When nature does many small experiments. *Evolution, 52*(6), 1537–1553.

van der Waerden, B. L. (1968). Mendel's experiments. *Centaurus, 12*(4), 275–288.

Wahlund, S. (1928). Zusammensetzung von Populationen und Korrelationserscheinungen vom Standpunkt der Vererbungslehre aus betrachtet (Eng. transl. in Wahlund (1975)). *Hereditas, 11*(1), 65–106.

Wahlund, S. (1932). *Demographic studies in the nomadic and the settled population of Northern Lapland*. Almqvist and Wiksell.

Wahlund, S. (1975). Composition of populations and of genotypic correlations from the viewpoint of population genetics (Eng. transl. by Marianne Langenbucher of Wahlund (1928)). In K. M. Weiss & P. A. Ballonoff (Eds.), *Demographic genetics* (pp. 224–263). Dowden, Hutchinson & Ross, Inc.

Waldeyer, W. (1888). Über Karyokinese und ihre Beziehungen zu den Befruchtungsvorgängen. *Archiv für Mikroskopische Anatomie, 32*(1), 1–122.

Wallace, A. R. (1853a). *A narrative of travels on the Amazon and Rio Negro: With an account of the native tribes, and observations on the climate, geology and natural history of the Amazon Valley*. Reeve and Co.

Wallace, A. R. (1853b). *Palm trees of the Amazon and their uses*. John Van Voorst.

Wallace, A. R. (1855). On the law which has regulated the introduction of new species. *Annals and Magazine of Natural History, 16*(93), 184–196.

Wallace, A. R. (1856a). A new kind of baby. *Chambers's Journal (ser. 3), 6*(151), 325–327.

Wallace, A. R. (1856b). On the habits of the Orang-Utan of Borneo. *Annals and Magazine of Natural History, 18*(103), 26–32.

Wallace, A. R. (1860). The ornithology of northern Celebes. *Ibis, 2*(2), 140–147.

Wallace, A. R. (1864). The origin of human races and the antiquity of man deduced from the theory of" natural selection". *Journal of the Anthropological Society of London, 2*, clviii–clxxxvii.

Wallace, A. R. (1866). *The scientific aspect of the supernatural: Indicating the desirableness of an experimental enquiry by men of science into the alleged powers of clairvoyants and mediums*. F. Farrah.

Wallace, A. R. (1869). Sir Charles Lyell on geological climates and the origin of species. *Quarterly Review, 126*, 359–394.

Wallace, A. R. (1870). Natural selection—Mr. Wallace's reply to Mr. Bennett. *Nature, 3*, 49–50.

Wallace, A. R. (1876). Review [of lessons from nature as manifested in mind and matter by St. George Mivart, 1876]. *Academy, 9*, 562–563.

Wallace, A. R. (1889). *Darwinism; an exposition of the theory of natural selection*. Macmillan.

Wallace, A. R. (1895). The method of organic evolution. *Fortnightly, 57*(338), 211–224.

Wallace, A. R. (1898). *The wonderful century: Its successes and its failures*. Dodd, Mead.

Wallace, A. R. (1903). The dawn of a great discovery: "My relations with Darwin in reference to the theory of natural selection". *Black and White, 25*, 78–79.

Wallace, A. R. (1905). *My life; a record of events and opinions*. Dodd, Mead & Company.
Wallace, B. (1968). *Topics in population genetics*. W W Norton Co.
Wallace, B. (1970). *Genetic load: Its biological and conceptual aspects*. Prentice-Hall.
Wallace, B. (1991). *Fifty years of genetic load: An odyssey*. Cornell University Press.
Walsh, B., & Lynch, M. (2018). *Evolution and selection of quantitative traits*. Oxford University Press.
Walsh, D. M. (2015). *Organisms, agency, and evolution*. Cambridge University Press.
Ward, P. (2018). *Lamarck's revenge: How epigenetics is revolutionizing our understanding of evolution's past and present*. Bloomsbury Publishing.
Warming, E., & Johannsen, W. (1900). *Den almindelige Botanik* (4th ed.). Nordiske Forlag.
Watson, H. W., & Galton, F. (1875). On the probability of the extinction of families. *The Journal of the Anthropological Institute of Great Britain and Ireland, 4*, 138–144.
Watterson, G. A. (1962). Some theoretical aspects of diffusion theory in population genetics. *The Annals of Mathematical Statistics, 33*(3), 939–957.
Watterson, G. A. (1975). On the number of segregating sites in genetical models without recombination. *Theoretical Population Biology, 7*(2), 256–276.
Weale, M. E. (2015). Patrick Matthew's law of natural selection. *Biological Journal of the Linnean Society, 115*(4), 785–791.
Weber, B. H. (2013). Complex systems dynamics in evolution and emergent processes. In B. Henning & A. Scarfe (Eds.), *Beyond mechanism: Putting life back into biology* (pp. 67–74). Lexington Books.
Weiling, F. (1966). Hat J. G. Mendel bei seinen Versuchen "zu genau" gearbeitet? — Der χ2-Test und seine Bedeutung für die Beurteilung genetischer Spaltungsverhältnisse (Eng. transl. in Weiling (1980)). *Der Züchter, 36*(8), 359–365.
Weiling, F. (1980). *Has J. G. Mendel been 'too accurate' in his experiments? The χ2-test and its significance to the evaluation of genetic segregation* (Eng. transl. by W. Piegorsch of Weiling (1966)).
Weiling, F. (1986). What about R.A. Fisher's statement of the "too good" data of J.G. Mendel's Pisum paper? *Journal of Heredity, 77*(4), 281–283.
Weiling, F. (1989). Which points are incorrect in RA Fisher's statistical conclusion: Mendel's experimental data agree too closely with his expectations? *Angewandte Botanik, 63*(1–2), 129–143.
Weiling, F. (1991). Historical study: Johann Gregor Mendel 1822–1884. *American Journal of Medical Genetics, 40*(1), 1–25.
Weinberg, W. (1901). Beiträge zur physiologie und pathologie der mehrlingsgeburten beim menschen. *Archiv für die gesamte Physiologie des Menschen und der Tiere, 88*(6), 346–430.
Weinberg, W. (1908a). Über den nachweis der vererbung beim menschen (Eng. transl. in Weinberg (1977)). *Jahreshefte des Vereins für vaterländische Naturkunde in Württemberg, 64*, 368–382.
Weinberg, W. (1908b). Über vererbungsgesetze beim menschen. I. *Zeitschrift für induktive Abstammungs-und Vererbungslehre, 1*(1), 377–392.
Weinberg, W. (1909). Über Vererbungsgesetze beim Menschen. II. *Zeitschrift für induktive Abstammungs-und Vererbungslehre, 2*(1), 276–330.
Weinberg, W. (1910). Weitere Beitrage zur Theorie der Vererbung (Eng. transl. of except in Weinberg (1984)). *Archiv für Rassen- und Gesellschafts-Biologie, 7*, 35–49.
Weinberg, W. (1977). On the demonstration of inheritance in man (Eng. transl. by Roy A Jameson of Weinberg (1908)). *Evolutionary Genetics (Benchmark Papers in Genetics, 8)*. Edited by D.L. Jameson, 115–125.
Weinberg, W. (1984). Further contributions to the theory of inheritance (Eng. transl. by Karen Meyer of excerpt of Weinberg (1910)). In W. G. Hill (Ed.), *Quantitative genetics: Explanation and analysis of continuous variation (Part I)* (pp. 42–57). Van Nostrand.
Weinstein, A. (1980). Cytology in the T. H. Morgan School. In E. Mayr & W. B. Provine (Eds.), *The evolutionary synthesis: Perspectives on the unification of biology* (pp. 80–86). Harvard University Press.
Weismann, A. (1875). *Studien zur Descendenz-Theorie. I. Über den Saison-Dimorphismus der Schmetterlinge* (Eng. transl. by R. Meldola of Weismann (1875)). Wilhelm Engelmann.

Weismann, A. (1882a). *Studies in the theory of descent. I. On the seasonal dimorphism of butterflies (Eng. transl. by R. Meldola of Vol. I of Weismann (1875))*. S. Low, Marston, Searle, & Rivington.
Weismann, A. (1882b). *Ueber die dauer des lebens* (Eng. transl. by A. E. Shipley in Weismann (1989)). G. Fischer.
Weismann, A. (1882c). Ueber die Vererbung. In *Aufsätze über Vererbung und verwandte biologische Fragen*. Gustav Fischer.
Weismann, A. (1883a). *Die entstehung der sexualzellen bei den hydromedusen: zugleich ein Betrag zur Kenntniss des Baues und der Lebenserscheinungen dieser Gruppe*. Fischer.
Weismann, A. (1883b). On heredity (Eng. transl. by A.E. Shipley of Weismann (1882c)). In *Essays upon heredity and kindred problems* (Vol. I, pp. 67–106). Clarendon Press.
Weismann, A. (1885a). *Die kontinuitat des Keimplasmas als grundlage einer theorie der vererbung* (Eng. transl. by E.B. Poulton et al. in Weismann (1885b)). Fischer.
Weismann, A. (1885b). The continuity of the germ-plasm as the foundation of a theory of heredity (Eng. transl. by E.B. Poulton et al. of Weismann (1885a)). In *Essays upon heredity and kindred biological problems* (pp. 163–254). Clarendon Press.
Weismann, A. (1886). *Die Bedeutung der sexuellen Fortpflanzung für die Selectionstheorie, Deutsche Medizinische Wochenschrift*. Gustav Fischer.
Weismann, A. (1887). On the number of polar bodies and their significance in heredity. In *Essays upon heredity and kindred problems* (pp. 343–396). Clarendon Press.
Weismann, A. (1889). The duration of life (Eng. transl. by A. E. Shipley of Weismann (1982b)). In *Essays upon heredity and kindred biological problems* (Vol. I, pp. 5–66). Clarendon Press.
Weismann, A. (1892). *Das Keimplasma: eine theorie der Vererbung* (Eng. transl. by W. Newton Parker in Weismann (1893)). G. Fischer.
Weismann, A. (1893). *The germ-plasm: A theory of heredity (Eng. transl. by W. Newton Parker of Weismann (1892))*. Charles Scribner's Sons.
Weismann, A. (1894). *The effect of external influences upon development*. Henry Frowde.
Weldon, W. F. R. (1883). Note on the early development of Lacerta muralis. *Quarterly Journal of Microscopical Science, 23*(89), 134–144.
Weldon, W. F. R. (1890). The variations occurring in certain Decapod Crustacea.-I. Crangon vulgaris. *Proceedings of the Royal Society, 47*, 445–453.
Weldon, W. F. R. (1892). Certain correlated variations in Crangon vulgaris. *Proceedings of the Royal Society, 51*, 2–21.
Weldon, W. F. R. (1893). On certain correlated variations in Carcinus maenas. *Proceedings of the Royal Society, 54*, 318–329.
Weldon, W. F. R. (1894a). II. Remarks on variation in animals and plants. To accompany the first report of the committee for conducting statistical inquiries into the measurable characteristics of plants and animals. *Proceedings of the Royal Society of London, 57*(340–346), 379–382.
Weldon, W. F. R. (1894b). Report of the committee. Part I. An attempt to measure the death-rate due to the selective destruction of Carcinus Moenas with respect to a particular dimension. *Proceedings of the Royal Society of London, 57*(340–346), 360–378.
Weldon, W. F. R. (1894c). The study of animal variation. *Nature, 50*(1280), 25–26.
Weldon, W. F. R. (1895a). Letter to the Editor: The origin of the cultivated Cineraria. *Nature, 52*(1333), 54.
Weldon, W. F. R. (1895b). The origin of cultivated Cineraria (Note on Bateson's article). *Nature, 52*(1335), 104.
Weldon, W. F. R. (1895c). The origin of the cultivated Cineraria. *Nature, 52*(1336), 129.
Weldon, W. F. R. (1902a). Mendel's laws of alternative inheritance in peas. *Biometrika, 1*(2), 228–254.
Weldon, W. F. R. (1902b). On the ambiguity of Mendel's categories. *Biometrika, 2*(1), 44–55.
Weldon, W. F. R. (1902c). Professor de Vries on the origin of species. *Biometrika, 1*(3), 365–374.
Weldon, W. F. R. (1903a). Letter to Editor: Mendel's principles of heredity in mice. *Nature, 67*(1744), 512.
Weldon, W. F. R. (1903b). Letter to Editor: Mendel's principles of heredity in mice. *Nature, 67*(1748), 610.

Weldon, W. F. R. (1903c). Letter to Editor: Mendel's principles of heredity in mice. *Nature, 68*(1750), 34.
Weldon, W. F. R. (1903d). Mr Bateson's revisions of Mendel's theory of heredity. *Biometrika, 2*(3), 286–298.
Weldon, W. F. R. (1895). Variations in animals and plants. *Nature, 51*, 449–450.
Weldon, W. F. R., & Pearson, K. (1903). Inheritance in Phaseolus vulgaris. *Biometrika, 2*, 499–503.
Wells, W. C. (1818). *Two essays: One upon single vision with two eyes; the other on dew*. Archibald Constable & Co.
Wentworth, E. N., & Remick, B. L. (1916). Some breeding properties of the generalized Mendelian population. *Genetics, 1*, 608–616.
Whewell, W. (1832). Review of volume 2 of Charles Lyell's principles of geology. *Quarterly Review, 47*, 103–132.
Whewell, W. (1833). *Astronomy and general physics considered with reference to natural theology*. Carey, Lea & Blanchard.
Whewell, W. (1834). On the connexion of the physical sciences: By Mrs. Somerville. *Quarterly Review, 51*, 54–68.
Whewell, W. (1837). *History of the inductive sciences: From the earliest times to the present* (Vol. 3). Parker.
Whewell, W. (1840). *The philosophy of the inductive sciences: Founded upon their history*. Parker.
White, A. D. (1897). *A history of the warfare of science with theology in Christendom*. D. Appleton.
Whiting, P. W. (1914). Heredity of bristles in the common greenbottle fly, Lucilia Sericata Meig. A study of factors governing distribution. *The American Naturalist, 48*(570), 339–355.
Wichler, G. (1961). *Charles Darwin: The founder of evolution and natural selection*. Pergamon Press.
Widmer, E. (1891). *Die europäischen Arten der Gattung*. Primula.
Wilberforce, S. (1860). [Review of] On the origin of species, by means of natural selection; or the preservation of favoured races in the struggle for life. *Quarterly Review, 108*, 225–264.
Wilkie, J. S. (1956a). The idea of evolution in the writings of Buffon.—I. *Annals of Science, 12*(1), 48–62.
Wilkie, J. S. (1956b). The idea of evolution in the writings of Buffon.—II. *Annals of Science, 12*(3), 212–227.
Wilkie, J. S. (1956c). The idea of evolution in the writings of Buffon.—III. *Annals of Science, 12*(4), 255–266.
Wilkie, J. S. (1960a). Nägeli's work on the fine structure of living matter—I. *Annals of Science, 16*(1), 11–41.
Wilkie, J. S. (1960b). Nägeli's work on the fine structure of living matter—II: The immediate reception of Nägeli's work. *Annals of Science, 16*(3), 171–207.
Wilkie, J. S. (1960c). Nageli's work on the fine structure of living matter—IIIa: Studies tending to confirm, correct and develop Nägeli's views I: Optical studies. *Annals of Science, 16*(4), 209–239.
Wilkie, J. S. (1961). Nägeli's work on the fine structure of living matter—IIIb: Studies tending to confirm, correct and develop Nägeli's views II: X-ray studies. *Annals of Science, 17*(1), 27–62.
Williams, G. (2019). *Unravelling the double helix: The lost heroes of DNA*. Weidenfeld & Nicolson.
Williams, G. C. (1992). *Natural selection: Domains, levels, and challenges*. Oxford University Press.
Willis, J. C. (1940). *The course of evolution*. Cambridge University Press.
Wills, C. (1980). *Genetic variability*. Oxford University Press.
Wilson, E. B. (1905). The chromosomes in relation to the determination of sex in insects. *Science, 22*(564), 500–502.
Wilson, E. B. (1925). *The cell in development and heredity* (3rd ed.). Macmillan.
Wilson, L. G. (1970). *Sir Charles Lyell's scientific journals on the species question*. Yale University Press.

Wilson, L. G. (1972). *Charles Lyell – The years to 1841: The revolution in geology*. Yale University Press.

Winkler, H. (1906). Ueber Parthenogenesis bei Wikstroemia indica (L.) CA Mey. *Annales du Jardin botanique de Buitenzorg, 20*, 208–276.

de Winter, W. (1997). The beanbag genetics controversy: Towards a synthesis of opposing views of natural selection. *Biology and Philosophy, 12*, 149–184.

Winther, R. G. (2001). August Weismann on germ-plasm variation. *Journal of the History of Biology, 34*, 517–555.

Wolff, C. (1728). *Philosophia rationalis: sive logica: methodo scientifica pertractata*. Officina libraria Rengeriana.

Wolff, C. F. (1759). *Theoria Generationis*. Hendelianis.

Wolfle, L. M. (1999). Sewall Wright on the method of path coefficients: An annotated bibliography. *Structural Equation Modeling, 6*(3), 280–291.

Wollaston, W. (1724). *Religion of nature delineated* (2nd ed.). Samuel Palmer.

Wood, T. B. (1905). Note on the inheritance of horns and face colour in sheep. *The Journal of Agricultural Science, 1*(3), 364–365.

Wool, D. (2020). *Milestones in the evolving theory of evolution*. Taylor and Francis.

Wright, C. (1871). The genesis of species. *North American Review, 113*, 63–104.

Wright, S. (1916). An intensive study of the inheritance of color and of other coat characters in guinea pigs with especial reference to graded variations. *Carnegie Institution of Washington Publication, 241*, 59–160.

Wright, S. (1918). On the nature of size factors. *Genetics, 3*(4), 367–374.

Wright, S. (1920). The relative importance of heredity and environment in determining the piebald pattern of guinea-pigs. *Proceedings of the National Academy of Sciences, 6*(6), 320–332.

Wright, S. (1921a). Correlation and causation. *Journal of Agricultural Research, 20*(7), 557–585.

Wright, S. (1921b). Systems of mating. I. The biometric relations between parent and offspring. *Genetics, 6*(2), 111–123.

Wright, S. (1921c). Systems of mating. II. The effects of inbreeding on the genetic composition of a population. *Genetics, 6*(2), 124–143.

Wright, S. (1921d). Systems of mating. III. Assortative mating based on somatic resemblance. *Genetics, 6*(2), 144–161.

Wright, S. (1921e). Systems of mating. IV. The effects of selection. *Genetics, 6*(2), 162–166.

Wright, S. (1921f). Systems of mating. V. General considerations. *Genetics, 6*(2), 168–178.

Wright, S. (1922a). Coefficients of inbreeding and relationship. *The American Naturalist, 56*(645), 330–338.

Wright, S. (1922b). The effects of inbreeding and crossbreeding on guinea-pigs: III. Crosses between highly inbred families. *United States Department of Agriculture Bulletin, 1121*, 1–61.

Wright, S. (1927). The effects in combination of the major color-factors of the guinea pig. *Genetics, 12*(6), 530–569.

Wright, S. (1929a). Evolution in a Mendelian population. *Anatomical Record, 44*, 287.

Wright, S. (1929b). Fisher's theory of dominance. *The American Naturalist, 63*(686), 274–279.

Wright, S. (1929c). The evolution of dominance: Comment on Dr Fisher's reply. *The American Naturalist, 63*(689), 556–561.

Wright, S. (1930). The genetical theory of natural selection: A review. *Journal of Heredity, 21*(8), 349–356.

Wright, S. (1931). Evolution in Mendelian populations. *Genetics, 16*(2), 97–159.

Wright, S. (1932). The roles of mutation, inbreeding, crossbreeding, and selection in evolution. *Proceedings of the Sixth International Congress of Genetics, 1*, 356–365.

Wright, S. (1934a). Physiological and evolutionary theories of dominance. *The American Naturalist, 68*(714), 24–53.

Wright, S. (1934b). Professor Fisher on the theory of dominance. *The American Naturalist, 68*(719), 562–565.

Wright, S. (1934c). The method of path coefficients. *Annals of Mathematical Statistics, 5*(3), 161–215.
Wright, S. (1935a). Evolution in populations in approximate equilibrium. *Journal of Genetics, 30*(2), 257–266.
Wright, S. (1935b). The analysis of variance and the correlations between relatives with respect to deviations from an optimum. *Journal of Genetics, 30*, 243–256.
Wright, S. (1937). The distribution of gene frequencies in populations. *Proceedings of the National Academy of Sciences, 23*(6), 307–320.
Wright, S. (1938). The distribution of gene frequencies under irreversible mutation. *Proceedings of the National Academy of Sciences, 24*(7), 253–259.
Wright, S. (1948). Genetics of populations. *Encyclopedia Britannica, 10*, 111.
Wright, S. (1952). The theoretical variance within and among subdivisions of a population that is in a steady state. *Genetics, 37*(3), 312–321.
Wright, S. (1960a). Genetics and twentieth century Darwinism—A review and discussion. *American Journal of Human Genetics, 12*(3), 365–372.
Wright, S. (1960b). Path coefficients and path regressions: Alternative or complementary concepts? *Biometrics*, 189–202.
Wright, S. (1969). *Evolution and the genetics of populations, volume 2: The theory of gene frequencies*. University of Chicago press.
Wright, S. (1970). Random drift and the shifting balance theory of evolution. In K. Kojima (Ed.), *Mathematical topics in population genetics* (pp. 1–31). Springer.
Wright, S. (1977). *Evolution and the genetics of populations, volume 3: Experimental results and evolutionary deductions*. University of Chicago press.
Wright, S. (1978). The relation of livestock breeding to theories of evolution. *Journal of Animal Science, 46*(5), 1192–1200.
Wright, S. (1980). Genic and organismic selection. *Evolution*, 825–843.
Wright, S. (1986). *Evolution: Selected papers*. University of Chicago Press.
van Wyhe, J. (2013). *Dispelling the darkness: Voyage in the Malay Archipelago and the discovery of evolution by Wallace and Darwin*. World Scientific.
Yule, G. U. (1900). VII. On the association of attributes in statistics: With illustrations from the material of the childhood society, &c. *Philosophical Transactions of the Royal Society of London. Series A, Containing Papers of a Mathematical or Physical Character, 194*(252–261), 257–319.
Yule, G. U. (1902). Mendel's laws and their probable relations to intra-racial heredity. *New Phytologist, 1*(9), 193–207, 222–238.
Yule, G. U. (1904). Prof. Johannsen on heredity. *Nature, 69*(1784), 223–224.
Yule, G. U. (1906a). On a property which holds good for all groupings of a normal distribution of frequency for two variables, with applications to the study of contingency-tables for the inheritance of unmeasured qualities. *Proceedings of the Royal Society of London A, 77*(517), 324–336.
Yule, G. U. (1906b). On the influence of bias and of personal equation in statistics of ill-defined qualities: An experimental study. *Proceedings of the Royal Society of London A, 77*(517), 337–339.
Yule, G. U. (1906c). On the theory of inheritance of quantitative compound characters on the basis of Mendel's laws. In *Report of the third international conference on genetics* (pp. 140–142).
Zhang, H., Chen, W., & Sun, K. (2017). Mendelism: New insights from Gregor Mendel's lectures in Brno. *Genetics, 207*(1), 1–8.
Zimmermann, F. (1975). Actualité de la pensée de Haldane. *Population, 30*, 285–302.
Zirkle, C. (1946). The early history of the idea of the inheritance of acquired characters and of pangenesis. *Transactions of the American Philosophical Society, 35*(2), 91–151.
Zirkle, C. (1951). Gregor Mendel and his precursors. *Isis, 42*(2), 97–104.
Zirkle, C. (1964). Some oddities in the delayed discovery of Mendelism. *Journal of Heredity, 55*(2), 65–72.

Index

P. Gorroochurn, *The Development of Evolutionary Genetics*,
https://doi.org/10.1007/978-3-031-69374-8

C

G

N

O

MIX
Papier aus verantwortungsvollen Quellen
Paper from responsible sources
FSC® C105338

If you have any concerns about our products,
you can contact us on
ProductSafety@springernature.com

In case Publisher is established outside the EU,
the EU authorized representative is:
Springer Nature Customer Service Center GmbH
Europaplatz 3, 69115 Heidelberg, Germany

Printed by Libri Plureos GmbH
in Hamburg, Germany